Principles and Practice
of Big Data

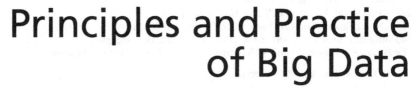

Principles and Practice of Big Data

Preparing, Sharing, and Analyzing Complex Information

Second Edition

Jules J. Berman

ELSEVIER

ACADEMIC PRESS

An imprint of Elsevier

Academic Press is an imprint of Elsevier
125 London Wall, London EC2Y 5AS, United Kingdom
525 B Street, Suite 1650, San Diego, CA 92101, United States
50 Hampshire Street, 5th Floor, Cambridge, MA 02139, United States
The Boulevard, Langford Lane, Kidlington, Oxford OX5 1GB, United Kingdom

Notices
Knowledge and best practice in this field are constantly changing. As new research and experience broaden our understanding, changes in research methods, professional practices, or medical treatment may become necessary.

Practitioners and researchers must always rely on their own experience and knowledge in evaluating and using any information, methods, compounds, or experiments described herein. In using such information or methods they should be mindful of their own safety and the safety of others, including parties for whom they have a professional responsibility.

To the fullest extent of the law, neither the Publisher nor the authors, contributors, or editors, assume any liability for any injury and/or damage to persons or property as a matter of products liability, negligence or otherwise, or from any use or operation of any methods, products, instructions, or ideas contained in the material herein.

Library of Congress Cataloging-in-Publication Data
A catalog record for this book is available from the Library of Congress

British Library Cataloguing-in-Publication Data
A catalogue record for this book is available from the British Library

ISBN: 978-0-12-815609-4

For information on all Academic Press publications
visit our website at https://www.elsevier.com/books-and-journals

Working together
to grow libraries in
developing countries

www.elsevier.com • www.bookaid.org

Publisher: Mara Conner
Acquisition Editor: Mara Conner
Editorial Project Manager: Mariana L. Kuhl
Production Project Manager: Punithavathy Govindaradjane
Cover Designer: Matthew Limbert

Typeset by SPi Global, India

Other Books by Jules J. Berman

Taxonomic Guide to Infectious Diseases
Understanding the Biologic Classes of Pathogenic Organisms (2012)
9780124158955

Principles of Big Data
Preparing, Sharing, and Analyzing Complex Information (2013)
9780124045767

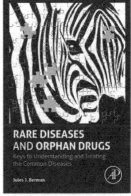

Rare Diseases and Orphan Drugs
Keys to Understanding and Treating the Common Diseases (2014)
9780124199880

Repurposing Legacy Data
Innovative Case Studies (2015)
9780128028827

Data Simplification
Taming Information with Open Source Tools (2016)
9780128037812

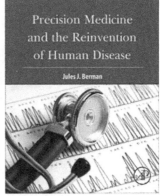

Precision Medicine and The Reinvention of Human Disease (2018)
9780128143933

Dedication

To my wife, Irene, who reads every day, and who understands why books are important.

Contents

About the Author

Jules J. Berman received two baccalaureate degrees from MIT; in Mathematics, and in Earth and Planetary Sciences. He holds a PhD from Temple University, and an MD, from the University of Miami. He was a graduate student researcher in the Fels Cancer Research Institute, at Temple University, and at the American Health Foundation in Valhalla, New York. His postdoctoral studies were completed at the US National Institutes of Health, and his residency was completed at the George Washington University Medical Center in Washington, DC. Dr. Berman served as Chief of Anatomic Pathology, Surgical Pathology, and Cytopathology at the Veterans Administration Medical Center in Baltimore, Maryland, where he held joint appointments at the University of Maryland Medical Center and at the Johns Hopkins Medical Institutions. In 1998, he transferred to the US National Institutes of Health, as a Medical Officer, and as the Program Director for Pathology Informatics in the Cancer Diagnosis Program at the National Cancer Institute. Dr. Berman is a past president of the Association for Pathology Informatics, and the 2011 recipient of the Association's Lifetime Achievement Award. He has first-authored over 100 scientific publications and has written more than a dozen books in the areas of data science and disease biology. Several of his most recent titles, published by Elsevier, include:

Taxonomic Guide to Infectious Diseases: Understanding the Biologic Classes of Pathogenic Organisms (2012)
Principles of Big Data: Preparing, Sharing, and Analyzing Complex Information (2013)
Rare Diseases and Orphan Drugs: Keys to Understanding and Treating the Common Diseases (2014)
Repurposing Legacy Data: Innovative Case Studies (2015)
Data Simplification: Taming Information with Open Source Tools (2016)
Precision Medicine and the Reinvention of Human Disease (2018)

Author's Preface to Second Edition

Everything has been said before, but since nobody listens we have to keep going back and beginning all over again.
Andre Gide

Good science writers will always jump at the chance to write a second edition of an earlier work. No matter how hard they try, that first edition will contain inaccuracies and misleading remarks. Sentences that seemed brilliant when first conceived will, with the passage of time, transform into examples of intellectual overreaching. Points too trivial to include in the original manuscript may now seem like profundities that demand a full explanation. A second edition provides rueful authors with an opportunity to correct the record.

When the first edition of Principles of Big Data was published in 2013 the field was very young and there were few scientists who knew what to do with Big Data. The data that kept pouring in was stored, like wheat in silos, throughout the planet. It was obvious to data managers that none of that stored data would have any scientific value unless it was properly annotated with metadata, identifiers, timestamps, and a set of basic descriptors. Under these conditions, the first edition of the Principles of Big Data stressed the proper and necessary methods for collecting, annotating, organizing, and curating Big Data. The process of preparing Big Data comes with its own unique set of challenges, and the First Edition was peppered with warnings and exhortations intended to steer readers clear of disaster.

It is now five years since the first edition was published and there have since been hundreds of books written on the subject of Big Data. As a scientist, it is disappointing to me that the bulk of Big Data, today, is focused on issues of marketing and predictive analytics (e.g., "Who is likely to buy product x, given that they bought product y two weeks previously?"); and machine learning (e.g., driverless cars, computer vision, speech recognition). Machine learning relies heavily on hyped up techniques such as neural networks and deep learning; neither of which are leading to fundamental laws and principles that simplify and broaden our understanding of the natural world and the physical universe. For the most part, these techniques use data that is relatively new (i.e., freshly collected), poorly annotated (i.e., provided with only the minimal information required for one particular analytic process), and not deposited for public evaluation or for re-use. In short, Big Data has followed the path of least resistance, avoiding most of the tough issues raised in the first edition of this book; such as the importance of sharing data with the public, the value of finding relationships (not similarities) among data objects, and the heavy, but inescapable, burden of creating robust, immortal, and well-annotated data.

It was certainly my hope that the greatest advances from Big Data would come as fundamental breakthroughs in the realms of medicine, biology, physics, engineering, and chemistry. Why has the focus of Big Data shifted from basic science over to machine learning? It may have something to do with the fact that no book, including the first edition of this book, has provided readers with the methods required to put the principles of Big Data into practice. In retrospect, it was not sufficient to describe a set of principles and then expect readers to invent their own methodologies.

Consequently, in this second edition, the publisher has changed the title of the book from "The Principles of Big Data," to "The Principles AND PRACTICE of Big Data." Henceforth and herein, recommendations are accompanied by the methods by which those recommendations can be implemented. The reader will find that all of the methods for implementing Big Data preparation and analysis are really quite simple. For the most part, computer methods require some basic familiarity with a programming language, and, despite misgivings, Python was chosen as the language of choice. The advantages of Python are:

- Python is a no-cost, open source, high-level programming language that is easy to acquire, install, learn, and use, and is available for every popular computer operating system.
- Python is extremely popular, at the present time, and its popularity seems to be increasing.
- Python distributions (such as Anaconda) come bundled with hundreds of highly useful modules (such as numpy, matplot, and scipy).
- Python has a large and active user group that has provided an extraordinary amount of documentation for Python methods and modules.
- Python supports some object-oriented techniques that will be discussed in this new edition

As everything in life, Python has its drawbacks:

- The most current versions of Python are not backwardly compatible with earlier versions. The scripts and code snippets included in this book should work for most versions of Python 3.x, but may not work with Python versions 2.x and earlier, unless the reader is prepared to devote some time to tweaking the code. Of course, these short scripts and snippets are intended as simplified demonstrations of concepts, and must not be construed as application-ready code.
- The built-in Python methods are sometimes maximized for speed by utilizing Random Access Memory (RAM) to hold data structures, including data structures built through iterative loops. Iterations through Big Data may exhaust available RAM, leading to the failure of Python scripts that functioned well with small data sets.
- Python's implementation of object orientation allows multiclass inheritance (i.e., a class can be the subclass of more than one parent class). We will describe why this is problematic, and the compensatory measures that we must take, whenever we use our Python programming skills to understand large and complex sets of data objects.

The core of every algorithm described in the book can be implemented in a few lines of code, using just about any popular programming language, under any operating system, on any modern computer. Numerous Python snippets are provided, along with descriptions of free utilities that are widely available on every popular operating system. This book stresses the point that most data analyses conducted on large, complex data sets can be achieved with simple methods, bypassing specialized software systems (e.g., parallelization of computational processes) or hardware (e.g., supercomputers). Readers who are completely unacquainted with Python may find that they can read and understand Python code, if the snippets of code are brief, and accompanied by some explanation in the text. In any case, readers who are primarily concerned with mastering the principles of Big Data can skip the code snippets without losing the narrative thread of the book.

This second edition has been expanded to stress methodologies that have been overlooked by the authors of other books in the field of Big Data analysis. These would include:

– **Data preparation.**

How to annotate data with metadata and how to create data objects composed of triples. The concept of the triple, as the fundamental conveyor of meaning in the computational sciences, is fully explained.

– **Data structures of particular relevance to Big Data**

Concepts such as triplestores, distributed ledgers, unique identifiers, timestamps, concordances, indexes, dictionary objects, data persistence, and the roles of one-way hashes and encryption protocols for data storage and distribution are covered.

– **Classification of data objects**

How to assign data objects to classes based on their shared relationships, and the computational roles filled by classifications in the analysis of Big Data will be discussed at length.

– **Introspection**

How to create data objects that are self-describing, permitting the data analyst to group objects belonging to the same class and to apply methods to class objects that have been inherited from their ancestral classes.

– **Algorithms that have special utility in Big Data preparation and analysis**

How to use one-way hashes, unique identifier generators, cryptographic techniques, timing methods, and time stamping protocols to create unique data objects that are immutable (never changing), immortal, and private; and to create data structures that facilitate a host of useful functions that will be described (e.g., blockchains and distributed ledgers, protocols for safely sharing confidential information, and methods for reconciling identifiers across data collections without violating privacy).

– **Tips for Big Data analysis**

How to overcome many of the analytic limitations imposed by scale and dimensionality, using a range of simple techniques (e.g., approximations, so-called back-of-the-envelope tricks, repeated sampling using a random number generator, Monte Carlo simulations, and data reduction methods).

– **Data reanalysis, data repurposing, and data sharing**

Why the first analysis of Big Data is almost always incorrect, misleading, or woefully incomplete, and why data reanalysis has become a crucial skill that every serious Big Data analyst must acquire. The process of data reanalysis often inspires repurposing of Big Data resources. Neither data reanalysis nor data repurposing can be achieved unless and until the obstacles to data sharing are overcome. The topics of data reanalysis, data repurposing, and data sharing are explored at length.

Comprehensive texts, such as the second edition of the Principles and Practice of Big Data, are never quite as comprehensive as they might strive to be; there simply is no way to fully describe every concept and method that is relevant to a multi-disciplinary field, such as Big Data. To compensate for such deficiencies, there is an extensive Glossary section for every chapter, that defines the terms introduced in the text, providing some explanation of the relevance of the terms for Big Data scientists. In addition, when techniques and methods are discussed, a list of references that the reader may find useful, for further reading on the subject, is provided. Altogether, the second edition contains about 600 citations to outside references, most of which are available as free downloads. There are over 300 glossary items, many of which contain short Python snippets that readers may find useful.

As a final note, this second edition uses case studies to show readers how the principles of Big Data are put into practice. Although case studies are drawn from many fields of science, including physics, economics, and astronomy, readers will notice an overabundance of examples drawn from the biological sciences (particularly medicine and zoology). The reason for this is that the taxonomy of all living terrestrial organisms is the oldest and best Big Data classification in existence. All of the classic errors in data organization, and in data analysis, have been committed in the field of biology. More importantly, these errors have been documented in excruciating detail and most of the documented errors have been corrected and published for public consumption. If you want to understand how Big Data can be used as a tool for scientific advancement, then you must look at case examples taken from the world of biology, a well-documented field where everything that can happen has happened, is happening, and will happen. Every effort has been made to limit Case Studies to the simplest examples of their type, and to provide as much background explanation as non-biologists may require.

Principles and Practice of Big Data, Second Edition, is devoted to the intellectual conviction that the primary purpose of Big Data analysis is to permit us to ask and answer a wide range of questions that could not have been credibly approached with small sets of data. There is every reason to hope that the readers of this book will soon achieve scientific breakthroughs that were beyond the reach of prior generations of scientists. Good luck!

Author's Preface to First Edition

We can't solve problems by using the same kind of thinking we used when we created them.
Albert Einstein

Data pours into millions of computers every moment of every day. It is estimated that the total accumulated data stored on computers worldwide is about 300 exabytes (that's 300 billion gigabytes). Data storage increases at about 28% per year. The data stored is peanuts compared to data that is transmitted without storage. The annual transmission of data is estimated at about 1.9 zettabytes or 1,900 billion gigabytes [1]. From this growing tangle of digital information, the next generation of data resources will emerge.

As we broaden our data reach (i.e., the different kinds of data objects included in the resource), and our data timeline (i.e., accruing data from the future and the deep past), we need to find ways to fully describe each piece of data, so that we do not confuse one data item with another, and so that we can search and retrieve data items when we need them. Astute informaticians understand that if we fully describe everything in our universe, we would need to have an ancillary universe to hold all the information, and the ancillary universe would need to be much larger than our physical universe.

In the rush to acquire and analyze data, it is easy to overlook the topic of data preparation. If the data in our Big Data resources are not well organized, comprehensive, and fully described, then the resources will have no value. The primary purpose of this book is to explain the principles upon which serious Big Data resources are built. All of the data held in Big Data resources must have a form that supports search, retrieval, and analysis. The analytic methods must be available for review, and the analytic results must be available for validation.

Perhaps the greatest potential benefit of Big Data is its ability to link seemingly disparate disciplines, to develop and test hypothesis that cannot be approached within a single knowledge domain. Methods by which analysts can navigate through different Big Data resources to create new, merged data sets, will be reviewed.

What exactly, is Big Data? Big Data is characterized by the three V's: volume (large amounts of data), variety (includes different types of data), and velocity (constantly accumulating new data) [2]. Those of us who have worked on Big Data projects might suggest throwing a few more v's into the mix: vision (having a purpose and a plan), verification (ensuring that the data conforms to a set of specifications), and validation (checking that its purpose is fulfilled).

Many of the fundamental principles of Big Data organization have been described in the "metadata" literature. This literature deals with the formalisms of data description (i.e., how to describe data); the syntax of data description (e.g., markup languages such as eXtensible Markup Language, XML); semantics (i.e., how to make computer-parsable statements that convey meaning); the syntax of semantics (e.g., framework specifications such as Resource Description Framework, RDF, and Web Ontology Language, OWL); the creation of data objects that hold data values and self-descriptive information; and the deployment of ontologies, hierarchical class systems whose members are data objects.

The field of metadata may seem like a complete waste of time to professionals who have succeeded very well, in data-intensive fields, without resorting to metadata formalisms. Many computer scientists, statisticians, database managers, and network specialists have no trouble handling large amounts of data, and they may not see the need to create a strange new data model for Big Data resources. They might feel that all they really need is greater storage capacity, distributed over more powerful computers that work in parallel with one another. With this kind of computational power, they can store, retrieve, and analyze larger and larger quantities of data. These fantasies only apply to systems that use relatively simple data or data that can be represented in a uniform and standard format. When data is highly complex and diverse, as found in Big Data resources, the importance of metadata looms large. Metadata will be discussed, with a focus on those concepts that must be incorporated into the organization of Big Data resources. The emphasis will be on explaining the relevance and necessity of these concepts, without going into gritty details that are well covered in the metadata literature.

When data originates from many different sources, arrives in many different forms, grows in size, changes its values, and extends into the past and the future, the game shifts from data computation to data management. I hope that this book will persuade readers that faster, more powerful computers are nice to have, but these devices cannot compensate for deficiencies in data preparation. For the foreseeable future, universities, federal agencies, and corporations will pour money, time, and manpower into Big Data efforts. If they ignore the fundamentals, their projects are likely to fail. On the other hand, if they pay attention to Big Data fundamentals, they will discover that Big Data analyses can be performed on standard computers. The simple lesson, that data trumps computation, will be repeated throughout this book in examples drawn from well-documented events.

There are three crucial topics related to data preparation that are omitted from virtually every other Big Data book: identifiers, immutability, and introspection.

A thoughtful identifier system ensures that all of the data related to a particular data object will be attached to the correct object, through its identifier, and to no other object. It seems simple, and it is, but many Big Data resources assign identifiers promiscuously, with the end result that information related to a unique object is scattered throughout the resource, attached to other objects, and cannot be sensibly retrieved when needed. The concept of object identification is of such overriding importance that a Big Data resource can be usefully envisioned as a collection of unique identifiers to which complex data is attached.

Immutability is the principle that data collected in a Big Data resource is permanent, and can never be modified. At first thought, it would seem that immutability is a ridiculous and impossible constraint. In the real world, mistakes are made, information changes, and the methods for describing information changes. This is all true, but the astute Big Data manager knows how to accrue information into data objects without changing the pre-existing data. Methods for achieving this seemingly impossible trick will be described in detail.

Introspection is a term borrowed from object-oriented programming, not often found in the Big Data literature. It refers to the ability of data objects to describe themselves when interrogated. With introspection, users of a Big Data resource can quickly determine the content of data objects and the hierarchical organization of data objects within the Big Data resource. Introspection allows users to see the types of data relationships that can be analyzed within the resource and clarifies how disparate resources can interact with one another.

Another subject covered in this book, and often omitted from the literature on Big Data, is data indexing. Though there are many books written on the art of the science of so-called back-of-the-book indexes, scant attention has been paid to the process of preparing indexes for large and complex data resources. Consequently, most Big Data resources have nothing that could be called a serious index. They might have a Web page with a few links to explanatory documents, or they might have a short and crude "help" index, but it would be rare to find a Big Data resource with a comprehensive index containing a thoughtful and updated list of terms and links. Without a proper index, most Big Data resources have limited utility for any but a few cognoscenti. It seems odd to me that organizations willing to spend hundreds of millions of dollars on a Big Data resource will balk at investing a few thousand dollars more for a proper index.

Aside from these four topics, which readers would be hard-pressed to find in the existing Big Data literature, this book covers the usual topics relevant to Big Data design, construction, operation, and analysis. Some of these topics include data quality, providing structure to unstructured data, data deidentification, data standards and interoperability issues, legacy data, data reduction and transformation, data analysis, and software issues. For these topics, discussions focus on the underlying principles; programming code and mathematical equations are conspicuously inconspicuous. An extensive Glossary covers the technical or specialized terms and topics that appear throughout the text. As each Glossary term is "optional" reading, I took the liberty of expanding on technical or mathematical concepts that appeared in abbreviated form in the main text. The Glossary provides an explanation of the practical relevance of each term to Big Data, and some readers may enjoy browsing the Glossary as a stand-alone text.

The final four chapters are non-technical; all dealing in one way or another with the consequences of our exploitation of Big Data resources. These chapters will cover legal, social, and ethical issues. The book ends with my personal predictions for the future of Big Data, and its impending impact on our futures. When preparing this book, I debated whether these four chapters might best appear in the front of the book, to whet the reader's appetite for the more technical chapters. I eventually decided that some readers would be unfamiliar with some of the technical language and concepts included in the final chapters, necessitating their placement near the end.

Readers may notice that many of the case examples described in this book come from the field of medical informatics. The healthcare informatics field is particularly ripe for discussion because every reader is affected, on economic and personal levels, by the Big Data policies and actions emanating from the field of medicine. Aside from that, there is a rich literature on Big Data projects related to healthcare. As much of this literature is controversial, I thought it important to select examples that I could document from reliable sources. Consequently, the reference section is large, with over 200 articles from journals, newspaper articles, and books. Most of these cited articles are available for free Web download.

Who should read this book? This book is written for professionals who manage Big Data resources and for students in the fields of computer science and informatics. Data management professionals would include the leadership within corporations and funding agencies who must commit resources to the project, the project directors who must determine a feasible set of goals and who must assemble a team of individuals who, in aggregate, hold the requisite skills for the task: network managers, data domain specialists, metadata specialists, software programmers, standards experts, interoperability experts, statisticians, data analysts, and representatives from the intended user community. Students of informatics, the computer sciences, and statistics will discover that the special challenges attached to Big Data, seldom discussed in university classes, are often surprising; sometimes shocking.

By mastering the fundamentals of Big Data design, maintenance, growth, and validation, readers will learn how to simplify the endless tasks engendered by Big Data resources. Adept analysts can find relationships among data objects held in disparate Big Data resources if the data is prepared properly. Readers will discover how integrating Big Data resources can deliver benefits far beyond anything attained from stand-alone databases.

References

[1] Martin Hilbert M, Lopez P. The world's technological capacity to store, communicate, and compute information. Science 2011;332:60–5.

[2] Schmidt S. Data is exploding: the 3V's of Big Data. Business Computing World; 2012. May 15.

1

Introduction

OUTLINE

Section 1.1. Definition of Big Data

It's the data, stupid.

Jim Gray

Back in the mid 1960s, my high school held pep rallies before big games. At one of these rallies, the head coach of the football team walked to the center of the stage carrying a large box of printed computer paper; each large sheet was folded flip-flop style against the next sheet and they were all held together by perforations. The coach announced that the athletic abilities of every member of our team had been entered into the school's computer (we were lucky enough to have our own IBM-360 mainframe). Likewise, data on our rival team had also been entered. The computer was instructed to digest all of this information and to produce the name of the team that would win the annual Thanksgiving Day showdown. The computer spewed forth the aforementioned box of computer paper; the very last output sheet revealed that we were the pre-ordained winners. The next day, we sallied forth to yet another ignominious defeat at the hands of our long-time rivals.

Fast-forward about 50 years to a conference room at the National Institutes of Health (NIH), in Bethesda, Maryland. A top-level science administrator is briefing me. She explains that disease research has grown in scale over the past decade. The very best research initiatives are now multi-institutional and data-intensive. Funded investigators are using high-throughput molecular methods that produce mountains of data for every tissue sample in a matter of minutes. There is only one solution; we must acquire supercomputers and a staff of talented programmers who can analyze all our data and tell us what it all means!

The NIH leadership believed, much as my high school coach believed, that if you have a really big computer and you feed it a huge amount of information, then you can answer almost any question.

Principles and Practice of Big Data. https://doi.org/10.1016/B978-0-12-815609-4.00001-7

1

That day, in the conference room at the NIH, circa 2003, I voiced my concerns, indicating that you cannot just throw data into a computer and expect answers to pop out. I pointed out that, historically, science has been a reductive process, moving from complex, descriptive data sets to simplified generalizations. The idea of developing an expensive supercomputer facility to work with increasing quantities of biological data, at higher and higher levels of complexity, seemed impractical and unnecessary. On that day, my concerns were not well received. High performance supercomputing was a very popular topic, and still is. [Glossary Science, Supercomputer]

Fifteen years have passed since the day that supercomputer-based cancer diagnosis was envisioned. The diagnostic supercomputer facility was never built. The primary diagnostic tool used in hospital laboratories is still the microscope, a tool invented circa 1590. Today, we augment microscopic findings with genetic tests for specific, key mutations; but we do not try to understand all of the complexities of human genetic variations. We know that it is hopeless to try. You can find a lot of computers in hospitals and medical offices, but the computers do not calculate your diagnosis. Computers in the medical workplace are relegated to the prosaic tasks of collecting, storing, retrieving, and delivering medical records. When those tasks are finished, the computer sends you the bill for services rendered.

Before we can take advantage of large and complex data sources, we need to think deeply about the meaning and destiny of Big Data.

Big Data is defined by the three V's:

1. Volume—large amounts of data;.
2. Variety—the data comes in different forms, including traditional databases, images, documents, and complex records;.
3. Velocity—the content of the data is constantly changing through the absorption of complementary data collections, the introduction of previously archived data or legacy collections, and from streamed data arriving from multiple sources.

It is important to distinguish Big Data from "lotsa data" or "massive data." In a Big Data Resource, all three V's must apply. It is the size, complexity, and restlessness of Big Data resources that account for the methods by which these resources are designed, operated, and analyzed. [Glossary Big Data resource, Data resource]

The term "lotsa data" is often applied to enormous collections of simple-format records. For example: every observed star, its magnitude and its location; the name and cell phone number of every person living in the United States; and the contents of the Web. These very large data sets are sometimes just glorified lists. Some "lotsa data" collections are spreadsheets (2-dimensional tables of columns and rows), so large that we may never see where they end.

Big Data resources are not equivalent to large spreadsheets, and a Big Data resource is never analyzed in its totality. Big Data analysis is a multi-step process whereby data is extracted, filtered, and transformed, with analysis often proceeding in a piecemeal, sometimes recursive, fashion. As you read this book, you will find that the gulf between "lotsa data" and Big Data is profound; the two subjects can seldom be discussed productively within the same venue.

Section 1.2. Big Data Versus Small Data

Actually, the main function of Big Science is to generate massive amounts of reliable and easily accessible data.... Insight, understanding, and scientific progress are generally achieved by 'small science.'
Dan Graur, Yichen Zheng, Nicholas Price, Ricardo Azevedo, Rebecca Zufall, and Eran Elhaik [1].

Big Data is not small data that has become bloated to the point that it can no longer fit on a spreadsheet, nor is it a database that happens to be very large. Nonetheless, some professionals who customarily work with relatively small data sets, harbor the false impression that they can apply their spreadsheet and database know-how directly to Big Data resources without attaining new skills or adjusting to new analytic paradigms. As they see things, when the data gets bigger, only the computer must adjust (by getting faster, acquiring more volatile memory, and increasing its storage capabilities); Big Data poses no special problems that a supercomputer could not solve. [Glossary Database]

This attitude, which seems to be prevalent among database managers, programmers, and statisticians, is highly counterproductive. It will lead to slow and ineffective software, huge investment losses, bad analyses, and the production of useless and irreversibly defective Big Data resources.

Let us look at a few of the general differences that can help distinguish Big Data and small data.

- **Goals**

small data—Usually designed to answer a specific question or serve a particular goal.

Big Data—Usually designed with a goal in mind, but the goal is flexible and the questions posed are protean. Here is a short, imaginary funding announcement for Big Data grants designed "to combine high quality data from fisheries, coast guard, commercial shipping, and coastal management agencies for a growing data collection that can be used to support a variety of governmental and commercial management studies in the Lower Peninsula." In this fictitious case, there is a vague goal, but it is obvious that there really is no way to completely specify what the Big Data resource will contain, how the various types of data held in the resource will be organized, connected to other data resources, or usefully analyzed. Nobody can specify, with any degree of confidence, the ultimate destiny of any Big Data project; it usually comes as a surprise.

- **Location**

small data—Typically, contained within one institution, often on one computer, sometimes in one file.

Big Data—Spread throughout electronic space and typically parceled onto multiple Internet servers, located anywhere on earth.

- **Data structure and content**

small data—Ordinarily contains highly structured data. The data domain is restricted to a single discipline or sub-discipline. The data often comes in the form of uniform records in an ordered spreadsheet.

Big Data—Must be capable of absorbing unstructured data (e.g., such as free-text documents, images, motion pictures, sound recordings, physical objects). The subject matter of the resource may cross multiple disciplines, and the individual data objects in the resource may link to data contained in other, seemingly unrelated, Big Data resources. [Glossary Data object]

– **Data preparation**

small data—In many cases, the data user prepares her own data, for her own purposes.

Big Data—The data comes from many diverse sources, and it is prepared by many people. The people who use the data are seldom the people who have prepared the data.

– **Longevity**

small data—When the data project ends, the data is kept for a limited time (seldom longer than 7 years, the traditional academic life-span for research data); and then discarded.

Big Data—Big Data projects typically contain data that must be stored in perpetuity. Ideally, the data stored in a Big Data resource will be absorbed into other data resources. Many Big Data projects extend into the future and the past (e.g., legacy data), accruing data prospectively and retrospectively. [Glossary Legacy data]

– **Measurements**

small data—Typically, the data is measured using one experimental protocol, and the data can be represented using one set of standard units. [Glossary Protocol]

Big Data—Many different types of data are delivered in many different electronic formats. Measurements, when present, may be obtained by many different protocols. Verifying the quality of Big Data is one of the most difficult tasks for data managers. [Glossary Data Quality Act]

– **Reproducibility**

small data—Projects are typically reproducible. If there is some question about the quality of the data, the reproducibility of the data, or the validity of the conclusions drawn from the data, the entire project can be repeated, yielding a new data set. [Glossary Conclusions]

Big Data—Replication of a Big Data project is seldom feasible. In general, the most that anyone can hope for is that bad data in a Big Data resource will be found and flagged as such.

– **Stakes**

small data—Project costs are limited. Laboratories and institutions can usually recover from the occasional small data failure.

Big Data—Big Data projects can be obscenely expensive [2,3]. A failed Big Data effort can lead to bankruptcy, institutional collapse, mass firings, and the sudden disintegration of all the data held in the resource. As an example, a United States National Institutes of Health Big Data project known as the "NCI cancer biomedical informatics grid" cost at least $350 million for fiscal years 2004–10. An ad hoc committee reviewing the resource found that despite the intense efforts of hundreds of cancer researchers and information specialists, it had accomplished so little and at so great an expense that a project moratorium was called [4]. Soon thereafter, the resource was terminated [5]. Though the costs of failure can be high, in terms of money, time, and labor, Big Data failures may have some redeeming value. Each failed effort lives on as intellectual remnants consumed by the next Big Data effort. [Glossary Grid]

– **Introspection**

small data—Individual data points are identified by their row and column location within a spreadsheet or database table. If you know the row and column headers, you can find and specify all of the data points contained within. [Glossary Data point]

Big Data—Unless the Big Data resource is exceptionally well designed, the contents and organization of the resource can be inscrutable, even to the data managers. Complete access to data, information about the data values, and information about the organization of the data is achieved through a technique herein referred to as introspection. Introspection will be discussed at length in Chapter 6. [Glossary Data manager, Introspection]

– **Analysis**

small data—In most instances, all of the data contained in the data project can be analyzed together, and all at once.

Big Data—With few exceptions, such as those conducted on supercomputers or in parallel on multiple computers, Big Data is ordinarily analyzed in incremental steps. The data are extracted, reviewed, reduced, normalized, transformed, visualized, interpreted, and re-analyzed using a collection of specialized methods. [Glossary Parallel computing, MapReduce]

Section 1.3. Whence Comest Big Data?

All I ever wanted to do was to paint sunlight on the side of a house.

Edward Hopper

Often, the impetus for Big Data is entirely ad hoc. Companies and agencies are forced to store and retrieve huge amounts of collected data (whether they want to or not). Generally, Big Data come into existence through any of several different mechanisms:

– An entity has collected a lot of data in the course of its normal activities and seeks to organize the data so that materials can be retrieved, as needed.

The Big Data effort is intended to streamline the regular activities of the entity. In this case, the data is just waiting to be used. The entity is not looking to discover anything or to do anything new. It simply wants to use the data to accomplish what it has always been doing;

only better. The typical medical center is a good example of an "accidental" Big Data resource. The day-to-day activities of caring for patients and recording data into hospital information systems results in terabytes of collected data, in forms such as laboratory reports, pharmacy orders, clinical encounters, and billing data. Most of this information is generated for a one-time specific use (e.g., supporting a clinical decision, collecting payment for a procedure). It occurs to the administrative staff that the collected data can be used, in its totality, to achieve mandated goals: improving quality of service, increasing staff efficiency, and reducing operational costs. [Glossary Binary units for Big Data, Binary atom count of universe]

- An entity has collected a lot of data in the course of its normal activities and decides that there are many new activities that could be supported by their data.

Consider modern corporations; these entities do not restrict themselves to one manufacturing process or one target audience. They are constantly looking for new opportunities. Their collected data may enable them to develop new products based on the preferences of their loyal customers, to reach new markets, or to market and distribute items via the Web. These entities will become hybrid Big Data/manufacturing enterprises.

- An entity plans a business model based on a Big Data resource.

Unlike the previous examples, this entity starts with Big Data and adds a physical component secondarily. Amazon and FedEx may fall into this category, as they began with a plan for providing a data-intense service (e.g., the Amazon Web catalog and the FedEx package tracking system). The traditional tasks of warehousing, inventory, pick-up, and delivery, had been available all along, but lacked the novelty and efficiency afforded by Big Data.

- An entity is part of a group of entities that have large data resources, all of whom understand that it would be to their mutual advantage to federate their data resources [6].

An example of a federated Big Data resource would be hospital databases that share electronic medical health records [7].

- An entity with skills and vision develops a project wherein large amounts of data are collected and organized, to the benefit of themselves and their user-clients.

An example would be a massive online library service, such as the U.S. National Library of Medicine's PubMed catalog, or the Google Books collection.

- An entity has no data and has no particular expertise in Big Data technologies, but it has money and vision.

The entity seeks to fund and coordinate a group of data creators and data holders, who will build a Big Data resource that can be used by others. Government agencies have been the major benefactors. These Big Data projects are justified if they lead to important discoveries that could not be attained at a lesser cost with smaller data resources.

Section 1.4. The Most Common Purpose of Big Data Is to Produce Small Data

If I had known what it would be like to have it all, I might have been willing to settle for less.

Lily Tomlin

Imagine using a restaurant locater on your smartphone. With a few taps, it lists the Italian restaurants located within a 10-block radius of your current location. The database being queried is big and complex (a map database, a collection of all the restaurants in the world, their longitudes and latitudes, their street addresses, and a set of ratings provided by patrons, updated continuously), but the data that it yields is small (e.g., five restaurants, marked on a street map, with pop-ups indicating their exact address, telephone number, and ratings). Your task comes down to selecting one restaurant from among the five, and dining thereat.

In this example, your data selection was drawn from a large data set, but your ultimate analysis was confined to a small data set (i.e., five restaurants meeting your search criteria). The purpose of the Big Data resource was to proffer the small data set. No analytic work was performed on the Big Data resource; just search and retrieval. The real labor of the Big Data resource involved collecting and organizing complex data, so that the resource would be ready for your query. Along the way, the data creators had many decisions to make (e.g., Should bars be counted as restaurants? What about take-away only shops? What data should be collected? How should missing data be handled? How will data be kept current? [Glossary Query, Missing data]

Big Data is seldom, if ever, analyzed *in toto*. There is almost always a drastic filtering process that reduces Big Data into smaller data. This rule applies to scientific analyses. The Australian Square Kilometre Array of radio telescopes [8], WorldWide Telescope, CERN's Large Hadron Collider and the Pan-STARRS (Panoramic Survey Telescope and Rapid Response System) array of telescopes produce petabytes of data every day. Researchers use these raw data sources to produce much smaller data sets for analysis [9]. [Glossary Raw data, Square Kilometer Array, Large Hadron Collider, World-Wide Telescope]

Here is an example showing how workable subsets of data are prepared from Big Data resources. Blazars are rare super-massive black holes that release jets of energy that move at near-light speeds. Cosmologists want to know as much as they can about these strange objects. A first step to studying blazars is to locate as many of these objects as possible. Afterwards, various measurements on all of the collected blazars can be compared, and their general characteristics can be determined. Blazars seem to have a gamma ray signature that is not present in other celestial objects. The WISE survey collected infrared data on the entire observable universe. Researchers extracted from the Wise data every celestial body associated with an infrared signature in the gamma ray range that was suggestive of blazars; about 300 objects. Further research on these 300 objects led the researchers to believe that about half were blazars [10]. This is how Big Data research often works; by constructing small data sets that can be productively analyzed.

Because a common role of Big Data is to produce small data, a question that data managers must ask themselves is: "Have I prepared my Big Data resource in a manner that helps it become a useful source of small data?"

Section 1.5. Big Data Sits at the Center of the Research Universe

Physics is the universe's operating system.

Steven R Garman

In the past, scientists followed a well-trodden path toward truth: hypothesis, then experiment, then data, then analysis, then publication. The manner in which a scientist analyzed his or her data was crucial because other scientists would not have access to the same data and could not re-analyze the data for themselves. Basically, the results and conclusions described in the manuscript was the scientific product. The primary data upon which the results and conclusion were based (other than one or two summarizing tables) were not made available for review. Scientific knowledge was built on trust. Customarily, the data would be held for 7 years, and then discarded. [Glossary Results]

In the Big data paradigm the concept of a final manuscript has little meaning. Big Data resources are permanent, and the data within the resource is immutable (See Chapter 6). Any scientist's analysis of the data does not need to be the final word; another scientist can access and re-analyze the same data over and over again. Original conclusions can be validated or discredited. New conclusions can be developed. The centerpiece of science has moved from the manuscript, whose conclusions are tentative until validated, to the Big Data resource, whose data will be tapped repeatedly to validate old manuscripts and spawn new manuscripts. [Glossary Immutability, Mutability]

Today, hundreds or thousands of individuals might contribute to a Big Data resource. The data in the resource might inspire dozens of major scientific projects, hundreds of manuscripts, thousands of analytic efforts, and millions or billions of search and retrieval operations. The Big Data resource has become the central, massive object around which universities, research laboratories, corporations, and federal agencies orbit. These orbiting objects draw information from the Big Data resource, and they use the information to support analytic studies and to publish manuscripts. Because Big Data resources are permanent, any analysis can be critically examined using the same set of data, or re-analyzed anytime in the future. Because Big Data resources are constantly growing forward in time (i.e., accruing new information) and backward in time (i.e., absorbing legacy data sets), the value of the data is constantly increasing.

Big Data resources are the stars of the modern information universe. All matter in the physical universe comes from heavy elements created inside stars, from lighter elements. All data in the informational universe is complex data built from simple data. Just as stars can exhaust themselves, explode, or even collapse under their own weight to become black holes; Big Data resources can lose funding and die, release their contents and burst into nothingness, or collapse under their own weight, sucking everything around them into a dark void. It is an interesting metaphor. In the following chapters, we will see how a Big Data resource can be designed and operated to ensure stability, utility, growth, and permanence; features you might expect to find in a massive object located in the center of the information universe.

Glossary

Big Data resource A Big Data collection that is accessible for analysis. Readers should understand that there are collections of Big Data (i.e., data sources that are large, complex, and actively growing) that are not designed to support analysis; hence, not Big Data resources. Such Big Data collections might include some of the older hospital information systems, which were designed to deliver individual patient records upon request; but could not support projects wherein all of the data contained in all of the records were opened for selection and analysis. Aside from privacy and security issues, opening a hospital information system to these kinds of analyses would place enormous computational stress on the systems (i.e., produce system crashes). In the late 1990s and the early 2000s data warehousing was popular. Large organizations would collect all of the digital information created within their institutions, and these data were stored as Big Data collections, called data warehouses. If an authorized person within the institution needed some specific set of information (e.g., emails sent or received in February, 2003; all of the bills paid in November, 1999), it could be found somewhere within the warehouse. For the most part, these data warehouses were not true Big Data resources because they were not organized to support a full analysis of all of the contained data. Another type of Big Data collection that may or may not be considered a Big Data resource are compilations of scientific data that are accessible for analysis by private concerns, but closed for analysis by the public. In this case a scientist may make a discovery based on her analysis of a private Big Data collection, but the research data is not open for critical review. In the opinion of some scientists, including myself, if the results of a data analysis are not available for review, then the analysis is illegitimate. Of course, this opinion is not universally shared, and Big Data professionals hold various definitions for a Big Data resource.

Binary atom count of universe There are estimated to be about 10^{80} atoms in the universe. Log2(10) is 3.32192809, so the number of atoms in the universe is $2^{80*3.32192809}$ or 2^{266} atoms.

Binary units for Big Data Binary sizes are named in 1000-fold intervals: 1 bit = binary digit (0 or 1); 1 byte = 8 bits (the number of bits required to express an ascii character); 1000 bytes = 1 kilobyte; 1000 kilobytes = 1 megabyte; 1000 megabytes = 1 gigabyte; 1000 gigabytes = 1 terabyte; 1000 terabytes = 1 petabyte; 1000 petabytes = 1 exabyte; 1000 exabytes = 1 zettabyte; 1000 zettabytes = 1 yottabyte.

Conclusions Conclusions are the interpretations made by studying the results of an experiment or a set of observations. The term "results" should never be used interchangeably with the term "conclusions." **Remember, results are verified. Conclusions are validated** [11].

Data Quality Act In the United States the data upon which public policy is based must have quality and must be available for review by the public. Simply put, public policy must be based on verifiable data. The Data Quality Act of 2002 requires the Office of Management and Budget to develop government-wide standards for data quality [12].

Data manager This book uses "data manager" as a catchall term, without attaching any specific meaning to the name. Depending on the institutional and cultural milieu, synonyms and plesionyms (i.e., near-synonyms) for data manager would include: technical lead, team liaison, data quality manager, chief curator, chief of operations, project manager, group supervisor, and so on.

Data object As used in this book, a data object consists of a unique object identifier along with all of the data/metadata pairs that rightly belong to the object identifier, and that includes one data/metadata pair that tells us the object's class.

```
75898039563441
   name              G. Willikers
   gender            male
   age               35
   is_a_class_member cowboy
```

In this example, the object identifier, 75898039563441, is followed by its data/metadata pairs, including the one pair that tells us that the object (a 35-year-old man named G. Willikers) belongs to the class of individuals known as "cowboy."

The utility of data objects, in the field of Big Data, is discussed in Section 6.2.

Data point The singular form of data is datum. Strictly speaking, the term should be datum point or datumpoint. Most information scientists, myself included, have abandoned consistent usage rules for the word "data." In this book, the term "data" always refers collectively to information, numeric or textual, structured or unstructured, in any quantity.

Data resource A collection of data made available for data retrieval. The data can be distributed over servers located anywhere on earth or in space. The resource can be static (i.e., having a fixed set of data), or in flux. Plesionyms for data resource are: data warehouse, data repository, data archive, and data store.

Database A software application designed specifically to create and retrieve large numbers of data records (e.g., millions or billions). The data records of a database are persistent, meaning that the application can be turned off, then on, and all the collected data will be available to the user.

Grid A collection of computers and computer resources (typically networked servers) that is coordinated to provide a desired functionality. In the most advanced Grid computing architecture, requests can be broken into computational tasks that are processed in parallel on multiple computers and transparently (from the client's perspective) assembled and returned. The Grid is the intellectual predecessor of Cloud computing. Cloud computing is less physically and administratively restricted than Grid computing.

Immutability Immutability is the principle that data collected in a Big Data resource is permanent and can never be modified. At first thought, it would seem that immutability is a ridiculous and impossible constraint. In the real world, mistakes are made, information changes, and the methods for describing information changes. This is all true, but the astute Big Data manager knows how to accrue information into data objects without changing the pre-existing data. Methods for achieving this seemingly impossible trick are described in Chapter 8.

Introspection Well-designed Big Data resources support introspection, a method whereby data objects within the resource can be interrogated to yield their properties, values, and class membership. Through introspection the relationships among the data objects in the Big Data resource can be examined and the structure of the resource can be determined. Introspection is the method by which a data user can find everything there is to know about a Big Data resource without downloading the complete resource.

Large Hadron Collider The Large Hadron Collider is the world's largest and most powerful particle accelerator and is expected to produce about 15 petabytes (15 million gigabytes) of data annually [13].

Legacy data Data collected by an information system that has been replaced by a newer system, and which cannot be immediately integrated into the newer system's database. For example, hospitals regularly replace their hospital information systems with new systems that promise greater efficiencies, expanded services, or improved interoperability with other information systems. In many cases, the new system cannot readily integrate the data collected from the older system. The previously collected data becomes a legacy to the new system. In such cases, legacy data is simply "stored" for some arbitrary period of time in case someone actually needs to retrieve any of the legacy data. After a decade or so the hospital may find itself without any staff members who are capable of locating the storage site of the legacy data, or moving the data into a modern operating system, or interpreting the stored data, or retrieving appropriate data records, or producing a usable query output.

MapReduce A method by which computationally intensive problems can be processed on multiple computers, in parallel. The method can be divided into a mapping step and a reducing step. In the mapping step a master computer divides a problem into smaller problems that are distributed to other computers. In the reducing step the master computer collects the output from the other computers. Although MapReduce is intended for Big Data resources, and can hold petabytes of data, most Big Data problems do not require MapReduce.

Missing data Most complex data sets have missing data values. Somewhere along the line data elements were not entered, records were lost, or some systemic error produced empty data fields. Big Data, being large, complex, and composed of data objects collected from diverse sources, is almost certain to have missing data. Various mathematical approaches to missing data have been developed; commonly involving assigning values on a statistical basis; so-called imputation methods. The underlying assumption for such methods is that missing data arises at random. When missing data arises nonrandomly, there is no satisfactory statistical fix. The Big Data curator must track down the source of the errors and somehow rectify the situation. In either case the issue of missing data introduces a potential bias and it is crucial to fully document the method by which missing data is handled. In the realm of clinical trials, only a minority of data analyses bothers to describe their chosen method for handling missing data [14].

Mutability Mutability refers to the ability to alter the data held in a data object or to change the identity of a data object. Serious Big Data is not mutable. Data can be added, but data cannot be erased or altered. Big Data resources that are mutable cannot establish a sensible data identification system, and cannot support verification and validation activities. The legitimate ways in which we can record the changes that occur in unique data objects (e.g., humans) over time, without ever changing the key/value data attached to the unique object, is discussed in Section 8.2.

For programmers, it is important to distinguish data mutability from object mutability, as it applies in Python and other object-oriented programming languages. Python has two immutable objects: strings and tuples. Intuitively, we would probably guess that the contents of a string object cannot be changed, and the contents of a tuple object cannot be changed. This is not the case. Immutability, for programmers, means that there are no methods available to the object by which the contents of the object can be altered. Specifically, a Python tuple object would have no methods it could call to change its own contents. However, a tuple may contain a list, and lists are mutable. For example, a list may have an append method that will add an item to the list object. You can change the contents of a list contained in a tuple object without violating the tuple's immutability.

Parallel computing Some computational tasks can be broken down and distributed to other computers, to be calculated "in parallel." The method of parallel programming allows a collection of desktop computers to complete intensive calculations of the sort that would ordinarily require the aid of a supercomputer. Parallel programming has been studied as a practical way to deal with the higher computational demands brought by Big Data. Although there are many important problems that require parallel computing, the vast majority of Big Data analyses can be easily accomplished with a single, off-the-shelf personal computer.

Protocol A set of instructions, policies, or fully described procedures for accomplishing a service, operation, or task. Protocols are fundamental to Big Data. Data is generated and collected according to protocols. There are protocols for conducting experiments, and there are protocols for measuring the results. There are protocols for choosing the human subjects included in a clinical trial, and there are protocols for interacting with the human subjects during the course of the trial. All network communications are conducted via protocols; the Internet operates under a protocol (TCP-IP, Transmission Control Protocol-Internet Protocol).

Query The term "query" usually refers to a request, sent to a database, for information (e.g., Web pages, documents, lines of text, images) that matches a provided word or phrase (i.e., the query term). More generally a query is a parameter or set of parameters that are submitted as input to a computer program that searches a data collection for items that match or bear some relationship to the query parameters. In the context of Big Data the user may need to find classes of objects that have properties relevant to a particular area of interest. In this case, the query is basically introspective, and the output may yield metadata describing individual objects, classes of objects, or the relationships among objects that share particular properties. For example, "weight" may be a property, and this property may fall into the domain of several different classes of data objects. The user might want to know the names of the classes of objects that have the "weight" property and the numbers of object instances in each class. Eventually the user might want to select several of these classes (e.g., including dogs and cats, but excluding microwave ovens) along with the data object instances whose weights fall within a specified range (e.g., 20–30 pound). This approach to querying could work with any data set that has been well specified with metadata, but it is particularly important when using Big Data resources.

Raw data Raw data is the unprocessed, original data measurement, coming straight from the instrument to the database with no intervening interference or modification. In reality, scientists seldom, if ever, work with raw data. When an instrument registers the amount of fluorescence emitted by a hybridization spot on a gene array, or the concentration of sodium in the blood, or virtually any of the measurements that we receive as numeric quantities, the output is produced by an algorithm executed by the measurement instrument. Pre-processing of data is commonplace in the universe of Big Data, and data managers should not labor under the false impression that the data received is "raw," simply because the data has not been modified by the person who submits the data.

Results The term "results" is often confused with the term "conclusions." Interchanging the two concepts is a source of confusion among data scientists. In the strictest sense, "results" consist of the full set of experimental data collected by measurements. In practice, "results" are provided as a small subset of data distilled from the raw, original data. In a typical journal article, selected data subsets are packaged as a chart or graph that emphasizes some point of interest. Hence, the term "results" may refer, erroneously, to subsets of the original data, or to visual graphics intended to summarize the original data. Conclusions are the inferences drawn from the results. Results are verified; conclusions are validated.

Science Of course, there are many different definitions of science, and inquisitive students should be encouraged to find a conceptualization of science that suits their own intellectual development. For me, science is all about finding general relationships among objects. In the so-called physical sciences the most important relationships are expressed as mathematical equations (e.g., the relationship between force, mass and acceleration; the relationship between voltage, current and resistance). In the so-called natural sciences, relationships are often expressed through classifications (e.g., the classification of living organisms). Scientific advancement is the discovery of new relationships or the discovery of a generalization that applies to objects hitherto confined within disparate scientific realms (e.g., evolutionary theory arising from observations of organisms and geologic strata). Engineering would be the area of science wherein scientific relationships are exploited to build new technology.

Square Kilometer Array The Square Kilometer Array is designed to collect data from millions of connected radio telescopes and is expected to produce more than one exabyte (1 billion gigabytes) every day [8].

Supercomputer Computers that can perform many times faster than a desktop personal computer. In 2015 the top supercomputers operate at about 30 petaflops. A petaflop is 10 to the 15 power floating point operations per second. By my calculations a 1 petaflop computer performs about 250,000 operations in the time required for my laptop to finish one operation.

WorldWide Telescope A Big Data effort from the Microsoft Corporation bringing astronomical maps, imagery, data, analytic methods, and visualization technology to standard Web browsers. More information is available at: http://www.worldwidetelescope.org/Home.aspx

References

[1] Graur D, Zheng Y, Price N, Azevedo RB, Zufall RA, Elhaik E. On the immortality of television sets: "function" in the human genome according to the evolution-free gospel of ENCODE. Genome Biol Evol 2013;5:578–90.

[2] Whittaker Z. UK's delayed national health IT programme officially scrapped. ZDNet, September 22, 2011.

[3] Kappelman LA, McKeeman R, Lixuan Zhang L. Early warning signs of IT project failure: the dominant dozen. Information Systems Management 2006;23:31–6.

[4] An assessment of the impact of the NCI cancer Biomedical Informatics Grid (caBIG). Report of the Board of Scientific Advisors Ad Hoc Working Group. National Cancer Institute; March 2011.

[5] Komatsoulis GA. Program announcement to the CaBIG community. National Cancer Institute. https://cabig.nci.nih.gov/program_announcement [viewed August 31, 2012].

[6] Freitas A, Curry E, Oliveira JG, O'Riain S. Querying heterogeneous datasets on the linked data web: challenges, approaches, and trends. IEEE Internet Computing 2012;16:24–33.

[7] Drake TA, Braun J, Marchevsky A, Kohane IS, Fletcher C, Chueh H, et al. A system for sharing routine surgical pathology specimens across institutions: the Shared Pathology Informatics Network (SPIN). Hum Pathol 2007;38:1212–25.

[8] Francis M. Future telescope array drives development of exabyte processing. Ars Technica; 2012. April 2.

[9] Markoff J. A deluge of data shapes a new era in computing. The New York Times; 2009. December 15.

[10] Harrington JD, Clavin W. NASA's WISE mission sees skies ablaze with blazars. NASA Release 12-109; 2002. April 12.

[11] Committee on Mathematical Foundations of Verification, Validation, and Uncertainty Quantification; Board on Mathematical Sciences and Their Applications, Division on Engineering and Physical Sciences, National Research Council. Assessing the reliability of complex models: mathematical and statistical foundations of verification, validation, and uncertainty quantification. National Academy Press; 2012. Available from: http://www.nap.edu/catalog.php?record_id=13395. [viewed January 1, 2015].

[12] Data Quality Act. 67 Fed. Reg. 8,452, February 22, 2002, addition to FY 2001 Consolidated Appropriations Act (Pub. L. No. 106-554. Codified at 44 U.S.C. 3516).

[13] Worldwide LHC Computing Grid. European Organization for Nuclear Research. Available from: http://public.web.cern.ch/public/en/lhc/Computing-en.html; 2008 [viewed September 19, 2012].

[14] Carpenter JR, Kenward MG. Missing data in randomised control trials: a practical guide. November 21. Available from:http://www.hta.nhs.uk/nihrmethodology/reports/1589.pdf; 2007.

2

Providing Structure to Unstructured Data

Section 2.1. Nearly All Data Is Unstructured and Unusable in Its Raw Form

I was working on the proof of one of my poems all the morning, and took out a comma. In the afternoon I put it back again.

Oscar Wilde

In the early days of computing, data was always highly structured. All data was divided into fields, the fields had a fixed length, and the data entered into each field was constrained to a pre-determined set of allowed values. Data was entered into punch cards with pre-configured rows and columns. Depending on the intended use of the cards, various entry and read-out methods were chosen to express binary data, numeric data, fixed-size text, or programming instructions. Key-punch operators produced mountains of punch cards. For many analytic purposes, card-encoded data sets were analyzed without the assistance of a computer; all that was needed was a punch card sorter. If you wanted the data card on all males, over the age of 18, who had graduated high school, and had passed their physical exam, then the sorter would need to make 4 passes. The sorter would pull every card listing a male, then from the male cards it would pull all the cards of people over the age of 18, and from this double-sorted sub-stack, it would pull cards that met the next criterion, and so on.

Principles and Practice of Big Data. https://doi.org/10.1016/B978-0-12-815609-4.00002-9

As a high school student in the 1960s, I loved playing with the card sorters. Back then, all data was structured data, and it seemed to me, at the time, that a punch-card sorter was all that anyone would ever need to analyze large sets of data. [Glossary Binary data]

How wrong I was! Today, most data entered by humans is unstructured in the form of free-text. The free-text comes in email messages, tweets, and documents. Structured data has not disappeared, but it sits in the shadows cast by mountains of unstructured text. Free-text may be more interesting to read than punch cards, but the venerable punch card, in its heyday, was much easier to analyze than its free-text descendant. To get much informational value from free-text, it is necessary to impose some structure. This may involve translating the text to a preferred language; parsing the text into sentences; extracting and normalizing the conceptual terms contained in the sentences; mapping terms to a standard nomenclature; annotating the terms with codes from one or more standard nomenclatures; extracting and standardizing data values from the text; assigning data values to specific classes of data belonging to a classification system; assigning the classified data to a storage and retrieval system (e.g., a database); and indexing the data in the system. All of these activities are difficult to do on a small scale and virtually impossible to do on a large scale. Nonetheless, every Big Data project that uses unstructured data must deal with these tasks to yield the best possible results with the resources available. [Glossary Parsing, Nomenclature, Nomenclature mapping, Thesaurus, Indexes, Plain-text]

Section 2.2. Concordances

The limits of my language are the limits of my mind. All I know is what I have words for. (Die Grenzen meiner Sprache bedeuten die Grenzen meiner Welt.)
                                                                    Ludwig Wittgenstein

A concordance is a list of all the different words contained in a text with the locations in the text where each word appears. Concordances have been around for a very long time, painstakingly constructed from holy scriptures thought to be of such immense value that every word deserved special attention. Creating a concordance has always been a straightforward operation. You take the first word in the text and you note its location (i.e., word 1, page 1); then onto the second word (word 2 page 1), and so on. When you come to a word that has been included in the nascent concordance, you add its location to the existing entry for the word. Continuing thusly, for a few months or so, you end up with a concordance that you can be proud of. Today a concordance for the Bible can be constructed in a small fraction of a second. [Glossary Concordance]

Without the benefit of any special analyses, skimming through a book's concordance provides a fairly good idea of the following:

- The topic of the text based on the words appearing in the concordance. For example, a concordance listing multiple locations for "begat" and "anointed" and "thy" is most likely to be the Old Testament.

- The complexity of the language. A complex or scholarly text will have a larger vocabulary than a romance novel.
- A precise idea of the length of the text, achieved by adding all of the occurrences of each of the words in the concordance. Knowing the number of items in the concordance, multiplied by the average number of locations of concordance items, provides a rough estimate of the total number of words in the text.
- The care with which the text was prepared, achieved by counting the misspelled words.

Here, in a short Python script, concord_gettysbu.py, that builds a concordance for the Gettysburg address, located in the external file "gettysbu.txt": [Glossary Script]

```python
import re, string
word_list=[];word_dict={};key_list=[]
count=0; word=""
in_text_string = open('gettysbu.txt', "r").read().lower()
word_list = re.split(r'[^a-zA-z\_\-]+',in_text_string)
for word in word_list:
    count = count + 1
    if word in word_dict:
      word_dict[word] = word_dict[word] + ',' + str(count)
    else:
      word_dict[word] = str(count)
key_list = list(word_dict)
key_list.sort()
for key in key_list:
    print(key + " " + word_dict[key])
```

The first few lines of output are shown:

```
a 14,36,59,70,76,104,243
above 131
add 136
advanced 185
ago 6
all 26
altogether 93
and 3,20,49,95,122,248
any 45
are 28,33,56
as 75
battlefield 61
be 168,192
before 200
birth 245
```

```
brave 119
brought 9
but 102,151
by 254
can 52,153
cannot 108,111,114
```

The numbers that follow each item in the concordance correspond to the locations (expressed as the nth words of the Gettysburg address) of each word in the text.

At this point, building a concordance may appear to an easy, but somewhat pointless exercise. Does the concordance provide any functionality beyond that provided by the ubiquitous "search" box. There are five very useful properties of concordances that you might not have anticipated.

- You can use a concordance to rapidly search and retrieve the locations where single-word terms appear.
- You can always reconstruct the original text from the concordance. Hence, after you've built your concordance, you can discard the original text.
- You can merge concordances without forfeiting your ability to reconstruct the original texts, provided that you tag locations with some character sequence that identifies the text of origin.
- With a little effort a dictionary can be transformed into a universal concordance (i.e., a merged dictionary/concordance of every book in existence) by attaching the book identifier and its concordance entries to the corresponding dictionary terms.
- You can easily find the co-locations among words (i.e., which words often precede or follow one another).
- You can use the concordance to retrieve the sentences and paragraphs in which a search word or a search term appears, without having access to the original text. The concordance alone can reconstruct and retrieve the appropriate segments of text, on-the-fly, thus bypassing the need to search the original text.
- A concordance provides a profile of the book and can be used to compute a similarity score among different books.

There is insufficient room to explore all of the useful properties of concordances, but let us examine a script, concord_reverse.py, that reconstructs the original text, in lowercase, from the concordance. In this case, we have pasted the output from the concord_get-tysbu.py script (vida supra) into the external file, "concordance.txt".

```
import re, string
concordance_hash = {} ; location_array = []
in_text = open('concordance.txt', "r")
for line in in_text:
    line = line.replace("\n","")
    location_word, separator, location_positions = line.partition(" ")
    location_array = location_positions.split(",")
```

```
   location_array = [int(x) for x in location_array]
   for location in location_array:
      concordance_hash[location] = location_word
for n in range(300):
   if n in concordance_hash:
      print((concordance_hash[n]), end = " ")
```

Here is the familiar output:

> *four score and seven years ago our fathers brought forth on this continent a new nation conceived in liberty and dedicated to the proposition that all men are created equal now we are engaged in a great civil war testing whether that nation or any nation so conceived and so dedicated can long endure we are met on a great battlefield of that war we have come to dedicate a portion of that field as a final resting-place for those who here gave their lives that that nation might live it is altogether fitting and proper that we should do this but in a larger sense we cannot dedicate we cannot consecrate we cannot hallow this ground the brave men living and dead who struggled here have consecrated it far above our poor power to add or detract the world will little note nor long remember what we say here but it can never forget what they did here it is for us the living rather to be dedicated here to the unfinished work which they who fought here have thus far so nobly advanced it is rather for us to be here dedicated to the great task remaining before us–that from these honored dead we take increased devotion to that cause for which they gave the last full measure of devotion–that we here highly resolve that these dead shall not have died in vain that this nation under god shall have a new birth of freedom and that government of the people by the people for the people shall not perish from the earth*

Had we wanted to write a script that produces a merged concordance, for multiple documents, we could have simply written a loop that repeated the concordance-building process for each text. Within the loop, we would have tagged each word location with a short notation indicating the particular source book. For example, locations from the Gettysburg address could have been prepended with "G:" and locations from the Bible might have been prepended with a "B:".

We have not finished with the topic of concordances. Later in this chapter (Section 2.8), we will show how concordances can be transformed to speed-up search and retrieval operations on large bodies of text.

Section 2.3. Term Extraction

> *There's a big difference between knowing the name of something and knowing something.*
>
> *Richard Feynman*

One of my favorite movies is the parody version of "Hound of the Baskervilles," starring Peter Cooke as Sherlock Holmes and Dudley Moore as his faithful hagiographer, Dr. Watson. Sherlock, preoccupied with his own ridiculous pursuits, dispatches Watson to the Baskerville family manse, in Dartmoor, to undertake urgent sleuth-related activities. The hapless Watson, standing in the great Baskerville Hall, has no idea how to proceed with the investigation. After a moment of hesitation, he turns to the incurious maid and commands, "Take me to the clues!"

Building an index is a lot like solving a fiendish crime; you need to know how to find the clues. For informaticians, the terms in the text are the clues upon which the index is built. Terms in a text file do not jump into your index file; you need to find them. There are several available methods for finding and extracting index terms from a corpus of text [1], but no method is as simple, fast, and scalable as the "stop word" method [2]. [Glossary Term extraction algorithm, Scalable]

The "stop word" method presumes that text is composed of terms that are somehow connected into sequences known as sentences. [Glossary Sentence]

Consider the following:

The diagnosis is chronic viral hepatitis.

This sentence contains two very specific medical concepts: "diagnosis" and "chronic viral hepatitis." These two concepts are connected to form a sentence, using grammatical bric-a-brac such as "the" and "is", and the sentence delimiter, ".". These grammatical bric-a-brac are found liberally sprinkled in every paragraph you are likely to read.

A term can be defined as a sequence of one or more uncommon words that are demarcated (i.e., bounded on one side or another) by the occurrence of one or more very common words (e.g., "and", "the", "a", "of") and phrase delimiters (e.g., ".", ",", and ";")

Consider the following:

An epidural hemorrhage can occur after a lucid interval.

The medical concepts "epidural hemorrhage" and "lucid interval" are composed of uncommon words. These uncommon word sequences are bounded by common words (i.e., "the", "an", "can", "a") or a sentence delimiter (i.e., ".").

If we had a list of all the words that were considered common, we could write a program that extracts the all the concepts found in any text of any length. The concept terms would consist of all sequences of uncommon words that are uninterrupted by common words. Here is an algorithm for extracting terms from a sentence:

1. Read the first word of the sentence. If it is a common word, delete it. If it is an uncommon word, save it.
2. Read the next word. If it is a common word, delete it, and place the saved word (from the prior step, if the prior step saved a word) into our list of terms found in the text. If it

is an uncommon word, concatenate it with the word we saved in step one, and save the 2-word term. If it is a sentence delimiter, place any saved term into our list of terms, and stop the program.

3. Repeat step two.

This simple algorithm, or something much like it, is a fast and efficient method to build a collection of index terms. The following list of common words might be useful: "about, again, all, almost, also, although, always, among, an, and, another, any, are, as, at, be, because, been, before, being, between, both, but, by, can, could, did, do, does, done, due, during, each, either, enough, especially, etc, for, found, from, further, had, has, have, having, here, how, however, i, if, in, into, is, it, its, itself, just, kg, km, made, mainly, make, may, mg, might, ml, mm, most, mostly, must, nearly, neither, no, nor, obtained, of, often, on, our, overall, perhaps, pmid, quite, rather, really, regarding, seem, seen, several, should, show, showed, shown, shows, significantly, since, so, some, such, than, that, the, their, theirs, them, then, there, therefore, these, they, this, those, through, thus, to, upon, use, used, using, various, very, was, we, were, what, when, which, while, with, within, without, would."

Such lists of common words are sometimes referred to as "stop word" lists or "barrier word" lists, as they demarcate the beginnings and endings of extraction terms. Let us look at a short Python script (terms.py) that uses our list of stop words (contained in the file stop.txt) and extracts the terms from the sentence: "Once you have a method for extracting terms from sentences the task of creating an index associating a list of locations with each term is child's play for programmers"

```
import re, string
stopfile = open("stop.txt",'r')
stop_list = stopfile.readlines()
stopfile.close()
item_list = []
line = "Once you have a method for extracting terms from \
sentences the task of creating an index associating a list \
of locations with each term is child's play for programmers"
for stopword in stop_list:
    stopword = re.sub(r'\n', ", stopword)
    line = re.sub(r' *\b' + stopword + r'\b *', '\n', line)
item_list.extend(line.split("\n"))
item_list = sorted(set(item_list))
for item in item_list:
    print(item)
```

Here is the output:

```
Once
child's play
creating
```

```
extracting terms
index associating
list
locations
method
programmers
sentences
task
term
```

Extracting terms is the first step in building a very crude index. Indexes built directly from term extraction algorithms always contain lots of unnecessary terms having little or no informational value. For serious indexers, the collection of terms extracted from a corpus, along with their locations in the text, is just the beginning of an intellectual process that will eventually lead to a valuable index.

Section 2.4. Indexing

Knowledge can be public, yet undiscovered, if independently created fragments are logically related but never retrieved, brought together, and interpreted.

Donald R. Swanson [3]

Individuals accustomed to electronic media tend to think of the Index as an inefficient or obsolete method for finding and retrieving information. Most currently available e-books have no index. It is far easier to pull up the "Find" dialog box and enter a word or phrase. The e-reader can find all matches quickly, providing the total number of matches, and bringing the reader to any or all of the pages containing the selection. As more and more books are published electronically, the book Index, as we have come to know it, may cease to be.

It would be a pity if indexes were to be abandoned by computer scientists. A well-designed book index is a creative, literary work that captures the content and intent of the book and transforms it into a listing wherein related concepts are collected under common terms, and keyed to their locations. It saddens me that many people ignore the book index until they want something from it. Open a favorite book and read the index, from A to Z, as if you were reading the body of the text. You will find that the index refreshes your understanding of the concepts discussed in the book. The range of page numbers after each term indicates that a concept has extended its relevance across many different chapters. When you browse the different entries related to a single term, you learn how the concept represented by the term applies itself to many different topics. You begin to understand, in ways that were not apparent when you read the book as a linear text, the versatility of the ideas contained in the book. When you have finished reading the index, you will notice that the indexer exercised great restraint when selecting terms.

Most indexes are under 20 pages. The goal of the indexer is not to create a concordance (i.e., a listing of every word in a book, with its locations), but to create a keyed encapsulation of concepts, sub-concepts and term relationships.

The indexes we find in today's books are generally alphabetized terms. In prior decades and prior centuries, authors and editors put enormous effort into building indexes, sometimes producing multiple indexes for a single book. For example, a biography might contain a traditional alphabetized term index, followed by an alphabetized index of the names of the people included in the text. A zoology book might include an index specifically for animal names, with animals categorized according to their taxonomic order. A geography index might list the names of localities sub-indexed by country, with countries sub-indexed by continent. A single book might have 5 or more indexes. In nineteenth century books, it was not unusual to publish indexes as stand-alone volumes. [Glossary Taxonomy, Systematics, Taxa, Taxon]

You may be thinking that all this fuss over indexes is quaint, but it cannot apply to Big Data resources. Actually, Big Data resources that lack a proper index cannot be utilized to their full potential. Without an index, you never know what your queries are missing. Remember, in a Big Data resource, it is the relationship among data objects that are the keys to knowledge. Data by itself, even in large quantities, tells only part of a story. The most useful Big Data resources have electronic indexes that map concepts, classes, and terms to specific locations in the resource where data items are stored. An index imposes order and simplicity on the Big Data resource. Without an index, Big Data resources can easily devolve into vast collections of disorganized information. [Glossary Class]

The best indexes comply with international standards (ISO 999) and require creativity and professionalism [4]. Indexes should be accepted as another device for driving down the complexity of Big Data resources. Here are a few of the specific strengths of an index that cannot be duplicated by "find" operations on terms entered into a query box:

- An index can be read, like a book, to acquire a quick understanding of the contents and general organization of the data resource.
- Index lookups (i.e., searches and retrievals) are virtually instantaneous, even for very large indexes (see Section 2.6 of this chapter, for explanation).
- Indexes can be tied to a classification. This permits the analyst to know the relationships among different topics within the index, and within the text. [Glossary Classification]
- Many indexes are cross-indexed, providing relationships among index terms that might be extremely helpful to the data analyst.
- Indexes from multiple Big Data resources can be merged. When the location entries for index terms are annotated with the name of the resource, then merging indexes is trivial, and index searches will yield unambiguously identified locators in any of the Big Data resources included in the merge.
- Indexes can be created to satisfy a particular goal; and the process of creating a made-to-order index can be repeated again and again. For example, if you have a Big Data

resource devoted to ornithology, and you have an interest in the geographic location of species, you might want to create an index specifically keyed to localities, or you might want to add a locality sub-entry for every indexed bird name in your original index. Such indexes can be constructed as add-ons, as needed. [Glossary Ngrams]

- Indexes can be updated. If terminology or classifications change, there is nothing stopping you from re-building the index with an updated specification. In the specific context of Big Data, you can update the index without modifying your data. [Glossary Specification]

- Indexes are created after the database has been created. In some cases, the data manager does not envision the full potential of the Big Data resource until after it is created. The index can be designed to facilitate the use of the resource in line with the observed practices of users.

- Indexes can serve as surrogates for the Big Data resource. In some cases, all the data user really needs is the index. A telephone book is an example of an index that serves its purpose without being attached to a related data source (e.g., caller logs, switching diagrams).

Section 2.5. Autocoding

The beginning of wisdom is to call things by their right names.

Chinese proverb

Coding, as used in the context of unstructured textual data, is the process of tagging terms with an identifier code that corresponds to a synonymous term listed in a standard nomenclature. For example, a medical nomenclature might contain the term renal cell carcinoma, a type of kidney cancer, attaching a unique identifier code for the term, such as "C9385000." There are about 50 recognized synonyms for "renal cell carcinoma." A few of these synonyms and near-synonyms are listed here to show that a single concept can be expressed many different ways, including: adenocarcinoma arising from kidney, adenocarcinoma involving kidney, cancer arising from kidney, carcinoma of kidney, Grawitz tumor, Grawitz tumour, hypernephroid tumor, hypernephroma, kidney adenocarcinoma, renal adenocarcinoma, and renal cell carcinoma. All of these terms could be assigned the same identifier code, "C9385000". [Glossary Coding, Identifier]

The process of coding a text document involves finding all the terms that belong to a specific nomenclature, and tagging each term with the corresponding identifier code.

A nomenclature is a specialized vocabulary, usually containing terms that comprehensively cover a knowledge domain. For example, there may be a nomenclature of diseases, of celestial bodies, or of makes and models of automobiles. Some nomenclatures are ordered alphabetically. Others are ordered by synonymy, wherein all synonyms and plesionyms (near-synonyms) are collected under a canonical (i.e., best or preferred) term. Synonym indexes are always corrupted by the inclusion of polysemous terms (i.e., terms with multiple meanings). In many nomenclatures, grouped synonyms are collected under a so-called code (i.e., a unique alphanumeric string) assigned to all of the terms in the group.

Nomenclatures have many purposes: to enhance interoperability and integration, to allow synonymous terms to be retrieved regardless of which specific synonym is entered as a query, to support comprehensive analyses of textual data, to express detail, to tag information in textual documents, and to drive down the complexity of documents by uniting synonymous terms under a common code. Sets of documents held in more than one Big Data resource can be harmonized under a nomenclature by substituting or appending a nomenclature code to every nomenclature term that appears in any of the documents. [Glossary Interoperability, Data integration, Plesionymy, Polysemy, Vocabulary, Uniqueness, String]

In the case of "renal cell carcinoma," if all of the 50+ synonymous terms, appearing anywhere in a medical text, were tagged with the code "C938500," then a search engine could retrieve documents containing this code, regardless of which specific synonym was queried (e.g., a query on Grawitz tumor would retrieve documents containing the word "hypernephroid tumor"). To do so the search engine would simply translate the query word, "Grawitz tumor" into its nomenclature code "C938500" and would pull every record that had been tagged by the code.

Traditionally, nomenclature coding, much like language translation, has been considered a specialized and highly detailed task that is best accomplished by human beings. Just as there are highly trained translators who will prepare foreign language versions of popular texts, there are highly trained coders, intimately familiar with specific nomenclatures, who create tagged versions of documents. Tagging documents with nomenclature codes is serious business. If the coding is flawed the consequences can be dire. In 2009 the Department of Veterans Affairs sent out hundreds of letters to veterans with the devastating news that they had contracted Amyotrophic Lateral Sclerosis, also known as Lou Gehrig's disease, a fatal degenerative neurologic condition. About 600 of the recipients did not, in fact, have the disease. The VA retracted these letters, attributing the confusion to a coding error [5]. Coding text is difficult. Human coders are inconsistent, idiosyncratic, and prone to error. Coding accuracy for humans seems to fall in the range of 85%–90% [6]. [Glossary Accuracy versus precision]

When dealing with text in gigabyte and greater quantities, human coding is simply out of the question. There is not enough time or money or talent to manually code the textual data contained in Big Data resources. Computerized coding (i.e., autocoding) is the only practical solution.

Autocoding is a specialized form of machine translation, the field of computer science wherein meaning is drawn from narrative text. Not surprisingly, autocoding algorithms have been adopted directly from the field of machine translation, particularly algorithms for natural language processing. A popular approach to autocoding involves using the natural rules of language to find words or phrases found in text and matching them to nomenclature terms. Ideally the terms found in text are correctly matched to their equivalent nomenclature terms, regardless of the way that the terms were expressed in the text.

For instance, the term "adenocarcinoma of lung" has much in common with alternate terms that have minor variations in word order, plurality, inclusion of articles, terms split by a word inserted for informational enrichment, and so on. Alternate forms would be "adenocarcinoma of the lung," "adenocarcinoma of the lungs," "lung adenocarcinoma," and "adenocarcinoma found in the lung." A natural language algorithm takes into account grammatical variants, allowable alternate term constructions, word roots (i.e., stemming), and syntax variation. Clever improvements on natural language methods might include string similarity scores, intended to find term equivalences in cases where grammatical methods come up short. [Glossary Algorithm, Syntax, Machine translation, Natural language processing]

A limitation of the natural language approach to autocoding is encountered when synonymous terms lack etymologic commonality. Consider the term "renal cell carcinoma." Synonyms include terms that have no grammatical relationship with one another. For example, hypernephroma, and Grawitz tumor are synonyms for renal cell carcinoma. It is impossible to compute the equivalents among these terms through the implementation of natural language rules or word similarity algorithms. The only way of obtaining adequate synonymy is through the use of a comprehensive nomenclature that lists every synonym for every canonical term in the knowledge domain.

Setting aside the inability to construct equivalents for synonymous terms that share no grammatical roots, the best natural language autocoders are pitifully slow. The reason for the slowness relates to their algorithm, which requires the following steps, at a minimum: parsing text into sentences; parsing sentences into grammatical units; re-arranging the units of the sentence into grammatically permissible combinations; expanding the combinations based on stem forms of words; allowing for singularities and pluralities of words, and matching the allowable variations against the terms listed in the nomenclature. A typical natural language autocoder parses text at about 1 kilobyte per second, which is equivalent to a terabyte of text every 30 years. Big Data resources typically contain many terabytes of data; thus, natural language autocoding software is unsuitable for translating Big Data resources. This being the case, what good are they?

Natural language autocoders have value when they are employed at the time of data entry. Humans type sentences at a rate far less than 1 kilobyte per second, and natural language autocoders can keep up with typists, inserting codes for terms, as they are typed. They can operate much the same way as auto-correct, auto-spelling, look-ahead, and other commonly available crutches intended to improve or augment the output of plodding human typists.

– Recoding and speed

It would seem that by applying the natural language parser at the moment when the data is being prepared, all of the inherent limitations of the algorithm can be overcome. This belief, popularized by developers of natural language software, and perpetuated by a generation of satisfied customers, ignores two of the most important properties that must be preserved in Big Data resources: longevity, and curation. [Glossary Curator]

Nomenclatures change over time. Synonymous terms and the codes will vary from year to year as new versions of old nomenclature are published and new nomenclatures are developed. In some cases, the textual material within the Big Data resource will need to be annotated using codes from nomenclatures that cover informational domains that were not anticipated when the text was originally composed.

Most of the people who work within an information-intensive society are accustomed to evanescent data; data that is forgotten when its original purpose is served. Do we really want all of our old e-mails to be preserved forever? Do we not regret our earliest blog posts, Facebook entries, and tweets? In the medical world, a code for a clinic visit or a biopsy diagnosis, or a reportable transmissible disease will be used in a matter of minutes or hours; maybe days or months. Few among us place much value on textual information preserved for years and decades. Nonetheless, it is the job of the Big Data manager to preserve resource data over years and decades. When we have data that extends back, over decades, we can find and avoid errors that would otherwise reoccur in the present, and we can analyze trends that lead us into the future.

To preserve its value, data must be constantly curated, adding codes that apply to currently available nomenclatures. There is no avoiding the chore; the entire corpus of textual data held in the Big Data resource needs to be recoded again and again, using modified versions of the original nomenclature, or using one or more new nomenclatures. This time, an autocoding application will be required to code huge quantities of textual data (possibly terabytes), quickly. Natural language algorithms, which depend heavily on regex operations (i.e., finding word patterns in text) are too slow to do the job. [Glossary RegEx]

A faster alternative is so-called lexical parsing. This involves parsing text, word by word, looking for exact matches between runs of words and entries in a nomenclature. When a match occurs, the words in the text that matched the nomenclature term are assigned the nomenclature code that corresponds to the matched term. Here is one possible algorithmic strategy for autocoding the sentence: "Margins positive malignant melanoma." For this example, you would be using a nomenclature that lists all of the tumors that occur in humans. Let us assume that the terms "malignant melanoma," and "melanoma" are included in the nomenclature. They are both assigned the same code, for example "Q5673013," because the people who wrote the nomenclature considered both terms to be biologically equivalent.

Let us autocode the diagnostic sentence, "Margins positive malignant melanoma":

1. Begin parsing the sentence, one word at a time. The first word is "Margins." You check against the nomenclature, and find no match. Save the word "margins." We will use it in step 2.
2. You go to the second word, "positive" and find no matches in the nomenclature. You retrieve the former word "margins" and check to see if there is a 2-word term, "margins positive." There is not. Save "margins" and "positive" and continue.
3. You go to the next word, "malignant." There is no match in the nomenclature. You check to determine whether the 2-word term "positive malignant" and the 3-word term "margins positive malignant" are in the nomenclature. They are not.

4. You go to the next word, "melanoma." You check and find that melanoma is in the nomenclature. You check against the two-word term "malignant melanoma," the three-word term "positive malignant melanoma," and the four-word term "margins positive malignant melanoma." There is a match for "malignant melanoma" but it yields the same code as the code for "melanoma."

5. The autocoder appends the code, "Q5673013" to the sentence, and proceeds to the next sentence, where it repeats the algorithm.

The algorithm seems like a lot of work, requiring many comparisons, but it is actually much more efficient than natural language parsing. A complete nomenclature, with each nomenclature term paired with its code, can be held in a single variable, in volatile memory. Look-ups to determine whether a word or phrase is included in the nomenclature are also fast. As it happens, there are methods that will speed things along. In Section 2.7, we will see a 12-line autocoder algorithm that can parse through terabytes of text at a rate that is much faster than commercial-grade natural language autocoders [7]. [Glossary Variable]

Another approach to the problem of recoding large volumes of textual data involves abandoning the attempt to autocode the entire corpus, in favor of on-the-fly autocoding, when needed. On-the-fly autocoding involves parsing through a text of any size, and searching for all the terms that match one particular concept (i.e., the search term).

Here is a general algorithm on-the-fly coding [8]. This algorithm starts with a query term and seeks to find every synonym for the query term, in any collection of Big Data resources, using any convenient nomenclature.

1. The analyst starts with a query term submitted by a data user. The analyst chooses a nomenclature that contains his query term, as well as the list of synonyms for the term. Any vocabulary is suitable, so long as the vocabulary consists of term/code pairs, where a term and its' synonyms are all paired with the same code.

2. All of the synonyms for the query term are collected together. For instance the 2004 version of a popular medical nomenclature, the Unified Medical Language System, had 38 equivalent entries for the code C0206708, nine of which are listed here:

```
C0206708|Cervical Intraepithelial Neoplasms
C0206708|Cervical Intraepithelial Neoplasm
C0206708|Intraepithelial Neoplasm, Cervical
C0206708|Intraepithelial Neoplasms, Cervical
C0206708|Neoplasm, Cervical Intraepithelial
C0206708|Neoplasms, Cervical Intraepithelial
C0206708|Intraepithelial Neoplasia, Cervical
C0206708|Neoplasia, Cervical Intraepithelial
C0206708|Cervical Intraepithelial Neoplasia
```

If the analyst had chosen to search on "Cervial Intraepithelial Neoplasia," his term will be attached to the 38 synonyms included in the nomenclature.

3. One-by-one, the equivalent terms are matched against every record in every Big Data resource available to the analyst.
4. Records are pulled that contain terms matching any of the synonyms for the term selected by the analyst.

In the case of the example, this would mean that all 38 synonymous terms for "Cervical Intraepithelial Neoplasms" would be matched against the entire set of data records. The benefit of this kind of search is that data records that contain any search term, or its nomenclature equivalent, can be extracted from multiple data sets in multiple Big Data resources, as they are needed, in response to any query. There is no pre-coding, and there is no need to match against nomenclature terms that have no interest to the analyst. The drawback of this method is that it multiplies the computational task by the number of synonymous terms being searched, 38-fold in this example. Luckily, there are published methods for conducting simple and fast synonym searches, using precompiled concordances [8].

Section 2.6. Case Study: Instantly Finding the Precise Location of Any Atom in the Universe (Some Assembly Required)

There's as many atoms in a single molecule of your DNA as there are stars in the typical galaxy. We are, each of us, a little universe.

Neil deGrasse Tyson, Cosmos

If you have sat through an introductory course in Computer Science, you are no doubt familiar with three or four sorting algorithms. Indeed, most computer science books devote a substantial portion of their texts to describing sorting algorithms. The reason for this infatuation with sorting is that all sorted lists can be searched nearly instantly, regardless of the size of the list. The so-called binary algorithm for searching a sorted list is incredibly simple. For the sake of discussion, let us consider an alphabetically sorted list of 1024 words. I want to determine if the word "kangaroo" is in the list; and, if so, its exact location in the list. Here is how a binary search would be conducted.

1. Go to the middle entry of the list.
2. Compare the middle entry to the word "kangaroo." If the middle entry comes earlier in the alphabet than "kangaroo," then repeat step 1, this time ignoring the first half of the list and using only the second half of the list (i.e., going to the middle entry of the second half of the file). Otherwise, go to step 1, this time ignoring the second half of the list and using only the first half.

These steps are repeated until you come to the location where kangaroo resides, or until you have exhausted the list without finding your kangaroo.

Each cycle of searching cuts the size of the list in half. Hence, a search through a sorted list of 1024 items would involve, at most, 10 cycles through the two-step algorithm (because $1024 = 2^{10}$).

Every computer science student is expected to write her own binary search script. Here is a simple script, binary.py, that does five look-ups through a sorted numeric list, reporting on which items are found, and which items are not.

```
def Search(search_list, search_item):
  first_item = 0
  last_item = len(search_list)-1
  found = False
  while (first_item <= last_item) and not found:
    middle = (first_item + last_item)//2
    if search_list[middle] == search_item:
      found = True
    else:
      if search_item < search_list[middle]:
        last_item = middle - 1
      else:
        first_item = middle + 1
  return found
sorted_list = [4, 5, 8, 15, 28, 29, 30, 45, 67, 82, 99, 101, 1002]
for item in [3, 7, 28, 31, 45, 1002]:
  print(Search(sorted_list, item))
```

```
output:
False
False
True
False
True
True
```

Let us say, just for fun, we wanted to search through a sorted list of every atom in the universe. First we would take each atom in the universe and assign it a location. Then we would sort the locations based on their distances from the center of the center of the universe, which is apparently located at the tip of my dog's left ear. We could then substitute the sorted atom list for the sorted_list in the binary.py script, shown above.

How long would it take to search all the atoms of the universe, using the binary.py script. As it happens, we could find the list location for any atom in the universe, almost instantly. The reason is that there are only about 2^260 atoms in the known universe. This means that the algorithm would required, at the very most, 260 2-step cycles. Each cycle is very fast, requiring only that we compare the search atom's distance from my dog's ear, against the middle atom of the list.

Of course, composing the list of atom locations may pose serious difficulties, and we might need another universe, much larger than our own, to hold the sorted list that we create. Nonetheless, a valid point emerges; that binary searches are fast, and the time to completion of a binary search is not significantly lengthened by any increase in the number of items in the list. Had we chosen, we could have annotated the items of sorted_list with any manner of information (e.g., locations in a file, nomenclature code, links to web addresses, definitions of the items, metadata), so that our binary searches would yield something more useful than the location of the item in the list.

Section 2.7. Case Study (Advanced): A Complete Autocoder (in 12 Lines of Python Code)

Software is a gas; it expands to fill its container.

Nathan Myhrvold

This script requires two external files:

1. The nomenclature file that will be converted into a Python dictionary, wherein each term is a dictionary key, and each nomenclature code is a value assigned to a term. [Glossary Dictionary]

   Here are a few sample lines from the nomenclature file (nomenclature_dict.txt, in this case):

```
oropharyngeal adenoid cystic adenocarcinoma , C6241000
peritoneal mesothelioma , C7633000
benign tumour arising from the exocrine pancreas , C4613000
basaloid penile squamous cell cancer , C6980000
cns malignant soft tissue tumor , C6758000
digestive stromal tumour of stomach , C5806000
bone with malignancy , C4016000
benign mixed tumor arising from skin , C4474000
```

2. The file containing a corpus of sentences that will be autocoded by the script. Here are a few sample lines from the corpus file (tumorabs.txt, in this case):

```
local versus diffuse recurrences of meningiomas factors correlated
to the extent of the recurrence

the effect of an unplanned excision of a soft tissue sarcoma on
prognosis

obstructive jaundice associated burkitt lymphoma mimicking
pancreatic carcinoma

efficacy of zoledronate in treating persisting isolated tumor
cells in bone marrow in patients with breast cancer a phase ii pilot
study
```

metastatic lymph node number in epithelial ovarian carcinoma does it have any clinical significance

extended three dimensional impedance map methods for identifying ultrasonic scattering sites

aberrant expression of connexin 26 is associated with lung metastasis of colorectal cancer

The 19-line python script, autocode.txt, produces a sentence-by-sentence list of extracted autocoded terms:

```
outfile = open("autocoded.txt", "w")
literalhash = {}
with open("nomenclature_dict.txt") as f:
    for line in f:
        (key, val) = line.split(" , ")
        literalhash[key] = val
corpus_file = open("tumorabs.txt", "r")
for line in corpus_file:
    sentence = line.rstrip()
    outfile.write("\n" + sentence[0].upper() + sentence[1:] + "." +
    "\n")
    sentence_array = sentence.split(" ")
    length = len(sentence_array)
    for i in range(length):
        for place_length in range(len(sentence_array)):
            last_element = place_length + 1
            phrase = ' '.join(sentence_array[0:last_element])
            if phrase in literalhash:
                outfile.write(phrase + " " + literalhash[phrase])
        sentence_array.pop(0)
```

The first seven lines of code are housekeeping chores, in which the external nomenclature is loaded into a Python dictionary (literalhash, in this case), and an external file composed of lines, with one sentence on each line, is opened and prepared for reading, and which another external file, autocoded.txt, is created to accept the script's output. We will not count these first seven lines as belonging to our autocoder because, in all fairness, they are not doing any of the work of autocoding. The meat of the script is the next twelve lines, beginning with "for line in corpus_file."

Here is a sample of the output:

Obstructive jaundice associated burkitt lymphoma mimicking pancreatic carcinoma.
    burkitt lymphoma C7188000

```
    lymphoma C7065000
    pancreatic carcinoma C3850000
    carcinoma C2000000

Littoral cell angioma of the spleen.
    littoral cell angioma C8541100
    littoral cell angioma of the spleen C8541100
    angioma C3085000
    angioma of the spleen C8541000

Isolated b cell lymphoproliferative disorder at the dura mater with b
cell chronic lymphocytic leukemia immunophenotype.
    lymphoproliferative disorder C4727100
    b cell chronic lymphocytic leukemia C3163000
    chronic lymphocytic leukemia C3163000
    lymphocytic leukemia C7539000
    leukemia C3161000
```

By observing a few samples of autocoded lines of text, we can see that the autocoder extracts all cancer terms, and supplies its nomenclature code, regardless of whether a term is contained within a longer term.

For example, the autocoder managed to find four terms within the sentence "Littoral cell angioma of the spleen," these being: littoral cell angioma, littoral cell angioma of the spleen, angioma, and angioma of the spleen. The ability to extract every valid term, even when they are subsumed by larger terms, guarantees that a query term and all its synonyms will always be retrieved, if the query term happens to be a valid nomenclature term.

This short autocoding script comes with a few advantages that are of particular interest to Big Data professionals:

− Scalable to any size

All nomenclatures are small. Most of us have a working vocabulary of a few thousand words. Most dictionaries are smaller, containing maybe 60,000 words. The most extreme case of verbiage about verbiage is The 20-volume Oxford English Dictionary, which contains about 170,000 entries. Even in this case, slurping the entire list of Oxford English dictionary items would be a simple matter for any modern computer.

Most importantly, the autocoding algorithm imposes no limits on the size of the Big Data corpus. The software proceeds line-by-line until the task is complete. Memory requirements and other issues of scalability are not a problem.

− Fast

On my modest desktop computer, the 12-line autocoding algorithm processes text at the rate of 1 megabyte every two seconds. A fast and powerful computer, using the same algorithm, would be expected to parse at rates of 1 gigabytes of text per second, or greater.

–   Repeatable

Code a gigabyte of data in the morning. Do it all over again in the afternoon. Use another version of the nomenclature, or use a different nomenclature, entirely. Recoding is not a problem.

–   Simple and adaptable, with easily maintained code

The larger the program, the more difficult it is to find bugs, or to recover from errors produced when the code is modified. It is nearly impossible to inflict irreversible damage upon a simple, 12-line script. As a general rule, tiny scripts are seldom a problem if you maintain records of where the scripts are located, how the scripts are used, and how the scripts are modified over time.

–   Reveals the dirty little secret that every programmer knows, but few are willing
    to admit.

Virtually all useful algorithms can be implemented in a few lines of code; autocoders are no exception. The thousands, or millions, of lines of code in just about any commercial software application are devoted, in one way or another, to the graphic user interface.

Section 2.8. Case Study: Concordances as Transformations of Text

Interviewer: Is there anything from home that you brought over with you to set up for yourself? Creature comforts?
Hawkeye: I brought a book over.
Interviewer: What book?
Hawkeye: The dictionary. I figure it's got all the other books in it.
Interview with the character Hawkeye, played by Alan Alda, from television show M*A*S*H

A transform is a mathematical operation that takes a function, a signal, or a set of data and changes it into something else, that is easier to work with than the original data. The concept of the transform is a simple but important idea that has revolutionized many scientific fields including electrical engineering, digital signal processing, and data analysis. In the field of digital signal processing, data in the time domain (i.e., wherein the amplitude of a measurement varies over time, as in a signal), is commonly transformed into the frequency domain (i.e., wherein the original data can be assigned to amplitude values for a range of frequencies). There are dozens, possibly hundreds, of mathematical transforms that enable data analysts to move signal data between forward transforms (e.g., time domain to frequency domain), and their inverse counterparts (e.g., frequency domain to time domain). [Glossary Transform, Signal, Digital signal, Digital Signal Processing, DSP, Fourier transform, Burrows-Wheeler transform]

A concordance is transform, for text. A concordance takes a linear text and transforms it a word-frequency distribution list; which can reversed as needed. Like any good transform, we can expect to find circumstances when it is easier to perform certain types of operations on the transformed data than on the original data. [Glossary Concordance]

Here is an example, from the Python script proximate_words.py, where we use a concordance to list the words in close proximity to the concordance entries (i.e., the words contained in the text). In this script, we use the previously constructed (vida supra) concordance of the Gettysburg address.

```python
import string
infile = open ("concordance.txt", "r")
places = []
word_array = []
concordance_hash = {}
words_hash = {}
for line in infile:
    line = line.rstrip()
    line_array = line.split(" ")
    word = line_array[0]
    places = line_array[1]
    places_array = places.split(",")
    words_hash[word] = places_array
    for word_position in places_array:
        concordance_hash[word_position] = word
for k, v in words_hash.items():
    print(k, end=" - \n")
    for items in v:
        n = 0
        while n < 5:
            nextone = str(int(items) + n)
            if nextone in concordance_hash:
                print(concordance_hash[nextone], end=" ")
            n = n+1
        print()
    print()
```

The script produces a list of the words from the Gettysburg address, along with short sequences of the text that follow each occurrence of the word in the text, as shown in this sampling from the output file:

```
to -
to the proposition that all
to dedicate a portion of
```

```
to add or detract. The
to be dedicated here to
to the unfinished work which
to be here dedicated to
to the great task remaining
to that cause for which

dedicated -
dedicated to the proposition that
dedicated can long endure. We
dedicated here to the unfinished
dedicated to the great task
```

Inspecting some of the output, we see that the word "to" appears 8 times in the Gettysburg address. We used the concordance to reconstruct four words that follow the word "to" wherever it occurs in the text. Likewise we see that the word "dedicated" occurs 4 times in the text, and the concordance tells us the four words that follow at each of the locations where "dedicated" appears. We can construct these proximity phrases very quickly, because the concordance tells us the exact location of the words in the text. If we were working from the original text, instead of its transform (i.e., the concordance), then our algorithm would run much more slowly, because each word would need to be individually found and retrieved, by parsing every word in the text, sequentially.

Section 2.9. Case Study (Advanced): Burrows Wheeler Transform (BWT)

All parts should go together without forcing. You must remember
that the parts you are reassembling were disassembled by you. Therefore,
if you can't get them together again, there must be a reason. By all
means, do not use a hammer.

IBM Manual, 1925

One of the most ingenious transforms in the field of data science is the Burrows Wheeler transform. Imagine an algorithm that takes a corpus of text and creates an output string consisting of a transformed text combined with its own word index, in a format that can be compressed to a smaller size than the compressed original file. The Burrows Wheeler Transform does all this, and more [9,10]. A clever informatician may find many ways to use the BWT transform in search and retrieval algorithms and in data merging projects [11]. Using the BWT file, you can re-compose the original file, or you can find any portion of a file preceding or following any word from the file [12]. [Glossary Data merging, Data fusion]

Excellent discussions of the algorithm are available, along with implementations in several languages [9,10,13]. The Python script, bwt.py, shown here, is a modification of a script available on Wikipedia [13]. The script executes the BWT algorithm in just three lines of code. In this example, the input string is a excerpt from Lincoln's Gettysburg address [12].

```
input = "four score and seven years ago our fathers brought forth upon"
input = input + " this continent a new nation conceived in liberty and"
input = input + "\0"
table = sorted(input[i:] + input[:i] for i in range(len(input)))
last_column = [row[-1:] for row in table]
print("".join(last_column))
```

Here is the transformed output:

```
dtsyesnsrtdnwaordnhn  efni n snenryvcvnhbsn  uatttgl tthe oioe oaai
eogipccc
fr fuuuobaeoerri nhra naro ooieet
```

Admittedly, the output does not look like much. Let us juxtapose our input string and our BWT's transform string:

```
four score and seven years ago our fathers brought forth upon this
continent a new nation conceived in liberty and
dtsyesnsrtdnwaordnhn  efni n snenryvcvnhbsn  uatttgl tthe oioe oaai
eogipcccfr fuuuobaeoerri nhra naro ooieet
```

We see that the input string and the transformed output string both have the same length, so there doesn't seem to be any obvious advantage to the transform. If we look a bit closer, though, we see that the output string consists largely of runs of repeated individual characters, repeated substrings, and repeated spaces (e.g., "ttt" "uuu"). These frequent repeats in the transform facilitate compression algorithms that hunt for repeat patterns. BWT's facility for creating runs of repeated characters accounts for its popularity in compression software (e.g., the Bunzip compression utility).

The Python script, bwt_inverse.py, computes the inverse BWT to re-construct the original input string. Notice that the inverse algorithm is implemented in just the last four lines of the python code (the first five lines re-created the forward BWT transform) [12]

```
input = "four score and seven years ago our fathers brought forth upon"
input = input + " this continent a new nation conceived in liberty and"
input = input + "\0"
table = sorted(input[i:] + input[:i] for i in range(len(input)))
last_column = [row[-1:] for row in table]
#The first lines re-created the bwt transform

#The next four lines compute the inverse transform
table = [""] * len(last_column)
for i in range(len(last_column)):
    table = sorted(last_column[i] + table[i] for i in range(len(input)))
print([row for row in table if row.endswith("\0")][0])
```

As we would expect, the output of the bwt_inverse.py script, is our original input string:

```
four score and seven years ago our fathers brought forth upon this
continent a new nation conceived in liberty and
```

The charm of the BWT transform is demonstrated when we create an implementation that parses the input string word-by-word; not character-by-character.

Here is the Python script, bwt_trans_inv.py, that transforms an input string, word-by-word, producing its transform; then reverses the process to yield the original string, as an array of words. As an extra feature, the script produces the first column, as an array, of the transform table [12]. [Glossary Numpy]

```
import numpy as np
input = "\0 four score and seven years ago our fathers brought forth upon"
input = input + " this continent a new nation conceived in liberty and"
word_list = input.rsplit()
table = sorted(word_list[i:] + word_list[:i] for i in range(len
(word_list)))
last_column = [row[-1:] for row in table]
first_column = [row[:1] for row in table]
print("First column of the transform table:\n" + str(first_column) +
"\n")
table = [""] * len(last_column)
for i in range(len(last_column)):
    table = sorted(str(last_column[i]) + " " + str(table[i]) for i in
range(len(word_list)))
original = [row for row in table][0]
print("Inverse transform, as a word array:\n" + str(original))
```

Here is the output of the bwt_trans_inv.py script. Notice once more that the word-by-word transform was implemented in 3 lines of code, and the inverse transform was implemented in four lines of code.

```
First column of the transform table:
[['\x00'], ['a'], ['ago'], ['and'], ['and'], ['brought'],
['conceived'], ['continent'], ['fathers'], ['forth'], ['four'],
['in'], ['liberty'], ['nation'], ['new'], ['our'], ['score'],
['seven'], ['this'], ['upon'], ['years']]

Inverse transform, as a word array:
['\x00'] ['four'] ['score'] ['and'] ['seven'] ['years'] ['ago']
['our'] ['fathers'] ['brought'] ['forth'] ['upon'] ['this']
['continent'] ['a'] ['new'] ['nation'] ['conceived'] ['in']
['liberty'] ['and']
```

The first column of the transform, created in the forward BWT, is a list of the words in the input string, in alphabetic order. Notice that words that occurred more than one time in the input text were repeated in the first column of the transform table (i.e., [and], [and] in the example sentence). Hence, the transform yields all the words from the original input, along with their frequency of occurrence in the text. As expected, the inverse of the transform yields our original input string.

Glossary

Accuracy versus precision Accuracy measures how close your data comes to being correct. Precision provides a measurement of reproducibility (i.e., whether repeated measurements of the same quantity produce the same result). Data can be accurate but imprecise. If you have a 10 pound object, and you report its weight as 7.2376 pounds, on every occasion when the object is weighed, then your precision is remarkable, but your accuracy is dismal.

Algorithm An algorithm is a logical sequence of steps that lead to a desired computational result. Algorithms serve the same function in the computer world as production processes serve in the manufacturing world and as pathways serve in the world of biology. Fundamental algorithms can be linked to one another, to create new algorithms (just as biological pathways can be linked). Algorithms are the most important intellectual capital in computer science. In the past half century, many brilliant algorithms have been developed for the kinds of computation-intensive work required for Big Data analysis [14,15].

Binary data Computer scientists say that there are 10 types of people. Those who think in terms of binary numbers, and those who do not. Pause for laughter and continue. All digital information is coded as binary data. Strings of 0s and 1s are the fundamental units of electronic information. Nonetheless, some data is more binary than other data. In text files, 8-bite sequences are converted into decimals in the range of 0–256, and these decimal numbers are converted into characters, as determined by the ASCII standard. In several raster image formats (i.e., formats consisting of rows and columns of pixel data), 24-bit pixel values are chopped into red, green and blue values of 8-bits each. Files containing various types of data (e.g., sound, movies, telemetry, formatted text documents), all have some kind of low-level software that takes strings of 0s and 1s and converts them into data that has some particular meaning for a particular use. So-called plain-text files, including HTML files and XML files are distinguished from binary data files and referred to as plain-text or ASCII files. Most computer languages have an option wherein files can be opened as "binary," meaning that the 0s and 1s are available to the programmer, without the intervening translation into characters or stylized data.

Burrows-Wheeler transform Abbreviated as BWT, the Burrows-Wheeler transform produces a compressed version of an original file, along with a concordance to the contents of the file. Using a reverse BWT, you can reconstruct the original file, or you can find any portion of a file preceding or succeeding any location in the file. The BWT transformation is an amazing example of simplification, applied to informatics. A detailed discussion of the BWT is found in Section 2.9, "Case Study (Advanced): Burrows Wheeler Transform."

Class A class is a group of objects that share a set of properties that define the class and that distinguish the members of the class from members of other classes. The word "class," lowercase, is used as a general term. The word "Class," uppercase, followed by an uppercase noun (e.g., Class Animalia), represents a specific class within a formal classification.

Classification A system in which every object in a knowledge domain is assigned to a class within a hierarchy of classes. The properties of superclasses are inherited by the subclasses. Every class has one immediate superclass (i.e., parent class), although a parent class may have more than one immediate subclass (i.e., child class). Objects do not change their class assignment in a classification, unless there was a mistake in the assignment. For example, a rabbit is always a rabbit, and does not change into a tiger. Classifications can be thought of as the simplest and most restrictive type of ontology, and serve to reduce the complexity of a knowledge domain [16].

Classifications can be easily modeled in an object-oriented programming language and are non-chaotic (i.e., calculations performed on the members and classes of a classification should yield the same output, each time the calculation is performed). A classification should be distinguished from an ontology. In an ontology a class may have more than one parent class and an object may be a member of more than one class. A classification can be considered a special type of ontology wherein each class is limited to a single parent class and each object has membership in one and only one class.

Coding The term "coding" has three very different meanings depending on which branch of science influences your thinking. For programmers, coding means writing the code that constitutes a computer programmer. For cryptographers, coding is synonymous with encrypting (i.e., using a cipher to encode a message). For medics, coding is calling an emergency team to handle a patient in extremis. For informaticians and library scientists, coding involves assigning a alphanumeric identifier, representing a concept listed in a nomenclature, to a term. For example, a surgical pathology report may includes the diagnosis, "Adenocarcinoma of prostate." A nomenclature may assign a code C4863000 that uniquely identifies the concept "Adenocarcinoma." Coding the report may involve annotating every occurrence of the work "Adenocarcinoma" with the "C4863000" identifier. For a detailed explanation of coding, and its importance for searching and retrieving data, see the full discussion in Section 3.4, "Autoencoding and Indexing with Nomenclatures."

Concordance A concordance is an index consisting of every word in the text, along with every location wherein each word can be found. It is computationally trivial to reconstruct the original text from the concordance. Before the advent of computers, concordances fell into the provenance of religious scholars, who painstakingly recorded the locations of the all words appearing in the Bible, ancient scrolls, and any texts whose words were considered to be divinely inspired. Today, a concordance for a Bible-length book can be constructed in about a second. Furthermore, the original text can be reconstructed from the concordance, in about the same time.

Curator The word "curator" derives from the latin, "curatus," the same root for "curative," indicating that curators "take care of" things. A data curator collects, annotates, indexes, updates, archives, searches, retrieves, and distributes data. Curator is another of those somewhat arcane terms (e.g., indexer, data archivist, lexicographer) that are being rejuvenated in the new millennium. It seems that if we want to enjoy the benefits of a data-centric world, we will need the assistance of curators, trained in data organization.

DSP Abbreviation for Digital Signal Processing.

Data fusion Data fusion is very closely related to data integration. The subtle difference between the two concepts lies in the end result. Data fusion creates a new and accurate set of data representing the combined data sources. Data integration is an on-the-fly usage of data pulled from different domains and, as such, does not yield a residual fused set of data.

Data integration The process of drawing data from different sources and knowledge domains in a manner that uses and preserves the identities of data objects and the relationships among the different data objects. The term "integration" should not be confused with a closely related term, "interoperability." An easy way to remember the difference is to note that **integration applies to data; interoperability applies to software.**

Data merging A nonspecific term that includes data fusion, data integration, and any methods that facilitate the accrual of data derived from multiple sources.

Dictionary In general usage a dictionary is a word list accompanied by a definition for each item. In Python a dictionary is a data structure that holds an unordered list of key/value pairs. A dictionary, as used in Python, is equivalent to an associative array, as used in Perl.

Digital Signal Processing Digital Signal Processing (DSP) is the field that deals with creating, transforming, sending, receiving, and analyzing digital signals. Digital signal processing began as a specialized

subdiscipline of signal processing, another specialized subdiscipline. For most of the twentieth century, many technologic advances came from converting non-electrical signals (temperature, pressure,
sound, and other physical signals) into electric signals that could be carried via electromagnetic waves,
and later transformed back into physical actions. Because electromagnetic waves sit at the center of so
many transform process, even in instances when the input and outputs are non-electrical in nature,
the field of electrical engineering and signal processing have paramount importance in every field of
engineering. In the past several decades the intermediate signals have been moved from the analog
domain (i.e., waves) into the digital realm (i.e., digital signals expressed as streams of 0s and 1s). Over
the years, as techniques have developed by which any kind of signal can be transformed into a digital
signal, the subdiscipline of digital signal processing has subsumed virtually all of the algorithms once
consigned to its parent discipline. In fact, as more and more processes have been digitized (e.g., telemetry, images, audio, sensor data, communications theory), the field of digital signal processing has
come to play a central role in data science.

Digital signal A signal is a description of how one parameter varies with some other parameter. The most
familiar signals involve some parameter varying over time (e.g., sound is air pressure varying over
time). When the amplitude of a parameter is sampled at intervals, producing successive pairs of values,
the signal is said to be digitized.

Fourier transform A transform is a mathematical operation that takes a function or a time series (e.g.,
values obtained at intervals of time) and transforms it into something else. An inverse transform takes
the transform function and produces the original function (Fig. 2.1). Transforms are useful when there
are operations that can be more easily performed on the transformed function than on the original
function. Possibly the most useful transform is the Fourier transform, which can be computed with
great speed on modern computers, using a modified form known as the fast Fourier Transform. Periodic functions and waveforms (periodic time series) can be transformed using this method. Operations on the transformed function can sometimes eliminate repeating artifacts or frequencies that
occur below a selected threshold (e.g., noise). The transform can be used to find similarities between
two signals. When the operations on the transform function are complete, the inverse of the transform
can be calculated and substituted for the original set of data (Fig. 2.2).

Identifier A string that is associated with a particular thing (e.g., person, document, transaction, data
object), and not associated with any other thing [17]. In the context of Big Data, identification usually
involves permanently assigning a seemingly random sequence of numeric digits (0–9) and alphabet
characters (a-z and A-Z) to a data object. The data object can be a class of objects.

Indexes Every writer must search deeply into his or her soul to find the correct plural form of "index". Is it
"indexes" or is it "indices"? Latinists insist that "indices" is the proper and exclusive plural form. Gram-
marians agree, reserving "indexes" for the third person singular verb form; "The student indexes his
thesis." Nonetheless, popular usage of the plural of "index," referring to the section at the end of a
book, is almost always "indexes," the form used herein.

$$\hat{f}(\xi) = \int_{-\infty}^{\infty} f(x)\; e^{-2\pi i x \xi}\; dx$$

$$f(x) = \int_{-\infty}^{\infty} \hat{f}(\xi)\; e^{2\pi i x \xi}\; d\xi$$

FIG. 2.1 The Fourier transform and its inverse. In this representation of the transform, *x* represents time in seconds and
the transform variable zeta represents frequency in hertz.

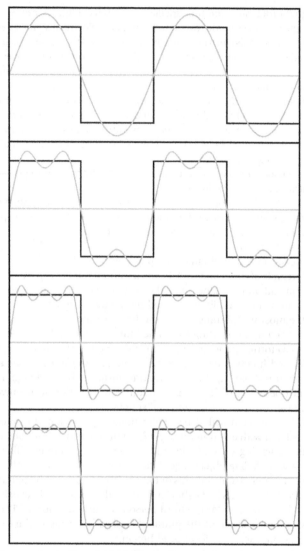

FIG. 2.2 A square wave is approximated by a single sine wave, the sum of two sine waves, three sine waves, and so on. As more components are added, the representation of the original signal or periodic set of data, is more closely approximated. *From Wikimedia Commons.*

Interoperability It is desirable and often necessary to create software that operates with other software, regardless of differences in hardware, operating systems and programming language. Interoperability, though vital to Big Data science, remains an elusive goal.

Machine translation Ultimately, the job of machine translation is to translate text from one language into another language. The process of machine translation begins with extracting sentences from text, parsing the words of the sentence into grammatical parts, and arranging the grammatical parts into an order that imposes logical sense on the sentence. Once this is done, each of the parts can be translated by a dictionary that finds equivalent terms in a foreign language, then re-assembled as a foreign language sentence by applying grammatical positioning rules appropriate for the target language. Because these steps apply the natural rules for sentence constructions in a foreign language, the process is often referred to as natural language machine translation. It is important to note that nowhere in the process of machine translation is it necessary to find meaning in the source text, or to produce meaning in the output. Good machine translation algorithms preserve ambiguities, without attempting to impose a meaningful result.

Natural language processing A field broadly concerned with how computers interpret human language (i.e., machine translation). At its simplest level this may involve parsing through text and organizing the grammatical units of individual sentences (i.e., tokenization). For example, we might assign the following tokens to the grammatical parts of a sentence: A = adjective, D = determiner, N = noun, P = preposition, V = main verb. A determiner is a word such as "a" or "the", which specifies the noun [18]. Consider the sentence, "The quick brown fox jumped over lazy dogs." This sentence can be grammatically tokenized as:

the::D
quick::A
brown::A
fox::N
jumped::V
over::P
the::D
lazy::A
dog::N

We can express the sentence as the sequence of its tokens listed in the order of occurrence in the sentence: DAANVPDAN. This does not seem like much of a breakthrough, but imagine having a large collection of such token sequences representing every sentence from a large text corpus. With such a data set, we could begin to understand the rules of sentence structure. Commonly recurring sequences, like DAANVPDAN, might be assumed to be proper sentences. Sequences that occur uniquely in a large text corpus are probably poorly constructed sentences. Before long, we might find ourselves constructing logic rules for reducing the complexity of sentences by dropping subsequences which, when removed, yield a sequence that occurs more commonly than the original sequence. For example, our table of sequences might indicate that we can convert DAANVPDAN into NVPAN (i.e., "Fox jumped over lazy dog"), without sacrificing too much of the meaning from the original sentence and preserving a grammatical sequence that occurs commonly in the text corpus.

This short example serves as an overly simplistic introduction to natural language processing. We can begin to imagine that the grammatical rules of a language can be represented by sequences of tokens that can be translated into words or phrases from a second language, and re-ordered according to grammatical rules appropriate to the target language. Many natural language processing projects involve transforming text into a new form, with desirable properties (e.g., other languages, an index, a collection of names, a new text with words and phrases replaced with canonical forms extracted from a nomenclature) [18]. When we use natural language rules to autocode text, the grammatical units are trimmed, reorganized, and matched against concept equivalents in a nomenclature.

Ngrams Ngrams are subsequences of text, of length n words. A complete collection of ngrams consists of all of the possible ordered subsequences of words in a text. Because sentences are the basic units of statements and ideas, when we speak of ngrams, we are confining ourselves to ngrams of sentences. Let us examine all the ngrams for the sentence, "Ngrams are ordered word sequences."

```
Ngrams (1-gram)
are (1-gram)
ordered (1-gram)
word (1-gram)
```

```
sequences (1-gram)
Ngrams are (2-gram)
are ordered (2-gram)
ordered word (2-gram)
word sequences (2-gram)
Ngrams are ordered (3-gram)
are ordered word (3-gram)
ordered word sequences (3-gram)
Ngrams are ordered word (4-gram)
are ordered word sequences (4-gram)
Ngrams are ordered word sequences (5-gram)
```

Here is a short Python script, ngram.py, that will take a sentence and produce a list of all the contained ngrams.

```python
import string
text = "ngrams are ordered word sequences"
partslist = []
ngramlist = {}
text_list = text.split(" ")
while(len(text_list) > 0):
   partslist.append(" ".join(text_list))
   del text_list[0]
for part in partslist:
   previous = ""
   wordlist = part.split(" ")
   while(len(wordlist) > 0):
      ngramlist[(" ".join(wordlist))] = ""
      firstword = wordlist[0]
      del wordlist[0]
      ngramlist[firstword] = ""
      previous = previous + " " + firstword
      previous = previous.strip()
      ngramlist[previous] = ""
for key in sorted(ngramlist):
      print(key)
exit
```

```
output:
are
are ordered
are ordered word
are ordered word sequences
ngrams
ngrams are
ngrams are ordered
ngrams are ordered word
ngrams are ordered word sequences
ordered
ordered word
ordered word sequences
sequences
```

word
word sequences

The ngram.py script can be easily modified to parse through all the sentences of any text, regardless of
length, building the list of ngrams as it proceeds.

Google has collected ngrams from scanned literature dating back to 1500. The public can enter their own
ngrams into Google's ngram viewer, and receive a graph of the published occurrences of the phrase,
through time [18]. We can use the Ngram viewer to find trends (e.g., peaks, valleys and periodicities) in
data. Consider the Google Ngram Viewer results for the two-word ngram, "yellow fever" (Fig. 2.3).

We see that the term "yellow fever" (a mosquito-transmitted hepatitis) appeared in the literature beginning about 1800, with several subsequent peaks. The dates of the peaks correspond roughly to outbreaks of yellow fever in Philadelphia (epidemic of 1793), New Orleans (epidemic of 1853), with
United States construction efforts in the Panama Canal (1904–14), and with well-documented WWII
Pacific outbreaks (about 1942). Following the 1942 epidemic an effective vaccine was available, and the
incidence of yellow fever, as well as the literature occurrences of the "yellow fever" n-gram, dropped
precipitously. In this case, a simple review of n-gram frequencies provides an accurate chart of historic
yellow fever outbreaks [19,18].

Nomenclature   A nomenclatures is a listing of terms that cover all of the concepts in a knowledge domain.
A nomenclature is different from a dictionary for three reasons: 1) the nomenclature terms are not annotated with definitions, 2) nomenclature terms may be multi-word, and 3) the terms in the nomenclature
are limited to the scope of the selected knowledge domain. In addition, most nomenclatures group synonyms under a group code. For example, a food nomenclature might collect submarine sandwich,
hoagie, po' boy, grinder, hero, and torpedo under an alphanumeric code such as "F63958." Nomenclatures simplify textual documents by uniting synonymous terms under a common code. Documents that
have been coded with the same nomenclature can be integrated with other documents that have been
similarly coded, and queries conducted over such documents will yield the same results, regardless of
which term is entered (i.e., a search for either hoagie, or po' boy will retrieve the same information, if both
terms have been annotated with the synonym code, "F63948"). Optimally, the canonical concepts listed
in the nomenclature are organized into a hierarchical classification [20,21,12].

Nomenclature mapping   Specialized nomenclatures employ specific names for concepts that are
included in other nomenclatures, under other names. For example, medical specialists often preserve
their favored names for concepts that cross into different fields of medicine. The term that pathologists
use for a certain benign fibrous tumor of the skin is "fibrous histiocytoma," a term spurned by

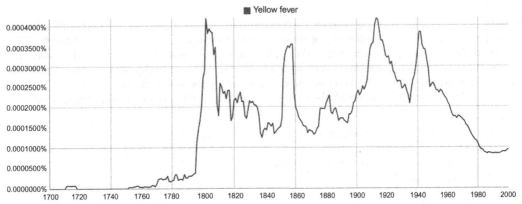

FIG. 2.3 Google Ngram for the phrase "yellow fever," counting occurrences of the term in a large corpus, from
the years 1700–2000. Peaks roughly correspond to yellow fever epidemics. *Source: Google Ngram viewer, with
permission from Google.*

dermatologists, who prefer to use "dermatofibroma" to describe the same tumor. As another horrifying example, the names for the physiologic responses caused by a reversible cerebral vasoconstricitve event include: thunderclap headache, Call-Fleming syndrome, benign angiopathy of the central nervous system, postpartum angiopathy, migrainous vasospasm, and migraine angiitis. The choice of term will vary depending on the medical specialty of the physician (e.g., neurologist, rheumatologist, obstetrician). To mitigate the discord among specialty nomenclatures, lexicographers may undertake a harmonization project, in which nomenclatures with overlapping concepts are mapped to one another.

Numpy Numpy (Numerical Python) is an open source extension to Python that supports matrix operations, as well as a rich assortment of mathematical functions. Numpy can be easily downloaded from sourceforge.net: http://sourceforge.net/projects/numpy/. Here is a short Python script, numpy_dot.py, that creates a 3x3 matrix, inverts the matrix, and calculates the dot produce of the matrix and its inverted counterpart.

```
import numpy
from numpy.linalg import inv
a = numpy.array([[1,4,6], [9,15,55], [62,-5, 4]])
print(a)
print(inv(a))
c = numpy.dot(a, inv(a))
print(numpy.round_(c))
```

The numpy_dot.py script employs numpy, numpy's linear algebra module, and numpy's matrix inversion method, and the numpy dot product method. Here is the output of the script, displaying the original matrix, its inversion, and the dot product, which happens to be the unity matrix:

```
c:\ftp\py>numpy_dot.py
[[ 1 4 6]
 [ 9 15 55]
 [62 -5 4]]
[[ 4.19746899e-02 -5.76368876e-03 1.62886856e-02]
 [ 4.22754041e-01 -4.61095101e-02 -1.25297582e-04]
 [ -1.22165142e-01 3.17002882e-02 -2.63124922e-03]]
[[1. 0. 0.]
 [0. 1. 0.]
 [0. 0. 1.]]
```

Parsing Much of computer programming involves parsing; moving sequentially through a file or some sort of data structure and performing operations on every contained item, one item at a time. For files, this might mean going through a text file line by line, or sentence by sentence. For a data file, this might mean performing an operation on each record in the file. For in-memory data structures, this may mean performing an operation on each item in a list or a tuple or a dictionary.

The parse_directory.py script prints all the file names and subdirectory names in a directory tree.

```
import os
for root, dirs, files in os.walk(".", topdown=False):
   for filename in files:
      print(os.path.join(root, filename))
   for dirname in dirs:
      print(os.path.join(root, dirname))
```

Plain-text Plain-text refers to character strings or files that are composed of the characters accessible to a typewriter keyboard. These files typically have a ".txt" suffix to their names. Plain-text files are sometimes referred to as 7-bit ascii files because all of the familiar keyboard characters have

ASCII vales under 128 (i.e., can be designated in binary with, just seven 0s and 1s. In practice, plain-text files exclude 7-bit ascii symbols that do not code for familiar keyboard characters. To further confuse the issue, plain-text files may contain ascii characters above 7 bits (i.e., characters from 128 to 255) that represent characters that are printable on computer monitors, such as accented letters.

Plesionymy Nearly synonymous words, or pairs of words that are sometimes synonymous; other times not. For example, the noun forms of "smell" and "odor" are synonymous. As verb forms, "smell" applies, but odor does not. You can small a fish, but you cannot odor a fish. Smell and odor are plesionyms. Plesionymy is another challenge for machine translators.

Polysemy Occurs when a word has more than one distinct meaning. The intended meaning of a word can sometimes be determined by the context in which the word is used. For example, "She rose to the occasion," and "Her favorite flower is the rose." Sometimes polysemy cannot be resolved. For example, "Eats shoots and leaves."

RegEx Short for Regular Expressions, RegEx is a syntax for describing patterns in text. For example, if I wanted to pull all lines from a text file that began with an uppercase "B" and contained at least one integer, and ended with the a lowercase x, then I might use the regular expression: " B.*[0-9].*x$". This syntax for expressing patterns of strings that can be matched by pre-built methods available to a programming language is somewhat standardized. This means that a RegEx expression in Perl will match the same pattern in Python, or Ruby, or any language that employs RegEx. The relevance of RegEx to Big Data is several-fold. RegEx can be used to build or transform data from one format to another; hence creating or merging data records. It can be used to convert sets of data to a desired format; hence transforming data sets. It can be used to extract records that meet a set of characteristics specified by a user; thus filtering subsets of data or executing data queries over text-based files or text-based indexes. The big drawback to using RegEx is speed: operations that call for many RegEx operations, particularly when those operations are repeated for each parsed line or record, will reduce software performance. RegEx-heavy programs that operate just fine on megabyte files may take hours, days or months to parse through terabytes of data.

A 12-line python script, file_search.py, prompts the user for the name of a text file to be searched, and then prompts the user to supply a RegEx pattern. The script will parse the text file, line by line, displaying those lines that contain a match to the RegEx pattern.

```
import sys, string, re
print("What is file would you like to search?")
filename = sys.stdin.readline()
filename = filename.rstrip()
print("Enter a word, phrase or regular expression to search.")
word_to_search = (sys.stdin.readline()).rstrip()
infile = open (filename, "r")
regex_object = re.compile(word_to_search, re.I)
for line in infile:
  m= regex_object.search(line)
  if m:
    print(line)
```

Scalable Software is scalable if it operates smoothly, whether the data is small or large. Software programs that operate by slurping all data into a RAM variable (i.e., a data holder in RAM memory) are not scalable, because such programs will eventually encounter a quantity of data that is too large to store in RAM. As a rule of thumb, programs that process text at speeds less than a megabyte per second are not scalable, as they cannot cope, in a reasonable time frame, with quantities of data in the gigabyte and higher range.

Script A script is a program that is written in plain-text, in a syntax appropriate for a particular programming language, that needs to be parsed through that language's interpreter before it can be compiled and executed. Scripts tend to run a bit slower than executable files, but they have the advantage that they can be understood by anyone who is familiar with the script's programming language.

Sentence Computers parse files line by line, not sentence by sentence. If you want a computer to perform operations on a sequence of sentences found in a corpus of text, then you need to include a subroutine in your scripts that list the sequential sentences. One of the simplest ways to find the boundaries of sentences is to look for a period followed by one or more spaces, followed by an uppercase letter. Here's a simple Python demonstration of a sentence extractor, using a few famous lines from the Lewis Carroll poem, Jabberwocky.

```
import re
all_text =\
"And, has thou slain the Jabberwock? Come \
to my arms, my beamish boy! O frabjous \
day! Callooh! Callay! He chortled in his \
joy. Lewis Carroll, excerpted from \
Jabberwocky";
sentence_list = re.split(r'[\.\!\?] +(?=[A-Z])', all_text)
print("\n".join(sentence_list))
```

Here is the output:

```
And, has thou slain the Jabberwock
Come to my arms, my beamish boy
O frabjous day
Callooh
Callay
He chortled in his joy
Lewis Carroll, excerpted from Jabberwocky
```

The meat of the script is the following line of code, which splits lines of text at the boundaries of sentences:

```
sentence_list = re.split(r'[\.\!\?] +(?=[A-Z])',in_text_string)
```

This algorithm is hardly foolproof, as periods are used for many purposes other than as sentence terminators. But it may suffice for most purposes.

Signal In a very loose sense a signal is a way of gauging how measured quantities (e.g., force, voltage, or pressure) change in response to, or along with, other measured quantities (e.g., time). A sound signal is caused by the changes in pressure, exerted on our eardrums, over time. A visual signal is the change in the photons impinging on our retinas, over time. An image is the change in pixel values over a two-dimensional grid. Because much of the data stored in computers consists of discrete quantities of describable objects, and because these discrete quantities change their values, with respect to one another, we can appreciate that a great deal of modern data analysis is reducible to digital signal processing.

Specification A specification is a method for describing objects (physical objects such as nuts and bolts or symbolic objects such as numbers). Specifications do not require specific types of information, and do not impose any order of appearance of the data contained in the document. Specifications do not generally require certification by a standards organization. They are generally produced by special interest organizations, and their legitimacy depends on their popularity. Examples of specifications are RDF (Resource Description Framework) produced by the W3C (WorldWide Web Consortium), and TCP/IP (Transfer Control Protocol/Internet Protocol), maintained by the Internet Engineering Task Force.

String A string is a sequence of characters. Words, phrases, numbers, and alphanumeric sequences (e.g., identifiers, one-way hash values, passwords) are strings. A book is a long string. The complete sequence of the human genome (3 billion characters, with each character an A,T,G, or C) is a very long string. Every subsequence of a string is another string.

Syntax Syntax is the standard form or structure of a statement. What we know as English grammar is equivalent to the syntax for the English language. If I write, "Jules hates pizza," the statement would be syntactically valid, but factually incorrect. If I write, "Jules drives to work in his pizza," the statement would be syntactically valid but nonsensical. For programming languages, syntax refers to the enforced structure of command lines. In the context of triplestores, syntax refers to the arrangement and notation requirements for the three elements of a statement (e.g., RDF format or N3 format). Charles Mead distinctly summarized the difference between syntax and semantics: "Syntax is structure; semantics is meaning" [22].

Systematics The term "systematics" is, by tradition, reserved for the field of biology that deals with taxonomy (i.e., the listing of the distinct types of organisms) and with classification (i.e., the classes of organisms and their relationships to one another). There is no reason why biologists should lay exclusive claim to the field of systematics. As used herein, systematics equals taxonomics plus classification, and this term applies just as strongly to stamp collecting, marketing, operations research, and object-oriented programming as it does to the field of biology.

Taxa Plural of taxon.

Taxon A taxon is a class. The common usage of "taxon" is somewhat inconsistent, as it sometimes refers to the class name, and at other times refers to the instances (i.e., members) of the class. In this book, the term "taxon" is abandoned in favor of "class," the plesionym used by computer scientists. Hence, the term "class" is used herein in the same manner that it is used in modern object oriented programming languages.

Taxonomy When we write of "taxonomy" as an area of study, we refer to the methods and concepts related to the science of classification, derived from the ancient Greek taxis, "arrangement," and nomia, "method." When we write of "a taxonomy," as a construction within a classification, we are referring to the collection of named instances (class members) in the classification. To appreciate the difference between a taxonomy and a classification, it helps to think of taxonomy as the scientific field that determines how different members of a classification are named. Classification is the scientific field that determines how related members are assigned to classes, and how the different classes are related to one another. A taxonomy is similar to a nomenclature; the difference is that in a taxonomy, every named instance must have an assigned class.

Term extraction algorithm Terms are phrases, most often noun phrases, and sometimes individual words, that have a precise meaning within a knowledge domain. For example, "software validation," "RDF triple," and "WorldWide Telescope" are examples of terms that might appear in the index or the glossary of this book. The most useful terms might appear up to a dozen times in the text, but when they occur on every page, their value as a searchable item is diminished; there are just too many instances of the term to be of practical value. Hence, terms are sometimes described as noun phrases that have low-frequency and high information content. Various algorithms are available to extract candidate terms from textual documents. The candidate terms can be examined by a curator who determines whether they should be included in the index created for the document from which they were extracted. The curator may also compare the extracted candidate terms against a standard nomenclature, to determine whether the candidate terms should be added to the nomenclature. For additional discussion, see Section 2.3, "Term Extraction."

Thesaurus A vocabulary that groups together synonymous terms. A thesaurus is very similar to a nomenclature. There are two minor differences. Nomenclatures do not always group terms by synonymy; and nomenclatures are often restricted to a well-defined topic or knowledge domain (e.g., names of stars, infectious diseases, etc.).

Transform (noun form) There are three truly great conceptual breakthroughs that have brought with them great advances to science and to civilization. The first two to be mentioned are well known to everyone: equations and algorithms. Equations permit us to relate variable quantities in a highly specific and repeatable way. Algorithms permit us to follow a series of steps that always produce the same results. The third conceptual breakthrough, less celebrated but just as important, is the transformation; a way of changing things to yield a something new, with properties that provide an advantage over the original item. In the case of reversible transformation, we can return the transformed item to its original form, and often in improved condition, when we have completed our task.

It should be noted that this definition applies only to the noun form of "transform." The meaning of the verb form of transform is to change or modify, and a transformation is the closest noun form equivalent of the verb form, "to transform."

Uniqueness Uniqueness is the quality of being separable from every other thing in the universe. For data scientists, uniqueness is achieved when data is bound to a unique identifier (i.e., a randomly chosen string of alphanumeric characters) that has not, and will never be, assigned to any data. The binding of data to a permanent and inseparable identifier constitutes the minimal set of ingredients for a data object. Uniqueness can apply to two or more indistinguishable objects, if they are assigned unique identifiers (e.g., unique product numbers stamped into identical auto parts).

Variable In algebra, a variable is a quantity, in an equation, that can change; as opposed to a constant quantity, that cannot change. In computer science, a variable can be perceived as a container that can be assigned a value. If you assign the integer 7 to a container named "x," then "x" equals 7, until you re-assign some other value to the container (i.e., variables are mutable). In most computer languages, when you issue a command assigning a value to a new (undeclared) variable, the variable automatically comes into existence to accept the assignment. The process whereby an object comes into existence, because its existence was implied by an action (such as value assignment), is called reification.

Vocabulary A comprehensive collection of the words used in a general area of knowledge. The term "vocabulary" and the term "nomenclature" are nearly synonymous. In common usage, a vocabulary is a list of words and typically includes a wide range of terms and classes of terms. Nomenclatures typically focus on a class of terms within a vocabulary. For example, a physics vocabulary might contain the terms "quark, black hole, Geiger counter, and Albert Einstein"; a nomenclature might be devoted to the names of celestial bodies.

References

[1] Krauthammer M, Nenadic G. Term identification in the biomedical literature. J Biomed Inform 2004;37:512–26.

[2] Berman JJ. Methods in medical informatics: fundamentals of healthcare programming in Perl, Python, and Ruby. Boca Raton: Chapman and Hall; 2010.

[3] Swanson DR. Undiscovered public knowledge. Libr Q 1986;56:103–18.

[4] Wallis E, Lavell C. Naming the indexer: where credit is due. The Indexer 1995;19:266–8.

[5] Hayes A. VA to apologize for mistaken Lou Gehrig's disease notices. CNN; 2009. August 26. Available from: http://www.cnn.com/2009/POLITICS/08/26/veterans.letters.disease [viewed September 4, 2012].

[6] Hall PA, Lemoine NR. Comparison of manual data coding errors in 2 hospitals. J Clin Pathol 1986;39:622–6.

[7] Berman JJ. Doublet method for very fast autocoding. BMC Med Inform Decis Mak 2004;4:16.

[8] Berman JJ. Nomenclature-based data retrieval without prior annotation: facilitating biomedical data integration with fast doublet matching. In Silico Biol 2005;5:0029.

[9] Burrows M, Wheeler DJ. a block-sorting lossless data compression algorithm. SRC Research Report 124, May 10, 1994.

[10] Berman JJ. Perl programming for medicine and biology. Sudbury, MA: Jones and Bartlett; 2007.

[11] Healy J, Thomas EE, Schwartz JT, Wigler M. Annotating large genomes with exact word matches. Genome Res 2003;13:2306–15.

[12] Berman JJ. Data simplification: taming information with open source tools. Waltham, MA: Morgan Kaufmann; 2016.

[13] Burrows-Wheeler transform. Wikipedia. Available at: https://en.wikipedia.org/wiki/Burrows%E2%80%93Wheeler_transform [viewed August 18, 2015].

[14] Cipra BA. The best of the 20th century: editors name top 10 algorithms. SIAM News May 2000;33(4).

[15] Wu X, Kumar V, Quinlan JR, Ghosh J, Yang Q, Motoda H, et al. Top 10 algorithms in data mining. Knowl Inf Syst 2008;14:1–37.

[16] Patil N, Berno AJ, Hinds DA, Barrett WA, Doshi JM, Hacker CR, et al. Blocks of limited haplotype diversity revealed by high-resolution scanning of human chromosome 21. Science 2001;294:1719–23.

[17] Paskin N. Identifier interoperability: a report on two recent ISO activities. D-Lib Mag 2006;12:1–23.

[18] Berman JJ. Repurposing legacy data: innovative case studies. Waltham, MA: Morgan Kaufmann; 2015.

[19] Berman JJ. Principles of big data: preparing, sharing, and analyzing complex information. Waltham, MA: Morgan Kaufmann; 2013.

[20] Berman JJ. Tumor classification: molecular analysis meets Aristotle. BMC Cancer 2004;4:10.

[21] Berman JJ. Tumor taxonomy for the developmental lineage classification of neoplasms. BMC Cancer 2004;4:88.

[22] Mead CN. Data interchange standards in healthcare IT–computable semantic interoperability: now possible but still difficult, do we really need a better mousetrap? J Healthc Inf Manag 2006;20:71–8.

3

Identification, Deidentification, and Reidentification

Section 3.1. What Are Identifiers?

Where is the 'any' key?

Homer Simpson, in response to his computer's instruction to "Press any key"

Let us begin this chapter with a riddle. "Is the number 5 a data object?" If you are like most people, you will answer "yes" because "5" is an integer and therefore it is represents numeric data, and "5" is an object because it exists and is different from all the other numbers. Therefore "5" is a data object. This line of reasoning happens to be completely erroneous. Five is not a data object. As a pure abstraction with nothing binding it to a physical object (e.g., 5 pairs of shoes, 5 umbrellas), it barely qualifies as data.

When we speak of a data object, in computer science, we refer to something that is identified and described. Consider the following statements:

```
<f183136d-3051-4c95-9e32-66844971afc5><name><Baltimore>
<f183136d-3051-4c95-9e32-66844971afc5><class><city>
<f183136d-3051-4c95-9e32-66844971afc5><population><620,961>
```

Without knowing much about data objects (which we will be discussing in detail in Section 6.2), we can start to see that these three statements are providing information about Baltimore. They tell us that Baltimore is a city of population 620,961, and that

Principles and Practice of Big Data. https://doi.org/10.1016/B978-0-12-815609-4.00003-0

53

Baltimore has been assigned an alphanumeric sequence, "f183136d-3051-4c95-9e32-66844971afc5," to which all our available information about Baltimore has been attached. Peeking ahead into Chapter 6, we can now surmise that a data object consists of a unique alphanumeric sequence (the object identifier) plus the descriptive information associated with the identifier (e.g., name, population number, class). We will see that there are compelling reasons for storing all information contained in Big Data resources within uniquely identified data objects. Consequently, one of the most important tasks for data managers is the creation of a dependable identifier system [1]. In this chapter, we will be focusing our attention on the unique identifier and how it is created and utilized in the realm of Big Data.

Identification issues are often ignored by data managers who are accustomed to working on small data projects. It is worthwhile to list, up front, the most important ideas described in this chapter, many of which are counterintuitive and strange to those whose careers are spent outside the confusing realm of Big Data.

- All Big Data resources can be imagined as identifier systems to which we attach our data.
- Without an adequate identification system, a Big Data resource has no value. In this case, the data within the resource cannot be sensibly analyzed.
- Data deidentification is a process whereby links to the public name of the subject of the record are removed.
- Deidentification should not be confused with the act of stripping a record of an identifier. A deidentified record, like any valid data object, must always have an associated identifier.
- Deidentification should not be confused with data scrubbing. Data scrubbers remove unwanted information from a data record, including information of a personal nature, and any information that is not directly related to the purpose of the data record. [Glossary Data cleaning, Data scrubbing]
- Reidentification is a concept that specifically involves personal and private data records. It involves ascertaining the name of the individual who is associated with a deidentified record. Reidentification is sometimes necessary to verify the contents of a record, or to provide information that is necessary for the well-being of the subject of a deidentified data record. Ethical reidentification always requires approval and oversight.
- Where there is no identification, there can be no deidentification and no reidentification.
- When a deidentified data set contains no unique records (i.e., every record has one or more additional records from which it cannot be distinguished, aside from its assigned identifier sequence), then it becomes impossible to maliciously uncover a deidentified record's public name.

Section 3.2. Difference Between an Identifier and an Identifier System

Many errors, of a truth, consist merely in the application the wrong names of things.
Baruch Spinoza

Data identification is among the most underappreciated and least understood Big Data issue. Measurements, annotations, properties, and classes of information have no informational meaning unless they are attached to an identifier that distinguishes one data object from all other data objects, and that links together all of the information that has been or will be associated with the identified data object. The method of identification and the selection of objects and classes to be identified relates fundamentally to the organizational model of the Big Data resource. If data identification is ignored or implemented improperly, the Big Data resource cannot succeed. [Glossary Annotation]

This chapter will describe, in some detail, the available methods for data identification, and the minimal properties of identified information (including uniqueness, exclusivity, completeness, authenticity, and harmonization). The dire consequences of inadequate identification will be discussed, along with real-world examples. Once data objects have been properly identified, they can be deidentified and, under some circumstances, reidentified. The ability to deidentify data objects confers enormous advantages when issues of confidentiality, privacy, and intellectual property emerge. The ability to reidentify deidentified data objects is required for error detection, error correction, and data validation. [Glossary Deidentification, Re-identification, Privacy versus confidentiality, Intellectual property]

Returning to the title of this section, let us ask ourselves, "What is the difference between an identifier and an identifier system?" To answer, by analogy, it is like the difference between having a $100 dollar bill in your pocket and having a savings account with $100 credited to the account. In the case of the $100 bill, anyone in possession of the bill can use it to purchase items. In the case of the $100 credit, there is a system in place for uniquely assigning the $100 to one individual, until such time as that individual conducts an account transaction that increases or decreases the account value. Likewise, an identifier system creates a permanent environment in which the identifiers are safely stored and used.

Every good information system is, at its heart, an identification system: a way of naming data objects so that they can be retrieved by their name, and a way of distinguishing each object from every other object in the system. If data managers properly identified their data, and did absolutely nothing else, they would be producing a collection of data objects with more informational value than many existing Big Data resources.

The properties of a good identifier system are the following:

- **Completeness**

Every unique object in the big data resource must be assigned an identifier.

– **Uniqueness**

Each identifier is a unique sequence.

– **Exclusivity**

Each identifier is assigned to a unique object, and to no other object.

– **Authenticity**

The objects that receive identification must be verified as the objects that they are intended to be. For example, if a young man walks into a bank and claims to be Richie Rich, then the bank must ensure that he is, in fact, who he says he is.

– **Aggregation**

The Big Data resource must have a mechanism to aggregate all of the data that is properly associated with the identifier (i.e., to bundle all of the data that belongs to the uniquely identified objected). In the case of a bank, this might mean collecting all of the transactions associated with an account holder. In a hospital, this might mean collecting all of the data associated with a patient's identifier: clinic visit reports, medication transactions, surgical procedures, and laboratory results. If the identifier system performs properly, aggregation methods will always collect all of the data associated with an object and will never collect any data that is associated with a different object.

– **Permanence**

The identifiers and the associated data must be permanent. In the case of a hospital system, when the patient returns to the hospital after 30 years of absence, the record system must be able to access his identifier and aggregate his data. When a patient dies, the patient's identifier must not perish.

– **Reconciliation**

There should be a mechanism whereby the data associated with a unique, identified object in one Big Data resource can be merged with the data held in another resource, for the same unique object. This process, which requires comparison, authentication, and merging is known as reconciliation. An example of reconciliation is found in health record portability. When a patient visits a hospital, it may be necessary to transfer her electronic medical record from another hospital. Both hospitals need a way of confirming the identity of the patient and combining the records. [Glossary Electronic medical record]

– **Immutability**

In addition to being permanent (i.e., never destroyed or lost), the identifier must never change (see Chapter 6) [2]. In the event that two Big Data resources are merged, or that legacy data is merged into a Big Data resource, or that individual data objects from two different Big Data resources are merged, a single data object will be assigned two identifiers; one from each of the merging systems. In this case, the identifiers must be preserved as they are, without modification. The merged data object must be provided with annotative information specifying the origin of each identifier (i.e., clarifying which identifier came from which Big Data resource).

– **Security**

The identifier system is vulnerable to malicious attack. A Big Data resource with an identifier system can be irreversibly corrupted if the identifiers are modified. In the case of human-based identifier systems, stolen identifiers can be used for a variety of malicious activities directed against the individuals whose records are included in the resource.

– **Documentation and Quality Assurance**

A system should be in place to find and correct errors in the identifier system. Protocols must be written for establishing the identifier system, for assigning identifiers, for protecting the system, and for monitoring the system. Every problem and every corrective action taken must be documented and reviewed. Review procedures should determine whether the errors were corrected effectively; and measures should be taken to continually improve the identifier system. All procedures, all actions taken, and all modifications of the system should be thoroughly documented. This is a big job.

– **Centrality**

Whether the information system belongs to a savings bank, an airline, a prison system, or a hospital, identifiers play a central role. You can think of information systems as a scaffold of identifiers to which data is attached. For example, in the case of a hospital information system, the patient identifier is the central key to which every transaction for the patient is attached.

– **Autonomy**

An identifier system has a life of its own, independent of the data contained in the Big Data resource. The identifier system can persist, documenting and organizing existing and future data objects even if all of the data in the Big Data resource were to suddenly vanish (i.e., when all of the data contained in all of the data objects are deleted).

In theory, identifier systems are incredibly easy to implement. Here is exactly how it is done:

1. Generate a unique character sequence, such as UUID, or a long random number. [Glossary UUID, Randomness]
2. Assign the unique character sequence (i.e., identifier) to each new object, at the moment that the object is created. In the case of a hospital a patient chart is created at the moment he or she is registered into the hospital information system. In the case of a bank a customer record is created at the moment that he or she is provided with an

account number. In the case of an object-oriented programming language, such as Ruby, this would be the moment when the "new" method is sent to a class object, instructing the class object to create a class instance. [Glossary Object-oriented programming, Instance]

3. Preserve the identifier number and bind it to the object. In practical terms, this means that whenever the data object accrues new data, the new data is assigned to the identifier number. In the case of a hospital system, this would mean that all of the lab tests, billable clinical transactions, pharmacy orders, and so on, are linked to the patient's unique identifier number, as a service provided by the hospital information system. In the case of a banking system, this would mean that all of the customer's deposits and withdrawals and balances are attached to the customer's unique account number.

Section 3.3. Generating Unique Identifiers

A UUID is 128 bits long, and can guarantee uniqueness across space and time.

P. Leach, M. Mealling and R. Salz [3]

Uniqueness is one of those concepts that everyone intuitively understands; explanations would seem unnecessary. Actually, uniqueness in the computational sciences is a some-what different concept than uniqueness in the natural world. In computational sciences, uniqueness is achieved when a data object is associated with an unique identifier (i.e., a character string that has not been assigned to any other data object). Most of us, when we think of a data object, are probably thinking of a data record, which may consist of the name of a person followed by a list of feature values (height, weight, and age), or a sample of blood followed by laboratory values (e.g., white blood cell count, red cell count, and hematocrit). For computer scientists a data object is a holder for data values (the so-called encapsulated data), descriptors of the data, and properties of the holder (i.e., the class of objects to which the instance belongs). Uniqueness is achieved when the data object is permanently bound to its own identifier sequence. [Glossary Encapsulation]

Unique objects have three properties:

- A unique object can be distinguished from all other unique objects.
- A unique object cannot be distinguished from itself.
- Uniqueness may apply to collections of objects (i.e., a class of instances can be unique).

UUID (Universally Unique IDentifier) is an example of one type of algorithm that creates unique identifiers, on command, at the moment when new objects are created (i.e., during the run-time of a software application). A UUID is 128 bits long and reserves 60 bits for a string computed directly from a computer time stamp, and is usually represented by a sequence of alphanumeric ASCII characters [3]. UUIDs were originally used in the Apollo

Network Computing System and were later adopted in the Open Software Foundation's Distributed Computing Environment [4]. [Glossary Time stamp, ASCII]

Linux systems have a built-in UUID utility, "uuidgen.exe," that can be called from the system prompt.

Here are a few examples of output values generated by the "uuidgen.exe" utility: [Glossary Command line utility, Utility]

```
$ uuidgen.exe
312e60c9-3d00-4e3f-a013-0d6cb1c9a9fe
$ uuidgen.exe
822df73c-8e54-45b5-9632-e2676d178664
$ uuidgen.exe
8f8633e1-8161-4364-9e98-fdf37205df2f
$ uuidgen.exe
83951b71-1e5e-4c56-bd28-c0c45f52cb8a
$ uuidgen -t
e6325fb6-5c65-11e5-b0e1-0ceee6e0b993
$ uuidgen -r
5d74e36a-4ccb-42f7-9223-84eed03291f9
```

Notice that each of the final two examples has a parameter added to the "uuidgen" command (i.e., "-t" and "-r"). There are several versions of the UUID algorithm that are available. The "-t" parameter instructs the utility to produce a UUID based on the time (measured in seconds elapsed since the first second of October 15, 1582, the start of the Gregorian calendar). The "-r" parameter instructs the utility to produce a UUID based on the generation of a pseudorandom number. In any circumstance, the UUID utility instantly produces a fixed length character string suitable as an object identifier. The UUID utility is trusted and widely used by computer scientists. Independent-minded readers can easily design their own unique object identifiers, using pseudorandom number generators, or with one-way hash generators. [Glossary One-way hash, Pseudorandom number generator]

Python has its own UUID generator. The uuid module is included in the standard python distribution and can be called directly from the script.

```
import uuid
print(uuid.uuid4())
```

When discussing UUIDs the question of duplicates (so-called collisions, in the computer science literature) always arises. How can we be certain that a UUID is unique? Isn't it possible that the algorithm that we use to create a UUID may, at some point, produce the same sequence on more than one occasion? Yes, but the odds are small. It has been estimated that duplicate UUIDs are produced, on average, once every 2.71 quintillion (i.e., $2.71 * 10^{18}$) executions [5]. It seems that reports of UUID collisions, when investigated, have been attributed to defects in the implementation of the UUID algorithms. The general consensus seems to be that UUID collisions are not worth worrying about, even in the realm of Big Data.

Section 3.4. Really Bad Identifier Methods

I always wanted to be somebody, but now I realize I should have been more specific.

Lily Tomlin

Names are poor identifiers. First off, we can never assume that any name is unique. Surnames such as Smith, Zhang, Garcia, Lo, and given names such as John and Susan are very common. In Korea, five last names account for nearly 50% of the population [6]. Moreover, if we happened to find an individual with a truly unique name (e.g., Mr. Mxyzptlk), there would be no guarantee that some other unique individual might one day have the same name. Compounding the non-uniqueness of names, there is the problem of the many variant forms of a single name. The sources for these variations are many. Here is a partial listing:

1. Modifiers to the surname (du Bois, DuBois, Du Bois, Dubois, Laplace, La Place, van de Wilde, Van DeWilde, etc.).
2. Accents that may or may not be transcribed onto records (e.g., acute accent, cedilla, diacritical comma, palatalized mark, hyphen, diphthong, umlaut, circumflex, and a host of obscure markings).
3. Special typographic characters (the combined "ae").
4. Multiple "middle names" for an individual, that may not be transcribed onto records. Individuals who replace their first name with their middle name for common usage, while retaining the first name for legal documents.
5. Latinized and other versions of a single name (Carl Linnaeus, Carl von Linne, Carolus Linnaeus, Carolus a Linne).
6. Hyphenated names that are confused with first and middle names (e.g., Jean-Jacques Rousseau, or Jean Jacques Rousseau; Louis-Victor-Pierre-Raymond, 7th duc de Broglie, or Louis Victor Pierre Raymond Seventh duc deBroglie).
7. Cultural variations in name order that are mistakenly rearranged when transcribed onto records. Many cultures do not adhere to the Western European name order (e.g., given name, middle name, surname).
8. Name changes; through marriage or other legal actions, aliasing, pseudonymous posing, or insouciant whim.

Aside from the obvious consequences of using names as record identifiers (e.g., corrupt database records, forced merges between incompatible data resources, impossibility of reconciling legacy record), there are non-obvious consequences that are worth considering. Take, for example, accented characters in names. These word decorations wreak havoc on orthography and on alphabetization. Where do you put a name that contains an umlauted character? Do you pretend the umlaut is not there, and alphabetize it according to its plain characters? Do you order based on the ASCII-numeric assignment for the character, in which the umlauted letter may appear nowhere near the plain-lettered words in an alphabetized list. The same problem applies to every special character. [Glossary American Standard Code for Information Interchange, ASCII]

A similar problem exists for surnames with modifiers. Do you alphabetize de Broglie under "D" or under "d" or under "B"? If you choose B, then what do you do with the concatenated form of the name, "deBroglie"? When it comes down to it, it is impossible to satisfactorily alphabetize a list of names. This means that searches based on proximity in the alphabet will always be prone to errors.

I have had numerous conversations with intelligent professionals who are tasked with the responsibility of assigning identifiers to individuals. At some point in every conversation, they will find it necessary to explain that although an individual's name cannot serve as an identifier, the combination of name plus date of birth provides accurate identification in almost every instance. They sometimes get carried away, insisting that the combination of name plus date of birth plus social security number provides perfect identification, as no two people will share all three identifiers: same name, same date of birth, same social security number. This argument rises to the height of folly and completely misses the point of identification. As we will see, it is relatively easy to assign unique identifiers to individuals and to any data object, for that matter. For managers of Big Data resources, the larger problem is ensuring that each unique individual has only one identifier (i.e., denying one object multiple identifiers). [Glossary Social Security Number]

Let us see what happens when we create identifiers from the name plus the birthdate. We will examine name + birthdate + social security number later in this section.

Consider this example. Mary Jessica Meagher, born June 7, 1912 decided to open a separate bank account in each of 10 different banks. Some of the banks had application forms, which she filled out accurately. Other banks registered her account through a teller, who asked her a series of questions and immediately transcribed her answers directly into a computer terminal. Ms. Meagher could not see the computer screen and could not review the entries for accuracy.

Here are the entries for her name plus date of birth:

1. Marie Jessica Meagher, June 7, 1912 (the teller mistook Marie for Mary).
2. Mary J. Meagher, June 7, 1912 (the form requested a middle initial, not name).
3. Mary Jessica Magher, June 7, 1912 (the teller misspelled the surname).
4. Mary Jessica Meagher, Jan 7, 1912 (the birth month was constrained, on the form, to three letters; Jun, entered on the form, was transcribed as Jan).
5. Mary Jessica Meagher, 6/7/12 (the form provided spaces for the final two digits of the birth year. Through a miracle of modern banking, Mary, born in 1912, was re-born a century later).
6. Mary Jessica Meagher, 7/6/2012 (the form asked for day, month, year, in that order, as is common in Europe).
7. Mary Jessica Meagher, June 1, 1912 (on the form, a 7 was mistaken for a 1).
8. Mary Jessie Meagher, June 7, 1912 (Marie, as a child, was called by the informal form of her middle name, which she provided to the teller).
9. Mary Jesse Meagher, June 7, 1912 (Marie, as a child, was called by the informal form of her middle name, which she provided to the teller, and which the teller entered as the male variant of the name).

10. Marie Jesse Mahrer, 1/1/12 (an underzealous clerk combined all of the mistakes on the form and the computer transcript, and added a new orthographic variant of the surname).

For each of these ten examples, a unique individual (Mary Jessica Meagher) would be assigned a different identifier at each of 10 banks. Had Mary re-registered at one bank, ten times, the outcome may have been the same.

If you toss the social security number into the mix (name + birth date + social security number) the problem is compounded. The social security number for an individual is anything but unique. Few of us carry our original social security cards. Our number changes due to false memory ("You mean I've been wrong all these years?"), data entry errors ("Character transpositoins, I mean transpositions, are very common"), intention to deceive ("I don't want to give those people my real number"), or desperation ("I don't have a number, so I'll invent one"), or impersonation ("I don't have health insurance, so I'll use my friend's social security number"). Efforts to reduce errors by requiring patients to produce their social security cards have not been entirely beneficial.

Beginning in the late 1930s, the E. H. Ferree Company, a manufacturer of wallets, promoted their product's card pocket by including a sample social security card with each wallet sold. The display card had the social security number of one of their employees. Many people found it convenient to use the card as their own social security number. Over time, the wallet display number was claimed by over 40,000 people. Today, few institutions require individuals to prove their identity by showing their original social security card. Doing so puts an unreasonable burden on the honest patient (who does not happen to carry his/her card) and provides an advantage to criminals (who can easily forge a card).

Entities that compel individuals to provide a social security number have dubious legal standing. The social security number was originally intended as a device for validating a person's standing in the social security system. More recently, the purpose of the social security number has been expanded to track taxable transactions (i.e., bank accounts, salaries). Other uses of the social security number are not protected by law. The Social Security Act (Section 208 of Title 42 U.S. Code 408) prohibits most entities from compelling anyone to divulge his/her social security number.

Considering the unreliability of social security numbers in most transactional settings, and considering the tenuous legitimacy of requiring individuals to divulge their social security numbers, a prudently designed medical identifier system will limit its reliance on these numbers. The thought of combining the social security number with name and date of birth will virtually guarantee that the identifier system will violate the strict one-to-a-customer rule.

Most identifiers are not purely random numbers; they usually contain some embedded information that can be interpreted by anyone familiar with the identification system. For example, they may embed the first three letters of the individual's family name in the identifier. Likewise, the last two digits of the birth year are commonly embedded in many types of identifiers. Such information is usually included as a crude "honesty" check by people "in the know." For instance, the nine digits of a social security number are divided into an area code (first three digits), a group number (the next two digits), followed by a serial number (last four digits). People with expertise in the social security numbering system can pry considerable information from a social security number, and can determine whether certain numbers are bogus, based on the presence of excluded sub-sequences.

Seemingly inconsequential information included in an identifier can sometimes be used to discover confidential information about individuals. Here is an example. Suppose every client transaction in a retail store is accessioned under a unique number, consisting of the year of the accession, followed by the consecutive count of accessions, beginning with the first accession of the new year. For example, accession 2010-3518582 might represent the 3,518,582nd purchase transaction in the year 2010. Because each number is unique, and because the number itself says nothing about the purchase, it may be assumed that inspection of the accession number would reveal nothing about the transaction.

Actually, the accession number tells you quite a lot. The prefix (2010) tells you the year of the purchase. If the accession number had been 2010-0000001, then you could safely say that accession represented the first item sold on the first day of business in the year 2010. For any subsequent accession number in 2010, simply divide the suffix number (in this case 3,518,582) by the last accession number of the year, and multiply by 365 (the number of days in a non-leap year), and you have the approximate day of the year that the transaction occurred. This day can easily be converted to a calendar date.

Unimpressed? Consider this scenario. You know that a prominent member of the President's staff had visited a Washington, D.C. Hospital on February 15, 2005, for the purpose of having a liver biopsy. You would like to know the results of that biopsy. You go to a Web site that lists the deidentified pathology records for the hospital, for the years 2000–2010. Though no personal identifiers are included in these public records, the individual records are sorted by accession numbers. Using the aforementioned strategy, you collect all of the surgical biopsies performed on or about February 15, 2010. Of these biopsies, only three are liver biopsies. Of these three biopsies, only one was performed on a person whose gender and age matched the President's staff member. The report provides the diagnosis. You managed to discover some very private information without access to any personal identifiers.

The alphanumeric character string composing the identifier should not expose the patient's identity. For example, a character string consisting of a concatenation of the patient's name, birth date, and social security number might serve to uniquely identify an individual, but it could also be used to steal an individual's identity. The safest identifiers are random character strings containing no information whatsoever.

Section 3.5. Registering Unique Object Identifiers

It isn't that they can't see the solution. It's that they can't see the problem.

G. K. Chesterton

Registries are trusted services that provide unique identifiers to objects. The idea is that everyone using the object will use the identifier provided by the central registry. Unique object registries serve a very important purpose, particularly when the object identifiers are persistent. It makes sense to have a central authority for Web addresses, library acquisitions, and journal abstracts. Such registries include:

- DOI, Digital object identifier
- PMID, PubMed identification number
- LSID (Life Science Identifier)
- HL7 OID (Health Level 7 Object Identifier)
- DICOM (Digital Imaging and Communications in Medicine) identifiers
- ISSN (International Standard Serial Numbers)
- Social Security Numbers (for United States population)
- NPI, National Provider Identifier, for physicians
- Clinical Trials Protocol Registration System
- Office of Human Research Protections FederalWide Assurance number
- Data Universal Numbering System (DUNS) number
- International Geo Sample Number
- DNS, Domain Name Service
- URL, Unique Resource Locator [Glossary URL]
- URN, Unique Resource Name [Glossary URN]

In some cases the registry does not provide the full identifier for data objects. The registry may provide a general identifier sequence that will apply to every data object in the resource. Individual objects within the resource are provided with a non-unique registry number. A unique suffix sequence is appended locally (i.e., not by a central registrar). Life Science Identifiers (LSIDs) serve as a typical example of a registered identifier. Every LSIDs is composed of the following 5 parts: Network Identifier, root DNS name of the issuing authority, name chosen by the issuing authority, a unique object identifier assigned locally, and an optional revision identifier for versioning information.

In the issued LSID identifier, the parts are separated by a colon, as shown:

urn:lsid:pdb.org:1AFT:1

This identifies the first version of the 1AFT protein in the Protein Data Bank. Here are a few LSIDs:

urn:lsid:ncbi.nlm.nih.gov:pubmed:12571434

This identifies a PubMed citation

urn:lsid:ncbi.nlm.nig.gov:GenBank:T48601:2

This refers to the second version of an entry in GenBank

An OID, short for Object Identifier, is a hierarchy of identifier prefixes. Successive numbers in the prefix identify the descending order of the hierarchy. Here is an example of an OID from HL7, an organization that deals with health data interchanges:

1.3.6.1.4.1.250

Each node is separated from the successor by a dot, Successively finer registration detail leads to the institutional code (the final node). In this case the institution identified by the HL7 OID happens to be the University of Michigan.

The final step in creating an OID for a data object involves placing a unique identifier number at the end of the registered prefix. OID organizations leave the final step to the institutional data managers.

The problem with this approach is that the final within-institution data object identifier is sometimes prepared thoughtlessly, corrupting the OID system [7]. Here is an example. Hospitals use an OID system for identifying images, part of the DICOM (Digital Imaging and Communications in Medicine) image standard. There is a prefix consisting of a permanent, registered code for the institution and the department, and a suffix consisting of a number generated for an image as it is created.

A hospital may assign consecutive numbers to its images, appending these numbers to an OID that is unique for the institution and the department within the institution. For example, the first image created with a CT-scanner might be assigned an identifier consisting of the OID (the assigned code for institution and department) followed by a separator such as a hyphen, followed by "1."

In a worst-case scenario, different instruments may assign consecutive numbers to images, independently of one another. This means that the CT-scanner in room A may be creating the same identifier (OID + image number) as the CT-scanner in Room B; for images on different patients. This problem could be remedied by constraining each CT-scanner to avoid using numbers assigned by any other CT-scanner. This remedy can be defeated if there is a glitch anywhere in the system that accounts for image assignments (e.g., if the counters are re-set, broken, replaced or simply ignored).

When image counting is done properly, and the scanners are constrained to assign unique numbers (not previously assigned by other scanners in the same institution), each image may indeed have a unique identifier (OID prefix + image number suffix). Nonetheless, the use of consecutive numbers for images will create havoc over time. Problems arise when the image service is assigned to another department in the institution, or when departments or institutions merge. Each of these shifts produces a change in the OID (the institutional and departmental prefix) assigned to the identifier. If a consecutive numbering system is used, then you can expect to create duplicate identifiers if institutional prefixes are replaced after the merge. The old records in both of the merging institutions will be assigned the same prefix and will contain replicate (consecutively numbered) suffixes (e.g., image 1, image 2, etc.).

Yet another problem may occur if one unique object is provided with multiple different unique identifiers. A software application may be designed to ignore any previously assigned unique identifier and to generate its own identifier, using its own assignment method. Doing so provides software vendors with a strategy that insulates the vendors from bad identifiers created by their competitor's software, and locks the customer to a vendor's software, and identifiers, forever.

In the end the OID systems provide a good set of identifiers for the institution, but the data objects created within the institution need to have their own identifier systems. Here is the HL7 statement on replicate OIDs:

Though HL7 shall exercise diligence before assigning an OID in the HL7 branch to third parties, given the lack of a global OID registry mechanism, one cannot make absolutely certain that there is no preexisting OID assignment for such third-party entity [8].

It remains to be seen whether any of the registration identifier systems will be used and supported with any serious level of permanence (e.g., over decades and centuries).

Section 3.6. Deidentification and Reidentification

Never answer an anonymous letter.

Yogi Berra

For scientists, deidentification serves two purposes:

- To protect the confidentiality and the privacy of the individual (when the data concerns a particular human subject), and
- To remove information that might bias the experiment (e.g., to blind the experimentalist to patient identities).

Deidentification involves stripping information from a data record that might link the record to the public name of the record's subject. In the case of a patient record, this would involve stripping any information from the record that would enable someone to connect the record to the name of the patient. The most obvious item to be removed in the deidentification process is the patient's name. Other information that should be removed would be the patient's address (which could be linked to the name), the patient's date of birth (which narrows down the set of individuals to whom the data record might pertain), and the patient's social security number. In the United States, patient privacy regulations include a detailed discussion of record deidentification and this discussion recommends 18 patient record items for exclusion from deidentified records [9].

Before going any further, it is important to clarify that deidentification is not achieved by removing an identifier from a data object. In point of fact, nothing good is ever achieved by simply removing an identifier from a data object; doing so simply invalidates the data object (i.e., every data object, identified or deidentified, must have an identifier). Deidentification involves removing information contained in the data object that reveals something about the publicly known name of the data object. This kind of information is often referred to as identifying information, but it would be much less confusing if we used another term for such data, such as "name-linking information." The point here is that we do not want to confuse the identifier of a data object with information contained in a data object that can link the object to its public name.

It may seem counterintuitive, but there is very little difference between an identifier and a deidentifier; under certain conditions the two concepts are equivalent. Here is how a dual identification/deidentification system might work:

1. Collect data on unique object. "Joe Ferguson's bank account contains $100."
2. Assign a unique identifier. "Joe Ferguson's bank account is 7540038947134."
3. Substitute name of object with its assigned unique identifier: "754003894713 contains $100.".

4. Consistently use the identifier with data.

5. Do not let anyone know that Joe Ferguson owns account "754003894713."

The dual use of an identifier/deidentifier is a tried and true technique. Swiss bank accounts are essentially unique numbers (identifiers) assigned to a person. You access the bank account by producing the identifier number. The identifier number does not provide information about the identity of the bank account holder (i.e., it is a deidentifier and an identifier).

The purpose of an identifier is to tell you that whenever the identifier is encountered, it refers to the same unique object, and whenever two different identifiers are encountered, they refer to different objects. The identifier, by itself, should contain no information that links the data object to its public name.

It is important to understand that the process of deidentification can succeed only when each record is properly identified (i.e., there can be no deidentification without identification). Attempts to deidentify a poorly identified data set of clinical information will result in replicative records (multiple records for one patient), mixed-in records (single records composed of information on multiple patients), and missing records (unidentified records lost in the deidentification process).

The process of deidentification is best understood as an algorithm performed on-the-fly, in response to a query from a data analyst. Here is how such an algorithm might proceed.

1. The data analyst submits a query requesting a record from a Big Data resource. The resource contains confidential records that must not be shared, unless the records are deidentified.

2. The Big Data resource receives the query and retrieves the record.

3. A copy of the record is parsed and any of the information within the data record that might link the record to the public name of the subject of the record (usually the name of an individual) is deleted from the copy. This might include the aforementioned name, address, date of birth, and social security number.

4. A pseudo-identifier sequence is prepared for the deidentified record. The pseudo-identifier sequence might be generated by a random number generator, by encrypting the original identifier, through a one-way hash algorithm, or by other methods chosen by the Big Data manager. [Glossary Encryption]

5. A transaction record is attached to the original record that includes the pseudo-identifier, the deidentified record, the time of the transaction, and any information pertaining to the requesting entity (e.g., the data analyst who sent the query) that is deemed fit and necessary by the Big Data resource data manager.

6. A record is sent to the data analyst that consists of the deidentified record (i.e., the record stripped of its true identifier and containing no data that links the record to a named person) and the unique pseudo-identifier created for the record.

Because the deidentified record, and its unique pseudo-identifier are stored with the original record, subsequent requests for the pseudo-identified record can be retrieved and provided, at the discretion of the Big Data manager. This general approach to data deidentification will apply to requests for a single record or to millions of records.

At this point, you might be asking yourself the following question, "What gives the data manager the right to distribute parts of a confidential record, even if it happens to be deidentified?" You might think that if you tell someone a secret, under the strictest confidence, then you would not want any part of that secret to be shared with anyone else. The whole notion of sharing confidential information that has been deidentified may seem outrageous and unacceptable.

We will discuss the legal and ethical issues of Big Data in Chapters 18 and 19. For now, readers should know that there are several simple and elegant principles that justify sharing deidentified data.

Consider the statement "Jules Berman has a blood glucose level of 85." This would be considered a confidential statement because it tells people something about my medical condition.

Consider the phrase, "Blood glucose 85."

When the name "Jules Berman" is removed, we are left with a disembodied piece of data. "Blood glucose 85" is no different from "Temperature 98.6" or "Apples 2" or "Terminator 3." They are simply raw data belonging to nobody in particular. The act of removing information linking data to a person renders the data harmless. Because the use of properly deidentified data poses no harm to human subjects, United States Regulations allow the unrestricted use of such data for research purposes [9,10]. Other countries have similar provisions.

– Reidentification

Because confidentiality and privacy concerns always apply to human subject data, it would seem imperative that deidentification should be an irreversible process (i.e., the names of the subjects and samples should be held a secret, forever).

Scientific integrity does not always accommodate irreversible deidentification. On occasion, experimental samples are mixed-up; samples thought to come from a certain individual, tissue, record, or account, may in fact come from another source. Sometimes major findings in science need to be retracted when a sample mix-up has been shown to occur [11,12,13,14,15]. When samples are submitted, without mix-up, the data is sometimes collected improperly. For example, reversing electrodes on an electrocardiogram may yield spurious and misleading results. Sometimes data is purposefully fabricated and otherwise corrupted, to suit the personal agendas of dishonest scientists. When data errors occur, regardless of reason, it is important to retract the publications [16,17]. To preserve scientific integrity, it is sometimes necessary to discover the identity of deidentified records.

In some cases, deidentification stops the data analyst from helping individuals whose confidentiality is being protected. Imagine you are conducting an analysis on a collection of deidentified data, and you find patients with a genetic marker for a disease that is curable, if treated at an early stage; or you find a new biomarker that determines which patients would benefit from surgery and which patients would not. You would be compelled to contact the subjects in the database to give them information that could potentially save their lives. Having an irreversibly deidentified data sets precludes any intervention with subjects; nobody knows their identities.

Deidentified records can, under strictly controlled circumstances, be reidentified. Reidentification is typically achieved by entrusting a third party with a confidential list that maps individuals to their deidentified records. Obviously, reidentification can only occur if the Big Data resource keeps a link connecting the identifiers of their data records to the identifiers of the corresponding deidentified record (what we've been calling pseudo-identifiers). The act of assigning a public name to the deidentified record must always involve strict oversight. The data manager must have in place a protocol that describes the process whereby approval for reidentification is obtained. Reidentification provides an opportunity whereby confidentiality can be breached and human subjects can be harmed. Consequently, stewarding the reidentification process is one of the most serious responsibilities of Big Data managers [18].

Section 3.7. Case Study: Data Scrubbing

It is a sin to believe evil of others but it is seldom a mistake.

Garrison Keillor

The term "data scrubbing" is sometimes used, mistakenly, as a synonym for deidentification. It is best to think of data scrubbing as a process that begins where deidentification ends. A data scrubber will remove unwanted information from a data record, including information of a personal nature and any information that is not directly related to the purpose of the data record. For example, in the case of a hospital record a data scrubber might remove the names of physicians who treated the patient; the names of hospitals or medical insurance agencies; addresses; dates; and any textual comments that are inappropriate, incriminating, irrelevant, or potentially damaging. [Glossary Data munging, Data scraping, Data wrangling]

In medical data records, there is a concept known as "minimal necessary" that applies to shared confidential data [9]. It holds that when records are shared, only the minimum necessary information should be released. Any information not directly relevant to the intended purposes of the data analyst should be withheld. The process of data scrubbing gives data managers the opportunity to render a data record that is free of information that would link the record to its subject and free of extraneous information that the data analyst does not actually require. [Glossary Minimal necessary]

There are many methods for data scrubbing. Most of these methods require that data managers develop an exception list of items that should not be included in shared records (e.g., cities, states, zip codes, and names of people). The scrubbing application moves through the records, extracting unnecessary information along the way. The end product is cleaned, but not sterilized. Though many undesired items can be successfully removed, this approach never produces a perfectly scrubbed set of data. In a Big Data resource, it is simply impossible for the data manager to anticipate every objectionable item and to include it in an exception list. Nobody is that smart.

There is, however, a method whereby data records can be cleaned, without error. This method involves creating a list of data (often in the form of words and phrases) that is acceptable for inclusion in a scrubbed and deidentified data set. Any data that is not in the list of acceptable information is automatically deleted. Whatever is left is the scrubbed data. This method can be described as a reverse scrubbing method. Everything is in the data set is automatically deleted, unless it is an approved "exception."

This method of scrubbing is very fast and can produce an error-free deidentified and scrubbed output [4,19,20]. An example of the kind of output produced by such a scrubber is shown:

> *Since the time when * * * * * * * * his own * and the * * * *, the anomalous * * have been * and persistent * * *; and especially * true of the construction and functions of the human *, indeed, it was the anomalous that was * * * in the * the attention, * * that were * to develop into the body * * which we now * *. As by the aid * * * * * * * * * our vision into the * * * has emerged *, we find * * and even evidence of *. To the highest type of * * it is the * the ordinary * * * * *. * to such, no less than to the most *, * * * is of absorbing interest, and it is often * * that the * * the most * into the heart of the mystery of the ordinary. * * been said, * * * * *. * * dermoid cysts, for example, we seem to * * * the secret * of Nature, and * out into the * * of her clumsiness, and * of her * * * *, *, * tell us much of * * * used by the vital * * * * even the silent * * * upon the * * *.*

The reverse-scrubber requires the preexistence of a set of approved terms. One of the simplest methods for generating acceptable terms involves extracting them from a nomenclature that comprehensively covers the terms used in a knowledge domain. For example, a comprehensive listing of living species will not contain dates or zip codes or any of the objectionable language or data that should be excluded from a scrubbed data set. In a method that I have published a list of approved doublets (approximately 200,000 two-word phrases collected from standard nomenclatures) are automatically collected for the scrubbing application [4]. The script is fast, and its speed is not significantly reduced by the size of the list of approved terms.

Here is a short python script. scrub.py, that will take any line of text and produce a scrubbed output. It requires an external file, doublets.txt, containing an approved list of doublet terms.

```python
import sys, re, string
doub_file = open("doublets.txt", "r")
doub_hash = {}
for line in doub_file:
    line = line.rstrip()
    doub_hash[line] = " "
```

```
doub_file.close()
print("What would you like to scrub?")
line = sys.stdin.readline()
line = line.lower()
line = line.rstrip()
linearray = re.split(r' +', line)
lastword = "*"
for i in range(0, len(linearray)):
  doublet = " ".join(linearray[i:i+2])
  if doublet in doub_hash:
    print(" " + linearray[i], end="")
    lastword = " " + linearray[i+1]
  else:
    print(lastword, end="")
    lastword = " *"
  if (i == len(linearray) + 1):
    print(lastword, end="")
```

Section 3.8. Case Study (Advanced): Identifiers in Image Headers

Plus ca change, plus c'est la meme chose.
> *Old French saying ("The more things change, the more things stay the same.")*

As it happens, nothing is ever as simple as it ought to be. In the case of an implementation of systems that employ long sequence generators to produce unique identifiers, the most common problem involves indiscriminate reassignment of additional unique identifiers to the same data object, thus nullifying the potential benefits of the unique identifier systems.

Let us look at an example wherein multiple identifiers are redundantly assigned to the same image, corrupting the identifier system. In Section 4.3, we discuss image headers, and we provide examples wherein the ImageMagick "identify" utility could extract the textual information included in the image header. One of the header properties created, inserted, and extracted by ImageMagick's "identify" is an image-specific unique string. [Glossary ImageMagick]

When ImageMagick is installed in our computer, we can extract any image's unique string, using the "identify" utility and the "-format" attribute, on the following system command line: [Glossary Command line]

```
c:\>identify -verbose -format "%#" eqn.jpg
```

Here, the image file we are examining is "eqn.jpg". The "%#" character string is ImageMagick's special syntax indicating that we would like to extract the image identifier from the image header. The output is shown.

```
219e41b4c761e4bb04fbd67f71cc84cd6ae53a26639d4bf33155a5f62ee36e33
```

We can repeat the command line whenever we like, for this image; and the same image-specific unique sequence of characters will be produced.

Using ImageMagick, we can insert text into the "comment" section of the header, using the "-set" attribute. Let us add the text, "I'm modifying myself":

```
c:\ftp>convert eqn.jpg -set comment "I'm modifying myself" eqn.jpg
```

Now, let us extract the comment that we just added, to satisfy ourselves that the "-set" attribute operated as we had hoped. We do this using the "-format" attribute and the "%c" character string, which is ImageMagick's syntax for extracting the comment section of the header.

```
c:\ftp>identify -verbose -format "%c" eqn.jpg
```

The output of the command line is:

```
I'm modifying myself
```

Now, let us run, one more time, the command line that produces the unique character string that is unique for the eqn.jpg image file

```
c:\ftp>identify -verbose -format "%#" eqn.jpg
```

The output is:

```
cb448260d6eeeb2e9f2dcb929fa421b474021584e266d486a6190067a278639f
```

What just happened? Why has the unique character string specific for the eqn.jpg image changed? Has our small modification of the file, which consisted of adding a text comment to the image header, resulted in the production of a new image object, worthy of a new unique identifier?

Before answering these very important questions, let us pose the following gedanken question. Imagine you have a tree. This tree, like every living organism, is unique. It has a unique history, a unique location, and a unique genome (i.e., a unique sequence of nucleotides composing its genetic material). In ten years, its leaves drop off and are replaced ten times. Its trunk expands in size and its height increases. In the ten years of its existence, has the identify of the tree changed? [Glossary Gedanken]

You would probably agree that the tree has changed, but that it has maintained its identity (i.e., it is still the same tree, containing the descendants of the same cells that grew within the younger version of itself).

In informatics, a newly created object is given an identifier, and this identifier is immutable (i.e., cannot be changed), regardless of how the object is modified. In the case of the unique string assigned to an image by ImageMagick, the string serves as an authenticator, not as an identifier. When the image is modified a new unique string is created. By comparing the so-called identifier string in copies of the image file, we can determine whether any modifications have been made. That is to say, we can authenticate the file.

Getting back to the image file in our example, when we modified the image by inserting a text comment, ImageMagick produced a new unique string for the image. The identity of the image had not changed, but the image was different from the original image (i.e., no longer authentic). It seems that the string that we thought to be an identifier string was actually an authenticator string. [Glossary Authentication]

If we want an image to have a unique identifier that does not change when the image is modified, we must create our own identifier that persists when the image is modified.

Here is a short Python script, image_id.py, that uses Python's standard UUID method to create an identifier, which is inserted into the comment section of the image's header, and flanking the identifier with XML tags. [Glossary XML, HTML]

```
import sys, os, uuid
my_id = "<image_id>" + str(uuid.uuid4()) + "</image_id>"
in_command = "convert leaf.jpg -set comment \"" + my_id + "\" leaf.jpg"
os.system(in_command)
out_command = "identify -verbose -format \"%c\" leaf.jpg"
print ("\nHere's the unique identifier:")
os.system(out_command)
print ("\nHere's the unique authenticator:")
os.system("identify -verbose -format \"%#\" leaf.jpg")
os.system("convert leaf.jpg -resize 325x500! leaf.jpg")
print ("\nHere's the new authenticator:")
os.system("identify -verbose -format \"%#\" leaf.jpg")
print ("\nHere's the unique identifier:")
os.system(out_command)
```

Here is the output of the image_id.py script:

```
Here's the unique identifier:
<image_id>b0836a26-8f0e-4a6b-842d-9b0dde2b3f59</image_id>

Here's the unique authenticator:
98c9fe07e90ce43f49961ab6226cd1ccffee648edd1a456a9d06a53ad6d3215a

Here's the new authenticator:
017e401d80a41aafa289ae9c2a1adb7c00477f7a943143141912189499d69ad2

Here's the unique identifier:
<image_id>b0836a26-8f0e-4a6b-842d-9b0dde2b3f59</image_id>
```

What did the script do and what does it teach us? It employed the UUID utility to create a unique and permanent identifier for the image (leaf.jpg, in this case), and inserted the unique identifier into the image header. This identifier, "b0836a26-8f0e-4a6b-842d-9b0dde2b3f59," did not change when the image was subsequently modified. A new authenticator string was automatically inserted into the image header, by ImageMagick, when the image was modified. Hence, we achieved what we needed to achieve: a unique identifier that never changes, and a unique authenticator that changes when the image is modified in any way.

If you have followed the logic of this section, then you are prepared for the following question posed as an exercise for Zen Buddhists. Imagine you have a hammer. Over the years, you have replaced its head, twice, and its handle, thrice. In this case, with nothing remaining of the original hammer, has it maintained its identity (i.e., is it still the same hammer?). The informatician would answer "Yes," the hammer has maintained its unique identity, but it is no longer authentic (i.e., it is what it must always be, though it has become something different).

Section 3.9. Case Study: One-Way Hashes

I live on a one-way street that's also a dead end. I'm not sure how I got there.

Steven Wright

A one-way hash is an algorithm that transforms a string into another string is such a way that the original string cannot be calculated by operations on the hash value (hence the term "one-way" hash). Popular one-way hash algorithms are MD5 and Standard Hash Algorithm (SHA). A one-way hash value can be calculated for any character string, including a person's name, or a document, or even another one-way hash. For a given input string, the resultant one-way hash will always be the same.

Here are a few examples of one-way hash outputs performed on a sequential list of input strings, followed by their one-way hash (md5 algorithm) output.

```
Jules Berman => Ri0oaVTIAilwnS8+nvKhfA
"Whatever" => n2YtKKG6E4MyEZvUKyGWrw
Whatever => OkXaDVQFYjwkQ+MOC8dpOQ
jules berman => SlnuYpmyn8VXLsxBWwO57Q
Jules J. Berman => i74wZ/CsIbxt3goH2aCS+A
Jules J Berman => yZQfJmAf4dIYO6Bd0qGZ7g
Jules Berman => Ri0oaVTIAilwnS8+nvKhfA
```

The one-way hash values are a seemingly random sequence of ASCII characters (the characters available on a standard keyboard). Notice that a small variation among input strings (e.g., exchanging an uppercase for a lowercase character, adding a period or quotation mark) produces a completely different one-way hash output. The first and the last entry (Jules Berman) yield the same one-way hash output (Ri0oaVTIAilwnS8+nvKhfA) because the two input strings are identical. A given string will always yield the same hash value, so long as the hashing algorithm is not altered. Each one-way hash has the same length (22 characters for this particular md5 algorithm) regardless of the length of the input term. A one-way hash output of the same length (22 characters) could have been produced for a string or file or document of any length. Once produced, there is no feasible mathematical algorithm that can reconstruct the input string from its one-way hash output. In our example, there is no way of examining the string "Ri0oaVTIAilwnS8+nvKhfA" and computing the name Jules Berman.

We see that the key functional difference between a one-way hash and a UUID sequence is that the one-way hash algorithm, performed on a unique string, will always yield the same random-appearing alphanumeric sequence. A UUID algorithm has no input string; it simply produces unique alphanumeric output, and never (almost never) produces the same alphanumeric output twice.

One-way hashes values can serve as ersatz identifiers, permitting Big Data resources to accrue data, over time, to a specific record, even when the record is deidentified (e.g., even when its UUID identifier has been stripped from the record). Here is how it works [18]:

1. A data record is chosen, before it is deidentified, and a one-way hash is performed on its unique identifier string.
2. The record is deidentified by removing the original unique identifier. The output of the one-way hash (from step 1) is substituted for the original unique identifier.
3. The record is deidentified because nobody can reconstruct the original identifier from the one-way hash that has replaced it.
4. The same process is done for every record in the database.
5. All of the data records that were associated with the original identifier will now have the same one-way hash identifier and can be collected under this substitute identifier, which cannot be computationally linked to the original identifier.

Implementation of one-way hashes carry certain practical problems. If anyone happens to have a complete listing of all of the original identifiers, then it would be a simple matter to perform one-way hashes on every listed identifier. This would produce a look-up table that can match deidentified records back to the original identifier, a strategy known as a dictionary attack. For deidentification to work, the original identifier sequences must be kept secret.

One-way hash protocols have many practical uses in the field of information science [21,18,4]. It is very easy to implement one-way hashes, and most programming languages and operating systems come bundled with one or more implementations of one-way hash algorithms. The two most popular one-way hash algorithms are md5 (message digest version 5) and SHA (Secure Hash Algorithm). [Glossary HMAC, Digest, Message digest, Check digit]

Here we use Cygwin's own md5sum.exe utility on the command line to produce a one-way hash for an image file, named dash.png:

```
c:\ftp>c:\cygwin64\bin\md5sum.exe dash.png
```

Here is the output:

```
db50dc33800904ab5f4ac90597d7b4ea *dash.png
```

We could call the same command line from a Python script:

```
import sys, os
os.system("c:/cygwin64/bin/md5sum.exe dash.png")
```

The output will always be the same, as long as the input file, dash.png, does not change:

```
db50dc33800904ab5f4ac90597d7b4ea *dash.png
```

OpenSSL contains several one-way hash implementations, including both md5 and several variants of SHA.

One-way hashes on files are commonly used as a quick and convenient authentication tool. When you download a file from a Web site, you are likely to see that the file distributor has posted the file's one-way hash value. When you receive the file, it is a good idea to calculate the one-way hash on the file that you have received. If the one-way hash value is equal to the posted one-way hash value, then you can be certain that the file received is an exact copy of the file that was intentionally sent. Of course, this does not ensure that the file that was intentionally sent was a legitimate file or that the website was an honest file broker. We will be using our knowledge of one-way hashes when we discuss trusted time stamps (Section 8.5), blockchains (Section 8.6) and data security protocols (Section 18.3).

Glossary

ASCII ASCII is the American Standard Code for Information Interchange, ISO-14962-1997. The ASCII standard is a way of assigning specific 8-bit strings (a string of 0s and 1s of length 8) to the alphanumeric characters and punctuation. Uppercase letters are assigned a different string of 0s and 1s than their matching lowercase letters. There are 256 ways of combining 0s and 1s in strings of length 8. This means that that there are 256 different ASCII characters, and every ASCII character can be assigned a number-equivalent, in the range of 0–255. The familiar keyboard keys produce ASCII characters that happen to occupy ASCII values under 128. Hence, alphanumerics and common punctuation are represented as 8-bits, with the first bit, "0", serving as padding. Hence, keyboard characters are commonly referred to as 7-bit ASCII, and files composed exclusively of common keyboard characters are referred to as plain-text files or as 7-bit ASCII files.
These are the classic ASCII characters:

```
!"#$%&'()*+,-./0123456789:;<=>
?@ABCDEFGHIJKLMNOPQRSTUVWXYZ
[\]^_`abcdefghijklmnopqrstuvwxyz{|}~
```

Python has several methods for removing non-printable characters from text, including the "printable" method, as shown in this short script, printable.py.

```
# -*- coding: iso-8859-15 -*-

import string
in_string = "prinüéâäàtable"
out_string = "".join(s for s in in_string if s in string.printable)
print(out_strung)
output:
printable
```

It is notable that the first line of code violates a fundamental law of Python programming; that the pound sign signifies that a comment follows, and that the Python interpreter will ignore the pound sign and any characters that follow the pound sign on the line in which they appear. For obscure reasons, the top line of the snippet is a permitted exception to the rule. In nonpythonic language, the top line conveys to the Python compiler that it may expect to find non-ASCII characters encoded in the iso-8859-15 standard.

The end result of this strange snippet is that non-ASCII characters are stripped from input strings; a handy script worth saving.

American Standard Code for Information Interchange Long form of the familiar acronym, ASCII.

Annotation Annotation involves describing data elements with metadata or attaching supplemental information to data objects.

Authentication A process for determining if the data object that is received (e.g., document, file, image) is the data object that was intended to be received. The simplest authentication protocol involves one-way hash operations on the data that needs to be authenticated. Suppose you happen to know that a certain file, named temp.txt will be arriving via email and that this file has an MD5 hash of "a0869a42609af6c712caeba454f47429". You receive the temp.txt file, and you perform an MD5 one-way hash operation on the file.

In this example, we will use the md5 hash utility bundled into the CygWin distribution (i.e., the Linux emulator for Windows systems). Any md5 implementation would have sufficed.

```
c:\cygwin64\bin>openssl md5 temp.txt
MD5(temp.txt)= a0869a42609af6c712caeba454f47429
```

We see that the md5 hash value generated for the received file is identical to the md5 hash value produced on the file, by the file's creator, before the file was emailed. This tells us that the received, temp.txt, is authentic (i.e., it is the file that you were intended to receive) because no other file has the same MD5 hash. Additional implementations of one-way hashes are described in **Section 3.9.**

The authentication process, in this example, does not tell you who sent the file, the time that the file was created, or anything about the validity of the contents of the file. These would require a protocol that included signature, time stamp, and data validation, in addition to authentication. In common usage, authentication protocols often include entity authentication (i.e., some method by which the entity sending the file is verified). Consequently, authentication protocols are often confused with signature verification protocols. An ancient historical example serves to distinguish the concepts of authentication protocols and signature protocols. Since earliest of recorded history, fingerprints were used as a method of authentication. When a scholar or artisan produced a product, he would press his thumb into the clay tablet, or the pot, or the wax seal closing a document. Anyone doubting the authenticity of the pot could ask the artisan for a thumbprint. If the new thumbprint matched the thumbprint on the tablet, pot, or document, then all knew that the person creating the new thumbprint and the person who had put his thumbprint into the object were the same individual. Hence, ancient pots were authenticated. Of course, this was not proof that the object was the creation of the person with the matching thumbprint. For all anyone knew, there may have been a hundred different pottery artisans, with one person pressing his thumb into every pot produced. You might argue that the thumbprint served as the signature of the artisan. In practical terms, no. The thumbprint, by itself, does not tell you whose print was used. Thumbprints could not be read, at least not in the same way as a written signature. The ancients needed to compare the pot's thumbprint against the thumbprint of the living person who made the print. When the person died, civilization was left with a bunch of pots with the same thumbprint, but without any certain way of knowing whose thumb produced them. In essence, because there was no ancient database that permanently associated thumbprints with individuals, the process of establishing the identity of the pot-maker became very difficult once the artisan died. A good signature protocol permanently binds an authentication code to a unique entity (e.g., a person). Today, we can find a fingerprint at the scene of a crime; we can find a matching signature in a database; and we can link the fingerprint to one individual. Hence, in modern times, fingerprints are true "digital" signatures, no pun intended. Modern uses of fingerprints include keying (e.g., opening locked devices based on an authenticated fingerprint), tracking (e.g., establishing the path and whereabouts of an individual by following a trail of fingerprints or other identifiers), and body part identification (i.e., identifying the remains of individuals recovered from mass graves or from the sites of catastrophic events based on fingerprint matches). Over the past decade, flaws in the vaunted process of fingerprint identification have been documented, and the improvement of the science of identification is an active area of investigation [22].

Check digit A checksum that produces a single digit as output is referred to as a check digit. Some of the common identification codes in use today, such as ISBN numbers for books, come with a built-in check digit. Of course, when using a single digit as a check value, you can expect that some transmitted errors will escape the check, but the check digit is useful in systems wherein occasional mistakes are tolerated; or wherein the purpose of the check digit is to find a specific type of error (e.g., an error produced by a substitution in a single character or digit), and wherein the check digit itself is rarely transmitted in error.

Command line Instructions to the operating system, that can be directly entered as a line of text from the a system prompt (e.g., the so-called C prompt, "c:\>", in Windows and DOS operating systems; the so-called shell prompt, "$", in Linux-like systems).

Command line utility Programs lacking graphic user interfaces that are executed via command line instructions. The instructions for a utility are typically couched as a series of arguments, on the command line, following the name of the executable file that contains the utility.

Data cleaning More correctly, data cleansing, and synonymous with data fixing or data correcting. Data cleaning is the process by which errors, spurious anomalies, and missing values are somehow handled. The options for data cleaning are: correcting the error, deleting the error, leaving the error unchanged, or imputing a different value [23]. Data cleaning should not be confused with data scrubbing.

Data munging Refers to a multitude of tasks involved in preparing data for some intended purpose (e.g., data cleaning, data scrubbing, and data transformation). Synonymous with data wrangling.

Data scraping Pulling together desired sections of a data set or text by using software.

Data scrubbing A term that is very similar to data deidentification and is sometimes used improperly as a synonym for data deidentification. Data scrubbing refers to the removal of unwanted information from data records. This may include identifiers, private information, or any incriminating or otherwise objectionable language contained in data records, as well as any information deemed irrelevant to the purpose served by the record.

Data wrangling Jargon referring to a multitude of tasks involved in preparing data for eventual analysis. Synonymous with data munging [24].

Deidentification The process of removing all of the links in a data record that can connect the information in the record to an individual. This usually includes the record identifier, demographic information (e.g., place of birth), personal information (e.g., birthdate), and biometrics (e.g., fingerprints). The deidentification strategy will vary based on the type of records examined. Deidentifying protocols exist wherein deidentificated records can be reidentified, when necessary.

Digest As used herein, "digest" is equivalent to a one-way hash algorithm. The word "digest" also refers to the output string produced by a one-way hash algorithm.

Electronic medical record Abbreviated as EMR, or as EHR (Electronic Health Record). The EMR is the digital equivalent of a patient's medical chart. Central to the idea of the EMR is the notion that all of the documents, transactions, and all packets of information containing test results and other information on a patient are linked to the patient's unique identifier. By retrieving all data linked to the patient's identifier, the EMR (i.e., the entire patient's chart) can be assembled instantly.

Encapsulation The concept, from object oriented programming, that a data object contains its associated data. Encapsulation is tightly linked to the concept of introspection, the process of accessing the data encapsulated within a data object. Encapsulation, Inheritance, and Polymorphism are available features of all object-oriented languages.

Encryption A common definition of encryption involves an algorithm that takes some text or data and transforms it, bit-by-bit, into an output that cannot be interpreted (i.e., from which the contents of the source file cannot be determined). Encryption comes with the implied understanding that there exists some reverse transform that can be applied to the encrypted data, to reconstitute the original source. As used herein, the definition of encryption is expanded to include any protocols by which files can be shared, in such a way that only the intended recipients can make sense of the received documents. This would include protocols that divide files into pieces that can only be reassembled into the original file using a password. Encryption would also include protocols that alter parts of a file while retaining the original text in other parts of the file. As described in Chapter 5, there are instances when some data in a file should be shared, while only specific parts need to be encrypted. The protocols that accomplish these kinds of file transformations need not always employ classic encryption algorithms (e.g., Winnowing and Chaffing [25], threshold protocols [21]).

Gedanken Gedanken is the German word for "thought." A gedanken experiment is one in which the scientist imagines a situation and its outcome, without resorting to any physical construction of a scientific trial. Albert Einstein, a consummate theoretician, was fond of inventing imaginary scenarios, and his use of the term "gedanken trials" has done much to popularize the concept. The scientific literature contains multiple descriptions of gedanken trials that have led to fundamental breakthroughs in our understanding of the natural world and of the universe [26].

HMAC Hashed Message Authentication Code. When a one-way hash is employed in an authentication protocol, it is often referred to as an HMAC.

HTML HyperText Markup Language is an ASCII-based set of formatting instructions for web pages. HTML formatting instructions, known as tags, are embedded in the document, and double-bracketed, indicating the start point and end points for instruction. Here is an example of an HTML tag instructing the web browser to display the word "Hello" in italics: <i> Hello </i>. All web browsers conforming to the HTML specification must contain software routines that recognize and implement the HTML instructions embedded within in web documents. In addition to formatting instructions, HTML also includes linkage instructions, in which the web browsers must retrieve and display a listed web page, or a web resource, such as an image. The protocol whereby web browsers, following HTML instructions, retrieve web pages from other Internet sites, is known as HTTP (HyperText Transfer Protocol).

ImageMagick An open source utility that supports a huge selection of robust and sophisticated image editing methods. ImageMagick is available for download at: https://www.imagemagick.org/script/download.php

Instance An instance is a specific example of an object that is not itself a class or group of objects. For example, Tony the Tiger is an instance of the tiger species. Tony the Tiger is a unique animal and is not itself a group of animals or a class of animals. The terms instance, instance object, and object are sometimes used interchangeably, but the special value of the "instance" concept, in a system wherein everything is an object, is that it distinguishes members of classes (i.e., the instances) from the classes to which they belong.

Intellectual property Data, software, algorithms, and applications that are created by an entity capable of ownership (e.g., humans, corporations, and universities). The entity holds rights over the manner in which the intellectual property can be used and distributed. Protections for intellectual property may come in the form of copyrights and patent. Copyright applies to published information. Patents apply to novel processes and inventions. Certain types of intellectual property can only be protected by being secretive. For example, magic tricks cannot be copyrighted or patented; this is why magicians guard their intellectual property so closely. Intellectual property can be sold outright, essentially transferring ownership to another entity; but this would be a rare event. In other cases, intellectual property is retained by the creator who permits its limited use to others via a legal contrivance (e.g., license, contract, transfer agreement, royalty, and usage fee). In some cases, ownership of the intellectual property is retained, but the property is freely shared with the world (e.g., open source license, GNU license, FOSS license, and Creative Commons license).

Message digest Within the context of this book, "message digest", "digest", "HMAC", and "one-way hash" are equivalent terms.

Minimal necessary In the field of medical informatics, there is a concept known as "minimal necessary" that applies to shared confidential data [9]. It holds that when records are shared, only the minimum necessary information should be released. Information not directly relevant to the intended purposes of the study should be withheld.

Object-oriented programming In object-oriented programming, all data objects must belong to one of the classes built into the language or to a class created by the programmer. Class methods are subroutines that belong to a class. The members of a class have access to the methods for the class. There is a hierarchy of classes (with superclasses and subclasses). A data object can access any method from any superclass of its class. All object-oriented programming languages operate under this general strategy. The two most important differences among the object oriented programming languages relate to syntax (i.e., the required style in which data objects call their available methods) and content (the built-in classes and methods available to objects). Various esoteric issues, such as types of polymorphism offered by the language, multi-parental inheritance, and non-Boolean logic operations may play a role in how expert programmer's choose a specific object-oriented language for the job at-hand.

One-way hash A one-way hash is an algorithm that transforms one string into another string (a fixed-length sequence of seemingly random characters) in such a way that the original string cannot be calculated by operations on the one-way hash value (i.e., the calculation is one-way only). One-way hash values can be calculated for any string, including a person's name, a document, or an image. For any given input string, the resultant one-way hash will always be the same. If a single byte of the input string is modified, the resulting one-way hash will be changed, and will have a totally different sequence than the one-way hash sequence calculated for the unmodified string.

Most modern programming languages have several methods for generating one-way hash values. Regardless of the language we choose to implement a one-way hash algorithm (e.g., md5, SHA), the output value will be identical. One-way hash values are designed to produce long fixed-length output strings (e.g., 256 bits in length). When the output of a one-way hash algorithm is very long, the chance of a hash string collision (i.e., the occurrence of two different input strings generating the same one-way hash output value) is negligible. Clever variations on one-way hash algorithms have been repurposed as identifier systems [27,28,29,30]. A detailed discussion of one-way hash algorithms can be found in Section 3.9, "Case Study: One-Way Hashes."

Privacy versus confidentiality The concepts of confidentiality and of privacy are often confused, and it is useful to clarify their separate meanings. Confidentiality is the process of keeping a secret with which you have been entrusted. You break confidentiality if you reveal the secret to another person. You violate privacy when you use the secret to annoy the person whose confidential information was acquired. If you give a friend your unlisted telephone number in confidence, then your fried is expected to protect this confidentiality by never revealing the number to other persons. In addition, your friend may be expected to protect your privacy by resisting the temptation to call you in the middle of the night, complain about a mutual acquaintance. In this case, the same information object (unlisted telephone number) is encumbered by separable confidentiality and privacy obligations.

Pseudorandom number generator It is impossible for computers to produce an endless collection of truly random numbers. Eventually, algorithms will cycle through their available variations and begins to repeat themselves, producing the same set of "random" numbers, in the same order; a phenomenon referred to as the generator's period. Because algorithms that produce seemingly random numbers are imperfect, they are known as pseudorandom number generators. The Mersenne Twister algorithm, which has an extremely long period, is used as the default random number generator in Python. This algorithm performs well on most of the tests that mathematicians have devised to test randomness.

Randomness Various tests of randomness are available [31]. One of the easiest to implement takes advantage of the property that random strings are uncompressible. If you can show that if a character string, a series of numbers, or a column of data cannot be compressed by gzip, then it is pretty safe to conclude that the data is randomly distributed, and without any informational value.

Reidentification A term casually applied to any instance whereby information can be linked to a specific person after the links between the information and the person associated with the information were removed. Used this way, the term reidentification connotes an insufficient deidentification process. In the healthcare industry, the term "reidentification" means something else entirely. In the United States, regulations define "reidentification" under the "Standards for Privacy of Individually Identifiable Health Information". Reidentification is defined therein as a legally valid process whereby deidentified records can be linked back to the respective human subjects, under circumstances deemed compelling by a privacy board. Reidentification is typically accomplished via a confidential list of links between human subject names and deidentified records, held by a trusted party. As used by the healthcare industry, reidentification only applies to the approved process of re-establishing the identity of a deidentified record. When a human subject is identified through fraud, trickery, or through the deliberate use of computational methods to break the confidentiality of insufficiently deidentified records, the term "reidentification" would not apply.

Social Security Number The common strategy, in the United States, of employing social security numbers as identifiers is often counterproductive, owing to entry error, mistaken memory, or the intention to deceive. Efforts to reduce errors by requiring individuals to produce their original social security cards puts an unreasonable burden on honest individuals, who rarely carry their cards, and provides an advantage to dishonest individuals, who can easily forge social security cards. Institutions that compel patients to provide a social security number have dubious legal standing. The social security number was originally intended as a device for validating a person's standing in the social security system. More recently, the purpose of the social security number has been expanded to track taxable transactions (i.e., bank accounts, salaries). Other uses of the social security number are not protected by law. The Social Security Act (Section 208 of Title 42 U.S. Code 408) prohibits most entities from compelling anyone to divulge his/her social security number. Legislation or judicial action may one day stop healthcare institutions from compelling patients to divulge their social security numbers as a condition for providing medical care. Prudent and forward-thinking institutions will limit their reliance on social security numbers as personal identifiers.

Time stamp Many data objects are temporal events and all temporal events must be given a time stamp indicating the time that the event occurred, using a standard measurement for time. The time stamp must be accurate, persistent, and immutable. The Unix epoch time (equivalent to the Posix epoch time) is available for most operating systems and consists of the number of seconds that have elapsed since January 1, 1970, midnight, Greenwhich mean time. The Unix epoch time can easily be converted into any other standard representation of time. The duration of any event can be easily calculated by subtracting the beginning time from the ending time. Because the timing of events can be maliciously altered, scrupulous data managers employ a trusted time stamp protocol by which a time stamp can be verified. A trusted time stamp must be accurate, persistent, and immutable. Trusted time stamp protocols are discussed in Section 8.5, "Case Study: The Trusted Time stamp."

URL Unique Resource Locator. The Web is a collection of resources, each having a unique address, the URL. When you click on a link that specifies a URL, your browser fetches the page located at the unique location specified in the URL name. If the Web were designed otherwise (i.e., if several different web pages had the same web address, or if one web address were located at several different locations), then the web could not function with any reliability.

URN Unique Resource Name. Whereas the URL identifies objects based on the object's unique location in the Web, the URN is a system of object identifiers that are location-independent. In the URN system, data objects are provided with identifiers, and the identifiers are registered with, and subsumed by, the URN.
For example:

```
urn:isbn-13:9780128028827
```

Refers to the unique book, "Repurposing Legacy Data: Innovative Case Studies," by Jules Berman

```
urn:uuid:e29d0078-f7f6-11e4-8ef1-e808e19e18e5
```

Refers to a data object tied to the UUID identifier e29d0078-f7f6-11e4-8ef1-e808e19e18e5.
In theory, if every data object were assigned a registered URN, and if the system were implemented as intended, the entire universe of information could be tracked and searched.

UUID UUID, the abbreviation for Universally Unique IDentifiers, is a protocol for assigning identifiers to data objects, without using a central registry. UUIDs were originally used in the Apollo Network Computing System [3].

Utility In the context of software, a utility is an application that is dedicated to performing one specific task, very well, and very fast. In most instances, utilities are short programs, often running from the command line, and thus lacking any graphic user interface. Many utilities are available at no cost, with open source code. In general, simple utilities are preferable to multi-purpose software applications [32]. Remember, an application that claims to do everything for the user is, most often, an application that requires the user to do everything for the application.

XML Abbreviation for eXtensible Markup Language. A syntax for marking data values with descriptors (metadata). The descriptors are commonly known as tags. In XML, every data value is enclosed by a start-tag, indicating that a value will follow, and an end-tag, indicating that the value had preceded the tag. For example: < name > Tara Raboomdeay < /name >. The enclosing angle brackets, "<>", and the end-tag marker, "/", are hallmarks of XML markup. This simple but powerful relationship between metadata and data allows us to employ each metadata/data pair as though it were a small database that can be combined with related metadata/data pairs from any other XML document. The full value of metadata/data pairs comes when we can associate the pair with a unique object, forming a so-called triple.

References

[1] Reed DP. Naming and synchronization in a decentralized computer system. Doctoral Thesis, MIT; 1978.

[2] Joint NEMA/COCIR/JIRA Security and Privacy Committee (SPC). Identification and allocation of basic security rules in healthcare imaging systems. Available from: http://www.medicalimaging. org/wp-content/uploads/2011/02/Identification_and_Allocation_of_Basic_Security_Rules_In_ Healthcare_Imaging_Systems-September_2002.pdf; September 2002 [viewed January 10, 2013].

[3] Leach P, Mealling M, Salz R. A Universally Unique IDentifier (UUID) URN Namespace. Network Working Group, Request for Comment 4122, Standards Track. Available from: http://www.ietf.org/rfc/ rfc4122.txt [viewed November 7, 2017].

[4] Berman JJ. Methods in medical informatics: fundamentals of healthcare programming in Perl, Python, and Ruby. Boca Raton: Chapman and Hall; 2010.

[5] Mathis FH. A generalized birthday problem. SIAM Rev 1991;33:265–70.

[6] Dimitropoulos LL. Privacy and security solutions for interoperable health information exchange perspectives on patient matching: approaches, findings, and challenges. RTI International, Indianapolis, June 30, 2009.

[7] Kuzmak P, Casertano A, Carozza D, Dayhoff R, Campbell K. Solving the Problem of Duplicate Medical Device Unique Identifiers High Confidence Medical Device Software and Systems (HCMDSS) Workshop, Philadelphia, PA, June 2–3. Available from: http://www.cis.upenn.edu/hcmdss/Papers/submissions/; 2005 [viewed August 26, 2012].

[8] Health Level 7 OID Registry. Available from: http://www.hl7.org/oid/frames.cfm [viewed August 26, 2012].

[9] Department of Health and Human Services. 45 CFR (code of federal regulations), parts 160 through 164. Standards for privacy of individually identifiable health information (final rule). Fed Regist 2000;65(250):82461–510. December 28.

[10] Department of Health and Human Services. 45 CFR (Code of Federal Regulations), 46. Protection of Human Subjects (Common Rule). Fed Regist 1991;56:28003–32. June 18.

[11] Knight J. Agony for researchers as mix-up forces retraction of ecstasy study. Nature 2003;425:109.

[12] Sainani K. Error: What biomedical computing can learn from its mistakes. Biomed Comput Rev 2011;12–9. Fall.

[13] Palanichamy MG, Zhang Y. Potential pitfalls in MitoChip detected tumor-specific somatic mutations: a call for caution when interpreting patient data. BMC Cancer 2010;10:597.

[14] Bandelt H, Salas A. Contamination and sample mix-up can best explain some patterns of mtDNA instabilities in buccal cells and oral squamous cell carcinoma. BMC Cancer 2009;9:113.

[15] Harris GUS. Inaction lets look-alike tubes kill patients. The New York Times; August 20, 2010.

[16] Flores G. Science retracts highly cited paper: study on the causes of childhood illness retracted after author found guilty of falsifying data. The Scientist; June 17, 2005.

[17] Gowen LC, Avrutskaya AV, Latour AM, Koller BH, Leadon SA. Retraction of: Gowen LC, Avrutskaya AV, Latour AM, Koller BH, Leadon SA. Science. 1998 Aug 14;281(5379):1009–12. Science 2003;300:1657.

[18] Berman JJ. Confidentiality issues for medical data miners. Artif Intell Med 2002;26:25–36.

[19] Berman JJ. Concept-match medical data scrubbing: how pathology datasets can be used in research. Arch Pathol Lab Med Arch Pathol Lab Med 2003;127:680–6.

[20] Berman JJ. Comparing de-identification methods, March 31. Available from: http://www.biomedcentral.com/1472-6947/6/12/comments/comments.htm; 2006 [viewed Jan. 1, 2015].

[21] Berman JJ. Threshold protocol for the exchange of confidential medical data. BMC Med Res Methodol 2002;2:12.

[22] A review of the FBI's handling of the Brandon Mayfield case. U. S. Department of Justice, Office of the Inspector General, Oversight and Review Division; March 2006.

[23] Van den Broeck J, Cunningham SA, Eeckels R, Herbst K. Data cleaning: detecting, diagnosing, and editing data abnormalities. PLoS Med 2005;2:e267.

[24] Lohr S. For big-data scientists, 'janitor work' is key hurdle to insights. The New York Times; August 17, 2014.

[25] Rivest RL. Chaffing and winnowing: confidentiality without encryption. MIT Lab for Computer Science; March 18, 1998. (rev. April 24, 1998). Available from: http://people.csail.mit.edu/rivest/chaffing-980701.txt [viewed January 10, 2017].

[26] Berman JJ. Armchair science: no experiments, just deduction. Amazon Digital Services, Inc. (Kindle Book); 2014.

[27] Faldum A, Pommerening K. An optimal code for patient identifiers. Comput Methods Prog Biomed 2005;79:81–8.

[28] Rivest R, Request for comments: 1321, the MD5 message-digest algorithm. Network Working Group. https://www.ietf.org/rfc/rfc1321.txt [viewed January 1, 2015].

[29] Bouzelat H, Quantin C, Dusserre L. Extraction and anonymity protocol of medical file. Proc AMIA Annu Fall Symp 1996;1996:323–7.

[30] Quantin CH, Bouzelat FA, Allaert AM, Benhamiche J, Faivre J, Dusserre L. Automatic record hash coding and linkage for epidemiological followup data confidentiality. Methods Inf Med 1998;37:271–7.

[31] Marsaglia G, Tsang WW. Some difficult-to-pass tests of randomness. J Stat Softw 2002;7:1–8.

[32] Brooks FP. No silver bullet: essence and accidents of software engineering. Computer 1987;20:10–9.

4

Metadata, Semantics, and Triples

Section 4.1. Metadata

Life is a concept.

Patrick Forterre [1]

When you think about it, numbers are meaningless. The number "8" has no connection to anything in the physical realm until we attach some information to the number (e.g., 8 candles, 8 minutes). Some numbers, like "0" or "−5" have no physical meaning under any set of circumstances. There really is no such thing as "0 dollars"; it is an abstraction indicating the absence of a positive number of dollars. Likewise, there is no such thing as "−5 walnuts"; it is an abstraction that we use to make sense of subtractions $(5 - 10 = -5)$.

When we write "8 walnuts," "walnuts" is the metadata that tells us what is being referred to by the data, in this case the number "8."

When we write "8 o'clock", "8" is the data and "o'clock" is the metadata.

Section 4.2. eXtensible Markup Language

The purpose of narrative is to present us with complexity and ambiguity.

Scott Turow

XML (eXtensible Markup Language) is a syntax for attaching descriptors (so-called metadata) to data values. [Glossary Metadata]

In XML, descriptors are commonly known as tags.

Principles and Practice of Big Data. https://doi.org/10.1016/B978-0-12-815609-4.00004-2

85

XML has its own syntax; a set of rules for expressing data/metadata pairs. Every data value is flanked by a start-tag and an end-tag. Enclosing angle brackets, "<>", and the end-tag marker, "/", are hallmarks of XML markup. For example:

```
< name >Tara Raboomdeay</name>
```

This simple but powerful relationship between metadata and data allows us to employ every metadata/data pair as a miniscule database that can be combined with related metadata/data pairs from the same XML document or from different XML documents.

It is impossible to overstate the importance of XML (eXtensible Markup Language) as a data organization tool. With XML, every piece of data tells us something about itself. When a data value has been annotated with metadata, it can be associated with other, related data, even when the other data is located in a seemingly unrelated database. [Glossary Integration].

When all data is flanked by metadata, it is relatively easy to port the data into spreadsheets, where the column headings correspond to the metadata tags, and the data values correspond to the value found in the cells of the spreadsheet. The rows correspond to the record number.

A file that contains XML markup is considered a proper XML document only if it is well formed. Here are the properties of a well-formed XML document.

- The document must have a proper XML header. The header can vary somewhat, but it usually looks something like:
  ```
  <?xml version="1.0" ?>
  ```
- XML files are ASCII files consisting of characters available to a standard keyboard.
- Tags in XML files must conform to composition rules (e.g., spaces are not permitted within a tag, and tags are case-sensitive).
- Tags must be properly nested (i.e., no overlapping). For example, the following is properly nested XML.

```
< chapter ><chapter_title >Introspection</chapter_title ></chapter >
```

Compare the previous example, with the following, improperly nested XML.

```
< chapter ><chapter_title >Introspection</chapter ></chapter_title >
```

Web browsers will not display XML files that are not well formed.

The actual structure of an XML file is determined by another XML file known as an XML Schema. The XML Schema file lists the tags and determines the structure for those XML files that are intended to comply with a specific Schema document. A valid XML file conforms to the rules of structure and content defined in its assigned XML Schema.

Every XML file that is valid under a particular Schema will contain data that is described using the same tags that are listed in that same XML schema, permitting data integration among those files. This is one of the strengths of XML.

The greatest drawback of XML is that data/metadata pairs are not assigned to a unique object. XML describes its data, but it does not tell us the object of the data. This gaping hole in XML was filled by RDF (Resource Description Framework), a modified XML syntax designed to associate every data/metadata pair with a unique data object. Before we can begin to understand RDF, we need to understand the concept of "meaning," in the context of information science.

Section 4.3. Semantics and Triples

Supplementary bulletin from the Office of Fluctuation Control, Bureau of Edible Condiments, Soluble and Indigestible Fats and Glutinous Derivatives, Washington, D.C. Correction of Directive #943456201, . . . the quotation on ground-hog meat should read 'ground hog meat.'

Bob Elliot and Ray Goulding, comedy routine

Metadata gives structure to data values, but it does not tell us anything about how the data value relates to anything else. For example,

```
<height_in_feet_inches>5'11"</height_in_feet_inches>
```

What does it mean to know that 5'11" is a height attribute, expressed in feet and inches? Nothing really. The metadata/data pair has no meaning, as it stands, because it does not describe anything in particular. If we were to assert that John Harrington has a height of 5'11", then we would be making a meaningful statement. This brings us to ask ourselves: What is the meaning of meaning? This question sounds like another one of those Zen mysteries that has no answer. In informatics, "meaningfulness" is achieved when described data (i.e., a metadata/data pair) is bound to the unique identifier of a data object.

Let us look once more at our example:

```
"John Harrington's height is five feet eleven inches."
```

This sentence has meaning because there is data (five feet eleven inches), and it is described (person's height), and it is bound to a unique individual (John Harrington). Let us generate a unique identifier for John Harrington using our UUID generator (discussed in Section 3.3) and rewrite our assertion in a format in which metadata/data pairs are associated with a unique identifier:

```
9c7bfe97-e637-461f-a30b-d931b97907fe   name    John Harrington
9c7bfe97-e637-461f-a30b-d931b97907fe   height  5'11"
```

We now have two meaningful assertions: one that associates the name "John Harrington" with a unique identifier (9c7bfe97-e637-461f-a30b-d931b97907fe); and one that tells us that the object associated with the unique identifier (i.e., John Harrington) is 5'11" tall. We could insert these two assertions into a Big Data resource, knowing that both assertions fulfill our definition of meaning. Of course, we would need to have some process in place to ensure that any future information collected on our unique John Harrington (i.e., the John Harrington assigned the identifier 9c7bfe97-e637-461f-a30b-d931b97907fe) will be assigned the same identifier.

A statement with meaning does not need to be a true statement (e.g., The height of John Harrington was not 5 feet 11 inches when John Harrington was an infant). That is to say, an assertion can be meaningful but false.

Semantics is the study of meaning. In the context of Big Data, semantics is the technique of creating meaningful assertions about data objects. All meaningful assertions, without exception, can be structured as a 3-item list consisting of an identified data object, a data value, and a descriptor for the data value. These 3-item assertions are referred to as "triples." Just as sentences are the fundamental informational unit of spoken languages, the triple is the fundamental unit of computer information systems.

In practical terms, semantics involves making assertions about data objects (i.e., making triples), combining assertions about data objects (i.e., aggregating triples), and assigning data objects to classes; hence relating triples to other triples. As a word of warning, few informaticians would define semantics in these terms, but I would suggest that all legitimate definitions for the term "semantics" are functionally equivalent to the definition offered here. For example every cell in a spreadsheet is a data value that has a descriptor (the column header), and a subject (the row identifier). A spreadsheet can be pulled apart and re-assembled as a set of triples (known as a triplestore) equal in number to the number of cells contained in the original spreadsheet. Each triple would be an assertion consisting of the following:

```
< row identifier > < column header > < content of cell >
```

Likewise, any relational database, no matter how many relational table are included, can be decomposed into a triplestore. The primary keys of the relational tables would correspond to the identifier of the RDF triple. Column header and cell contents complete the triple.

If spreadsheets and relational databases are equivalent to triplestores, then is there any special advantage to creating triplestores? Yes. A triple is a stand-alone unit of meaning. It does not rely on the software environment (e.g., excel spreadsheet or SQL database engine) to convey its meaning. Hence, triples can be merged without providing any additional structure. Every triple on the planet could be concatenated to create the ultimate superduper triplestore, from which all of the individual triples pertaining to any particular unique identifier, could be collected. This is something that could not be done with spreadsheets and database engines. Enormous triplestores can serve as native databases or as a large relational table, or as pre-indexed tables. Regardless, the final products have all the functionality of any popular database engine [2].

Section 4.4. Namespaces

It is once again the vexing problem of identity within variety; without a solution to this disturbing problem there can be no system, no classification.

Roman Jakobson

A namespace is the metadata realm in which a metadata tag applies. The purpose of a namespace is to distinguish metadata tags that have the same name, but different meaning. For example, within a single XML file, the metadata term "date" may be used to signify a calendar date, or the fruit, or the social engagement. To avoid confusion, the metadata term is given a prefix that is associated with a Web document that defines the term within an assigned Web location. [Glossary Namespace]

For example, an XML page might contain three date-related values, and their metadata descriptors:

```
<calendar:date>June 16, 1904</caldendar:date>
<agriculture:date>Thoory</agriculture:date>
<social:date>Pyramus and Thisbe<social:date>
```

At the top of the XML document you would expect to find declarations for the namespaces used in the XML page. Formal XML namespace declarations have the syntax:

```
xmlns:prefix="URI"
```

In the fictitious example used in this section, the namespace declarations might appear in the "root" tag at the top of the XML page, as shown here (with fake web addresses):

```
<root xmlns:calendar="http://www.calendercollectors.org/"
xmlns:agriculture="http://www.farmersplace.org/"
xmlns:social="http://hearts_throbbing.com/">
```

The namespace URIs are the web locations that define the meanings of the tags that reside within their namespace.

The relevance of namespaces to Big Data resources relates to the heterogeneity of information contained in or linked to a resource. Every description of a value must be provided a unique namespace. With namespaces, a single data object residing in a Big Data resource can be associated with assertions (i.e., object-metadata-data triples) that include descriptors of the same name, without losing the intended sense of the assertions. Furthermore, triples held in different Big Data resources can be merged, with their proper meanings preserved.

Here is an example wherein two resources are merged, with their data arranged as assertion triples.

```
Big Data resource 1

29847575938125        calendar:date        February 4, 1986
83654560466294        calendar:date        June 16, 1904

Big Data resource 2

57839109275632        social:date          Jack and Jill
83654560466294        social:date          Pyramus and Thisbe
```

```
Merged Big Data Resource 1 + 2

29847575938125      calendar:date      February 4, 1986
57839109275632      social:date        Jack and Jill
83654560466294      social:date        Pyramus and Thisbe
83654560466294      calendar:date      June 16, 1904
```

There you have it. The object identified as 83654560466294 is associated with a "date" metadata tag in both resources. When the resources are merged, the unambiguous meaning of the metadata tag is conveyed through the appended namespaces (i.e., social: and calendar:)

Section 4.5. Case Study: A Syntax for Triples

I really do not know that anything has ever been more exciting than diagramming sentences.

Gertrude Stein

If you want to represent data as triples, you will need to use a standard grammar and syntax. RDF (Resource Description Framework) is a dialect of XML designed to convey triples. Providing detailed instruction in RDF syntax, or its dialects, lies far outside the scope of this book. However, every Big Data manager must be aware of those features of RDF that enhance the value of Big Data resources. These would include:

1. The ability to express any triple in RDF (i.e., the ability to make RDF statements).
2. The ability to assign the subject of an RDF statement to a unique, identified, and defined class of objects (i.e., that ability to assign the object of a triple to a class).

RDF is a formal syntax for triples. The subjects of triples can be assigned to classes of objects defined in RDF Schemas and linked from documents composed of RDF triples. RDF Schemas will be described in detail in Section 5.9.

When data objects are assigned to classes, the data analysts can discover new relationships among the objects that fall into a class, and can also determine relationships among different related classes (i.e., ancestor classes and descendant classes, also known as superclasses and subclasses). RDF triples plus RDF Schemas provide a semantic structure that supports introspection and reflection. [Glossary Child class, Subclass, RDF Schema, RDFS, Introspection, Reflection]

3. The ability for all data developers to use the same publicly available RDF Schemas and namespace documents with which to describe their data, thus supporting data integration over multiple Big Data resources.

This last feature allows us to turn the Web into a worldwide Big Data resource composed of RDF documents.

We will briefly examine each of these three features in RDF. First, consider the following triple:

```
pubmed:8718907    creator    Bill Moore
```

Every triple consists of an identifier (the subject of the triple), followed by metadata, followed by a value. In RDF syntax the triple is flanked by metadata indicating the beginning and end of the triple. This is the <rdf:description> tag and its end-tag </rdf:description). The identifier is listed as an attribute within the <rdf:description> tag, and is described with the rdf:about tag, indicating the subject of the triple. There follows a metadata descriptor, in this case <author>, enclosing the value, "Bill Moore."

```
<rdf:description rdf:about="urn:pubmed:8718907">
  <creator>Bill Moore</creator>
</rdf:description>
```

The RDF triple tells us that Bill Moore wrote the manuscript identified with the PubMed number 8718907. The PubMed number is the National library of Medicine's unique identifier assigned to a specific journal article. We could express the title of the article in another triple.

```
pubmed:8718907, title, "A prototype Internet autopsy database. 1625
consecutive fetal and neonatal autopsy facesheets spanning 20 years."
```

In RDF, the same triple is expressed as:

```
<rdf:description rdf:about="urn:pubmed:8718907">
  <title>A prototype Internet autopsy database. 1625 consecutive
fetal and neonatal autopsy facesheets spanning 20 years</title>
</rdf:description>
```

RDF permits us to nest triples if they apply to the same unique object.

```
<rdf:description rdf:about="urn:pubmed:8718907">
  <author>Bill Moore</author>
  <title>A prototype Internet autopsy database. 1625 consecutive
fetal and neonatal autopsy facesheets spanning 20 years</title>
</rdf:description>
```

Here we see that the PubMed manuscript identified as 8718907 was written by Bill Moore (the first triple) and is titled "A prototype Internet autopsy database. 1625 consecutive fetal and neonatal autopsy facesheets spanning 20 years" (a second triple).

What do we mean by the metadata tag "title"? How can we be sure that the metadata term "title" refers to the name of a document and does not refer to an honorific (e.g., The Count of Monte Cristo or the Duke of Earl). We append a namespace to the metadata. Namespaces were described in Section 4.4.

```
<rdf:description rdf:about="urn:pubmed:8718907">
  <dc:creator>Bill Moore</dc:creator>
  <dc:title>A prototype Internet autopsy database. 1625 consecutive
fetal and neonatal autopsy facesheets spanning 20 years</dc:title>
</rdf:description>
```

In this case, we appended "dc:" to our metadata. By convention, "dc:" refers to the Dublin Core metadata set at: http://dublincore.org/documents/2012/06/14/dces/.

We will be describing the Dublin Core in more detail, in Section 4.6. [Glossary Dublin Core metadata].

RDF was developed as a semantic framework for the Web. The object identifier system for RDF was created to describe Web addresses or unique resources that are available through the Internet. The identification of unique addresses is done through the use of a Uniform Resource Name (URN) [3]. In many cases the object of a triple designed for the Web will be a Web address. In other cases the URN will be an identifier, such as the PubMed reference number in the example above. In this case, we appended the "urn:" prefix to the PubMed reference in the "about" declaration for the object of the triple.

```
<rdf:description rdf:about="urn:pubmed:8718907">
```

Let us create an RDF triple whose subject is an actual Web address.

```
<rdf:Description rdf:about="http://www.usa.gov/">
  <dc:title>USA.gov: The U.S. Government's Official Web Portal</dc:
title>
</rdf:Description>
```

Here we created a triple wherein the object is uniquely identified by the unique Web address http://www.usa.gov/, and the title of the Web page is "USA.gov: The U.S. Government's Official Web Portal." The RDF syntax for triples was created for the purpose of identifying information with its URI (Unique Resource Identifier). The URI is a string of characters that uniquely identifies a Web resource (such as a unique Web address, or some unique location at a Web address, or some unique piece of information that can be ultimately reached through the Worldwide Web). In theory, using URIs as identifiers for triples will guarantee that all triples will be accessible through the so-called "Semantic Web" (i.e., the Web of meaningful assertions) [3]. Using RDF, Big Data resources can design a scaffold for their information that can be understood by humans, parsed by computers, and shared by other Big Data resources. This solution transforms every RDF-compliant Web page into a an accessible database whose contents can be searched, extracted, aggregated, and integrated along with all the data contained in every existing Big Data resource.

In practice, the RDF syntax is just one of many available formats for packaging triples, and can be used with identifiers that have invalid URIs (i.e., that do not relate in any way to Web addresses or Web resources). The point to remember is that Big Data resources that employ triples can port their data into RDF syntax, or into any other syntax for triples, as needed. [Glossary Notation 3, Turtle]

Section 4.6. Case Study: Dublin Core

For myself, I always write about Dublin, because if I can get to the heart of Dublin I can get to the heart of all the cities of the world. In the particular is contained the universal.

James Joyce

James Joyce believed that Dublin held the meaning of every city in the world. In a similar vein, the Dublin Core metadata descriptors hold the meaning of every document in the world. The principle difference between the two Dublin-centric philosophies is that James Joyce hailed from Dublin, Ireland, while the Dublin Core metadata descriptors hailed from Dublin, Ohio, United States. For it was in Dublin, Ohio, in 1995, that a coterie of interested Internet technologists and librarians met for the purpose of identifying a core set of descriptive data elements that every electronic document should contain.

The specification resulting from this early workshop came to be known as the Dublin Core [4]. The Dublin Core elements include such information as the date that the file was created, the name of the entity that created the file, and a general comment on the contents of the file. The Dublin Core elements aid in indexing and retrieving electronic files, and should be included in every electronic document, including every image file. The Dublin Core metadata specification is found at:

http://dublincore.org/documents/dces/

Some of the most useful Dublin Core elements are [5]:

- Contributor—the entity that contributes to the document
- Coverage—the general area of information covered in the document
- Creator—the entity primarily responsible for creating the document
- Date—a time associated with an event relevant to the document
- Description—description of the document
- Format—file format
- Identifier—a character string that uniquely and unambiguously identifies the document
- Language—the language of the document
- Publisher—the entity that makes the resource available
- Relation—a pointer to another, related document, typically the identifier of the related document
- Rights—the property rights that apply to the document
- Source—an identifier linking to another document from which the current document was derived
- Subject—the topic of the document
- Title—title of the document
- Type—genre of the document

An XML syntax for expressing the Dublin Core elements is available [6,7].

Glossary

Child class The direct or first generation subclass of a class. Sometimes referred to as the daughter class or, less precisely, as the subclass.

Dublin Core metadata The Dublin Core is a set of metadata elements developed by a group of librarians who met in Dublin, Ohio. It would be very useful if every electronic document were annotated with the Dublin Core elements. The Dublin Core Metadata is discussed in detail in Chapter 4. The syntax for including the elements is found at: http://dublincore.org/documents/dces/

Integration Occurs when information is gathered from multiple data sets, relating diverse data extracted from different data sources. Integration can broadly be categorized as pre-computed or computed on-the fly. Pre-computed integration includes such efforts as absorbing new databases into a Big Data resource or merging legacy data from with current data. On-the-fly integration involves merging data objects at the moment when the individual objects are parsed. This might be done during a query that traverses multiple databases or multiple networks. On-the-fly data integration can only work with data objects that support introspection. The two closely related topics of integration and interoperability are often confused with one another. An easy way to remember the difference is to note that integration refers to data; interoperability refers to software.

Introspection Well-designed Big Data resources support introspection, a method whereby data objects within the resource can be interrogated to yield their properties, values, and class membership. Through introspection the relationships among the data objects in the Big Data resource can be examined and the structure of the resource can be determined. Introspection is the method by which a data user can find everything there is to know about a Big Data resource without downloading the complete resource.

Metadata Data that describes data. For example in XML, a data quantity may be flanked by a beginning and an ending metadata tag describing the included data quantity. < age > 48 years < /age >. In the example, < age > is the metadata and 48 years is the data.

Namespace A namespace is the metadata realm in which a metadata tag applies. The purpose of a namespace is to distinguish metadata tags that have the same name, but a different meaning. For example, within a single XML file, the metadata term "date" may be used to signify a calendar date, or the fruit, or the social engagement. To avoid confusion the metadata term is given a prefix that is associated with a Web document that defines the term within the document's namespace.

Notation 3 Also called n3. A syntax for expressing assertions as triples (unique subject + metadata + data). Notation 3 expresses the same information as the more formal RDF syntax, but n3 is compact and easy for humans to read. Both n3 and RDF can be parsed and equivalently tokenized (i.e., broken into elements that can be re-organized in a different format, such as a database record).

RDF Schema Resource Description Framework Schema (RDFS). A document containing a list of classes, their definitions, and the names of the parent class(es) for each class (e.g., Class Marsupiala is a subclass of Class Metatheria). In an RDF Schema, the list of classes is typically followed by a list of properties that apply to one or more classes in the Schema. To be useful, RDF Schemas are posted on the Internet, as a Web page, with a unique Web address. Anyone can incorporate the classes and properties of a public RDF Schema into their own RDF documents (public or private) by linking named classes and properties, in their RDF document, to the web address of the RDF Schema where the classes and properties are defined.

RDFS Same as RDF Schema.

Reflection A programming technique wherein a computer program will modify itself, at run-time, based on information it acquires through introspection. For example, a computer program may iterate over a collection of data objects, examining the self-descriptive information for each object in the collection (i.e., object introspection). If the information indicates that the data object belongs to a particular class of objects, the program might call a method appropriate for the class. The program executes in a manner determined by descriptive information obtained during run-time; metaphorically reflecting upon the purpose of its computational task. Because introspection is a property of well-constructed Big Data resources, reflection is an available technique to programmers who deal with Big Data.

Subclass A class in which every member descends from some higher class (i.e., a superclass) within the class hierarchy. Members of a subclass have properties specific to the subclass. As every member of a subclass is also a member of the superclass, the members of a subclass inherit the properties and methods of the ancestral classes. For example, all mammals have mammary glands because mammary glands are a defining property of the mammal class. In addition, all mammals have vertebrae because the class of mammals is a subclass of the class of vertebrates. A subclass is the immediate child class of its parent class.

Turtle Another syntax for expressing triples. From RDF came a simplified syntax for triples, known as Notation 3 or N3 [8]. From N3 came Turtle, thought to fit more closely to RDF. From Turtle came an even more simplified form, known as N-Triples.

References

[1] Forterre P. The two ages of the RNA world, and the transition to the DNA world: a story of viruses and cells. Biochimie 2005;87:793–803.

[2] Neumann T, Weikum G. xRDF3X: Fast querying, high update rates, and consistency for RDF databases. Proceedings of the VLDB Endowment 2010;3:256–63.

[3] Berners-Lee T. Linked data—design issues. July 27. Available at: https://www.w3.org/DesignIssues/LinkedData.html; 2006 [viewed December 20, 2017].

[4] Kunze J. Encoding Dublin Core Metadata in HTML. Dublin Core Metadata Initiative. Network Working Group Request for Comments 2731. The Internet Society; 1999. December. Available at: http://www.ietf.org/rfc/rfc2731.txt [viewed August 1, 2015].

[5] Berman JJ. Principles of big data: preparing, sharing, and analyzing complex information. Waltham, MA: Morgan Kaufmann; 2013.

[6] Dublin Core Metadata Initiative. Available from: http://dublincore.org/ The Dublin Core is a set of basic metadata that describe XML documents. The Dublin Core were developed by a forward-seeing group library scientists who understood that every XML document needs to include self-describing metadata that will allow the document to be indexed and appropriately retrieved.

[7] Dublin Core Metadata Element Set, Version 1.1: Reference Description. Available from: http://dublincore.org/documents/1999/07/02/dces/ [viewed January 18, 2018].

[8] Primer: Getting into RDF & Semantic Web using N3. Available from: http://www.w3.org/2000/10/swap/Primer.html [viewed September 17, 2015].

5

Classifications and Ontologies

Section 5.1. It's All About Object Relationships

Order and simplification are the first steps toward the mastery of a subject.

Thomas Mann

Information has limited value unless it can take its place within our general understanding of the world. When a financial analyst learns that the price of a stock has suddenly dropped, he cannot help but wonder if the drop of a single stock reflects conditions in other stocks in the same industry. If so, the analyst may check to ensure that other industries are following a downward trend. He may wonder whether the downward trend represents a shift in the national or global economies. There is a commonality to all of the questions posed by the financial analyst. In every case, the analyst is asking a variation on a single question: "How does this thing relate to that thing?"

Big Data resources are complex. When data is simply stored in a database, without any general principles of organization, it becomes impossible to find the relationships among the data objects. To be useful the information in a Big Data resource must be divided into classes of data. Each data object within a class shares a set of properties chosen to enhance our ability to relate one piece of data with another.

Relationships are the fundamental properties of an object that determine the class in which it is placed. Every member of a class shares these same fundamental properties. A core set of relational properties is found in all the ancestral classes of an object and in all the descendant classes of an object. Similarities are just features that one or more

Principles and Practice of Big Data. https://doi.org/10.1016/B978-0-12-815609-4.00005-4

objects have in common, but they are not fundamental relationships upon which classes can be organized. Related objects tend to be similar to one another, but these similarities occur as the consequence of their relationships; not vice versa. For example, you may have many similarities to your father. If so, you are similar to your father because you are related to him; you are not related to him because you are similar to him.

The distinction between grouping data objects by similarity and grouping data objects by relationship is sometimes lost on computer scientists. I have had numerous conversations with intelligent scientists who refuse to accept that grouping by similarity (e.g., clustering) is fundamentally different from grouping by relationship (i.e., building a classification). [Glossary Cluster analysis]

Consider a collection of 300 objects. Each object belongs to one of two classes, marked by an asterisk or by an empty box. The three hundred objects naturally cluster into three groups. It is tempting to conclude that the graph shows three classes of objects that can be defined by their similarities, but we know from the outset that the objects fall into two classes, and we see from the graph that objects from both classes are distributed in all three clusters (Fig. 5.1).

Is this graph far-fetched? Not really. Suppose you have a collection of felines and canines. The collection of dogs might include Chihuahuas, St. Bernards, and other breeds. The collection of cats might include housecats, lions, and other species, and the data collected on each animal might include weight, age, and hair length. We do not know what to expect when we cluster the animals by similarities (i.e., weight, age, and hair length) but we can be sure that short-haired cats and short-haired Chihuahuas of the same age will probably fall into one cluster. Cheetahs and greyhounds, having similar size and build,

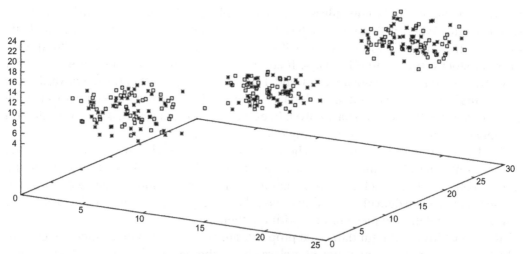

FIG. 5.1 The spatial distribution of 300 objects represented by data points in three dimensions. Each data object falls into one of two classes, represented by an asterisk or an empty box. The data naturally segregates into three clusters. Objects of type asterisk and type box are distributed throughout each cluster.

might fall into another cluster. The similarity clusters will mix together unrelated animals and will separate related animals.

OK, similarities are different from relationships; but how do we know when we are dealing with a similarity and when we are dealing with a true relationship? Here are two stories that may clarify the functional differences between the two concepts:

1. You look up at the clouds, and you begin to see the shape of a lion. The cloud has a tail, like a lion's tail, and a fluffy head, like a lion's mane. With a little imagination, the mouth of the lion seems to roar down from the sky. You have succeeded in finding similarities between the cloud and a lion. If you look at a cloud and you imagine a teakettle producing a head of steam, then you are establishing a relationship between the physical forces that create a cloud and the physical forces that produce steam from a heated kettle, and you understand that clouds are composed of water vapor.
2. You look up at the stars and you see the outline of a flying horse, Pegasus, or the soup ladle, the Big Dipper. You have found similarities upon which to base the names of celestial landmarks, the constellations. The constellations help you orient yourself to the night sky, but they do not tell you much about the physical nature of the twinkling objects. If you look at the stars and you see the relationship between the twinkling stars in the night sky, and the round sun in the daylight sky, then you can begin to understand how the universe operates.

For taxonomists, the importance of grouping by relationship, not by similarity, is a lesson learned the hard way. Literally two thousand years of mis-classifications, erroneous biological theorizations, impediments to progress in medicine and agriculture, have occurred whenever similarities were confused with relationships. Early classifications of animals were based on similarities (e.g., beak shape, color of coat, or number of toes). These kinds of classifications led to the erroneous conclusion that the various juvenile forms of holometabolous insects (i.e., insects that undergo metamorphosis) were distinct organisms, unrelated to the adult form into which they would mature. The vast field of animal taxonomy was a useless mess until taxonomists began to think very deeply about classes of organisms and the fundamental properties that accounted for the relationships among the classes. [Glossary Classification system versus identification system, Classification versus index, Phenetics]

Geneticists have learned that sequence similarities among genes may bear no relationship to their functionalities, their inheritance from higher organisms, their physical locations, or to any biological process whatsoever. Geneticists use the term homology to describe the relationship among sequences that can be credited to descent from a common ancestral sequence. Similarity among different sequences can be non-homologous, developing randomly in non-related organisms, or developing by convergence, through selection for genes that have common functionality. Sequence similarity that is not acquired from a common ancestral sequence seldom relates to the shared fundamental cellular properties that characterize inherited relationships. Biological inferences drawn from gene analyses are more useful when they are built upon phylogenetic relationships, rather than on superficial genetic or physiologic similarities [1]. [Glossary Nonphylogenetic property]

The distinction between classification by similarity and classification by relationship is vitally important to the field of computer science and to the future of Big Data analysis. I have discussed this point with many of my colleagues, who hold the opposite view: that the distinction between similarity classification and relationship classification is purely semantic. There is no practical difference between the two methods. Regardless of which side you may choose, the issue is worth pondering for a few moments.

Two arguments support the opinion that classification should be based on similarity measures. The first argument is that classification by similarity is the standard method by which relational classifications are built. The second argument is that relational properties are always unknown at the time that the classification is built. The foundation of every classification must be built on measurable features and the only comparison we have for measurable features is similarity. This argument has no scientific merit insofar as comparisons by relationship are always feasible, though not always readily computable.

The second argument, that classification by relationship requires access to unobtainable knowledge is a clever observation that hits on a weakness in the relational theory of classification. To build a classification, you must first know the relational properties that define classes, superclasses, and subclasses; but if you want to know the relationships among the classes, you must refer to the classification. It is another bootstrapping problem. [Glossary Bootstrapping]

Building a classification is an iterative process wherein you hope that your tentative selection of relational properties and your class assignments will be validated by the test of time. You build a classification by guessing which properties are fundamental and relational and by guessing which system of classes will make sense when all of the instances of the classes are assigned. A classification is often likened to a hypothesis that must be tested again and again as the classification grows.

Is it ever possible to build a classification using a hierarchical clustering algorithm based on measuring similarities among objects? The answer is a qualified yes, assuming that the object features that you have measured happen to be the relational properties that define the classes. A good example of this process is demonstrated by the work of Carl Woese and his coworkers in the field of the classification of terrestrial organisms [2]. Woese compared ribosomal RNA sequences among organisms. Ribosomal RNA is involved in the precise synthesis of proteins according to instructions coded in genes. According to Woese, the genes coding for ribosomal RNA mutate more slowly than other genes, because ribosomal RNA has scarcely any leeway in its functionality. Changes in the sequence of ribosomal RNA act like a chronometer for evolution. Using sequence similarities Woese developed a brilliant classification of living organisms that has revolutionized evolutionary genetics. Woese's analysis is not perfect and where there are apparent mistakes in his classification, disputations focus on the limitations of using similarity as a substitute for fundamental relational properties [3,4]. [Glossary Non-living organism]

The field of medical genetics has been embroiled in a debate, lasting well over a decade, on the place of race in science. Some would argue that when the genomes of humans from different races are compared, there is no sensible way to tell one genome from another, on the basis of assigned race. The genes of a tall man and the short man are more different than the genes of an African-American man and a white man. Judged by genetic similarity, race has no scientific meaning [5]. On the other hand, every clinician understands that various diseases, congenital and acquired, occur at different rates in the African-American population than in the white population. Furthermore, the clinical symptoms, clinical outcome, and even the treatment of these diseases in African-American and white individuals will sometimes differ among ethnic or racial groups. Hence, many medical epidemiologists and physicians perceive race as a clinical reality [6]. The discord stems from a misunderstanding of the meanings of similarity and of relationship. It is quite possible to have a situation wherein similarities are absent, while relationships pertain. The lack of informative genetic similarities that distinguish one race from another does not imply that race does not exist. The basis for race is the relationship created by shared ancestry. The morphologic and clinical by-product of the ancestry relationship may occur as various physical features and epidemiologic patterns found by clinicians. [Glossary Cladistics]

Fundamentally, all analysis is devoted to finding relationships among objects or classes of objects. All we ever know about the universe, and the processes that play out in our universe, can be reduced to simple relationships. In many cases the process of finding and establishing relationships often begins with finding similarities; but it must never end there.

Section 5.2. Classifications, the Simplest of Ontologies

Consciousness is our awareness of our own awareness.

Descartes

The human brain is constantly processing visual and other sensory information collected from the environment. When we walk down the street, we see images of concrete and asphalt and millions of blades of grass, birds, dogs, and other persons. Every step we take conveys a new world of sensory input. How can we process it all? The mathematician and philosopher Karl Pearson (1857–1936) has likened the human mind to a "sorting machine" [7]. We take a stream of sensory information and sort it into objects; we then collect the individual objects into general classes. The green stuff on the ground is classified as "grass," and the grass is subclassified under some larger grouping, such as "plants." A flat stretch of asphalt and concrete may be classified as a "road" and the road might be subclassified under "man-made constructions." If we lacked a culturally determined classification of objects for our world, we would be overwhelmed by sensory input, and we would have no way to remember what we see, and no way to draw general inferences about anything. Simply put, without our ability to classify, we would not be human [8].

Every culture has some particular way to impose a uniform way of perceiving the environment. In English-speaking cultures, the term "hat" denotes a universally recognized object. Hats may be composed of many different types of materials, and they may vary greatly in size, weight, and shape. Nonetheless, we can almost always identify a hat when we see one, and we can distinguish a hat from all other types of objects. An object is not classified as a hat simply because it shares a few structural similarities with other hats. A hat is classified as a hat because it has a class relationship; all hats are items of clothing that fit over the head. Likewise, all biological classifications are built by relationships, not by similarities [9,8].

Aristotle was one of the first experts in classification. His greatest insight came when he correctly identified a dolphin as a mammal. Through observation, he knew that a large group of animals was distinguished by a gestational period in which a developing embryo is nourished by a placenta, and the offspring are delivered into the world as formed, but small versions of the adult animals (i.e., not as eggs or larvae), and the newborn animals feed from milk excreted from nipples, overlying specialized glandular organs (mammae). Aristotle knew that these features, characteristic of mammals, were absent in all other types of animals. He also knew that dolphins had all these features; fish did not. He correctly reasoned that dolphins were a type of mammal, not a type of fish. Aristotle was ridiculed by his contemporaries for whom it was obvious that dolphins were a type of fish. Unlike Aristotle, they based their classification on similarities, not on relationships. They saw that dolphins looked like fish and dolphins swam in the ocean like fish, and this was all the proof they needed to conclude that dolphins were indeed fish. For about two thousand years following the death of Aristotle, biologists persisted in their belief that dolphins were a type of fish. For the past several hundred years, biologists have acknowledged that Aristotle was correct after all; dolphins are mammals. Aristotle discovered and taught the most important principle of classification; that classes are built on relationships among class members; not by counting similarities [8].

Today, the formal systems that assign data objects to classes, and that relate classes to other classes, are known as ontologies. When the data within a Big Data resource is classified within an ontology, data analysts can determine whether observations on a single object will apply to other objects in the same class.

A classification is a very simple form of ontology, in which each class is allowed to have only one parent class. To build a classification, the ontologist must do the following: (1) define classes (i.e., find the properties that define a class and extend to the subclasses of the class); (2) assign instances to classes; (3) position classes within the hierarchy; and (4) test and validate all the above. [Glossary Parent class]

The constructed classification becomes a hierarchy of data objects conforming to a set of principles:

1. The classes (groups with members) of the hierarchy have a set of properties or rules that extend to every member of the class and to all of the subclasses of the class, to the exclusion of unrelated classes. A subclass is itself a type of class wherein the members have the defining class properties of the parent class plus some additional property(ies) specific for the subclass.

2. In a hierarchical classification, each subclass may have no more than one parent class. The root (top) class has no parent class. The biological classification of living organisms is a hierarchical classification.
3. At the bottom of the hierarchy is the class instance. For example, your copy of this book is an instance of the class of objects known as "books."
4. Every instance belongs to exactly one class.
5. Instances and classes do not change their positions in the classification. As examples, a horse never transforms into a sheep, and a book never transforms into a harpsichord. [Glossary Intransitive property]
6. The members of classes may be highly similar to one another, but their similarities result from their membership in the same class (i.e., conforming to class properties), and not the other way around (i.e., similarity alone cannot define class inclusion).

Classifications are always simple; the parental classes of any instance of the classification can be traced as a simple, non-branched list, ascending through the class hierarchy. As an example, here is the lineage for the domestic horse (Equus caballus), from the classification of living organisms:

Equus caballus
Equus subg. Equus
Equus
Equidae
Perissodactyla
Laurasiatheria
Eutheria
Theria
Mammalia
Amniota
Tetrapoda
Sarcopterygii
Euteleostomi
Teleostomi
Gnathostomata
Vertebrata
Craniata
Chordata
Deuterostomia
Coelomata
Bilateria
Eumetazoa
Metazoa
Fungi/Metazoa group
Eukaryota
cellular organisms

The words in this zoological lineage may seem strange to laypersons, but taxonomists who view this lineage instantly grasp the place of domestic horses in the classification of all living organisms.

A classification is a list of every member class along with their relationships to other classes. Because each class can have only one parent class, a complete classification can be provided when we list all the classes, adding the name of the parent class for each class on the list. For example, a few lines of the classification of living organisms might be:

```
Craniata, subclass of Chordata
Chordata, subclass of Duterostomia
Deuterostomia, subclass of Coelomata
Coelomata, subclass of Bilateria
Bilateria, subclass of Eumetazoa
```

Given the name of any class a programmer can compute (with a few lines of code), the complete ancestral lineage for the class, by iteratively finding the parent class assigned to each ascending class [10]. [Glossary Iterator]

A taxonomy is a classification with the instances "filled in." This means that for each class in a taxonomy, all the known instances (i.e., member objects) are explicitly listed. For the taxonomy of living organisms the instances are named species. Currently, there are several million named species of living organisms, and each of these several million species is listed under the name of some class included in the full classification.

Classifications drive down the complexity of their data domain because every instance in the domain is assigned to a single class and every class is related to the other classes through a simple hierarchy.

It is important to distinguish a classification system from an identification system. An identification system puts a data object into its correct slot within the classification. For example, a fingerprint matching system may look for a set of features that puts a fingerprint into a special subclass of all fingerprint, but the primary goal of fingerprint matching is to establish the identity of an instance (i.e., to determine whether two sets of fingerprints belong to the same person). In the realm of medicine, when a doctor renders a diagnosis on a patient's diseases, she is not classifying the disease; she is finding the correct slot, within the preexisting classification of diseases, that holds her patient's diagnosis.

Section 5.3. Ontologies, Classes With Multiple Parents

...science is in reality a classification and analysis of the contents of the mind...

Karl Pearson [7]

Ontologies are constructions that permit an object to be a direct subclass of more than one classes. In an ontology, the class "horse" might be a subclass of Equu, a zoological term; as well as a subclass of "racing animals" and "farm animals," and "four-legged animals." The class "book" might be a subclass of "works of literature," as well as a subclass of "wood-pulp materials," and "inked products." Ontologies are unrestrained classifications. Hence, all classifications are ontologies, but not all ontologies are classifications. Ontologies are predicated on the belief that a single object or class of objects might have multiple different fundamental identities, and that these different identities will often place one class of objects directly under more than one superclass. [Glossary Multiclass classification, Multiclass inheritance]

Data analysts sometimes prefer ontologies to classifications because they permit the analyst to find relationships among classes of objects that would have been impossible to find under a classification. For example, a data analyst might be interested in determining the relationships among groups of flying animals, such as butterflies, birds, and bats. In the classification of living organisms, these animals occupy classes that are not closely related to one another; no two of the different types of flying animals share a single parent class. Because classifications follow relationships through a lineage, they cannot connect instances of classes that fall outside the line of descent.

Ontologies are not subject to the analytic limitations imposed by classifications. In an ontology, a data object can be an instance of many different kinds of classes; thus, the class does not define the essence of the object, as it does in a classification. In an ontology the assignment of an object to a class and the behavior of the members of the objects of a class, are determined by rules. An object belongs to a class when it behaves like the other members of the class, according to a rule created by the ontologist. Every class, subclass, and superclass is defined by rules; and rules can be programmed into software.

Classifications were created and implemented at a time when scientists did not have powerful computers that were capable of handling the complexities of ontologies. For example, the classification of all living organisms on earth was created over a period of two millennia. Several million species have been assigned to date to the classification. It is currently estimated that we will need to add another 10–50 million species before we come close to completing the taxonomy of living organisms. Prior generations of scientists could cope with a simple classification, wherein each class of organisms falls under a single superclass; they could not hope to cope with a complex ontology of organisms.

The advent of powerful and accessible computers has spawned a new generation of computer scientists who have developed powerful methods for building complex ontologies. It is the goal of these computer scientists to analyze data in a manner that allows us to find and understand ontologic relationships among data objects.

In simple data collections, such as spreadsheets, data is organized in a very specific manner that preserves the relationships among specific types of data. The rows of the spreadsheet are the individual data objects (i.e., people, experimental samples, and class of information). The left-hand field of the row is typically the name assigned to the data object and the cells of the row are the attributes of the data object (e.g., quantitative measurements, categorical data, and other information). Each cell of each row occurs in a specific order and the order determines the kind of information contained in the cell. Hence, every column of the spreadsheet has a particular type of information in each spreadsheet cell. [Glossary Categorical data, Observational data]

Big Data resources are much more complex than spreadsheets. The set of features belonging to an object (i.e., the values, sometimes called variables, belonging to the object, and corresponding to the cells in a spreadsheet row) will be different for different classes of objects. For example, a member of Class Automobile may have a feature such as "average miles per gallon in city driving," while a member of Class Mammal would not. Every data object must be assigned membership in a class (e.g., Class Persons, Class Tissue Samples, and Class Bank Accounts), and every class must be assigned a set of class properties. In Big Data resources that are based on class models, the data objects are not defined by their location in a rectangular spreadsheet; they are defined by their class membership. Classes, in turn, are defined by their properties and by their relations to other classes. [Glossary Properties versus classes]

The question that should confront every Big Data manager is, "Should I model my data as a classification, wherein every class has one direct parent class; or should I model the resource as an ontology, wherein classes may have multiparental inheritance?"

Section 5.4. Choosing a Class Model

Taxonomy is the oldest profession practiced by people with their clothes on.
Quentin Wheeler, referring to the belief that Adam was assigned the task of naming all the
creatures.

The simple, and fundamental question, "Can a class of objects have more than one parent class?" lies at the heart of several related fields: database management, computational informatics, object oriented programming, semantics, and artificial intelligence. Computer scientists are choosing sides, often without acknowledging the problem or fully understanding the stakes. For example, when a programmer builds object libraries in the Python or the Perl programming languages, he is choosing to program in a permissive environment that supports multiclass object inheritance. In Python and Perl, any object can have as many parent classes as the programmer prefers. When a programmer chooses to program in the Ruby programming language, he shuts the door on multiclass inheritance. A Ruby object can have only one direct parent class. Many programmers are totally unaware of the liberties and restrictions imposed by their choice of programming language, until they start to construct their own object libraries, or until they begin to use class libraries prepared by another programmer. [Glossary Artificial intelligence]

In object oriented programming the programming language provides a syntax whereby a named method is "sent" to data objects and a result is calculated. The named methods are functions and short programs contained in a library of methods created for a class. For example, a "close" method, written for file objects, typically shuts a file so that it cannot be accessed for read or write operations. In object-oriented languages a "close" method is sent to an instance of class "File" when the programmer wants to prohibit access to the file. The programming language, upon receiving the "close" method, will look for a method named "close" somewhere in the library of methods prepared for the "File" class.

If it finds the "close" method in the "File" class library, it will apply the method to the object to which the method was sent. In simplest terms the specified file would be closed.

If the "close" method were not found among the available methods for the "File" class library, the programming language would automatically look for the "close" method in the parent class of the "File" class. In some languages the parent class of the "File" class is the "Input/Output" class. If there were a "close" method in the "Input/Output" class, the method would be sent to the "File" Object. If not, the process of looking for a "close" method would be repeated for the parent class of the "Input/Output" class. You get the idea. Object oriented languages search for methods by moving up the lineage of ancestral classes for the object instance that receives the method.

In object oriented programming, every data object is assigned membership to a class of related objects. Once a data object has been assigned to a class, the object has access to all of the methods available to the class in which it holds membership, and to all of the methods in all the ancestral classes. This is the beauty of object oriented programming. If the object oriented programming language is constrained to single parental inheritance, as happens in the Ruby programming language, then the methods available to the programmer are restricted to a tight lineage. When the object oriented language permits multiparental inheritance, as happens in the Perl and Python programming languages, a data object can have many different ancestral classes spread horizontally and vertically through the class libraries. [Glossary Beauty]

Freedom always has its price. Imagine what happens in a multiparental object oriented programming language when a method is sent to a data object, and the data object's class library does not contain the method. The programming language will look for the named method in the library belonging to a parent class. Which parent class library should be searched? Suppose the object has two parent classes, and each of those two parent classes has a method of the same name in their respective class libraries? The functionality of the method will change depending on its class membership (i.e., a "close" method may have a different function within class File than it may have within class Transactions or class Boxes). There is no way to determine how a search for a named method will traverse its ancestral class libraries; hence, the output of a software program written in an object oriented language that permits multiclass inheritance is unpredictable.

The rules by which ontologies assign class relationships can become computationally difficult. When there are no restraining inheritance rules, a class within the ontology might be an ancestor of a child class that is an ancestor of its parent class (e.g., a single class might be a grandfather and a grandson to the same class). An instance of a class might be an instance of two classes, at once. The combinatorics and the recursive options can become impossible to compute. [Glossary Combinatorics]

Those who use ontologies that allow multiclass inheritance will readily acknowledge that they have created a system that is complex and unpredictable. The ontology expert justifies his complex and unpredictable model on the observation that reality itself is complex and unpredictable. A faithful model of reality cannot be created with a simple-minded classification. With time and effort, modern approaches to complex systems will isolate and eliminate computational impedimenta; these are the kinds of problems that computer scientists are trained to solve. For example, recursion within an ontology can be avoided if the ontology is acyclic (i.e., class relationships are not permitted to cycle back onto themselves). For every problem created by an ontology an adept computer scientist will find a solution. Basically, many modern ontologists believe that the task of organizing and understanding information cannot reside within the ancient realm of classification.

For those non-programmers who believe in the supremacy of classifications, over ontologies, their faith may have nothing to do with the computational dilemmas incurred with multiclass parental inheritance. They base their faith on epistemological grounds; on the nature of objects. They hold that an object can only be one thing. You cannot pretend that one thing is really two or more things simply because you insist that it is so. One thing can only belong to one class. Once class can only have one ancestor class; otherwise, it would have a dual nature. For classical taxonomists, assigning more than one parental class to an object indicates that you have failed to grasp the essential nature of the object. The classification expert believes that ontologies (i.e., classifications that permit one class to have more than one parent classes and that permit one object to hold membership in more than one class), do not accurately represent reality.

At the heart of traditional classifications is the notion that everything in the universe has an essence that makes it one particular thing and nothing else. This belief is justified for many different kinds of systems. When an engineer builds a radio, he knows that he can assign names to components, and these components can be relied upon to behave in a manner that is characteristic of its type. A capacitor will behave like a capacitor, and a resistor will behave like a resistor. The engineer need not worry that the capacitor will behave like a semiconductor or an integrated circuit.

What is true for the radio engineer may not hold true for the Big Data analyst. In many complex systems the object changes its function depending on circumstances. For example, cancer researchers discovered an important protein that plays a very important role in the development of cancer. This protein, p53, was, at one time, considered to be the primary cellular driver for human malignancy. When p53 mutated, cellular regulation was disrupted and cells proceeded down a slippery path leading to cancer. In the past few decades, as more information was obtained, cancer researchers have learned that p53 is just one of many proteins that play some role in carcinogenesis, but the role changes depending on the species, tissue type, cellular microenvironment, genetic background of the cell, and many other factors. Under one set of circumstances, p53 may play a role in DNA repair; under another set of circumstances, p53 may cause cells to arrest the growth cycle [11,12]. It is difficult to classify a protein that changes its primary function based on its biological context.

As someone steeped in the ancient art of classification, and as someone who has written extensively on object oriented programming, I am impressed, but not convinced, by arguments on both sides of the ontology/classification debate. As a matter of practicality, complex ontologies are nearly impossible to implement in Big Data projects. The job of building and operating a Big Data resource is always difficult. Imposing a complex ontology framework onto a Big Data resource tends to transform a tough job into an impossible job. Ontologists believe that the Big Data resources must match the complexity of their data domain. They would argue that the dictum "Keep it simple, stupid!" only applies to systems that are simple at the outset. I would comment here that one of the problems with ontology builders is that they tend to build ontologies that are much more complex than our reality. They do so because it is actually quite easy to add layers of abstraction to an ontology without incurring any immediate penalty. [Glossary KISS]

Without stating a preference for single-class inheritance (classifications) or multi-class inheritance (ontologies), I would suggest that when modeling a complex system, you should always strive to design a model that is as simple as possible. The wise ontologist will settle for a simplified approximation of the truth. Regardless of your personal preference, you should learn to recognize when an ontology has become too complex for its own good.

Here are the danger signs of an overly-complex ontology:

- You realize that the ontology makes no sense. The solutions obtained by data analysts contradict direct observations. The ontologists perpetually tinker with the model in an effort to achieve a semblance of reality and rationality. Meanwhile, the data analysts tolerate the flawed model because they have no choice in the matter.
- For a given problem, no two data analysts seem able to formulate the query the same way and no two query results are ever equivalent.
- The time spent on ontology design and improvement exceeds the time spent on collecting the data that populates the ontology.
- The ontology lacks modularity. It is impossible to remove a set of classes within the ontology without reconstructing the entire ontology. When anything goes wrong the entire ontology must be fixed or redesigned.
- The ontology cannot be fitted into a higher level ontology or a lower-level ontology.
- The ontology cannot be debugged when errors are detected.
- Errors occur without anyone knowing where the error has occurred.
- Nobody, even the designers, fully understands the ontology model.

Simple classifications are not flawless. Here are a few danger signs of an overly-simple classifications.

1. The classification is too granular.

You find it difficult to associate observations with particular instances within a class or to particular classes within the classification.

2. The classification excludes important relationships among data objects.

For example, dolphins and fish both live in water. As a consequence, dolphins and fish will both be subject to some of the same influences (e.g., ocean pollutants and water-borne infectious agents). In this case, relationships that are not based on species ancestry are simply excluded from the classification of living organisms and cannot be usefully examined.

3. The classes in the classification lack inferential competence.

Competence in the ontology field is the ability to infer answers based on the rules for class membership. For example, in an ontology you can subclass wines into white wines and red wines and you can create a rule that specifies that the two subclasses are exclusive. If you know that a wine is white, then you can infer that the wine does not belong to the subclass of red wines. Classifications are built by understanding the essential features of an object that make it what it is; they are not generally built on rules that might serve the interests of the data analyst or the computer programmer. Unless a determined effort has been made to build a rule-based classification, the ability to draw logical inferences from observations on data objects will be sharply limited.

4. The classification contains a "miscellaneous" class.

A formal classification requires that every instance belongs to a class with well-defined properties. A good classification does not contain a "miscellaneous" class that includes objects that are difficult to assign. Nevertheless, desperate taxonomists will occasionally assign objects of indeterminate nature to a temporary class, waiting for further information to clarify the object's correct placement. In the field of biological taxonomy, the task of creating and assigning the correct classes for the members of these unnatural and temporary groupings, has frustrated biologists over many decades, and is still a source of some confusion [13]. [Glossary Unclassifiable objects]

5. The classification is unstable.

Simplistic approaches may yield a classification that serves well for a limited number of tasks, but fails to be extensible to a wider range of activities or fails to integrate well with classifications created for other knowledge domains. All classifications require review and revision, but some classifications are just awful and are constantly subjected to major overhauls.

It seems obvious that in the case of Big Data, a computational approach to data classification is imperative, but a computational approach that consistently leads to failure is not beneficial. Many of the ontologies that have been created for data collected in many of the fields of science have been ignored or abandoned by their intended beneficiaries. Ontologies, due to their multi-lineage ancestries, are simply too difficult to understand and too difficult to implement.

Section 5.5. Class Blending

It ain't what you don't know that gets you into trouble. It's what you know for sure that just ain't so.

Mark Twain

A blended class, also known as a noisy class, results when the taxonomist assigns unrelated objects to the same class. This almost always leads to errors in data analysis whose cause is nearly impossible to find. As an example of class blending, suppose you were testing the effectiveness of an antibiotic on a group of subjects all having a specific type of bacterial pneumonia. In this case, the accuracy of your results will be forfeit when your study population includes subjects with viral pneumonia, smoking-related lung damage, or a pneumonia produced by some bacteria other than the bacteria that is known to be sensitive to the antibiotic under study. Basically, a classification has no value if its classes contain unrelated members.

Errors induced by blending classes are often overlooked by data analysts who incorrectly assume that the experiment was designed to ensure that each data group is composed of a uniform and representative population. Sometimes class blending occurs when an incompetent curator misplaces data objects into the wrong class. For example, you would not want to hire an astronomer who cannot distinguish a moon from a planet. More commonly, however, the problem lies within the classification itself. It is not uncommon for the formal class definition (which includes objective criteria for including or excluding objects from the class) to be ill-conceived.

One caveat. Efforts to eliminate class blending can be counterproductive if undertaken with excessive zeal. For example, in an effort to reduce class blending, a researcher may choose groups of subjects who are uniform with respect to every known observable property. For example, suppose you want to actually compare apples with oranges. To avoid class blending, you might want to make very sure that your apples do not include any cumquats or persimmons. You should be certain that your oranges do not include any limes or grapefruits. Imagine that you go even further, choosing only apples and oranges of one variety (e.g., Macintosh apples and Navel oranges), size (e.g., 10 cm), and origin (e.g., California). How will your comparisons apply to the varieties of apples and oranges that you have excluded from your study? You may actually reach conclusions that are invalid and irreproducible for more generalized populations within each class. In this case, you have succeeded in eliminated class blending at the expense of losing representative subpopulations of the classes. Some days, the more you try, the more you lose. [Glossary Representation bias, Confounder]

Section 5.6. Common Pitfalls in Ontology Development

The hallmark of good science is that it uses models and theory but never believes them.

Martin Wilk

Do ontologies serve a necessary role in the design and development of Big Data resources? Yes. Because every Big Data resource is composed of many different types of information, it becomes important to assign types of data into groups that have similar properties: images, music, movies, documents, and so forth. The data manager needs to distinguish one type of data object from another, and must have a way of knowing the set of properties that apply to the members of each class. When a query comes in asking for a list of songs written by a certain composer, or performed by a particular musician, the data manager will need to have a software implementation wherein the features of the query are matched to the data objects for which those features apply. The ontology that organizes the Big Data resource may be called by many other names (class systems, tables, data typing, database relationships, object model), but it will always come down to some way of organizing information into groups that share a set of properties.

Despite the importance of ontologies to Big Data resources the process of building an ontology is seldom undertaken wisely. There is a rich and animated literature devoted to the limitations and dangers of ontology-building [14,15]. Here are just a few pitfalls that you should try to avoid:

– **Do not build transitive classes.**

Class assignment is permanent. If you assign your pet beagle to the "dog" class, you cannot pluck him from this class and reassign him to the "feline" class. Once a dog, always a dog. This may seem like an obvious condition for an ontology, but it can be very tempting to make a class known as "puppy." This practice is forbidden because a dog assigned to class "puppy" will grow out of his class when he becomes an adult. It is better to assign "puppy" as a property of Class Dog, with a property definition of "age less than one year."

– **Do not build miscellaneous classes.**

As previously mentioned, even experienced ontologists will stoop to creating a "miscellaneous" class, as an act of desperation. The temptation to build a "miscellaneous" class arises when you have an instance (of a data object) that does not seem to fall into any of the well-defined classes. You need to assign the instance to a class, but you do not know enough about the instance to define a new class for the instance. To keep the project moving forward, you invent a "miscellaneous" class to hold the object until a better class can be created. When you encounter another object that does not fit into any of the defined classes, you simply assign it to the "miscellaneous" class. Now you have two objects in the "miscellaneous" class. Their only shared property is that neither object can be readily assigned to any of the defined classes. In the classification of living organisms, Class Protoctista was invented in the mid-nineteenth century to hold, temporarily, some of the organisms that could not be classified as animal, plant, or fungus. It has taken a century for taxonomists to rectify the oversight, and it may take another century for the larger scientific community to fully adjust to the revisions. Likewise, mycologists (fungus experts) have accumulated a large group of unclassifiable fungi. A pseudoclass of fungi, deuteromycetes (spelled with a lowercase "d", signifying its questionable validity as a true biologic class) was created to hold these indeterminate organisms until definitive classes can be assigned. At present, there are several thousand such fungi, sitting in taxonomic limbo, until they can be placed into a definitive taxonomic class [16]. [Glossary Negative classifier]

Sometimes, everyone just drops the ball and miscellaneous classes become permanent [17]. Successive analysts, unaware that the class is illegitimate, assumed that the "miscellaneous" objects were related to one another (i.e., related through their "miscellaneousness"). Doing so led to misleading interpretations (e.g., finding similarities among unrelated data objects, and failing to see relationships that would have been obvious had the objects been assigned to their correct classes). The creation of an undefined "miscellaneous" class is an example of a general design flaw known as "ontological promiscuity" [14]. When an ontology is promiscuous the members of one class cannot always be distinguished from members of other classes.

– **Do not confuse properties with classes.**

Whenever I lecture on the topic of classifications and ontologies, I always throw out the following question: "Is a leg a subclass of the human body?" Most people answer yes. They reason that the normal human body contains a leg; hence leg is a subclass of the human body. They forget that a leg is not a type of human body, and is therefore not a subclass of the human body. As a part of the human body, "leg" is a property of a class. Furthermore, lots of different classes of things have legs (e.g., dogs, cows, tables). The "leg" property can be applied to many different classes and is usually asserted with a "has_a" descriptor (e.g., "Fred has_a leg"). The fundamental difference between classes and properties is one of the more difficult concepts in the field of ontology.

– **Do not invent classes and properties that have already been invented** [18].

Time-pressured ontologists may not wish to search, find, and study the classes and properties created by other ontologists. It is often easier to invent classes and properties as you need them, defining them in your own Schema document. If your ambitions are limited to using your own data for your own purposes, there really is no compelling reason to hunt for external ontologies. Problems will surface only if you need to integrate your data objects with the data objects held in other Big Data resources. If every resource invented its own set of classes and properties, then there could be no sensible comparisons among classes, and the relationships among the data objects from the different resources could not be explored.

Most data records, even those that are held in seemingly unrelated databases, contain information that applies to more than one type of class of data. A medical record, a financial record and a music video may seem to be heterogeneous types of data, but each is associated with the name of a person, and each named person might have an address. The classes of information that deal with names and addresses can be integrated across resources is they all fit into the same ontology, and if they all have the same intended meanings in each resource. [Glossary Heterogeneous data]

– **Do not use a complex data description language.**

If you decide to represent your data objects as triples, you will have a choice of languages, each with their own syntax, with which to describe your data objects. Examples of "triple" languages, roughly listed in order of increasing complexity, are: Notation 3, Turtle, RDF, DAML/OIL, and OWL. Experience suggests that syntax languages start out simple; complexity is added as users demand additional functionalities. The task of expressing triples in DAML/OIL or OWL has gradually become a job for highly trained specialists who work in the obscure field of descriptive logic. As the complexity of the descriptive language increases the number of people who can understand and operate the resource tends to diminish. In general, complex descriptive languages should only be used by well-staffed and well-funded Big Data resources capable of benefiting from the added bells and whistles. [Glossary RDF, Triple]

Section 5.7. Case Study: An Upper Level Ontology

An idea can be as flawless as can be, but its execution will always be full of mistakes.
                                                            Brent Scowcroft

Knowing that ontologies reach into higher ontologies, ontologists have endeavored to create upper level ontologies to accommodate general classes of objects, under which the lower ontologies may take their place. Once such ontology is SUMO, the Suggested Upper Merged Ontology, created by a group of talented ontologists [19]. SUMO is owned by IEEE (Institute of Electrical and Electronics Engineers), and is freely available, subject to a usage license [14]. [Glossary RDF Ontology]

As an upper level ontology, SUMO contains classes of objects that other ontologies can refer to as their superclasses. SUMO permits multiple class inheritance. For example, in SUMO, the class of humans is assigned to two different parent classes: Class Hominid and Class CognitiveAgent. "HumanCorpse," another SUMO class, is defined in SUMO as "A dead thing that was formerly a Human." Human corpse is a subclass of Class OrganicObject; not of Class Human. This means that a human, once it ceases to live, transits to a class that is not directly related to the class of humans. Consequently, members of Class Human, in the SUMO ontology, will change their class and their ancestral lineage, at different moments in time, thus violating the non-transitive rule of classification. [Glossary Superclass]

What went wrong?

- Class HumanCorpse was not created as a subclass of Class Human. This was a mistake, as all humans will eventually die. If we were to create two classes, one called Class Living Human and one called Class Deceased Human, we would certainly cover all possible human states of being, but we would be creating a situation where members of a class are forced to transition out of their class and into another (violating the intransitive rule of classification). The solution, in this case, is simple. Life and death are properties of organisms, and all organisms can and will have both properties, but never at the same time. Assign organisms the properties of life and of death, and stop there.

One last quibble. Consider these two classes from the SUMO ontology, both of which happen to be subclasses of Class Substance.

```
Subclass NaturalSubstance
Subclass SyntheticSubstance
```

It would seem that these two subclasses are mutually exclusive. However, diamonds occur naturally, and diamonds can be synthesized. Hence, diamond belongs to Subclass NaturalSubstance and to Subclass SyntheticSubstance. The ontology creates two mutually exclusive classes that contain members of the same objects. This is problematic, because it violates the uniqueness rule of classifications. We cannot create sensible inference rules for objects that occupy mutually exclusive classes.

What went wrong?

– At first glance, the concepts "NaturalSubstance" and "SyntheticSubstance" would appear to be subclasses of "Substance." Are they really? Would it not be better to think that being "natural" or being "synthetic" are just properties of substances; not types of substances. If we agree that diamonds are a member of class substance, we can say that any specific diamond may have occurred naturally or through synthesis. We can eliminate two subclasses (i.e., "NaturalSubstance" and "SyntheticSubstance") and replace them with two properties of class "Substance": synthetic and natural. By assigning properties to a class of objects, we simplify the ontology (by reducing the number of subclasses), and we eliminate problems created when a class member belongs to two mutually exclusive subclasses. We will discuss the role of properties in classifications in Section 5.9.

As ontologies go, SUMO is one of the best, serving a useful purpose as an upper level repository of classes that can be used freely by Big Data scientists who are trying to simplify how they classify their data objects. Nonetheless, SUMO is not perfect and we are reminded that all ontologies are works-in-progress that must be critically examined, tested, and improved, in perpetuity. [Glossary Data scientist]

Section 5.8. Case Study (Advanced): Paradoxes

Owners of dogs will have noticed that, if you provide them with food, water, shelter, and affection, they will think you are god. Whereas owners of cats are compelled to realize that, if you provide them with food, water, shelter, and affection, they draw the conclusion that they are gods.

Christopher Hitchens

The rules for constructing classifications seem obvious and simplistic. Surprisingly, the task of building a logical, self-consistent classification is extremely difficult. Most classifications are rife with logical inconsistencies and paradoxes. Let us look at a few examples.

In 1975, while touring the Bethesda, Maryland, campus of the National Institutes of Health, I was informed that their Building 10 was the largest all-brick building in the world, providing a home to over 7 million bricks. Soon thereafter, an ambitious construction project was undertaken to greatly expand the size of Building 10. When the work was finished, building 10 was no longer the largest all-brick building in the world. What happened? The builders used material other than brick, and Building 10 lost its classification as an all-brick building, violating the immutability rule of class assignments.

Apparent paradoxes that plague any formal conceptualization of classifications are not difficult to find. Let us look at a few more examples.

Consider the geometric class of ellipses; planar objects in which the sum of the distances to two focal points is constant. Class Circle is a child of Class Ellipse, for which the two focal points of instance members occupy the same position, in the center, producing a radius of constant size. Imagine that Class Ellipse is provided with a class method called "stretch," in which the foci are moved further apart, thus producing flatter objects. When the parent class "stretch" method is applied to members of the Class Circle the circle stops being a circle and becomes an ordinary ellipse. Hence the inherited "stretch" method forces members of Class Circle to transition out of their assigned class, violating the intransitive rule of classifications. [Glossary Method]

Let us look at the "Bag" class of objects. A "Bag" is a collection of objects and the Class Bag is included in most object oriented programming languages. A "Set" is also a collection of objects (i.e., a subclass of Bag), with the special feature that duplicate instances are not permitted. For example, if Kansas is a member of the set of United States states, then you cannot add a second state named "Kansas" to the set. If Class Bag were to have an "increment" method, that added "1" to the total count of objects in the bag, whenever an object is added to Class Bag, then the "increment" method would be inherited by all of the subclasses of Class Bag, including Class Set. But Class Set cannot increase in size when duplicate items are added. Hence, inheritance creates a paradox in the Class Set. [Glossary Inheritance]

How does a data scientist deal with class objects that disappear from their assigned class and reappear elsewhere? In the examples discussed here, we saw the following:

1. Building 10 at NIH was defined as the largest all-brick building in the world. Strictly speaking, Building 10 was a structure; it had a certain weight and dimensions, and it was constructed of brick. "Brick" is an attribute or property of buildings and properties cannot form the basis of a class of building, if they are not a constant feature shared by all members of the class (i.e., some buildings have bricks; others do not). Had we not conceptualized an "all-brick" class of building, we would have avoided any confusion.

2. Class Circle qualified as a member of Class Ellipse, because a circle can be imagined as an ellipse whose two focal points happen to occupy the same location. Had we defined Class Ellipse to specify that class members must have two separate focal points, we could have excluded circles from class Ellipse. Hence, we could have safely included the stretch method in Class Ellipse without creating a paradox.

3. Class Set was made a subset of Class Bag, but the increment method of class Bag could not apply to Class Set. We created Class Set without taking into account the basic properties of Class Bag, which must apply to all its subclasses. Perhaps it would have been better if Class Set and Class Bag were created as children of Class Collection; each with its own set of properties.

Section 5.9. Case Study (Advanced): RDF Schemas and Class Properties

It's OK to figure out murder mysteries, but you shouldn't need to figure out code. You should be able to read it.

Steve McConnell

In Section 4.5, "Case Study: A Syntax for Triples," we introduced the topic of RDF Schemas, and defined them as web-accessible documents that contain the definitions of classes. How does the RDF schema know how to describe the classes in such a way that computers can understand the class definitions and determine the properties that convey to all the members of a class, and to every member of every subclass of a class? Without moving too far beyond the scope of this book, we can discuss here the marvelous "trick" that RDF Schema employs that solves many of the complexity problems of ontologies and many of the over-simplification issues associated with classifications. It does so by introducing the new concept of class property. The class property permits the developer to assign features that can be associated with a class and its members. A property can apply to more than one class, and may apply to classes that are not directly related (i.e., neither an ancestor class nor a descendant class). The concept of the assigned class property permits developers to create simple ontologies, by reducing the need to create classes to account for every feature of interest to the developer. Moreover, the concept of the assigned property gives classification developers the ability to relate instances belonging to unrelated classes through their shared property features. The RDF Schema permits developers to build class structures that preserve the best qualities of both complex ontologies and simple classifications.

How do the Class and Property definitions of RDF Schema work? The RDF Schema is a file that defines Classes and Properties. When an RDF Schema is prepared, it is simply posted onto the Internet, as a public Web page, with a unique Web address.

An RDF Schema contains a list of classes, their definition, and the names of the parent class(es). This is followed by a list of properties that apply to one or more classes in the Schema. The following is an example of an RDF Schema written in plain English, without formal RDF syntax.

```
Class: Fungi
Definition: Contains all fungi
Subclass of: Class Opisthokonta (described in another RDF Schema)

Class Plantae
Definition: Includes multicellular organisms such as flowering plants,
conifers, ferns and mosses.
Subclass of: Class Archaeplastida (described in another RDF Schema)
```

```
Property: Stationary existence
Definition: Adult organism does not ambulate under its own power.
Range of classes: Class Fungi, Class Plantae

Property: Soil-habitation
Definition: Lives in soil.
Range of classes: Class Fungi, Class Plantae

Property: Chitinous cell wall
Definition: Chitin is an extracellular material often forming part of
the matrix surrounding cells.
Range of classes: Class Opisthokonta

Property: Cellulosic cell wall
Definition: Cellulose is an extracellular material often forming part
of the matric surrounding cells.
Range of classes: Class Archaeplastida
```

This Schema defines two classes: Class Fungi, containing all fungal species, and Class Plantae containing the flowering plants, conifers and mosses. The Schema defines four properties. Two of the properties (Property Stationary existence and Property Soil-habitation apply to two different classes. Two of the properties (Property Chitinous cell wall and Property Cellulosic cell wall) apply to only one class.

By assigning properties that apply to several unrelated classes, we keep the class system small, but we permit property comparisons among unrelated classes. In this case, we defined Property Stationary growth and we indicated that the property applied to instances of Class Fungi and Class Plantae. This schema permits databases that contain data objects assigned to Class Fungi or data objects assigned to Class Plantae to include data object values related to Property Stationary Growth. Data analysts can collect data from any plant or fungus data object and examine these objects for data values related to Stationary Growth.

Property Soil-habitation applies to Class Fungi and to Class Plantae. Objects of either class may include soil-habitation data values. Data objects from two unrelated classes (Class Fungi and Class Plantae) can be analyzed by a shared property.

The schema lists two other properties, Property Chitinous cell wall and Property Cellulosic cell wall. In this case each property is assigned to one class only. Property Chitinous cell wall applies to Class Opisthokonta. Property Cellulosic cell wall applies to Class Archaeplastidae. These two properties are exclusive to their class. If a data object is described as having a cellulosic cell wall, it cannot be a member of Class Opisthokonta. If a data object is described as having a chitinous cell wall, then it cannot be a member of Class Archaeplastidae.

A property assigned to a class will extend to every member of every descendant class. Class Opisthokonta includes Class Fungi and it also includes Class Animalia, the class of all animals. This means that all animals may have the property of chitinous cell wall. In point of fact, chitin is distributed widely through the animal kingdom, but is not found in mammals.

As the name implies, RDF Schema are written in RDF syntax. In practice, many of the so-called RDF Schema documents found on the web are prepared in alternate formats. They are nominally RDF syntax because they create a namespace for classes and properties referred by triples listed in RDF documents.

Here is a short schema, written as Turtle triples, and held in a fictitious web site, "http://www.fictitious_site.org/schemas/life#" [Glossary Turtle]

```
@prefix rdf: <http://www.w3.org/1999/02/22-rdf-syntax-ns#>
@prefix rdfs: <http://www.w3.org/2000/01/rdf-schema#>
@base <http://www.fictitious_site.org/schemas/life#>
:Homo instance_of rdfs:Class.
:HomoSapiens instance_of rdfs:Class;
    rdfs:subClassOf :Homo.
```

Turtle triples have a somewhat different syntax than N-triples or N3 triples. As you can see, the turtle triple resembles RDF syntax in form, allowing for nested metadata/data pairs assigned to the same object. Nonetheless, turtle triples use less verbiage than RDF, but convey equivalent information. In this minimalist RDF Schema, we specify two classes that would normally be included in the much larger classification of living organisms: Homo and HomoSapiens.

A triple that refers to our "http://www.fictitious_site.org/schemas/life#" Schema might look something like this:

```
:Batman instance_of <http://www.fictitious_site.org/schemas/
life#>:HomoSapiens.
```

The triple asserts that Batman is an instance of Homo Sapiens. The data "HomoSapiens" links us to the RDF Schema, which in turn tells us that HomoSapiens is a class and is the subclass of Class Homo.

One of the many advantages of triples is their fungibility. Once you have created your triple list, you can port them into spreadsheets, or databases, or morph them into alternate triple dialects, such as RDF or N3. Triples in any dialect can be transformed into any other dialect with simple scripts using your preferred programming language.

RDF documents can be a pain to create, but they are very easy to parse. Even in instances when an RDF file is composed of an off-kilter variant of RDF, it is usually quite easy to write a short script that will parse through the file, extracting triples, and using the components of the triples to serve the programmer's goals. Such goals may include: counting occurrences of items in a class, finding properties that apply to specific subsets of items in specific classes, or merging triples extracted from various triplestore databases. [Glossary Triplestore]

RDF seems like a panacea for ontologists, but it is seldom used in Big Data resources. The reasons for its poor acceptance are largely due to its strangeness. Savvy data mangers who have led successful careers using standard database technologies are understandably reluctant to switch over to an entirely new paradigm of information management. Realistically, a novel and untested approach to data description, such as RDF, will take decades to catch on. Whether RDF emerges as the data description standard for Big Data resources is immaterial. The fundamental principles upon which RDF is built are certain to dominate the world of Big Data.

Section 5.10. Case Study (Advanced): Visualizing Class Relationships

The ignoramus is a leaf who doesn't know he is part of a tree

Attributed to Michael Crichton

When working with classifications or ontologies, it is useful to have an image that represents the relationships among the classes. GraphViz is an open source software utility that produces graphic representations of object relationships.

The GraphViz can be downloaded from:

http://www.graphviz.org/

GraphViz comes with a set of applications that generate graphs of various styles. Here is an example of a GraphViz dot file, number.dot, constructed in GraphViz syntax [20]. Aside from a few lines that provide instructions for line length and graph size the dot file is a list of classes and their child classes.

```
digraph G {
  size="7,7";
  Object -> Numeric;
  Numeric -> Integer;
  Numeric -> Float;
  Integer -> Fixnum
  Integer -> Bignum
}
```

After the GraphViz exe file (version graphviz-2.14.1.exe, on my computer) is installed, you can launch the various GraphViz methods as command lines from its working directory, or through a system call from within a script. [Glossary Exe file, System call]

```
c:\ftp\dot>dot -Tpng number.dot -o number.png
```

The command line tells GraphViz to use the dot method to produce a rendering of the number.dot text file, saved as an image file, with filename number.png. The output file contains a class hierarchy, beginning with the highest class and branching until it reaches the lowest descendant class.

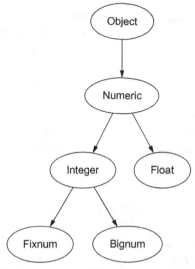

FIG. 5.2 A class hierarchy, described by the number.dot file and converted to a visual file, using GraphViz.

With a glance, we see that the highest class is Class Object (Fig. 5.2). Class Object has one child class, Class Numeric. Numeric has two child classes, Class Integer and Class Float. Class Integer has two child classes, Class Fixnum and Class Bignum. You might argue that a graphic representation of classes was unnecessary; the textual listing of class relationships was all that you needed. Maybe so, but when the class structure becomes complex, visualization can greatly simplify your understanding of the relationships among classes.

Here is a visualization of a classification of human neoplasms (Fig. 5.3). It was produced by GraphViz, from a .dot file containing a ranking of classes and their subclasses, and rendered with the "twopi" method, shown: [Glossary Object rank]

```
c:\ftp>twopi -Tpng neoplasms.dot -o neoplasms_classes.png
```

We can look at the graphic version of the classification and quickly make the following observations:

1. The root class (i.e., the ancestor to every class) is Class Neoplasm. The GraphViz utility helped us find the root class, by placing it in the center of the visualization.
2. Every class is connected to other classes. There are no classes sitting out in space, unrelated to other classes.
3. Every class that has a parent class has exactly one parent class.
4. There are no recursive branches to the graph (e.g., the ancestor of a class cannot also be a descendant of the class).

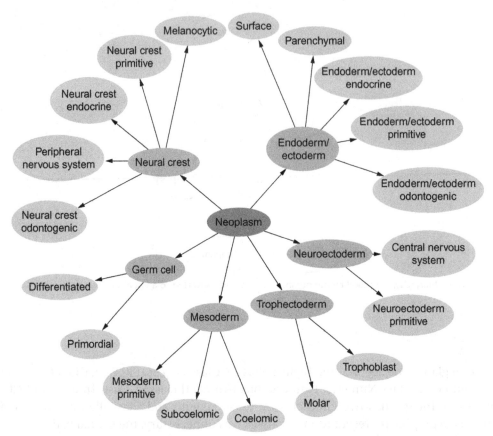

FIG. 5.3 A visualization of relationships in a classification of tumors. The image was rendered with the GraphViz utility, using the twopi method, which produced a radial classification, with the root class in the center.

If we had only the textual listing of class relationships, without benefit of a graphic visualization, it would be very difficult for a human to verify, at a glance, the internal logic of the classification.

With a few tweaks to the neo.dot GraphViz file, we can create a nonsensical graphic visualization:

Notice that one cluster of classes is unconnected to the other, indicating that class Endoderm/Ectoderm has no parent classes (Fig. 5.4). Elsewhere, Class Mesoderm is both child and parent to Class Neoplasm. Class Melanocytic and Class Molar are each the child class to two different parent classes. At a glance, we have determined that the classification is highly flawed. The visualization simplified the relationships among classes, and allowed us to see where the classification went wrong. Had we only looked at the textual listing of classes and subclasses, we may have missed some or all of the logical flaws in our classification.

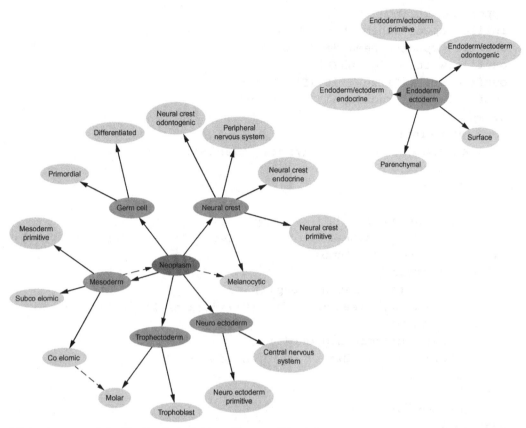

FIG. 5.4 A corrupted classification that might qualify as a valid ontology.

At this point, you may be thinking that visualizations of class relationships are nice, but who has the time and energy to create the long list of classes and subclasses, in GraphViz syntax, that are the input files for the GraphViz methods? Now comes one of the great payoffs of data specifications. You must remember that good data specifications are fungible. A modestly adept programmer can transform a specification into whatever format is necessary to do a particular job. In this case, the classification of neoplasms had been specified as an RDF Schema (vida supra). An RDF Schema includes the definitions of classes and properties, with each class provided with the name of its parent class and each property provided with its range (i.e., the classes to which the property applies). Because class relationships in an RDF Schema are specified, it is easy to transform an RDF Schema into a .dot file suitable for Graphviz.

Here is a short RDF python script, dot.py that parses an RDF Schema (contained in the plain-text file, schema.txt) and produces a GraphViz .dot file, named schema.dot. [Glossary Metaprogramming]

```
import re, string
infile = open('schema.txt', "r")
outfile = open("schema.dot", "w")
outfile.write("digraph G {\n")
outfile.write("size=\"15,15\";\n")
outfile.write("ranksep=\"3.00\";\n")
clump = ""
for line in infile:
  namematch = re.match(r'\<\/rdfs\:Class>', line)
  if (namematch):
    father = ""
    child = ""
    clump = re.sub(r'\n', ' ', clump)
    fathermatch = re.search(r'\:resource\=\"[a-zA-Z0-9\:\/\_\.\-]*
\#([a-zA-Z\_]+)\"', clump)
    if fathermatch:
      father = fathermatch.group(1)
    childmatch = re.search(r'rdf\:ID\=\"([a-zA-Z\_]+)\"', clump)
    if childmatch:
      child = childmatch.group(1)
    outfile.write(father + " -> " + child + ";\n")
    clump = ""
  else:
    clump = clump + line
outfile.write("}\n")
```

The first 15 lines of output of the dot.pl script:

```
digraph G {
size="15,15";
ranksep="2.00";
Class -> Tumor_classification;
Tumor_classification -> Neoplasm;
Tumor_classification -> Unclassified;
Neural_tube -> Neural_tube_parenchyma;
Mesoderm -> Sub_coelomic;
Neoplasm -> Endoderm_or_ectoderm;
Unclassified -> Syndrome;
Neoplasm -> Neural_crest;
Neoplasm -> Germ_cell;
Neoplasm -> Pluripotent_non_germ_cell;
Sub_coelomic -> Sub_coelomic_gonadal;
Trophectoderm -> Molar;
```

The full schema.dot file, not shown, is suitable for use as an input file for the GraphViz utility.

Glossary

Artificial intelligence Artificial intelligence is the field of computer science that seeks to create machines and computer programs that seem to have human intelligence. The field of artificial intelligence some-times includes the related fields of machine learning and computational intelligence. Over the past few decades the term "artificial intelligence" has taken a battering from professionals inside and outside the field, for good reasons. First and foremost is that computers do not think in the way that humans think. Though powerful computers can now beat chess masters at their own game, the algorithms for doing so do not simulate human thought processes. Furthermore, most of the predicted benefits from artificial intelligence have not come to pass, despite decades of generous funding. The areas of neural networks, expert systems, and language translation have not met expectations. Detractors have sug-gested that artificial intelligence is not a well-defined subdiscipline within computer science as it has encroached into areas unrelated to machine intelligence, and has appropriated techniques from other fields, including statistics and numerical analysis. Some of the goals of artificial intelligence have been achieved (e.g., speech-to-text translation), and the analytic methods employed in Big Data analysis should be counted among the enduring successes of the field.

Beauty To mathematicians, beauty and simplicity are virtually synonymous, both conveying the idea that someone has managed to produce something of great meaning or value from a minimum of material. Euler's identity, relating e, i, pi, 0, and 1 in a simple equation, is held as an example of beauty in math-ematics. When writing this book, I was tempted to give it the title, "The Beauty of Data," but I feared that a reductionist flourish, equating data simplification with beauty, was just too obscure.

Bootstrapping The act of self-creation, from nothing. The term derives from the ludicrous stunt of pulling oneself up by one's own bootstraps. Its shortened form, "booting" refers to the startup process in com-puters in which the operating system is somehow activated via its operating system, which has not been activated. The absurd and somewhat surrealistic quality of bootstrapping protocols serves as one of the most mysterious and fascinating areas of science. As it happens, bootstrapping processes lie at the heart of some of the most powerful techniques in data simplification (e.g., classification, object oriented programming, resampling statistics, and Monte Carlo simulations).

It is worth taking a moment to explore the philosophical and the pragmatic aspects of bootstrap-ping. Starting from the beginning, how was the universe created? For believers, the universe was cre-ated by an all-powerful deity. If this were so, then how was the all-powerful deity created? Was the deity self-created, or did the deity simply bypass the act of creation altogether? The answers to these questions are left as an exercise for the reader, but we can all agree that there had to be some kind of bootstrapping process, if something was created from nothing. Otherwise, there would be no universe, and this book would be much shorter than it is. Getting back to our computers, how is it possible for any computer to boot its operating system, when we know that the process of managing the startup process is one of the most important functions of the fully operational operating system? Basically, at startup, the operating system is non-functional. A few primitive instructions hardwired into the computer's processors are sufficient to call forth a somewhat more complex process from memory, and this newly activated process calls forth other processes, until the operating system is eventually up and running. The cascading rebirth of active processes takes time and explains why booting your computer may seem to be a ridiculously slow process.

What is the relationship between bootstrapping and classification? The ontologist creates a clas-sification based on a worldview in which objects hold specific relationships with other objects. Hence, the ontologist's perception of the world is based on preexisting knowledge of the classification of things; which presupposes that the classification already exists. Essentially, you cannot build a classification without first having the classification. How does an ontologist bootstrap a classification into existence? She may begin with a small assumption that seems, to the best of her knowledge, unassailable. In the case of the classification of living organisms, she may assume that the first organisms were primitive, consisting of a few self-replicating molecules and some physiologic actions, confined to a small space, capable of a self-sustaining system. Primitive viruses and prokaryotes (i.e., bacteria) may have started the ball rolling. This first assumption might lead to observations and deductions, which eventually yield the classification of living organisms that we know today. Every thoughtful ontologist will admit that a classification is, at its best, a hypothesis-generating machine; not a factual representation of reality. We use the classification to create new hypotheses about the world and about the classification itself. The process of testing hypotheses may reveal that the classification is flawed; that our early assumptions were incorrect. More often, testing hypotheses will reassure us that our assumptions were consistent with new observations, adding to our understanding of the relations between the classes and instances within the classification.

Categorical data Non-numeric data in which objects are assigned categories, with categories having no numeric order. Yes or no, male or female, heads or tails, snake-eyes or boxcars, are types of unordered categorical data. Traditional courses in mathematics and statistics stress the analysis of numeric data, but data scientists soon learn that much of their work involves the collection and analysis of non-numeric data.

Cladistics The technique of producing a hierarchy of clades, wherein each branch includes a parent species and all its descendant species, while excluding species that did not descend from the parent (Fig. 5.5). If a subclass of a parent class omits any of the descendants of the parent class, then the parent class is said to be paraphyletic. If a subclass of a parent class includes organisms that did not descend from the parent, then the parent class is polyphyletic. A class can be paraphyletic and polyphyletic, if it excludes organisms that were descendants of the parent and if it includes organisms that did not descend from the parent. The goal of cladistics is to create a hierarchical classification that consists exclusively of monophyletic classes (i.e., no paraphyly, no polyphyly). Classifications of the kinds described in this chapter, are monophyletic.

Classification system versus identification system It is important to distinguish a classification system from an identification system. An identification system matches an individual organism with its assigned object name (or species name, in the case of the classification of living organisms). Identification is based on finding several features that, taken together, can help determine the name of an organism. For example, if you have a list of characteristic features: large, hairy, strong, African, jungle-dwelling, knuckle-walking; you might correctly identify the organisms as a gorilla. These identifiers are different from the phylogenetic features that were used to classify gorillas within the hierarchy of

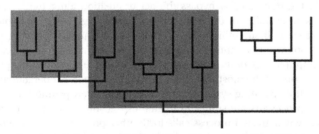

FIG. 5.5 Schematic (cladogram) of all the descendant branches of a common ancestor (stem at bottom of image). The left and the right groups represent clades insofar as they contain all their descendants and exclude classes that are not descendants of the group root. The middle group is not a valid clade because it does not contain all of the descendants of its group root (i.e., it is paraphyletic). Specifically, it excludes the left-most group in the diagram. *From Wikimedia Commons, author "Life of Riley".*

organisms (Animalia: Chordata: Mammalia: Primates: Hominidae: Homininae: Gorillini: Gorilla). Specifically, you can identify an animal as a gorilla without knowing that a gorilla is a type of mammal. You can classify a gorilla as a member of Class Gorillini without knowing that a gorilla happens to be large. One of the most common mistakes in science is to confuse an identification system with a classification system. The former simply provides a handy way to associate an object with a name; the latter is a system of relationships among objects.

Classification versus index   In practice, an index is an alphabetized listing of the important terms in a work (e.g., book), with the locations of each term within the work. Ideally, though, an index should be much more than that. An idealized index is a conceptualization of a corpus of work that enables users to locate the concepts that are discussed and created within the work. How does an idealized index differ from a classification? A classification is a way of organizing concepts in classes, wherein the relationships of the concepts are revealed. The classification can incorporate all of the information held in an index by encapsulating the concept locations together with the names of the concepts. Because the relationships among the objects in a classification can be used to draw inferences about the objects, we can think of a classification as an index that can help us think.

Cluster analysis   Clustering algorithms provide a way of taking a large set of data objects that seem to have no relationship to one another, and to produce a visually simple collection of clusters wherein each cluster member is similar to every other member of the same cluster. The algorithmic methods for clustering are simple. One of the most popular clustering algorithms is the k-means algorithm, which assigns any number of data objects to one of k clusters [21]. The number k of clusters is provided by the user. The algorithm is easy to describe and to understand, but the computational task of completing the algorithm can be difficult when the number of dimensions in the object (i.e., the number of attributes associated with the object), is large. There are some serious drawbacks to the algorithm: (1) The final set of clusters will sometimes depend on the initial, random choice of k data objects. This means that multiple runs of the algorithm may produce different outcomes; (2) The algorithms are not guaranteed to succeed. Sometimes, the algorithm does not converge to a final, stable set of clusters; (3) When the dimensionality is very high, the distances between data objects (i.e., the square root of the sum of squares of the measured differences between corresponding attributes of two objects) can be ridiculously large and of no practical meaning. Computations may bog down, cease altogether, or produce meaningless results. In this case, the only recourse may require eliminating some of the attributes (i.e., reducing dimensionality of the data objects); (4) The clustering algorithm may succeed, producing a set of clusters of similar objects, but the clusters may have no practical value. They may miss important relationships among the objects, or they might group together objects whose similarities are totally non-informative. The biggest drawback associated with cluster analyses is that researchers may mistakenly believe that that the groupings produced by the method constitute a valid biological classification. This is not the case because biological entities (genes, proteins, cells, organs, organisms) may share many properties and still be fundamentally different. For example, two genes may have the same length and share some sub-sequences, but both genes may have no homology with one another (i.e., no shared ancestry) and may have no common or similar expressed products. Another set of genes may be structurally dissimilar but may belong to the same family. The groupings produced by cluster analysis should never be equated with a classification. At best, cluster analysis produces groups that can be used to start piecing together a biological classification.

Combinatorics   The analysis of complex data often involves combinatorics; the evaluation, on some numeric level, of combinations of things. Often, combinatorics involves pairwise comparisons of all possible combinations of items. When the number of comparisons becomes large, as is the case with virtually all combinatoric problems involving large data sets, the computational effort becomes massive. For this reason, combinatorics research has become a subspecialty in applied mathematics and data science. There are four "hot" areas in combinatorics. The first involves building increasingly powerful computers capable of solving complex combinatoric problems. The second involves developing methods whereby combinatoric problems can be broken into smaller problems that can be distributed to many computers, to provide relatively fast solutions to problems that could not otherwise be solved in any reasonable length of time The third area of research involves developing new algorithms for solving combinatoric problems quickly and efficiently. The fourth area, perhaps the most promising area, involves developing innovative non-combinatoric solutions for traditionally combinatoric problems, a golden opportunity for experts in the field of data simplification.

Confounder Unanticipated or ignored factor that alters the outcome of a data analysis. Confounders are particularly important in Big Data analytics, because most analyses are observational; based on collected parameters from large numbers of data records, and there is very little control over confounders. Confounders are less of a problem in controlled prospective experiments, in which a control group and a treated group are alike, to every extent feasible; only differing in their treatment. Differences between the control group and the treated group are presumed to be caused by the treatment, as all of the confounders have been eliminated. One of the greatest challenges of Big Data analytics involves developing new analytic protocols that reduce the effect of confounders in observational studies.

Data scientist Anyone who practices data science and who has some expertise in a field subsumed by data science (i.e., informatics, statistics, data analysis, programming, and computer science).

Exe file Short for executable file and also known as application file. A file containing a program, in binary code, that can be executed when the name of the file is invoked on the command line.

Heterogeneous data Sets of data that are dissimilar with regard to content, purpose, format, organization, and annotations. One of the purposes of Big Data is to discover relationships among heterogeneous data sources. For example, epidemiologic data sets may be of service to molecular biologists who have gene sequence data on diverse human populations. The epidemiologic data is likely to contain different types of data values, annotated and formatted in a manner that is completely different from the data and annotations in a gene sequence database. The two types of related data, epidemiologic and genetic, have dissimilar content; hence they are heterogeneous to one another.

Inheritance The method by which a child is endowed with features of the parent. In object oriented programming, inheritance is passed from parent class to child class, meaning that the child class has access to all of the methods and properties that are held in the parent class.

Intransitive property One of the criteria for a classification is that every object (sometimes referred to as member or as instance) belongs to exactly one class. From this criteria comes the intransitive property of classifications; namely, an object cannot change its class. Otherwise an object would belong to more than one class at different times. It is easy to apply the intransitive rule under most circumstances. A cat cannot become a dog and a horse cannot become a sheep. What do we do when a caterpillar becomes a butterfly? In this case, we must recognize that caterpillar and butterfly represent phases in the development of one particular instance of a species, and do not belong to separate classes.

Iterator Iterators are simple programming shortcuts that call functions that operate on consecutive members of a data structure, such as a list, or a block of code. Typically, complex iterators can be expressed in a single line of code. Perl, Python and Ruby all have iterator methods. In Ruby, the iterator methods are each, find, collect, and inject. In Python, there are types of objects that are iterable (not to be confused with "irritable"), and these objects accept implicit or scripted iteration methods.

KISS Acronym for Keep It Simple Stupid. With respect to Big Data, there are basically two schools of thought. This first is that reality is quite complex, and the advent of powerful computers and enormous data collections allows us to tackle important problems, despite their inherent size and complexity. KISS represents a second school of thought; that Big Problems are just small problems that are waiting to be simplified.

Metaprogramming A metaprogram is a program that creates or modifies other programs. Metaprogramming is a particularly powerful feature of languages that are modifiable at runtime. Perl, Python, and Ruby are all metaprogramming languages. There are several techniques that facilitate metaprogramming features, including introspection and reflection.

Method Roughly equivalent to functions, subroutines, or code blocks. In object-oriented languages, a method is a subroutine available to an object (class or instance). In Ruby and Python, instance methods are declared with a "def" declaration followed by the name of the method, in lowercase. Here is an example, in Ruby, for the "hello" method, is written for the Salutations class.

```
class Salutations
  def hello
      puts "hello there"
  end
end
```

Multiclass classification A misnomer imported from the field of machine translation, and indicating the assignment of an instance to more than one class. Classifications, as defined in this book, impose one-class classification (i.e., an instance can be assigned to one and only one class). It is tempting to think that a ball should be included in class "toy" and in class "spheroids," but multiclass assignments create unnecessary classes of inscrutable provenance, and taxonomies of enormous size, consisting largely of replicate items.

Multiclass inheritance In ontologies, multiclass inheritance occurs when a child class has more than one parent class. For example, a member of Class House may have two different parent classes: Class Shelter, and Class Property. Multiclass inheritance is generally permitted in ontologies but is forbidden in classifications that restrict inheritance to a single parent class (i.e., each class can have at most one parent class, though it may have multiple child classes). When an object-oriented program language permits multiparental inheritance (e.g., Perl and Python programming languages), data objects may have many different ancestral classes spread horizontally and vertically through the class libraries. There are many drawbacks to multi-class inheritance in object oriented programming languages and these have been discussed at some length in the computer science literature [22]. Medical taxonomists should understand that when multi-class inheritance is permitted, a class may be an ancestor of a child class that is an ancestor of its parent class (e.g., a single class might be a grandfather and a grandson to the same class). An instance of a class might be an instance of two classes, at once. The combinatorics and the recursive options can become computationally difficult or impossible. Those who use taxonomies that permit multiclass inheritance will readily acknowledge that they have created a system that is complex. Ontology experts justify the use of multiclass inheritance on the observation that such ontologies provide accurate models of nature and that faithful models of reality cannot be created with simple, uniparental classification. Taxonomists who rely on simple, uniparental classifications base their model on epistemological grounds; on the nature of objects. They hold that an object can have only one nature and can belong to only one defining class, and can be derived from exactly one parent class. Taxonomists who insist upon uniparental class inheritance believe that assigning more than one parental class to an object indicates that you have failed to grasp the essential nature of the object [22–24].

Negative classifier One of the most common mistakes committed by ontologists involves classification by negative attribute. A negative classifier is a feature whose absence is used to define a class. An example is found in the Collembola, popularly known as springtails, a ubiquitous member of Class Hexapoda, and readily found under just about any rock. These organisms look like fleas (same size, same shape) and were formerly misclassified among the class of true fleas (Class Siphonaptera). Like fleas, springtails are wingless, and it was assumed that springtails, like fleas, lost their wings somewhere in evolution's murky past. However, true fleas lost their wings when they became parasitic. Springtails never had wings, an important taxonomic distinction separating springtails from fleas. Today, springtails (Collembola) are assigned to Class Entognatha, a separate subclass of Class Hexapoda. Alternately, taxonomists may be deceived by a feature whose absence is falsely conceived to be a fundamental property of a class of organisms. For example, all species of Class Fungi were believed to have a characteristic absence of a flagellum. Based on the absence of a flagellum, the fungi were excluded from Class Opisthokonta and were put in Class Plantae, which they superficially resembled. However, the chytrids, which have a flagellum, were have been shown to be a primitive member of Class Fungi. This finding places fungi among the true descendants of Class Opisthokonta (from which Class Animalia descended). This means that fungi are much more closely related to people than to plants, a shocking revelation [13]!

Non-living organism Herein, viruses and prions are referred to as non-living organisms. Viruses lack key features that distinguish life from non-life. They depend entirely on host cells for replication; they do not partake in metabolism, and do not yield energy; they cannot adjust to changes in their environment (i.e., no homeostasis), nor can they respond to stimuli. Most scientists consider viruses to be mobile genetic elements that can travel between cells (much as transposons are considered mobile genetic elements that travel within a cell). All viruses have a mechanism that permits them to infect cells and to use the host cell machinery to replicate. At minimum, viruses consist of a small RNA or DNA genome, encased by a protective protein coat, called a capsid. Class Mimiviridae, discovered in 1992, occupies a niche that seems to span the biological gulf separating living organisms from viruses. Members of Class Mimiviridae are complex, larger than some bacteria, with enormous genomes (by viral standards), exceeding a million base pairs and encoding upwards of 1000 proteins. The large size and complexity of Class Mimiviridae exemplifies the advantage of a double-stranded DNA genome. Class Megaviridae is a newly reported (October, 2011) class of viruses, related to Class Mimiviridae, but even larger [25]. Biologically, the life of a mimivirus is not very different from that of obligate intracellular bacteria (e.g., Rickettsia). The discovery of Class Mimiviridae inspires biologists to reconsider the "non-living" status relegated to viruses and compels taxonomists to examine the placement of viruses within the phylogenetic development of prokaryotic and eukaryotic organisms [13].

Nonphylogenetic property Properties that do not hold true for a class; hence, cannot be used by ontologists to create a classification. For example, we do not classify animals by height, or weight because animals of greatly different heights and weights may occupy the same biological class. Similarly, animals within a class may have widely ranging geographic habitats; hence, we cannot classify animals by locality. Case in point: penguins can be found virtually anywhere in the southern hemisphere, including hot and cold climates. Hence, we cannot classify penguins as animals that live in Antarctica or that prefer a cold climate. Scientists commonly encounter properties, once thought to be class-specific that prove to be uninformative, for classification purposes. For many decades, all bacteria were assumed to be small; much smaller than animal cells. However, the bacterium *Epulopiscium fishelsoni* grows to about 600 microns by 80 microns, much larger than the typical animal epithelial cell (about 35 microns in diameter) [26]. *Thiomargarita namibiensis*, an ocean-dwelling bacterium, can reach a size of 0.75 mm, visible to the unaided eye. What do these admittedly obscure facts teach us about the art of classification? Superficial properties, such as size, seldom inform us how to classify objects. The ontologist must think very deeply to find the essential defining features of classes.

Object rank A generalization of Page rank, the indexing method employed by Google. Object ranking involves providing objects with a quantitative score that provides some clue to the relevance or the popularity of an object. For the typical object ranking project, objects take the form of a key word phrase.

Observational data Data obtained by measuring existing things or things that occurred without the help of the scientist. Observational data needs to be distinguished from experimental data. In general, experimental data can be described with a Gaussian curve, because the experimenter is trying to measure what happens when a controlled process is performed on every member of a uniform population. Such experiments typically produce Gaussian (i.e., bell-shaped or normal) curves for the control population and the test population. The statistical analysis of experiments reduces to the chore of deciding whether the resulting Gaussian curves are different from one another. In observational studies, data is collected on categories of things, and the resulting data sets often follow a Zipf distribution, wherein a few types of data objects account for the majority of observations For this reason, many of the assumptions that apply to experimental data (i.e., the utility of parametric statistical descriptors including average, standard deviation and p-values), will not necessarily apply to observational data sets [24].

Parent class The immediate ancestor, or the next-higher class (i.e., the direct superclass) of a class. For example, in the classification of living organisms, Class Vertebrata is the parent class of Class Gnathostomata. Class Gnathostomata is the parent class of Class Teleostomi. In a classification, which imposes single class inheritance, each child class has exactly one parent class; whereas one parent class may have several different child classes. Furthermore, some classes, in particular the bottom class in the lineage, have no child classes (i.e., a class need not always be a superclass of other classes). A class can be defined by its properties, its membership (i.e., the instances that belong to the class), and by the name of its parent class. When we list all of the classes in a classification, in any order, we can always reconstruct the complete class lineage, in their correct lineage and branchings, if we know the name of each class's parent class [13].

Phenetics The classification of organisms by feature similarity, rather than through relationships. Starting with a set of feature data on a collection of organisms, you can write a computer program that will cluster the organisms into classes, according to their similarities. In theory, one computer program, executing over a large dataset containing measurements for every earthly organism, could create a complete biological classification. The status of a species is thereby reduced from a fundamental biological entity, to a mathematical construction.

There are a host of problems consequent to computational methods for classification. First, there are many different mathematical algorithms that cluster objects by similarity. Depending on the chosen algorithm, the assignment of organisms to one species or another would change. Secondly, mathematical algorithms do not cope well with species convergence. Convergence occurs when two species independently acquire identical or similar traits through adaptation; not through inheritance from a shared ancestor. Examples are: the wing of a bat and the wing of a bird; the opposable thumb of opossums and of primates; the beak of a platypus and the beak of a bird. Unrelated species frequently converge upon similar morphologic adaptations to common environmental conditions or shared physiological imperatives. Algorithms that cluster organisms based on similarity are likely to group divergent organisms under the same species.

It is often assumed that computational classification, based on morphologic feature similarities, will improve when we acquire whole-genome sequence data for many different species. Imagine an experiment wherein you take DNA samples from every organism you encounter: bacterial colonies cultured from a river, unicellular non-bacterial organisms found in a pond, small multicellular organisms found in soil, crawling creatures dwelling under rocks, and so on. You own a powerful sequencing machine, that produces the full-length sequence for each sampled organism, and you have a powerful computer that sorts and clusters every sequence. At the end, the computer prints out a huge graph, wherein all the samples are ordered and groups with the greatest sequence similarities are clustered together. You may think you have created a useful classification, but you have not really, because you do not know anything about the organisms that are clustered together. You do not know whether each cluster represents a species, or a class (a collection of related species), or whether a cluster may be contaminated by organisms that share some of the same gene sequences, but are phylogenetically unrelated (i.e., the sequence similarities result from chance or from convergence, but not by descent from a common ancestor). The sequences do not tell you very much about the biological properties of specific organisms, and you cannot infer which biological properties characterize the classes of clustered organisms. You have no certain knowledge whether the members of any given cluster of organisms can be characterized by any particular gene sequence (i.e., you do not know the characterizing gene sequences for classes of organisms). You do not know the genus or species names of the organisms included in the clusters, because you began your experiment without a presumptive taxonomy. Basically, you simply know what you knew before you started; that individual organisms have unique gene sequences that can be grouped by sequence similarity.

Taxonomists, who have long held that a species is a natural unit of biological life, and that the nature of a species is revealed through the intellectual process of building a consistent taxonomy [27], are opposed to the process of phenetics-based classification [27,13]. In the realm of big data, computational phenetics may create a complex web of self-perpetuating nonsense that cannot be sensibly analyzed. Over the next decade or two, the brilliance or the folly of computational phenetics will most likely be revealed.

Properties versus classes When creating classifications, the most common mistake is to assign class status to a property. When a property is inappropriately assigned as a class, then the entire classification is ruined. Hence, it is important to be very clear on the difference between these two concepts, and to understand why it is human nature to confuse one with the other. A class is a holder of related objects (e.g., items, records, categorized things). A property is a feature or trait that can be assigned to an item. When inclusion in a class requires items to have a specific property, we often name the class by its defining property. For example Class Rodentia, which includes rats, mice, squirrels, and gophers, are all gnawing mammals. The word rodent derives from the Latin roots rodentem, rodens, from rodere, "to gnaw." Although all rodents gnaw, we know that gnawing is not unique to rodents. Rabbits (Class Lagomorpha) also gnaw.

Objects from many different classes may have some of the same properties. Here's another example. Normal human anatomy includes two legs. This being the case, is "leg" a subclass of "human." The answer is no. A leg is not a type of human. Having a leg is just one of many properties associated with normal human anatomy. You would be surprised how many people can be tricked into thinking that a leg, which is itself an object, should be assigned as a subclass of the organisms to which it is attached. Some of this confusion comes from the way that we think about relationships between objects and properties. We say "He is hungry," using a term of equality, "is" to describe the relationship between "He" and "hungry." Technically, the sentence, "He is hungry" asserts that "He" and "hungry" are equivalent objects. We never bother to say "He has hunger," but other languages are more fastidious. A German might say "Ich habe Hunger," indicating that he has hunger, and avoiding any inference that he and hunger are equivalent terms (i.e., never "Ich bin Hunger"). It may seem like a trivial point, but mistaking classes for properties is a common error that nearly always leads to disaster.

RDF Resource Description Framework (RDF) is a syntax in XML notation that formally expresses assertions as triples. The RDF triple consists of a uniquely identified subject plus a metadata descriptor for the data plus a data element. Triples are necessary and sufficient to create statements that convey meaning. Triples can be aggregated with other triples from the same data set or from other data sets, so long as each triple pertains to a unique subject that is identified equivalently through the data sets. Enormous data sets of RDF triples can be merged or functionally integrated with other massive or complex data resources.

RDF Ontology A term that, in common usage, refers to the class definitions and relationships included in an RDF Schema document. The classes in an RDF Schema need not comprise a complete ontology. In fact, a complete ontology could be distributed over multiple RDF Schema documents.

Representation bias Occurs when the population sampled does not represent the population intended for study. For example, the population for which the normal range of prostate specific antigen (PSA) was based, was selected from a county in the state of Minnesota. The male population under study consisted almost exclusively of white men (i.e., virtually no African-Americans, Asians, Hispanics, etc.). It may have been assumed that PSA levels would not vary with race. It was eventually determined that the normal PSA ranges varied greatly by race [28]. The Minnesota data, though plentiful, did not represent racial subpopulations. A sharp distinction must drawn between Big-ness and Whole-ness [29].

Superclass Any of the ancestral classes of a subclass. For example, in the classification of living organisms, the class of vertebrates is a superclass of the class of mammals. The immediate superclass of a class is its parent class. In common parlance, when we speak of the superclass of a class, we are usually referring to its parent class.

System call Refers to a command, within a running script, that calls the operating system into action, momentarily bypassing the programming interpreter for the script. A system call can do essentially anything the operating system can do via a command line.

Triple In computer semantics, a triple is an identified data object associated with a data element and the description of the data element. In theory, all Big Data resources can be composed as collections of triples. When the data and metadata held in sets of triples are organized into ontologies consisting of classes of objects and associated properties (metadata), the resource can potentially provide intro-spection (the ability of a data object to be self-descriptive). An in-depth discussion of triples is found in Chapter 4, "Metadata, Semantics, and Triples."

Triplestore A list or database composed entirely of triples (statements consisting of an item identifier plus the metadata describing the item plus an item of data. The triples in a triplestore need not be saved in any particular order, and any triplestore can be merged with any other triplestore; the basic semantic meaning of the contained triples is unaffected. Additional discussion of triplestores can be found in Section 6.5, "Case Study: A Visit to the TripleStore."

Turtle Another syntax for expressing triples. From RDF came a simplified syntax for triples, known as Notation 3 or N3 [30]. From N3 came Turtle, thought to fit more closely to RDF. From Turtle came an even more simplified form, known as N-Triples.

Unclassifiable objects Classifications create a class for every object and taxonomies assign each and every object to its correct class. This means that a classification is not permitted to contain unclassified objects; a condition that puts fussy taxonomists in an untenable position. Suppose you have an object, and you simply do not know enough about the object to confidently assign it to a class. Or, suppose you have an object that seems to fit more than one class, and you can't decide which class is the correct class. What do you do?

Historically, scientists have resorted to creating a "miscellaneous" class into which otherwise unclassifiable objects are given a temporary home, until more suitable accommodations can be provided. I have spoken with numerous data managers, and everyone seems to be of a mind that "miscellaneous" classes, created as a stopgap measure, serve a useful purpose. Not so. Historically, the promiscuous application of "miscellaneous" classes has proven to be a huge impediment to the advancement of science. In the case of the classification of living organisms, the class of protozoans stands as a case in point. Ernst Haeckel, a leading biological taxonomist in his time, created the King-dom Protista (i.e., protozoans), in 1866, to accommodate a wide variety of simple organisms with superficial commonalities. Haeckel himself understood that the protists were a blended class that included unrelated organisms, but he believed that further study would resolve the confusion. In a sense, he was right, but the process took much longer than he had anticipated; occupying generations of taxonomists over the following 150 years.

Today, Kingdom Protista no longer exists. Its members have been reassigned to positions among the animals, plants, and fungi. Nonetheless, textbooks of microbiology still describe the protozoans, just as though this name continued to occupy a legitimate place among terrestrial organisms. In the meantime, therapeutic opportunities for eradicating so-called protozoal infections, using class-targeted agents, have no doubt been missed [13].

You might think that the creation of a class of living organisms, with no established scientific rela-tion to the real world, was a rare and ancient event in the annals of biology, having little or no chance of being repeated. Not so. A special pseudoclass of fungi, deuteromyctetes (spelled with a lowercase "d," signifying its questionable validity as a true biologic class) has been created to hold fungi of indeterminate speciation. At present, there are several thousand such fungi, sitting in a taxonomic limbo, waiting to be placed into a definitive taxonomic class [16,13].

References

[1] Wu D, Hugenholtz P, Mavromatis K, Pukall R, Dalin E, Ivanova NN, et al. A phylogeny-driven genomic encyclopaedia of Bacteria and Archaea. Nature 2009;462:1056–60.

[2] Woese CR, Fox GE. Phylogenetic structure of the prokaryotic domain: the primary kingdoms. PNAS 1977;74:5088–90.

[3] Mayr E. Two empires or three? PNAS 1998;95:9720–3.

[4] Woese CR. Default taxonomy: Ernst Mayr's view of the microbial world. PNAS 1998;95(19):11043–6.

[5] Bamshad MJ, Olson SE. Does race exist? Sci Am December, 2003;78–85.

[6] Wadman M. Geneticists struggle towards consensus on place for 'race'. Nature 2004;431:1026.

[7] Pearson K. The grammar of science. London: Adam and Black; 1900.

[8] Berman JJ. Racing to share pathology data. Am J Clin Pathol 2004;121:169–71.

[9] Scamardella JM. Not plants or animals: a brief history of the origin of Kingdoms Protozoa, Protista and Protoctista. Int Microbiol 1999;2:207–16.

[10] Berman JJ. Methods in medical informatics: fundamentals of healthcare programming in Perl, Python, and Ruby. Boca Raton: Chapman and Hall; 2010.

[11] Madar S, Goldstein I, Rotter V. Did experimental biology die? Lessons from 30 years of p53 research. Cancer Res 2009;69:6378–80.

[12] Zilfou JT, Lowe SW. Tumor suppressive functions of p53. Cold Spring Harb Perspect Biol 2009;00: a001883.

[13] Berman JJ. Taxonomic guide to infectious diseases: understanding the biologic classes of pathogenic organisms. Cambridge, MA: Academic Press; 2012.

[14] Suggested Upper Merged Ontology (SUMO). The Ontology Portal. Available from: http://www.ontologyportal.org [viewed August 14, 2012].

[15] de Bruijn J. Using ontologies: enabling knowledge sharing and reuse on the Semantic Web. Digital Enterprise Research Institute Technical Report DERI-2003-10-29. Available from: http://www.deri.org/fileadmin/documents/DERI-TR-2003-10-29.pdf; October 2003 [viewed August 14, 2012].

[16] Guarro J, Gene J, Stchigel AM. Developments in fungal taxonomy. Clin Microbiol Rev 1999; 12:454–500.

[17] Nakayama R, Nemoto T, Takahashi H, Ohta T, Kawai A, Seki K, et al. Gene expression analysis of soft tissue sarcomas: characterization and reclassification of malignant fibrous histiocytoma. Mod Pathol 2007;20:749–59.

[18] Cote R, Reisinger F, Martens L, Barsnes H, Vizcaino JA, Hermjakob H. The ontology lookup service: bigger and better. Nucleic Acids Res 2010;38:W155–160.

[19] Niles I, Pease A. In: Welty C, Smith B, editors. Towards a standard upper ontology. Proceedings of the 2nd international conference on formal ontology in information systems (FOIS-2001), Ogunquit, Maine, October 17-19; 2001.

[20] Gansner E, Koutsofios E. Drawing graphs with dot. January 26. Available at: http://www.graphviz.org/Documentation/dotguide.pdf; 2006 [viewed on June 29, 2015].

[21] Wu X, Kumar V, Quinlan JR, Ghosh J, Yang Q, Motoda H, et al. Top 10 algorithms in data mining. Knowl Inf Syst 2008;14:1–37.

[22] Berman JJ. Principles of big data: preparing, sharing, and analyzing complex information. Waltham, MA: Morgan Kaufmann; 2013.

[23] Berman JJ. Repurposing legacy data: innovative case studies. Waltham, MA: Morgan Kaufmann; 2015.

[24] Berman JJ. Data simplification: taming information with open source tools. Waltham, MA: Morgan Kaufmann; 2016.

[25] Arslan D, Legendre M, Seltzer V, Abergel C, Claverie J. Distant Mimivirus relative with a larger genome highlights the fundamental features of Megaviridae. PNAS 2011;108:17486–91.

[26] Angert ER, Clements KD, Pace NR. The largest bacterium. Nature 1993;362:239–41.

[27] DeQueiroz K. Ernst Mayr and the modern concept of species. PNAS 2005;102(suppl 1):6600–7.

[28] Sawyer R, Berman JJ, Borkowski A, Moore GW. Elevated prostate-specific antigen levels in black men and white men. Mod Pathol 1996;9:1029–32.

[29] Boyd D. Privacy and publicity in the context of big data. Raleigh, North Carolina: Open Government and the World Wide Web (WWW2010); 2010. April 29. Available from: http://www.danah.org/papers/talks/2010/WWW2010.html [viewed August 26, 2012].

[30] Primer: Getting into RDF & Semantic Web using N3. Available from: http://www.w3.org/2000/10/swap/Primer.html [viewed September 17, 2015].

Introspection

Section 6.1. Knowledge of Self

All science is description and not explanation.

Karl Pearson [1]

Not very long ago a cancer researcher sent me one of his published papers. For his study, he used a publicly available collection of gene micro-array data collected on tumors. He knew that I was a long-time proponent of open access scientific data sets and that I had been encouraging my colleagues to use these available data sources for various analytic projects. I read the paper with admiration, but the "methods" section of the paper did not provide much description of the human lung cancer tissues that were used to generate the micro-array data. [Glossary Open access]

I called the researcher and asked, perhaps a bit undiplomatically, the following question: "The methods section indicates that data on 12 lung cancer tissues, provided by the repository, were studied. How do you distinguish whether these were 12 lung cancer samples from 12 different patients, or 12 samples of tissue all taken from one lung cancer, in one patient?" If it were the former (12 lung cancers from each of 12 patients), his study conclusions would have applied to a sampling of different tumors and might reasonably apply to lung cancers in general. If it were the latter (12 samples of one tumor), then his generalized conclusion was unjustified.

There was a pause on the line, and I was told that he had neglected to include that information in the manuscript, but the paper included a link to the repository Web site, where the detailed sample information was stored.

After our conversation, I visited the Web site, and found that there was very little information describing the samples included in the database. There was a sample number, followed by the name of a type of cancer (lung cancer, in this case), and then there was the raw gene-array data. Were the multiple samples taken from multiple patients, or were all those samples taken from one tumor, in one patient? We could not say, because the data would not tell us.

I contacted the researcher again and reiterated the problem. He agreed that the people at the repository should have been more attentive to data annotation. It has been my experience that some data analysts believe that their professional responsibility begins with the received data. In their opinion, pre-analytic issues, such as those described above, do not fall under their professional jurisdiction. This approach to Big Data analysis is an invitation for disaster. Studies emanating from Big Data resources have no scientific value and the Big Data resources are all a waste of time and money, if data analysts cannot uncover all the information that fully describes their data.

The aforementioned story serves as an introduction to the concept of introspection, a term that is not commonly applied to Big Data resources; but should be. Introspection is a term taken from the field of object oriented programming, and it refers to the ability of data objects to describe themselves, when called upon. In object oriented programming languages, everything is objectified. Variables are objects, parameters are objects, methods are objects, and so on. Every object carries around its own data values, as well as an identifier, and self-descriptive information, including its class assignment (i.e., the name of the class of objects to which it belongs). An object can have its own methods (similar to subroutines), and it has access to a library of methods built for its class (i.e., class methods) and from the ancestor classes in its lineage (i.e., superclass methods).

Most object oriented programming languages have methods that can call an object to describe itself. To illustrate, let us see how Ruby, a popular object oriented programming language, implements introspection.

First, let us create a new object, "x"; we will assign "hello world" to the object.

```
x = "hello world"    yields "hello world"
```

Ruby knows that "hello world" is a string and automatically assigns "x" to Class String. We can check any object to see determine its class by sending the "class" method to the object, as shown.

```
x.class        yields String
```

When we send the "class" method to x, Ruby outputs its class assignment, "String." Every class (except the top level class in the hierarchy) has a single parent class, also known as a superclass. We can learn the name of the superclass of Class String, by sending the superclass method, as shown.

```
x.class.superclass    yields Object
```

Ruby tells us that the superclass of Class String is Class Object.

Ruby assigns a unique identifier to every created object. We can find the object identifier by sending "x" the object_id method.

```
x.object_id       yields 22502910
```

The unique object identifier for "x" is 22502910.

If we ever need to learn the contents of "x," we can send the inspect method to the object.

```
x.inspect        yields "hello world"
```

Ruby reminds us that "x" contains the string, "hello world."

Every data object in Ruby inherits dozens of class methods. We can generate a list of methods available to "x" by sending it the "methods" method.

```
x.methods
```

Ruby yields a list of dozens of methods, including a few that we can try out here: "length", "is_a?", "upcase", "downcase", "capitalize", and "reverse."

```
x.length         yields 11
```

The length method, sent to "x" yields the number of characters in "hello world."

```
x.is_a?(String)     yields true
```

When Ruby uses the is_a? method to determine if x is a member of Class String, it yields "true."

```
x.is_a?(Integer)    yields false
```

When Ruby uses the is_a? method to determine if x is a member of Class Integer, it yields "false."

```
x.upcase          yields "HELLO WORLD"
x.downcase        yields "hello world"
x.capitalize      yields "Hello world"
x.reverse         yields "dlrow olleh"
```

String methods sent to the "x" object return appropriate values, as shown above.

What happens when we send "x" a method from a library of methods built for some other class?

The "nonzero?" method tests to see whether an object of Class Integer is zero. This method is useful to avoid division by zero.

Let us see what happens when we send "x" the "nonzero?" method.

```
x.nonzero?        Yields NoMethodError: undefined method `nonzero?'
                  for "hello world":String
```

Ruby sends us an error message, indicating that "nonzero?" is an undefined method for an object of Class String.

How does introspection, a feature of object oriented programming languages, apply to Big Data? In principle, Big Data resources must have the same features of introspection that are automatically provided by object oriented programming languages. Specifically, all data pertaining to the object must be encapsulated within data objects to include the raw data, a description for the raw data (the so-called metadata), the name of the class to which the data object belongs, and a unique identifier that distinguishes the data object from all other data objects.

I must admit that most Big Data resources lack introspection. Indeed, most Big Data managers are unfamiliar with the concept of introspection as it applies to Big Data. When you speak to the people who manage and use these resources, you may be surprised to learn that they are happy, even ecstatic, about their introspection-free resource. As far as they are concerned, their resource functions just fine, without introspection. When pressed on the subject of data introspection, data managers may confess that their Big Data resources may fall short of perfection, but their users have accommodated themselves to minor deficiencies.

There is always a price to pay when Big Data resources lack introspection. Symptoms of an introspection-free Big Data resource include:

– The resource is used for a narrow range of activities, somewhat less than was originally proposed when the resource was initiated.
– The resource is used by a small number of domain experts; to all others, the resource is inscrutable.
– The data records for the resource cannot be verified. It is impossible to eliminate the possibility that records may have been duplicated or that data may have been mistakenly inserted into the wrong records (i.e., the data records may have been corrupted).
– The resource cannot merge its data with data contained in other Big Data resources.
– The resource cannot match its record identifiers with record identifiers from other resources. For example, if each of two Big Data resources has a record on the same individual, the resources cannot sensibly combine the two records into a single record.
– The resource cannot add legacy data, collected by their own institution on older software system, into the current Big Data resource.
– Despite employing a crew of professionals for the resource, only one individual seems to be privy to the properties of data objects in the largely undocumented system. When he is absent, the system tends to crash.

Introspection is not a feature that you can attach to a Big Data resource, as an afterthought. Introspection is a foundational component of any well-designed data resource. Most Big Data resources will never attain the level of introspection available in object oriented programming languages, but some introspective features would seem essential.

Section 6.2. Data Objects: The Essential Ingredient of Every Big Data Collection

Computer Science is no more about computers than astronomy is about telescopes.
Edsger W. Dijkstra

If you have been following the computer science literature, you may have noticed that the term "data object" has been slowly replacing the shorter, simpler, and more understandable term "data." Do we really need to clutter our minds with yet another example of dispensable technojargon?

Yes, we must. Despite every intention to minimize the use of jargon in this book, the term "data object" has already insinuated itself into this book dozens of times. Back in Section 3.1, we offered a loose definition of data object as "a collection of data that contains self-describing information, and one or more data values." In this section, we will expand the definition to indicate the role of data objects in Big Data construction and analyses.

Like everything else in this fledgling field of Big Data, there is no canonical definition for "data object." As you might expect, practitioners of subdisciplines of computer science provide definitions of data object that coincide with the way they happen to employ data objects in their work. For example, someone who works exclusively with relational databases will refer to data tables, indexes, and views as data objects. A programmer who uses assembly language might refer to a data object as any data that can be referenced from a particular address in memory. A programmer who works with a typed language, such as Ada, might think of a data object as being data that has been assigned a particular type (e.g., string, integer, float).

We can try to find a reasonable definition for data object that serves the imperatives of Big Data, but before we do, let us look at a few triples.

```
75898039563441  name            G. Willikers
75898039563441  gender          male
75898039563441  age             35
75898039563441  is_a_class_member human
```

These triples tell us a few things about a 35-year-old male named G. Willikers, who is a human. Without losing any information, we can rearrange this collection of triples under its identifier, as shown here:

```
75898039563441
name            G. Willikers
gender          male
age             35
is_a_class_member human
```

Now, we can begin to see how a collection of triples, all having the same identifier, might compose a single data object. What we have is one identifier followed by all the available meta/data data pairs that bind to the same identifier. Someone who prefers working with spreadsheets might interpret this as a row (with "75898039563441" as its key); having metadata as its column headers, and having the data as the contents of the row's cells. We can guess that the relational database programmer would recognize this as a table.

The assembly language programmer would look at the same collection and surmise that it represents the data culled from a referenced address in memory.

For our purposes, let us try to think of the collection as a data object, defined as an object identifier along with all of the data/metadata pairs that rightly belong to the object identifier, and that includes one data/metadata pair that tells us the object's class (i.e., "human" in this example).

By adding a triple that provides class membership, the data object immediately gains all of the properties associated with its class "Human," and this might include a birth date, a social security number, and an address. If a programmer were to write a set of computational methods for the class "Human," then every member of class "Human," including instance 75898039563441, would be qualified to access those methods. In the next section, we will describe how programmers use the information available within data objects to understand and to utilize the relationships among data objects. Before proceeding, there are a few properties of data objects that we should examine. Notice that there is no special order for the data/metadata pairs encapsulated within the data object. We could shuffle the data/metadata pairs any way we please. Furthermore, if each data/metadata pair was attached to its identifier (75898039563441, in this case), as a triple, then there would be no special reason to store the components of the data object in one particular memory location.

The following triple could be stored in a server in California:

```
75898039563441  name      G. Willikers
```

The next triple could be stored in a server in Iceland:

```
75898039563441  gender      male
```

The triples that compose the data object may exist anywhere and everywhere (i.e., stored as replicates), and we might have no knowledge of the number of object's data/metadata pairs that exist at any moment of time. Wherever the pieces of the data object may reside, they will forever have the same unique identifier, and will always belong to the same data object. It is best to think of a data object as an abstraction that is made practical by software created by programmers. Object oriented programming languages are designed to create data objects assigned to classes, and provide them with useful computational methods.

Section 6.3. How Big Data Uses Introspection

"Si sol deficit, respicit me nemo" ("If the sun's gone, nobody looks at me")

Latin motto

Let us look at how data objects are used to understand and explore Big Data. First, we must understand a few new concepts that have been developed for Object-Oriented programming languages, but which apply to all data that supports introspection: encapsulation, inheritance, polymorphism, and reflection.

Encapsulation refers to general property of a data object to contain the data pertaining to itself (i.e., its identifier and its data/metadata pairs). When we say "contain," we are not referring to a physical container. We are indicating that there is some way by which a programmer can access an object's identifier and its data/metadata pairs, through methods provided by a programming language. The data can be scattered on servers throughout the globe. So long as there are methods for retrieving the data/metadata pairs, and ascertaining that these pairs belong to a unique data object, and to no other data object, then we say that the data object encapsulates its data.

Inheritance refers to the ability of a data object to respond appropriately to methods created for its class, and for all of its ancestral classes. For example, all members of Class Document can respond to methods created for its class, such as a "screen_display" method, or a "printer_print" or a "copy_me" method. Furthermore, since all documents are composed of strings of alphanumeric sequences, we know that Class Document is a descendant of Class String. Hence, members of Class Document will inherit the methods created for Class String, such as a "lowercase" method, a "concatenate" method, or a "find_substring" method.

This object oriented concept of inheritance fits nicely with the concept of inheritance, as known to zoologists: every animal inherits the properties of its ancestors. For example, humans are descendants of the class of animals known as the vertebrates (i.e., Class Vertebrata). This means that every human, like all animal classes that descend from Class Vertebrata, contains a vertebra, and all such animals have shared properties inherited from their common ancestor (e.g., they all have anatomic structures derived from gill arches that appear in embryologic development and they all share genes and proteins that were included in the vertebrates from which they descended).

The key thing to understand, whether you are a computer programmer or a zoologist, is that inheritance only helps us if we have created a sensible classification (see the principles of classification in Section 5.2).

Polymorphism is the ability of an object to respond to a named method in a manner that is appropriate for its own class. For example, if I sent a "double" method to an object belonging to Class Integer, I might expect it to multiply its contained integer by itself (e.g., $5 * 2 = 10$). If I sent the "double" method to an object belonging to Class String, I might expect it to simply concatenate the contained string to itself (e.g., "3y228hw" would become "3y228hw3y228hw").

How does a data object know how to respond polymorphically (i.e., in a manner appropriate for its class) to a method? In object oriented programming, classes have methods that apply to every instance (i.e., member) belonging to the class. When you send the "double" method to an integer object the integer object knows the name of its own class and will look inside its class object for a class method named "double." If it finds the method, it will do whatever its class method tells it to do; in this case, it will multiply its integer contents by two. If the double method were sent to a member of Class String, the data object would pull the "double" method written for objects of the String class, and would respond appropriately; by concatenating its contained string to itself.

Polymorphism is achieved by having objects respond to different methods, written for different classes. The different methods may happen to have the same name (i.e., "double" in this example), but each data object only has access to the "double" method that was written for its class.

Now, suppose the data object looks within the collection of methods available to its class, and fails to find what it's looking for. In that case, it will look at the methods contained in the parent class of Class String. Remember, in a good classification, every class contains the name of its parent class; hence the ancestral lineage of any class object can be computationally traced, up to the root class of the classification. This means that a class object can search its entire ancestry, if need be, looking for a "double" method. When it finds the method, it stops and does whatever the method instructs it to perform. This is known as inheritance polymorphism.

Underlying all these methods (encapsulation, inheritance, and polymorphism) is a technique known as abstraction. Abstraction encompasses all techniques wherein data objects are unencumbered by the details of their operational repertoires. For example, the programmer who sends a method to an object does not need to create the program by which the method operates. Object methods can be chosen from class libraries. The object that receives the method does not need to contain the instructions for executing the method. The object simply needs to know the class to which it has been assigned membership. Objects will always pull the methods that are appropriate for their own classes. In object oriented languages, the class libraries subsume the nitty-gritty of programming, and the burden of holding all the information required by data objects is abstracted into the class structure of the data domain. Not surprisingly, programs written in object oriented programming languages are famously short, consisting mostly of one-word methods, sent to one-word names for complex data objects.

There is one more concept that we must discuss: reflection. If you were to take introspective data gleaned on-the-fly during the execution of a program and you used that data to modify the run-time instructions of the same program, then you would be achieving reflection. There are many situations when reflection might come in handy. For example, you might use introspection to determine that a data object was created prior to 2010; and then exclude that data object from subsequent computations intended to show the average value of measurements performed from 2010 to the present.

What are the benefits of these object-oriented concepts. For the purposes of Big Data, object-oriented approaches drive down the complexity of the system. Once the classification has been created, and all of the data objects are assigned to one and only one class within the classification, all of the wonderful concepts of object-oriented programming (encapsulation, inheritance, polymorphism, and reflection) come to us gratis. The methods in the class libraries can be written without knowing anything about the individual class objects. The instances of class objects can exist unencumbered by any information pertaining to the classification, other than the name of the class in which they belong.

Section 6.4. Case Study: Time Stamping Data

People change and forget to tell each other.

Lillian Hellman, playwright (1905–1984)

Consider the following assertions:

```
Alexander Goodboy, 34 inches height
Alexander Goodboy, 42 inches height
Alexander Goodboy, 46 inches height
Alexander Goodboy, 52 inches height
```

At first glance these assertions seem contradictory. How can Alexander Goodboy be 34, 42, 46, and 52 inches tall? The confusion is lifted when we add some timing information to the assertions:

```
Alexander Goodboy, age 3 years, 34 inches height
Alexander Goodboy, age 5 years, 42 inches height
Alexander Goodboy, age 7 years, 46 inches height
Alexander Goodboy, age 9 years, 52 inches height
```

All events, measurements and transactions occur at a particular time, and it is essential to annotate data objects with their moment of creation and with every moment when additional data is added to the data object (i.e., event times) [2]. It is best to think of data objects as chronicles of a temporal sequence of immutable versions of the object. In the case of Alexander Goodboy, the boy changes in height as he grows, but each annotated version of Alexander Goodboy (e.g., Alexander Goodboy, age 3 years, height 34 inches) is eternal and immutable. [Glossary Immutability]

Time stamping is nothing new. Ancient scribes were fastidious time stampers. It would be an unusual Sumerian, Egyptian, or Mayan document that lacked an inscribed date. In contrast, it is easy to find modern, Web-based news reports that lack any clue to the date that the Web page was created. Likewise, it is a shameful fact that most spreadsheet data lacks time stamps for individual data cells. Data sets that lack time stamps, unique identifiers, and metadata have limited value to anyone other than the individual who created the data and who happens to have personal knowledge of how the data was created and what the data means.

Fortunately, all computers have an internal clock. This means that all computer events can be time stamped. Most programming languages have a method for generating the epoch time; the number of seconds that have elapsed since a particular moment in time. On most systems the epoch is the first second of January 1, 1970. Perl, Python, and Ruby have methods for producing epoch time. For trivia-sake, we must observe that the UUID time stamp is generated for an epoch time representing the number of seconds elapsed since the first second of Friday, October 15, 1582 (See Section 5.1, "Unique Identifiers"). This moment marks the beginning of the Gregorian calendar. The end of the Julian calendar occurred on October 4, 1582. The 11 days intervening, from the end of the Julian calendar to the start of the Gregorian calendar, are lost somewhere in time and space.

From Python's interactive environment:

```
import time
print(time.time())
output:
1442353742.456994
```

if you would like the GMT (Greenwich Mean Time), try this gmtime.py script:

```
import time
print(time.gmtime())
output:
time.struct_time(tm_year=2017, tm_mon=10, tm_mday=12, tm_hour=14,
tm_min=3, tm_sec=0, tm_wday=3, tm_yday=285, tm_isdst=0)
```

It is very important to understand that country-specific styles for representing the date are a nightmare for data scientists. As an example, consider: "2/4/97." This date signifies February 4, 1997 in America; and April 2, 1997 in Great Britain and much of the world. There basically is no way of distinguishing with certainty 2/4/97 and 4/2/97.

It is not surprising that an international standard, the ISO-8601, has been created for representing date and time [3]. The international format for date and time is: YYYY-MM-DD hh:mm:ss.

The value "hh" is the number of complete hours that have passed since midnight. The upper value of hh is 24 (midnight). If hh = 24, then the minute and second values must be zero (think about it). An example of and ISO-8601-compliant data and time is:

```
1995-02-04 22:45:00
```

An alternate form, likewise ISO-8601-compliant, is:

```
1995-02-04T22:45:00Z
```

In the alternate form, a "T" replaces the space left between the date and the time, indicating that time follows date. A "Z" is appended to the string indicating that the time and date are computed for UTC (Coordinated Universal Time, formerly known as Greenwich Mean Time, and popularly known as Zulu time, hence the "Z").

Here is a Python script, format_time.py, that generates the date and time, compliant with ISO-8601.

```
import time, datetime
timenow = time.time()
print(datetime.datetime.fromtimestamp(timenow).strftime('%Y-%m-%d
%H:%M:%S'))
```

Here is the output of the format_time.py script:

```
2015-09-16 07:44:09
```

It is sometimes necessary to establish, beyond doubt, that a time stamp is accurate and has not been modified. Through the centuries, a great many protocols have been devised to prove that a time stamp is trustworthy. One common implementation of a trusted time stamp protocol involves sending a message digest (i.e., a one-way hash) of a confidential document to a time stamp authority. The timestamp authority adds a date to the received message digest and returns a time-annotated message, encrypted with the time stamp authority's private key, containing the original one-way hash plus the trusted date. The received message can be decrypted with the timestamp authority's public key to reveal the date/time and the message digest that is unique for the original document. It might seem as though the trusted time stamp process is a lot of work, but regular users of these services can routinely process hundreds of documents in seconds. We will be revisiting the subject of time stamps in Chapter 8, Immutability and Immortality. [Glossary Message digest, Symmetric key, Trusted time stamp]

Section 6.5. Case Study: A Visit to the TripleStore

Before I speak, I have something important to say.

Groucho Marx

Enormous benefits follow when data objects are expressed as triples and assigned to defined classes. All of the attributes of object oriented programming languages (i.e., inheritance, encapsulation, abstraction, and polymorphism) are available to well-organized collections of triples. Furthermore, desirable features in any set of data, including integration, interoperability, portability, and introspection are available to data scientists who analyze triplestore data. Most importantly, when triples are collected as a triplestore, a simple analysis of the triplestore yields all the relations among data objects, and all the information needed to assemble every data object,

Here is a small example of a triplestore:

```
9f0ebdf2^^object_name^^Class
9f0ebdf2^^property^^subclass_of
9f0ebdf2^^property^^property
9f0ebdf2^^property^^definition
9f0ebdf2^^property^^object_name
9f0ebdf2^^property^^instance_of
9f0ebdf2^^subclass_of^^Class
9f0ebdf2^^instance_of^^Class
701cb7ed^^object_name^^Property
701cb7ed^^subclass_of^^Class
```

```
701cb7ed^^definition^^^^the metadata class
77cb79d5^^object_name^^instance_of
77cb79d5^^instance_of^^Property
77cb79d5^^definition^^the name of the class to which the object is an
instance
a03fbc3b^^object_name^^object_name
a03fbc3b^^instance_of^^Property
a03fbc3b^^definition^^word equivalent of its predicate identifying
sequence
de0e5aa1^^object_name^^subclass_of
de0e5aa1^^instance_of^^Property
de0e5aa1^^definition^^the name of the parent class of the referred object
4b675067^^object_name^^property
4b675067^^instance_of^^Property
4b675067^^definition^^an identifier a for class property
c37529c5^^object_name^^definition
c37529c5^^instance_of^^Property
c37529c5^^definition^^the meaning of the referred object
a29c59c0^^object_name^^dob
a29c59c0^^instance_of^^Property
a29c59c0^^definition^^date of birth, as Day, Month, Year
a34a1e35^^object_name^^glucose_at_time
a34a1e35^^instance_of^^Property
a34a1e35^^definition^^glucose level in mg/Dl at time drawn (GMT)
03cc6948^^object_name^^Organism
03cc6948^^subclass_of^^Class
7d7ff42b^^object_name^^Hominidae
7d7ff42b^^subclass_of^^Organism
7d7ff42b^^property^^dob
a0ce8ec6^^object_name^^Homo
a0ce8ec6^^subclass_of^^Hominidae
a0ce8ec6^^property^^glucose_at_time
a1648579^^object_name^^Homo sapiens
a1648579^^subclass_of^^Homo
98495efc^^object_name^^Andy Muzeack
98495efc^^instance_of^^Homo sapiens
98495efc^^dob^^1 January, 2001
98495efc^^glucose_at_time^^87, 02-12-2014 17:33:09
```

Perusal of the triples provides the following observations:

1. Individual triples are easy to understand, consisting only of a unique
   identifier followed by a metadata/data pair. We could have used any

separator, but in this example, we chose to separate the parts of the triple by a double caret, """.

```
7d7ff42b^^subclass_of^^Organism
```

As noted, the individual parts of the triple are:

```
7d7ff42b is the identifier
subclass_of is the metadata
Organism is the data
```

Notice that these triples are expressed in a format different from RDF, Notation3, or Turtle. Do we care? Not at all. We know that with a few lines of code, we could convert our triples-tore into any alternate format we might prefer. Furthermore, our triplestore could be converted into a spreadsheet, in which the identifiers are record keys, the metadata are column headings, and the data occupy cells. We could also port our triples into a database, if we so desired.

2. Using triples, we have defined various classes and properties. For example:

```
03cc6948^^object_name^^Organism
03cc6948^^subclass_of^^Class
```

With one triple, we create a new object, with name Organism, and we associate it with a unique identifier (03cc6948). With another triple, we establish that the Organism object is a class that happens to be the child class of the root class, Class. Because Organism is a subclass of Class, it will inherit all of the properties of its parent class.

Let's skip down to the bottom of the file:

```
98495efc^^object_name^^Andy Muzeack
98495efc^^instance_of^^Homo sapiens
98495efc^^dob^^1 January, 2001
98495efc^^glucose_at_time^^87, 02-12-2014 17:33:09
```

Here we create a few triples that provide information about a person named Andy Muzeack. First, we assign a unique identifier to our new object, named Andy Muzeack. We learn, from the next triple that Andy Muzeack is a member of class Homo Sapiens. As such, we infer that Andy Muzeack inherits all the properties contained in class Homo (the parent class of class Homo Sapiens) and all the ancestors of class Homo, leading to the top, or root ancestor, class Class. We learn that Andy Muzeack has a "dob" of January 1, 2001. By ascending the list of triples, we learn that "dob" is a property, with a unique identifier (a29c59c0), and a definition, "date of birth, as Day, Month, Year." Finally, we learn that Andy Muzeack has a glucose_at_time of "87, 02-12-2014 17:33:09." Elsewhere in the triplestore, we find that the "glucose_at_time" metadata is defined as the glucose level in mg/Dl at time drawn, in Greenwich Mean Time.

If we wished, we could simply concatenate our triplestore with other triplestores that contain triples relevant to Andy Muzeack. It would not make any difference how the triples are ordered. If Andy Muzeack's identifier is reconcilable, the metadata is defined, and each triple is assigned to a class, then we will be able to fully understand and analyze the data held in the triplestore. [Glossary Reconciliation]

Of course, when we have millions and billions of triples, we could not perform our analyses by reading through the file. We would need scripts and/or a database application. Here is a simple Python script, nested.py, that loads a triplestore into a nested dictionary, and reads the dictionary items:

```python
import collections, sys, re, string, os
from collections import defaultdict
def make_dictionary():
    return defaultdict(make_dictionary)
tripledictionary=defaultdict(make_dictionary)
triple_file = open("triple_2.txt", "r")
for line in triple_file:
  line = line.rstrip()
  triple_items = line.split("^^")
  tripledictionary[triple_items[0]][triple_items[1]][triple_items
[2]] = ""
triple_file.close()
def iter_all(tripledictionary,depth=0):
    for key,value in tripledictionary.items():
        if (depth == 0):
            print("\nidentifier " + key)
        else:
          print("-"*(depth) + key)
        if type(value) is defaultdict:
            iter_all(value,depth+1)
iter_all(tripledictionary)
```

Here is the partial output of the nested.py script:

```
identifier a34a1e35
-definition
--glucose level in mg/Dl at time drawn (GMT)
-object_name
--glucose_at_time
-instance_of
--Property
identifier a29c59c0
-definition
--date of birth, as Day, Month, Year
-object_name
--dob
```

```
-instance_of
--Property
identifier 7d7ff42b
-object_name
--Hominidae
-subclass_of
--Organism
-property
--dob
identifier a0ce8ec6
-object_name
--Homo
-subclass_of
--Hominidae
-property
--glucose_at_time
identifier 98495efc
-object_name
--Andy Muzeack
-instance_of
--Homo sapiens
-glucose_at_time
--87, 02-12-2014 17:33:09
-dob
--1 January, 2001
```

The first listed data object, followed by its nested metadata/data pairs, is "a34a1e35."

```
identifier a34a1e35
-definition
--glucose level in mg/Dl at time drawn (GMT)
-object_name
--glucose_at_time
-instance_of
--Property
```

The triples belonging to "a34a1e35" tell us that the data object is a Property. The property's name is "glucose_at_time," and the object is defined as the "glucose level in mg/Dl at time drawn (GMT)." Had we examined all of the output of the nested.py script, we would have learned that "glucose_at_time" is a property of Class Homo, the subclass of Class Hominidae.

The last listed data object is "98495efc."

```
identifier 98495efc
-object_name
--Andy Muzeack
-instance_of
--Homo sapiens
-glucose_at_time
--87, 02-12-2014 17:33:09
-dob
--1 January, 2001
```

The triples are telling us what we have previously learned; that "98495efc" is a member of Class Homo sapiens, named Andy Muzeack, and that he has a glucose_at_time of 87 mg/Dl drawn at December 2, 2014, at 5:33 Greenwich Mean Time. He was born on January 1, 2001.

Triplestores can be difficult to understand, at first, owing to the seemingly convoluted self-definitions of the highest-level classes and properties. For example, must we really know that a property is a member of Class Property and that Class Property is a subclass of Class Class? Yes and no. These preliminary triples must exist somewhere, but they need not appear in every triplestore. Ideally, the high level triples would be stored, for reference, in an upper level ontology. Most triplestores would have the appearance of a list of spreadsheet cells with row and column headers attached. The power of a well-designed triplestore comes from the ease with which they can be merged, integrated, and introspected.

Section 6.6. Case Study (Advanced): Proof That Big Data Must Be Object-Oriented

The worst form of inequality is to try to make unequal things equal.

Aristotle

Everyone knows what the meaning of the following equation (or do we?):

x = y

Does it mean that x and y are the same thing? Certainly, if x is equal to zero, and x equals y, then y must also equal zero. But what if x is the truck blocking my view in traffic? Must I assume that y is the same truck, also blocking my view in traffic? Or does it mean that y is the same kind of truck as x, but not blocking my view?

Perhaps the equation is an assignment function, indicating that the value of y is being assigned to x. In this case, if y is 5, then x is assigned the value of 5. In that case, what happens when y is incremented by 1, to become y + 1, or 6. Does x, being equal to y, also become equal to 6, or does it keep its assigned value of 5?

Suppose x is a global variable (i.e., a variable that persists for the life of the executing program) and y is a local variable (a variable that persists only for the life of the subroutine in which it is created). Then what happens to x when y's subroutine ends?

Maybe the equivalency between x and y indicates that x and y happen to contain the same types and quantities of data objects. For example x is equivalent to y if x contains an orange and an apple and y contains an orange and an apple. If y gives x its apple and its orange, then does x become 2x? If so, then how do we describe the result of a transaction where y gives x its apple but retains its orange. Do we then have 1.5x and 0.5y. Have we become guilty of falsely comparing apples and oranges?

What is the point of all this annoying sophistry? The "=" sign is an example of a polymorphic method. Equality can indicate the assignment of a variable, the establishment of identity, a property of belonging to sets of objects, or any number of alternate meanings. Each meaning of "=" is determined by the class of objects it operates upon.

You can see that if the simple "=" sign is polymorphic, then other methods that operate on objects of different types can also be polymorphic. For example, a "rounding" method applied to a geometric object would be quite different from a "rounding" method applied to a floating point number.

How does this relate to Big Data? Remember that Big Data is complex, meaning that it contains heterogeneous data types. When Big Data contains may different types of data (i.e., may different classes of data objects), we must be prepared to accommodate polymorphic methods. The only way to accommodate polymorphic methods is with object-oriented rules. Doing so guarantees that an object will respond to a method based on the method's defined functionality within the object's class.

Hence, Big Data must be object oriented.

Glossary

Immutability Immutability is the principle that data collected in a Big Data resource is permanent, and can never be modified. At first thought, it would seem that immutability is a ridiculous and impossible constraint. In the real world, mistakes are made, information changes, and the methods for describing information changes. This is all true, but the astute Big Data manager knows how to accrue information into data objects without changing the pre-existing data. Methods for achieving this seemingly impossible trick are described in Chapter 8.

Message digest Within the context of this book, "message digest", "digest", "HMAC", and "one-way hash" are equivalent terms.

Open access A document is open access if its complete contents are available to the public. Open access applies to documents in the same manner as open source applies to software.

Reconciliation Usually refers to identifiers, and involves verifying an object that is assigned a particular identifier in one information system will be provided the same identifier in some other system. For example, if you were assigned identifier 967bc9e7-fea0-4b09-92e7-d9327c405d78 in a legacy record system, you should like to be assigned the same identifier in the new record system. If that were the case, your records in both systems could be combined. If you were assigned an identifier in one system that is different from your assigned identifier in another system, then the two identifiers must be reconciled to determine that they both refer to the same unique data object (i.e., yourself). This may involve creating a link between the two identifiers. Despite claims to the contrary, there is no possible way by which information systems with poor identifier systems can be sensibly reconciled. Consider this example. A hospital has two separate registry systems: one for dermatology cases and another for psychiatry cases. The hospital would like to merge records from the two services. Because of sloppy identifier and registration protocols, a single patient has been registered 10 times in the dermatology system, and 6 times in the psychiatry system, each time with different addresses, social security numbers, birthdates and spellings of the name. A reconciliation algorithm is applied, and one of the identifiers from the dermatology service is matched positively against one of the records from the psychiatry service. Performance studies on the algorithm indicate that the merged records have a 99.8% chance of belonging to the same patient. So what? Though the two merged identifiers correctly point to the same patient, there are 14 (9 + 5) residual identifiers for the patient still unmatched. The patient's merged record will not contain his complete clinical history. Furthermore, in this hypothetical instance, analyses of patient population data will mistakenly attribute one patient's clinical findings to as many as 15 different patients, and the set of 15 records in the corrupted deidentified dataset may contain mixed-in information from an indeterminate number of additional patients! If the preceding analysis seems harsh, consider these words, from the Healthcare Information and Management Systems Society, "A local system with a poorly maintained or 'dirty' master person index (MPI) will only proliferate and contaminate all of the other systems to which it links" [4].

Symmetric key A key (i.e., a password) that can be used to encrypt and decrypt the same file. AES is an encryption/decryption algorithm that employs a symmetric key.

For example, you may wish to use the AES protocol to encrypt the file myfile.txt, using the following command line code:

```
openssl.exe aes128 -in myfile.txt -out myfile.aes -pass pass:12345
```

In this example, the encrypted output file is myfile.aes, and the password is "12345".
To decrypt the encrypted file, you would use the same password that you used to encrypt the file, and a decrypt instruction ("-d" in this case):

```
openssl aes128 -d -in myfile.aes -out myfiledecrypted.txt -pass pass:12345
```

Trusted time stamp It is sometimes necessary to establish, beyond doubt, that a time stamp is accurate and has not been modified. Through the centuries, a great many protocols have been devised to prove that a time stamp is trustworthy. One of the simplest methods, employed in the late twentieth century, involved creating a digest of a document (e.g., a concatenated sequence consisting of the first letter of each line in the document) and sending the sequence to a newspaper for publication in the "Classifieds" section. After publication of the newspaper, anyone in possession of the original document could extract the same sequence from the document, thus proving that the document had existed on the date that the sequence appeared in the newspaper's classified advertisements.

Near the end of the twentieth century, one-way hash values become the sequences of choice for trusted time stamp protocols. Today, newspapers are seldom used to establish trust in time stamps. More commonly, a message digest of a confidential document is sent to a time stamp authority that adds a date to the digest and returns a message, encrypted with the time stamp authority's private key, containing the original one-way hash plus the trusted date. The received message can be decrypted with the time stamp authority's public key, to reveal the data/time and the message digest that is unique for the original document. It seems like the modern trusted time stamp protocol is a lot of work, but those who use these services can quickly and automatically process huge batches of documents.

References

[1] Pearson K. The grammar of science. London: Adam and Black; 1900.

[2] Reed DP. Naming and synchronization in a decentralized computer system. Doctoral Thesis, MIT; 1978.

[3] Klyne G. Newman C. Date and time on the Internet: time stamps. Network Working Group Request for Comments RFC:3339, Available from: http://tools.ietf.org/html/rfc3339 [viewed on September 15, 2015].

[4] Patient Identity Integrity. A White Paper by the HIMSS Patient Identity Integrity Work Group. Available from: http://www.himss.org/content/files/PrivacySecurity/PIIWhitePaper.pdf; December 2009 [viewed September 19, 2012].

7

Standards and Data Integration

OUTLINE

Section 7.1. Standards

The nice thing about standards is that you have so many to choose from.

Andrew S. Tanenbaum

Everyone is taught, at an early age, the standard composition of a written letter. You start with the date, then you include the name and address of your correspondent, then you write a salutation (e.g., "Dear Abby,"), then comes the body of the letter, followed by a closing (e.g., "Best wishes,") and your name and signature on the next lines. It is all rather rigid and anyone can recognize a page of correspondence, from across the room, just by the configuration of lines and paragraphs on the page.

Now, consider the reminder notes that you leave for yourself. You might jot a thought down on a Post-it and hang your Post-it notes on your refrigerator, you might use a small paper notepad, or you might write something on your computer or your smartphone. You might carry a little voice recorder for this purpose. The point is that there are an endless variety of methods whereby people leave notes for themselves, yet there is only one format for writing a letter to a friend.

The reason for this disparity in available options relates to the important distinction between self and non-self. When you write a note to yourself, you are free to do as you please. When you write a note to another person, you must conform to a standard.

The entire concept of data integration, and software interoperability draws from the same basic rule. If you intend to create your own data to serve your own purposes, then you need not be concerned with data integration and software interoperability. Everyone else must toe the line.

Until the last decade or two, most data systems were created for use within one organization or corporation. The last thing on anyone's minds was providing access to outsiders. All this has changed. Today, data means very little if it cannot be integrated with

Principles and Practice of Big Data. https://doi.org/10.1016/B978-0-12-815609-4.00007-8

related data sources. Today's software protocols operate with standard application programming interfaces that mediate the interchange of data over operating systems and networks. [Glossary Data interfaces]

In small data projects, a single standard will often support the successful interchange of data. In a Big Data project, data integration and system interoperability might involve the use of multiple standards with data conforming to multiple versions of each standard. Sharing the data across networks may involve the use of many different interchange protocols. The purpose of this chapter is familiarize data managers with standards issues that are important to Big Data resources.

Standards are sometimes touted as the solution to every data integration issue [1]. When implemented as intended, they can support the exchange of data between heterogeneous systems [2]. Such exchanges may involve non-equivalent databases (i.e., databases with different data models, different software holding different types of data). Exchanges may also involve information transfer between humans and databases, or between software agents and mechanical devices. Any exchanges between one data source and another data source can benefit from standards for describing the data and standards for transferring the data.

Whereas a single, all-purpose, unchanging, and perpetual standard is a blessing for Big Data managers, an assortment of incompatible standards can be a curse. The utility of data standards has been undermined by the proliferation of standards, the frequent versioning of data standards, the intellectual property encumbrances placed upon standards, the tendency for standards to increase in complexity over time, the abandonment of unpopular standards, and the idiosyncratic ways in which standards are implemented by data managers.

Look at the field of information science and the growing role of Big Data in science and society; it is tempting to believe that the profusion of standards that we see today is the result of rapid growth in a new field. As the field matures, there will be a filtering-out process wherein the weaker standards are replaced by the strongest, most useful standards, until we reach a point when a few tested and stable standards dominate. This scenario will probably never happen. To the contrary, there is every indication that the number of standards will increase, that the newly created standards will not serve their intended purposes, and that future revisions of these early and inadequate standards will be more complex and less useful than their inadequate predecessors.

Of course, the future need not be so dreary, but it is worth taking a look at some of the scientific, economic, legal, and social forces that push us to create more and more standards of lower and lower quality.

1. **There is no guiding force that has either the authority or the popular support to limit the flood of new standards. They just keep coming**

Today, there are thousands of organizations that develop standards; these are called Standards Development Organizations (SDOs). The development of standards has become part of the established culture of technology. SDOs may become members of a Standards Activities Organization, such as the American National Standards Institute

(ANSI), which coordinates between Standards Development Organizations and Standards Organizations, providing guidance and procedures to attain certified new standards. Above the Standards Activities Organizations are the standards certifying agencies. The two most important are ISO (International Organization for Standardization) and IEC (International Electrochemical Commission).

Aside from SDOs, there are independent-minded groups that create their own standards without following the aforementioned route. These groups often develop standards for their members of for their own private consumption. They see no reason to make the effort to follow a path to the ISO or the IEC. There is no way to count the number of independent standards that are being created.

In addition to the standards produced by SDOs and independent groups, there are the *de facto* standards that seem to arise out of thin air and rapidly gain in popularity. These represent the "better mousetraps" that somebody builds and to which the world beats a path. In the long run, de facto standards such as TCP/IP, QWERTY keyboards, PDF files, and Microsoft Word DOC documents, will have a much greater influence than any official standards.

2. Standards can be easy to create, especially if they are narrowly focused.

Many standards are created for a niche audience. When the topic is very narrow, a standard can be developed in under a month, through the part-time efforts of a few motivated individuals. The time-consuming component of the standards process is vetting; getting your committee members and your user community to read, understand, approve, support, and use the finished product. For the technically-minded, the vetting process can be an insurmountable obstacle. The creators of the standard may not have the stamina, social skills, money, or influence to produce a popular and widely implemented standard. Nonetheless, it is relatively easy to write a standards document and publish it as a journal article or as a Web posting, vetted or not.

3. Standards are highly profitable, with many potential revenue streams.

When there is no industry standard for data representation, then each vendor may prepare his or her own proprietary data model to establish "vendor lock-in." The customer's data is held in the format provided by the vendor. Because the format is proprietary, competing vendors cannot appropriate the format in their own hardware and software. The customer becomes locked into the vendor's original system, upgrades, and add-ons. Proprietary systems provide vendors with an opportunity to gain customer loyalty, without necessarily producing a superior product.

One of the purposes of industry-wide standards is to abolish proprietary systems. The hope is that if every vendor's software, hardware and data models were equivalent, then buyers could break away from locked-in systems; the free market would prevail.

Who sits on standards development committees? Who has the time to invest in the vetting process? Who has the expertise to write a standard? Who can afford to send representatives across the globe to attend committee meetings? Vendors; vendors write the standards, vendors vet the standards, and vendors implement the standards.

Large corporations can afford to send a delegation of standards experts to a standards committee. Furthermore, the corporation that sends delegates will pay for membership in the committee. Consequently, the standards committee becomes dependent on corporations that finance the standards process, and this dependence strengthens the corporation's influence. The corporation will work to create a standard that can be technically supported by the products in place or under development. The standards-making corporations secure an advantage over competitors who do not participate in the standards committee meetings and who cannot anticipate the outcome of the standards process or who cannot comply with finalized rulings for reasons of system incompatibility or simply because the proposed standard is technically beyond the capacity of their staff.

It is one of the great ironies of informatics that standards are written by the very same people who are the standard's intended targets of restraint. Vendors are clever and have learned to benefit from the standards-making process. In some cases, a member of a standards committee may knowingly insert a fragment of patented property into the standard. After the standard is released and implemented in many different vendor systems, the patent holder rises to assert the hidden patent. In this case, all those who implemented the standard may find themselves required to pay a royalty for the use of some intellectual property sequestered within the standard. The practice of hiding intellectual property within a standard or device is known as patent farming or patent ambushing [3]. The patent farmer plants seeds in the standard and harvests his crop when the standard has grown to maturity; a rustic metaphor for some highly sophisticated and cynical behavior.

Savvy standards committees take measures to reduce patent farming. This often takes the form of an agreement, signed by all members of the standards committee, to refrain from asserting patent claims on the users of the standards. There are several ways to circumvent these agreements. If a corporation holds patents on components of a standard, the corporation can sell their patents to a third party. The third party would be a so-called patent holding company that buys patents in selected technologies with the intention of eventually asserting patents over an array of related activities [4]. If the patent holder asserts the patent, the corporation might profit from patent farming, through their sale of the patent, without actually breaking the agreement. [Glossary Patent farming]

Corporations can profit from standards indirectly by obtaining patents on the uses of the standard; not on the patent itself. For example, an open standard may have been created that can be obtained at no cost, is popular among its intended users, and contains no hidden intellectual property. An interested corporation or individual may discover a use for the standard that is non-obvious, novel, and useful; these are the three criteria for awarding patents. The corporation or individual can patent the use of the standard, without needing to patent the standard itself. The patent holder will have the legal right to assert the patent over anyone who uses the standard for the purpose claimed by the patent. This patent protection will apply even when the standard is free and open.

The world of standards is a very strange place. Big Data managers are particularly vulnerable to the legal trappings associated with standards because Big Data is complex and diverse and requires different standards for different types of data and for different types of software.

4. Standards are popular (everyone wants one of their own).

Having your own standard is somewhat of a status symbol. Whenever a team of scientists develops a new method, a variant of an old method, and an organized way of collecting the data produced by the method, there will be a natural urge to legitimize and aggrandize their efforts with a new standard. The standard will dictate how the method is used, and how the data is collected, labeled, and stored. In the late 1990s the most favored way to represent data was through a new markup language; basically a list of specialized XML tags and a Schema that dictated the nesting hierarchy of the tags. In almost every case, these niche markup languages were self-contained constructs that did not re-use tags from related markup languages. For example, many different markup languages contained an equivalent tag that described the sample name or the sample identifier, but these mark-up languages did not refer to pre-existing equivalent tags in other Schemas (i.e., they did not make use of established namespaces). Consequently, a Babel of markup languages sprang into existence, with no attempt at harmonizing the languages or sharing content among the different languages. Thankfully, the markup language fad has passed, but a basic problem persists. Deep down, scientists believe that their way of organizing their own data should be the standard for the entire world. This irrational belief accounts for much of the unchecked proliferation of personalized standards.

5. Most standards are created for reasons that applied in the past, but which do not apply in the Big Data era.

For the last half century the purpose of a standard was to ensure that everyone who created a particular type of object (e.g., a physical object, or a document, or a collection of a specific type of information) would do so in the same way, so that the objects could be compared and categorized.

For example, imagine an international standard for death certificates. You would naturally want each certificate to contain the same information, including the name of the deceased, identifying information (e.g., date of birth, gender, race), causes of death and contributing factors, all coded in accordance with a standard nomenclature. With the cause of death, you would want to find details of the circumstances of the death (e.g., date and time of death, time at which the certificate was signed). Regarding format, you might want every country to list the contents of the document in the same order, numbered identically, so that item 4 in a Portuguese death certificate would correspond to item 4 in an Australian certificate. You may want the layout of the documents to be identical (e.g., name of deceased in the upper left, date of birth of deceased in the upper right). These restrictions are intended to facilitate comparisons among death certificates worldwide. This detailed approach to layout is terribly outdated and largely irrelevant to the purposes of standards in Big Data resources.

In the Big Data universe the purpose of a standard is not to compare one document with another document of the same kind; the purpose of a standard is to enable data analysts to relate data objects within a document to data objects contained in documents of a different kind.

This last point is the most difficult for people to accept, particularly those people who have been supporters of data standards and who have used them to good effect in their work. It is a near-impossible task to convince someone to abandon a paradigm that has served him or her well. But it is worth a try!

Let us reexamine the role of the standard death certificate in Big Data analysis. The "cause of death" section will contain the primary cause of death plus any diseases that contributed to the primary cause of death. Another database, in a hospital information system, might list various diseases that co-exist in living patients. By comparing data in the database of death certificates with data in a hospital information system, it may be possible to find sets of diseases that co-occur with a high risk of death. By comparing the average age at which a particular disease is associated with death, it may be possible to predict when a disease under treatment is likely to lead to death. The occurrence of diseases in particular racial groups included in death certificate data may lead to disparities found in the occurrence of the same diseases in a living population. These are extremely simple examples wherein data values included in one standard data set (death certificates) are compared with data values in another standard data set (Electronic Health Records). The comparisons are made between selected data values in heterogeneous data sets; the comparisons are not made between two documents that conform to the same standard.

The phenomenon of data integration over heterogeneous sources is repeated in virtually every Big Data effort. A real estate property with a known address is matched against crime statistics collected for its listed zip code. A planting chart based on a list of popular flowers and vegetables within a locality is matched against a climate zone dataset matched to geographic region. A data set of personal buying preferences for a population of individuals is matched against a list of previously sold items, and their features, and a list of items-for-sale and their features. In each case, the comparisons are made for data values held in heterogeneous data sets.

In an earlier era, standards served to create data homogeneity. In the Big Data era, standards should help us find the data relationships in heterogeneous data sources.

Section 7.2. Specifications Versus Standards

Good specifications will always improve programmer productivity far better than any programming tool or technique.

Milt Bryce

The two terms, "standards" and "specifications" are used interchangeably in the informatics literature, but they are different from one another in very important ways. A "standard" is a set of construction rules that tells you how to represent a required set of information.

For a given subject (i.e., an image, a movie, a legal certificate, a programming language), the standard tells you exactly how the contents must be organized, from top to bottom, and the contents that must be included, and how those contents are expressed. For a standard to have value, it generally requires approval from a standards-certifying organization (such as the ISO), or from some large and influential industry group.

A specification is a general way of describing objects (i.e., physical objects such as nuts and bolts or symbolic objects such as numbers) so that anyone can fully understand your intended meaning. Specifications do not force you to include specific types of information, and do not impose a specific order on the data contained in the document. Specifications are not generally certified by a standards organization. Their legitimacy depends on their popularity. Examples of specifications are RDF (Resource Description Framework) produced by the W3C (WorldWide Web Consortium), and TCP/IP (Transfer Control Protocol/Internet Protocol), maintained by the Internet Engineering Task Force.

The strength of a standard is that it imposes uniformity; the weakness of a standard is that it has no flexibility and impedes innovation. An engineer might want to manufacture a cup with a very wide bottom rim and a narrow top rim; or with no handle; or with three handles; or with an attached computer chip. If the standard prohibits the bottom rim diameter to exceed the top rim diameter, or requires exactly one handle, or has no method for describing ancillary attachments, then the innovator cannot comply with the standard.

The strength of the specification is that it is highly flexible; the weakness of the specification that its flexibility allows designers to omit some of the information required to fully specify the object. In practice, proper implementation of specifications is ensured by usability tests. If everyone seems to understand your implementation of a specification, and if your implementation functions adequately, and operates with other systems without problems, then the specification has served its intended purpose.

Both standards and specifications suffer from the following:

1. **New versions may appear, without much notice, and the new versions may not be fully compatible with older versions.**

For example, Python 3.x has a somewhat different syntax than Python 2.x. Your Python 2.x programs will not necessarily run in a Python 3.x environment, and your Python 3.x programs may not run in a Python 2.x environment. Incompatible programs may run for a while, and then stop when a conflict arises. Because the glitch is caused by a language incompatibility, not a programming error, you may find the debugging process exasperating.

2. **Both standards and specifications may be overly complex.**

It is easy for a standards committee to create a complex standard or for an organization to develop a specification language that contains thousands of metadata tags. A complex standard or specification can easily exceed human comprehension. Data managers may be hesitant to stake their resource on tools that they cannot understand.

3. There are too many standards and specifications from which to choose.

Big Data managers would like to stay in conformance with industry standards. The problem is that Big Data serves many different purposes and must comply with many different standards, all at the same time.

After a standard has been created, there follows a Darwinian struggle for supremacy. Standards committees sometimes display group behavior that can be described as antisocial or even sociopathic. They want their standard to be the only standard used in a data domain. If there are other standards in the data domain, they sometimes use coercive methods to force everyone to use their standard.

The most common coercive argument involves threatening colleagues with the inflated claim that everyone will be using the standard; failing to switch to the standard will result in loss of business opportunities. The proponents of a standard may suggest that those who fail to adopt the standard will be ostracized and marginalized by their colleagues. I have personally heard coercive arguments from some of my colleagues who, in every other respect, are decent and altruistic individuals. The reason for their nastiness often comes down to economics. Vendors and Big Data managers select a standard in the full knowledge that a poor choice may bring financial ruin. If the vendor builds a data model to fit a standard, and their market does not adapt the standard, then they will not be able to sell their software. If a Big Data manager annotates terabytes of data in conformance with an ontology that is soon-to-be-abandoned by its user community, then the value of the resource will plummet. Nevertheless, there can be no excuses for bad behavior; coercion should not be tolerated.

A few commonsense measures might help the data manager:

- Learn how to decompose the standard document into an organized collection of data objects that can be merged with other data object collections or inserted into a preferred data model.
- If feasible, avoid using any standard as your data object model for the resource. It is often best to model your own data in a simple but flexible format that can be ported into any selected standard, as needed.
- Know the standards you use. Read the license agreements. Keep your legal staff apprised of your pending decisions.
- Try your best to use standards that are open source or that belong to the public domain. [Glossary Public domain]

Section 7.3. Versioning

I visited the Sage of reverend fame
And thoughtful left more burden'd than I came.
I went- - and ere I left his humble door
The busy World had quite forgot his name.

Ecclesiastes

In the year 2000 I attended a workshop in San Diego whose purpose was to introduce pathologists to new, standardized protocols for describing different types of cancer specimens (e.g., cancer of the adrenal gland, cancer of the prostate, cancer of the lung, etc.) This was not the first such standardization effort. Over the past decade, several groups had been pushing for standards that would ensure that pathology reports prepared in any United States hospital would contain the same kind of information for a given type of specimen. Having a reporting standard seemed like a good idea, but as I looked at the protocols I saw lots of problems. Lists of required items seemed incomplete and many of the descriptors were poorly defined. Some of the descriptors were non-qualitative and required subjective data. The final reports would not be highly reproducible between laboratories or within a single laboratory. These deficiencies are par for the course in any standards effort. I asked the chairman how she planned to deal with producing and controlling new versions of the standard. She replied that because the standards had been prepared by experts and thoroughly tested by a panel of implementers, there would be no need to develop new versions of the standard. She was telling me that the new standard had been born perfect! Eighteen years have passed, during which time the standards have been subjected to unceasing modifications. [Glossary Reproducibility]

For most types of standards and specifications, versioning is a requirement. Nomenclatures in every area of science and technology are constantly being updated. Every year, the Medical Subject Headings comes out with an updated version. Some nomenclatures are actually named according to the version (e.g., ICD-10 is the tenth version of the International Classification of Diseases). New versions of nomenclatures are not simple expansions of older versions. Aside from the addition of new terms, old terms must be retired, and new coding sequences are sometimes created. The relationships among terms (i.e., the class or classes to which a term belongs) might change.

Without exception, all large nomenclatures are unstable. Changes in a nomenclature may have a ripple effect, changing the meaning of terms that are not included in the nomenclature. Here is an example from the world of mycology (the study of fungi). When the name of a fungus changes, so must the name of the associated infection. Consider "Allescheria boydii," People infected with this organisms were said to suffer from the disease known as allescheriasis. When the organism's name was changed to *Petriellidium boydii*, the disease name was changed to petriellidosis. When the fungal name was changed, once more, to *Pseudallescheria boydii*, the disease name was changed to pseudallescheriasis [5]. All three names appear in the literature (past and present). In this case, changes in the fungal nomenclature necessitate reciprocal changes in every disease nomenclature. Such changes may require months, years, and even decades to adjudicate and finalize in the newer version of the nomenclature. Within this period, the term may change again and the corrected version of the disease nomenclature may be obsolete on its release date.

We discussed classifications and ontologies in Chapter 5. Classifications have a very strong advantage over ontologies with regard to the ease of versioning. Because each class in a classification is restricted to a single parent, the hierarchical tree of a classification is simple. When a class needs to be repositioned in the classification tree, it is a simple matter to move the class, with its intact branches, to another node on the tree. We do this from time to time with the classification of living organisms.

Unlike the case with uniparental classifications, it is virtually impossible to make sweeping changes in multiparental ontologies. In every complex ontology, we can expect to encounter class branches insinuated across multiple classes. A class cannot simply be cut and repositioned elsewhere. The more complex the ontology, the more difficult it is to modify its structure.

Section 7.4. Compliance Issues

It's not worth doing something unless someone, somewhere, would much rather you weren't doing it.

Terry Pratchett.

When it comes to complex standards, compliance is in the eye of the beholder. One vendor's concept of standard-compliant software might be entirely different from another vendor's concept. Standards organizations seldom have the time, manpower, money, or energy to ensure compliance with their standards; consequently, the implementations of standards are often non-standard and incompatible with one another.

In large part, non-compliance is caused by the instability of modern standards. As we have seen, standards themselves may contain flaws related to the complexity of the technologies they model. When a technology outpaces the standard built for the technology, it may be impossible for businesses to adequately model all of their data and processes within the standard.

Small businesses may not have the manpower to keep abreast of every change in a complex standard. Large businesses may have the staff and the expertise to stay compliant; but they may lack the incentive. If they produce a product that works well, and is believed, wrongly or not, to be compliant with a standard, then it may be in the best interest of the business to purposefully introduce a bit of non-compliance. The expectation being that small deviations from the standard will create incompatibilities between their products and their competitors; thus achieving vendor lock-in. Their customers will be loath to switch to another vendor's products if they fear that their original system will not support software or hardware produced by rival companies.

Compliance with specifications is, in general, much easier than compliance with standards. Data specifications provide a syntax and a general method for describing data objects, without demanding much in the way of structuring the data. In most cases, it is relatively easy to produce a program that determines whether a file conforms to a specification.

When a file conforms to the syntax of a specification, it is said to be well formed. When a file conforms to a document that describes how certain types of objects should be annotated (e.g., which tags should be used, the relationships among tags, the data value properties that can be assigned to tags, the inclusion of all required tags), then the file is said to be valid. A file that is fully compliant with a specification is said to be well formed and valid.

In the case of RDF (as discussed in Section 4.5), a well-formed document would comply with RDF syntax rules. A valid file would conform to the classes and properties found in the RDF Schemas linked from within the RDF statements contained in the file.

Section 7.5. Case Study: Standardizing the Chocolate Teapot

History doesn't repeat itself, but it rhymes.
Attributed variously to Mark Twain and to Joseph Anthony Wittreich

Malcolm Duncan has posted an insightful and funny essay entitled "The Chocolate Teapot (Version 2.3)" [6]. In this essay, he shows how new versions of nomenclatures may unintentionally alter the meanings of classes of terms contained in earlier versions, making it impossible to compare or sensibly aggregate and interpret terms and concepts contained in any of the versions. The essay is a must-read for anyone seriously interested in terminologies, but we can examine a few of the points raised by Duncan.

Suppose you have a cooking-ware terminology with a single "teapot" item. We will call this Version 1.0. Early teapots were made of porcelain and porcelain came in two colors; white and blue. Version 2 of the terminology might accommodate the two sub-types: blue teapot and white teapot. If a teapot were neither blue nor white, it would presumably be coded under the parent term, "teapot." Suppose version 3 expands to accommodate some new additions to the teapot pantheon: chocolate teapot, ornamental teapot, china teapot, and industrial teapot. Now the teapot world is shaken by a tempest of monumental proportions. The white and the blue teapots, implicitly considered to be made of porcelain, like all china teapots, stand divided across the subtypes. How does one deal with a white porcelain teapot that is not a china teapot? If we had previously assumed that a teapot was an item in which tea is made, how do we adjust, conceptually, to the new term "ornamental teapot?" If the teapot is ornamental, then it has no tea-making functionality, and if it cannot be used to make tea, how can it be a teapot? Must we change our concept of the teapot to include anything that looks like a teapot? If so, how can we deal with the new term "industrial teapot," which is likely to be a big stainless steal vat that has more in common, structurally, with a microbrewery fermenter than with an ornamental teapot? What is the meaning of a chocolate teapot? Is it something made of chocolate, is it chocolate-colored, or does it brew chocolate-flavored tea? Suddenly we have lost the ability to map terms in version 3 to terms in versions 1 and 2. We no longer understand the classes of objects (i.e., teapots) in the various versions of our cookware nomenclature. We cannot unambiguously attach nomenclature terms to objects in our data collection (e.g., blue china teapot). We no longer have a precise definition of a teapot or of the subtypes of teapot.

Glossary

Data interfaces Interfaces to Big Data resources often come in one of several types including:

Direct user interfaces. These interfaces permit individuals to submit simple queries, constructed within a narrow range of options, producing an output that is truncated to produce a manageable visual display. Google is an example. You never know what information is excluded from the indexed resource, or exactly how the search is conducted, and the output may or may not have the results you actually need. Regarding the actual query, it is limited to words and phrases entered into a box, and although it permits some innovative methods to specify the query, it does not permit you to enter hundreds of items at once, or to search based on a user-invented algorithms, or to download the entire search output into a file. Basically, Google gives users the opportunity to enter a query according to a set of Google-specified query rules, and Google provides an output. What happens in the very short moment that begins when the query has been launched, and ends when the reply is displayed, is something that only Google fully understands. For most users, the Google reply may as well be conjured by magic.

Programmer or software interfaces. These are standard commands and instructions that a data service releases to the public, and that individual developers can use to link to and interact with the service. The usual term applied to these interfaces is API (Application Programming Interface), but other related terms, including SaaS (Software as a Service) might also apply. Amazon is an example of a company that provides an API. Web developers can use the Amazon API to link to information related to specific Amazon products. Current information for the product can be displayed on the third party Web site, and a buyer's link can broker a purchase. The API enables transactions to be completed through interactions between the developer's software and the company's software.

Autonomous agent interfaces. These are programs that are launched into a network of communicating computers, carrying a query. The program contains communication and interface protocols that enable it to interrogate various databases. The response from a database is stored and examined. Depending on the information received, the autonomous agent might proceed to another database or may modify its interrogation of the first database. The agent continues to collect and process information, traveling to different networked databases in the process. At some point, the software program returns to the client (the user who initiated the query) with its collected output. Web crawlers, familiar to anyone who reviews Internet server logs, are somewhat primitive examples of partly autonomous software agents. They use an interface (Internet protocols) to visit servers, conducting an inventory of the contents, and visiting other servers based on the addresses of links listed on Web pages. If a Big Data resource opens its data to programs that employ a compatible communications protocol (such as a Web services language), then the problem of constructing a software agent becomes relatively straightforward. Opening a system to autonomous agents comes with risk. The consequences of opening a system to complex interactions with innumerable agents, each operating under its own set of instructions, is difficult, or impossible, to predict and control [7].

Patent farming Also known as patent ambushing [3]. The practice of hiding intellectual property within a standard or device, at the time of its creation, is known as patent farming. After the property is marketed, the patent farmer announces the presence of his or her hidden patented material and presses for royalties; metaphorically harvesting his crop.

Public domain Data that is not owned by an entity. Public domain materials include documents whose copyright terms have expired, materials produced by the federal government, materials that contain no creative content (i.e., materials that cannot be copyrighted), or materials donated to the public domain by the entity that holds copyright. Public domain data can be accessed, copied, and re-distributed without violating piracy laws. It is important to note that plagiarism laws and rules of ethics apply to public domain data. You must properly attribute authorship to public domain documents. If you fail to attribute authorship or if you purposefully and falsely attribute authorship to the wrong person (e.g., yourself), then this would be an unethical act and an act of plagiarism.

Reproducibility Reproducibility is achieved when repeat studies produce the same results, over and over. Reproducibility is closely related to validation, which is achieved when you draw the same conclusions, from the data, over and over again. Implicit in the concept of "reproducibility" is that the original research must somehow convey the means by which the study can be reproduced. This usually requires the careful recording of methods, algorithms, and materials. In some cases, reproducibility requires access to the data produced in the original studies. If there is no feasible way for scientists to undertake a reconstruction of the original study, or if the results obtained in the original study cannot be obtained in subsequent attempts, then the study is irreproducible. If the work is reproduced, but the results and the conclusions cannot be repeated, then the study is considered invalid.

References

[1] National Committee on Vital and Health Statistics. Report to the Secretary of the U.S. Department of Health and Human Services on Uniform Data Standards for Patient Medical Record Information. July 6. Available from: http://www.ncvhs.hhs.gov/hipaa000706.pdf; 2000.

[2] Kuchinke W, Ohmann C, Yang Q, Salas N, Lauritsen J, Gueyffier F, et al. Heterogeneity prevails: the state of clinical trial data management in Europe—results of a survey of ECRIN centres. Trials 2010;11:79.

[3] Gates S. Qualcomm v. Broadcom—The Federal Circuit Weighs in on "Patent Ambushes", December 5. Available from: http://www.mofo.com/qualcomm-v-broadcom—the-federal-circuit-weighs-in-on-patent-ambushes-12-05-2008; 2008 [viewed January 22, 2013].

[4] Cahr D, Kalina I. Of pacs and trolls: how the patent wars may be coming to a hospital near you. ABA Health Lawyer 2006;19:15–20.

[5] Guarro J, Gene J, Stchigel AM. Developments in fungal taxonomy. Clin Microbiol Rev 1999;12:454–500.

[6] Duncan M. Terminology version control discussion paper: the chocolate teapot. Medical Object Oriented Software Ltd.; 2009. September 15. Available from: http://www.mrtablet.demon.co.uk/chocolate_teapot_lite.htm [viewed August 30, 2012].

[7] Jennings N. On agent-based software engineering. Artif Intell 2000;117:277–96.

8

Immutability and Immortality

Section 8.1. The Importance of Data That Cannot Change

Cheese is milk's leap toward immortality
Clifton Fadiman (editor of Mathematical Magpie)

Immutability is one of those issues, like identifiers and introspection, that seem unimportant, until something goes terribly wrong. Then, in the midst of the problem, you realize that your entire information system was designed incorrectly, and there really is nothing you can do to cope.

Here is an example of a immutability problem. You are a pathologist working in a university hospital that has just installed a new, $600 million information system. On Tuesday, you released a report on a surgical biopsy, indicating that it contained cancer. On Friday morning, you showed the same biopsy to your colleagues, who all agreed that the biopsy was not malignant, and contained a benign condition that simulated malignancy (looked a little like a cancer, but was not). Your original diagnosis was wrong, and now you must rectify the error. You return to the computer, and access the prior report, changing the wording of the diagnosis to indicate that the biopsy is benign. You can do this, because pathologists are granted "edit" access for pathology reports. Now, everything seems to have been set right. The report has been corrected, and the final report in the computer is the official diagnosis.

Unknown to you, the patient's doctor read the incorrect report on Wednesday, the day after the incorrect report was issued, and two days before the correct report replaced the incorrect report. Major surgery was scheduled for the following Wednesday (five days after the corrected report was issued). Most of the patient's liver was removed. No cancer was

Principles and Practice of Big Data. https://doi.org/10.1016/B978-0-12-815609-4.00008-X

found in the excised liver. Eventually, the surgeon and patient learned that the original report had been altered. The patient sued the surgeon, the pathologist, and the hospital.

You, the pathologist, argued in court that the computer held one report issued by the pathologist (following the deletion of the earlier, incorrect report) and that report was correct and available to the surgeon prior to the surgery date. Therefore, you said, you made no error. The patient's lawyer had access to a medical chart in which paper versions of the diagnosis had been kept. The lawyer produced, for the edification of the jury, two reports from the same pathologist, on the same biopsy: one positive for cancer, the other negative for cancer. The hospital, conceding that they had no credible defense, settled out of court for a very large quantity of money. Meanwhile, back in the hospital, a fastidious intern is deleting an erroneous diagnosis, and substituting her improved rendition.

One of the most important features of serious Big Data resources (such as the data collected in hospital information systems) is immutability. The rule is simple. Data is immortal and cannot change. You can add data to the system, but you can never alter data and you can never erase data. Immutability is counterintuitive to most people, including most data analysts. If a patient has a glucose level of 100 on Monday, and the same patient has a glucose level of 115 on Tuesday, then it would seem obvious that his glucose level changed between Monday and Tuesday. Not so. Monday's glucose level remains at 100. For the end of time, Monday's glucose level will always be 100. On Tuesday, another glucose level was added to the record for the patient. Nothing that existed prior to Tuesday was changed. [Glossary Serious Big Data]

Section 8.2. Immutability and Identifiers

People change and forget to tell each other.

Lillian Hellman

Immutability applies to identifiers. In a serious Big Data resource, data objects never change their identity (i.e., their identifier sequences). Individuals never change their names. A person might add a married name, but the married name does not change the maiden name. The addition of a married name might occur as follows:

```
18843056488   is_a          patient
18843056488   has_a         maiden_name
18843056488   has_a         married_name
9937564783    is_a          maiden_name
4401835284    is_a          married_name
18843056488   maiden_name   Karen Sally Smith
18843056488   married_name  Karen Sally Smythe
```

Here, we have a woman named Karen Sally Smith. She has a unique, immutable identifier, "18843056488." Her patient record has various metadata/data pairs associated with her unique identifier. Karen is a patient, Karen has a maiden name, and Karen has a married name. The metadata tags that describe the data that is associated with Karen include "maiden_name" and "married_name." These metadata tags are themselves data objects. Hence, they must be provided with unique, immutable identifiers. Though metadata tags are themselves unique data objects, each metadata tag can be applied to many other data objects. In the following example, the unique maiden_name and married_name tags are associated with two different patients.

```
9937564783  is_a           maiden_name
4401835284  is_a           married_name
18843056488 is_a           patient
18843056488 has_a          maiden_name
18843056488 has_a          married_name
18843056488 maiden_name    Karen Sally Smith
18843056488 married_name   Karen Sally Smythe
73994611839 is_a           patient
73994611839 has_a          maiden_name
73994611839 has_a          married_name
73994611839 maiden_name    Barbara Hay Wire
73994611839 married_name   Barbara Haywire
```

The point here is that patients may acquire any number of names over the course of their lives, but the Big Data resource must have a method for storing, and describing each of those names and associating them with the same unique patient identifier. Everyone who uses a Big Data resource must be confident that all the data objects in the resource are unique, identified, and immutable.

By now, you should be comfortable with the problem confronted by the pathologist who changed his mind. Rather than simply replacing one report with another, the pathologist might have issued a modification report, indicating that the new report supercedes the earlier report. In this case, the information system does not destroy or replace the earlier report, but creates a companion report. As a further precaution the information system might flag the early report with a link to the ensuant entry. Alternately, the information system might allow the pathologist to issue an addendum (i.e., add-on text) to the original report. The addendum could have clarified that the original diagnosis is incorrect, stating the final diagnosis is the diagnosis in the addendum. Another addendum might indicate that the staff involved in the patient's care was notified of the updated diagnosis. The parts of the report (including any addenda) could be dated and authenticated with the electronic signature of the pathologist. Not one byte in the original report is ever changed. Had these procedures been implemented, the unnecessary surgery, the harm inflicted on the patient, the lawsuit, and the settlement, might have all been avoided. [Glossary Digital signature]

The problem of updating diagnoses may seem like a problem that is specific for the healthcare industry. It is not. The content of Big Data resources is constantly changing; the trick is to accommodate all changes by the addition of data, not by the deletion or modification of data. For example, suppose a resource uses an industry standard for catalog order numbers assigned to parts of an automobile. These 7-digit numbers are used whenever a part needs to be purchased. The resource may inventory millions of different parts, each with an order number annotation. What happens when the standard suddenly changes, and 12-digit numbers replace all of the existing 7-digit numbers? A well-managed resource will preserve all of the currently held information, including the metadata tag that describe the 7-digit standard and the 7-digit order number for each part in the resource inventory. The new standard, containing 12-digit numbers, will have a different metadata tag from the prior standard, and the new metadata/data pair will be attached to the internal identifier for the part. This operation will work if the resource maintains its own unique identifiers for every data object held in the resource and if the data objects in the resource are associated with metadata/data pairs. All of these actions involve adding information to data objects, not deleting information.

In the days of small data, this was not much of a problem. The typical small data scenario would involve creating a set of data, all at once, followed soon thereafter by a sweeping analytic procedure applied against the set of data, culminating in a report that summarized the conclusions. If there was some problem with the study, a correction would be made, and everything would be repeated. A second analysis would be performed in the new and improved data set. It was all so simple.

A procedure for replicative annotations to accommodate the introduction of new standards and nomenclatures as well as new versions of old standards and nomenclatures is one of the more onerous jobs of the Big Data curator. Over the years, dozens of new or additional annotations could be required. It should be stressed that replicative annotations for nomenclatures and standards can be avoided if the data objects in the resource are not tied to any specific standard. If the data objects are well specified (i.e., providing adequate and uniform descriptions), queries can be matched against any standard nomenclature on-the-fly (i.e., as needed, in response to queries), as previously discussed in Section 2.5, "Autocoding" [1]. [Glossary Curator]

Why is it always bad to change the data objects held in a Big Data resource? Though there are many possible negative repercussions to deleting and modifying data, most of the problems come down to data verification, and time stamping. All Big Data resources must be able to verify that the data held in the resource conforms to a set of protocols for preparing data objects and measuring data values. When you change pre-existing data, all of your efforts at resource verification are wasted, because the resource that you once verified no longer exists. The resource has become something else. Aside from producing an unverifiable resource, you put the resource user into the untenable position of deciding which data to believe; the old data or the new data. Time stamping is another component of data objects. Events (e.g., a part purchased, a report issued, a file opened) have no meaning unless you know when they occurred. Timestamps applied to data objects must be unique and immutable. A single event cannot occur at two different times. [Glossary Time stamp, Verification and validation]

– Immortal Data Objects

In Section 6.2, we defined the term "data object." To review, a data object is a collection of triples that have the same identifier. A respectable data object should always encapsulate two very specific triples: one that tells us the class to which the data object holds membership, and another that tells us the name of the parent class from which the data object descends. When these two triples are included in the data object, we can apply the logic and the methods of object-oriented programming to Big Data objects.

In addition, we should note that if the identifier and the associated metadata/data pairs held by the data object are immutable (as they must be, vida supra), and if all the data held in the Big Data resource is preserved indefinitely (as it should be), then the data objects achieve immortality. If every data object has metadata/data pairs specifying its class and parent class, then all of the relationships among every data object in the Big Data resource will apply forever. In addition, all the class-specific methods can be applied to objects belonging to its class and its subclass descendants, can always be applied; and all of the encapsulated data can always be reconstructed. This would hold true, even if the data objects were reduced to their individual triples, scattered across the planet, and deposited into countless data clouds. The triples could, in theory, reassemble into data objects under their immortal identifier.

Big Data should be designed to last forever. Hence, Big Data managers must do what seems to be impossible; they must learn how to modify data without altering the original content. The rewards are great.

Section 8.3. Coping With the Data That Data Creates

The chief problem in historical honesty isn't outright lying. It is omission or de-emphasis of important data.

Howard Zinn

Imagine this scenario. A data analyst extracts a large set of data from a Big Data resource. After subjecting the data to several cycles of the usual operations (data cleaning, data reduction, data filtering, data transformation, and the creation of customized data metrics), the data analyst is left with a new set of data, derived from the original set. The data analyst has imbued this new set of data with some added value, not apparent in the original set of data.

The question becomes, "How does the data analyst insert her new set of derived data back into the original Big Data resource, without violating immutability?" The answer is simple but disappointing; re-inserting the derived data is impossible, and should not be attempted. The transformed data set is not a collection of original measurements; the data manager of the Big Data Resource can seldom verify it. Data derived from other data (e.g., age-adjustments, normalized data, averaged data values, and filtered data) will not sensibly fit into the data object model upon which the resource was created. There simply is no substitute for the original and primary data.

The data analyst should make her methods and her transformed data available for review by others. Every step involved in creating the new data set needs to be carefully recorded and explained, but the transformed set of data should not be absorbed back into the resource. The Big Data resource may provide a link to sources that hold the modified data sets. Doing so provides the public with an information trail leading from the original data to the transformed data prepared by the data analyst. [Glossary Raw data]

Section 8.4. Reconciling Identifiers Across Institutions

Mathematics is the art of giving the same name to different things.

Henri Poincare

In math, we are taught that variables are named "x" or "y," or sometimes "n," (if you are sure the variable is an integer). Using other variable names, such as "h" or "s," is just asking for trouble. Computer scientists have enlarged their list of familiar variables to include "foo" and "bar." A long program with hundreds of different local variables, all named "foo" is unreadable, even to the person who wrote the code. The sloppiness with which mathematicians and programmers assign names has carried over into the realm of Big Data. Sometimes, it seems that data professionals just don't care much about how we name our data records, just so long as we have lots of them to play with. Consequently, we must deal with the annoying problem that arises when multiple data records, for one unique object, are assigned different identifiers (e.g., when identifier x and identifier y and identifier foo all refer to the same unique data object). The process of resolving identifier replications is known as reconciliation. [Glossary Metasyntactic variable]

In many cases, the biggest obstacle to achieving Big Data immutability is data record reconciliation [2]. When different institutions merge their data systems, it is crucial that no data is lost, and all identifiers are sensibly preserved. Cross-institutional identifier reconciliation is the process whereby institutions determine which data objects, held in different resources, are identical (i.e., the same data object). The data held in reconciled identical data objects can be combined in search results, and the identical data objects themselves can be merged (i.e., all of the encapsulated data can be combined into one data object), when Big Data resources are integrated, or when legacy data is absorbed into a Big data resource.

In the absence of successful reconciliation, there is no way to determine the unique identity of records (i.e., duplicate data objects may exist across institutions and data users will be unable to rationally analyze data that relates to or is dependent upon the distinctions among objects in a data set). For all practical purposes, without data object reconciliation, there is no way to understand data received from multiple sources.

Reconciliation is particularly important for healthcare agencies. Some countries provide citizens with a personal medical identifier that is used in every medical facility in the nation. Hospital A can send a query to Hospital B for medical records pertaining to a patient sitting Hospital A's emergency room. The national patient identifier insures that the cross-institutional query will yield all of Hospital B's data on the patient, and will not include data on other patients. [Glossary National Patient Identifier]

Consider the common problem of two institutions trying to reconcile personal records (e.g., banking records, medical charts, dating service records, credit card information). When both institutions are using the same identifiers for individuals in their resources, then reconciliation is effortless. Searches on an identifier will retrieve all the information attached to the identifier, if the search query is granted access to the information systems in both institutions. However, universal identifier systems are rare. If any of the institutions lack an adequate identifier system, the data from the systems cannot be sensibly reconciled. Data pertaining to a single individual may be unattached to any identifier, attached to one or more of several different identifiers, or mixed into the records of other individuals. The merging process would fail, at this point.

Assuming both institutions have adequate identifiers, then the two institutions must devise a method whereby a new identifier is created, for each record, that will be identical to the new identifier created for the same individual's record, in the other institution. For example, suppose each institution happens to store biometric data (e.g., retinal scan, DNA sequences, fingerprints), then the institutions might agree on a way to create a new identifier validated against these unique markers. With some testing, they could determine whether the new identifier works as specified (i.e., either institution will always create the same identifier for the same individual, and the identifier will never apply to any other individual). Once testing is finished, the new identifiers can be used for cross-institutional searches.

Lacking a unique biometric for individuals, reconciliation between institutions is feasible, but difficult. Some combination of identifiers (e.g., date of birth, social security number, name) might be developed. Producing an identifier from a combination of imperfect attributes has its limitations (as discussed in detail in Section 3.4, "Really Bad Identifier Methods"), but it has the advantage that if all the pre-conditions of the identifier are met, errors in reconciliation will be uncommon. In this case, both institutions will need to decide how they will handle the set of records for which there is no identifier match in the other institution. They may assume that some individuals will have records in both institutions, but their records were not successfully reconciled by the new identifier. They may also assume that unmatched group contains individuals that actually have no records in the other institution. Dealing with unreconciled records is a nasty problem. In most cases, it requires a curator to slog through individual records, using additional data from records or new data supplied by individuals, to make adjustments, as needed. This issue will be explored further, in Section 18.5, "Case Study: Personal Identifiers."

Section 8.5. Case Study: The Trusted Timestamp

Time is what keeps everything from happening at once.
Ray Cummings in his 1922 novel, "The Girl in the Golden Atom"

Time stamps are not tamper-proof. In many instances, changing a recorded time residing in a file or data set requires nothing more than viewing the data on your computer screen and substituting one date and time for another. Dates that are automatically recorded, by your computer system, can also be altered. Operating systems permit users to reset the system date and time. Because the timing of events can be altered, scrupulous data managers employ a trusted timestamp protocol by which a timestamp can be verified.

Here is a description of how a trusted time stamp protocol might work. You have just created a message, and you need to document that the message existed on the current date. You create a one-way hash on the message (a fixed-length sequence of seemingly random alphanumeric characters). You send the one-way hash sequence to your city's newspaper, with instructions to publish the sequence in the classified section of that day's late edition. You are done. Anyone questioning whether the message really existed on that particular date can perform their own one-way hash on the message and compare the sequence with the sequence that was published in the city newspaper on that date. The sequences will be identical to each other. [Glossary One-way hash]

Today, newspapers are seldom used in trusted time stamp protocols. A time authority typically receives the one-way hash value on the document, appends a time, and encrypts a message containing the one-way hash value and the appended time, using a private key. Anyone receiving this encrypted message can decrypt it using the time authority's public key. The only messages that can be decrypted with the time authority's public key are messages that were encrypted using the time authority's private key; hence establishing that the message had been sent by the time authority. The decrypted message will contain the one-way hash (specific for the document) and the time that the authority received the document. This time stamp protocol does not tell you when the message was created; it tells you when the message was stamped.

Section 8.6. Case Study: Blockchains and Distributed Ledgers

It's worse than tulip bulbs.
JP Morgan CEO Jamie Dimon, referring to Bitcoin, a currency exchange system based on blockchains

Today, no book on the subject of Big Data would be complete without some mention of blockchains, which are likely to play an important role in the documentation and management of data transactions for at least the next decade, or until something better comes along. Fortunately, blockchains are built with two data structures that we have already introduced: one-way hashes and triples. All else is mere detail, determined by the user's choice of implementation.

At its simplest, a blockchain is a collection of short data records, with each record consisting of some variation on the following:

```
<head>-<message>-<tail>
```

Here are the conditions that the blockchain must accommodate:

1. The head (i.e., first field) in each blockchain record consists of the tail of the preceding data record.
2. The tail of each data record consists of a one-way hash of the head of the record concatenated with the record message.
3. Live copies of the blockchain (i.e., a copy that grows as additional blocks are added) are maintained on multiple servers.
4. A mechanism is put in place to ensure that every copy of the blockchain is equivalent to one another, and that when a blockchain record is added, it is added to every copy of the blockchain, in the same sequential order, and with the same record contents.

We will soon see that conditions 1 through 3 are easy to achieve. Condition 4 can be problematic, and numerous protocols have been devised, with varying degrees of success, to ensure that the blockchain is updated identically, at every site. Most malicious attacks on blockchains are targeted against condition 4, which is considered to be the most vulnerable point in every blockchain enterprise.

By convention, records are real-time transactions, acquired sequentially, so that we can usually assume that the nth record was created at a moment in time prior to the creation of the $n+1$th record.

Let us assume that the string that lies between the head and the tail of each record is a triple. This assumption is justified because all meaningful information can be represented as a triple or as a collection of triples.

Here is our list of triples that we will be blockchaining.

```
a0ce8ec6^^object_name^^Homo
a0ce8ec6^^subclass_of^^Hominidae
a0ce8ec6^^property^^glucose_at_time
a1648579^^object_name^^Homo sapiens
a1648579^^subclass_of^^Homo
98495efc^^object_name^^Andy Muzeack
98495efc^^instance_of^^Homo sapiens
98495efc^^dob^^1 January, 2001
98495efc^^glucose_at_time^^87, 02-12-2014 17:33:09
```

Let us create our own blockchain using these nine triples as our messages.

Each blockchain record will be of the form:

```
<tail of prior blockchain link—the current record's triple—md5 hash
of the current triple concatenated with the header>
```

For example, to compute the tail of the second link, we would perform an md5 hash on:

```
ufxOaEaKfw7QBrgsmDYtIw—a0ce8ec6^^subclass_of^^Hominidae
```

Which yields:

```
=> PhjBvwGf6dk9oUK/+yxrCA
```

The resulting blockchain is shown here.

```
                a0ce8ec6^^object_name^^Homo—ufxOaEaKfw7QBrgsmDYtIw
ufxOaEaKfw7QBrgsmDYtIw—a0ce8ec6^^subclass_of^^Hominidae—
PhjBvwGf6dk9oUK/+yxrCA
PhjBvwGf6dk9oUK/+yxrCA—a0ce8ec6^^property^^glucose_at_time—
P40p5GHp4hElgsstKbrFPQ
P40p5GHp4hElgsstKbrFPQ—a1648579^^object_name^^Homo sapiens—
2wAF1kWPFi35f6jnGOecYw
2wAF1kWPFi35f6jnGOecYw—a1648579^^subclass_of^^Homo—
N2y3fZgiOgRcqfx86rcpwg
N2y3fZgiOgRcqfx86rcpwg—98495efc^^object_name^^Andy Muzeack—
UXSrchXFR457g4JreErKiA
UXSrchXFR457g4JreErKiA—98495efc^^instance_of^^Homo sapiens—
5wDuJUTLWBJjQIu0Av1guw
5wDuJUTLWBJjQIu0Av1guw—98495efc^^glucose_at_time^^87, 02-12-2014
17:33:09—Y1jCYB7YyRBVIhm4PUUbaA
```

Whether you begin with a list of triples that you would like to convert into a blockchain data structure, or whether you are creating a blockchain one record at a time, through transactions that occur over time, it is easy to write a short script that will generate the one-way hashes and attach them to the end of the nth triple and the beginning of the $n+1$th triple, as needed.

Looking back at our blockchain, we can instantly spot an anomaly, in the header of the very first record. The header to the record is missing. Whenever we begin to construct a new blockchain, the first record will have no antecedent record from which a header can be extracted. This poses another computational bootstrap paradox. In this instance, we cannot begin until there is a beginning. The bootstrap paradox is typically resolved with the construction of a root record (record 0). The root record is permitted to break the rules.

Now that we have a small blockchain, what have we achieved? Here are the properties of a blockchain

- Every blockchain header is built from the values in the entire succession of preceding blockchain links

- The blockchain is immutable. Changing any of the messages contained in any of the blockchain links, would produce a totally different blockchain. Dropping any of the links of the blockchain or inserting any new links (anywhere other than as an attachment to the last validated link) will produce an invalid blockchain.
- The blockchain is recomputable. Given the same message content, the entire blockchain, with all its headers and tails, can be rebuilt. If it cannot recompute, then the blockchain is invalid.
- The blockchain, in its simplest form, is a trusted "relative time" stamp. Our blockchain does not tell us the exact time that a record was created, but it gives its relative time of creation compared with the preceding and succeeding records.

With a little imagination, we can see that a blockchain can be used as a true time stamp authority, if the exact time were appended to each of the records in the container at the moment when the record was added to the blockchain. The messages contained in blockchain records could be authenticated by including data encrypted with a private key. Tampering of the blockchain data records could be prevented by having multiple copies of the blockchain at multiple sites, and routinely checking for discrepancies among the different copies of the data.

We might also see that the blockchain could be used as a trusted record of documents, legal transactions (e.g., property deals), monetary exchanges (e.g., Bitcoin). Blockchains may also be used for authenticating voters, casting votes, and verifying the count. The potential value of blockchains in the era of Big Data is enormous, but the devil hides in the details. Every implementation of a blockchain comes with its own vulnerabilities and much has been written on this subject [3,4].

Section 8.7. Case Study (Advanced): Zero-Knowledge Reconciliation

Experience is what you have after you've forgotten her name.

Milton Berle

Though record reconciliation across institutions is always difficult, the task becomes truly Herculean when it must be done blindly, without directly comparing records. This awkward situation occurs quite commonly whenever confidential data records from different institutions must be checked to see if they belong to the same person. In this case, neither institution is permitted to learn anything about the contents of records in the other institutions. Reconciliation, if it is to occur, must implement a zero-knowledge protocol; a protocol that does not reveal any information concerning the reconciled records [5].

We will be describing a protocol for reconciling identifiers without exchanging information about the contents of data records. Because the protocol is somewhat abstract and unintuitive, a physical analogy may clarify the methodology. Imagine two people each holding a box containing an item. Neither person knows the contents of the box that they are holding or of the box that the other person is holding. They want to determine whether they are holding identical items, but they don't want to know anything about the items. They work together to create two identical imprint stamps, each covered by a complex random collection of raised ridges. With eyes closed, each one pushes his imprint stamp against his item. By doing so, the randomly placed ridges in the stamp are compressed in a manner characteristic of the object's surface. The stamps are next examined to determine if the compression marks on the ridges are distributed identically in both stamps. If so, the items in the two boxes, whatever they may be, are considered to be identical. Not all of the random ridges need to be examined-just enough of them to reach a high level of certainty. It is theoretically possible for two different items to produce the same pattern of compression marks, but it is highly unlikely. After the comparison is made, the stamps are discarded.

The physical analogy demonstrates the power of a zero-knowledge protocol. Neither party knows the identity of his own item. Neither party learns anything about his item or the other party's item during the transaction. Yet, somehow, the parties can determine whether the two items are identical.

Here is how the zero-knowledge protocol to reconcile confidential records across institutions [5]:

1. Both institutions generate a random number of a pre-determined length and each institution sends the random number to the other institution.
2. Each institution sums their own random number with the random number provided by the other institution. We will refer to this number as Random_A. In this way, both institutions have the same final random number and neither institution has actually transmitted this final random number. The splitting of the random number was arranged as a security precaution.
3. Both institutions agree to create a composite representation of information contained in the record that could establish the human subject of the record. The composite might be a concatenation of the social security number, the date of birth, the first initial of the surname.
4. Both institutions create a program that automatically creates the composite numeric representation of the record (which we will refer to as the record signature) and immediately sums the signature with Random_A, the random number that was negotiated between the two institutions (steps 1 and 2). The sum of the composite representation of the record plus Random_A is a random number that we will call Random_B.
5. If the two records being compared across institutions belong to the same human subject, then Random_B will the identical in both institutions. At this point, the two institutions must compare their respective versions of Random_B in such a way that they do not actually transmit Random_B to the other institution. If they were to transmit Random_B to the other institution, then the receiving institution could subtract Random_A from Random B and produce the signature string for a confidential record contained in the other institution. This would be a violation of the requirement to share zero knowledge during the transaction.

6. The institutions take turns sending consecutive characters of their versions of Random_B. For example, the first institution sends the first character to the second institution. The second institution sends the second character to the first institution. The first institution sends the third character to the second institution. The exchange of characters proceeds until the first discrepancy occurs, or until the first 8 characters of the string match successfully. If any of the characters do not match, both institutions can assume that the records belong to different human subjects (i.e., reconciliation failed). If the first 8 characters match, then it is assumed that both institutions are holding the same Random_B string, and that the records are reconciled.

At the end, both institutions learn whether their respective records belong to the same individual; but neither institution has learned anything about the records held in the other institution. Anyone eavesdropping on the exchange would be treated to a succession of meaningless random numbers.

Glossary

Curator The word "curator" derives from the Latin, "curatus," and the same root for "curative," indicating that curators "take care of" things. A data curator collects, annotates, indexes, updates, archives, searches, retrieves and distributes data. Curator is another of those somewhat arcane terms (e.g., indexer, data archivist, lexicographer) that are being rejuvenated in the new millennium. It seems that if we want to enjoy the benefits of a data-centric world, we will need the assistance of curators, trained in data organization.

Digital signature As it is used in the field of data privacy a digital signature is an alphanumeric sequence that could only have been produced by a private key owned by one particular person. Operationally, a message digest (e.g., a one-way hash value) is produced from the document that is to be signed. The person "signing" the document encrypts the message digest using her private key, and submits the document and the encrypted message digest to the person who intends to verify that the document has been signed. This person decrypts the encrypted message digest with her public key (i.e., the public key complement to the private key) to produce the original one-way hash value. Next, a one-way hash is performed on the received document. If the resulting one-way hash is the same as the decrypted one-way hash, then several statements hold true: the document received is the same document as the document that had been "signed." The signer of the document had access to the private key that complemented the public key that was used to decrypt the encrypted one-way hash. The assumption here is that the signer was the only individual with access to the private key. Digital signature protocols, in general, have a private method for encrypting a hash, and a public method for verifying the signature. Such protocols operate under the assumption that only one person can encrypt the hash for the message, and that the name of that person is known; hence, the protocol establishes a verified signature. It should be emphasized that a digital signature is quite different from a written signature; the latter usually indicates that the signer wrote the document or somehow attests to agreement with the contents of the document. The digital signature merely indicates that the document was received from a particular person, contingent on the assumption that the private key was available only to that person. To understand how a digital signature protocol may be maliciously deployed, imagine the following scenario: I contact you and tell you that I am Elvis Presley and would like you to have a copy of my public key plus a file that I have encrypted using my private key. You receive the file and the public key; and you use the public key to decrypt the file. You conclude that the file was indeed sent by Elvis Presley. You read the decrypted file and learn that Elvis advises you to invest all your money in a company that manufactures concrete guitars; which, of course, you do. Elvis knows guitars. The problem here is that the signature was valid, but the valid signature was not authentic.

Metasyntactic variable A variable name that imports no specific meaning. Popular metasyntactic variables are x, y, n, foo, bar, foobar, spam, eggs, norf, wubble, and blah. Dummy variables are often used in iterating loops. For example:

```
for($i=0;$i<1000;$i++)
```

Good form dictates against the liberal use of metasyntactic variables. In most cases, programmers should create variable names that describe the purpose of the variable (e.g., time_of_day, column_sum, current_line_from_file).

National Patient Identifier Many countries employ a National Patient Identifier (NPI) system. In these cases, when a citizen receives treatment at any medical facility in the country, the transaction is recorded under the same permanent and unique identifier. Doing so enables the data collected on individuals, from multiple hospitals, to be merged. Hence, physicians can retrieve patient data that was collected anywhere in the nation. In countries with NPIs, data scientists have access to complete patient records and can perform healthcare studies that would be impossible to perform in countries that lack NPI systems. In the United States, where a system of NPIs has not been adopted, there is a perception that such a system would constitute an invasion of privacy and would harm citizens.

One-way hash A one-way hash is an algorithm that transforms one string into another string (a fixed-length sequence of seemingly random characters) in such a way that the original string cannot be calculated by operations on the one-way hash value (i.e., the calculation is one-way only). One-way hash values can be calculated for any string, including a person's name, a document, or an image. For any given input string, the resultant one-way hash will always be the same. If a single byte of the input string is modified, the resulting one-way hash will be changed, and will have a totally different sequence than the one-way hash sequence calculated for the unmodified string.

Most modern programming languages have several methods for generating one-way hash values. Regardless of the language we choose to implement a one-way hash algorithm (e.g., md5, SHA), the output value will be identical. One-way hash values are designed to produce long fixed-length output strings (e.g., 256 bits in length). When the output of a one-way hash algorithm is very long, the chance of a hash string collision (i.e., the occurrence of two different input strings generating the same one-way hash output value) is negligible. Clever variations on one-way hash algorithms have been repurposed as identifier systems [6–9]. A detailed discussion of one-way hash algorithms can be found in Section 3.9, "Case Study: One-Way Hashes."

Raw data Raw data is the unprocessed, original data measurement, coming straight from the instrument to the database, with no intervening interference or modification. In reality, scientists seldom, if ever, work with raw data. When an instrument registers the amount of fluorescence emitted by a hybridization spot on a gene array, or the concentration of sodium in the blood, or virtually any of the measurements that we receive as numeric quantities, an algorithm executed by the measurement instrument produces the output. Pre-processing of data is commonplace in the universe of Big Data, and data managers should not labor under the false impression that the data received is "raw," simply because the data has not been modified by the person who submits the data.

Serious Big Data 3 V's (data volume, data variety and data velocity) plus "seriousness." Seriousness is a tongue-in-cheek term that the author applies to Big Data resources whose objects are provided with an adequate identifier and a trusted timestamp and provide data users with introspection, including pointers to the protocols that produced the data objects. The metadata in Big Data resources are appended with namespaces. Serious Big Data resources can be merged with other serious Big Data resources. In the opinion of the author, Big Data resources that lack seriousness should not be used in science, legal work, banking, and in the realm of public policy.

Time stamp Many data objects are temporal events and all temporal events must be given a time stamp indicating the time that the event occurred, using a standard measurement for time. The time stamp must be accurate, persistent, and immutable. The Unix epoch time (equivalent to the Posix epoch time) is available for most operating systems and consists of the number of seconds that have elapsed since January 1, 1970, midnight, Greenwhich mean time. The Unix epoch time can easily be converted into any other standard representation of time. The duration of any event can be easily calculated by subtracting the beginning time from the ending time. Because the timing of events can be maliciously altered, scrupulous data managers employ a trusted time stamp protocol by which a time stamp can be verified. A trusted time stamp must be accurate, persistent, and immutable. Trusted time stamp protocols are discussed in Section 8.5, "Case Study: The Trusted Time stamp."

Verification and validation As applied to data resources, verification is the process that ensures that data conforms to a set of specifications. Validation is the process that checks whether the data can be applied in a manner that fulfills its intended purpose. This often involves showing that correct conclusions can be obtained from a competent analysis of the data. For example, a Big Data resource might contain position, velocity, direction, and mass data for the earth and for a meteor that is traveling sunwards. The data may meet all specifications for measurement, error tolerance, data typing, and data completeness. A competent analysis of the data indicates that the meteor will miss the earth by a safe 50,000 miles, plus or minus 10,000 miles. If the asteroid smashes into the earth, destroying all planetary life, then an extraterrestrial observer might conclude that the data was verified, but not validated.

References

[1] Berman JJ. Nomenclature-based data retrieval without prior annotation: facilitating biomedical data integration with fast doublet matching. In Silico Biol 2005;5:0029.

[2] Beaudoin J. National experts at odds over patient identifiers. Healthcare IT News; 2004. October 18. Available at: http://www.healthcareitnews.com/news/national-experts-odds-over-patient-identifiers [viewed October 23, 2017].

[3] Ugarte H. A more pragmatic Web 3.0: linked blockchain data. 2017. https://doi.org/10.13140/RG.2.2.10304.12807/1 Available at: https://www.researchgate.net/publication/315619465_A_more_pragmatic_Web_30_Linked_Blockchain_Data [viewed October 27, 2017].

[4] Blockchains: the great chain of being sure about things. The Economist; 2015. October 31. Available at: https://www.economist.com/news/briefing/21677228-technology-behind-bitcoin-lets-people-who-do-not-know-or-trust-each-other-build-dependable [viewed October 27, 2017].

[5] Berman JJ. Zero-check: a zero-knowledge protocol for reconciling patient identities across institutions. Arch Pathol Lab Med 2004;128:344–6.

[6] Faldum A, Pommerening K. An optimal code for patient identifiers. Comput Methods Prog Biomed 2005;79:81–8.

[7] Rivest R. Request for comments: 1321, the MD5 message-digest algorithm. Network Working Group. https://www.ietf.org/rfc/rfc1321.txt [viewed January 1, 2015].

[8] Bouzelat H, Quantin C, Dusserre L. Extraction and anonymity protocol of medical file. Proc AMIA Annu Fall Symp 1996;1996:323–7.

[9] Quantin CH, Bouzelat FA, Allaert AM, Benhamiche J, Faivre J, Dusserre L. Automatic record hash coding and linkage for epidemiological followup data confidentiality. Methods Inf Med 1998;37:271–7.

9

Assessing the Adequacy of a Big Data Resource

Section 9.1. Looking at the Data

discovery is ".....seeing what others have seen, but thinking what others have not."

Albert Szent-Gyorgyi

Big Data must not be a Big Waste of time. Looking at the data will tell you immediately if you can use the data. Moving forward with calculations before looking at the data is inexcusable. Before you choose and apply analytic methods to data sets, you should spend time studying your raw data. The following steps may be helpful:

1. Find a free ASCII editor.

When I encounter a large data file, in plain ASCII format, the first thing I do is open the file and take a look at its contents. Unless the file is small (i.e., under about 20 megabytes), most commercial word processors will fail at this task. They simply cannot open really large files (in the Gigabyte range). You will want to use an editor designed to work with large ASCII files. Two of the more popular, freely available editors are Emacs and vi (also available under the name vim). Downloadable versions are available for Linux, Windows, and Macintosh systems. On most computers, these editors will open files in the range of a Gigabyte. For even larger files, there are operating system utilities that can do the job. These will be discussed in Section 9.4, "Case Study: Utilities for Viewing and Searching Large Files." [Glossary Text editor]

Principles and Practice of Big Data. https://doi.org/10.1016/B978-0-12-815609-4.00009-1

2. Download and study the "readme" or index files, or their equivalent.

In prior decades, large collections of data were often assembled as files within subdirectories and these files could be downloaded in part or *in toto*, via ftp (file transfer protocol). Traditionally, a "readme" file would be included with the files, and the "readme" file would explain the purpose, contents, and organization of all the files. In some cases, an index file might be available, providing a list of terms covered in the files and their locations in the various files. When such files are prepared thoughtfully, they are of great value to the data analyst. It is always worth a few minutes time to open and browse the "readme" file. I think of "readme" files as treasure maps. The data files contain great treasure, but you are unlikely to find anything of value unless you study and follow the map.

In the past few years, data resources have grown in size and complexity. Today, Big Data resources are often collections of resources, housed on multiple servers. New and innovative access protocols are continually being developed, tested, released, updated, and replaced. Still, some things remain the same. There will always be documents to explain how the Big Data resource "works" for the user. It behooves the data analyst to take the time to read and understand this prepared material. If there is no prepared material, or if the prepared material is unhelpful, then you may want to reconsider using the resource.

3. Assess the number of records in the Big Data resource.

There is a tendency among some data managers to withhold information related to the number of records held in the resource. In many cases, the number of records says a lot about the inadequacies of the resource. If the total number of records is much smaller than the typical user might have expected or desired, then the user might seek their data elsewhere. Data managers, unlike data users, sometimes dwell in a perpetual future that never merges into the here and now. They think in terms of the number of records they will acquire in the next 24 hours, the next year, or the next decade. To the data manager, limitations in the present are often irrelevant.

Data managers may be reluctant to divulge the number of records held in the Big Data resource when the number is so large as to defy credibility. Consider this example. There are about 5700 hospitals in the United States serving a population of about 313 million people. If each hospital served a specific subset of the population with no overlap in service between neighboring hospitals, then each would provide care for about 54,000 people. In practice, there is always some overlap in catchment population and a popular estimate for the average (overlapping) catchment for United States hospitals is 100,000. The catchment population for any particular hospital can be estimated by factoring in a parameter related to its size. For example, if a hospital hosts twice the number of beds than the average United States hospital, then one would guess that its catchment population would be about 200,000. The catchment population represents the approximate number of electronic medical records for living patients served by the hospital (one living individual, one hospital record). If you are informed that a hospital, of average size, contains 10 million records (when you are expecting about 100,000), then you can infer that something is very wrong. Most likely, the hospital is creating multiple records for individual patients. In general, institutions do not voluntarily provide users with information that casts doubt on the quality of their information systems. Hence, the data analyst, ignorant of the total number of records in the system, might proceed under the false assumption that each patient is assigned one and only one hospital record. Suffice it to say that the data user must know the number of records available in a resource, and the manner in which records are identified and internally organized.

A related issue of particular importance is the sample number/sample dimension dichotomy. Some resources with enormous amounts of data may have very few data records. This occurs when individual records contain mountains of data (e.g., sequences, molecular species, images), but the number of individual records is woefully low (e.g., hundreds or thousands). This problem, falling under the curse of dimensionality, will be further discussed in Section 14.6, "Case Study (Advanced): Curse of Dimensionality."

4. Determine how data objects are identified and classified.

As discussed in previous chapters, if you know the identifier for a data object, then you can collect all of the information associated with the object, regardless of its location in the resource. If other Big Data resources use the same identifier for the data object, you can integrate all of the data associated with the data object, regardless of its location in external resources. Furthermore, if you know the class that holds a data object, you can combine objects of a class and study all of the members of the class. Consider the following example.

```
Big Data resource 1

75898039563441    name                G. Willikers
75898039563441    gender              male

Big Data resource 2

75898039563441    age                 35
75898039563441    is_a_class_member   cowboy
94590439540089    name                Hopalong Tagalong
94590439540089    is_a_class_member   cowboy

Merged Big Data Resource 1 + 2

75898039563441    name                G. Willikers
75898039563441    gender              male
75898039563441    is_a_class_member   cowboy
75898039563441    age                 35
94590439540089    name                Hopalong Tagalong
94590439540089    is_a_class_member   cowboy
```

The merge of two Big Data resources combines data related to identifier 75898039563441 from both resources. We now know a few things about this data object that we did not know before the merge. The merge also tells us that the two data objects identified as 75898039563441 and 94590439540089 are both members of class cowboy. We now have two instance members from the same class, and this gives us information related to the types of instances contained in the class.

The consistent application of standard methods for object identification and for class assignments, using a standard classification or ontology, greatly enhances the value of a Big Data resource. A savvy data analyst will quickly determine whether the resource provides these important features. [Glossary Identification]

5. Determine whether data objects contain self-descriptive information.

Data objects should be well specified. All values should be described with metadata, all metadata should be defined, and the definitions for the metadata should be found documents whose unique names and locations are provided. The data should be linked to protocols describing how the data was obtained and measured. [Glossary ISO metadata standard]

6. Assess whether the data is complete and representative.

You must be prepared to spend hours reading through the records; otherwise, you will never really understand the data. After you have spent a few weeks of your life browsing through Big Data resources, you will start to appreciate the value of the process. Nothing comes easy. Just as the best musicians spend thousands of hours practicing and rehearsing their music, the best data analysts must devote thousands of hours to studying their data sources. It is always possible to run sets of data through analytic routines that summarize the data, but drawing insightful observations from the data requires thoughtful study.

An immense Big Data resource may contain spotty data. On one occasion, I was given a large hospital-based data set, with assurances that the data was complete (i.e., containing all necessary data relevant to the project). After determining how the records and the fields were structured, I looked at the distribution frequency of diagnostic entities contained in the data set. Within a few minutes I had the frequencies of occurrence of the different diseases, categorized under broad diagnostic categories. I spent another few hours browsing through the list, and before long I noticed that there were very few skin diseases included in the data. I am not a dermatologist, but I knew that skin diseases are among the most common conditions encountered in medical clinics. Where were the missing skin diseases? I asked one of the staff clinicians assigned to the project. He explained that the skin clinic operated somewhat autonomously from the other hospital departments. The dermatologists maintained their own information system, and their cases were not integrated into the general disease data set. I inquired as to why I had been assured that the data set was complete, when everyone other than myself knew full well that the data set lacked skin cases. Apparently, the staff had become so accustomed to ignoring the field of dermatology that it never crossed their minds to mention the matter.

It is a quirk of human nature to ignore anything outside one's own zone of comfort and experience. Otherwise fastidious individuals will blithely omit relevant information from Big Data resources if they consider the information to be inconsequential, irrelevant, or insubstantial. I have had conversations with groups of clinicians who requested that the free-text information in radiology and pathology reports (the part of the report containing descriptions of findings and other comments) be omitted from the compiled electronic records on the grounds that it is all unnecessary junk. Aside from the fact that "junk" text can serve as important analytic clues (e.g., measurements of accuracy, thoroughness, methodological trends), the systematic removal of parts of data records produces a biased and incomplete Big Data resource. In general, data managers should not censor data. It is the job of the data analyst to determine what data should be included or excluded from analysis; and to justify his or her decision. If the data is not available to the data analyst, then there is no opportunity to reach a thoughtful and justifiable determination.

On another occasion, I was given an anonymized set of clinical data from an undisclosed hospital. As I always do, I looked at the frequency distributions of items on the reports. In a few minutes, I noticed that germ cell tumors, rare tumors that arise from a cell lineage that includes oocytes and spermatocytes, were occurring in high numbers. At first, I thought that I might have discovered an epidemic of germ cell tumors in the hospital's catchment population. When I looked more closely at the data, I noticed that the increased incidence occurred in virtually every type of germ cell tumor, and there did not seem to be any particular increase associated with gender, age, or ethnicity. Cancer epidemics raise the incidence of one or maybe two types of cancer and may involve a particular at-risk population. A cancer epidemic would not be expected to raise the incidence of all types of germ cell tumors, across ages and genders. It seemed more likely that the high numbers of germ cell tumors were explained by a physician or specialized care unit that concentrated on treating patients with germ cell tumors, receiving referrals from across the nation. Based on the demographics of the data set (the numbers of patients of different ethnicities), I could guess the geographic region of the hospital. With this information and knowing that the institution probably had a prestigious germ cell clinic, I guessed the name of the "undisclosed" hospital. My suspicions were eventually confirmed. [Glossary Anonymization versus deidentification]

It sometimes helps to compare the distribution of data in a new collection against the distribution in data in a known and trusted population. For example, you may want to stratify data records by the age of individuals and compare it with the distribution of ages in a control or normal population of individuals. You might also create a word list or index of terms extracted from the data to determine if the frequency of occurrences of the included words or terms are similar to what you have come to expect from comparable data sets. If you find that there are too many kinds of data that are missing from your new collection of data, then you may need to abandon the project. You may find that the information contained in the new collection is similar in kind, but dissimilar in frequency to other populations. For example, if you encounter a population of men and women of all ages, but with a woman:male ration of 5:1 and with very few men over the age of 70 included in the population, then you might want to normalize your population against a control population. [Glossary Age-adjusted incidence]

The point here is that if you take the time to study raw data, you can spot systemic deficiencies or excesses in the data, if they exist, and you may gain deep insights that would not be obtained by mathematical techniques.

7. Plot some of the data.

Plotting data is quick, easy, and surprisingly productive. Within minutes, the data analyst can assess long-term trends, short-term and periodic trends, the general shape of data distribution and general notions of the kinds of functions that might represent the data (e.g., linear, exponential, power series). Simply knowing that the data can be expressed as a graph is immeasurably reassuring to the data analyst.

There are many excellent data visualization tools that are widely available. Without making any recommendation, I mention that graphs produced for this book were made with Matplotlib, a plotting library for the Python programming language; and Gnuplot, a graphing utility available for a variety of operating systems. Both Matplotlib and Gnuplot are open source applications that can be downloaded, at no cost, and are available at sourceforge.net. [Glossary Open source]

Gnuplot is extremely easy to use, either as stand-alone scripts containing gnuplot commands, or from the system command line. Most types of plots can be created with a single gnuplot command line. Gnuplot can fit a mathematically expressed curve to a set of data using the nonlinear least-squares Marquardt-Levenberg algorithm [1,2]. Gnuplot can also provide a set of statistical descriptors (e.g., median, mean, and standard deviation) for plotted sets of data.

Gnuplot operates from data held in tab-delimited ASCII files. Typically, data extracted from a Big Data resource is ported into a separate ASCII file, with column fields separated with a tab character, and rows separated by a newline character. In most cases, you will want to modify your raw data, readying it for plotting. Use your favorite programming language to normalize, shift, transform, covert, filter, translate, or munge your raw data, as you see fit. Export the data as a tab-delimited file, named with a .dat suffix.

It takes about a second to generate a plot for 10,000 data points (Fig. 9.1).

One command line in Gnuplot produced the graph, from the data.

```
splot 'c:\ftp\xyz_rand.dat'
```

It is very easy to plot data, but one of the most common mistakes of the data analyst is to assume that the available data actually represents the full range of data that may occur. If the data under study does not include the full range of the data, the data analyst will often reach a completely erroneous explanation for the observed data distribution.

Data distributions will almost always appear to be linear at various segments of their range. An oscillating curve that reaches equilibrium may look like a sine wave early in its course, and a flat-line later on. In the larger oscillations, it may appear linear along the length of a half-cycle. Any of these segmental interpretations of the data will miss observations that would lead to a full explanation of the data (Fig. 9.2).

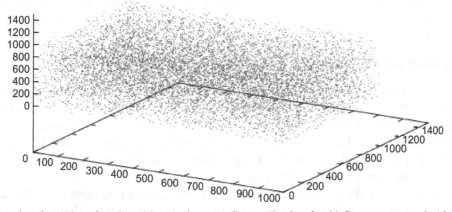

FIG. 9.1 A plot of 10,000 random data points, in three coordinates. The data for this figure was created with a 7 line script using the Perl programming language, but any scripting language would have been sufficient [3]. Ten thousand data points were created, with the *x*, *y*, and *z* coordinates for each point produced by a random number generator. The point coordinates were put into a file named xyz_rand.dat.

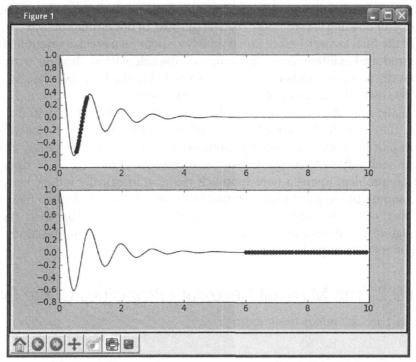

FIG. 9.2 An oscillating wave reaching equilibrium. The top graph uses circle-points to emphasize a linear segment for a half-cycle oscillation. The bottom graph of the same data emphasizes a linear segment occurring at equilibrium.

An adept data analyst can eyeball a data distribution and guess the kind of function that might model the data. For example, a symmetric bell-shaped curve is probably a normal or Gaussian distribution. A curve with an early peak and a long, flat tail is often a power law distribution. Curves that are simple exponential or linear can also be assayed by visual inspection. Distributions that may be described by a Fourier series or a power series, or that can be segmented into several different distributions, can also be assessed. [Glossary Power law, Power series, Fourier series]

8. Estimate the solution to your multi-million dollar data project, on day 1.

This may seem difficult to accept, and there will certainly be exceptions to the rule, but the solution to almost every multi-million dollar analytic problem can usually be estimated in just a few hours, sometimes minutes, at the outset of the project. If an estimate cannot be attained fairly quickly, then there is a good chance that the project will fail. If you do not have the data for a quick and dirty estimate, then you will probably not have the data needed to make a precise determination.

The past several decades have witnessed a profusion of advanced mathematical techniques for analyzing large data sets. It is important that we have these methods, but in most cases, newer methods serve to refine and incrementally improve older methods that do not rely on powerful computational techniques or sophisticated mathematical algorithms. As someone who was raised prior to the age of hand-held calculators and personal computers, I was taught quick-and-dirty estimation methods for adding, subtracting, multiplying, and dividing lists of numbers. The purpose of the estimation was to provide a good idea of the final answer, before much time was spent on a precise solution. If no mistake was introduced in either the estimate or the long calculation, then the two numbers would come close to one another. Conversely, mistakes in the long calculations could be detected if the two calculations yielded different numbers.

If data analysts go straight to the complex calculations before they perform a simple estimation, they will find themselves accepting wildly ridiculous calculations. For comparison purposes, there is nothing quite like a simple, intuitive estimate to pull a overly-eager analyst back to reality. Often, the simple act of looking at a stripped-down version of the problem opens a new approach that can drastically reduce computation time [4]. In some situations, analysts will find that a point is reached when higher refinements in methods yield diminishing returns. When everyone has used their most advanced algorithms to make an accurate prediction, they may sometimes find that their best effort offers little improvement over a simple estimator.

Section 9.2. The Minimal Necessary Properties of Big Data

In God we trust, all others bring data.

William Edwards Deming (1900–1993)

Many of today's statisticians and scientists came of age in the world of small data. When you are working with a few hundred measurements, most of the issues discussed in this book have almost no relevance. Small data does not need to be dressed up with identifiers and metadata. Scientists did not worry very much about creating self-explanatory data; each scientist understood their own data, and that was usually good enough.

Big Data, with its volume, complexity, velocity, and permanence, requires a remarkable amount of annotation and curation. For the most part, the issues raised in this book are unknown to the bulk of individuals who collect Big Data. Hence, most of the Big Data that has been collected and stored has no scientific value; it is simply incomprehensible and unusable [5–8]. This may seem like an outrageous claim, particularly when you consider how much of the world's activities are data-driven. If you speak with scientists who collect and analyze data, and that would include just about every scientist you are likely to encounter, you will hear them tell you that their data is just fine, and perfectly suitable for their own scientific studies. The point that must be made is that the scientists who collected and analyzed the data cannot judge the value of scientific data. The true value of data must be assessed by the scientists who verify, validate, and re-analyze the data that was collected by other scientists. If the original data cannot be obtained and analyzed by the scientific community, now and in the future, then the original assertions cannot be confirmed, and the data cannot be usefully merged with other data sets, extended, and repurposed. [Glossary Abandonware, Dark data, Universal and perpetual, Data versus datum, Identifier, Data repurposing]

For data to be useful to the scientific community, it must have a set of basic properties, and, unfortunately, these properties are seldom taught or utilized. Here are the universal properties of good data that has lasting scientific value.

– Data that has been annotated with metadata
– Data that establishes uniqueness or identity
– Time stamped data that accrues over time [Glossary Time, Time stamp]
– Data that resides within a data object
– Data that has membership in a defined class
– Introspective data—data that explains itself
– Immutable data
– Data that has been simplified

Let us take a moment to examine each of these data features:

– **Data that has been annotated with metadata**

Metadata, the data that explains data, was discussed in Sections 4.1 through 4.3. The modern specification for metadata is the eXtensible Markup Language (XML). The importance of XML to data scientists cannot be overstated. As a data-organizing technology, it is as important as the invention of written language (circa 3000 bc) or the appearance of mass-printed books (circa 1450 ad). Markup allows us to convey any message as XML (a pathology report, a radiology image, a genome database, a workflow process, a software program, or an e-mail). [Glossary Data annotation, Annotation, Data sharing]

– **Data that establishes uniqueness or identity**

The most useful data establishes the identity of objects. In many cases, objects have their own, natural identifiers that come very close to establishing uniqueness.

Examples include fingerprints, iris patterns, and the sequence of nucleotides in an organism's genetic material.

In regard to identifying data objects, we need not depend on each data object having its own naturally occurring identifier. As discussed in Section 3.1, we can simply generate and assign unique identifiers to our data objects [3,8–10]. Identifiers are data simplifiers, when implemented properly. They allow us to collect all of the data associated with a unique object, while ensuring that we exclude that data that should be associated with some other object.

– **Time stamped data that accrues over time**

When a data set contains data records that collect over time, it becomes possible to measure how the attributes of data records may change as the data accumulates. Signals analysts use the term time series to refer to attribute measurements that change over time. The shape of the time series can be periodic (i.e., repeating over specific intervals), linear, non-linear, Gaussian, or multimodal (i.e., having multiple peaks and troughs), or chaotic. A large part of data science is devoted to finding trends in data, determining simple functions that model the variation of data over time, or predicting how data will change in the future. All these analytic activities require data that is annotated with the time that a measurement is made, or the time that a record is prepared, or the time that an event has occurred. [Glossary Data science, Waveform]

You may be shocked to learn that many, if not most, web pages lack a time stamp to signify the date and time when the page's textual content was created. This oversight applies to news reports, announcements from organizations and governments, and even scientific papers; all being instances for which a time stamp would seem to be an absolute necessity. When a scientist publishes an undated manuscript, how would anyone know if the results are novel? If a news article describes an undated event, how would anyone know whether the report is current? For the purposes of data analysis, undated documents and data records are useless [5].

Whereas undated documents have very little value, all transactions, statements, documents and data points that are annotated with reliable time stamps will always have some value, particularly if the information continues to collect over time. Today, anyone with a computer can easily time stamp his or her data, with the date and the time, accurate to within a second. As discussed in Section 6.4, "Case Study: Time stamping Data," every operating system and every programming language has access to the time, and can easily annotate any data point with the time that it was created. Time data can be formatted in any of dozens of ways, all of which can be instantly converted to an international standard [11].

It's human nature to value newly collected data and to dismiss old data as being outdated or irrelevant. Nothing could be further from the truth. New data, in the absence of old data, has little value. All historical events develop through time, and the observations made at any given moment in time are always influenced by events that transpired at earlier times. Whenever we speak of "new" data, alternately known as prospectively acquired data, we must think in terms that relate the new data to the "old" data that preceded it. Old data can be used to analyze trends over time and to predict the data values into the future. Essentially, old data provides the opportunity to see the past, the present, and the future. The dependence of new data on old data can be approached computationally. The autocorrelation function is a method for producing a type of measurement indicating the dependence of data elements on prior data elements. Long-range dependence occurs when a value is dependent on many prior values. Long-range dependence is determined when the serial correlation (i.e., the autocorrelation over multiple data elements) is high when the number of sequential elements is large [12]. These are nifty tools for data analysis, but they cannot be employed if the data is not time stamped [6]. [Glossary Correlation distance]

– **Data that is held in a data object**

In Section 6.2, "Data Objects: The Essential Ingredient of Every Big Data Collection," we defined a data object as an object identifier plus all of the data/metadata pairs that rightly belong to the object identifier, including a data/metadata pair that tells us the object's class. Lucky for us, some of the most common data creations (e.g., emails and photographic images) are automatically composed as data objects by our software (i.e., email clients and digital cameras).

When you send a message, your email client automatically creates a data object that holds the contents of your message, descriptive information about the message, a message identifier, and a time stamp. Here is a sample email header, obtained by selecting the email client's long or detailed version of the message. The actual message contents would normally follow, but are omitted here for brevity.

- `MIME-Version: 1.0`
- `Received: by 10.36.165.75 with HTTP; Tue, 2 May 2017 14:46:47 -0700 (PDT)`
- `Date: Tue, 2 May 2017 17:46:47 -0400`
- `Delivered-To: you@gmail.com`
- `Message-ID: < CALVNVe-kk7fqYJ82MfsV6a4kFKW4v57c4y9BLpOUYf1cBHq9pQ@mail.gmail.com >`
- `Subject: tiny fasts`
- `From: Anybody <me@gmail.com>`
- `To: Anybody Else <you@gmail.com>`
- `Content-Type: multipart/alternative; boundary=94eb2c07ab4c054062054e917a03`

Notice that each line of the header consists of a colon ":" flanked to the right by metadata (e.g., Subject, From, To) and on the left by the described data. There is a line for a time stamp and a line for an identifier assigned by the email client.

- `Date: Tue, 2 May 2017 17:46:47 -0400`
- `Message-ID: < CALVNVe-kk7fqYJ82MfsV6a4kFKW4v57c4y9BLpOUYf1cBHq9pQ@mail.gmail.com >`

Email messages are an example of data objects that are automatically created when you push the "send" button. When we read about the remarkable results achieved by forensic data analysts, who gather time stamped, immutable, and identified evidence from millions of stored messages, we must give credit to the power of data objects.

– **Data that has membership in a defined class**

In Chapter 5, we discussed classifications and ontologies and explained the importance of assigning instances (e.g., diseases, trucks, investments) to classes wherein every instance shares a set of features typical of the class. All good classifications have a feature that is known as competence; the ability to draw inferences about data objects, and their relationships to other data objects, based on class definitions. Data that is unclassified may have some immediate observational or experimental value to scientists, but such data cannot be used to draw inferences from classes of data objects obtained from Big Data resources.

– **Introspective data (data that explains itself)**

Introspection, as previously discussed in Chapter 6, refers to the ability of data (e.g., data records, documents, and all types of data objects) to describe itself when interrogated. Introspection gives data users the opportunity to see relationships among the individual data records that are distributed in different data sets, and is one of the most useful features of data objects, when implemented properly.

Modern programming languages allow us to interrogate data, and learn everything there is to know about the information contained in data objects. Information about data objects, acquired during the execution of a program, can be used to modify a program's instructions, during run-time, a useful feature known as "reflection". Detailed information about every piece of data in a data set (e.g., the identifier associated with the data object, the class of objects to which the data object belongs, the metadata and the data values that are associated with the data object), permit data scientists to integrate data objects collected from multiple Big Data resources.

It should be noted that the ability to perform introspection is not limited to object oriented programming languages. Introspection is provided by the data, and any programming language will suffice, so long as the data itself is organized as data objects assigned to classes within a sensibly structured classification.

– **Immutable data**

When you are permitted to change preexisting data, all of your collected data becomes tainted. None of the analyses performed on the data in the database can be verified, because the data that was originally analyzed no longer exists. It has become something else, which you cannot fully understand. Aside from producing an unverifiable data collection, you put the data analyst in the impossible position of deciding which data to believe; the old data or the new data.

- **Data that has been simplified**

Big Data is complex data, and complex data is difficult to understand and analyze. As it happens, all of the properties that we consider the minimal necessary for Big Data preparation happen to be simplifying. Metadata, identifiers, data objects, and classifications all work to drive down the complexity of data and render the data understandable to man or machine.

It is easy for data managers to shrug off the data requirements described in this section as high-tech nuisances. Big Data requires an enormous amount of fussy work that was simply not necessary when data was small. Nonetheless, it is necessary, if we hope to use more than an insignificant fraction of the data that is being collected every day.

Section 9.3. Data That Comes With Conditions

This site has been moved.
We'd tell you where, but then we'd
have to delete you.

Computer-inspired haiku by Charles Matthews

I was involved in one project where the data holders could not be deterred from instituting a security policy wherein data access would be restricted to pre-approved users. Anyone wishing to query the database would first submit an application, which would include detailed information about themselves and their employer. The application required users to explain how they intended to use the resource, providing a description of their data project. Supplying this information was a warm-up exercise for the next step.

A screening committee composed primarily of members of the Big Data team would review the submitted application. A statistician would be consulted to determine if the applicant's plan was feasible. The committee would present their findings to an executive committee that would compare each application's merits against those of the other applicants. The very best applications would be approved for data access.

The data team could not seem to restrain their enthusiasm for adding layers of complexity to the security system. They decided that access to data would be tiered. Some users would be given less access to data than other users. No users would be given free access to the entire set of data. No user would have access to individual deidentified records; only aggregate views of record data would be released. A system would be designed to identify users and to restrict data access based on the identity and assigned access status.

These security measures were unnecessary. The data in the system had been rendered harmless via deidentification and could be distributed without posing any risk to the data subjects or to the data providers. The team seemed oblivious to the complexities engendered by a tiered access system. Bruce Schneier, a widely cited security expert, wrote an essay entitled, "A plea for simplicity: you can't secure what you don't understand" [13].

In this essay, he explained that as you add complexity to a system, the system becomes increasingly difficult to secure. I doubted that the team had the resources or the expertise to implement a complex, multi-tiered access system for a Big Data resource. I suspected that if the multi-tiered access system were actually put into place, the complexity of the system would render the resource particularly vulnerable to attack. In addition, the difficulty of accessing the system would discourage potential users and diminish the scientific value of the Big Data resource.

Many data holders believe that their job, as responsible stewards of data, is to deny data access to undeserving individuals and to ensure that any incorrect conclusions drawn from their data will never see the light of day. I have seen examples wherein the data holders require data users to sign an agreement indicating that the results of their analyses must be submitted back to the data holders before being released to the public in the form of manuscripts, public announcements, or conference presentations. The data holders typically reserve the right to forbid releasing results with which they disapprove. It is easy to see that a less-than-saintly committee might disapprove results that cast their Big Data resource in a bad light, or results that compete in any way with the products of their own research, or results that they hold in disfavor for any capricious reason whatsoever.

Aside from putting strict restrictions on who gets access to data, and which results are permitted to be published, it is commonplace to impose strict restrictions on how the data can be viewed. Anyone who has visited online databases is familiar with the query box. The idea is that the user enters a query and waits for some output to appear on the screen. The assumption here is that the user knows how the query must be composed to produce the most complete output. Of course, this is never the case. When a user enters a query, she cannot know, in advance, whether some other query term might have yielded a better output. Such query boxes almost never return details about the data set or the algorithm employed in responding to the query. It is difficult, under these circumstances, to imagine any scenario wherein these kinds of queries have any scientific merit.

If Big Data resources are to add significantly to the advancement of science, the kinds of complex and stingy data sharing practices that have evolved over the past few decades must face extinction.

Section 9.4. Case Study: Utilities for Viewing and Searching Large Files

It isn't that they can't see the solution. It's that they can't see the problem.

G. K. Chesterton

In Section 9.1, we discussed the importance of looking at your data. Several free and open source text editors were suggested (Open Office, vi, emacs). These text editors can open immense files (gigabytes in length and longer), but they have their limits. Files much larger than a gigabyte may be slow to load, or may actually be unloadable on systems with small memory capacity. In such cases, your computer's operating system may offer a convenient alternative to text editors.

In the Windows operating system, then you can read any text file, one screen at a time, with the "more" command.

For example, on Windows systems, at the prompt:

    c:\>type huge_file.txt |more

The first lines from the huge_file.txt file will fill the screen, and you can proceed through the file by pressing and holding the <Enter> key. [Glossary Line]

Using this simple command, you can assess the format and general organization of any file. For the uninitiated, ASCII data files are inscrutable puzzles. For those who take a few moments to learn the layout of the record items, ASCII records can be read and understood, much like any book.

In contrarian Unix and Linux systems the "less" command functions much like the Windows "more" command, but offers many additional options. At the Unix (or Linux) system prompt, type the following command (substituting your preferred file for "huge_file.txt"):

```
$ less huge_file.txt
```

This will load a screen-sized chunk of huge_file.txt onto your monitor. By pressing the "enter" key, or the "arrow down" key, additional lines will scroll onto the monitor, one line at a time. For fast screen scrolls, keep your finger on the "Page Down" key. The "Page Up" key lets you back the screens.

The less command accommodates various options.

```
$ less -S huge_file.txt
```

the use of the -S switch cuts off line wrap so that the lines are truncated at the edge of the screen. In general, this speeds up the display.

When you use the Unix "less" command, you will find that the last line at the bottom of the screen is a ":". The ":" is a prompt for additional instructions. If you were to enter a slash character ("/") followed by a word or phrase or regex pattern, you would immediately see the line in which the first occurrence of your search term appeared. If you typed "&" and the pattern, at the ":" prompt, you would see all the lines from the file, in which your search pattern appears.

The Unix "less" command is a versatile and fast utility for viewing and searching very large files. If you do not use Unix systems, do not despair. Windows users can install Cygwin, a free Unix-like interface. Cygwin, and supporting documentation, can be downloaded from:

    http://www.cygwin.com/

Cygwin opens in a window that produces a shell prompt (equivalent to Windows C prompt) from which Unix programs can be launched. For myself, I use Cygwin primarily as a source of Unix and Linux utilities, of which there are hundreds. In addition, Cygwin comes bundled with some extremely useful applications, such as Perl, Python, OpenSSL, and Gnuplot.

Windows users are not restricted to launching Unix and Linux applications from within the Cygwin shell prompt. A command line from the Windows C prompt will launch Cygwin utilities. For example:

```
c:\cygwin64\bin>wc temp.txt
  11587 217902 1422378 temp.txt
```

The command "wc temp.txt," launched the Unix/Linux word counter utility ("wc") from the Windows C prompt, yielding a count of the lines, words, and bytes in the temp.txt file. Likewise, a system call from a Python script can invoke Cygwin utilities and applications.

Big Data scientists eventually learn that there are some tasks that are best left to Unix/Linux. Having Cygwin installed on your Windows system will make life easier for you, and for your collaborators, who may prefer to work in Linux.

Section 9.5. Case Study: Flattened Data

Everything should be made as simple as possible, but not simpler.

Albert Einstein

Data flattening is a term that is used differently by data analysts, database experts, and informaticians. Though the precise meaning changes from subfield to subfield, the term always seems to connote a simplification of the data and the elimination of unnecessary structural restraints.

In the field of informatics, data flattening is a popular but ultimately counter-productive method of data organization and data reduction. Data flattening involves removing data annotations that are not needed for the interpretation of data [5].

Imagine, for the sake of illustration, a drastic option that was seriously considered by a large medical institution. This institution, that shall remain nameless, had established an excellent Electronic Medical Record (EMR) system. The EMR assigns a unique and permanent identifier string to each patient, and attaches the identifier string to every hospital transaction involving the patient (e.g., biopsy reports, pharmacy reports, nursing notes, laboratory reports). All of the data relevant to a patient, produced anywhere within the hospital system is linked by the patient's unique identifier. The patient's EMR can be assembled, instantly, whenever needed, via a database query.

Over time, the patient records in well-designed information systems accrue a huge number of annotations (e.g., time stamped data elements, object identifiers, linking elements, metadata). The database manager is saddled with the responsibility of maintaining the associations among all of the annotations. For example, an individual with a particular test, conducted at a particular time, on a particular day, will have annotations that link the test to a test procedure protocol, an instrument identifier, a test code, a laboratory name, a test sample, a sample accession time, and so on. If data objects could be stripped of most of their annotations, after some interval of time, then it would reduce the overall data management burden on the hospital information system. This can be achieved by composing simplified reports and deleting the internal annotations. For example, all of the data relevant to a patient's laboratory test could be reduced to the patient's name, the date, the name of the test, and the test result. All of the other annotations can be deleted. This process is called data flattening.

Should a medical center, or any entity that collects data, flatten their data? The positive result would be a streamlining of the system, with a huge reduction in annotation overhead. The negative result would be the loss of the information that connects well-defined data objects (e.g., test result with test protocol, test instrument with test result, name of laboratory technician with test sample, name of clinician with name of patient). Because the fundamental activity of the data scientist is to find relationships among data objects, data flattening will reduce the scope and value of data repurposing projects. Without annotations and metadata, the data from different information systems cannot be sensibly merged. Furthermore, if there is a desire or a need to reanalyze flattened data, then the data scientist will not be able to verify the data and validate the conclusions drawn from the data [5]. [Glossary Verification and validation, Validation]

Glossary

Abandonware Software that that is abandoned (e.g., no longer updated, supported, distributed, or sold) after its economic value is depleted. In academic circles, the term is often applied to software that is developed under a research grant. When the grant expires, so does the software. Most of the software in existence today is abandonware.

Age-adjusted incidence An age-adjusted incidence is the crude incidence of disease occurrence within an age category (e.g., age 0–10 years, age 70–80 years), weighted against the proportion of persons in the age groups of a standard population. When we age-adjust incidence, we cancel out the changes in the incidence of disease occurrence, in different populations, that result from differences in the proportion of people in different age groups. For example, suppose you were comparing the incidence of childhood leukemia in two populations. If the first population has a large proportion of children, then it will likely have a higher number of childhood leukemia in its population, compared with another population with a low proportion of children. To determine whether the first population has a true, increased rate of leukemia, we need to adjust for the differences in the proportion of young people in the two populations [14].

Annotation Annotation involves describing data elements with metadata or attaching supplemental information to data objects.

Anonymization versus deidentification Anonymization is a process whereby all the links between an individual and the individual's data record are irreversibly removed. The difference between anonymization and deidentification is that anonymization is irreversible. There is no method for re-establishing the identity of the patient from anonymized records. Deidentified records can, under strictly controlled circumstances, be reidentified. Reidentification is typically achieved by entrusting a third party with a confidential list that maps individuals to deidentified records. Obviously, reidentification opens another opportunity of harming individuals, if the confidentiality of the reidentification list is breached. The advantages of reidentification is that suspected errors in a deidentified database can be found, and corrected, if permission is obtained to reidentify individuals. For example, if the results of a study based on blood sample measurements indicate that the original samples were mislabeled, it might be important to reidentify the samples and conduct further tests to resolve the issue. In a fully anonymized data set, the opportunities for verifying the quality of data are highly limited.

Correlation distance Also known as correlation score. The correlation distance provides a measure of similarity between two variables. Two similar variables will rise and fall together [15,16]. The Pearson correlation score is popular, and can be easily implemented [3,17]. It produces a score that varies from -1 to 1. A score of 1 indicates perfect correlation; a score of -1 indicates perfect anti-correlation (i.e., one variable rises while the other falls). A Pearson score of 0 indicates lack of correlation. Other correlation measures can be applied to Big Data sets [15,16].

Dark data Unstructured and ignored legacy data, presumed to account for most of the data in the "infoverse". The term gets its name from "dark matter" which is the invisible stuff that accounts for most of the gravitational attraction in the physical universe.

Data annotation The process of supplementing data objects with additional data, often providing descriptive information about the data (i.e., metadata, identifiers, time information, and other forms of information that enhances the utility of the data object.

Data repurposing Involves using old data in new ways, that were not foreseen by the people who originally collected the data. Data repurposing comes in the following categories: (1) Using the preexisting data to ask and answer questions that were not contemplated by the people who designed and collected the data; (2) Combining preexisting data with additional data, of the same kind, to produce aggregate data that suits a new set of questions that could not have been answered with any one of the component data sources; (3) Reanalyzing data to validate assertions, theories, or conclusions drawn from the original studies; (4) Reanalyzing the original data set using alternate or improved methods to attain outcomes of greater precision or reliability than the outcomes produced in the original analysis; (5) Integrating heterogeneous data sets (i.e., data sets with seemingly unrelated types of information), for the purpose an answering questions or developing concepts that span diverse scientific disciplines; (6) Finding subsets in a population once thought to be homogeneous; (7) Seeking new relationships among data objects; (8) Creating, on-the-fly, novel data sets through data file linkages; (9) Creating new concepts or ways of thinking about old concepts, based on a reexamination of data; (10) Fine-tuning existing data models; and (11) Starting over and remodeling systems [5].

Data science A vague term encompassing all aspects of data collection, organization, archiving, distribution, and analysis. The term has been used to subsume the closely related fields of informatics, statistics, data analysis, programming, and computer science.

Data sharing Providing one's own data to another person or entity. This process may involve free or purchased data, and it may be done willingly, or under coercion, as in compliance with regulations, laws, or court orders.

Data versus datum The singular form of data is datum, but the word "datum" has virtually disappeared from the computer science literature. The word "data" has assumed both a singular and plural form. In its singular form, it is a collective noun that refers to a single aggregation of many data points. Hence, current usage would be "The data is enormous," rather than "These data are enormous."

Fourier series Periodic functions (i.e., functions with repeating trends in the data, including waveforms and periodic time series data) can be represented as the sum of oscillating functions (i.e., functions involving sines, cosines, or complex exponentials). The summation function is the Fourier series.

ISO metadata standard ISO 11179 is the standard produced by the International Standards Organization (ISO) for defining metadata, such as XML tags. The standard requires that the definitions for metadata used in XML (the so-called tags) be accessible and should include the following information for each tag: Name (the label assigned to the tag), Identifier (the unique identifier assigned to the tag), Version (the version of the tag), Registration Authority (the entity authorized to register the tag), Language (the language in which the tag is specified), Definition (a statement that clearly represents the concept and essential nature of the tag), Obligation (indicating whether the tag is required), Datatype (indicating the type of data that can be represented in the value of the tag), Maximum Occurrence (indicating any limit to the repeatability of the tag), and Comment (a remark describing how the tag might be used).

Identification The process of providing a data object with an identifier, or the process of distinguishing one data object from all other data objects on the basis of its associated identifier.

Identifier A string that is associated with a particular thing (e.g., person, document, transaction, data object), and not associated with any other thing [18]. In the context of Big Data, identification usually involves permanently assigning a seemingly random sequence of numeric digits (0–9) and alphabet characters (a–z and A–Z) to a data object. The data object can be a class of objects.

Line A line in a non-binary file is a sequence of characters that terminate with an end-of-line character. The end-of-line character may differ among operating systems. For example, the DOS end of line character is ASCII 13 (i.e., the carriage return character) followed by ASCII 10 (i.e., the line feed character), simulating the new line movement in manual typewriters. The Linux end-of-line character is ASCII 10 (i.e., the line feed character only). When programming in Perl, Python or Ruby, the newline character is represented by "\n" regardless of which operating system or file system is used. For most purposes, use of "\n" seamlessly compensates for discrepancies among operating systems with regard to their preferences for end-of-line characters. Binary files, such as image files or telemetry files, have no designated end-of-line characters. When a file is opened as a binary file, any end-of-line characters that happen to be included in the file are simply ignored as such, by the operating system.

Open source Software is open source if the source code is available to anyone who has access to the software.

Power law A mathematical formula wherein a particular value of some quantity varies as an inverse power of some other quantity [19,20]. The power law applies to many natural phenomena and describes the Zipf distribution or Pareto's principle. The power law is unrelated to the power of a statistical test.

Power series A power series of a single variable is an infinite sum of increasing powers of x, multiplied by constants. Power series are very useful because it is easy to calculate the derivative or the integral of a power series, and because different power series can be added and multiplied together. When the high exponent terms of a power series are small, as happens when x is less than one, or when the constants associated with the higher exponents all equal 0, the series can be approximated by summing only the first few terms. Many different kinds of distributions can be represented as a power series. Distributions that cannot be wholly represented by a power series may sometimes by segmented by ranges of x. Within a segment, the distribution might be representable as a power series. A power series should not be confused with a power law distribution.

Text editor A text editor (also called ASCII editor) is a software application designed to create, modify, and display simple unformatted text files. Text editors are different from word processes that are designed to include style, font, and other formatting symbols. Text editors are much faster than word processors because they display the contents of files without having to interpret and execute formatting instructions. Unlike word processors, text editors can open files of enormous size (e.g., gigabyte range).

Time A large portion of data analysis is concerned, in one way or another, with the times that events occur or the times that observations are made, or the times that signals are sampled. Here are three examples demonstrate why this is so: (1) most scientific and predictive assertions relate how variables change with respect to one another, over time; and (2) a single data object may have many different data values, over time, and only timing data will tell us how to distinguish one observation from another; (3) computer transactions are tracked in logs, and logs are composed of time-annotated descriptions of the transactions. Data objects often lose their significance if they are not associated with an accurate time measurement. Because modern computers easily capture accurate time data, there is not annotating all data points with the time when they are measured.

Time stamp Many data objects are temporal events and all temporal events must be given a time stamp indicating the time that the event occurred, using a standard measurement for time. The time stamp must be accurate, persistent, and immutable. The Unix epoch time (equivalent to the Posix epoch time) is available for most operating systems and consists of the number of seconds that have elapsed since January 1, 1970, midnight, Greenwhich mean time. The Unix epoch time can easily be converted into any other standard representation of time. The duration of any event can be easily calculated by subtracting the beginning time from the ending time. Because the timing of events can be maliciously altered, scrupulous data managers employ a trusted time stamp protocol by which a time stamp can be verified. A trusted time stamp must be accurate, persistent, and immutable. Trusted time stamp protocols are discussed in Section 8.5, "Case Study: The Trusted Time Stamp."

Universal and perpetual Wherein a set of data or methods can be understood and utilized by anyone, from any discipline, at any time. It is a tall order, but a worthy goal. Much of the data collected over the centuries of recorded history is of little value because it was never adequately described when it was recorded (e.g., unknown time of recording, unknown source, unfamiliar measurements, unwritten protocols). Efforts to resuscitate large collections of painstakingly collected data are often abandoned simply because there is no way of verifying, or even understanding, the original data [5]. Data scientists who want their data to serve for posterity should use simple specifications, and should include general document annotations such as the Dublin Core. The importance of creating permanent data is discussed elsewhere [6].

Validation Involves demonstrating that the conclusions that come from data analyses fulfill their intended purpose and are consistent [21]. You validate a conclusion (which my appear in the form of an hypothesis, or a statement about the value of a new laboratory test, or a therapeutic protocol) by showing that you draw the same conclusion repeatedly whenever you analyze relevant data sets, and that the conclusion satisfies some criteria for correctness or suitability. Validation is somewhat different from reproducibility. Reproducibility involves getting the same measurement over and over when you perform the test. Validation involves drawing the same conclusion over and over.

Verification and validation As applied to data resources, verification is the process that ensures that data conforms to a set of specifications. Validation is the process that checks whether the data can be applied in a manner that fulfills its intended purpose. This often involves showing that correct conclusions can be obtained from a competent analysis of the data. For example, a Big Data resource might contain position, velocity, direction, and mass data for the earth and for a meteor that is traveling sunwards. The data may meet all specifications for measurement, error tolerance, data typing, and data completeness. A competent analysis of the data indicates that the meteor will miss the earth by a safe 50,000 miles, plus or minus 10,000 miles. If the asteroid smashes into the earth, destroying all planetary life, then an extraterrestrial observer might conclude that the data was verified, but not validated.

Waveform A graph showing a signal's amplitude over time. By convention, the amplitude of the signal is shown on the y-axis, while the time is shown on the x-axis. A .wav file can be easily graphed as a waveform, in python.

The waveform.py script graphs a sample .wav vile, alert.wav, but any handy .wav file should suffice (Fig. 9.3).

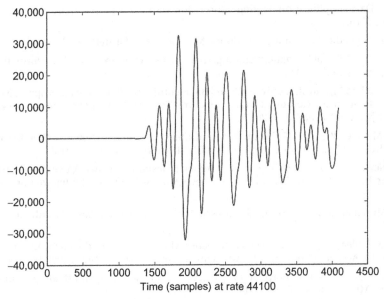

FIG. 9.3 The plotted waveform of a .wav file, alert.wav.

```
from scipy.io.wavfile import read
import matplotlib.pyplot as plt
input_data = read("alert.wav")
# returns a two-item tuple with sampling rate as
#the 0th item and audio samples as the 1st item
audio = input_data[1]
# we'll plot the first 4096 samples
plt.plot(audio[0:4096])
plt.xlabel("time (samples) at rate " + str(input_data[0]))
plt.show()
```

References

[1] Levenberg K. A method for the solution of certain non-linear problems in least squares. Q Appl Math 1944;2:164–8.

[2] Marquardt DW. An algorithm for the least-squares estimation of nonlinear parameters. J SIAM Appl Math 1963;11:431–41.

[3] Berman JJ. Methods in medical informatics: fundamentals of healthcare programming in Perl, Python, and Ruby. Boca Raton: Chapman and Hall; 2010.

[4] Lee J, Pham M, Lee J, Han W, Cho H, Yu H, et al. Processing SPARQL queries with regular expressions in RDF databases. BMC Bioinf 2011;12:S6.

[5] Berman JJ. Repurposing legacy data: innovative case studies. Waltham, MA: Morgan Kaufmann; 2015.

[6] Berman JJ. Data simplification: taming information with open source tools. Waltham, MA: Morgan Kaufmann; 2016.

[7] Berman JJ. Biomedical informatics. Sudbury, MA: Jones and Bartlett; 2007.

[8] Berman JJ. Principles of big data: preparing, sharing, and analyzing complex information. Waltham, MA: Morgan Kaufmann; 2013.

[9] Leach P, Mealling M, Salz R. A universally unique identifier (UUID) URN namespace. Network Working Group, Request for Comment 4122, Standards Track. Available from: http://www.ietf.org/rfc/rfc4122.txt [viewed November 7, 2017].

[10] Mealling M. RFC 3061. A URN namespace of object identifiers. Network Working Group; 2001. Available from: https://www.ietf.org/rfc/rfc3061.txt [viewed January 1, 2015].

[11] Klyne G. Newman C. Date and time on the internet: timestamps. Network Working Group Request for Comments RFC:3339, Available from: http://tools.ietf.org/html/rfc3339 [viewed on September 15, 2015].

[12] Downey AB, Think DSP. Digital signal processing in python, Version 0.9.8. Needham, MA: Green Tea Press; 2014.

[13] Schneier B. A plea for simplicity: you can't secure what you don't understand. Information Security November 19. Available from: http://www.schneier.com/essay-018.html; 1999 [viewed July 1, 2015].

[14] Berman JJ. Rare diseases and orphan drugs: keys to understanding and treating common diseases. Cambridge, MD: Academic Press; 2014.

[15] Reshef DN, Reshef YA, Finucane HK, Grossman SR, McVean G, Turnbaugh PJ, et al. Detecting novel associations in large data sets. Science 2011;334:1518–24.

[16] Szekely GJ, Rizzo ML. Brownian distance covariance. Ann Appl Stat 2009;3:1236–65.

[17] Lewis PD. R for medicine and biology. Sudbury: Jones and Bartlett Publishers; 2009.

[18] Paskin N. Identifier interoperability: a report on two recent ISO activities. D-Lib Mag 2006;12:1–23.

[19] Newman MEJ. Power laws, Pareto distributions and Zipf's law. Contemp Phys 2005;46:323–51.

[20] Clauset A, Shalizi CR, Newman MEJ. Power-law distributions in empirical data. SIAM Rev 2009;51:661–703.

[21] Committee on Mathematical Foundations of Verification, Validation, and Uncertainty Quantification; Board on Mathematical Sciences and Their Applications, Division on Engineering and Physical Sciences, National Research Council. Assessing the reliability of complex models: mathematical and statistical foundations of verification, validation, and uncertainty quantification. National Academy Press; 2012. Available from: http://www.nap.edu/catalog.php?record_id=13395 [viewed January 1, 2015].

10

Measurement

OUTLINE

Section 10.1. Accuracy and Precision

Get your facts first, then you can distort them as you please.

Mark Twain

Precision is the degree of exactitude of a measurement and is verified by its reproducibility (i.e., whether repeated measurements of the same quantity produce the same result). Accuracy measures how close your data comes to being correct. Data can be accurate but imprecise or precise but inaccurate. If you have a 10-pound object, and you report its weight as 7.2376 pounds, every time you weigh the object, then your precision is remarkable, but your accuracy is dismal.

What are the practical limits of precision measurements? Let us stretch our imaginations, for a moment, and pretend that we have just found an artifact left by an alien race that excelled in the science of measurement. As a sort of time capsule for the universe, their top scientists decided to collect the history of their civilization, and encoded it in binary. Their story looked something like "001011011101000..." extended to about 5 million places. Rather than print the sequence out on a piece of paper or a computer disc, these aliens simply converted the sequence to a decimal length (i.e., .001011011101000...") and marked the length on a bar composed of a substance that would never change its size. To decode the bar and recover the history of the alien race, one would simply need to have a highly precise measuring instrument that would yield the original binary sequence. Computational linguists could translate the sequence to text, and the recorded history of the alien race would be revealed! Of course, the whole concept is built on an impossible premise. Nothing

Principles and Practice of Big Data. https://doi.org/10.1016/B978-0-12-815609-4.00010-8

can be measured accurately to 5 million places. We live in a universe with practical limits (i.e., the sizes of atomic particles, the speed of light, the Heisenberg uncertainty principle, the maximum mass of a star, the second law of thermodynamics, the unpredictability of highly complex systems). There are many things that we simply cannot do, no matter how hard we try. The most precise measurement achieved by modern science has been in the realm of atomic clocks, where accuracy of 18 decimal places has been claimed [1]. Nonetheless, many scientific disasters are caused by our ignorance of our own limitations, and our persistent gullibility, leading us to believe that precision claimed is precision obtained.

It is quite common for scientists to pursue precision when they should be seeking accuracy. For an example, we need look no further than the data-intensive field of "Precision Medicine" [2]. One of the goals of precision medicine is to determine the specific genetic alterations that account for human disease. In many cases, this means finding a change in a single nucleotide (from among the 3 billion nucleotides in the DNA sequence that accounts for the human genome) responsible for the development of a disease. Precision Medicine has had tremendous success for a variety of rare diseases and for rare subtypes of common diseases, but has had less luck with common diseases such as type 2 diabetes and adult onset hypertension. Why is this? The diagnosis of diabetes and hypertension are based on a cut-off measurement. Above a certain glucose level in the blood, the patient is said to have diabetes. Above a certain pressure, the patient is said to have hypertension. It is not much different from being overweight (i.e., above a certain weight) or tall (i.e., above a certain height). Theory, strengthened by empiric observations, informs us that quantitative traits have multiple genetic and environmental influences, a phenomenon recognized since the early studies of RA Fisher, in 1919 [3–5]. Hence, we would expect that hypertension and diabetes would not be amenable to precise diagnosis [2].

At this point, we can determine, with credible accuracy, whether a person is diabetic, hypertensive, obese, tall, able to hold his breath for a long time, or able to run 100m in record time. We cannot determine, with any precision, the precise genes that are necessary for the development of any acquired human traits. Should we be devoting time and money to attain higher and higher precision in the genetic diagnosis of common, polygenic diseases, if increasing precision brings us no closer to a practical cure for these diseases? Put plainly, shouldn't we be opting for "Accurate Medicine" rather than "Precision Medicine"?

The conflict between seeking accuracy and seeking precision is a common dilemma in the universe of Big Data, wherein access to highly precise measurements is ridiculously abundant.

– **Steganography: using imprecision to your advantage**

You look at them every day, the ones that others create, and that you create your own, that you share with your friends or with the world. They are part of your life, and you would feel a deep sense of loss if you lost them. I am referring to high resolution digital images. We love them, but we give them more credit than they deserve. When you download a

16-megapixel image of your sister's lasagna, you can be certain that most of the pixel information is padded with so-called empty resolution; pixel precision that is probably inaccurate and certainly exceeding the eye's ability to meaningfully resolve. Most images in the megabyte size range can safely be reduced to the kilobyte size range, without loss of visual information. Steganography is an encryption technique that takes advantage of the empty precision in pixel data by inserting secret text messages into otherwise useless bits of pseudodata.

Steganography is one of several general techniques in which a message is hidden within an object, such as a book or a painting. The forerunners of modern steganography have been around for centuries and were described as early as AD 1500 by Trithemious [6]. Watermarking is closely related to steganography. Digital watermarking is a way of secretly insinuating the name of the owner or creator of a digital object into the object, as a mechanism of rights management [7]. [Glossary Steghide]

Section 10.2. Data Range

> *Many an object is not seen, though it falls within the range of our visual ray, because it does not come within the range of our intellectual ray, i.e., we are not looking for it. So, in the largest sense, we find only the world we look for.*
>
> *Henry David Thoreau*

Always determine the highest and the lowest observed values in your data collection. These two numbers are often the most important numbers in any set of data; even more important than determining the average or the standard deviation. There is always a compelling reason, relating to the measurement of the data or to the intrinsic properties of the data set, to explain the high and the low of data.

Here is an example. You are looking at human subject data that includes weights. The minimum weight is a pound (the round-off weight of a viable but premature newborn infant). You find that the maximum weight in the data set is 300 pounds, exactly. There are many individuals in the data set who have a weight of 300 pounds, but no individuals with a weight exceeding 300 pounds. You also find that the number of individuals weighing 300 pounds is much greater than the number of individuals weighting 290 pounds or 280 pounds. What does this tell you? Obviously, the people included in the data set have been weighed on a scale that tops off at 300 pounds. Most of the people whose weight was recorded as 300 will have a false weight measurement. Had we not looked for the maximum value in the data set, we would have assumed, incorrectly, that the weights were always accurate.

It would be useful to get some idea of how weights are distributed in the population exceeding 300 pounds. One way of estimating the error is to look at the number of people weighing 295 pounds, 290 pounds, 285 pounds, etc. By observing the trend, and knowing the total number of individuals whose weight is 300 pounds or higher, you can estimate the number of people falling into weight categories exceeding 300 pounds.

Here is another example where knowing the maxima for a data set measurement is useful. You are looking at a collection of data on meteorites. The measurements include weights. You notice that the largest meteorite in the large collection weighs 66 tons (equivalent to about 60,000 kg), and has a diameter of about 3 m. Small meteorites are more numerous than large meteorites, but one or more meteorites account for almost every weight category up to 66 tons. There are no meteorites weighing more than 66 tons. Why do meteorites have a maximum size of about 66 tons?

A little checking tells you that meteors in space can come in just about any size, from a speck of dust to a moon-sized rock. Collisions with earth have involved meteorites much larger than 3 m. You check the astronomical records and you find that the meteor that may have caused the extinction of large dinosaurs about 65 million years ago was estimated at 6–10 km (at least 2000 times the diameter of the largest meteorite found on earth).

There is a very simple reason why the largest meteorite found on earth weighs about 66 tons, while the largest meteorites to impact the earth are known to be thousands of time heavier. When meteorites exceed 66 tons, the impact energy can exceed the energy produced by an atom bomb blast. Meteorites larger than 66 tons leave an impact crater, but the meteor itself disintegrates on impact.

As it turns out, much is known about meteorite impacts. The kinetic energy of the impact is determined by the mass of the meteor and the square of the velocity. The minimum velocity of a meteor at impact is about 11 km/s (equivalent to the minimum escape velocity for sending an object from earth into space). The fastest impacts occur at about 70 km/s. From this data, the energy released by meteors, on impact with the earth, can be easily calculated.

By observing the maximum weight of meteors found on earth we learn a great deal about meteoric impacts. When we look at the distribution of weights, we can see that small meteorites are more numerous than larger meteorites. If we develop a simple formula that relates the size of a meteorite with its frequency of occurrence, we can predict the likelihood of the arrival of a meteorite on earth, for every weight of meteorite, including those weighing more than 66 tons, and for any interval of time.

Here is another profound example of the value of knowing the maximum value in a data distribution. If you look at the distance from the earth to various cosmic objects (e.g., stars, black holes, nebulae) you will quickly find that there is a limit for the distance of objects from earth. Of the many thousands of cataloged stars and galaxies, none of them have a distance that is greater than 13 billion light years. Why? If astronomers could see a star that is 15 billion light years from earth, the light that is received here on earth must have traveled 15 billion light years to reach us. The time required for light to travel 15 billion light years is 15 billion years; by definition. The universe was born in a big bang about 14 billion years ago. This would imply that the light from the star located 15 billion miles from earth must have begun its journey about a billion years before the universe came into existence. Impossible!

By looking at the distribution of distances of observed stars and noting that the distances never exceed about 13 billion years, we can infer that the universe must be at least 13 billion years old. You can also infer that the universe does not have an infinite age and size; otherwise, we would see stars at a greater distance than 13 billion light years.

If you assume that stars popped into the universe not long after its creation, then you can infer that the universe has an age of about 13 or 14 billion years. All of these deductions, confirmed independently by theoreticians and cosmologists, were made without statistical analysis, simply by noting the maximum number in a distribution of numbers.

Section 10.3. Counting

On two occasions I have been asked, 'If you put into the machine wrong figures, will the right answers come out?' I am not able rightly to apprehend the kind of confusion of ideas that could provoke such a question.

Charles Babbage

For the bulk of Big Data projects, analysis begins with counting. If you cannot count the data held in a Big Data resource, then you will derive little benefit from the resource. Systemic counting errors account for irreproducible or misleading results. Surprisingly, there is very little written about this issue in the Big Data literature. Presumably, the subject is considered too trivial for serious study. To rectify this oversight, this section describes, in some depth, the surprising intellectual challenges of Big Data counting.

Most people would agree that the simple act of counting data is something that can be done accurately and reproducibly, from laboratory to laboratory. Actually, this is not the case. Counting is fraught with the kinds of errors previously described in this chapter, plus many other hidden pitfalls. Consider the problem of counting words in a paragraph. It seems straightforward, until you start asking yourself how you might deal with hyphenated words. "Deidentified" is certainly one word. "Under-represented" is probably one word, but sometimes the hyphen is replaced by a space, and then it is certainly two words. How about the term "military-industrial," which seems as though it should be two words? When a hyphen occurs at the end of a line, should we force a concatenation between the syllables at the end of one line and the start of the next?

Slashes are a tougher nut to crack than hyphens. How should we count terms that combine two related words by a slash, such as "medical/pharmaceutical"; one word or two words? If we believe that the slash is a word separator (i.e., slashes mark the end of one word and the beginning of another), then we would need to parse Web addresses into individual words. For example:

www.science.com/stuff/neat_stuff/super_neat_stuff/balloons.htm

The Web address could be broken into a string of words, if the "." and "_" characters could be considered valid word separators. In that case, the single Web address would consist of 11 words: www, science, com, stuff, neat, stuff, super, neat, stuff, balloons, htm. If you were only counting words that match entries in a standard dictionary, then the split Web address would contain 8 words: science, stuff, neat, stuff, super, neat, stuff, balloons. If we defined a word as a string bounded by a space or a part-of-sentence separator (e.g., period, comma, colon, semicolon, question mark, exclamation mark, end of line character), then the unsplit Web address would count as 1 word. If the word must match a dictionary term, then the unsplit Web address would count as zero words. So, which is it: 11 words, 8 words, 1 word, or 0 words?

This is just the start of the problem. How shall we deal with abbreviations [8,9]? Should all abbreviations be counted as one word, or as the sum of words represented by the abbreviation? Is "U.S." One word or two words? Suppose, before counting words, the text is pre-processed to expand abbreviations. All the abbreviated terms (i.e., every instance of "U.S." becomes an instance of United States, and UCLA would count as 4 words). This would yield an artificial increase in the number of words in the document. How would a word counter deal with abbreviations that look like words, such as "mumps" which could be the name of a viral disease of childhood, or it could be an abbreviation for a computer language used by medical informaticians and expanded as "Massachusetts General Hospital Utility Multi-Programming System."

How would we deal with numeric sequences appearing in the text? Should each numeric sequence be counted as a word? If not, how do we handle Roman numbers? Should "IV" be counted as a word, because it is composed of alphabetic characters, or should it be omitted as a word, because it is equivalent to the numeric value, "4"? When we encounter "IV," how can we be certain that we are parsing a Roman numeral? Could "IV," within the context of our document, represent the abbreviation for "intravenous"?

It is obvious that the number of words in a document will depend on the particular method used to count the words. If we use a commercial word counting application, how can we know which word counting rules are applied? In the field of informatics, the total number of words is an important feature of a document. The total word count often appears in the denominator of common statistical measurements. Counting words seems to be a highly specialized task. My favorite estimator of the number of words in any text file is simply the size of the file divided by 6.5, the average number of characters in a word plus one separator character.

The point here is that a simple counting task, such as word counting, can easily become complex. A complex counting task, involving subjective assessments of observations, seldom yields accurate results. When the criteria for counting change over time, then results that were merely inaccurate may devolve even further, into irreproducibility. An example of a counting task that is complex and objective is the counting of hits and errors in baseball. The rules for counting errors are subjective and based on the scorer's judgment of the intended purpose of the hit (e.g., sacrifice fly) and the expected number of bases reached in the absence of the error. The determination of an error sometimes depends on the outcome of the play after the presumptive error has occurred (i.e., on events that are not controlled or influenced by the error). Counting is also complex with rules covering specific instances of play. For example, passed balls and wild pitches are not scored as errors; they are assigned to another category of play. Plays involving catchers are exempt from certain rules for errors that apply to fielders. It would be difficult to find an example of a counting task that is more complex than counting baseball errors.

Sometimes counting criteria inadvertently exclude categories of items that should be counted. The diagnoses that appear on death certificates are chosen from a list of causes of death included in the International Classification of Diseases (ICD). Diagnoses collected from all of the death certificates issued in the United States are aggregated by the CDC (Centers for Disease Control and Prevention) and published in the National Vital Statistics Report [10]. As it happens, "medical error" is not included as a cause of death in the ICD; hence, United States casualties of medical errors are not counted as such in the official records. Official tally notwithstanding, it is estimated that about one of every six deaths in the United States result from medical error [10].

Big Data is particularly prone to counting errors, as data is typically collected from multiple sources, each with its own method for annotating data. In addition, Big Data may extend forwards and backwards in time; constantly adding new data and merging with legacy data sets. The criteria for counting data may change over time, producing misleading results. Here are a few examples of counts that changed radically when the rules for counting changed. [Glossary Meta-analysis]

– Beachy Head is a cliff in England with a straight vertical drop and a beautiful sea-view. It is a favorite jumping off point for suicides. The suicide rate at Beachy Head dropped as sharply as the cliff when the medical examiner made a small policy change. From a certain moment onward, bodies found at the cliff bottom would be counted as suicides only if their post-mortem toxicology screen was negative for alcohol. Intoxicated subjects were pronounced dead by virtue of accident (i.e., not suicide) [11].
– Sudden Infant Death Syndrome (SIDS, also known as crib death) was formerly considered to be a disease of unknown etiology that caused infants to stop breathing, and die, often during sleep. Today, most SIDS deaths are presumed to be due to unintentional suffocation from bedclothes, often in an overheated environment, and aggravated by a prone (i.e., face down) sleeping position. Consequently, some infant deaths that may have been diagnosed as SIDS in past decades are now diagnosed as unintentional suffocations. This diagnostic switch has resulted in a trend characterized by increasing numbers of infant suffocations and a decreasing number of SIDS cases [12]. This trend is, in part, artifactual, arising from changes in reporting criteria.
– In the year 2000, nearly a half-century after the Korean War, the United States Department of State downsized its long-standing count of United States military war deaths; to 36,616 down from an earlier figure of about 54,000. The drop of 17,000 deaths resulted from the exclusion of United States military deaths that occurred during the Korean War, in countries outside Korea [13]. The old numbers reflected deaths during the Korean War; the newer number reflects deaths occurring due to the Korean War. Aside from historical interest, the alteration indicates how collected counts may change retroactively.
– Human life is flanked by two events, birth and death; both events are commemorated with a certificate. Death certificates are the single most important gauge of public health. They tell us the ages at which deaths occur, and the causes of those deaths. With this information, we can determine the most common causes of death in the population, changes in the frequency of occurrences of the different causes of death, and the effect of interventions intended to increase overall life expectancy and

reduce deaths caused by particular causes. Death certificates are collected from greater than 99% of individuals who die in the United States [14]. This data, vital to the health of every nation, is highly error prone, and the problems encountered in the U.S. seem to apply everywhere [15,16]. A survey of 49 national and international health atlases has shown that there is virtually no consistency in the way that death data are prepared [17]. Within the United States there is little consistency among states in the manner in which the causes of death are listed [18]. Death data is Big Data, as it is complex (i.e., containing detailed, non-standard information within death certificates), comes from many sources (i.e., every municipality), arrives continually (i.e., deaths occur every minute), with many records (i.e., everyone dies eventually). The rules for annotating the data change regularly (i.e., new versions of the International Classification of Diseases contain different new terms and codes). The consistency of the data decreases as the Big Data grows in size and in time. Our basic understanding of how humans die, and our ability to measure the effect of potentially life-saving public health interventions, is jeopardized by our inability to count the causes of death.

– Dealing with Negations

A large portion of Big Data is categorical, not quantitative. Whenever counting categorical features, you need to know whether a feature is present or absent. Unstructured text has no specific format for negative assertions (i.e., statements indicating that a feature is absent or that an assertion is false). Negations in unstructured data come into play during parsing routines wherein various features need to be counted.

If a computer program is seeking to collect, count, or annotate occurrences of a particular diagnosis included in a pathology report, or a particular type of "buy" order on a financial transaction, or the mention of a particular species of frog on a field report, there should be some way to distinguish a positive occurrence of the term (e.g., Amalgamated Widget is traded), from a negation statement (e.g., Amalgamated Widget is not traded."). Otherwise, counts of the positive occurrences of trades would include cases that are demonstrably negative. Informaticians have developed a variety of techniques that deal with negations occurring in textual data [19].

In general, negation techniques rely on finding a negation term (e.g., not present, not found, not seen) in proximity with an annotation term (e.g., a term that matches some term in a standard nomenclature, or a term that has been cataloged or otherwise indexed for the data set, onto which a markup tag is applied). A negated term would not be collected or counted as a positive occurrence of the annotation.

Examples of negation terms included in sentences are shown here:

- He cannot find evidence for the presence of a black hole.
- We cannot find support for the claim.
- A viral infection is not present.
- No presence of Waldo is noted.
- Bigfoot is not in evidence in this footprint analysis.

It is easy to exclude terms that are accompanied by an obvious negation term. When terms are negated or otherwise nullified by terms that are not consistently characterized by a negative inference, the problem becomes complex.

Here is a short list of implied negations, each lacking an unambiguous negation term, followed by the re-written sentence that contains an unambiguous negation term (i.e., "not").

- "Had this been a tin control processor, the satellite would have failed."—The satellite did not fail.
- "There is a complete absence of fungus."—Fungus is not present
- "We can rule out the presence of invasive cancer."—Invasive cancer is not present.
- "Hurricanes failed to show."—Hurricanes did not show.
- "The witness fails to make an appearance."—The witness did not appear.
- "The diagnosis is incompatible with psoriasis."—Psoriasis is not present.
- "Drenching rain is inconsistent with drought."—Drought does not occur with drenching rain.
- "There is insufficient evidence for a guilty verdict."—Not guilty.
- "Accidental death is excluded."—Not an accidental death.
- "A drug overdose is ruled out."—Not a drug overdose.
- "Neither fish nor foul."—Not fish. Not foul.
- There is zero evidence for aliens in Hoboken."—Aliens have not been found in Hoboken.

In addition to lacking outright negations, sentences may contain purposefully ambiguating terms, intended to prohibit readers from drawing any conclusion, positive or negative. For example, "The report is inconclusive for malignancy." How would this report be counted? Was a malignancy present, or was it not?

The point here is that, like everything else in the field of Big Data, the individuals who prepare and use resources must have a deep understanding of the contained data. They must also have a realistic understanding of the kinds of questions that can be sensibly asked and answered with the available data. They must have an understanding of the limits of their own precision.

Section 10.4. Normalizing and Transforming Your Data

Errors have occurred.
We won't tell you where or why.
Lazy programmers.

Computer-inspired haiku by Charlie Gibbs

When extracting data from multiple sources, recorded at different times, and collected for different purposes, the data values may not be directly comparable. The Big Data analyst must contrive a method to normalize or harmonize the data values.

– **Adjusting for population differences.**

Epidemiologists are constantly reviewing large data sets on large populations (e.g., local, national, and global data). If epidemiologists did not normalize their data, they would be in a constant state of panic. Suppose you are following long-term data on the incidence of a rare childhood disease in a state population. You notice that the number of people with the disease has doubled in the past decade. You are about to call the New York Times with the shocking news when one of your colleagues taps you on the shoulder and explains that the population of the state has doubled in the same time period. The incidence, described as cases per 100,000 population, has remained unchanged. You calm yourself down and continue your analysis to find that the reported cases of the disease have doubled in a different state that has had no corresponding increase in state population. You are about to alert the White House with the news when your colleague taps you on the shoulder and explains that the overall population of the state has remained unchanged, but the population of children in the state has doubled. The incidence as expressed as cases occurring in the affected population, has remained unchanged.

An age-adjusted rate is the rate of a disease within an age category, weighted against the proportion of persons in the age groups of a standard population. When we age-adjust rates, we cancel out the changes in the rates of disease that result from differences in the proportion of people in different age groups.

Some of the most notorious observations on non-adjusted data come from the field of baseball. In 1930 Bill Terry maintained a batting average of 0.401, the best batting average in the National league. In 1968 Carl Yastrzemski led his league with a batting average of 0.301. You would think that the facts prove that Terry's lead over his fellow players was greater than Yastrzemski's. Actually, both had averages that were 27% higher than the average of their fellow ballplayers of the year. Normalized against all the players for the year in which the data was collected, Terry and Yastrzemski tied.

– **Rendering data values dimensionless.**

Histograms express data distributions by binning data into groups and displaying the bins in a bar graph. A histogram of an image may have bins (bars) whose heights consist of the number of pixels in a black and white image that fall within a certain gray-scale range (Fig. 10.1).

When comparing images of different sizes, the total number of pixels in the images is different, making it impossible to usefully compare the heights of bins. In this case, the number of pixels in each bin can be divided by the total number of pixels in the image, to produce a number that corresponds to the fraction of the total image pixels that are found in the bin. The normalized value (now represented as a fraction), can be compared between two images. Notice that by representing the bin size as a fraction, we have stripped the dimension from the data (i.e., a number expressed as pixels), and rendered a dimensionless data item (i.e., a purely numeric fraction).

FIG. 10.1 An image of the author, left, converted into a histogram representing the number of pixels that have a gray-scale value of 0, 1, 2, 3 and so on up to the top gray-scale value of 256. Each gray-scale value is a bin.

– Converting one data type to another, more useful, data type.

A zip code is an example of data formed by numeric digits that lack numeric properties. You cannot add two zip codes and expect to get any useful information from the process. However, every zip code has been mapped to a specific latitude and longitude at the center of the zip code region, and these values can be used as spherical coordinates from which distances between locations can be computed. It is often useful to assign geographic coordinates to every zip code entry in a database.

– Converting to a (0, 1) interval.

Any set of data values can be converted into an interval between 0 and 1, wherein the original data values maintain their relative positions in the new interval. There are several simple ways to achieve the result. The most straightforward is to compute the range of the data by subtracting the smallest data value in your data set from the largest data value. To determine the location of any data value in the 0, 1 range, simply subtract from it the smallest value in the data set and then divide the result by the range of the data (Fig. 10.2). This tells you where your value is located, in a 0, 1 interval, as a fraction of the range of the data.

$$y = \frac{x - x_{min}}{x_{max} - x_{min}}$$

FIG. 10.2 A formula that will convert any value to a fraction between 0 and 1 by dividing the distance of the value from the smallest value of the attribute in the population by the full data range of the value in the population [20].

Another popular method for converting data sets to a standard interval is to subtract the mean from any data value and divide by the standard deviation. This gives you the position of the data value expressed as its deviation from the mean as a fraction of the standard deviation. The resulting value is called the z-score.

When comparing different data sets, it is frequently important to normalize all of the data points to a common interval. In the case of multi-dimensional data it is usually necessary to normalize the data in every dimension using some sensible scaling computation. This may include the methods just described (i.e., dividing by range or by standard deviation, or by substituting data with a dimensionless transformed value, such as a correlation measure).

– **Weighting.**

Weighting is a method whereby the influence of a value is moderated by some factor intended to yield an improved value. In general, when a data value is replaced by the sum of weighted factors, the weights are chosen to add to 1. For example, if you are writing your own smoothing function, in which each value in a data set is replaced by a value computed by summing contributions from itself and its immediate neighbor on the left and the right, you might multiply each number by one-third, so that the final number is scaled to a magnitude similar to your original number. Alternately, you might multiply the number to the left and to the right by one-quarter and the original by one-half, to provide a summed number weighted to favor the original number.

It is a shame that Big Data never comes with instructions. Data analysts are constantly required to choose a normalization method, and the choice will always depend on their intended use of the data. Here is an example. Three sources of data provide records on children that include an age attribute. Each source measures age in the same dimension; years. You would think that because the ages are all recorded in years, not months or decades, you can omit a normalization step. When you study the data, you notice that one source contains children up to the year 14, while another is cut off at age 12, and another stops at age 16. Suddenly, you are left with a difficult problem. Can your ignore the differences in the cut-off age in the three data sets? Should you truncate all of the data above age 12? Should you use all of the data, but weigh the data differently for the different sources? Should you divide by the available ranges for the data? Should you compute z-scores? It all depends on what you are trying to learn from the data.

Section 10.5. Reducing Your Data

There is something fascinating about science. One gets such a wholesale return of conjecture out of such a trifling investment of fact.

Mark Twain

At first glance, it seems obvious that gravitational attraction is a Big Data problem. We know that gravitation between any two bodies is proportional to the product of their masses and inversely proportional to the square of the distance between them. If we want to predict the gravitational forces on an object, we would need to know the position and mass of every body in the universe. With this data, we would compute a force vector, from which we could determine the net gravitational influence of the universe upon the mass. Of course, this is absurd. If we needed all that data for our computation, physicists would be forever computing the orbit of the earth around the sun. We are lucky to live in a universe wherein gravity follows an inverse square distance rule, as this allows us to neglect the influences of heavenly bodies that are far away from earth and sun, and of nearby bodies that have small masses compared with the sun. Any high school student can compute the orbits of the earth around the sun, predicting their relative positions millennia into the future.

Likewise, if we can see two galaxies in space and we notice that they are similar in shape, size, and have a similar star density, then we can assume that they both produce about the same amount of light. If the light received on Earth from one of those galaxies is four times that received by the other galaxy, we can apply the inverse square law for light intensity and infer that the dimmer galaxy is probably twice as far from earth as the brighter galaxy. In this short analysis, we start with our observations on every visible galaxy in the universe. Next, we compare just two galaxies and from this comparison we can develop general methods of analysis that may apply to the larger set of data.

The point here is that when Big Data is analyzed it is seldom necessary to include every point of data in your system model. In the Big Data field the most successful analysts will often be those individuals who are adept at simplifying the system model; thus eliminating unnecessary calculations.

Because Big Data is complex, you will often find that your data objects have high dimensionality; each data object is annotated with a large number of values. The types of values that are shared among all the different data objects are usually referred to as parameters. It is very difficult to make much sense of high dimensional data. It is always best to develop a filtering mechanism that expunges useless parameters. A useless parameter will often have one of these two properties:

1. Redundancy. If a parameter correlates perfectly with some other parameter, you know that you can safely drop one of the two parameters. For example, you may have some physiologic data on a collection of people, and the data may include weight, waist size, body fat index, weight adjusted by height, and density. These measurements seem to

be measuring about the same thing; are they all necessary? If several attributes closely correlate with one another, you might want to drop a few.

Association scores provide a measure of similarity between two variables. Two similar variables will rise and fall together. The Pearson correlation score is popular and can be easily implemented [18,21]. It produces a score that varies from −1 to 1. A score of 1 indicates perfect correlation; a score of −1 indicates perfect anti-correlation (i.e., one variable rises while the other falls). A Pearson score of 0 indicates lack of correlation. Other correlation measures are readily available, as discussed in Section 11.3, "The Dot Product, a Simple and Fast Correlation Method" [22,23]. Big Data analysts should not demure from developing their own correlation scores, as needed, to ensure enhanced speed, or to provide a scoring measure that best serves their particular goals.

2. Randomness. If a parameter is totally random, then it cannot tell you anything meaningful about the data object, and you can drop the parameter. There are many tests that measure randomness; most were designed to measure the quality of random number generators [24]. They can also be used to determine the randomness of data sets.

Putting your set of parameter values into a file, and compressing the file can achieve a simple but useful test for randomness. If the values of the parameter are distributed randomly, the file will not compress well, whereas a set of data that has a regular distribution (e.g., a simple curve, or a Zipf-like distribution, or a distribution with a sharp peak), will compress down into a very small file.

As a small illustration, I wrote a short program that created three files, each 10,000 bytes in length. The first file consisted of the number 1, repeated 10,000 times (i.e., 11111111...). The second file consisted of the numbers 0 through 9, distributed as a sequence of 1000 zeros followed by 1000 ones, followed by 1000 twos, and so on, up to 1000 nines. The final file consisted of the numbers 0 through 9 repeated in a purely random sequence (e.g., 2859632222021860260840955273643117), extended to fill a file of 10,000 bytes. Each file was compressed with gunzip, which uses the DEFLATE compression algorithm, combining LZ77 and Huffman coding.

The uncompressed files (10,000 bytes) were compressed into the following file sizes:

```
compressed file size: 58 bytes for 10,000 consecutive "1"
compressed file size: 75 bytes for 1,000 consecutive values of 0 through 9
compressed file size: 5,092 bytes for a random sequence of 10,000 digits
```

In the third file, which consisted of a random sequence of digits, a small compression was achieved simply through the conversion from ASCII to binary representation. In general, though, a purely random sequence cannot be compressed. A data analyst can compare the compressibility of data values, parameter by parameter, to determine which parameters might be expunged, at least during the preliminary analysis of a large, multi-dimensional data set.

When random data are not omitted from the data parameters the unwary analyst may actually develop predictive models and classifiers based entirely on noise. This can occur because clustering algorithms and predictive methods, including neural networks, will produce an outcome from random input data. It has been reported that some published diagnostic tests have been developed from random data [25]. [Glossary Classifier, Neural network]

Aside from eliminating redundant or random parameters, you might want to review the data and eliminate parameters that do not contribute in any useful way toward your analysis. For example, if you have the zip code for an individual, you will probably not need to retain the street address. If you have the radiologic diagnosis for a patient's chest X-ray, you might not need to retain the file containing the X-ray image unless you are conducting an image analysis project.

The process of reducing parameters applies to virtually all of the fields of data analysis, including standard statistical analysis. Names for this activity include feature reduction or selection, variable reduction and variable subset reduction, and attribute selection. There is sometimes a fine line between eliminating useless data parameters and cherry-picking your test set. It is important to document the data attributes you have expunged and your reason for doing so. Your colleagues should be given the opportunity of reviewing all of your data, including the expunged parameters. [Glossary Cherry-picking, Second trial bias]

An example of a data elimination method is found in the Apriori algorithm. At its simplest, it expresses the observation that a collection of items cannot occur frequently unless each item in the collection also occurs frequently. To understand the algorithm and its significance, consider the items placed together in a grocery checkout cart. If the most popular combination of purchase items is a sack of flour, a stick of butter, and a quart of milk, then you can be certain that collections of each of these items individually, and all pairs of items from the list of 3, must also occur frequently. In fact, they must occur at least as often as the combination of all three, because each of these smaller combinations are subsets of the larger set and will occur with the frequency of the larger set plus the frequency of their occurrences in any other item sets. The importance of the apriori algorithm to Big Data relates to data reduction. If the goal of the analysis is to find association rules for frequently occurring combinations of items, then you can restrict your analysis to combinations composed of single items that occur frequently [26,20].

After a reduced data set has been collected, it is often useful to transform the data by any of a variety of methods that enhance our ability to find trends, patterns, clusters or relational properties that might be computationally invisible in the untransformed data set. The first step is data normalization, described in the next section. It is critical that data be expressed in a comparable form and measure. After the data is normalized, you can further reduce your data by advanced transformative methods.

As a final caveat, data analysts should be prepared to learn that there is never any guarantee that a collection of data will be helpful, even if it meets every criterion for accuracy and reproducibility. Sometimes the data you have is not the data you need. Data analysts should be aware that advanced analytic methods may produce a result that does not take you any closer to a meaningful answer. The data analyst must understand that there is an important difference between a result and an answer. [Glossary Support vector machine]

Section 10.6. Understanding Your Control

The purpose of computing is insight, not numbers.

Richard Hamming

In the small data realm the concept of "control" is easily defined and grasped. Typically, a group is divided into treatment and control sub-groups. Heterogeneity in the population (e.g., gender, age, health status) is randomly distributed into both groups, so that the treatment and the control subgroups are, to the extent possible, indistinguishable from one another. If the treatment group receives a drug administered by syringe suspended in a saline solution, then the control group might receive an injection of saline solution by syringe, without the drug. The idea is to control the experimental groups so that they are identical in every way, save for one isolated factor. Measurable differences in the control and the treatment groups that arise after treatment are potentially due to the action of the one treatment factor.

The concept of "control" does not strictly apply to Big Data; the data analyst never actually "controls" the data. We resign ourselves to doing our best with the "uncontrolled" data that is provided. In the absence of controlling an experiment, what can the data analyst do to exert some kind of data selection that simulates a controlled experiment? It often comes down to extracting two populations, from the Big Data resource, that are alike in every respect, but one: the treatment.

Let me relate a hypothetical situation that illustrates the special skills that Big Data analysts must master. An analyst is charged with developing a method for distinguishing endometriosis from non-diseased (control) tissue using gene expression data. By way of background, endometriosis is a gynecologic condition wherein endometrial tissue that is usually confined to the endometrium (the tissue that lines the inside cavity of the uterus) is found growing outside the uterus, on the surfaces of the ovaries, pelvis, and other organs found in the pelvis. He finds a public data collection that provides gene expression data on endometriosis tissue (five samples) and on control tissues (five samples). By comparing the endometriosis samples with the control samples, he finds a set of 1000 genes that are biomarkers for endometriosis (i.e., that have "significantly" different expression in the disease samples compared with the control samples).

Let us set aside the natural skepticism reserved for studies that generate 1000 new biomarkers from an analysis of 10 tissue samples. The analyst is asked the question, "What was your control tissue, and how was it selected and prepared for analysis?" The analyst indicates that he does not know anything about the control tissues. He points out that the selection and preparation of control tissues is a pre-analytic task (i.e., outside the realm of influence of the data analyst). In this case, the choice of the control tissue was not at all obvious. If the control tissue were non-uterine tissue, taken from the area immediately adjacent to the area from which the endometriosis was sampled, then the analysis would have been comparing endometriosis with the normal tissue that covers the surface of pelvic organs (i.e., a mixture of various types of connective tissue cells unlike endometrial cells). If the control consisted of samples of normal endometriotic tissue (i.e., the epithelium lining the endometrial canal), then the analysis would have been comparing endometriosis with its normal counterpart. In either case, the significance and rationale for the study would have been very different, depending on the choice of controls.

In this case, as in every case, the choice and preparation of the control is of the utmost importance to the analysis that will follow. In a "small data" controlled study, every system variable but one, the variable studied in the experiment, is "frozen"; an experimental luxury lacking in Big Data. The Big Data analyst must somehow invent a plausible control from the available data. This means that the data analyst, and his close co-workers, must delve into the details of data preparation and have a profound understanding of the kinds of questions that the data can answer. Finding the most sensible control and treatment groups from a Big Data resource can require a particular type of analytic mind that has been trained to cope with data drawn from many different scientific disciplines.

Section 10.7. Statistical Significance Without Practical Significance

The most savage controversies are those about matters as to which there is no good evidence either way.

Bertrand Russell

Big Data provides statistical significance without necessarily providing any practical significance. Here is an example. Suppose you have two populations of people and you suspect that the adult males in the first population are taller than the second population. To test your hypothesis, you measure the heights of a random sampling (100 subjects) from both groups. You find that the average height of group 1 is 172.7 cm, while the average height of the second group is 172.5 cm. You calculate the standard error of the mean (the standard deviation divided by the square root of the number of subjects in the sampled population), and you use this statistic to determine the range in which the mean is expected to fall. You find that the difference in the average height in the two sampled populations is not significant, and you cannot exclude the null hypothesis (i.e., that the two sampled groups are equivalent, height-wise).

This outcome really bugs you! You have demonstrated a 2 mm difference in the average heights of the two groups, but the statistical tests do not seem to care. You decide to up the ante. You use a sampling of one million individuals from the two populations and recalculate the averages and the standard errors of the means. This time, you get a slightly smaller difference in the heights (172.65 for group 1 and 172.51 for group 2). When you calculate the standard error of the mean for each population, you find a much smaller number, because you are dividing the standard deviation by the square root of one million (i.e., one thousand); not by the square root of 100 (i.e., 10) that you used for the first calculation. The confidence interval for the ranges of the averages is much smaller now, and you find that the differences in heights between group 1 and group 2 are sufficient to exclude the null hypothesis with reasonable confidence.

Your Big Data project was a stunning success; you have shown that group 1 is taller than group 2, with reasonable confidence. However, the average difference in their heights seems to be about a millimeter. There are no real life situations where a difference of this small magnitude would have any practical significance. You could not use height to distinguish individual members of group 1 from individual members of group 2; there is too much overlap among the groups, and height cannot be accurately measured to within a millimeter tolerance. You have used Big Data to achieve statistical significance, without any practical significance.

There is a tendency among Big Data enthusiasts to promote large data sets as a cure for the limited statistical power and frequent irreproducibility of small data studies. In general, if an effect is large, it can be evaluated in a small data project. If an effect is too small to confirm in a small data study, statistical analysis may benefit from a Big Data study, by increasing the sample size and reducing variances. Nonetheless, the final results may have no practical significance, or the results may be unrepeatable in a small-scale (i.e., real life) setting, or may be invalidated due to the persistence of biases that were not eliminated when the sample size was increased.

Section 10.8. Case Study: Gene Counting

There is a chasm
of carbon and silicon
the software can't bridge.

Computer-inspired haiku by Rahul Sonnad

The Human Genome Project is a massive bioinformatics project in which multiple laboratories helped to sequence the 3 billion base pair haploid human genome. The project began its work in 1990, a draft human genome was prepared in 2000, and a completed genome was finished in 2003, marking the start of the so-called post-genomics era. There are about 2 million species of proteins synthesized by human cells. If every protein had its own private gene containing its specific genetic code, then there would be about two million protein-coding genes contained in the human genome. As it turns out, this estimate is completely erroneous. Analysis of the human genome indicates that there are somewhere between 20,000 and 150,000 genes. The majority of estimates come in at the low end (about 25,000 genes). Why are the current estimates so much lower than the number of proteins, and why is there such a large variation in the lower and upper estimates (20,000 to 150,000)? [Glossary Human Genome Project]

Counting is difficult when you do not fully understand the object that you are counting. The reason that you are counting objects is to learn more about the object, but you cannot fully understand an object until you have learned what you need to know about the object. Perceived this way, counting is a bootstrapping problem. In the case of proteins a small number of genes can account for a much larger number of protein species because proteins can be assembled from combinations of genes, and the final form of a unique protein can be modified by so-called post-translational events (folding variations, chemical modifications, sequence shortening, clustering by fragments, etc.). The methods used to count protein-coding genes can vary [27]. One technique might look for sequences that mark the beginning and the end of a coding sequence; another method might look for segments containing base triplets that correspond to amino acid codons. The former method might count genes that code for cellular components other than proteins, and the later might miss fragments whose triplet sequences do not match known protein sequences [28]. Improved counting methods are being developed to replace the older methods, but a final number evades our grasp.

The take-home lesson is that the most sophisticated and successful Big Data projects can be stumped by the simple act of counting.

Section 10.9. Case Study: Early Biometrics, and the Significance of Narrow Data Ranges

The proper study of Mankind is Man.

Alexander Pope in "An Essay on Man," 1734.

It is difficult to determine the moment in history when we seriously began collecting biometric data. Perhaps it started with the invention of the stethoscope. Rene-Theophile-Hyacinthe Laennec (1781–1826) is credited with inventing this device, which provided us with the opportunity to listen to the sounds generated within our bodies. Laennec's 1816 invention was soon followed by his 900-page analysis of sounds, heard in health and disease, the Traite de l'Aascultation Mediate (1819). Laennec's meticulous observations were an early effort in Big Data medical science. A few decades later, in 1854, Karl Vierordt's 1854 sphygmograph was employed to routinely monitor the pulse of patients. Perhaps the first large monitoring project came in 1868 when Carl Wunderlich published Das Verhalten der Eigenwarme in Krankheiten, which collected body temperature data on approximately 25,000 patients [29]. Wunderlich associated peaks and fluctuations of body temperature with 32 different diseases. Not only did this work result in a large collection of patient data, it also sparked considerable debate over the best way to visualize datasets. Competing suggestions for the representation of thermometric data (as it was called) included time interval (discontinuous) graphs and oscillating realtime (continuous) charts. Soon thereafter, sphygmomanometry (blood pressure recordings) was invented (1896). With bedside recordings of pulse, blood pressure, respirations, and temperature (the so-called vital signs), the foundations of modern medical data collection were laid.

At the same time that surveillance of vital signs became commonplace, a vast array of chemical assays of blood and body fluids were being developed. By the third decade of the twentieth century, physicians had at their disposal most of the common blood tests known to modern medicine (e.g., electrolytes, blood cells, lipids, glucose, nitrogenous compounds). What the early twentieth century physicians lacked was any sensible way to interpret the test results. Learning how to interpret blood tests required examination of old data collected on many thousands of individuals, and it took considerable time and effort to understand the aggregated results.

The results of blood tests, measured under a wide range of physiologic and pathologic circumstances, produced a stunning conclusion. It was shown that nearly every blood test conducted on healthy individuals fell into a very narrow range, with very little change between individuals. This was particularly true for electrolytes (e.g., Sodium and Calcium) and to a somewhat lesser extent for blood cells (e.g., white blood cells, red blood cells). Furthermore, for any individual, multiple recordings at different times of the day and on different days, tended to produce consistent results (e.g., Sodium concentration in the morning was equivalent to Sodium concentration in the evening). These finding were totally unexpected at the time [30].

Analysis of the data also showed that significant deviations from the normal concentrations of any one of these blood chemicals is always an indicator of disease. Backed by data, but lacking any deep understanding of the physiologic role of blood components, physicians learned to associate deviations from the normal range with specific disease processes. The discovery of the "normal range" revolutionized the field of physiology. Thereafter physiologists concentrated their efforts toward understanding how the body regulates its blood constituents. These early studies led to nearly everything we now know about homeostatic control mechanisms, and the diseases thereof.

To this day, much of medicine consists of monitoring vital signs, blood chemistries, and hematologic cell indices (i.e., the so-called complete blood count), and seeking to find a cause and a remedy for deviations from the normal range.

Glossary

Cherry-picking The process whereby data objects are chosen for some quality that is intended to boost the likelihood that an experiment is successful, but which biases the study. For example, a clinical trial manager might prefer patients who seem intelligent and dependable, and thus more likely to comply with the rigors of a long and complex treatment plan. By picking those trial candidates with a set of desirable attributes, the data manager is biasing the results of the trial, which may no longer apply to a real-world patient population.

Classifier As used herein, refers to algorithms that assign a class (from a pre-existing classification) to an object whose class is unknown [26]. It is unfortunate that the term classifier, as used by data scientists, is often misapplied to the practice of classifying, in the context of building a classification. Classifier algorithms cannot be used to build a classification, as they assign class membership by similarity to other members of the class; not by relationships. For example, a classifier algorithm might assign a terrier to the same class as a housecat, because both animals have many phenotypic features in common (e.g., similar size and weight, presence of a furry tail, four legs, tendency to snuggle in a lap). A terrier is dissimilar to a wolf, and a housecat is dissimilar to a lion, but the terrier and the wolf are directly related to one another; as are the housecat and the lion. **For the purposes of creating a**

classification, relationships are all that are important. Similarities, when they occur, arise as a consequence of relationships; not the other way around. At best, classifier algorithms provide a clue to classification, by sorting objects into groups that may contain related individuals. Like clustering techniques, classifier algorithms are computationally intensive when the dimension is high, and can produce misleading results when the attributes are noisy (i.e., contain randomly distributed attribute values) or non-informative (i.e., unrelated to correct class assignment).

Human Genome Project The Human Genome Project is a massive bioinformatics project in which multiple laboratories contributed to sequencing the 3 billion base pair haploid human genome (i.e., the full sequence of human DNA). The project began its work in 1990, a draft human genome was prepared in 2000, and a completed genome was finished in 2003, marking the start of the so-called post-genomics era. All of the data produced for the Human Genome Project is freely available to the public.

Meta-analysis Meta-analysis involves combining data from multiple similar and comparable studies to produce a summary result. The hope is that by combining individual studies, the meta-analysis will carry greater credibility and accuracy than any single study. Three of the most recurring flaws in meta-analysis studies are selection bias (e.g., negative studies are often omitted from the literature), inadequate knowledge of the included sets of data (e.g., incomplete methods sections in the original articles), and non-representative data (e.g., when the published data are non-representative samples of the original data sets).

Neural network A dynamic system in which outputs are calculated by a summation of weighted functions operating on inputs. The weights for the individual functions are determined by a learning process, simulating the learning process hypothesized for human neurons. In the computer model, individual functions that contribute to a correct output (based on the training data) have their weights increased (strengthening their influence to the calculated output). Over the past ten or fifteen years, neural networks have lost some favor in the artificial intelligence community. They can become computationally complex for very large sets of multidimensional input data. More importantly, complex neural networks cannot be understood or explained by humans, endowing these systems with a "magical" quality that some scientists find unacceptable.

Second trial bias Can occur when a clinical trial yields a greatly improved outcome when it is repeated with a second group of subjects. In the medical field, second trial bias arises when trialists find subsets of patients from the first trial who do not respond well to treatment, thereby learning which clinical features are associated with poor trial response (e.g., certain pre-existing medical conditions, lack of a good home support system, obesity, nationality). During the accrual process for the second trial, potential subjects who profile as non-responders are excluded. Trialists may justify this practice by asserting that the purpose of the second trial is to find a set of subjects who will benefit from treatment. With a population enriched with good responders, the second trial may yield results that look much better than the first trial. Second trial bias can be considered a type of cherry-picking that is often justifiable.

Steghide Steghide is an open source GNU license utility that invisibly embeds data in image or audio files. Windows and Linux versions are available for download from SourceForge, at:

http://steghide.sourceforge.net/download.php

A Steghide manual is available at:

http://steghide.sourceforge.net/documentation/manpage.php

After installing, you can invoke steghide at the system prompt as a command line launched from the subdirectory in which steghide.exe resides.

Here is an example of a command line invocation of Steghide. Your chosen password can be inserted directly into the commandline. For example:

```
steghide embed -cf myphoto.jpg -ef mytext.txt -p hideme
```

The command line was launched from the subdirectory that holds the steghide executable files on my computer. The command instructs steghide to embed the text file, berman_author_bio.txt into the image file, berman_author_photo.jpg, under the password "hideme".

That is all there is to it. The image file, containing a photo of myself, now contains an embedded text file, containing my short biography. No longer need I keep track of both files. I can generate my biography file, from my image file, but I must remember the password.

I could have called Steghide from a script. Here is an example of an equivalent Python script that invokes steghide from a system call.

```
import os
command_string = "steghide embed -cf myphoto.jpg -ef mytext.txt -p hideme"
os.system(command_string)
```

You can see how powerful this method can be. With a bit of tweaking, you can write a short script that uses the Steghide utility to embed a hidden text message in thousands of images, all at once. Anyone viewing those images would have no idea that they contained a hidden message, unless and until you told them so.

Support vector machine A machine learning technique that classifies objects. The method starts with a training set consisting of two classes of objects as input. The support vector machine computes a hyperplane, in a multidimensional space, that separates objects of the two classes. The dimension of the hyperspace is determined by the number of dimensions or attributes associated with the objects. Additional objects (i.e., test set objects) are assigned membership in one class or the other, depending on which side of the hyperplane they reside.

References

[1] Bloom BJ, Nicholson TL, Williams JR, Campbell SL, Bishof M, Zhang X, et al. An optical lattice clock with accuracy and stability at the 10–18 level. Nature 2014;506:71–5.

[2] Berman J. Precision medicine, and the reinvention of human disease. Cambridge, MA: Academic Press; 2018.

[3] Fisher RA. The correlation between relatives on the supposition of Mendelian inheritance. Trans R Soc Edinb 1918;52:399–433.

[4] Ward LD, Kellis M. Interpreting noncoding genetic variation in complex traits and human disease. Nat Biotechnol 2012;30:1095–106.

[5] Visscher PM, McEvoy B, Yang J. From Galton to GWAS: quantitative genetics of human height. Genet Res 2010;92:371–9.

[6] Trithemius J. Steganographia (Secret Writing), by Johannes Trithemius. 1500.

[7] Berman JJ. Biomedical informatics. Sudbury, MA: Jones and Bartlett; 2007.

[8] Booker D, Berman JJ. Dangerous abbreviations. Hum Pathol 2004;35:529–31.

[9] Berman JJ. Pathology abbreviated: a long review of short terms. Arch Pathol Lab Med 2004;128:347–52.

[10] Patient Safety in American Hospitals. HealthGrades; 2004. July. Available from: http://www.healthgrades.com/media/english/pdf/hg_patient_safety_study_final.pdf [viewed September 9, 2012].

[11] Gordon R. Great medical disasters. New York: Dorset Press; 1986. p. 155–60.

[12] Vital Signs. Unintentional injury deaths among persons aged 0–19 years; United States, 2000–2009. In: Morbidity and Mortality Weekly Report (MMWR). vol. 61. Centers for Disease Control and Prevention; 2012. p. 1–7. April 16.

[13] Rigler T. DOD discloses new figures on Korean War dead. Army News Service; 2000. May 30.

[14] Frey CM, McMillen MM, Cowan CD, Horm JW, Kessler LG. Representativeness of the surveillance, epidemiology, and end results program data: recent trends in cancer mortality rate. JNCI 1992;84:872.

[15] Ashworth TG. Inadequacy of death certification: proposal for change. J Clin Pathol 1991;44:265.

[16] Kircher T, Anderson RE. Cause of death: proper completion of the death certificate. JAMA 1987;258:349–52.

[17] Walter SD, Birnie SE. Mapping mortality and morbidity patterns: an international comparison. Int J Epidemiol 1991;20:678–89.

[18] Berman JJ. Methods in medical informatics: fundamentals of healthcare programming in Perl, Python, and Ruby. Boca Raton: Chapman and Hall; 2010.

[19] Mitchell KJ, Becich MJ, Berman JJ, Chapman WW, Gilbertson J, Gupta D, et al. Implementation and evaluation of a negation tagger in a pipeline-based system for information extraction from pathology reports. Medinfo 2004;2004:663–7.

[20] Janert PK. Data Analysis with Open Source Tools. O'Reilly Media; 2010.

[21] Lewis PD. R for Medicine and Biology. Sudbury: Jones and Bartlett Publishers; 2009.

[22] Szekely GJ, Rizzo ML. Brownian distance covariance. Ann Appl Stat 2009;3:1236–65.

[23] Reshef DN, Reshef YA, Finucane HK, Grossman SR, McVean G, Turnbaugh PJ, et al. Detecting novel associations in large data sets. Science 2011;334:1518–24.

[24] Marsaglia G, Tsang WW. Some difficult-to-pass tests of randomness. J Stat Softw 2002;7:1–8.

[25] Venet D, Dumont JE, Detours V. Most random gene expression signatures are significantly associated with breast cancer outcome. PLoS Comput Biol 2011;7:e1002240.

[26] Wu X, Kumar V, Quinlan JR, Ghosh J, Yang Q, Motoda H, et al. Top 10 algorithms in data mining. Knowl Inf Syst 2008;14:1–37.

[27] Pennisi E. Gene counters struggle to get the right answer. Science 2003;301:1040–1.

[28] How Many Genes Are in the Human Genome? Human Genome Project Information. Available from: http://www.ornl.gov/sci/techresources/Human_Genome/faq/genenumber.shtml [viewed June 10, 2012].

[29] Wunderlich CR. Das Verhalten der Eigenwarme in Krankheiten (The behavior of the self-warmth in diseases). Leipzig: O. Wigand; 1868.

[30] Berman JJ. Repurposing legacy data: innovative case studies. Waltham, MA: Morgan Kaufmann; 2015.

11

Indispensable Tips for Fast and Simple Big Data Analysis

Section 11.1. Speed and Scalability

It's hardware that makes a machine fast. It's software that makes a fast machine slow.

Craig Bruce

Speed and scalability are the two issues that never seem to go away when discussions turn to Big Data. Will we be able to collect, organize, search, retrieve, and analyze Big Data at the same speed that we have grown accustomed to in small data systems? Will the same algorithms, software, protocols, and systems that work well with small data scale up to Big Data?

Let us turn the question around for a moment. Will Big Data solutions scale down to provide reliable and fast solutions for small data? It may seem as though the answer is obvious. If a solution works for large data it must also work for small data. Actually, this is not the case. Methods that employ repeated sampling of a population may produce meaningless results when the population consists of a dozen data points. Nonsensical results can be expected when methods that determine trends, or analyze signals come up against small data.

The point here is that computational success is always an artifact of the way that we approach data-related issues. By customizing solutions to some particular set of circumstances (number of samples, number of attributes, required response time, required precision), we make our products ungeneralizable. We tend to individualize

Principles and Practice of Big Data. https://doi.org/10.1016/B978-0-12-815609-4.00011-X

our work, at first, because we fail to see the downside of idiosyncratic solutions. Hence, we learn the hard way that solutions that work well under one set of circumstances will fail miserably when the situation changes. Desperate to adapt to the new circumstances, we often pick solutions that are expensive and somewhat short-sighted (e.g., "Let's chuck all our desktop computers and buy a supercomputer."; "Let's parallelize our problems and distribute the calculations to multiple computers", "Let's forget about trying to understand the system and switch to deep learning with neural networks."; "Let's purchase a new and more powerful information system and abandon our legacy data.")

In this book, which offers a low-tech approach to Big Data, we have been stressing the advantages of simple and general concepts that allow data to be organized at any level of size and complexity (e.g., identifiers, metadata/data pairs, triples, triplestores), and extremely simple algorithms for searching, retrieving, and analyzing data that can be accomplished with a few lines of code in any programming language. The solutions discussed in this book may not be suitable for everyone, but it is highly likely that many of the difficulties associated with Big Data could be eliminated or ameliorated, if all data were well organized.

Aside from issues of data organization, here are a few specific suggestions for avoiding some of the obstacles that get in the way of computational speed and scalability:

– **High-level programming languages (including Python) employ built-in methods that fail when the variables are large, and such methods must be avoided by programmers.**

Modern programming languages relieve programmers from the tedium of allocating memory to every variable. The language environment is tailored to the ample memory capacities of desktop and laptop computers and provides data structures (e.g., lists, dictionaries, strings) that are intended to absorb whatever data they are provided. By doing so, two problems are created for Big Data users. First, the size of data may easily exceed the loading capacity of variables (e.g., don't try to put a terabyte of data into a string variable). Second, the built-in methods that work well with small data will fail when dealing with large, multidimensional data objects, such as enormous matrices. This means that good programmers may produce terrific applications, using high-level programming languages that fail miserably in the Big Data realm.

Furthermore, the equivalent methods, in different versions of a high-level programming language, may deal differently with memory. Successive versions of a programming language, such as Python or Perl, may be written with the tacit understanding that the memory capacity of computers is increasing and that methods can be speeded assuming that memory will be available, as needed. Consequently, a relatively slow method may work quite well for a large load of data in an early version of the language. That same method, in a later version of the same language, written to enhance speed, may choke on the same data. Because the programmer is using the same named methods in her programs, run under different version of the language, the task of finding the source of consequent software failures may be daunting.

What is the solution? Should Big Data software be written exclusively in assembly language, or (worse) machine language? Probably not, although it is easy to see why programmers skilled in low-level programming languages will always be valued for their expertise. It is best to approach Big Data programming with the understanding that methods are not infinitely expandable to accommodate any size data, of any dimension. Sampling methods, such as those discussed in detail in Chapter 13, "Using Random Numbers to Knock Your Big Data Analytic Problems Down to Size" might be one solution.

– **Line-by-line reading is slow, but is always scalable.**

Years ago, I did a bit of programming in Mumps, a programming language developed in the 1960s (that will be briefly discussed in Section 11.8 of this chapter). Every variable had a maximum string size of 255 bytes. Despite this limitation, Mumps managed enormous quantities of data in hierarchical data structure (the so-called Mumps global). Efficient and reliable, Mumps was loved by a cult of loyal programmers. Every programmer of a certain age can regale younger generations with stories of magnificent software written for computers whose RAM memory was kilobytes (not Gigabytes).

A lingering residue of Mumps-like parsing is the line-by-line file read. Every file, even a file of Big Data enormity, can be parsed from beginning to end by repeated line feeds. Programmers who resist the urge to read whole files into a variable will produce software that scales to any size, but which will run slower than comparable programs that rely on memory. There are many trade-offs in life.

– **Make your data persistent.**

In Section 11.5 of this chapter, "Methods for Data Persistence" we will be discussing methods whereby data structures can be moved from RAM memory into external files, which can be retrieved in whole or in part, as needed. Doing so relieves many of the consequences of memory overload, and eliminates the necessity of rebuilding data structures each time a program or a process is called to execute.

– **Don't test your software on subsets of data**

Programs that operate with complex sets of input may behave unpredictably. It is important to use software that has been extensively tested by yourself and by your colleagues. After testing, it is important to constantly monitor the performance of software. This is especially important when using new sources of data, which may contain data values that have been formatted differently than prior data.

Here is one solution that everyone tries. If the test takes a lot of processing time, just reduce the size of the test data. Then you can quickly go through many test/debug cycles. Uh uh. When testing software, you cannot use a small subset of your data because the kinds of glitches you need to detect may only be detectable in large datasets. So, if you want to test software that will be used in large datasets, you must test the software on large datasets. How testing can be done, on Big Data, without crashing the live system, is a delicate problem. Readers are urged to consult publications in the large corpus of literature devoted to software testing.

- **Avoid turn-key applications.**

Vendors may offer Big Data solutions in the form of turn-key applications and systems that do everything for you. Often, such systems are opaque to users. When difficulties arise, including system crashes, the users are dependent upon the vendor to provide a fix. When the vendor is unreliable, or the version of the product that you are using is no longer supported, or when the vendor has gone out of business, or when the vendor simply cannot understand and fix the problem, the consequences to the Big Data effort can be catastrophic.

Everyone has his or her own opinions about vendor-provided solutions. In some cases, it may be reasonable to begin Big Data projects with an extremely simple, open source, database that offers minimal functionality. If the data is simple but identified (e.g., collections of triples), then a modest database application may suffice. Additional functionalities can be added incrementally, as needed.

- **Avoid proprietary software (when conducting scientific research)**

Proprietary software applications have their place in the realm of Big Data. Such software is used to operate the servers, databases, and communications involved in building and maintaining the Big Data resource. They are often well tested and dependable, faithfully doing the job they were designed to do. However, in the analysis phase, it may be impossible to fully explain your research methods if one of the steps is: "Install the vendor's software and mouse-click on the 'Run' button." Responsible scientists should not base their conclusion on software they cannot understand. [Glossary Black box]

- **Use small, efficient, and fast utilities**

Utilities are written to perform one task optimally. For data analysts, utilities fit perfectly with the "filter" paradigm of Big Data (i.e., that the primary purpose of Big Data is to provide a comprehensive source of small data). A savvy data analyst will have hundreds of small utilities, most being free and open source products, that can be retrieved and deployed, as needed. A utility can be tested with data sets of increasing size and complexity. If the utility scales to the job, then it can be plugged into the project. Otherwise, it can be replaced with an alternate utility or modified to suit your needs. [Glossary Undifferentiated software]

- **Avoid system calls from within iterative loops**

Many Big Data programs perform iterative loops operating on each of the elements of a large list, reading large text files line by line, or calling every key in a dictionary. Within these long loops, programmers must exercise the highest degree of parsimony, avoiding any steps that may unnecessarily delay the execution of the script, inasmuch as any delay will be multiplied by the number of iterations in the loop. System calls to external methods or utilities are always time consuming. In addition to the time spent executing the command, there is also the time spent loading and interpreting called methods, and this time is repeated at each iteration of the loop. [Glossary System call]

To demonstrate the point, let us do a little experiment. As discussed in Section 3.3, when we create new object identifiers with UUID, we have the choice of calling the Unix UUID method as a system call from a Python script; or, we can use Python's own uuid method. We will run two versions of a script. The first version will create 10,000 new UUID identifiers, using system calls to the external unix utility, uuid.gen. We will create another 10,000 UUID identifiers with Perl's own uuid method. We will keep time of how long each script runs.

Here is the script using system calls to uuidgen.exe

```
timenow = time.time()
for i in range(1, 10000):
    os.system("uuidgen.exe >uuid.out")
timenew = time.time()
print("Time for 10,000 uuid assignments: " + str(timenew - timenow))

output:
Time for 10,000 uuid assignments: 422.0238435268402
```

10,000 system calls to uuidgen.exe required 422 seconds to complete.

Here is the equivalent script using Python's built-in uuid method.

```
timenow = time.time()
for i in range(1, 10000):
    uuid.uuid4()
timenew = time.time()
print("Time for 10,000 uuid assignments: " + str(timenew - timenow))

output:
Time for 10,000 uuid assignments: 0.06850886344909668
```

Python's built-in method required 0.07 seconds to complete, a dramatic time savings.

– **Use look-up tables, and other pre-computed pointers**

Computers are very fast at retrieving information from look-up tables, and these would include concordances, indexes, color maps (for images), and even dictionary objects (known also as associative arrays). For example, the Google search engine uses a look-up table built upon the PageRank algorithm. PageRank (alternate form Page Rank) is a method, developed at Google, for displaying an ordered set of results (for a phrase search conducted over every page of the Web). The rank of a page is determined by two scores: the relevancy of the page to the query phrase; and the importance of the page. The relevancy of the page is determined by factors such as how closely the page matches the query phrase, and whether the content of the page is focused on the subject of the query. The importance of the page is determined by how many Web pages link to and from the page, and the importance of the Web pages involved in the linkages. It is easy to see that the methods for scoring relevance and importance are subject to many algorithmic variances, particularly with respect to the choice of measures (i.e., the way in which a page's focus on a particular topic is quantified), and the weights applied to each measurement. The reasons that PageRank query responses can be completed very rapidly is that the score of a page's importance can be pre-computed, and stored with the page's Web addresses. Word matches from the query phrase to Web pages are quickly assembled using a precomputed index of words, the pages containing the words, and the locations of the words in the pages [1]. [Glossary Associative array]

– **Avoid RegEx, especially in iterative processes**

RegEx (short for Regular Expressions) is a language for describing string patterns. The RegEx language is used in virtually every modern programming language to describe search, find, and substitution operations on strings. Most programmers love RegEx, especially those programmers who have mastered its many subtleties. There is a strong tendency to get carried away by the power and speed of Regex. I have personally reviewed software in which hundreds of RegEx operations are performed on every line read from files. In one such program the software managed to parse through text at the numbingly slow rate of 1000 bytes (about a paragraph) every 4 seconds. As this rate a terabyte of data would require a 4 billion seconds to parse (somewhat more than one century). In this particular case, I developed an alternate program that used a fast look-up table and did not rely upon RegEx filters. My program ran at a speed 1000 times faster than the RegEx intense program [2]. [Glossary RegEx]

– **Avoid unpredictable software.**

Everyone thinks of software as something that functions in a predetermined manner, as specified by the instructions in its code. It may seem odd to learn that software output may be unpredictable. It is easiest to understand the unpredictability of software when we examine how instructions are followed in software that employs class libraries (C++, Java) or that employs some features of object-oriented languages (Python) or is fully object oriented (Smalltalk, Ruby, Eiffel) [3]. When a method (e.g., an instruction to perform a function) is sent to an object, the object checks to see if the method belongs to itself (i.e., if the method is an instance method for the object). If not, it checks to see if the method belongs to its class (i.e., if the method is a class method for the object's class). If not, it checks its through the lineage of ancestral classes, searching for the method. When classes are per-mitted to have more than one parent class, there will be occasions when a named method exists in more than one ancestral class, for more than one ancestral lineage. In these cases, we cannot predict with any certainty which class method will be chosen to fulfill the method call. Depending on the object receiving the method call, and its particular ancestral line-ages, and the route taken to explore the lineages, the operation and output of the software will change.

In the realm of Big Data, you do not need to work in an object oriented environment to suffer the consequences of method ambiguity. An instruction can be sent over a network of servers as an RPC (Remote Procedure Call) that is executed in a different ways by the various servers that receive the call.

Unpredictability is often the very worst kind of programming bug because incorrect outputs producing adverse outcomes often go undetected. In cases where an adverse outcome is detected, it may be nearly impossible to find the glitch.

- **Avoid combinatorics.**

Much of Big Data analytics involves combinatorics; the evaluation, on some numeric level, of combinations of things. Often, Big Data combinatorics involves pairwise comparisons of all possible combinations of data objects, searching for similarities, or proximity (a distance measure) of pairs. The goal of these comparisons often involves clustering data into similar groups, finding relationships among data that will lead to classifying the data objects, or predicting how data objects will respond or change under a particular set of conditions. When the number of comparisons becomes large, as is the case with virtually all combinatoric problems involving Big Data, the computational effort may become massive. For this reason, combinatorics research has become somewhat of a subspecialty for Big Data mathematics. There are four "hot" areas in combinatorics. The first involves building increasingly powerful computers capable of solving combinatoric problems for Big Data. The second involves developing methods whereby combinatoric problems can be broken into smaller problems that can be distributed to many computers, to provide relatively fast solutions for problems that could not otherwise be solved in any reasonable length of time. The third area of research involves developing new algorithms for solving combinatoric problems quickly and efficiently. The fourth area, perhaps the most promising, involves finding innovative non-combinatoric solutions for traditionally combinatoric problems.

- **Pay for smart speed**

The Cleveland Clinic developed software that predicts disease survival outcomes from a large number of genetic variables. Unfortunately the time required for these computations was unacceptable. As a result, the Cleveland Clinic issued a Challenge "to deliver an efficient computer program that predicts cancer survival outcomes with accuracy equal or better than the reference algorithm, including 10-fold validation, in less than 15 hours of real world (wall clock) time" [4]. The Cleveland Clinic had its own working algorithm, but it was not scalable to the number of variables analyzed. The Clinic was willing to pay for faster service.

Section 11.2. Fast Operations, Suitable for Big Data, That Every Computer Supports

> *No one will need more than 637 kb of memory for a personal computer. 640K ought to be enough for anybody.*
>
> Bill Gates, founder of Microsoft Corporation, in 1981

- Random access to files

Most programming languages have a way of providing so-called random access to file locations. This means that if you want to retrieve the 2053th line of a file, you need not sequentially read lines 1 through 2052 before reaching your desired line. You can go directly to any line in the text, or any byte location, for that matter.

In Python, so-called random access to file locations is invoked by the seek() command. Here is a 9-line Python script that randomly selects twenty locations in a text file (the plain-text version of James Joyce's Ulysses in this example) for file access.

```
import os, sys, itertools, random
size = os.path.getsize("ulysses.txt")
infile = open("ulysses.txt", "r")
random_location = "0"
for i in range(20):
    random_location = random.uniform(0,size)
    infile.seek(random_location, 0)
    print(infile.readline(), end="\n")
infile.close()
```

Random access to files is a gift to programmers. When we have indexes, concordances, and other types of lookup tables, we can jump to the file locations we need, nearly instantaneously.

– Addition and multiplication

Some mathematical operations are easier than others. Addition and multiplication and the bitwise logic operations (e.g., XOR) are done so quickly that programmers can include these operations liberally in programs that loop through huge data structures.

– Time stamps

Computers have an intimate relationship with time. As discussed in Section 6.4, every computer has several different internal clocks that set the tempo for the processor, the motherboard, and for software operations. A so-called real-time clock (also known as system clock) knows the time internally as the number of seconds since the epoch. By convention, in Unix systems, the epoch begins at midnight on New Year's Eve, 1970. Prior dates are provided a negative time value. On most systems, the time is automatically updated 50–100 times per second, providing us with an extremely precise way of measuring the time between events.

Never hesitate to use built-in time functions to determine the time of events and to determine the intervals between times. Many important data analysis opportunities have been lost simply because the programmers who prepared the data neglected to annotate the times that the data was obtained, created, updated, or otherwise modified.

– One-way hashes

One-way hashes were discussed in Section 3.9. One-way hash algorithms have many different uses in the realm of Big Data, particularly in areas of data authentication and security. In some applications, one-way hashes are called iteratively, as in blockchains (Section 8.6) and in protocol for exchanging confidential information [5,6]. As previously discussed, various hash algorithms can be invoked via system calls from Python scripts to the openssl suite of data security algorithms. With few exceptions, a call to methods from within the running programming environment, is much faster than a system call to an external program. Python provides a suite of one-way hash algorithms in the hashlib module.

```
>>> import hashlib
>>> hashlib.algorithms_available
{'sha', 'SHA', 'SHA256', 'sha512', 'ecdsa-with-SHA1', 'sha256',
'whirlpool', 'sha1', 'RIPEMD160', 'SHA224', 'dsaEncryption',
'dsaWithSHA', 'sha384', 'SHA384', 'DSA-SHA', 'MD4', 'ripemd160',
'DSA', 'SHA512', 'md4', 'sha224', 'MD5', 'md5', 'SHA1'}
```

The python zlib module also provides some one-way hash functions, including adler32, with extremely fast algorithms, producing a short string output.

The following Python command lines imports Python's zlib module and calls the adler32 hash, producing a short one-way hash for "hello world." The bottom two command lines imports sha256 from Python's hashlib module and produces a much longer hash value.

```
>>> import zlib
>>> zlib.adler32("hello world".encode('utf-8'))
436929629
>>> import hashlib
>>> hashlib.sha256(b"hello world").hexdigest()
'b94d27b9934d3e08a52e52d7da7dabfac484efe37a5380ee9088f7
ace2efcde9'
```

Is there any difference in the execution time when we compare the adler32 and sha256 algorithms. Let's find out with the hash_speed.py script that repeats 10,000 one-way hash operations with each one-way hash algorithms, testing both algorithms on a short phrase ("hello world") and a long file (meshword.txt, in this example, which happens to be 1,901,912 bytes in length).

```
import time, zlib, hashlib
timenow = time.time()
for i in range(1, 10000):
    zlib.adler32("hello world".encode('utf-8'))
timenew = time.time()
print("Time for 10,000 adler32 hashes on a short string: " + str(timenew -
timenow))
timenow = time.time()
for i in range(1, 10000):
    hashlib.sha256(b"hello world").hexdigest()
timenew = time.time()
print("Time for 10,000 sha256 hashes on a short string: " + str(timenew -
timenow))
with open('meshword.txt', 'r') as myfile:
```

```
    data=myfile.read()
timenow = time.time()
for i in range(1, 10000):
   zlib.adler32(data.encode('utf-8'))
timenew = time.time()
print("Time for 10,000 adler32 hashes on a file: " + str(timenew -
timenow))
timenow = time.time()
for i in range(1, 10000):
   hashlib.sha256(data.encode('utf-8')).hexdigest()
timenew = time.time()
print("Time for 10,000 sha256 hashes on a file: " + str(timenew -
timenow))
```

Here is the output of the hash_speed.py script

```
Time for 10,000 adler32 hashes on a short string: 0.006000041961669922
Time for 10,000 sha256 hashes on a short string: 0.014998674392700195
Time for 10,000 adler32 hashes on a file: 20.740180492401123
Time for 10,000 sha256 hashes on a file: 76.80237746238708
```

Both adler32 and SHA256 took much longer (several thousand times longer) to hash a 2 Megabyte file than a short, 11-character string. This indicates that one-way hashes can be performed on individual identifiers and triples, at high speed.

The adler32 hash is several times faster than sha256. This difference may be insignificant under most circumstances, but would be of considerable importance in operations that repeat millions or billions of times. The adler 32 hash is less secure against attack than the sha256, and has a higher chance of collisions. Hence, the adler32 hash may be useful for projects where security and confidentiality are not at issue, and where speed is required. Otherwise, a strong hashing algorithm, such as sha256, is recommended.

– Pseudorandom number generators are fast.

As will be discussed in Chapter 13, "Using Random Numbers to Knock Your Big Data Analytic Problems Down to Size," random number generators have many uses in Big Data analyses. [Glossary Pseudorandom number generator]

Let us look at the time required to compute 10 million random numbers.

```
import random, time
timenow = time.time()
for iterations in range(10000000):
   random.uniform(0,1)
timenew = time.time()
print("Time for 10 million random numbers: " + str(timenew - timenow))
```

Here is the output from the ten_million_rand.py script:

```
output:
Time for 10 million random numbers: 6.093759775161743
```

Ten million random numbers were generated in just over 6 seconds, on my refurbished home desktop running at a CPU speed of 3.40 GHz. This tells us that, under most circumstances, the time required to generate random numbers will not be a limiting factor, even when we need to generate millions of numbers. Next, we need to know whether the random numbers we generate are truly random. Alas, it is impossible for computers to produce an endless collection of truly random numbers. Eventually, algorithms will cycle through their available variations and begin to repeat themselves, producing the same set of "random" numbers, in the same order; a phenomenon referred to as the generator's period. Because algorithms that produce seemingly random numbers are imperfect, they are known as pseudorandom number generators. The Mersenne Twister algorithm, which has an extremely long period, is used as the default random number generator in Python. This algorithm performs well on most of the tests that mathematicians have devised to test randomness [7].

– Be aware that calls to external cryptographic programs may slow your scripts.

In Section 18.3, we will be discussing cryptographic protocols. For now, suffice it to say that encryption protocols can be invoked from Python scripts with a system call to the openssl toolkit. Let us look at aes128, a strong encryption standard used by the United States government. We will see how long it takes to encrypt a nearly two megabyte file, 10,000 times, with the Python crypt_speed.py script. [Glossary AES]

```
#!/usr/bin/python
import time, os
os.chdir("c:/cygwin64/bin/")
timenow = time.time()
for i in range(1, 10000):
   os.system("openssl.exe aes128 -in c:/ftp/py/meshword.txt -out
meshword.aes -pass pass:12345")
timenew = time.time()
print("Time for 10,000 aes128 encryptions on a long file:
" + str(timenew - timenow))
exit
```

```
outputtp\py>crypt_speed.py
Time for 10,000 aes128 encryptions on a long file: 499.7346260547638
```

We see that it would take take about 500 seconds to encrypt a file 10,000 times (or 0.05 seconds per encryption); glacial in comparison to other functions (e.g., random number, time). Is it faster to encrypt a small file than a large file? Let us repeat the process, using the 11 byte helloworld.txt file.

```
import time, os
os.chdir("c:/cygwin64/bin/")
timenow = time.time()
for i in range(1, 10000):
   os.system("openssl.exe aes128 -in c:/ftp/py/helloworld.txt -out
meshword.aes -pass pass:12345")
timenew = time.time()
print("Time for 10,000 aes128 encryptions on a short file:
" + str(timenew - timenow))

c:\ftp\py>crypt_speed.py
Time for 10,000 aes128 encryptions on a short file: 411.52050709724426
```

Short files encrypt faster than longer files, but the savings is not great. As we have seen, calling an external program from within Python is always a time-consuming process, and we can presume that most of the loop time was devoted to finding the openssl.exe program, interpreting the entire program and returning a value."

– **Do not insist on precision when there is no practical value in precise answers.**

Approximation methods are often orders of magnitude faster than exact methods. Furthermore, algorithms that produce inexact answers are preferable to exact algorithms that crash under the load of a gigabyte of data. Real world data is never exact, so why must we pretend that we need exact solutions?

– **Write you scripts in such a way that calculations are completed in one pass through the data.**

Programmers commonly write iterative loops through their data, calculating some particular component of an equation, only to repeat the loop to calculate another piece of the puzzle. As an example, consider the common task of calculating the variance (square of the standard deviation) of a population. The typical algorithm involves calculating the population mean, by summing all the data values in the population and dividing by the number of values summed. After the population mean is calculated, the variance is obtained by a second pass through the population and applying the formula below (Fig. 11.1):

$$\sigma^2 = \frac{1}{N}\sum_{i=1}^{N}(x_i - \mu)^2$$

FIG. 11.1 Calculating the variance (square of standard deviation) after first calculating the population mean.

The variance can be calculated in a fast, single pass through the population, using the equivalent formula, below, which does not involve precalculating the population mean (Fig. 11.2).

$$\sigma^2 = \tfrac{1}{n}\sum x_i^2 - (\tfrac{1}{n}\sum x_i)^2$$

FIG. 11.2 Calculating the variance, in a single pass through the values of a population [8].

By using the one-pass equation, after any number of data values have been processed, the running values of the average and variance can be easily determined; a handy trick, especially applicable to signal processing [8,9].

Section 11.3. The Dot Product, a Simple and Fast Correlation Method

Our similarities are different.

Yogi Berra

Similarity scores are based on comparing one data object with another, attribute by attribute. Two similar variables will rise and fall together. A score can be calculated by summing the squares of the differences in magnitude for each attribute, and using the calculation to compute a final outcome, known as the correlation score. One of the most popular correlation methods is Pearson's correlation, which produces a score that can vary from −1 to +1. Two objects with a high score (near +1) are highly similar [10]. Two uncorrelated objects would have a Pearson score near zero. Two objects that correlated inversely (i.e., one falling when the other rises) would have a Pearson score near −1. [Glossary Correlation distance, Normalized compression distance, Mahalanobis distance]

The Pearson correlation for two objects, with paired attributes, sums the product of their differences from their object means and divides the sum by the product of the squared differences from the object means (Fig. 11.3).

Python's Scipy module offers a Pearson function. In addition to computing Pearson's correlation, the scipy function produces a two-tailed *P*-value, which provides some indication of the likelihood that two totally uncorrelated objects might produce a Pearson's correlation value as extreme as the calculated value. [Glossary *P* value, Scipy]

Let us look at a short python script, sci_pearson.py, that calculates the Pearson correlation on two lists.

$$r = \frac{\sum_{i=1}^{n}(x_i - \bar{x})(y_i - \bar{y})}{\sqrt{\sum_{i=1}^{n}(x_i - \bar{x})^2}\sqrt{\sum_{i=1}^{n}(y_i - \bar{y})^2}}$$

FIG. 11.3 Formula for Pearson's correlation, for two data objects, with paired sets of attributes, *x* and *y*.

```
from scipy.stats.stats import pearsonr
a = [1, 2, 3, 4]
b = [2, 4, 6, 8]
c = [1,4,6,9,15,55,62,-5]
d = [-2,-8,-9,-12,-80,14,15,2]
print("Correlation a with b: " + str(pearsonr(a,b)))
print("Correlation c with d: " + str(pearsonr(c,d)))
```

Here is the output of pearson.py

```
Correlation a with b: (1.0, 0.0)
Correlation c with d: (0.32893766587262174, 0.42628658412101167)
```

The Pearson correlation of a with b is 1 because the values of b are simply double the values of a; hence the values in a and b correlate perfectly with one another. The second number, "0.0", is the calculated P value.

In the case of c correlated with d, the Pearson correlation, 0.329, is intermediate between 0 and 1, indicating some correlation. How does the Pearson correlation help us to simplify and reduce data? If two lists of data have a Pearson correlation of 1 or of -1, this implies that one set of the data is redundant. We can assume the two lists have the same information content. If we were comparing two sets of data and found a Pearson correlation of zero, then we might assume that the two sets of data were uncorrelated, and that it would be futile to try to model (i.e., find a mathematical relationship for) the data. [Glossary Overfitting]

There are many different correlation measurements, and all of them are based on assumptions about how well-correlated sets of data ought to behave. A data analyst who works with gene sequences might impose a different set of requirements, for well-correlated data, than a data analyst who is investigating fluctuations in the stock market. Hence, there are many available correlation values that are available to data scientists, and these include: Pearson, Cosine, Spearman, Jaccard, Gini, Maximal Information Coefficient, and Complex Linear Pathway score. The computationally fastest of the correlation scores is the dot product (Fig. 11.4). In a recent paper comparing the performance of 12 correlation formulas the simple dot product led the pack [11].

Let us examine the various dot products that can be calculated for three sample vectors,

$$\sum_i x_i y_i$$

FIG. 11.4 The lowly dot product. For two vectors, the dot product is the sum of the products of the corresponding values. To normalize the dot product, we would divide the dot product by the product of the lengths of the two vectors.

```
a = [1,4,6,9,15,55,62,-5]
b = [-2,-8,-9,-12,-80,14,15,2]
c = [2,8,12,18,30,110,124,-10]
```

Notice that vector *c* has twice the value of each paired attribute in vector *a*. We'll use the Python script, numpy_dot.py to compute the lengths of the vectors *a*, *b*, and *c*; and we will calculate the simple dot products, normalized by the product of the lengths of the vectors.

```
from __future__ import division
import numpy
from numpy import linalg
a = [1,4,6,9,15,55,62,-5]
b = [-2,-8,-9,-12,-80,14,15,2]
c = [2,8,12,18,30,110,124,-10]
a_length = linalg.norm(a)
b_length = linalg.norm(b)
c_length = linalg.norm(c)
print(numpy.dot(a,b) / (a_length * b_length))
print(numpy.dot(a,a) / (a_length * a_length))
print(numpy.dot(a,c) / (a_length * c_length))
print(numpy.dot(b,c) / (b_length * c_length))
```

Here is the commented output:

```
0.0409175385118    (Normalized dot product of a with b)
1.0                (Normalized dot product of a with a)
1.0                (Normalized dot product of a with c)
0.0409175385118    (Normalized dot product of b with c)
```

Inspecting the output, we see that the normalized dot product of a vector with itself is 1. The normalized dot product of a and c is also 1, because *c* is perfectly correlated with *a*, being twice its value, attribute by attribute. We also see that the normalized dot product of *a* and *b* is equal to the normalized dot product of *b* and *c* (0.0409175385118); because *c* is perfectly correlated with a and because dot products are transitive.

Section 11.4. Clustering

Reality is merely an illusion, albeit a very persistent one.

Albert Einstein

Clustering algorithms are currently very popular. They provide a way of taking a large set of data objects that seem to have no relationship to one another and to produce a visually simple collection of clusters wherein each cluster member is similar to every other member of the same cluster.

The algorithmic methods for clustering are simple. One of the most popular clustering algorithms is the k-means algorithm, which assigns any number of data objects to one of k clusters [12]. The number k of clusters is provided by the user. The algorithm is easy to describe and to understand, but the computational task of completing the algorithm can be difficult when the number of dimensions in the object (i.e., the number of attributes associated with the object), is large. [Glossary K-means algorithm, K-nearest neighbor algorithm]

Here is how the algorithm works for sets of quantitative data:

1. The program randomly chooses k objects from the collection of objects to be clustered. We will call each of these these k objects a focus.
2. For every object in the collection, the distance between the object and all of randomly chosen k objects (chosen in step 1) is computed.
3. A round of k-clusters are computed by assigning every object to its nearest focus.
4. The centroid focus for each of the k clusters is calculated. The centroid is the point that is closest to all of the objects within the cluster. Another way of saying this is that if you sum the distances between the centroid and all of the objects in the cluster, this summed distance will be smaller than the summed distance from any other point in space.
5. Steps 2, 3, and 4 are repeated, using the k centroid foci as the points for which all distances are computed.
6. Step 5 is repeated until the k centroid foci converge on a non-changing set of values (or until the program slows to an interminable crawl).

There are some serious drawbacks to the algorithm:

- The final set of clusters will sometimes depend on the initial, random choice of k data objects. This means that multiple runs of the algorithm may produce different outcomes.
- The algorithms are not guaranteed to succeed. Sometimes, the algorithm does not converge to a final, stable set of clusters.
- When the dimensionality is very high the distances between data objects (i.e., the square root of the sum of squares of the measured differences between corresponding attributes of two objects) can be ridiculously large and of no practical meaning. Computations may bog down, cease altogether, or produce meaningless results. In this case the only recourse may require eliminating some of the attributes (i.e., reducing dimensionality of the data objects). Subspace clustering is a method wherein clusters are found for computationally manageable subsets of attributes. If useful clusters are found using this method, additional attributes can be added to the mix to see if the clustering can be improved. [Glossary Curse of dimensionality]
- The clustering algorithm may succeed, producing a set of clusters of similar objects, but the clusters may have no practical value. They may miss important relationships among the objects, or they might group together objects whose similarities are totally non-informative.

At best, clustering algorithms should be considered a first step toward understanding how attributes account for the behavior of data objects.

Classifier algorithms are different from clustering algorithms. Classifiers assign a class (from a pre-existing classification) to an object whose class is unknown [12]. The k-nearest neighbor algorithm (not to be confused with the k-means clustering algorithm) is a simple and popular classifier algorithm. From a collection of data objects whose class is known, the algorithm computes the distances from the object of unknown class to the objects of known class. This involves a distance measurement from the feature set of the objects of unknown class to every object of known class (the test set). The distance measure uses the set of attributes that are associated with each object. After the distances are computed, the k classed objects with the smallest distance to the object of unknown class are collected. The most common class in the nearest k classed objects is assigned to the object of unknown class. If the chosen value of k is 1, then the object of unknown class is assigned the class of its closest classed object (i.e., the nearest neighbor).

The k-nearest neighbor algorithm is just one among many excellent classifier algorithms, and analysts have the luxury of choosing algorithms that match their data (e.g., sample size, dimensionality) and purposes [13]. Classifier algorithms differ fundamentally from clustering algorithms and from recommender algorithms in that they begin with an existing classification. Their task is very simple; assign an object to its proper class within the classification. Classifier algorithms carry the assumption that similarities among class objects determine class membership. This may not be the case. For example, a classifier algorithm might place cats into the class of small dogs because of the similarities among several attributes of cats and dogs (e.g., four legs, one tail, pointy ears, average weight about 8 pounds, furry, carnivorous, etc.). The similarities are impressive, but irrelevant. No matter how much you try to make it so, a cat is not a type of dog. The fundamental difference between grouping by similarity and grouping by relationship has been discussed in Section 5.1. [Glossary Recommender, Modeling]

Like clustering techniques, classifier techniques are computationally intensive when the dimension is high, and can produce misleading results when the attributes are noisy (i.e., contain randomly distributed attribute values) or non-informative (i.e., unrelated to correct class assignment).

Section 11.5. Methods for Data Persistence (Without Using a Database)

A file that big?
It might be very useful.
But now it is gone.

Haiku by David J. Liszewski

Your scripts create data objects, and the data objects hold data. Sometimes, these data objects are transient, existing only during a block or subroutine. At other times, the data objects produced by scripts represent prodigious amounts of data, resulting from complex and time-consuming calculations. What happens to these data structures when the script finishes executing? Ordinarily, when a script stops, all the data structures produced by the script simply vanish.

Persistence is the ability of data to outlive the program that produced it. The methods by which we create persistent data are sometimes referred to as marshalling or serializing. Some of the language specific methods are called by such colorful names as data dumping, pickling, freezing/thawing, and storable/retrieve. [Glossary Serializing, Marshalling, Persistence]

Data persistence can be ranked by level of sophistication. At the bottom is the exportation of data to a simple flat-file, wherein records are each one line in length, and each line of the record consists of a record key, followed by a list of record attributes. The simple spreadsheet stores data as tab delimited or comma separated line records. Flat-files can contain a limitless number of line records, but spreadsheets are limited by the number of records they can import and manage. Scripts can be written that parse through flat-files line by line (i.e., record by record), selecting data as they go. Software programs that write data to flat-files achieve a crude but serviceable type of data persistence.

A middle-level technique for creating persistent data is the venerable database. If nothing else, databases can create, store, and retrieve data records. Scripts that have access to a database can achieve persistence by creating database records that accommodate data objects. When the script ends, the database persists, and the data objects can be fetched and reconstructed for later use.

Perhaps the highest level of data persistence is achieved when complex data objects are saved in toto. Flat-files and databases may not be suited to storing complex data objects, holding encapsulated data values. Most languages provide built-in methods for storing complex objects, and a number of languages designed to describe complex forms of data have been developed. Data description languages, such as YAML (Yet Another Markup Language) and JSON (JavaScript Object Notation), can be adopted by any programming language.

Let us review some of the techniques for data persistence that are readily accessible to Python programmers.

Python pickles its data. Here, the Python script, pickle_up.py, pickles a string variable, in the save.p file.

```
import pickle
pumpkin_color = "orange"
pickle.dump( pumpkin_color, open("save.p","wb"))
```

The Python script, pickle_down.py, loads the pickled data, from the "save.p" file, and prints it to the screen.

```
import pickle
pumpkin_color = pickle.load(open("save.p","rb"))
print(pumpkin_color)
```

The output of the pickle_down.py script is shown here:

```
orange
```

Python has several database modules that will insert database objects into external files that persist after the script has executed. The database objects can be quickly called from the external module, with a simple command syntax [10]. Here is the Python script, lucy. py, that creates a tiny external database, using Python's most generic dbm.dumb module.

```
import dbm.dumb
lucy_hash = dbm.dumb.open('lucy', 'c')
lucy_hash["Fred Mertz"] = "Neighbor"
lucy_hash["Ethel Mertz"] = "Neighbor"
lucy_hash["Lucy Ricardo"] = "Star"
lucy_hash["Ricky Ricardo"] = "Band leader"
lucy_hash.close()
```

Here is the Python script, lucy_untie.py, that reads all of the key/value pairs held in the persistent database created for the lucy_hash dictionary object.

```
import dbm.dumb
lucy_hash = dbm.dumb.open('lucy')
for character in lucy_hash.keys():
  print(character.decode('utf-8') + " " + lucy_hash[character].
  decode('utf-8'))
lucy_hash.close()
```

Here is the output produced by the Python script, lucy_untie.py script.

```
Ethel Mertz Neighbor
Lucy Ricardo Star
Ricky Ricardo Band leader
Fred Mertz Neighbor
```

Persistence is a simple and fundamental process ensuring that data created in your scripts can be recalled by yourself or by others who need to verify your results. Regardless of the programming language you use, or the data structures you prefer, you will need to familiarize with at least one data persistence technique.

Section 11.6. Case Study: Climbing a Classification

But - once I bent to taste an upland spring
And, bending, heard it whisper of its Sea.

Ecclesiastes

Classifications are characterized by a linear ascension through a hierarchy. The parental classes of any instance of the classification can be traced as a simple, non-branched, and non-recursive, ordered, and uninterrupted list of ancestral classes.

In a prior work [10], I described how a large, publicly available, taxonomy data file could be instantly queried to retrieve any listed organism, and to compute its complete class lineage, back to the "root" class, the primordial origin the classification of living organisms [10]. Basically, the trick to climbing backwards up the class lineage involves building two dictionary objects, also known as associative arrays. One dictionary object (which we will be calling "namehash") is composed of key/value pairs wherein each key is the identifier code of a class (in the nomenclature of the taxonomy data file), and each value is its name or label. The second dictionary object (which we'll be calling "parenthash") is composed of key/value pairs wherein each key is the identifier code of a class, and each value is the identifier code of the parent class. Once you have prepared the namehash dictionary and the parenthash dictionary the entire ancestral lineage of every one of the many thousands of organisms included in the taxonomy of living species (contained in the taxonomy.dat file) can be reconstructed with just a few lines of Python code, as shown here:

```
for i in range(30):
    if id_name in namehash:
      outtext.write(namehash[id_name] + "\n")
    if id_name in parenthash:
      id_name = parenthash[id_name]
```

The parts of the script that build the dictionary objects are left as an exercise for the reader. As an example of the script's output, here is the lineage for the Myxococcus bacteria:

```
Myxococcus
Myxococcaceae
Cystobacterineae
Myxococcales
Deltaproteobacteria
delta/epsilon subdivisions
Proteobacteria
Bacteria
cellular organisms
root
```

The words in this lineage may seem strange to laypersons, but taxonomists who view this lineage instantly grasp the place of organism within the classification of all living organisms. Every large and complex knowledge domain should have its own taxonomy, complete with a parent class for every child class. The basic approach to reconstructing lineages from the raw taxonomy file would apply to every field of study. For those interested in the taxonomy of living organisms, possibly the best documented classification of any kind, the taxonomy.dat file (exceeding 350 Mbytes) is available at no cost via ftp at:

ftp://ftp.ebi.ac.uk/pub/databases/taxonomy/

Section 11.7. Case Study (Advanced): A Database Example

Experts often possess more data than judgment.

Colin Powell

For industrial strength persistence, providing storage for millions or billions of data objects, database applications are a good choice. SQL (Systems Query Language, pronounced like "sequel") is a specialized language used to query relational databases. SQL allows programmers to connect with large, complex server-based network databases. A high level of expertise is needed to install and implement the software that creates server-based relational databases responding to multi-user client-based SQL queries. Fortunately, Python provides access to SQLite, a free, and widely available spin-off of SQL [10]. The source code for SQLite is public domain. [Glossary Public domain]

SQLite is bundled into the newer distributions of Python, and can be called from Python scripts with an "import sqlite3" command. Here is a Python script, sqlite.py, that reads a very short dictionary into an SQL database.

```python
import sqlite3, itertools
from sqlite3 import dbapi2 as sqlite
import string, re, os
mesh_hash = {}
entry = ()
mesh_hash["Fred Mertz"] = "Neighbor"
mesh_hash["Ethel Mertz"] = "Neighbor"
mesh_hash["Lucy Ricardo"] = "Star"
mesh_hash["Ricky Ricardo"] = "Band leader"
con = sqlite.connect('test1.db')
cur = con.cursor()
cur.executescript("""
    create table lucytable
    (
      name      varchar(64),
      term      varchar(64)
    );
    """)
for key, value in mesh_hash.items():
  entry = (key, value)
  cur.execute("insert into lucytable (name, term) values (?, ?)",
  entry)
con.commit()
```

Once created, entries in the SQL database file, test1.db, can be retrieved, as shown in the Python script, sqlite_read.py:

```
import sqlite3
from sqlite3 import dbapi2 as sqlite
import string, re, os
con=sqlite.connect('test1.db')
cur=con.cursor()
cur.execute("select * from lucytable")
for row in cur:
  print(row[0], row[1])
```

Here is the output of the sqlite_read.py script

```
Fred Mertz Neighbor
Ethel Mertz Neighbor
Lucy Ricardo Star
Ricky Ricardo Band leader
```

Databases, such as SQLite, are a great way to achieve data persistence, if you are adept at programming in SQL, and if you need to store millions of simple data objects. You may be surprised to learn that built-in persistence methods native to your favorite programming language may provide a simpler, more flexible option than proprietary database applications, when dealing with Big Data.

Section 11.8. Case Study (Advanced): NoSQL

The creative act is the defeat of habit by originality

George Lois

Triples are the basic commodities of information science. Every triple represents a meaningful assertion, and collections of triples can be automatically integrated with other triples. As such all the triples that share the same identifier can be collected to yield all the available information that pertains to the unique object. Furthermore, all the triples that pertain to all the members of a class of objects can be combined to provide information about the class, and to find relationships among different classes of objects. This being the case, it should come as no surprise that databases have been designed to utilize triples as their data structure; dedicating their principal functionality to the creation, storage, and retrieval of triples.

Triple databases, also known as triplestores, are specialized variants of the better-known NoSQL databases,; databases that are designed to store records consisting of nothing more than a key and value. In the case of triplestores, the key is the identifier of a data object, and the value is a metadata/data pair belonging associated with the identifier belonging to the data object.

Today, large triplestores exist, holding trillions of triples. At the current time, software development for triplestore databases is in a state of volatility. Triplestore databases are dropping in an out of existence, changing their names, being incorporated into other systems, or being redesigned from the ground up.

At the risk of showing my own personal bias, as an unapologetic Mumps fan, I would suggest that readers may want to investigate the value of using native Mumps as a triplestore database. Mumps, also known as the M programming language, is one of a small handful of ANSI-standard (American National Standard Institute) languages that includes C, Ada, and Fortran. It was developed in the 1960s and is still in use, primarily in hospital information systems and large production facilities [14]. The simple, hierarchical database design of Mumps lost favor through the last decades of the twentieth century, as relational databases gained popularity. In the past decade, with the push toward NoSQL databases holding massive sets of simplified data, Mumps has received renewed interest. As it happens, Mumps can be implemented as a powerful and high performance Triplestore database.

Versions of Mumps are available as open source, free distributions [15,16]. but the Mumps installation process can be challenging for those who are unfamiliar with the Mumps environment. Stalwarts who successfully navigate the Mumps installation process may find that Mumps' native features render it suitable for storing triples and exploring their relationships [17].

Glossary

AES The Advanced Encryption Standard (AES) is the cryptographic standard endorsed by the United States government as a replacement for the old government standard, DES (Data Encryption Standard). AES was chosen from among many different encryption protocols submitted in a cryptographic contest conducted by the United States National Institute of Standards and Technology, in 2001. AES is also known as Rijndael, after its developer. It is a symmetric encryption standard, meaning that the same password used for encryption is also used for decryption.

Associative array A data structure consisting of an unordered list of key/value data pairs. Also known as hash, hash table, map, symbol table, dictionary, or dictionary array. The proliferation of synonyms suggests that associative arrays, or their computational equivalents, have great utility. Associative arrays are used in Perl, Python, Ruby and most modern programming languages.

Black box In physics, a black box is a device with observable inputs and outputs, but what goes on inside the box is unknowable. The term is used to describe software, algorithms, machines, and systems whose inner workings are inscrutable.

Correlation distance Also known as correlation score. The correlation distance provides a measure of similarity between two variables. Two similar variables will rise rise and fall together [18,19]. The Pearson correlation score is popular, and can be easily implemented [10,20]. It produces a score that varies from −1 to 1. A score of 1 indicates perfect correlation; a score of −1 indicates perfect anti-correlation (i.e., one variable rises while the other falls). A Pearson score of 0 indicates lack of correlation. Other correlation measures can be applied to Big Data sets [18,19].

Curse of dimensionality As the number of attributes for a data object increases, the distance between data objects grows to enormous size. The multidimensional space becomes sparsely populated, and the distances between any two objects, even the two closest neighbors, becomes absurdly large. When you have thousands of dimensions, the space that holds the objects is so large that distances between objects become difficult or impossible to compute, and computational products become useless for most purposes.

K-means algorithm The k-means algorithm assigns any number of data objects to one of k clusters [12]. The algorithm is described fully in Chapter 9. The k-means algorithm should not be confused with the k-nearest neighbor algorithm.

K-nearest neighbor algorithm The k-nearest neighbor algorithm is a simple and popular classifier algorithm. From a collection of data objects whose class is known, the algorithm computes the distances from the object of unknown class to the objects of known class. This involves a distance measurement from the feature set of the objects of unknown class to every object of known class (the test set). After the distances are computed, the k classed objects with the smallest distance to the object of unknown class are collected. The most common class (i.e., the class with the most objects) among the nearest k classed objects is assigned to the object of unknown class. If the chosen value of k is 1, then the object of unknown class is assigned the class of its closest classed object (i.e., the nearest neighbor).

Mahalanobis distance A distance measure based on correlations between variables; hence, it measures the similarity of the objects whose attributes are compared. As a correlation measure, it is not influenced by the relative scale of the different attributes. It is used routinely in clustering and classifier algorithms.

Marshalling Marshalling, like serializing, is a method for achieving data persistence (i.e., saving variables and other data structures produced in a program, after the program has stopped running). Marshalling methods preserve data objects, with their encapsulated data and data structures.

Modeling Modeling involves explaining the behavior of a system, often with a formula, sometimes with descriptive language. The formula for the data describes the distribution of the data and often predicts how the different variables will change with one another. Consequently, modeling often provides reasonable hypotheses to explain how the data objects within a system will influence one another. Many of the great milestones in the physical sciences have arisen from a bit of data modeling supplemented by scientific genius (e.g., Newton's laws of mechanics and optics, Kepler's laws of planetary orbits, Quantum mechanics). The occasional ability to relate observation with causality endows modeling with greater versatility and greater scientific impact than the predictive techniques (e.g., recommenders, classifiers and clustering methods). Unlike the methods of predictive analytics, which tend to rest on a few basic assumptions about measuring similarities among data objects, the methods of data modeling are selected from every field of mathematics and are based on an intuitive approach to data analysis. In many cases, the modeler simply plots the data and looks for familiar shapes and patterns that suggest a particular type of function (e.g., logarithmic, linear, normal, Fourier series, Power law).

Normalized compression distance String compression algorithms (e.g., zip, gzip, bunzip) should yield better compression from a concatenation of two similar strings than from a concatenation of two highly dissimilar strings. The reason being that the same string patterns that are employed to compress a string (i.e., repeated runs of a particular pattern) are likely to be found in another, similar string. If two strings are completely dissimilar, then the compression algorithm would fail to find repeated patterns that enhance compressibility. The normalized compression distance is a similarity measure based on the enhanced compressibility of concatenated strings of high similarity [21]. A full discussion, with examples, is found in the Open Source Tools section of Chapter 4.

Overfitting Overfitting occurs when a formula describes a set of data very closely, but does not lead to any sensible explanation for the behavior of the data, and does not predict the behavior of comparable data sets. In the case of overfitting the formula is said to describe the noise of the system rather than the characteristic behavior of the system. Overfitting occurs frequently with models that perform iterative approximations on training data, coming closer and closer to the training data set with each iteration. Neural networks are an example of a data modeling strategy that is prone to overfitting.

***P* value** The *P* value is the probability of getting a set of results that are as extreme or more extreme as the set of results you observed, assuming that the null hypothesis is true (that there is no statistical difference between the results). The *P*-value has come under great criticism over the decades, with a growing consensus that the *P*-value is often misinterpreted, used incorrectly, or used in situations wherein it does not apply [22]. In the realm of Big Data, repeated samplings of data from large data sets will produce small *P*-values that cannot be directly applied to determining statistical significance. It is best to think of the *P*-value as just another piece of information that tells you something about how sets of observations compare with one another; and not as a test of statistical significance.

Persistence Persistence is the ability of data to remain available in memory or storage after the program in which the data was created has stopped executing. Databases are designed to achieve persistence. When the database application is turned off, the data remains available to the database application when it is re-started at some later time.

Pseudorandom number generator It is impossible for computers to produce an endless collection of truly random numbers. Eventually, algorithms will cycle through their available variations and begins to repeat themselves, producing the same set of "random" numbers, in the same order; a phenomenon referred to as the generator's period. Because algorithms that produce seemingly random numbers are imperfect, they are known as pseudorandom number generators. The Mersenne Twister algorithm, which has an extremely long period, is used as the default random number generator in Python. This algorithm performs well on most of the tests that mathematicians have devised to test randomness.

Public domain Data that is not owned by an entity. Public domain materials include documents whose copyright terms have expired, materials produced by the federal government, materials that contain no creative content (i.e., materials that cannot be copyrighted), or materials donated to the public domain by the entity that holds copyright. Public domain data can be accessed, copied, and re-distributed without violating piracy laws. It is important to note that plagiarism laws and rules of ethics apply to public domain data. You must properly attribute authorship to public domain documents. If you fail to attribute authorship or if you purposefully and falsely attribute authorship to the wrong person (e.g., yourself), then this would be an unethical act and an act of plagiarism.

Recommender A collection of methods for predicting the preferences of individuals. Recommender methods often rely on one or two simple assumptions: (1) If an individual expresses a preference for a certain type of product, and the individual encounters a new product that is similar to a previously preferred product, then he is likely to prefer the new product; (2) If an individual expresses preferences that are similar to the preferences expressed by a cluster of individuals, and if the members of the cluster prefer a product that the individual has not yet encountered, then the individual will most likely prefer the product.

RegEx Short for Regular Expressions, RegEx is a syntax for describing patterns in text. For example, if I wanted to pull all lines from a text file that began with an uppercase "B" and contained at least one integer, and ended with the a lowercase x, then I might use the regular expression: " B.*[0-9].*x$". This syntax for expressing patterns of strings that can be matched by pre-built methods available to a programming language is somewhat standardized. This means that a RegEx expression in Perl will match the same pattern in Python, or Ruby, or any language that employs RegEx. The relevance of Regex to Big Data is several-fold. Regex can be used to build or transform data from one format to another; hence creating or merging data records. It can be used to convert sets of data to a desired format; hence transforming data sets. It can be used to extract records that meet a set of characteristics specified by a user; thus filtering subsets of data or executing data queries over text-based files or text-based indexes. The big drawback to using RegEx is speed: operations that call for many Regex operations, particularly when those operations are repeated for each parsed line or record, will reduce software performance. Regex-heavy programs that operate just fine on megabyte files may take hours, days or months to parse through terabytes of data.

A 12-line python script, file_search.py, prompts the user for the name of a text file to be searched, and then prompts the user to supply a RegEx pattern. The script will parse the text file, line by line, displaying those lines that contain a match to the RegEx pattern.

```
import sys, string, re
print("What is file would you like to search?")
filename = sys.stdin.readline()
filename = filename.rstrip()
print("Enter a word, phrase or regular expression to search.")
word_to_search = (sys.stdin.readline()).rstrip()
infile = open (filename, "r")
regex_object = re.compile(word_to_search, re.I)
for line in infile:
  m= regex_object.search(line)
  if m:
    print(line)
```

Scipy Scipy, like numpy, is an open source extension to Python [23]. It includes many very useful mathematical routines commonly used by scientists, including: integration, interpolation, Fourier transforms, signal processing, linear algebra, and statistics. Scipy can be downloaded as a suite of modules from: http://www.scipy.org/scipylib/download.html. You can spare yourself the trouble of downloading individual installations of numpy and scipy by downloading Anaconda, a free distribution that bundles hundreds of python packages, along with a recent version of Python. Anaconda is available at: https://store.continuum.io/cshop/anaconda/

Serializing Serializing is a plesionym (i.e., near-synonym) for marshalling and is a method for taking data produced within a script or program, and preserving it in an external file, that can be saved when the program stops, and quickly reconstituted as needed, in the same program or in different programs. The difference, in terms of common usage, between serialization and marshalling is that serialization usually involves capturing parameters (i.e., particular pieces of information), while marshalling preserves all of the specifics of a data object, including its structure, content, and code content). As you might imagine, the meaning of terms might change depending on the programming language and the intent of the serializing and marshalling methods.

System call Refers to a command, within a running script, that calls the operating system into action, momentarily bypassing the programming interpreter for the script. A system call can do essentially anything the operating system can do via a command line.

Undifferentiated software Intellectual property disputes have driven developers to divide software into two categories: undifferentiated software and differentiated software. Undifferentiated software comprises the fundamental algorithms that everyone uses whenever they develop a new software application. It is in nobody's interest to assign patents to basic algorithms and their implementations. Nobody wants to devote their careers to prosecuting or defending tenuous legal claims over the ownership of the fundamental building blocks of computer science. Differentiated software comprises customized applications that are sufficiently new and different from any preceding product that patent protection would be reasonable.

References

[1] Brin S, Page L. The anatomy of a large-scale hypertextual Web search engine. Comput Netw ISDN Syst 1998;33:107–17.

[2] Berman JJ. Doublet method for very fast autocoding. BMC Med Inform Decis Mak 2004;4:16.

[3] Goldberg A, Robson D, Harrison MA. Smalltalk-80: the language and its implementation. Boston, MA: Addison-Wesley; 1983.

[4] Cleveland Clinic. Build an efficient pipeline to find the most powerful predictors. In: Innocentive September 8, 1911. https://www.innocentive.com/ar/challenge/9932794 [viewed September 25, 2012].

[5] Berman JJ. Threshold protocol for the exchange of confidential medical data. BMC Med Res Methodol 2002;2:12.

[6] Rivest RL. Chaffing and winnowing: confidentiality without encryption. MIT Lab for Computer Science; March 18, 1998. (rev. April 24, 1998). Available from: http://people.csail.mit.edu/rivest/chaffing-980701.txt [viewed January 10, 2017].

[7] Matsumoto M, Nishimura T. Mersenne twister: a 623-dimensionally equidistributed uniform pseudo-random number generator. ACM Trans Model Comput Simul 1998;8:3–30.

[8] Janert PK. Data analysis with open source tools. O'Reilly Media, 2010.

[9] Smith SW. The scientist and engineer's guide to digital signal processing. Concord, CA: California Technical Publishing; 1997.

[10] Berman JJ. Methods in medical informatics: fundamentals of healthcare programming in Perl, Python, and Ruby. Boca Raton: Chapman and Hall; 2010.

[11] Deshpande R, VanderSluis B, Myers CL. Comparison of profile similarity measures for genetic interaction networks. PLoS ONE 2013;8:e68664.

[12] Wu X, Kumar V, Quinlan JR, Ghosh J, Yang Q, Motoda H, et al. Top 10 algorithms in data mining. Knowl Inf Syst 2008;14:1–37.

[13] Zhang L, Lin X. Some considerations of classification for high dimension low-sample size data. Stat Methods Med Res November 23, 2011. http://smm.sagepub.com/content/early/2011/11/22/0962280211428387.long [viewed January 26, 2013].

[14] Epic CEO Faulkner tells why she wants to keep her company private. Modern Healthcare; March 14, 2015.

[15] GT.M High end TP database engine: industrial strength nosql application development platform. Available at: http://sourceforge.net/projects/fis-gtm/ [viewed on August 29, 2015].

[16] MUMPS database and language. ANSI standard MUMPS. Available at: http://sourceforge.net/projects/mumps/files/ [viewed August 29, 2015].

[17] Tweed R, James G. A universal NoSQL engine, using a tried and tested technology. 2010. Available at: http://www.mgateway.com/docs/universalNoSQL.pdf [viewed on August 29, 2015]

[18] Reshef DN, Reshef YA, Finucane HK, Grossman SR, McVean G, Turnbaugh PJ, et al. Detecting novel associations in large data sets. Science 2011;334:1518–24.

[19] Szekely GJ, Rizzo ML. Brownian distance covariance. Ann Appl Stat 2009;3:1236–65.

[20] Lewis PD. R for medicine and biology. Sudbury: Jones and Bartlett Publishers; 2009.

[21] Cilibrasi R, Vitanyi PMB. Clustering by compression. IEEE Trans Inform Theory 2005;51:1523–45.

[22] Cohen J. The earth is round (p < .05). Am Psychol 1994;49:997–1003.

[23] SciPy Reference Guide, Release 0.7. Written by the SciPy Community. December 07, 2008.

12

Finding the Clues in Large Collections of Data

Section 12.1. Denominators

The question is not what you look at, but what you see.

Henry David Thoreau in Journal, 5 August 1851

Denominators are the numbers that provide perspective to other numbers. If you are informed that 10,000 persons die each year in the United States, from a particular disease, then you might want to know the total number of deaths, from all causes. When you compare the death from a particular disease with the total number of deaths from all causes (the denominator), you learn something about the relative importance of your original count (e.g., an incidence of 10,000 deaths/350 million persons). Epidemiologists typically represent incidences as numbers per 100,000 population. An incidence of 10,000/350 million is equivalent to an incidence of 2.9 per 100,000.

Denominators are not always easy to find. In most cases the denominator is computed by tallying every data object in a Big Data resource. If you have a very large number of data objects, then the time required to reach a global tally may be quite long. In many cases a Big Data resource will permit data analysts to extract subsets of data, but analysts will be forbidden to examine the entire resource. In such cases the denominator will be computed for the subset of extracted data and will not accurately represent all of the data objects available to the resource.

If you are using Big Data collected from multiple sources, your histograms (i.e., graphic representations of the distribution of objects by some measureable attribute such as age, frequency, size) will need to be represented as fractional distributions for each source's

Principles and Practice of Big Data. https://doi.org/10.1016/B978-0-12-815609-4.00012-1

data; not as value counts. The reason for this is that a histogram from one source may not have the same total number of distributed values compared with the histogram created from another source. As discussed in Section 10.4, "Normalizing and Transforming Your Data," we achieve comparability among histograms by dividing the binned values by the total number of values in a distribution, for each data source. Doing so renders the bin value as a percentage of total, rather than a sum of data values.

Big Data managers should make an effort to supply information that summarizes the total set of data available at any moment in time, and should also provide information on the sources of data that contribute to the total collection. Here are some of the numbers that should be available to analysts: the number of records in the resource, the number of classes of data objects in the resource, the number of data objects belonging to each class in the resource, and the number of data values (preferably expressed as metadata/data pairs) that belong to data objects.

Section 12.2. Word Frequency Distributions

Poetry is everywhere; it just needs editing.

James Tate

There are two general types of data: quantitative and categorical. Quantitative data refers to measurements. Categorical data is simply a number that represents the number of items that have a feature. For most purposes the analysis of categorical data reduces to counting and binning.

Categorical data typically conforms to a Zipf distribution. George Kingsley Zipf (1902–50) was an American linguist who demonstrated that, for most languages, a small number of words account for the majority of occurrences of all the words found in prose. Specifically, he found that the frequency of any word is inversely proportional to its placement in a list of words, ordered by their decreasing frequencies in text. The first word in the frequency list will occur about twice as often as the second word in the list and three times as often as the third word in the list. [Glossary Word lists]

The Zipf distribution applied to languages is a special form of Pareto's principle, or the 80/20 rule. Pareto's principle holds that a small number of causes may account for the vast majority of observed instances. For example a small number of rich people account for the majority of wealth. Likewise, a small number of diseases account for the vast majority of human illnesses. A small number of children account for the majority of the behavioral problems encountered in a school. A small number of states hold the majority of the population of the United States. A small number of book titles, compared with the total number of publications, account for the majority of book sales. Much of Big Data is categorical and obeys the Pareto principle. Mathematicians often refer to Zipf distributions as Power law distributions. A short Python script for producing Zipf distribution's is found under its Glossary item. [Glossary Power law, Pareto's principle, Zipf distribution]

Let us take a look at the frequency distribution of words appearing in a book. Here is the list of the 30 most frequent words in a sample book and the number of occurrence of each word.

```
01 003977 the
02 001680 and
03 001091 class
04 000946 are
05 000925 chapter
06 000919 that
07 000884 species
08 000580 virus
09 000570 with
10 000503 disease
11 000434 for
12 000427 organisms
13 000414 from
14 000412 hierarchy
15 000335 not
16 000329 humans
17 000320 have
18 000319 proteobacteria
19 000309 human
20 000300 can
21 000264 fever
22 000263 group
23 000248 most
24 000225 infections
25 000219 viruses
26 000219 infectious
27 000216 organism
28 000216 host
29 000215 this
30 000211 all
```

As Zipf would predict, the most frequent word, "the" occurs 3977 times, roughly twice as often as the second most frequently occurring word, "and," which occurs 1689 times. The third most frequently occurring word "class" occurs 1091 times, or very roughly one-third as frequently as the most frequently occurring word.

What can we learn about the text from which these word frequencies were calculated? As discussed in Chapter 1 "stop" words are high frequency words that separate terms and tell us little or nothing about the informational content of text. Let us look at this same list with the "stop" words removed:

```
03 001091 class
05 000925 chapter
07 000884 species
08 000580 virus
10 000503 disease
12 000427 organisms
14 000412 hierarchy
16 000329 humans
18 000319 proteobacteria
19 000309 human
21 000264 fever
22 000263 group
24 000225 infections
25 000219 viruses
26 000219 infectious
27 000216 organism
28 000216 host
```

What kind of text could have produced this list? Could there be any doubt that the list of words and frequencies shown here came from a book whose subject is related to microbiology? As it happens, this word-frequency list came from a book that I previously wrote entitled "Taxonomic Guide to Infections Diseases: Understanding the Biologic Classes of Pathogenic Organisms" [1]. By glancing at a few words from a large text file, we gain a deep understanding of the subject matter of the text. The words with the top occurrence frequencies told us the most about the book, because these words are low-frequency in most books (e.g., words such as hierarchy, proteobacteria, organism). They occurred in high frequency because the text was focused on a narrow subject (e.g., infectious diseases).

A clever analyst will always produce a Zipf distribution for categorical data. A glance at the output reveals a great deal about the contents of the data.

Let us go one more step, and produce a cumulative index for the occurrence of words in the text, arranging them in order of descending frequency of occurrence.

```
01 003977 0.0559054232618291 the
02 001680 0.0795214934352948 and
03 001091 0.0948578818634204 class
04 000946 0.108155978520622 are
05 000925 0.121158874300655 chapter
06 000919 0.134077426972926 that
07 000884 0.146503978183249 species
08 000580 0.154657145266946 virus
09 000570 0.162669740504372 with
10 000503 0.169740504371784 disease
11 000434 0.175841322499930 for
```

```
12 000427 0.181843740335686 organisms
13 000414 0.187663414771290 from
14 000412 0.193454974837640 hierarchy
15 000335 0.198164131687706 not
16 000329 0.202788945430009 humans
17 000320 0.207287244510669 have
18 000319 0.211771486406702 proteobacteria
19 000309 0.216115156456465 human
20 000300 0.220332311844584 can
21 000264 0.224043408586128 fever
22 000263 0.227740448143046 group
23 000248 0.231226629930558 most
24 000225 0.234389496471647 infections
25 000219 0.237468019904973 viruses
26 000219 0.240546543338300 infectious
27 000216 0.243582895217746 organism
28 000216 0.246619247097191 host
29 000215 0.249641541792010 this
30 000211 0.252607607748320 all
   .
   .
   .
   .
   .
8957 000001 0.999873485338356 acanthaemoeba
8958 000001 0.999887542522984 acalculous
8959 000001 0.999901599707611 academic
8960 000001 0.999915656892238 absurd
8961 000001 0.999929714076865 abstract
8962 000001 0.999943771261492 absorbing
8963 000001 0.999957828446119 absorbed
8964 000001 0.999971885630746 abrasion
8965 000001 0.999985942815373 abnormalities
8966 000001 1.000000000000000 abasence
```

In this cumulative listing the third column is the fraction of the occurrences of the word along with the preceding words in the list as a fraction of all the occurrences of every word in the text.

The list is truncated after the thirtieth entry and picks up again at entry number 8957. There are a total of 8966 different, sometimes called unique, words in the text. The total number of words in the text happens to be 71,138. The last word on the list "abasence" has a cumulative fraction of 1.0, as all of the preceding words, plus the last word, account for

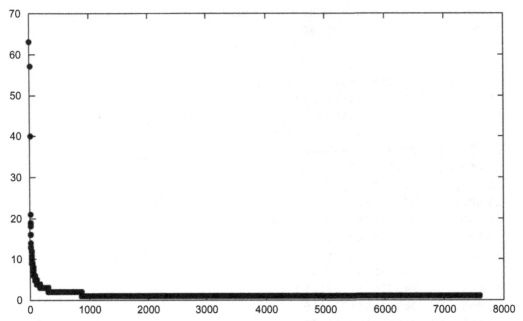

FIG. 12.1 A frequency distribution of word occurrences from a sample text. The bottom coordinates indicate that the entire text is accounted for by a list of about 9000 different words. The steep and early rise indicates that a few words account for the bulk of word occurrences. Graphs with this shape are sometimes referred to as Zipf distributions.

100% of word occurrences. The cumulative frequency distribution for the different words in the text is shown (Fig. 12.1). As an aside, the tail of the Zip distribution, which typically contains items occurring once only in a large data collection, are often "mistakes." In the case of text distributions, typographic errors can be found in the farthest and thinnest part of the tail. In this case the word "abasence" occurs just once, as the last item in the distribution. It is a misspelling for the word "absence."

Notice that though there are a total of 8957 unique words in the text, the first thirty words account for more than 25% of all word occurrences. The final ten words on the list occurred only once in the text. Common statistical measurements, such as the average of a population or the standard deviation, fail to provide any useful description of Zipf distributions. [Glossary Nonparametric statistics]

Section 12.3. Outliers and Anomalies

The mere formulation of a problem is far more essential than its solution, which may be merely a matter of mathematical or experimental skills. To raise new questions, new possibilities, to regard old problems from a new angle requires creative imagination and marks real advances in science.

Albert Einstein

On occasion the maxima or minima of a set of data will be determined by an outlier value; a value lying nowhere near any of the other values in the data set. If you could just eliminate the outlier, then you might enjoy a maxima and minima that were somewhat close to your other data values (i.e., the second-highest data value and the second-lowest data values would be close to the maxima and the minima, respectively). In these cases the data analyst must come to a decision, to drop or not to drop the outlier. There is no simple guideline for dealing with outliers, but it is sometimes helpful to know something about the dynamic range of the measurements. If a thermometer can measure temperature from −20 to 140°F, and your data outlier has a temperature of 390°F, then you know that the outlier must be an error; the thermometer does not measure above 140 degrees. The data analyst can drop the outlier, but it would be prudent to determine why the outlier was generated. [Glossary Dynamic range, Outlier, Case report, Dimensionality]

Outliers are extreme data values. The occurrence of outliers hinders the task of developing models, equations, or curves that closely fit all the available data. In some cases, outliers are simply mistakes; while in other cases, outliers may be the most important data in the data set. Examples of outliers that have advanced science are many, including: the observance of wobbly starts leading to the discovery of exoplanets; anomalous X-ray bursts from space leading to the discovery of magnetars, highly magnetized neutron stars; individuals with unusual physiological measurements leading to the discovery of rare diseases. The special importance of outliers to Big Data is that as the size of data sets increases, the number of outliers also increases.

True outliers (i.e., outliers not caused by experimental design error or errors in observation and measurement) obey the same physical laws as everything else in the universe. Therefore a valid outlier will always reveal something that is generally true about reality. Put another way, outliers are not exceptions to the general rules; outliers are the exceptions upon which the general rules are based. This assertion brings us to the sadly underappreciated and underutilized creation known as "the case report."

The case report, also known as the case study, is a detailed description of a single event or situation, often focused on a particular outlier, detail, or unique event. The concept of the case study is important in the field of Big Data because it highlights the utility of seeking general truths based on observations of outliers that can only be found in Big Data resources. Case reports are common in the medical literature, often beginning with a comment regarding the extreme rarity of the featured disease. You can expect to see phrases such as "fewer than a dozen have been reported in the literature" or "the authors have encountered no other cases of this lesion," or such and such a finding makes this lesion particularly uncommon and difficult to diagnose. The point that the authors are trying to convey is that the case report is worthy of publication specifically because the observation departs from normal experience. This is just wrong.

Too often, case reports serve merely as cautionary exercises, intended to ward against misdiagnosis. The "beware this lesion" approach to case reporting misses the most important aspect of this type of publication; namely that science, and most aspects of human understanding, involve generalizing from single observations to general rules. When Isaac

Newton saw an apple falling, he was not thinking that he could write a case report about how he once saw an apple drop, thus warning others not to stand under apple trees lest a rare apple might thump them upon the head. Newton generalized from the apple to all objects, addressing universal properties of gravity, and discovering the laws by which gravity interacts with matter. Case reports give us an opportunity to clarify the general way things work, by isolating one specific and rarely observed factor [2,3].

Section 12.4. Back-of-Envelope Analyses

Couldn't Prove, Had to Promise.

Book title, Wyatt Prunty

It is often assumed that Big Data resources are too large and complex for human comprehension. The analysis of Big Data is best left to software programs. Not so. When data analysts go straight to the complex calculations, before they perform a simple estimation, they will find themselves accepting wildly ridiculous calculations. For comparison purposes, there is nothing quite like a simple, and intuitive estimate to pull an overly-eager analyst back to reality. Often, the simple act of looking at a stripped-down version of the problem opens a new approach that can drastically reduce computation time. In some situations, analysts will find that a point is reached when higher refinements in methods yield diminishing returns. After the numerati have used their most advanced algorithms to make an accurate prediction, they may find that their best efforts offer little improvement over a simple estimator. This chapter reviews simple methods for analyzing big and complex data.

–   Estimation-only analyses

The sun is about 93 million miles from the Earth. At this enormous distance, the light hitting Earth arrives as near-parallel rays and the shadow produced by the earth is nearly cylindrical. This means that the shadow of the earth is approximately the same size as the Earth itself. If the Earth's circular shadow on the moon, as observed during a lunar eclipse, appears to be about 2.5 times the diameter of the moon itself, then the moon must have a diameter approximately 1/2.5 times that of the earth. The diameter of the earth is about 8000 miles, so the diameter of the moon must be about 8000/2.5 or about 3000 miles.

The true diameter of the moon is smaller, about 2160 miles. Our estimate is inaccurate because the Earth's shadow is actually conical, not cylindrical. If we wanted to use a bit more trigonometry, we'd arrive at a closer approximation. Still, we arrived at a fair approximation of the moon's size from one, simple division, based on a casual observation made during a lunar eclipse. The distance was not measured; it was estimated from a simple observation. Credit for the first astronomer to use this estimation goes to the Greek astronomer Aristarchus of Samos (310 BCE–230 BCE). In this particular case, a direct measurement of the moon's distance was impossible. Aristarchus' only option was the rough estimate. His predicament was not unique. Sometimes estimation is the only recourse for data analysts.

A modern-day example wherein measurements failed to help the data analyst is the calculation of deaths caused by heat waves. People suffer during heat waves, and municipalities need to determine whether people are dying from heat-related conditions. If heat-related deaths occur, then the municipality can justifiably budget for supportive services such as municipal cooling stations, the free delivery of ice, and increased staffing for emergency personnel. If the number of heat-related deaths is high, the governor may justifiably call a state of emergency.

Medical examiners perform autopsies to determine causes of death. During a heat wave the number of deceased individuals with a heat-related cause of death seldom rises as much as anyone would expect [4]. The reason for this is that stresses produced by heat cause death by exacerbating pre-existing non-heat-related conditions. The cause of death can seldom be pinned on heat. The paucity of autopsy-proven heat deaths can be relieved, somewhat, by permitting pathologists to declare a heat-related death when the environmental conditions at the site of death are consistent with hyperthermia (e.g., a high temperature at the site of death, and a high body temperature of the deceased measured shortly after death). Adjusting the criteria for declaring heat-related deaths is a poor remedy. In many cases the body is not discovered anytime near the time of death, invalidating the use of body temperature. More importantly, different municipalities may develop their own criteria for heat-related deaths (e.g., different temperature threshold measures, different ways of factoring night-time temperatures and humidity measurements). Basically, there is no accurate, reliable, or standard way to determine heat-related deaths at autopsy [4].

How would you, a data estimator, handle this problem? It is simple. You take the total number of deaths that occurred during the heat wave. Then you go back over your records of deaths occurring in the same period, in the same geographic region, over a series of years in which a heat wave did not occur. You average that number, giving you the expected number of deaths in a normal (i.e., without heat wave) period. You subtract that number from the number of deaths that occurred during the heat wave, and that gives you an estimate of the number of people who died from heat-related mortality. This strategy, applied to the 1995 Chicago heat wave, estimated that the number of heat-related deaths rose from 485 to 739 [5].

– **Mean-field averaging**

The average behavior of a collection of objects can be applied toward calculations that would exceed computational feasibility if applied to individual objects. Here is an example. Years ago, I worked on a project that involved simulating cell colony growth, using a Monte Carlo method [6]. Each simulation began with a single cell that divided, producing two cells, unless the cell happened to die prior to cell division. Each simulation applied a certain chance of cell death, somewhere around 0.5, for each cell, at each cell division. When you simulate colony growth, beginning with a single cell, the chance that the first cell will die on the first cell division would be about 0.5; hence, there is about a 50% chance that the colony will die out on the first cell division. If the cell survives the first cell division, the cell might go through several additional cell divisions before it dies, by chance. By that time, there are other progeny that are dividing, and these progeny cells might successfully divide, thus enlarging the size of the colony. A Monte Carlo simulation randomly assigned death or life at each cell division, for each cell in the colony. When the colony manages to reach a large size (e.g., ten million cells), the simulation slows down, as the Monte Carlo algorithm must parse through ten million cells, calculating whether each cell will live or die, and assigning two offspring cells for each simulated division, and removing cells that were assigned a simulated "death." When the computer simulation slowed to a crawl, I found that the whole population displayed an "average" behavior. There was no longer any need to perform a Monte Carlo simulation on every cell in the population. I could simply multiply the total number of cells by the cell death probability (for the entire population), and this would tell me the total number of cells that survived the cycle. For a large colony of cells, with a death probability of 0.5 for each cell, half the cells will die at each cell cycle, and the other half will live and divide, produce two progeny cells; hence the population of the colony will remain stable. When dealing with large numbers, it becomes possible to dispense with the Monte Carlo simulation and to predict each generational outcome with a pencil and paper. [Glossary Monte Carlo simulation]

Substituting the average behavior for a population of objects, rather than calculating the behavior of every single object, is called mean-field approximation. It uses a physical law telling us that large collections of objects can be characterized by their average behavior. Mean-field approximation has been used with great success to understand the behavior of gases, epidemics, crystals, viruses, and all manner of large population problems. [Glossary Mean-field approximation]

Section 12.5. Case Study: Predicting User Preferences

He has no enemies, but is intensely disliked by his friends.

Oscar Wilde

Imagine you have all the preference data for every user of a large movie subscriber service, such as Netflix. You want to develop a system whereby the preference of any subscriber, for any movie, can be predicted. Here are some analytic options, listed in order of increasing complexity; omitting methods that require advanced mathematical skills.

1. Ignore your data and use experts.

Movie reviewers are fairly good predictors of a movie's appeal. If they were not good predictors, they would have been replaced by people with better predictive skills. For any movie, go to the newspapers and magazines and collect about ten movie reviews. Average the review scores and use the average as the predictor for all of your subscribers.

You can refine this method a bit by looking at the average subscriber scores, after the movie has been released. You can compare the scores of the individual experts to the average score of the subscribers. Scores from experts that closely matched the scores from the subscribers can be weighted a bit more heavily than experts whose scores were nothing like the average subscriber score.

2. Use all of your data, as it comes in, to produce an average subscriber score.

Skip the experts; go straight to your own data. In most instances, you would expect that a particular user's preference will come close to the average preference of the entire population in the data set for any given movie.

3. Lump people into preference groups based on shared favorites.

If Ann's personal list of top-favored movies is the same as Fred's top-favored list, then it is likely that their preferences will coincide. For movies that Ann has seen but Fred has not, use Ann's score as a predictor.

In a large data set, find an individual's top ten movie choices and add the individual to a group of individuals who share the same top-ten list. Use the average score for members of the group, for any particular movie as that movie's predictor for each of the members of the group.

As a refinement, find a group of people who share the top-ten and the bottom-ten scoring movies. Everyone in this group shares a love of the same top movies and a loathing for the same bottom movies.

4. Focus your refined predictions.

For many movies, there really is not much of a spread in ratings. If just about everyone loves "Star Wars" and "Raiders of the Lost Arc" and "It's a Wonderful Life," then there really is no need to provide an individual prediction for such movies. Likewise, if a movie is universally loathed, or universally accepted as an "average" flick, then why would you want to use computationally intensive models for these movies?

Most data sets have a mixture of easy and difficult data. There is seldom any good reason to develop predictors for the easy data. In the case of movie predictors, if there is very little spread in a movie's score, then you can safely use the average rating as the predicted rating for all individuals. By removing all of the "easy" movies from your group-specific calculations, you reduce the total number of calculations for the data collection.

This method of eliminating the obvious has application in many different fields. As a program director at the National Cancer Institute, I was peripherally involved in efforts to predict cancer treatment options for patients diagnosed in different stages of disease. Traditionally, large numbers of patients, at every stage of disease, were included in a prediction model that employed a list of measurable clinical and pathological parameters (e.g., age and gender of patient, size of tumor, the presence of local or distant metastases, and so on). It turned out that early models produced predictions where none were necessary. If a patient had a tumor that was small, confined to its primary site of growth, and minimally invasive at its origin, then the treatment was always limited to surgical excision; there were no options for treatment, and hence no reason to predict the best option for treatment. If a tumor was widely metastatic to distant organs at the time of diagnosis, then there were no available treatments known, at that time, that could cure the patient. By focusing their analyses on the subset of patients who could benefit from treatment and for whom the best treatment option was not predetermined, the data analysts reduced the size and complexity of the data and simplified the problem.

The take-away lesson from this section is that predictor algorithms, so popular now among marketers, are just one of many different ways of determining how individuals and subgroups may behave, under different circumstances. Big Data analysts should not be reluctant to try several different analytic approaches, including approaches of their own invention. Sometimes the simplest algorithms, involving nothing more than arithmetic, are the best.

Section 12.6. Case Study: Multimodality in Population Data

What is essential is invisible to the eye.

Antoine de saint-exupery

Big Data distributions are sometimes multi-modal with several peaks and troughs. Multimodality always says something about the data under study. It tells us that the population is somehow non-homogeneous. Hodgkin lymphoma is an example of a cancer with a bimodal age distribution. There is a peak in occurrences at a young age, and another peak of occurrences at a more advanced age. This two-peak phenomenon can be found whenever Hodgkin Lymphoma is studied in large populations [7,8].

In the case of Hodgkin lymphoma, lymphomas occurring in the young may share diagnostic features with the lymphomas occurring in the older population, but the occurrence of lymphomas in two separable populations may indicate that some important distinction may have been overlooked: a different environmental cause, or different genetic alterations of lymphomas in the two age sets, or two different types of lymphomas that were mistakenly classified under one name, or there may be something wrong with the data (i.e., misdiagnoses, mix-ups during data collection). Big Data, by providing large numbers of cases, makes it easy to detect data incongruities (such as multimodality), when they are present. Explaining the causes for data incongruities is always a scientific challenge.

Multimodality in the age distribution of human diseases is an uncommon but well-known phenomenon. In the case of deaths resulting the Spanish flu of 1918, a tri-modal distribution was noticed (i.e., a high death rate in young, middle aged, and old individuals). In such cases, the observation of multimodality has provoked scientific interest, leading to fundamental discoveries in disease biology [9].

Section 12.7. Case Study: Big and Small Black Holes

If I didn't believe it, I would never have seen it.

Anon

The importance of inspecting data for multi-modality also applies to black holes. Most black holes have mass equivalents under 33 solar masses. Another set of black holes are supermassive, with mass equivalents of 10 or 20 billion solar masses. When there are objects of the same type, whose masses differ by a factor of a billion, scientists infer that there is something fundamentally different in the origin or development of these two variant forms of the same object. Black hole formation is an active area of interest, but current theory suggests that lower-mass black holes arise from pre-existing heavy stars. The supermassive black holes presumably grow from large quantities of matter available at the center of galaxies. The observation of bimodality inspired astronomers to search for black holes whose masses are intermediate between black holes with near-solar masses and the supermassive black holes. Intermediates have been found, and, not surprisingly, they come with a set of fascinating properties that distinguish them from other types of black holes. Fundamental advances in our understanding of the universe may sometimes follow from simple observations of multimodal data distributions.

Glossary

Case report The case report, also known as the case study, is a detailed description of a single event or situation, often devoted to an outlier, or a detail, or a unique occurrence of an observation. The concept of the case study is important in the field of data simplification because it highlights the utility of seeking general truths based on observations of rare events. Case reports are common in the biomedical literature, often beginning with a comment regarding the extreme rarity of the featured disease. You can expect to see phrases such as "fewer than a dozen have been reported in the literature" or "the authors have encountered no other cases of this lesion," or such and such a finding makes this lesion particularly uncommon and difficult to diagnose; and so on. The point that the authors are trying to convey is that the case report is worthy of publication specifically because the observation is rare. Too often, case reports serve merely as a cautionary exercise, intended to ward against misdiagnosis. The "beware this lesion" approach to case reporting misses the most important aspect of this type of publication; namely that science, and most aspects of human understanding, involve generalizing from the specific. When Isaac Newton saw an apple falling, he was not thinking that he could write a case report about how he once saw an apple drop, thus warning others not to stand under apple trees lest a rare apple might thump them upon the head. Newton generalized from the apple to all objects, addressing universal properties of gravity, and discovering the laws by which gravity interacts with matter. The case report gives us an opportunity to clarify the general way things work, by isolating one specific and rarely observed factor [2]. Data scientists must understand that rare cases are not exceptions to the general laws of reality; they are the exceptions upon which the general laws of reality are based.

Dimensionality The dimensionality of a data objects consists of the number of attributes that describe the object. Depending on the design and content of the data structure that contains the data object (i.e., database, array, list of records, object instance, etc.), the attributes will be called by different names, including field, variable, parameter, feature, or property. Data objects with high dimensionality create computational challenges, and data analysts typically reduce the dimensionality of data objects wherever possible.

Dynamic range Every measuring device has a dynamic range beyond which its measurements are without meaning. A bathroom scale may be accurate for weights that vary from 50 to 250 pounds, but you would not expect it to produce a sensible measurement for the weight of a mustard seed or an elephant.

Mean-field approximation A method whereby the average behavior for a population of objects substitutes for the behavior of each and every object in the population. This method greatly simplifies calculations. It is based on the observation that large collections of objects can be characterized by their average behavior. Mean-field approximation has been used with great success to understand the behavior of gases, epidemics, crystals, viruses, and all manner of large population phenomena.

Monte Carlo simulation This technique was introduced in 1946 by John von Neumann, Stan Ulam, and Nick Metropolis [10]. For this technique, the computer generates random numbers and uses the resultant values to simulate repeated trials of a probabilistic event. Monte Carlo simulations can easily simulate various processes (e.g., Markov models and Poisson processes) and can be used to solve a wide range of problems [6,11]. The Achilles heel of the Monte Carlo simulation, when applied to enormous sets of data, is that so-called random number generators may introduce periodic (non-random) repeats over large stretches of data [12]. What you thought was a fine Monte Carlo simulation, based on small data test cases, may produce misleading results for large data sets. The wise Big Data analyst will avail himself of the best possible random number generators, and will test his outputs for randomness. Various tests of randomness are available [13].

Nonparametric statistics Statistical methods that are not based on assumptions about the distribution of the sample population (e.g., not based on the assumption that the sample population fits a Gaussian distribution). Median, mode, and range are examples of common nonparametric statistics.

Outlier The term refers to a data point that lies far outside the value of the other data points in a distribution. The outlier may occur as the result of an error, or it may represent a true value that needs to be explained. When computing a line that is the "best fit" to the data, it is usually prudent to omit the outliers; otherwise, the best fit line may miss most of the data in your distribution. There is no strict rule for identifying outliers, but by convention, statisticians may construct a cut-off that lies 1.5 times the range of the lower quartile of the data, for small outliers, or 1.5 times the upper quartile range for large values.

Pareto's principle Also known as the 80/20 rule, Pareto's principle holds that a small number of items account for the vast majority of observations. For example, a small number of rich people account for the majority of wealth. Just 2 countries, India plus China, account for 37% of the world population. Within most countries, a small number of provinces or geographic areas contain the majority of the population of a country (e.g., East and West coastlines of the United States). A small number of books, compared with the total number of published books, account for the majority of book sales. Likewise, a small number of diseases account for the bulk of human morbidity and mortality. For example, two common types of cancer, basal cell carcinoma of skin and squamous cell carcinoma of skin account for about 1 million new cases of cancer each year in the United States. This is approximately the sum total of for all other types of cancer combined. We see a similar phenomenon when we count causes of death. About 2.6 million people die each year in the United States [14]. The top two causes of death account for 1,171,652 deaths (596,339 deaths from heart disease and 575,313 deaths from cancer [15]), or about 45% of all United States deaths. All of the remaining deaths are accounted for by more than 7000 conditions. Sets of data that follow Pareto's principle are often said to follow a Zipf distribution, or a power law distribution. These types of distributions are not tractable by standard statistical descriptors because they do not produce a symmetric bell-shaped curve. Simple measurements such as average and standard deviation have virtually no practical meaning when applied to Zipf distributions. Furthermore, the Gaussian distribution does not apply, and none of the statistical inferences built upon an assumption of a Gaussian distribution will hold on data sets that observe Pareto's principle [16].

Power law A mathematical formula wherein a particular value of some quantity varies as an inverse power of some other quantity [17,18]. The power law applies to many natural phenomena and describes the Zipf distribution or Pareto's principle. The power law is unrelated to the power of a statistical test.

Word lists Word lists are collections, usually in alphabetic order, of the different words that might appear in a corpus of text or a language dictionary. Such lists are easy to create. Here is a short Python script, words.py, that prepares an alphabetized list of the words occurring in a line of text. This script can be easily modified to create word lists from plain-text files.

```
import string
line = "a way a lone a last a loved a long the riverrun past eve and adam's from swerve of
    shore to bend of bay brings us by a commodius vicus"
linearray = sorted(set(line.split(" ")))
for item in linearray:
  print(item)
```

Here is the output:

```
a
adam's
and
bay
bend
brings
by
commodius
eve
from
last
lone
long
loved
of
past
riverrun
shore
swerve
the
to
us
vicus
way
```

Aside from word lists you create for yourself, there are a wide variety of specialized knowledge domain nomenclatures that are available to the public [19–24]. Linux distributions often bundle a wordlist, under filename "words," that is useful for parsing and natural language processing applications. A copy of the Linux wordlist is available at: http://www.cs.duke.edu/~ola/ap/linuxwords

Curated lists of terms, either generalized or restricted to a specific knowledge domain, are indispensable for a variety of applications (e.g., spell-checkers, natural language processors, machine translation, coding by term, indexing.) Personally, I have spent an inexcusable amount of time creating my own lists, when no equivalent public domain resource was available.

Zipf distribution George Kingsley Zipf (1902–50) was an American linguist who demonstrated that, for most languages, a small number of words account for the majority of occurrences of all the words found in prose. Specifically, he found that the frequency of any word is inversely proportional to its placement in a list of words, ordered by their decreasing frequencies in text. The first word in the frequency list will occur about twice as often as the second word in the list, three times as often as the third word in the list, and so on. Many Big Data collections follow a Zipf distribution (income distribution in a population, energy consumption by country, and so on). Zipf distributions within Big Data cannot be sensibly described by the standard statistical measures that apply to normal distributions. Zipf distributions are instances of Pareto's principle.

Here is a short Python script, zipf.py, that produces a Zipf distribution for a few lines of text.

```
import re, string
word_list=[];freq_list=[];format_list=[];freq={}
my_string = "Peter Piper picked a peck of pickled \
peppers. A peck of pickled peppers Peter Piper picked. \
If Peter Piper picked a peck of pickled peppers, \
Where is the peck of pickled peppers that Peter Piper \
picked?".lower()
word_list = re.findall(r'(\b[a-z]{1,}\b)', my_string)
for item in word_list:
  count = freq.get(item,0)
  freq[item] = count + 1
for key, value in freq.items():
  value = "000000" + str(value)
  value = value[-6:]
  format_list += [value + " " + key]
format_list = reversed(sorted(format_list))
print("\n".join(format_list))
```

Here is the output of the zipf,py script:

```
000004 piper
000004 pickled
000004 picked
000004 peter
000004 peppers
000004 peck
000004 of
000003 a
000001 where
000001 the
000001 that
000001 is
000001 if
```

References

[1] Berman JJ. Taxonomic guide to infectious diseases: understanding the biologic classes of pathogenic organisms. Cambridge, MA: Academic Press; 2012.

[2] Brannon AR, Sawyers CL. N of 1 case reports in the era of whole-genome sequencing. J Clin Invest 2013;123:4568–70.

[3] Subbiah IM, Subbiah V. Exceptional responders: in search of the science behind the miracle cancer cures. Future Oncol 2015;11:1–4.

[4] Perez-Pena R. New York's tally of heat deaths draws scrutiny. The New York Times; August 18, 2006.

[5] Chiang S. Heat waves, the "other" natural disaster: perspectives on an often ignored epidemic. Global Pulse, American Medical Student Association; 2006.

[6] Berman JJ, Moore GW. The role of cell death in the growth of preneoplastic lesions: a Monte Carlo simulation model. Cell Prolif 1992;25:549–57.

[7] Berman JJ. Methods in medical informatics: fundamentals of healthcare programming in Perl, Python, and Ruby. Boca Raton: Chapman and Hall; 2010.

[8] SEER. Surveillance Epidemiology End Results. National Cancer Institute. Available from: http://seer.cancer.gov/

[9] Berman J. Precision medicine, and the reinvention of human disease. Cambridge, MA: Academic Press; 2018.

[10] Cipra BA. The Best of the 20th century: editors name top 10 algorithms. SIAM News May 2000;33(4).

[11] Berman JJ, Moore GW. Spontaneous regression of residual tumor burden: prediction by Monte Carlo simulation. Anal Cell Pathol 1992;4:359–68.

[12] Sainani K. Error: What biomedical computing can learn from its mistakes. Biomed Comput Rev Fall 2011;12–9.

[13] Marsaglia G, Tsang WW. Some difficult-to-pass tests of randomness. J Stat Softw 2002;7:1–8.

[14] The World Factbook. Washington, DC: Central Intelligence Agency; 2009.

[15] Hoyert DL, Heron MP, Murphy SL, Kung H-C. Final data for 2003. Natl Vital Stat Rep 2006;54(13) April 19.

[16] Berman JJ. Rare diseases and orphan drugs: keys to understanding and treating common diseases. Cambridge, MD: Academic Press; 2014.

[17] Newman MEJ. Power laws, Pareto distributions and Zipf's law. Contemp Phys 2005;46:323–51.

[18] Clauset A, Shalizi CR, Newman MEJ. Power-law distributions in empirical data. SIAM Rev 2009;51:661–703.

[19] Medical Subject Headings. U.S. National Library of Medicine. Available at: https://www.nlm.nih.gov/mesh/filelist.html [viewed on July 29, 2015].

[20] Berman JJ. A tool for sharing annotated research data: the "Category 0" UMLS (unified medical language system) vocabularies. BMC Med Inform Decis Mak 2003;3:6.

[21] Berman JJ. Tumor taxonomy for the developmental lineage classification of neoplasms. BMC Cancer 2004;4:88.

[22] Hayes CF, O'Connor JC. English-Esperanto dictionary. London: Review of Reviews Office; 1906. Available at: http://www.gutenberg.org/ebooks/16967 [viewed July 29, 2105].

[23] Sioutos N, de Coronado S, Haber MW, Hartel FW, Shaiu WL, Wright LW. NCI Thesaurus: a semantic model integrating cancer-related clinical and molecular information. J Biomed Inform 2007;40:30–43.

[24] NCI Thesaurus. National Cancer Institute, U.S. National Institutes of Health. Bethesda, MD. Available at: ftp://ftp1.nci.nih.gov/pub/cacore/EVS/NCI_Thesaurus/ [viewed on July 29, 2015].

13

Using Random Numbers to Knock Your Big Data Analytic Problems Down to Size

Section 13.1. The Remarkable Utility of (Pseudo)Random Numbers

Chaos reigns within.
Reflect, repent, and reboot.
Order shall return.

Computer haiku by Suzie Wagner

As discussed in Section 11.1, much of the difficulty that of Big Data analysis comes down to combinatorics. As the number of data objects increases, along with the number of attributes that describe each object, it becomes computationally difficult, or impossible, to compute all the pairwise comparisons that would be necessary for analytics tasks (e.g., clustering algorithms, predictive algorithms). Consequently, Big Data analysts are always on the lookout for innovative, non-combinatoric approaches to traditionally combinatoric problems.

There are many different approaches to data analysis that help us to reduce the complexity of data (e.g., principal component analysis) or transform data into a domain that facilitates various types of analytic procedures (e.g., the Fourier transform). In this chapter, we will be discussing an approach that is easy to learn, easy to implement, and which is

Principles and Practice of Big Data. https://doi.org/10.1016/B978-0-12-815609-4.00013-3

particularly well-suited to enormous data sets. The techniques that we will be describing fall under different names, depending on how they are applied (e.g., resampling, Monte Carlo simulations, bootstrapping), but they all make use of random number generators, and they all involve repeated sampling from a large population of data, or from infinite points in a distribution. Together, these heuristic techniques permit us to perform nearly any type of Big Data analysis we might imagine in a few lines of Python code [1–4]. [Glossary Heuristic technique, Principal component analysis]

In this chapter, we will explore:

- Random numbers (strictly, pseudorandom numbers) [Glossary Pseudorandom number generator]
- General problems of probability
- Statistical tests
- Monte Carlo simulations
- Bayesian models
- Methods for determining whether there are multiple populations represented in a data set
- Determining the minimal sample size required to test a hypothesis [Glossary Power]
- Integration (calculus)

Let us look at how a random number generator works in Python, with a 3-line random.py script.

```
import random
for iterations in range(10):
    print(random.uniform(0,1))
```

Here is the sample output, listing 10 random numbers in the range 0–1:

```
0.594530508550135
0.289645594799927
0.393738321195123
0.648691742041396
0.215592023796071
0.663453594144743
0.427212189295081
0.730280586218356
0.768547788018729
0.906096189758145
```

Had we chosen, we could have rendered an integer output by multiplying each random number by 10 and rounding up or down to the closest integer.

Now, let us perform a few very simple simulations that confirm what we already know, intuitively. Imagine that you have a pair of dice and you would like to know how often you might expect each of the numbers (from one to six) to appear after you've thrown one die [5].

Let us simulate 60,000 throws of a die using the Python script, randtest.py:

```python
import random, itertools
one_of_six = 0
for i in itertools.repeat(None, 60000):
  if (int(random.uniform(1,7))) > 5:
    one_of_six = one_of_six + 1
print(str(one_of_six))
```

The script, randtest.pl, begins by setting a loop that repeats 60,000 times, each repeat simulating the cast of a die. With each cast of the die, Python generates a random integer, 1 through 6, simulating the outcome of a throw.

The script yields the total number die casts that would be expected to come up "6." Here is the output of seven consecutive runs of the randtest.py script

```
10020
10072
10048
10158
10000
9873
9899
```

As one might expect, a "6" came up about 1/6th of the time, or about 10,000 times in the 60,000 simulated roles. We could have chosen any of the other five outcomes of a die role (i.e., 1, 2, 3, 4, or 5), and the outcomes would have been about the same.

Let us use a random number generator to calculate the value of pi, without measuring anything, and without resorting to summing an infinite series of numbers. Here is a simple python script, pi.py, that does the job.

```python
import random, itertools
from math import sqrt
totr = 0; totsq = 0
for i in range(10000000):
  x = random.uniform(0,1)
  y = random.uniform(0,1)
  r = sqrt((x*x) + (y*y))
  if r < 1:
    totr = totr + 1
  totsq = totsq + 1
print(float(totr)*4.0/float(totsq))
```

The script returns a fairly close estimate of pi.

```
output: 3.1414256
```

The value of pi is the ratio of a circle of unit radius (pi * r^2) divided by the area of a square of unit radius (r ̂2). When we randomly select points within a unit distance in the x or the y dimension, we are filling up a unit square. When we count the number of randomly selected points within a unit radius of the origin, and compare this number with the number of points in the unit square, we are actually calculating a number that comes fairly close to pi. In this simulation, we are looking at one quadrant, but the results are equivalent, and we come up with a number that is a good approximation of pi (i.e., 3.1414256, in this simulation).

With a few extra lines of code, we can send the output to a to a data file, named pi.dat, that will helps us visualize how the script works. The data output (held in the pi.dat file) contains the x, y data points, generated by the random number generator, meeting the "if" statement's condition that the hypotenuse of the x, y coordinates must be less than one (i.e., less than a circle of radius 1). We can plot the output of the script with a few lines of Gnuplot code:

```
gnuplot > set size square
gnuplot > unset key
gnuplot > plot 'c:\ftp\pi.dat'
```

The resulting graph is a quarter-circle within a square (Fig. 13.1).

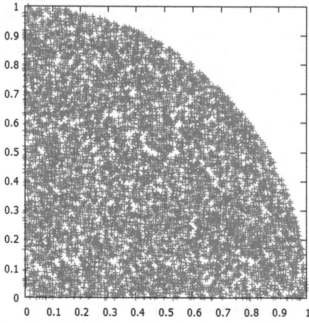

FIG. 13.1 The data points produced by 10,000 random assignments of x and y coordinates in a range of 0–1. Randomly assigned data points whose hypotenuse exceeds "1" are excluded.

   Let us see how we can simulate calculus operations, using a random number generator. We will integrate the expression f(x) = $x^3 - 1$. First, let us visualize the relationship between f(x) and x, by plotting the expression. We can use a very simple Python script, plot_funct.py, that can be trivially generalized to plot almost any expression we choose, over any interval.

```
import math, random, itertools
import numpy as np
import matplotlib.pyplot as plt
x=0.000

def f(x):
   return (x*x*x -1)

x = np.arange(0.7, 1.5, 0.01)
plt.plot(x, f(x))
plt.show()
```

The output is a graph, produced by Python's matplotlib module (Fig. 13.2).

   Let us use a random number generator, in the Python script integrator.py, to perform the integration and produce an approximate evaluation of the integral (i.e., the area under the curve of the equation).

```
import math, random, itertools
def f(x):
   return (x*x*x -1)
```

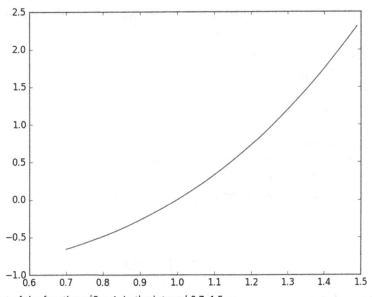

FIG. 13.2 The plot of the function $x^3 - 1$, in the interval 0.7–1.5.

```
range = 2       #let's evaluate the integral from x=1 to x=3
running_total = 0
for i in itertools.repeat(None, 1000000):
   x = float(random.uniform(1,3))
   running_total = running_total + f(x)
integral = (running_total / 1000000) * range
print(str(integral))
```

Here is the output:

```
18.001426696174935
```

The output comes very close to the exact integral, 18, calculated for the interval between 1 and 3. How did our short Python script calculate the integral? The integral of a function is equivalent to the area under of the curve of the function, in the chosen interval. The area under the cover is equal to the size of the interval (i.e., the x coordinate range) multiplied by the average value of the function in the interval. By calculating a million values of the function, from randomly chosen values of x in the selected interval, and taking the average of all of those values, we get the average value of the function. When we multiply this by the interval, we get the area, which is equal to the value of the integral.

Let us apply our newfound ability to perform calculus with a random number generator and calculate one of the most famous integrals in mathematics.

The integral of $1/x$ equals the natural logarithm of x (Fig. 13.3). This tells us that the integral of $(1/x)dx$ evaluated in the integral from 5 to 105 will be equal to ln(105) minus ln(5). Let us evaluate the integral in the range from $x = 5$ to $x = 105$, using our random number generator, alongside the calculation performed with Python's numpy (math) module, to see how close our approximation came. We will use the Python script, natural.py

```
import math, random, itertools
import numpy as np
def f(x):
   return (1 / x)
range = 100
running_total = 0
for i in itertools.repeat(None, 1000000):
   x = float(random.uniform(5,105))
   running_total = float(running_total + f(x))
```

$$\ln(x) = \int \tfrac{1}{x} dx$$

FIG. 13.3 The integral of the inverse of x is equal to the natural logarithm of x.

```
integral = (running_total / 1000000) * range
print("The estimate produced with a random number generator is: " + str
    (integral))
print("The value produced with Python's numpy function for ln(x) is: " +
    str(np.log(105) - np.log(5)))
```

Here is the output of the natural.py script.

```
The estimate produced with a random number generator
    is:    3.04209561069317
The value produced with Python's numpy function for ln(x)
    is: 3.04452243772
```

Not bad. Of course, python programmers know that they do not need to use a random number generator to solve calculus problems in python. The numpy and sympy modules seem to do the job nicely. Here is a short example wherein the a classic integral equation is demonstrated:

```
>>> import sympy as sy
>>> import numpy as np
>>> x = sy.Symbol('x')
>>> sy.integrate(1/x,x)
log(x)
```

Or, as any introductory calculus book would demonstrate: $\ln x = $ integral $1/x \, dx$

Why would anyone bother to integrate using a random number generator when they can produce an exact solution using elementary calculus? It happens that integration by repeated sampling using a random number generator comes in handy when dealing with multi-dimensional functions, particularly when the number of dimensions exceeds 8 (i.e., when there are 8 or more quantitative attributes describing each variable). In Big Data, where the variables have many attributes, standard computational approaches may fail. In these cases the methods described in this section may provide the most exact and the most practical solutions to a large set of Big Data computations. [Glossary Curse of dimensionality]

Section 13.2. Repeated Sampling

Every problem contains within itself the seeds of its own solution.

Stanley Arnold

In Big Data analytics, we can use a random number generator to solve problems in the areas of clustering, correlations, sample size determination, differential equations, integral equations, digital signals processing, and virtually every subdiscipline of physics [3,1,6,5]. Purists would suggest that we should be using formal, exact, and robust mathematical techniques for all these calculations. Maybe so, but there is one general set of problems for which the random number generator is the ideal tool; requiring no special assumptions about the data being explored, and producing answers that are as reliable as anything that can be produced with computationally intensive exact techniques. This set of problems typically consists of hypothesis testing on sets of data of uncertain distribution (i.e., not Gaussian) that are not strictly amenable to classic statistical analysis. The methods by which these problems are solved are the closely related techniques of resampling (from a population or data set) and permutating, both of which employ random number generators. [Glossary Resampling statistics, Permutation test, Resampling versus Repeated Sampling, Modified random sampling]

Resampling methods are not new. The resampling methods that are commonly used today by data analysts have been around since the early 1980s [3,1,2]. The underlying algorithms for these methods are so very simple that they have certainly been in use, using simple casts of dice, for thousands of years.

The early twentieth century saw the rise of mathematically rigorous statistical methods, that enabled scientists to test hypotheses and draw conclusions from small or large collections of data. These tests, which required nothing more than pencil and paper to perform, dominated the field of analysis and are not likely to be replaced anytime soon. Nonetheless, the advent of fast computers provides us with alternative methods of analysis that may lack the rigor of advanced statistics, but have the advantage of being easily comprehensible. Calculations that require millions of operations can be done essentially instantly, and can be programmed with ease. Never before, in the history of the world, has it been possible to design and perform resampling exercises, requiring millions or billions of iterative operations, in a matter of seconds, on computers that are affordable to a vast number of individuals in developed or developing countries. The current literature abounds with resources for scientists with rudimentary programming skills, who might wish to employ resampling techniques [7,8].

For starters, we need to learn a new technique: shuffling. Python's numpy module provides a simple method for shuffling the contents of a container (such as the data objects listed in an array) to produce a random set of objects. Here is Python's shuffle_100.py script, that produces a shuffled list of numbers ranging from 0 to 99:

```
import numpy as np
sample = np.arange(100)
gather = []
for i in sample:
  np.random.shuffle(sample)
print(sample)
```

Here is the output of the shuffle_100.py script:

```
[27 60 21 99 17 79 49 62 81  2 88 90 45 61 66 80 50 31 59 24 53 29 64 33 30
 41 13 23  0 67 78 70  1 35 18 86 25 93  6 98 97 84  9 12 56 48 74 96  4 32
```

```
44 11 19 38 26 52 87 77 39 91 92 76 65 75 63 57  8 94 51 69 71  7 73 34 20
40 68 22 82 15 37 72 28 47 95 54 55 58  5  3 89 46 85 16 42 36 43 10 83 14]
```

We will be using the shuffle method to help us answer a question that arises often, whenever we examine sets of data: "Does my data represent samples from one homogeneous population of data objects, or does my data represent samples from different classes of data objects that have been blended together in one collection?" The blending of distinctive classes of data into one data set is one of the most formidable biases in experimental design, and has resulted in the failures of many clinical trials, the misclassification of diseases, and the misinterpretation of the significance of outliers. How often do we fail to understand our data, simply because we cannot see the different populations lurking within? In Section 12.6, we discussed the importance of finding multimodal peaks in data distributions. Our discussion of multimodal peaks and separable subpopulations begged the question as to how we can distinguish peaks that represent subpopulations from peaks that represent random perturbations in our data. [Glossary Blended class]

Let us begin by using a random number generator to make two separate populations of data objects. The first population of data objects will have values that range uniformly between 1 and 80. The second population of data objects will have values that range uniformly between 20 and 100. These two populations will overlap (between 20 and 80), but they are different populations, with different population means, and different sets of properties that must account for their differences in values. We'll call the population ranging from 1 to 80 our low_array and the population ranging from 20 to 100 as high_array.

Here is a Python script, low_high.py, that generates the two sets of data, representing 50 randomly selected objects from each population:

```python
import numpy as np
from random import randint
low_array = [] ; high_array = []
for i in range(50):
  low_array.append(randint(1,80))
print("Here's the low-skewed data set " + str(low_array))
for i in range(50):
  high_array.append(randint(21,100))
print("\nHere's the high-skewed data set " + str(high_array))
av_diff = (sum(high_array)/len(high_array)) - (sum(low_array)/len
  (low_array))
print("\nThe difference in average value of the two arrays is:
  " + str(av_diff ))
```

Here is the output of the low_high.py script:

```
Here is the low-skewed data set [31, 8, 60, 4, 64, 35, 49, 80, 6, 9, 14, 15,
50, 45, 61, 77, 58, 24, 54, 45, 44, 6, 78, 59, 44, 61, 56, 8, 30, 34, 72, 33,
14, 13, 45, 10, 49, 65, 4, 51, 25, 6, 37, 63, 19, 74, 78, 55, 34, 22]
```

```
Here is the high-skewed data set [36, 87, 54, 98, 33, 49, 37, 35, 100, 48,
71, 86, 76, 93, 98, 99, 92, 68, 29, 34, 64, 30, 99, 76, 71, 32, 77, 32, 73,
54, 34, 44, 37, 98, 42, 81, 84, 56, 55, 85, 55, 22, 98, 72, 89, 24, 43, 76,
87, 61]
```

```
The difference in average value of the two arrays is: 23.919999999999995
```

The low-skewed data set consists of 50 random integers selected from the interval 1–80. The high-skewed data set consists of 50 random numbers selected from the interval between 20 and 100. Notice that not all possible outcomes in these two intervals are represented (i.e., there is no number 2 in the low-skewed data set and there is no number 25 in the high-skewed data set). If we were to repeat the low_high.py script, we would generate two different sets of numbers. Also, notice that the two populations have different average values. The difference between the average value of the low-skewed data population and the high-skewed data population is 23.9, in this particular simulation.

Now, we are just about ready to determine whether the two populations are statistically different. Let us repeat the simulation. This time, we will combine the two sets of data into one array that we will call "total_array," containing 100 data elements. Then we will shuffle all of the values in total_array and we will create two new arrays: a left array consisting of the first 50 values in the shuffled total_array and a right array consisting of the last 50 values in the total_array. Then we will find the difference between the average of the 50 values of the left array of the right array. We will repeat this 100 times, printing out the lowest five differences in averages and the highest five differences in averages. Then we will stop and contemplate what we have done.

Here is the Python script, pop_diff.py

```python
import numpy as np
from random import randint
low_array = []
high_array = []
gather = []
for i in range(50):
    low_array.append(randint(1,80))
for i in range(50):
    high_array.append(randint(21,100))
av_diff = (sum(high_array)/len(high_array)) - (sum(low_array)/len
    (low_array))
print("The difference in the average value of the high and low arrays is:
    " + str(av_diff))
sample = low_array + high_array
for i in sample:
    np.random.shuffle(sample)
    right = sample[50:]
```

```
    left = sample[:50]
    gather.append(abs((sum(left)/len(left)) - (sum(right)/len
    (right))))
sorted_gather = sorted(gather)
print("The 5 largest differences of averages of the shuffled arrays
    are:")
print(str(sorted_gather[95:]))

output:
The difference in the average value of the high and low arrays is:
    19.580000000000005
The 5 largest differences of averages of the shuffled arrays are:
    [8.780000000000001, 8.899999999999999, 9.46, 9.82,
    9.899999999999999]
```

Believe it or not, we just demonstrated that the two arrays that we began with (i.e., the array with data values randomly distributed between 0 and 80; and the array with data values randomly distributed between 20 and 100) represent two different populations of data (i.e., two separable classes of data objects). Here is what we did and how we reached our conclusion.

1. We recomputed two new arrays, with 50 items each, with data values randomly distributed between 0 and 80; and the array with data values randomly distributed between 20 and 100.
2. We calculated the difference of the average size of an item in the first array, compared with the average size of an item in the second array. This came out to 19.58 in this simulation.
3. We combined the two arrays into a new array of 100 items, and we shuffled these items 100 times, each time splitting the shuffle in half to produce two new arrays of 50 items each., We calculated the difference in the average value of the two arrays (produced by the shuffle).
4. We found the five sets of shuffled arrays (the two arrays produced by a shuffle of the combined array) that had the largest differences in their values (corresponding to the upper 5% of the combined and shuffled populations) and we printed these numbers.

The upper 5 percentile differences among the shuffled arrays (i.e., 8.78, 8.89, 9.46, 9.82, 9.89) came nowhere close to the difference of 19.58 we calculated for the original two sets of data. This tells us that whenever we shuffle the combined array, we never encounter differences anywhere near as great as what we observed in the original arrays. Hence, the original two arrays cannot be explained by random selection from one population (obtained when we combined the two original arrays). The two original arrays must represent two different populations of data objects.

A note of caution regarding scalability. Shuffling is not a particularly scalable function [9]. Shuffling a hundred items is a lot easier than shuffling a million items. Hence, when writing short programs that test whether two data arrays are statistically separable, it is best to impose a limit to the size of the arrays that you create when you sample from the combined array. If you keep the shuffling size small, you can compensate by repeating the shuffle nearly as often as you like.

– Sample Size and Power Estimates

In the prior exercise, we generated two populations, of 50 samples each, and we determined that the two populations were statistically separable from one another. Would we have been able to draw the same conclusion if we had performed the exercise using 25 samples in each population in each population? How about 10 samples? We would expect that as the population sizes of the two populations shrinks, the likelihood that we could reliably distinguish one population from another will fall. How do we determine the minimal population size necessary to perform our experiment?

The power of a trial is the likelihood of detecting a difference in two populations, if the difference actually exists. The power is related to the sample size. At a sufficiently large sample size, you can be virtually certain that the difference will be found, if the difference exists. Resampling permits the experimenter to conduct repeated trials, with different sample sizes, and under conditions that simulate the population differences that are expected. For example, an experimenter might expect that a certain drug produces a 15% difference in the measured outcomes in the treated population compared with the control population. By setting the conditions of the trials, and by performing repeated trials with increasing sizes of simulated populations, the data scientist can determine the minimum sampling size that consistently (e.g., in greater than 95% of the trials), demonstrates that the treated population and the control population are separable. Hence, using a random number generator and a few short scripts, the data scientist can determine the sampling size required to yield a power that is acceptable for a "real" experiment. [Glossary Sample size, Sampling theorem]

Section 13.3. Monte Carlo Simulations

One of the marks of a good model - it is sometimes smarter than you are.
Paul Krugman, Nobel prize-winning economist

Random number generators are well suited to Monte Carlo simulations, a technique described in 1946 by John von Neumann, Stan Ulam, and Nick Metropolis [10]. For these simulations, the computer generates random numbers and uses the resultant values to represent outcomes for repeated trials of a probabilistic event. Monte Carlo simulations can easily simulate various processes (e.g., Markov models and Poisson processes) and can be used to solve a wide range of problems [11,12].

For example, consider how biologists may want to model the growth of clonal colonies of cells. In the simplest case, wherein cell growth is continuous, a single cell divides, producing two cells. Each of the daughter cells divides, producing a total of four cells. The size of the colony increases as powers of 2. A single liver cancer cell happens to have a volume of about 30,000 cubic microns [13]. If the cell cycle time is one day, then the volume of a liver cell colony, grown for 45 days, and starting at day 1 with a single cell, would be $1\,m^3$. In 55 days, the volume would exceed $1000\,m^3$. If an unregulated tumor composed of malignant liver cells were to grow for the normal lifetime of a human, it would come to occupy much of the measured universe [14]. Obviously, unregulated cellular growth is unsustainable. In tumors, as in all systems that model the growth of cells and cellular organisms, the rate of cell growth is countered by the rate of cell death.

With the help of a random number generator, we can model the growth of colonies of cells by assigning each cell in the colony a probability of dying. If we say that the likelihood that a cell will die is 50%, then we are saying that its chance of dividing (i.e., producing two cells) is the same as its chance of dying (i.e., producing zero cells and thus eliminating itself from the population). We can create a Monte Carlo simulation of cell growth by starting with some arbitrary number of cells (let us say three), and assigning each cell an arbitrary chance of dying (let us say 49%). We can assign each imaginary cell to an array, and we can iterate through the array, cell by cell. As we iterate over each cell, we can use a random number generator to produce a number between 0 and 1. If the random number is less than 0.49, we say that the cell must die, dropping out of the array. If a cell is randomly assigned a number that is greater than 0.49, then we say that the cell can reproduce, to produce two cells that will take their place in the array. Every iteration over the cells in the array produces a new array, composed of the lucky winners and their offspring, from the prior array. In theory, we can repeat this process forever. More practically, we can repeat this process until the size of the colony reduces to zero, or until the colony becomes so large that additional iterations become tedious (even for a computer).

Here is the Python script, clone.py, that creates a Monte Carlo simulation for the growth of a colony, beginning with three cells, with a likelihood of cell death for all cells, during any generation, of 0.41. The script exits if the clone size dwindles to zero (i.e., dies out) or reaches a size exceeding 800 (presumably on the way to growing without limit).

```
import numpy.random as npr
import sys
death_chance = 0.41; cell_array = [1, 1, 1]; cell_array_incremented =
[1,1,1]
while(len(cell_array) > 0):
  for cell in cell_array:
      randnum = npr.randint(1,101)
      if randnum > 100 * death_chance:
        cell_array_incremented.append(1)
      else:
        cell_array_incremented.remove(1)
  if len(cell_array_incremented) < 1:
    sys.exit()
```

```
        if len(cell_array_incremented) > 800:
            sys.exit()
        cell_array = cell_array_incremented
        print(len(cell_array_incremented), end = ", ")
```

For each cell in the array, if the random number generated for the cell exceeds the likelihood of cell death, then the colony array is incremented by one. Otherwise, the colony array is decremented by 1. This simple step is repeated for every cell in the colony array, and is repeated for every clonal generation.

Here is a sample output when death_chance = 0.49

```
multiple outputs:

First trial:
2, 1, 1, 1,

Second trial:
4, 3, 3, 1,

Third trial:
6, 12, 8, 15, 16, 21, 30, 37, 47, 61, 64, 71, 91, 111, 141, 178, 216, 254,
310, 417, 497, 597, 712,

Fourth trial:
2, 4, 4, 4, 3, 3, 4, 4, 8, 5, 6, 4, 4, 6, 4, 10, 19, 17, 15, 32, 37, 53, 83, 96,
128, 167, 188, 224, 273, 324, 368, 416, 520, 642,

Fifth trial
4, 3, 7, 9, 19, 19, 26, 36, 45, 71, 88, 111, 119, 157, 214, 254, 319, 390,
480, 568, 675,

Sixth trial:
2, 1, 2, 2, 1,

Seventh trial:
2, 1,

Eighth trial:
2, 1, 1, 1, 4, 8, 10, 8, 10, 9, 10, 15, 11, 12, 16, 15, 21, 21, 35, 44, 43, 41,
47, 68, 62, 69, 90, 97, 121, 181, 229, 271, 336, 439, 501, 617, 786,
```

In each case the clone increases or decreases with each cell cycle until the clone reaches extinction or exceeds our cut off limit. These simulations indicate that clonal growth is precarious, under conditions when the cell death probability is approximately 50%. In these two simulations the early clones eventually die out. Had we repeated the simulation hundreds of times, we would have seen that most clonal simulations end in the extinction of the clone; while a few rare simulations yield a large, continuously expanding population, with virtually no chance of reaching extinction. In a series of papers by Dr. William Moore and myself, we developed Monte Carlo simulations of tumor cell growth. Those simulations suggested that very small changes in a tumor cell's death probability (per replication generation) profoundly affected the tumor's growth rate [11,12,15]. This simulation suggested that chemotherapeutic agents that can incrementally increase the death rate of tumor cells may have enormous treatment benefit. Furthermore, simulations showed that if you simulate the growth of a cancer from a single abnormal cancer cell, most simulations result in the spontaneous extinction of the tumor, an unexpected finding that helps us to understand the observed high spontaneous regression rate of nascent growths that precede the development of clinically malignant cancers [16,17].

This example demonstrates how simple Monte Carlo simulations, written in little more than a dozen lines of Python code, can simulate outcomes that would be difficult to compute using any other means.

Section 13.4. Case Study: Proving the Central Limit Theorem

The solution to a problem changes the nature of the problem.

John Peers ("Peer's Law" 1,001 Logical Laws)

The Central Limit Theorem is a key concept in probability theory and statistics. It asserts that when independent random variables are added, their sum tends toward a normal distribution (i.e., a "bell curve"). The importance of this theorem is that statistical methods designed for normal distributions will also apply in some situations wherein variables are chosen randomly from non-normal distributions.

For many years, the Central Limit Theorem was a personal stumbling block for me; I simply could not understand why it was true. It seemed to me that if you randomly select numbers in an interval, you would always get a random number, and if you summed and averaged two random numbers, you'll get another random number that is equally likely to by anywhere within the interval (i.e., not distributed along a bell curve with a central peak). According to the Central Limit Theorem, the sum of repeated random samplings will produce lots of numbers in the center of the interval, and very few or no numbers at the extremes.

As previously mentioned, repeated sampling allows us to draw inferences about a population, without examining every member of the population; a handy trick for Big Data analyses. In addition, repeated sampling allows us to test hypotheses that we are too dumb to understand (speaking for myself). In the case of the Central Limit Theorem, we can simulate a proof of the Central Limit Theorem by randomly selecting numbers in an interval (between 0 and 1) many times (say 10,000), and averaging their sum. We can repeat each of these trials 10,000 times, and then plot where the 10,000 values lie. If we get a Bell curve, then the Central Limit Theorem must be correct. Here is the Python script, central. py, that computes the results of the simulation, and plots the points as a graph.

```
import random, itertools
from matplotlib import numpy
import matplotlib.pyplot as plt
out_file = open('central.dat',"w")
randhash = {}
for i in itertools.repeat(None, 10000):
  product = 0
  for n in itertools.repeat(None, 10000):
    product = product + random.uniform(0,1)
  product = int(product)
  if product in randhash:
    randhash[product] = randhash[product] + 1
  else:
    randhash[product] = 1
lists = sorted(randhash.items())
x, y = zip(*lists)
plt.plot(x, y)
plt.show()
```

Here is the resultant graph (Fig. 13.4).

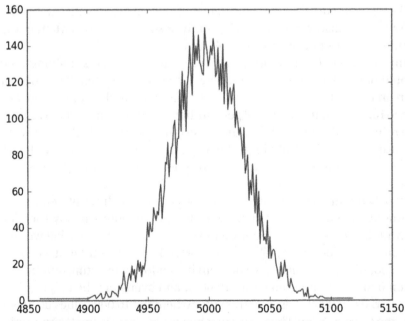

FIG. 13.4 The results of the simulation of the Central Limit Theorem (as executed in the Python script central.py), is a bell-shaped curve.

Without bludgeoning the point, after performing the simulation, and looking at the results, the Central Limit Theorem suddenly makes perfect sense to me. The reader can draw her own conclusion.

Section 13.5. Case Study: Frequency of Unlikely String of Occurrences

Luck is believing you're lucky.

Tennessee Williams

Imagine this scenario. A waitress drops a serving tray three times while serving each of three consecutive customers on the same day. Her boss tells her that she is incompetent, and indicates that he should fire her on the spot. The waitress objects, saying that the other staff drops trays all the time. Why should she be singled out for punishment simply because she had the bad luck of dropping three trays in a row. The manager and the waitress review the restaurant's records and find that there is a 2% drop rate, overall (i.e., a tray is dropped 2% of the time when serving customers), and that the waitress who dropped the three trays in one afternoon had had a low drop rate prior to this day's performance. She cannot explain why three trays dropped consecutively, but she supposes that if anyone works the job long enough, the day will come when they drop three consecutive trays.

We can resolve this issue, very easily, with the Python script runs.py, that simulates a million customer wait services and determines how often we are likely to see a string of three consecutive dropped trays.

```
import random, itertools
errorno = 0;
for i in itertools.repeat(None, 1000000): #let's do 1 million table
   services
   x = int(random.uniform(0,100))      #x any integer from 0 to 99
   if (x < 2):                #x simulates a 2% error rate
      errorno = errorno + 1
   else:
      errorno = 0
   if (errorno == 3):
      print("Uh oh. 3 consecutive errors")
      errorno = 0
```

Here is the output from one execution of runs.py

```
Uh oh. 3 consecutive errors
Uh oh. 3 consecutive errors
Uh oh. 3 consecutive errors
Uh oh. 3 consecutive errors
```

```
Uh oh. 3 consecutive errors
Uh oh. 3 consecutive errors
Uh oh. 3 consecutive errors
Uh oh. 3 consecutive errors
```

The 11-line Python script simulates 1 million table services. Each service is assigned a random number between 0 and 100. If the randomly assigned number is less than 2 the simulation counts as a dropped tray (because the drop rate is 2%). Now we watch to see the outcome of the next two served trays. If three consecutive trays are dropped, we print out "Uh oh. 3 consecutive errors") and we resume our simulations.

In this trial of 1 million table services, using a 2% error rate, the modeled waitress had eight runs of three consecutive tray drops. Since a million table services might possibly represent the total number of customers served by a waitress in her entire career, one can say that she should be allowed at least eight episodes of three consecutive tray drops, in her lifetime.

Section 13.6. Case Study: The Infamous Birthday Problem

I want to thank you for making this day necessary.

Yogi Berra

Let us use our random number generator to tackle the infamous birthday problem. It may seem unintuitive, but if you have a room holding 23 people, the odds are about even that two or more of the group will share the same birth date. The solution of the birthday problem has become a popular lesson in introductory probability courses. The answer involves an onerous calculation, involving lots of multiplied values, divided by an enormous exponential (Fig. 13.5).

If we wanted to know the probability of finding two or more individuals with the same birthday, in a group of 30 individuals, we could substitute 365 for n and 30 for k, and we would learn that the odds are about 70%. Or, we could design a simple little program, using a random number generator, to create an intuitively obvious simulation of the problem.

Here is the Python script, birthday.py, that conducts 10,000 random simulations of the birthday problem:

```
import random, itertools
success = 0
for i in itertools.repeat(None, 10000):
```

$$\frac{n(n-1)(n-2)(n-k+1)}{n^k}$$

FIG. 13.5 Calculating the general solution of the birthday problem. "n" is the number of days in a year. "k" is the number of people.

```
  date_hash = {}
  for j in itertools.repeat(None, 30):
    date = int(random.uniform(1,365))
    if date in date_hash:
      date_hash[date] = date_hash[date] + 1
    else:
      date_hash[date] = 1
    if (date_hash[date] == 2):
      success = success + 1
      break
print(str(success / 10000))
```

Here is the output from six consecutive runs of birthday.py, with 10,000 trials in each run:

```
0.7076
0.7083
0.7067
0.7101
0.7087
0.7102
```

The calculated probability is about 70%. The birthday.py script created variables, assigning each variable a birth date selected at random from a range of 1–365 (the number of days in the year). The script then checked among the 30 assigned variables, to see if any of them shared the same birthday (i.e., the same randomly assigned number). If so, then the set of 30 assigned variables was deemed a success. The script repeated this exercise 10,000 times, incrementing, by one, the number of successes whenever a match was found in the 30 assignments. At the end of it all, the total number of successes (i.e., instances where there is a birthday match in the group of 30) divided by the total number of simulations (10,000 in this case) provides the likelihood that any given simulation will find a birthday match. The answer, which happens to be about 70%, is achieved without the use of any knowledge of probability theory.

Section 13.7. Case Study (Advanced): The Monty Hall Problem

Of course I believe in luck. How otherwise to explain the success of some people you detest?

Jean Cocteau

This is the legendary Monty Hall problem, named after the host of a televised quiz show, where contestants faced a similar problem: "The player faces three closed containers, one containing a prize and two empty. After the player chooses, s/he is shown that one of the other two containers is empty. The player is now given the option of switching from her/his original choice to the other closed container. Should s/he do so? Answer: Switching will double the chances of winning."

Marilyn vos Savant, touted by some as the world's smartest person, correctly solved the Monty Hall problem in her newspaper column. When she published her solution, she received thousands of responses, many from mathematicians, disputing her answer.

Basically, this is one of those rare problems that seems to defy common sense. Personally, whenever I have approached this problem using an analytic approach based on probability theory, I come up with the wrong answer.

In desperation, I decided to forget everything I thought I knew about probability, in favor of performing the Monty Hall game, with a 10-line Python script, montyhall.py, that uses a random number generator to simulate outcomes.

```
import random, itertools
winner = 0; box_array = [1,2,3]
for i in itertools.repeat(None, 10000):
    full_box=int(random.uniform(1,4))   #randomly picks 1,2,or 3 as prize box
    guess_box=int(random.uniform(1,4))) #represents your guess, for prize box
    del box_array[full_box - 1]         #prize box deletes itself from array
    if guess_box in box_array:          #if your first guess is in the remaining array
                                        # (which excludes prize and includes the second
                                        #empty box), then you must have won
                                        #when you chose to switch your choice
        winner = winner + 1
    box_array = [1,2,3]
print(winner)
```

Here are the outputs of nine consecutive runs of montyhall.py script:

```
6710
6596
6657
6698
6653
6684
6661
6607
6674
```

The script simulates the Monty Hall strategy where the player takes the option of switching her selection. By taking the switch option, she wins nearly two thirds of the time (about 6600 wins in 10,000 simulations), twice as often as when the switch option is declined. The beauty of the resampling approach is that the programmer does not need to understand why it works. The programmer only needs to know how to use a random number generator to create an accurate simulation of the Monte Hall problem that can be repeated thousands and thousands of times.

How does the Monty Hall problem relate to Big Data? Preliminary outcomes of experimental trials are often so dramatic that the trialists choose to re-design their protocol mid-trial. For example, a drug or drug combination may have demonstrated sufficient effectiveness to justify moving control patients into the treated group. Or adverse reactions may necessitate switching patients off a certain trial arm. In either case, the decision to switch protocols based on mid-trial observations is a Monty Hall scenario.

Section 13.8. Case Study (Advanced): A Bayesian Analysis

It's hard to detect good luck - it looks so much like something you've earned.

Frank A. Clark

The Bayes theorem relates probabilities of events that are conditional upon one another (Fig. 13.6). Specifically, the probability of A occurring given that B has occurred multiplied by the probability that B occurs is equal to the probability of B occurring, given that A has occurred multiplied by the probability that A occurs. Despite all the hype surrounding Bayes theorem, it basically indicates the obvious: that if A and B are conditional, then A won't occur unless B occurs and B won't occur unless A occurs.

Bayesian inferences involve computing conditional probabilities, based on having information about the likelihood of underlying events. For example, the probability of rain would be increased if it were known that the sky is cloudy. The fundamentals of Bayesian analysis are deceptively simple. In practice, Bayesian analysis can easily evade the grasp of intelligent data scientists. By simulating repeated trials, using a random number generator, some of the toughest Bayesian inferences can be computed in a manner that is easily understood, without resorting to any statistical methods.

Here is a problem that was previously posed by William Feller, and adapted for resampling statistics by Julian L. Simon [1]. Imagine a world wherein there are two classes of drivers. One class, the good drivers, comprise 80% of the population, and the likelihood that a member of this class will crash his car is 0.06 per year. The other class, the bad drivers, comprise 20% of the population, and the likelihood that a member of this class will crash his car is 0.6 per year. An insurance company charges clients $100 times the likelihood of having an accident, as expressed as a percentage. Hence, a member of the good driver class would pay $600 per year; a member of the bad driver class would pay $6000 per year. The question is: If nothing is known about a driver other than that he had an accident in the prior year, then what should he be charged for his annual car insurance payment?

The Python script, bayes.py, calculates the insurance cost, based on 10,000 trial simulations:

$$P(A \mid B) * P(B) = P(B \mid A) * P(A)$$

FIG. 13.6 Bayes' theorem relating probabilities of event that are conditional upon one another.

```
import random, itertools
accidents_next_year = 0
no_accidents_next_year = 0
for i in itertools.repeat(None, 10000):
  group_likelihood = random.uniform(0,1)
  if (group_likelihood< 0.2):              #puts trial in poor-judgment group
    bad_likelihood=random.uniform(0,1) #roll the dice to see if accident occurs
    if (bad_likelihood< 0.6):              #an accident occurred, simulating an initial
                                           #condition of poor-judgment with accident
      next_bad_likelihood = random.uniform(0,1)
      if (next_bad_likelihood < 0.6):
        accidents_next_year = accidents_next_year + 1
      else:
        no_accidents_next_year = no_accidents_next_year + 1
  else: #othwerwise we bump the trial into the good-judgment group
    bad_likelihood=random.uniform(0,1) #simulates an accident accident
    if (bad_likelihood < 0.06):          #an accident with good-judgment
      next_bad_likelihood = random.uniform(0,1)
      if (next_bad_likelihood < 0.06):
        accidents_next_year = accidents_next_year + 1
      else:
        no_accidents_next_year = no_accidents_next_year + 1
cost = int(((accidents_next_year)/(accidents_next_year+no_accidents_next_year)
*100*100))

print("Insurance cost is $" + str(cost))

outputs of 7 executions of bayes.py script:
Insurance cost is 4352
Insurance cost is 4487
Insurance cost is 4406
Insurance cost is 4552
Insurance cost is 4454
Insurance cost is 4583
Insurance cost is 4471
```

In all eight executions of the script, each having 10,000 trials, we find that the insurance cost, based on initial conditions, should be about $4500.

What does our bayes.py do? First, it creates a loop for 10,000 trial simulations. Within each simulation, it begins by choosing a random number between 0 and 1. If the random number is less than 0.2, then this simulates an encounter with a member of the bad-driver class (i.e., the bottom 20% of the population). In this case the random number generator produces another number between 0 and 1. If this number is less than 0.6 (the annual likelihood of a bad driver having an accident), then this would be simulate a member of the bad-driver class who had an accident and who is applying for car insurance. Now, we run the random number generator one more time, to simulate whether the bad driver will have an accident during the insurance year. If the generated random number is less than 0.6, we will consider this a simulation of a bad-driver, who had an accident prior to asking for applying for car insurance, having an accident in the subsequent year. We will do the same for the simulations that apply to the good drivers (i.e., the trials for which our group likelihood random number was greater than 0.2). After the simulations have looped 10,000 times, all that remains is to use our tallies to calculate the likelihood of an accident, which in turn gives us the insurance cost. In this example, as in all our other examples, we really did not need to know any statistics. We only needed to know the conditions of the problem, and how to simulate those conditions as Monte Carlo trials.

Glossary

Blended class Also known as class noise. Blended class refers to inaccuracies (e.g., misleading results) introduced in the analysis of data due to errors in class assignments (e.g., inaccurate diagnosis). If you are testing the effectiveness of an antibiotic on a class of people with bacterial pneumonia, the accuracy of your results will be forfeit when your study population includes subjects with viral pneumonia, or smoking-related lung damage. Errors induced by blending classes are often overlooked by data analysts who incorrectly assume that the experiment was designed to ensure that each data group is composed of a uniform and representative population. A common source of class blending occurs when the classification upon which the experiment is designed is itself blended. For example, imagine that you are a cancer researcher and you want to perform a study of patients with malignant fibrous histiocytomas (MFH), comparing the clinical course of these patients with the clinical course of patients who have other types of tumors. Let us imagine that the class of tumors known as MFH does not actually exist; that it is a grab-bag term erroneously assigned to a variety of other tumors that happened to look similar to one another. This being the case, it would be impossible to produce any valid results based on a study of patients diagnosed as MFH. The results would be a biased and irreproducible cacophony of data collected across different, and undetermined, classes tumors. Believe it or not, this specific example, of the blended MFH class of tumors, is selected from the real-life annals of tumor biology [18–20]. The literature is rife with research of dubious quality, based on poorly designed classifications and blended classes. One caveat; efforts to reduce class blending can be counterproductive if undertaken with excess zeal. For example, in an effort to reduce class blending, a researcher may choose groups of subjects who are uniform with respect to every known observable property. For example, suppose you want to actually compare apples with oranges. To avoid class blending, you might want to make very sure that your apples do not included any cumquats, or persimmons. You should be certain that your oranges do not include any limes or grapefruits. Imagine that you go even further, choosing only apples and oranges of one variety (e.g., Macintosh apples and Navel oranges), size (e.g., 10 cm), and origin (e.g., California). How will your comparisons apply to the varieties of apples and oranges that you have excluded from your study? You may actually reach conclusions that are invalid and irreproducible for more generalized populations within each class. In this case, you have succeeded in eliminated class blending at the expense of having representative populations of the classes.

Curse of dimensionality As the number of attributes for a data object increases, the distance between data objects grows to enormous size. The multidimensional space becomes sparsely populated, and the distance between any two objects, even the two closest neighbors, becomes absurdly large. When you have thousands of dimensions, the space that holds the objects is so large that distances between objects become difficult or impossible to compute, and computational products become useless for most purposes.

Heuristic technique A way to solve problems with inexact but quick methods, sufficient for most practical purposes.

Modified random sampling When we think of random sampling, we envision a simple, unbiased random selection from all the data objects within a collection. Consider a population of 7 billion people, where the number of individuals aged 60 and above account for 75% of the population. A random sampling of this population would be skewed to select senior citizens, and might yield very few children in kindergarten. Depending on the study, the data analyst may want to change the sampling rules to attain a random sampling that produces a population that fits the study goals. In this example the population might be partitioned into age groups (by decade), with an individual in any partitioned group having the same chance of being randomly selected as an individual in any other partitioned age group.

There are many different ways in which the rules of random sampling can be modified to accommodate an analytic approach [21]. Here are a few:

- Event-based sampling, in which data are collected only at specific moments, when the data being received meet particular criteria or exceed a preset threshold
- Adaptive random sampling, in which the rules for selection are determined by prior observations on the selected samples
- Attribute-based sampling, in which the probability of selection is weighted by a feature attribute of a data object

Of course, when we introduce a modification to the simple process of random selection, we risk introducing new and unexpected biases and confounders, and we open ourselves to criticism and the possibility that our conclusions cannot be repeated in other populations. It is another of those damned if you do and damned if you don't scenarios that we can expect in nearly every Big Data analysis.

Permutation test A method whereby the null hypothesis is accepted or rejected after testing all possible outcomes under rearrangements of the observed data elements.

Power In statistics, power describes the likelihood that a test will detect an effect, if the effect actually exists. In many cases, power reflects sample size. The larger the sample size, the more likely that an experiment will detect a true effect; thus correctly rejecting the null hypothesis.

Principal component analysis One popular methods for transforming data to reduce the dimensionality of data objects is multidimensional scaling, which employs principal component analysis [22]. Without going into the mathematics, principal component analysis takes a list of parameters and reduces it to a smaller list, with each component of the smaller list constructed from variables in the longer list (as a sum of variables multiplied by weighted coefficients). Furthermore, principal component analysis provides an indication of which variables are least correlated with the other variables (as determined by the size of the coefficients). Principal component analysis requires operations on large matrices. Such operations are computationally intensive and can easily exceed the capacity of most computers [22].

Pseudorandom number generator It is impossible for computers to produce an endless collection of truly random numbers. Eventually, algorithms will cycle through their available variations and begins to repeat themselves, producing the same set of "random" numbers, in the same order; a phenomenon referred to as the generator's period. Because algorithms that produce seemingly random numbers are imperfect, they are known as pseudorandom number generators. The Mersenne Twister algorithm, which has an extremely long period, is used as the default random number generator in Python. This algorithm performs well on most of the tests that mathematicians have devised to test randomness.

Resampling statistics A technique whereby a sampling of observations is artifactually expanded by randomly selecting observations and adding them to the original data set; or by creating new sets of data by randomly selecting, without removing, data elements from the original data.

Resampling versus Repeated Sampling In resampling statistics, a limited number of data measurements is expanded by randomly selecting data and adding them back to the original data. In Big Data, there are so many data points that statistical tools cannot quickly evaluate them. Repeated sampling involves randomly selecting subpopulations of enormous data sets, over and over again, and performing statistical evaluations on these multiple samplings, to arrive at some reasonable estimate of the behavior of the entire set of data. Although random sampling is involved in resampling statistics and repeated sampling statistics, the latter does not resample the same data points.

Sample size The number of samples used in a study. Methods are available for calculating the required sample size to rule out the null hypothesis, when an effect is present at a specified significance level, in a population with a known population mean, and a known standard deviation [23]. The sample size formula should not be confused with the sampling theorem, which deals with the rate of sampling that would be required to adequately digitize an analog (e.g., physical or electromagnetic) signal.

Sampling theorem A foundational principle of digital signal processing, also known as the Shannon sampling theorem or the Nyquist sampling theorem. The theorem states that a continuous signal can be properly sampled, only if it does not contain components with frequencies exceeding one-half of the sampling rate. For example, if you want to sample at a rate of 4000 samples per second, you would prefer a signal containing no frequencies greater than 2000 cycles per second.

References

[1] Simon JL. Resampling: the new statistics. 2nd ed., 1997. Available at: http://www.resample.com/intro-text-online/ [viewed September 21, 2015].

[2] Efron B, Tibshirani RJ. An introduction to the bootstrap. Boca Raton: CRC Press; 1998.

[3] Diaconis P, Efron B. Computer-intensive methods in statistics. Sci Am 1983;116–30.

[4] Anderson HL. Metropolis, Monte Carlo and the MANIAC. Los Alamos Sci 1986;14:96–108.

[5] Berman JJ. Biomedical informatics. Sudbury, MA: Jones and Bartlett; 2007.

[6] Candes EJ, Wakin MB. An introduction to compressive sampling. IEEE Signal Process Mag March 2008;21:.

[7] Berman JJ. Methods in medical informatics: fundamentals of healthcare programming in Perl, Python, and Ruby. Boca Raton: Chapman and Hall; 2010.

[8] Berman JJ. Data simplification: taming information with open source tools. Waltham, MA: Morgan Kaufmann; 2016.

[9] Van Heel M. A fast algorithm for transposing large multidimensional image data sets. Ultramicroscopy 1991;38:75–83.

[10] Cipra BA. The best of the 20th century: editors name top 10 algorithms. SIAM News May 2000;33(4).

[11] Berman JJ, Moore GW. The role of cell death in the growth of preneoplastic lesions: a Monte Carlo simulation model. Cell Prolif 1992;25:549–57.

[12] Berman JJ, Moore GW. Spontaneous regression of residual tumor burden: prediction by Monte Carlo Simulation. Anal Cell Pathol 1992;4:359–68.

[13] Elias H, Sherrick JC. Morphology of the liver. Cambridge, MA: Academic Press; 1969.

[14] Berman JJ. Neoplasms: principles of development and diversity. Sudbury: Jones & Bartlett; 2009.

[15] Moore GW, Berman JJ. Cell growth simulations that predict polyclonal origins for 'monoclonal' tumors. Cancer Lett 1991;60:113–9.

[16] Berman JJ, Albores-Saavedra J, Bostwick D, Delellis R, Eble J, Hamilton SR, et al. Precancer: a conceptual working definition results of a consensus conference. Cancer Detect Prev 2006;30(5):387–94.

[17] Berman JJ. Precancer: the beginning and the end of cancer. Sudbury: Jones and Bartlett; 2010.

[18] Al-Agha OM, Igbokwe AA. Malignant fibrous histiocytoma: between the past and the present. Arch Pathol Lab Med 2008;132:1030–5.

[19] Nakayama R, Nemoto T, Takahashi H, Ohta T, Kawai A, Seki K, et al. Gene expression analysis of soft tissue sarcomas: characterization and reclassification of malignant fibrous histiocytoma. Mod Pathol 2007;20:749–59.

[20] Daugaard S. Current soft-tissue sarcoma classifications. Eur J Cancer 2004;40:543–8.

[21] National Research Council. Frontiers in massive data analysis. Washington, DC: The National Academies Press; 2013.

[22] Janert PK. Data analysis with open source tools. O'Reilly Media; 2010.

[23] How to determine sample size, determining sample size. Available at: http://www.isixsigma.com/tools-templates/sampling-data/how-determine-sample-size-determining-sample-size/ [viewed July 8, 2015].

14

Special Considerations in Big Data Analysis

Section 14.1. Theory in Search of Data

If triangles had a god, they would give him three sides.

Voltaire

Here is a riddle: "Which came first, the data, or the data analyst?" The intuitive answer would be that data precedes the data analyst. Without data, there really is no reason for the data analyst to exist. In the Big Data universe nothing is as it seems, and the data analyst commonly precedes the data. All too often the analyst develops a question or a hypothesis or a notion of what the facts "should be," and then goes about rummaging through the Big Data resource until he or she has created a data set that proves the point.

Several intrinsic flaws plague Big Data statistics. When the amount of data is sufficiently large, you can find almost anything you seek lurking somewhere within. Such findings may have statistical significance without having any practical significance. Also, whenever you select a subset of data from an enormous collection, you may have no way of knowing the relevance of the data that you excluded. Most importantly, Big Data resources cannot be designed to examine every conceivable hypothesis. Many types of analytic errors ensue when a Big Data resource is forced to respond to questions that it cannot possibly answer. The purpose of this chapter is to provide general recommendations for the responsible use of analytic methods, while avoiding some of the pitfalls in Big Data analysis.

We cannot escape the dangerous practice of imposing models on selected sets of data. Historians, who have the whole of human history to study, are just as guilty as technical data analysts in this regard. Consider this hypothetical example: the United States is on the

Principles and Practice of Big Data. https://doi.org/10.1016/B978-0-12-815609-4.00014-5

brink of a military intervention against entrenched and hostile revolutionaries on the other side of the globe. Two historians are asked to analyze the situation and render their opinions. The first historian compares the current crisis to the entrance of the United States into World War II. World War II worked out well for the United States. The first historian insists that World War II is a useful model for today's emergency and that we should engage our military against the current threat. The second historian says that the current crisis is very much like the crisis that preceded the Vietnam War. The Vietnam War did not turn out well for United States interests, and it would be best if we avoided direct military involvement in the current emergency. When you have all of history from which to choose, you can select any set of data that supports your biases. As humans, we do this all the time, whenever we make decisions.

Scientists have accused their peers of developing models for the purpose of reinforcing belief in their own favorite paradigms [1]. Big Data will not help advance science if analysts preferentially draw data to support their previously held biases. One of the important tasks for Big Data analysts will involve developing methods for creating unbiased models from Big Data. In the meantime, there is no practical way to validate conclusions drawn from Big Data, other than to test the hypothesis on additional data sets.

Section 14.2. Data in Search of Theory

Without highly specified a-priori hypotheses, there are hundreds of ways to analyse the dullest data set.

John P A Ioannidis [2]

In the prior section the point was made that data analysts can abuse Big Data if data is selected to confirm a hypothesis. In this section the point is made that scientists must enter their analysis with a model theory; otherwise they will choose a hypothesis to fit their data, even if the hypothesis makes no sense. [Glossary Multiple comparisons bias]

Here is a good example. Suppose I am at a shooting range and shoot ten shots at a bull's eye target. I can measure the distance of each bullet from the center of the target, from which I would develop some type of score with which I could compare my marksmanship against that of others. Now, imagine shooting ten shots at a wall that has no target. I may find that six of the bullets clustered very close together. I could then superimpose the bullet holes with a bull's eye target, placing the center of the target over the center of the tight clusters of bullet holes. A statistician analyzing the data might find that the six tightly clustered bullet holes at the center of the bull's eye indicated that I scored very well and that it was highly likely that I had better aim than others (who had actually aimed at the target). Scientists who troll large data sets will always find clusters of data upon which they can hang a bull's eye. Statisticians provided with such data can be tricked into confirming a ridiculous hypothesis that was contrived to fit the data. This deceptive practice is referred to as moving the target to the bullet hole.

Big Data analysts walk a thin line. If they start their project with a preconceived theory, then they run the risk of choosing a data set that confirms their bias. If they start their project without a theory, then they run the risk of favoring a false hypothesis that happens to fit their data. [Glossary Type errors]

Is there a correct approach to Big Data analysis? It is important to remember that a scientific theory is a plausible explanation of observations. Theories are always based on some set of pre-existing principles that are generally accepted as truth. When a scientist approaches a large set of data, he or she asks whether a set of commonly held principles will extend to the observations in the current set of data. Reconciling what is known with what is observed accounts for much of the activity of scientists.

For Big Data projects, holding an a prior theory or model is almost always necessary; otherwise, the scientist is overwhelmed by the options. Adequate analysis can be ensured if three conditions are met:

1. All of the available data is examined, or a subset is prepared from a random sampling (i.e., no cherry-picking data to fit the theory).
2. The analyst must be willing to modify or abandon the theory, if it does not fit the data.
3. The analyst must not believe that fitting the theory to the data validates the theory. Theories must be tested against multiple sets of data.
4. The analyst must accept that the theory may be wrong, even if it is validated. Validation is not proof that a theory is correct. It is proof that the theory is consistent with all of the observed data. A better theory may also be consistent with the observed data and may provide a true explanation of the observations.

One of the greatest errors of Big Data analysts is to believe that data models are tethered to reality; they seldom are. Models are made to express data sets as formulas or as a system that operates under a set of rules. When the data are numeric representations of physical phenomenon, it may sometimes be possible to link the model to a physical law. For example, repeated measurements of force, mass, and acceleration observed on moving bodies might produce a formula that applies consistently, at any time, any place, and with any object (i.e., $f = ma$). Most mathematical models are abstract, and cannot be ranked as physical laws. At best, they provide a quick glimpse of an ephemeral reality.

Section 14.3. Bigness Biases

Every increased possession loads us with new weariness.

John Ruskin

Because Big Data methods use enormous sets of data, there is a tendency to give the results more credence than would be given to a set of results produced from a small set of data. This is almost always a mistaken belief. In fact, Big Data is seldom a complete or accurate data collection. You can expect most Big Data resources to be selective, intentionally or not, for the data that is included and excluded from the resource. When dealing with Big Data, expect missing values, missing records, "noisy" data, huge variations in the quality of records, plus any and all of the inadequacies found in small data resources.

Nevertheless, the belief that Big Data is somehow more reliable, and more useful than smaller data is pervasive in the science community.

When a study is done on a very large number of human subjects (or with a very large number of samples), each annotated with a large number of observations, there is a tendency to accept the results, even when the results defy intuition. In 2007, a study using the enormous patient data set held by the U.S. Veterans Administration Medical Centers reported that the use of statins reduced the risk of developing lung cancer by about half [3]. The study, which involved nearly half a million patients, showed that the reduction in cancer risk held whether patients were smokers or non-smokers. The highest reduction in lung cancers (77%) occurred in people who had taken statins for four years or longer [3].

The potential importance of this study cannot be overestimated. Lung cancer is the most common cause of cancer deaths in the United States. A 77% reduction in lung cancer incidence would prevent the cancer deaths of about 123,000 U.S. residents each year. This number is equivalent to the total number of cancer deaths attributed each year to prostate cancer, breast cancer and colon cancer combined [4]!

As it happens these marvelous findings were as unintuitive as they were exciting. Statins are a widely used drugs that reduce the blood levels of cholesterol and various other blood lipids. There is absolutely nothing known about the biology of statins that would lead anyone to suspect that this drug would lower the incidence of lung cancer, or any other cancer, for that matter. It is always risky to accept a scientific conclusion without some sort of biological mechanism to explain the results.

In 2011, a second study, by another group of researchers, was published on the effect of statins on lung cancer incidence. This study was also big, using about 133,000 patients. The results failed to show any effect of statins on lung cancer incidence [5]. That same year, a third study, using a population of about 365,000 people, also failed to find any influence of statins on the incidence of lung cancer [6]. The authors of the negative studies blamed time-window bias on the misleading results of the first study.

To understand time-window bias, consider the undisputed observation that Nobel prize laureates live longer than other scientists. It would seem that scientists who want to live a long life should try their utmost to win a Nobel prize. Likewise, Popes live longer than other clergymen. If you are a priest, and you want to live long, aim for the Papacy. Both these biases are based on time-window conditions. The Nobel prize committee typically waits decades to determine whether a scientific work is worthy of the Nobel prize, and the prize is only awarded to living scientists. Would-be Nobelists who die before their scientific career begins, and accomplished scientists who die before their works are deemed Nobel-worthy, are omitted from the population of potential winners. Similarly, the Vatican seldom confers the Papacy on its junior clergy. The time-window surrounding Nobel winners and Popes skews their observed longevities upwards. Time-window bias is just one of a general class of biases wherein studies are invalidated by the pre-conditions imposed on the studies [7]. [Glossary Time-window bias]

Time-window bias affected the original large patient-based study because a population that had taken a statin for four years, without dying in the interim, was compared to a general population. Basically, the study imposed a cancer-free 4-year window for the treated population, artifactually conferring a lower cancer incidence on the statin-treated group.

The point here is simply that analytic errors occur just as easily in studies of large populations as they do in studies involving a small number of individuals. Because Big Data analysis tends to be complex, and difficult for anyone to thoroughly review, the chances of introducing Big Data errors is larger than the chances of introducing small data errors.

In the late 1990s, interest was growing in the medical research community and in biomarkers for cancers. It was believed then, as it is believed now, that the different types of cancers must contain biological markers to tell us everything we need to know about their clinical behaviors. Such biomarkers would be used to establish the presence of a cancer, the precise diagnosis of the cancer, the stage of the cancer (i.e., the size and the extent of spread of the cancer), and to predict the response of the cancer to any type of treatment. By the turn of the century, there was a sense that useful cancer biomarkers were not forthcoming; the pipeline for new biomarkers had apparently dried up [8–11]. What was the problem? A gnawing suspicion held that biomarkers failed because we weren't collecting enough data. A consensus had grown that we were wasting cancer research funds on small-scale studies that were irreproducible. What we needed, or so everyone thought, were Big Data studies, producing lots of data, yielding trustworthy results based on many observations. If researchers abandoned their small studies, in favor of large studies, then the field would surely move forward at a rapid pace.

In the past two decades, biomarker studies have seen enormous successes. Surprisingly, though, much of the recent progress has come from relatively small genomic studies, on very rare cancers, with limited numbers of specimens [2,12–14]. Why has Big Data not yielded the kind of progress that nearly everyone expected?

When a new potential biomarker is discovered using large and complex sets of data and advanced analytic tools, it needs to be validated; and validation involves repeating the original study, and drawing the same set of conclusions [15,16]. As a general rule the more complex the experiment, the data, and the analysis, the less likely that it can be reproduced. In addition to these basic limitations on conclusions drawn from Big Data, we must remember that it can be very difficult to analyze systems whose complexity exceeds our comprehension. We assume, quite incorrectly, that given sufficient data, we can understand complex systems. There is nothing to support this kind of self-confidence. Biological systems are highly complex, and we do not, at this time, have a deep understanding of their workings. For that matter, we have very little understanding of the kinds of data that ought to be collected. We are slowly learning that it seldom helps to throw Big Data at a problem, before we have a thorough understanding of what we need to find. In the case of cancer biomarkers, it was much easier to find the key mutations that accounted for rare tumors than it was to find common biomarkers in a general population [12,13].

Still unconvinced that Bigness bias is a real concern for Big Data studies? In the United States, our knowledge of the causes of death in the population is based on death certificate data collected by the Vital Statistics Program of the National Center for Health Statistics.

Death certificate data is notoriously faulty [17–19]. In most cases, the data in death certificates is supplied by clinicians, at or near the time of the patient's death, without benefit of autopsy results. In many cases, the clinicians who fill out the death certificate are not well trained for the task, often mistaking the mode of death (e.g., cardiac arrest, cardiopulmonary arrest), with cause of death (e.g., the disease process leading to cardiac arrest or cardiopulmonary arrest), thus nullifying the intended purpose of the death certificate. Thousands of instructional pages have been written on the proper way to complete a death certificate. Nonetheless, these certificates are seldom completed in a consistent manner. Clinicians become confused when there are multiple, sometimes unrelated, conditions that contribute to the patient's death. Though the death certificates are standardized throughout the United States, there are wide variations from state to state in the level of detail provided on the forms [20]. Despite all this the venerable death certificate is the bedrock of vital statistics. What we know, or think we know, about the causes of death in the United States population, is based on an enormous repository, collected since 1935, of many millions of death certificates.

Why do we believe death certificate data when we know that death certificates are highly flawed? Again, it is the bigness factor that prevails. There seems to be a belief, based on nothing but wishful thinking, that if you have a very large data set, bad measurements will cancel themselves out, leaving a final result that comes close to a fair representation of reality. For example, if a clinician forgets to list a particular condition as a cause of death, another physician will mistakenly include the condition on another death certificate, thus rectifying the error.

The cancel-out hypothesis puts forward the delightful idea that whenever you have huge amounts of data, systemic errors cancel out in the long run, yielding conclusions that are accurate. Sadly, there is neither evidence nor serious theory to support this hypothesis. If you think about it, you will see that it makes no sense. One of the most flagrant weaknesses is the fact that it is impossible to balance something that must always be positive. Every death certificate contains a cause of death. You cannot balance a false positive cause of death with a false negative cause of death (i.e., there is no such thing as a negative cause of death). The same applies to numeric databases. An incorrect entry for 5000 pairs of shoes cannot be balanced by a separate incorrect entry for negative 5000 pairs of shoes; there is no such thing as a negative shoe. [Glossary Negative study bias]

Perhaps the most prevalent type of bigness bias relates to the misplaced faith that complete data is representative data. Certainly, you might think that if a Big Data resource contains every measurement for a data domain, then biases imposed by insufficient sampling are eliminated. Danah Boyd, a social media researcher, draws a sharp distinction between Big-ness and Whole-ness [21]. She gives the example of a scientist who is exploring a huge data set of tweets collected by Twitter. If Twitter removes tweets containing expletives, or tweets composed of non-word character strings, or containing certain types of private information, then the resulting data set, no matter how large it may be, is not representative of the population of senders. If the tweets are available as a stripped-down set of messages, without any identifier for senders, then the compulsive tweeters (those who send hundreds or thousands of tweets) will be over-represented, and the one-time tweeters will be under-represented. If each tweet were associated with an account and all the tweets from a single account were collected as a unique record, then there would still be the problem created by tweeters who maintain multiple accounts. Basically, when you have a Big Data resource, the issue of sample representation does not disappear; it becomes more complex and less controlled. For Big Data resources lacking introspection and identifiers, data representation becomes an intractable problem.

– Too Much Data

Intuitively, you might think that the more data we have at our disposal, the more we can learn about the system that we are studying. This is not always the case. There are circumstances when more data simply takes you further and further from the solution you seek. As a trivial example, consider the perennial task of finding a needle in a haystack. As you add more hay, you make the problem harder to solve. You would be much better off if the haystack were small, consisting of a single straw, behind which lies your sought-after needle [22].

In the field of molecular biology the acquisition of whole genome sequencing on many individual organisms, representing hundreds of different species, has brought a flood of data, but many of the most fundamental questions cannot be answered when the data is complex and massive. Evolutionary biologists have invented a new term for a certain type of sequence data: "non-phylogenetic signal." The term applies to DNA sequences that cannot yield any useful conclusions related to the classification of an organism, or its evolutionary relationships to other organisms.

Evolutionary geneticists draw conclusions by comparing DNA sequences in organism, looking for similar, homologous regions (i.e., sequences that were inherited from a common ancestor). Because DNA mutations arise stochastically over time (i.e., at random locations in the gene, and at random times), unrelated organisms may attain the same sequence in a chosen stretch of DNA, without inheritance through a common ancestor. Such occurrences could lead to false inferences about the relatedness of different organisms. When mathematical phylogeneticists began modeling inferences for gene data sets, they assumed that most class assignment errors would be restricted to a narrow range of situations. This turned out not to be the case. In practice, errors due to non-phylogenetic signal occur due to just about any mechanism that causes DNA to change over time (e.g., random mutations, adaptive convergence) [23,24]. At the moment, there seems to be an excess of genetic information. The practical solution seems to involve moving away from purely automated data analyses and using a step-by-step approach involving human experts who take into account independently acquired knowledge concerning the relationships among organisms and their genes.

– Overfitting

Overfitting occurs when a formula describes a set of data very closely, but does not predict the behavior of comparable data sets. In overfitting, the formula is said to describe the noise of the system, rather than the characteristic behavior of the system. Overfitting commonly occurs with models that perform iterative approximations on training data. Neural networks are an example of a data modeling strategy that is prone to overfitting. In general, the bigger the data set, the easier it is to overfit the model.

Overfitting is discovered by testing your predictor or model on one or several new sets of data [25]. If the data is overfitted the model will fail with the new data. It can be heart-breaking to spend months or years developing a model that works like a charm for your training data and for your first set of test data (collected from the same data set as your training data), but fails completely for a new set of data.

Overfitting can sometimes be avoided by evaluating the model before it has been fitted to a mathematical formula, often during the data reduction stage. There are a variety of techniques that will produce a complex formula fitted to all your variables. It might be better to select just a few variables from your data that you think are most relevant to the model. You might try a few mathematical relationships that seem to describe the data plotted for the subset of variables. A formula built from an intuitive understanding of the relationships among variables may sometimes serve much better than a formula built to fit a multi-dimensional data set. [Glossary Data reduction]

Section 14.4. Data Subsets in Big Data: Neither Additive Nor Transitive

If you're told that a room has 3 people inside, and you count 5 people exiting the room, a mathematician would feel compelled to send in 2 people to empty it out.

Anon

It is often assumed that Big Data has one enormous advantage over small data: that sets of Big Data can be merged to create large populations that reinforce or validate conclusions drawn from small studies. This assumption is simply incorrect. In point of fact, it is possible to draw the same conclusion from two sets of data, only to draw an opposite conclusion when the two sets of data are combined. This phenomenon, well known to statisticians as Simpson's paradox, has particular significance when Big Data resources combine observations collected from multiple populations.

One of the most famous examples of Simpson's paradox was demonstrated in the 1973 Berkeley gender bias study [26]. A preliminary review of admissions data indicated that women had a lower admissions rate than men:

```
Men  Number of applicants..  8,442 Percent applicants admitted.. 44%
Women Number of applicants..  4,321 Percent applicants admitted.. 35%
```

A nearly 10% lower overall admission rate for women, compared with men, seemed significant, but what did it mean? Was the admissions office guilty of gender bias?

A look at admissions department-by-department (in distinction to admissions for the total number of applicants to the university, by gender) showed a very different story. Women were being admitted at higher rates than men, in almost every department.

The department-by-department data seemed incompatible with the data obtained when the admissions from all the departments were combined.

The explanation was simple. Women tended to apply to the most popular and oversubscribed departments, such as English and History, that had a high rate of admission denials. Men tended to apply to departments that the women of 1973 avoided, such as mathematics, engineering, and physics, that had high relatively few applicants and high acceptance rates. Though women had an equal footing with men in departmental admissions, the high rate of rejections in the large departments, accounted for an overall lower acceptance rate for women at Berkeley.

Simpson's paradox demonstrates that data is not additive. It also shows us that data is not transitive; you cannot make inferences based on subset comparisons. For example in randomized drug trials, you cannot assume that if drug A tests better than drug B, and drug B tests better than drug C, then drug A will test better than drug C [27]. When drugs are tested, even in well-designed trials, the test populations are drawn from a general population specific for the trial. When you compare results from different trials, you can never be sure whether the different sets of subjects are comparable. Each set may contain individuals whose responses to a third drug are unpredictable. Transitive inferences (i.e., if A is better than B, and B is better than C, then A is better than C), are unreliable.

Simpson's paradox has particular significance for Big Data research, wherein data samples are variously recombined and reanalyzed at different stages of the analytic process.

Section 14.5. Additional Big Data Pitfalls

Any problem in Computer Science can be solved with another level of indirection.
Butler Lampson

...except the problem of indirection complexity.

Bob Morgan

There is a large literature devoted to the pitfalls of data analysis. It would seem that all of the errors associated with small data analysis will apply to Big Data analysis. There are, however, a collection of Big Data errors that do not apply to small data, such as:

– **The misguided belief that Big Data is good data**

For decades, it was common for scientists to blame their failures on the paucity of their data. You would often here, at public meetings and in private, statements such as "It was a small study, using just a few samples and limited number of measurements on each sample. We really should not generalize at the moment. Let us wait for a definitive study based on a large group of samples."

There has always been the sense, based on nothing in particular, that a small study cannot be validated by another small study. A small study must be validated by a big study.

Anyone who has ever worked on a project that collects large, complex, quickly streaming data knows that such efforts are much more prone to systemic flaws in data collection than are smaller projects. In Section 16.1, "First Analysis (Nearly) Always Wrong," we will see why conclusions drawn from Big Data are notoriously misleading.

Big Data comes from many different sources, produced by many different protocols, and must undergo a series of tricky normalizations, transformation, and annotations, before it has any value whatsoever. Data analysts can never assume that Big Data is accurate. Competent analysts will always validate their conclusions based on alternate, independently collected data; big or small.

– Blending bias

If you are studying the properties of a class of records (e.g., records of individuals with a specific disease or data collected on a particular species of fish), then any analysis of the data, no matter how large the data set, will be biased if your class assignments are erroneous (e.g., if the disease was misdiagnosed, or if you mistakenly included other species of fish in your collection). Classifications can be deeply flawed when individual classes are poorly defined, or not based on a well-understood set of scientific principles, or are assembled through the use of poor analytic techniques.

Let us look at one example in some depth. Suppose you are a physician living in Southern Italy, in the year 1640, where people are dying in great number, from a mysterious disease characterized by recurring fevers, delirium, and pain. You are approached by an explorer who has just returned from a voyage to South America, in an area corresponding to modern-day Brazil. He holds a bag containing an herbal extract, and says "Give this to your patients, and they will quickly recover."

It happens that the drug is extracted from the bark of the Cinchona tree. It is a sure-fire cure for malaria. Unknown to you, many of your patients are suffering from malaria, and would benefit greatly from this miraculous drug. Nonetheless, you are skeptical and would like to test this new drug before subjecting your patients to any unanticipated horrors. Though you are not a statistician, you do know something about designing clinical trials. In short order, you collect 100 patients, all of whom have the symptoms of fever and delirium. You administer the cinchona powder, also known as quinine, to all the patients. A few improve, but most do not. Knowing that some patients recover without any medical assistance, you call the trial a wash-out. In the end, you decide not to administer quinine to your patients.

What happened? We know that quinine arrived as a miracle cure for malaria. It should have been effective in a population of 100 malarial patients. The problem with this hypothetical clinical trial is that the patients under study were assembled based on their mutual symptoms: fever and delirium. These same symptoms could have been accounted for by any of hundreds of other diseases that were prevalent in England at the time. The criterion employed at the time to classify diseases was imprecise, and the trial population was diluted with non-malarial patients who were guaranteed to be non-responders. Consequently, the trial failed, and you missed a golden opportunity to treat your malaria patients with quinine, a new, highly effective, miracle drug.

Back in Section 5.5, we discussed Class Blending, an insidious flaw found in many classifications, that virtually guarantees that any analysis will yield misleading results. Having lots and lots of data will not help you. The only way to overcome the bias introduced by class blending is to constantly test and refine your classification.

– **Complexity bias**

The data in Big Data resources comes from many different sources. Data from one source may not be strictly comparable to data from another source. The steps in data selection, including data filtering, and data transformation, will vary among analysts. Together, these factors create an error-prone analytic environment for all Big Data studies that does not apply to small data studies.

– **Statistical method bias**

Statisticians can apply different statistical methods to one set of data, and arrive at any of several different, even contradictory, conclusions. Statistical method biases are particularly dangerous for Big Data. The standard statistical tests that apply to small data and to data collected in controlled experiments, may not apply to Big Data. Analysts are faced with the unsatisfying option of applying standard methods to non-standard data, or of developing their own methodologies for their Big Data project. History suggests that given a choice, scientists will adhere to the analysis that reinforces their own scientific prejudices [28].

– **Ambiguity of system elements**

Big Data analysts want to believe that complex systems are composed of simple elements, having well-defined attributes and functions. Clever systems analysts, using advanced techniques, enjoy believing that algorithms can predict the behavior of complex systems, when the elements of the system are understood. We learn from biological systems that the components of complex systems have ambiguous functionalities, changing from one moment to the next, rendering our best predictions tentative, at best. [Glossary Deep analytics]

For example, living cells are complex systems in which many different metabolic pathways operate simultaneously. A metabolic pathway is a multi-step chemical process involving more than one enzyme and various additional substrate and non-substrate chemicals. Depending on the conditions within a cell, a single enzyme may participate in several different metabolic pathways; and any given pathway may exert any of a number of different biological effects [29–32]. As we learn more and more about cells, we are stunned by their complexities [33,13]. Big Data analysts, working with highly complex systems, cannot assume that any of the elements of their system have a single, defined function. This tells us that all Big Data analyses on living systems (e.g., all biomedical systems and all non-biomedical data that depends in any way on the predictability or reproducibility of biomedical data) may be intractable to the kinds of systems analysis techniques that we have come to understand.

Despite all the potential biases, at the very least Big Data offers us an opportunity to validate predictions based on small data studies. As a ready-made source of observations, Big Data resources may provide the fastest, most economical, and easiest method to "reality test" limited experimental studies. Testing against large, external data sets, on independently collected data, and coming up with the equivalent conclusions, is a reasonable way to validate scientific assertions [34,35].

Section 14.6. Case Study (Advanced): Curse of Dimensionality

As the number of spatial dimensions goes up, finding things or measuring their size and shape gets harder.
The Curse of Dimensionality, attributed to Richard Bellman, and sometimes called Bellman's curse

Any serious student of Big Data will eventually fall prey to the dreadful Curse of Dimensionality. This curse cannot be reversed, and cannot be fully fathomed by 3-dimensional entities. Luckily, we can see the tell-tale signs that indicate where the curse is strongest, and thus avoid the full force of its evil power.

First, let's understand what we mean when we talk about n-dimensional data objects. Each attribute of an object is a dimension. The object might have three attributes: height, width, and depth; and these three attributes would correspond to the familiar three dimensional measurements that we are taught in geometry. The object in a Big Data collection might have attributes of age, length of left foot, width of right foot, hearing acuity, time required to sprint 50 yards, and yearly income. In this case the object is described by 6 attributes and would occupy 6 dimensions of Big Data space.

Let us say that we have normalized the values of every attribute so that each attribute value lies between zero and two (i.e., the age is between 0 and 2; the length of the left foot is between 0 and 2; the width of the right foot is between 0 and 2, and so on for every dimension in the object.

The 6-dimensional cube that encloses the set of data objects with attributes measuring between 0 and 2 will have sides measuring 2 units in length. The general formula for the volume of an n-dimensional cube is the length of a side raised to the nth power. In the case of a 260 dimensional cube, this would give us a volume of 2^260. Just to give you some idea of the size of this number, 2^260 is roughly the estimated number of atoms contained in our universe. So the volume of the 260-dimensional cube, of side 2 units, is large enough to hold the total number of atoms in the universe, spaced one unit apart in every dimension. Because there are many more atoms in the universe than there are data objects in our Big Data resources, we can infer that all high-dimensional volumes of data will be sparsely populated (i.e., lots of space separating data objects from one another). In our physical universe, there is much more empty space than there is matter; in the infoverse, it's much the same thing, only moreso. [Glossary Euclidean distance]

So what? What does it matter that n-dimensional data space is mostly empty, so long as every data object has an n-coordinate location somewhere within the hypervolume?

Let us consider the problem of finding a data object that lies within one unit of a reference object located in the exact center of the data space. As an example, we will continue to use an n-dimensional data object composed of attributes with normalized values between 0 and 2. We will begin by looking at a two dimensional data space.

If the data objects in the 2-dimensional data space are uniformly distributed in the space, then the chances of finding a data object within one unit of the center of the space (i.e., at coordinate 1,1) will be the ratio of the circle of radius one unit around the center divided by the area of the square that contains the data space (i.e., a square whose sides have length of 2). This works out to pi/4, or 0.785. This tells us that in two dimensions, we'll have an excellent chance of finding an object within 1 unit of the center (Fig. 14.1).

We can easily imagine that as the number of the dimensions of our data space increases, with an exponentially increasing n-dimensional volume, so too will the volume of the hypersphere that accounts for all the objects lying within 1 radial unit from the center. Regardless of how fast the volume of the space is growing, our hypersphere will keep apace, and we will always be able to find data objects in a 1-radial unity vicinity. Actually, no. Here is where the Curse of Dimensionality truly kicks in.

The general formula for the volume of an n-dimensional sphere is shown in Fig. 14.2.

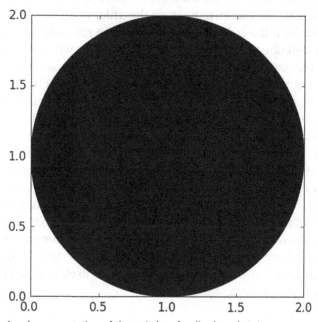

FIG. 14.1 A two-dimensional representation of the a circle, of radius length 1, in a square, of side length 2. The fraction of the square's area occupied by the circle is (pi * r^2)/4 or 3.1416/4 or 0.7854.

$$V_n(R) = \frac{\pi^{\frac{n}{2}}}{\Gamma(\frac{n}{2}+1)} R^n$$

FIG. 14.2 General formula for the volume of sphere of radius *R*, in *n* dimensions.

Let's not get distracted by the lambda function in the denominator. It suffices to know that the volume of a hypersphere in n dimensions is easily computable. Using the formula, here are the volumes of a 1 radial unit sphere in multiple dimensions [36].

```
Hypersphere volumes when radius = 1, in higher dimensions
n=1, V = 2
n=2, V = 3.1416
n=3, V = 4.1888
n=4, V = 4.9348
n=5, V = 5.2638
n=6, V = 5.1677
n=7, V = 4.7248
n=8, V = 4.0587
n=9, V = 3.2985
n=10, V = 2.5502
```

As the dimensionality increases, the volume of the sphere increases until we reach the fifth dimension. After that, the volumes of the 1-unit radius sphere begin to shrink. At 10 dimensions, the volume is down to 2.5502. From there on, the volume decreases faster and faster. The 20-dimension 1-radial unit sphere has a volume of only 0.0285, while the volume of the sphere in 100 dimensions is on the order of 10^{-40} [36].

How is this possible? If the central hypersphere has a radius of one unit, and the coordinate space is a hypercube that is 2 units on each side, then we know that, for any dimension, the hypersphere touches each and every face of the hypersphere at one point. In the two dimensional example shown above, the inside circle touches the enclosing square on all four sides: at points (1,0), (1, 2), (1, 2), and (0,1). If an n-dimensional sphere touches one point on every face of the enclosing hypercube, then how could the sphere be infinitesimally small while the hypercube is immensely large?

The secret of the curse is that as the dimensionality of the space increases, most of the volume of the hypercube comes to lie in the corners, outside the central hypersphere. The hypersphere misses the corners, just like the 2-dimensional circle misses the corners of the square. This means that as the dimensionality of data objects increases, the likelihood of finding similar objects (i.e., object at a close n-dimensional proximity from one another) drops to about zero. When you have thousands of dimensions, the space that holds the objects is so large that distances between objects become difficult or impossible to compute. Basically, you can't find similar objects if the likelihood of finding two objects in close proximity is always zero.

Glossary

Data reduction When a very large data set is analyzed, it may be impractical or counterproductive to work with every element of the collected data. In such cases, the data analyst may choose to eliminate some of the data, or develop methods whereby the data is approximated. Some data scientists reserve the term "data reduction" for methods that reduce the dimensionality of multivariate data sets.

Deep analytics Insipid jargon occasionally applied to the skill set needed for Big Data analysis. Statistics and machine learning are often cited as two of the most important areas of deep analytic expertise. In a recent McKinsey report, entitled "Big data: The next frontier for innovation, competition, and productivity," the authors asserted that the United States "faces a shortage of 140,000 to 190,000 people with deep analytical skills" [37].

Euclidean distance Two points, $(x1, y1)$, $(x2, y2)$ in Cartesian coordinates are separated by a hypotenuse distance, that being the square root of the sum of the squares of the differences between the respective *x*-axis and *y*-axis coordinates. In *n*-dimensional space, the Euclidean distance between two points is the square root of the sum of the squares of the differences in coordinates for each of the n dimensional coordinates. The significance of the Euclidean distance for Big Data is that data objects are often characterized by multiple feature values, and these feature values can be listed as though they were coordinate values for an *n*-dimensional object. The smaller the Euclidian distance between two objects, the higher the similarity to each other. Several of the most popular correlation and clustering algorithms involve pairwise comparisons of the Euclidean distances between data objects in a data collection.

Multiple comparisons bias When you compare a control group against a treated group using multiple hypotheses based on the effects of many different measured parameters, you will eventually encounter statistical significance, based on chance alone. For example, if you are trying to determine whether a population that has been treated with a particular drug is likely to suffer a serious clinical symptom, and you start looking for statistically significant associations (e.g., liver disease, kidney disease, prostate disease, heart disease, etc.), then eventually you will find an organ in which disease is more likely to occur in the treated group than in the untreated group. Because Big Data tends to have high dimensionality, biases associated with multiple comparisons must be carefully avoided. Methods for reducing multiple comparison bias are available to Big Data analysts. They include the Bonferroni correction, the Sidak correction and the Holm-Bonferroni correction.

Negative study bias When a project produces negative results (fails to confirm a hypothesis), there may be little enthusiasm to publish the work [38]. When statisticians analyze the results from many different published manuscripts (i.e., perform a meta-analysis), their work is biased by the pervasive absence of negative studies [39]. In the field of medicine, negative study bias creates a false sense that every kind of treatment yields positive results.

Time-window bias A bias produced by the choice of a time measurement. In medicine, survival is measured as the interval between diagnosis and death. Suppose a test is introduced that provides early diagnoses. Patients given the test will be diagnosed at a younger age than patients who are not given the test. Such a test will always produce improved survival simply because the interval between diagnosis and death will be lengthened. Assuming the test does not lead to any improved treatment, the age at which the patient dies is unchanged by the testing procedure. The bias is caused by the choice of timing interval (i.e., time from diagnosis to death). Survival is improved without a prolongation of life beyond what would be expected without the test. Some of the touted advantages of early diagnosis are the direct result of timing bias.

Type errors Statistical tests should not be confused with mathematical truths. Every statistician understands that conclusions drawn from statistical analyses are occasionally wrong. Statisticians, resigned to accept their own fallibilities, have classified their errors into five types: Type 1 error—Rejecting the null hypothesis when the null hypothesis is correct (i.e., seeing an effect when there was none). Type 2 error—Accepting the null hypotheses when the null hypothesis is false (i.e., seeing no effect when there was one). Type 3 error—Rejecting the null hypothesis correctly, but for the wrong reason, leading to an erroneous interpretation of the data in favor of an incorrect affirmative statement. Type 4 error—Erroneous conclusion based on performing the wrong statistical test. Type 5 error—Erroneous conclusion based on bad data.

References

[1] Sainani K. Meet the skeptics: why some doubt biomedical models, and what it takes to win them over. Biomed Comput Rev June 5, 2012.

[2] Ioannidis JP. Microarrays and molecular research: noise discovery? Lancet 2005;365:454–5.

[3] Khurana V, Bejjanki HR, Caldito G, Owens MW. Statins reduce the risk of lung cancer in humans: a large case-control study of US veterans. Chest 2007;131:1282–8.

[4] Jemal A, Murray T, Ward E, et al. Cancer statistics, 2005. CA Cancer J Clin 2005;55:10–30.

[5] Jacobs EJ, Newton CC, Thun MJ, Gapstur SM. Long-term use of cholesterol-lowering drugs and cancer incidence in a large United States cohort. Cancer Res 2011;71:1763–71.

[6] Suissa S, Dellaniello S, Vahey S, Renoux C. Time-window bias in case-control studies: statins and lung cancer. Epidemiology 2011;22:228–31.

[7] Shariff SZ, Cuerden MS, Jain AK, Garg AX. The secret of immortal time bias in epidemiologic studies. J Am Soc Nephrol 2008;19:841–3.

[8] Innovation or stagnation: challenge and opportunity on the critical path to new medical products. U. S. Department of Health and Human Services, Food and Drug Administration; 2004.

[9] Wurtman RJ, Bettiker RL. The slowing of treatment discovery, 1965–1995. Nat Med 1996;2:5–6.

[10] Saul S. Prone to error: earliest steps to find cancer. New York Times; July 19, 2010.

[11] Benowitz S. Biomarker boom slowed by validation concerns. J Natl Cancer Inst 2004;96:1356–7.

[12] Berman JJ. Rare diseases and orphan drugs: keys to understanding and treating common diseases. Cambridge, MD: Academic Press; 2014.

[13] Berman J. Precision medicine, and the reinvention of human disease. Cambridge, MA: Academic Press; 2018.

[14] Weigelt B, Reis-Filho JS. Molecular profiling currently offers no more than tumour morphology and basic immunohistochemistry. Breast Cancer Res 2010;12:S5.

[15] Begley S. In cancer science, many 'discoveries' don't hold up. Reuters; March 28, 2012.

[16] Abu-Asab MS, Chaouchi M, Alesci S, Galli S, Laassri M, Cheema AK, et al. Biomarkers in the age of omics: time for a systems biology approach. OMICS 2011;15:105–12.

[17] Ashworth TG. Inadequacy of death certification: proposal for change. J Clin Pathol 1991;44:265.

[18] Kircher T, Anderson RE. Cause of death: proper completion of the death certificate. JAMA 1987;258:349–52.

[19] Walter SD, Birnie SE. Mapping mortality and morbidity patterns: an international comparison. Int J Epidemiol 1991;20:678–89.

[20] Berman JJ. Methods in medical informatics: fundamentals of healthcare programming in Perl, Python, and Ruby. Boca Raton: Chapman and Hall; 2010.

[21] Boyd D. Privacy and publicity in the context of big data. Raleigh, North Carolina: Open Government and the World Wide Web (WWW2010); 2010. April 29, 2010. Available from: http://www.danah.org/papers/talks/2010/WWW2010.html [viewed August 26, 2012].

[22] Li W. The more-the-better and the less-the-better. Bioinformatics 2006;22:2187–8.

[23] Philippe H, Brinkmann H, Lavrov DV, Littlewood DT, Manuel M, Worheide G, et al. Resolving difficult phylogenetic questions: why more sequences are not enough. PLoS Biol 2011;9:e1000602.

[24] Bergsten J. A review of long-branch attraction. Cladistics 2005;21:163–93.

[25] Ransohoff DF. Rules of evidence for cancer molecular-marker discovery and validation. Nat Rev Cancer 2004;4:309–14.

[26] Bickel PJ, Hammel EA, O'Connell JW. Sex bias in graduate admissions: data from Berkeley. Science 1975;187:398–404.

[27] Baker SG, Kramer BS. The transitive fallacy for randomized trials: If A bests B and B bests C in separate trials, is A better than C? BMC Med Res Methodol 2002;2:13.

[28] Tatsioni A, Bonitsis NG, Ioannidis JP. Persistence of contradicted claims in the literature. JAMA 2007;298:2517–26.

[29] Wistow G. Evolution of a protein superfamily: relationships between vertebrate lens crystallins and microorganism dormancy proteins. J Mol Evol 1990;30:140–5.

[30] Waterham HR, Koster J, Mooyer P, van Noort G, Kelley RI, Wilcox WR, et al. Autosomal recessive HEM/Greenberg skeletal dysplasia is caused by 3-beta-hydroxysterol delta(14)-reductase deficiency due to mutations in the lamin B receptor gene. Am J Hum Genet 2003;72:1013–7.

[31] Madar S, Goldstein I, Rotter V. Did experimental biology die? Lessons from 30 years of p53 research. Cancer Res 2009;69:6378–80.

[32] Zilfou JT, Lowe SW. Tumor suppressive functions of p53. Cold Spring Harb Perspect Biol 2009;00: a001883.

[33] Rosen JM, Jordan CT. The increasing complexity of the cancer stem cell paradigm. Science 2009;324:1670–3.

[34] Mallett S, Royston P, Waters R, Dutton S, Altman DG. Reporting performance of prognostic models in cancer: a review. BMC Med 2010;30:21.

[35] Ioannidis JP. Is molecular profiling ready for use in clinical decision making? Oncologist 2007;12:301–11.

[36] Hayes B. An adventure in the nth dimension. Am Sci 2011;99:442–6.

[37] Manyika J, Chui M, Brown B, Bughin J, Dobbs R, Roxburgh C, et al. Big data: the next frontier for innovation, competition, and productivity. McKinsey Global Institute; June 2011.

[38] McGauran N, Wieseler B, Kreis J, Schuler Y, Kolsch H, Kaiser T. Reporting bias in medical research—a narrative review. Trials 2010;11:37.

[39] Dickersin K, Rennie D. Registering clinical trials. JAMA 2003;290:51.

15

Big Data Failures and How to Avoid (Some of) Them

Section 15.1. Failure Is Common

As a rule, software systems do not work well until they have been used, and have failed repeatedly, in real applications.

Dave Parnas

There are many ways in which a complex system can be broken. In 2000, a Concorde crashed on take-off from Charles de Gaulle Airport, Paris. The Concorde was a supersonic transport jet, one of the most advanced and complex planes ever built. Some debris left on the runway had flipped up and tore a tire and some of the underside of the hull. All passengers were killed.

Big Data resources are complex; they are difficult to build and easy to break. After they break, they cannot be easily fixed.

Most Big Data failures do not result from accidents. Most failures occur when the Big Data resource is never completed, or never attains an acceptable level of performance. What goes wrong? Let us run down the reasons for failure that have been published in blogs, magazine articles and books on the subject of Big Data disappointments: Inappropriate selection and use of human resources (wrong leadership, wrong team, wrong people, wrong direction, wrong milestones, wrong deadlines); Incorrect funding (too little funding, too much funding, incorrect allocation of resources, wrong pay scales, wrong incentives); Legal snags (patent infringements, copyright infringements, inept technology transfer, wrong legal staff, inadequate confidentiality and privacy measures, untenable consent forms, poor contracts with unhelpful non-compete clauses, non-compliance with applicable laws and regulations, inadequate financial records and poor documentation of protocol compliances); Bad data (inaccurate and imprecise data, data obtained without regard to established protocols, data that is not fully specified, un-representative data, data that is not germane to the purpose of the resource), poor data security (purposely corrupted data, data stolen by malevolent entities, data inadvertently copied and distributed by staff, non-compliance with internal security policies, poor internal security policies). The list goes on. Generally, we see failure in terms of our own weaknesses: funders see failure as the result of improper funding; managers see failure as the result of poor management; programmers see deficiencies in programming methods, informaticians see deficiencies in metadata annotations, and so on. The field of Big Data is still young; the most senior members of a Big Data team are little more than newbies, and there's plenty of room for self-doubt. [Glossary Fair use]

It may be useful to accept every imaginable defect in a Big Data project as a potential cause of failure. For convenience sake, these defects can be divided into two general categories: (1) failures due design and operation flaws in Big Data resource, and (2) failures due to improper analysis and interpretation of results. Analytic and interpretive errors were discussed in Chapter 14. This chapter deals with the problems that arise when Big Data resources are poorly planned and operated.

Big Data resources are new arrivals to the information world. With rare exceptions, database managers are not trained to deal with the layers of complexity of Big Data resources. It is hard to assemble a team with the composite skills necessary to build a really good Big Data resource. At this time, all data managers are reflexively acquiring new software applications designed to deal with Big Data collections. Far fewer data managers are coming to grips with the fundamental concepts discussed in earlier chapters (e.g., identifier systems, introspection, metadata annotation, immutability, and data triples). It may take several decades before these fundamental principles sink in, allowing Big Data resources to reach their highest potential.

In the field of hospital informatics, costs run very high. It is not unusual for large, academic medical centers to purchase information systems that cost in excess of $500 million. Bad systems are costly and failures are frequent [1–3]. About three quarters of hospital information systems are failures [4]. Successfully implemented electronic health record systems have not been shown to improve patient outcomes [5]. Based on a study of the kinds of failures that account for patient safety errors in hospitals, it has been suggested that hospital information systems will not greatly reduce the incidence of safety-related incidents [6]. Clinical decision support systems, built into electronic health record systems, have not had much impact on physician practice [7]. These systems tend to be too complex for the hospital staff to master and are not well utilized.

The United Kingdom's National Health Service embarked on a major overhaul of its information systems, with the goal of system-wide interoperability and data integration. After investing $17 billion dollars, the project was ditched when members of Parliament called the effort "unworkable" [8–10].

It is difficult to determine the failure rate of Big Data projects. Organizations herald their triumphs but hide their misadventures. There is no registry of Big Data projects that can be followed to determine which projects fail over time. There is no formal designation "Big Data project" that is bestowed on some projects and withheld from others. Furthermore, we have no definition for failure, as applied to Big Data. Would we require a project to be officially disbanded, with all funds withdrawn, before we say that it is defunct? Or would we say that a Big Data project has failed if it did not meet its original set of goals? If a Big Data resource is built, and operates as planned, can we say that it has failed if nobody actually uses the resource? With these caveats in mind, it is believed that the majority of information technology projects fail, and that failure is positively correlated with the size and cost of the projects [11]. We know that public projects costing hundreds of billions of dollars have failed quietly, without raising much attention [12]. Big Data projects are characterized by large size, high complexity, and novel technology, all of which aggravate any deficiencies in management, personnel, or process practices [11].

Section 15.2. Failed Standards

Don't be afraid of missing opportunities. Behind every failure is an opportunity somebody wishes they had missed.

Lily Tomlin

Most standards fail. Examples are easy to find. OSI (Open Systems Interconnection) was a seven-layer protocol intended as the Internet standard. It was backed by the U.S. government and approved by the ISO/IEC (International Organization for Standardization) and the International Electrotechnical Commission). It has been supplanted by TCP/IP (Transmission Control Protocol/Internet Protocol), preferred by Unix. Simply because a standard has been developed by experts, backed by the U.S. government, and approved by an international standards organization, there is no guarantee that it will be accepted by its intended users.

In the realm of format standards (e.g., for documents, images, sound, movies), there are hundreds of standards. Some of these standards were developed for specific devices (e.g., cameras, image grabbers, word processors), and served a specific purpose in a small window of time. Today, most of these standard formats are seldom used. A few dozen remain popular. There is an on-going effort to incorporate all of the various image and media formats under one standard and one horrifying term: the BLOB (Binary Large OBject). [Glossary BLOB]

Every new programming language is born with the hope that it will be popular and immortal. In the past half century, well over 2000 programming languages have been devised. Most of these languages are seldom used and often forgotten. In 1995, Ada 95 became an ANSI/ISO standard programming language, a distinction held by only a few programming languages. The U.S. National Institute of Standards announced that Ada would be used by Federal departments and agencies in software applications that involve control of real-time or parallel processes, very large systems, and systems with requirements for very high reliability [13]. The official announcement was entitled, Announcing the Standard for ADA. The programming language, Ada, was named for Ada Lovelace (1815–52), who wrote the first computer program (an algorithm for computing Bernoulli numbers) for Charles Babbage's prototype computer (the so-called analytic engine). Every Ada programmer knows that Ada is not an acronym; they bristle whenever Ada is spelled with all-uppercase letters. The federal government's announcement of the new "ADA" standard did not bode well. Ada is a fine programming language, but declaring it a government standard could not guarantee error-free implementations; nor could it guarantee its popularity among programmers. Following its ascension to standards status, the popularity of Ada declined rapidly. Today, it is rare to find an Ada programmer.

Even the best standards seldom meet expectations. Consider the metric system. It is used throughout the world, and it is generally acknowledged as a vast improvement over every preceding measurement standard. Nonetheless, in the United States our height is measured in feet and inches, not meters and centimeters, and our weight is measured in pounds, not kilograms. Here in the United States, it would be difficult to find a bathroom scale marked with metric graduations. The next time you look at your calendar, remember that about half the world uses a solar calendar. Most other earthlings follow a lunar calendar. Some base their calendars on a combination of solar and lunar observations.

When a standard is ignored, or improperly implemented, the results may be catastrophic. On June 4, 1996, the maiden flight of the French Ariane 5 exploded 37 seconds after launch. A software exception occurred during a data conversion from a 64-bit floating point to 16-bit signed integer value. The data conversion instructions (in Ada code) were not protected from causing an Operand Error [14].

On September 23, 1999, the United States launched the Mars Climate Orbiter, which crashed on impact on the red planet. An official investigation, by the Mars Climate Orbiter Mishap Investigation Board, concluded that the crash occurred due to a software glitch that arose when English units of measurement were used in the software when Metric units were supplied as input [15]. The flight software was coded in Ada.

The most successful standards are specifications that achieved popularity before they achieved the status of "standard." The best of these filled a need, enjoyed broad use, had few or no barriers to implementation (e.g., free and easy to use), and had the bugs ironed out (i.e., did not require excessive modifications and version updates). The most unsuccessful standards are those prepared by a committee of special interests who create the standard ab initio (i.e., without a pre-existing framework), without a user community, and without a proven need. The altogether worst standards seem to be those that only serve the interests of the standards committee members.

Robert Sowa has written a useful essay entitled "The Law of Standards" [16]. His hypothesis is, "Whenever a major organization develops a new system as an official standard for X, the primary result is the widespread adoption of some simpler system as a de facto standard for X." He gives many examples. The PL/I standard, developed by IBM was soon replaced by Fortran and COBOL. The Algol 68 standard was replaced by Pascal. Ada, promoted by the U.S. Department of Defense, was replaced by C. The OS/2 operating system produced by IBM was replaced by Windows.

For small data projects and for software applications, the instability of data standards is not a major problem. Small data projects are finite in length and will seldom extend beyond the life span of the standards implemented within the project. For software designers, a standard implemented within an application can be replaced in the next upgrade. Any costs are passed onto the licensed user. Instability in standards serves the interests of software developers by coercing customers to purchase upgrades that comply with the new versions of included standards.

For Big Data resources, instability in standards is always bad news. A failed standard may invalidate the data model for the resource; undoing years of work. How can the data manager cope with failed standards? Over twenty years ago, I was approached by a pathologist who was tasked with annotating his diagnostic reports with a standard vocabulary of diseases. The principal options at the time were ICD (International Classification of Diseases), SNOMED (Systematized Nomenclature of Medicine) and MeSH (Medical Subject Headings produced by the National Library of Medicine). The ICD seemed too granular (i.e., not enough names of diseases). SNOMED was constantly changing; newer versions were incompatible with older versions, and he worried that annotations under an old version of SNOMED could not be integrated into the newer hospital information systems. MeSH was a well-curated public nomenclature, analogous to the Dewey Decimal System for the health informatics community, but it was not widely adopted by the pathology community.

I suggested all of his options were untenable. His best bet, under the circumstances, was to write his reports in simple, declarative sentences, using canonical diagnostic terms (i.e., terms expressed in a form suitable for a nomenclature). Reduced sentences could be easily parsed into constituent parts and accurately translated or mapped to any chosen vocabulary, as needed [17].

Consider the following sentence:

The patient has an scc, and we see invasion to the subcutaneous tissue, all the way to the deep margins, but the lateral margins are clear.

This sentence, which is understandable to clinicians, would not be understandable to a computer program that parsed text. Among other impediments, a computer would not know that the abbreviation "scc" corresponds to the diagnostic term "squamous cell carcinoma." A computer that has an index list matching abbreviations to terms may falsely map the abbreviation to the wrong expansion term (e.g., small cell carcinoma rather than squamous cell carcinoma).

The complex sentence could be rewritten as six declarative statements:

Diagnosis: squamous cell carcinoma.
  Invasion is present.
  Invasion extends to subcutaneous tissue.
  Margin is involved.
  Tumor extends to deep margin.
  Tumor does not extend to lateral margins.

It would be relatively easy to write a computer program that could autocode these very simple sentences. Every surgical pathology case entered in the hospital information system could be coded again and again, using any new version of any nomenclature. [Glossary Autocoding, Autoencoding]

We could go one step further, expressing every statement as a triple consisting of an identifier, a metadata term, and a data value. As discussed in Section 4.3, "Semantics and Triples," if all of the data in the resource is available as simple triples, and if the model provides a method whereby data objects can be assigned to classes, then every data object can be fully specified. Specified data objects, expressed as a simple triples, can be ported into any old or new data standard, as needed. [Glossary Class-oriented programming]

It is best to keep in mind two general principles of data management:

1. Data objects can be well specified, without a standard. You do not need to store your data in a format that is prescribed by a standard.
2. Data standards are fungible. If you know the rules for standards, you can write a program that converts to the standard, as needed.

In many instances, a simple, generic data model may free the Big Data manager from the problems that ensue when a data standard becomes obsolete.

Section 15.3. Blaming Complexity

Complexity is the worst enemy of security.

Bruce Schneier [18]

Big Data is complex, and complexity is dangerous. It is easy to write software that attains a level of complexity that exceeds anything encountered in the physical realm. Likewise, there is no limit to the complexity of operational methods, security standards, data models, and virtually every component of a Big Data resource. When a Big Data resource somehow manages to cope with complexity, it can be just a matter of time before key personnel depart to follow other opportunities, errors are introduced into the system, and a once-great resource grinds to a halt.

When errors occur in complex systems, even catastrophic errors, they can be very difficult to detect. A case in point is the Toyota Lexus ES 350 Sedan. Thousands of vehicle owners experienced unintended vehicle acceleration; the complex electronic control system was the chief suspect [19]. Over the years, Toyota expended enormous resources trying to understand and solve the problem [20]. A host of agencies and authorities were involved in the investigation; first came the Department of Transportation and the National Highway Traffic Safety Administration. Then, owing to its expertise in software integrity, computer control systems, and electromagnetic interference, the National Aeronautics and Space Administration was called into the fray. Later, the National Academy of Sciences launched its own study of unintended acceleration in the auto industry. During these investigations, Toyota paid about $50 million in fines and recalled about 9 million cars. The dust may never settle completely on this problem, but it now appears that most, if not all, problems were due to sticky pedals, driver error, or improperly placed floor mats; no errors were uncovered in the complex electronic control system.

The most vexing problems in software engineering involve "sometimes" errors; software that runs perfectly under most circumstances, but fails at apparently random intervals. Finding the source of the problem is virtually impossible, because the most thorough evaluations will indicate that everything is working well. Sometimes the mistake occurs because of the chaotic and unpredictable quality of complex systems. Sometimes mistakes occur because the numbers get too big, or too small, or too impossible (division by zero); sometimes the order by which events occur are unexpected, causing the system to behave oddly. In all these situations, finding the problem is very difficult and could have been avoided if the system had been less complex. Knowing this, you might expect that data managers try their best to reduce the complexity of their resources. Actually, no. For most resources, increasing complexity is the normal state because it is easier to solve problems with more complexity than with more simplicity.

Every Big Data project should be designed for simplicity. The design team should constantly ask "Can we achieve this functionality with less complexity?" When the complexity cannot be reduced for a desired level of functionality, a trade-off might be reached. The team is justified in asking, "Do we need this level of functionality? Might we achieve a reduced but adequate level of functionality with a less complex system?" After the design phase, every addition, and every modification to the system should be examined for complexity. If complexity needs to be added to the system, then the team must analyze the consequences of the increased complexity.

– **When does complexity help?**

There are times when complexity is necessary. Think of the human mind and the human body. Our achievements as individuals and as a species come as the result of our complexity. This complexity was achieved over 4 billion of years of evolution, during which time disadvantageous traits were lost and advantageous traits were retained. The entire process was done incrementally. The complexity of a Big Data resource is created in a moment. We do not have four billion years to debug the system. When can complexity be permitted in a Big Data resource? There are several scenarios:

– When approximate or locally accurate solutions are not acceptable.

In the case of weather forecasting, the purpose is to attain predictions of ever-increasing accuracy. Each new forecasting model contains more parameters than the prior model, requires more computing power, and is expected to provide accurate forecasts that extend further and further into the future. Complex models that do not yield greater accuracy than simpler models are abandoned. The whole process mimics evolution.

– When complexity is achieved incrementally.

Many of the most important devices implemented in society are complex (televisions, computers, smartphones, jet airplanes, magnetic resonance imaging devices). They all started as simpler devices, with complexity added incrementally. These complex devices did not require 4 billion years to evolve, but they did require the intense participation of many different individuals, teams, corporations, and users to attain their current utility.

The venerable U-2 U.S. spy plane is an example of incrementally achieved complexity. The U-2 was born in the 1950s and designed as a cold war spy plane. Despite its advanced age, the U-2 has stubbornly resisted obsolescence by incrementally increasing in complexity and utility. Today, it is still in service, with a functionality far greater than anything imaginable when it was created [21]. The value of incremental complexity has been emulated in some modern Big Data technologies [22].

– When your model really needs to match, item by item, the complexity of the real system that it is modeling.

Biologists have learned, in the past decade, that cellular processes are much more complex than they had originally imagined. Genes are controlled by interactions with other genes, with RNA, with proteins, and with chemical modifications to DNA. Complex chemical interactions that occur in the cell's nucleus and the cytoplasm have made it impossible to find simple genetic variants that account for many biologic processes. The entire field of gene research has shifted to accommodate previously unanticipated complexities that have thwarted our earlier analyses [23]. Our progress in disease biology, developmental biology, and aging seems to hinge on our willingness to accept that life is complex, and cannot be reduced to a sequence of nucleotides in a strand of DNA.

There are occasions when we cannot "wish away" complexity. The best we can do is to prepare a model that does not amplify the irreducible complexity that exists in reality.

Section 15.4. An Approach to Big Data That May Work for You

The most likely way for the world to be destroyed, most experts agree, is by accident. That's where we come in; we're computer professionals. We cause accidents.
                                                                        Nathaniel Borenstein

The old saying, "You can bring a horse to water, but you can't make it drink," aptly describes the situation for many Big Data resources. Sometimes Big Data projects fail because the intended users simply do not know what to do with the data. They have no approach that matches the available data, and they blame the Resource for their inability to succeed. After a few would-be data miners give up, the Resource gets a reputation of being useless. Funding is withdrawn, and the Resource dies.

At this point, you may feel completely overwhelmed by the complexities of Big Data resources. It may seem that analysis is humanly impossible. The best way to tackle a large and complex project is to divide it into a smaller, less intimidating tasks. The approach to data analysis described in this chapter involves nine sequential steps.

– **Step 1. A question is formulated**

It takes a certain talent to ask a good question. Sometimes, a question, even a brilliant question, cannot be answered until it is phrased in a manner that clarifies the methods by which the question can be solved. For example, suppose I am interested in how much money is spent, each year, on military defense, in the United States. I could probably search the Internet and find the budget for the Department of Defense, in the year 2011. The budget for the Department of Defense would not reflect the costs associated with other agencies that have a close relationship with the military, such as intelligence agencies and the State Department. The Department of Defense budget would not reflect the budget of the Veterans Administration (an agency that is separate from the Department of Defense). The budget for the Department of Defense might include various items that have no obvious relationship to military defense. Because I am asking for the "annual" budget, I might need to know how to deal with projects whose costs are annualized over 5, 10, or 15 years. If large commitments were made, in 2005, to pay for long-term projects, with increasing sums of money paid out over the next decade, then the 2018 annual budget may reflect payouts on 2010 commitments. A 2018 budget may not provide a meaningful assessment of costs incurred by 2018 activities. After a little thought, it becomes obvious that the question: "How much money is spent, each year, on military defense, in the United States?" is complex, and probably cannot be answered by any straightforward method.

At this point, it may be best to table the question for a while, and to think deeply about what you can reasonably expect from Big Data. Many analysts start with the following general question: "How can this Big Data resource provide the answer to my question?" A more fruitful approach may be: "What is the data in this resource trying to tell me?" The two approaches are quite different, and I would suggest that data analysts begin their analyses with the second question.

– **Step 2. Resource evaluation**

Every good Big Data resource provides users with a detailed description of its data contents. This might be done through a table of contents or an index, or through a detailed "readme" file, or a detailed user license. It all depends on the type of resource and its intended purposes. Resources should provide detailed information on their methods for collecting and verifying data, and their protocols supporting outsider queries and data extractions. Big Data resources that do not provide such information generally fall into two categories: (1) highly specialized resources with a small and devoted user base who are thoroughly familiar with every aspect of the resource and who do not require guidance; or (2) bad resources.

Before developing specific queries related to your research interest, data analysts should develop queries designed to evaluate the range of information contained in the resource (discussed in detail in Chapter 9, "Assessing the Adequacy of a Big Data Resource". Even the best Big Data resources may contain systemic biases. For example, PubMed contains abstracted data on about 20 million research articles. Research articles are published on positive findings. It is very difficult for a scientist to publish a paper that reports on the absence of an effect or the non-occurrences of a biological phenomenon. PubMed has a positive result bias. The preferential exclusion or inclusion of specific types of data is very common, and data analysts must try to identify such biases.

Every Big Data resource has its blind spots; areas in which data is missing or scarce, or otherwise unrepresentative of the data domain. Often, the Big Data managers are unaware of such deficiencies. In some cases, Big Data managers blame the data analyst for "inventing" a deficiency that pertains exclusively to unauthorized uses of the resource. When a data analyst wishes to use a Big Data resource for something other than its intended purposes (e.g., using PubMed to predict NIH funding priorities over the next decade, using the Netflix query box to determine what kinds of actors appear in zombie movies), then the Big Data manager may be reluctant to respond to the analyst's complaints.

Simply having access to large amounts of subject data does not guaranteed that you have all the data you would need to draw a correct conclusion.

- **Step 3. A question is re-formulated**

"If you can dream - and not make dreams your master"

From If (poem), by Rudyard Kipling

The available data cannot always answer the exact question you started with. After you have assessed the content and design of your Big Data resource(s), you will want to calibrate your question to your available data sources. In the case of our original question, from Step 1, we wanted to know how much money is spent, each year, on military defense, in the United States. If we are unable to answer this question, we may be able to answer questions related to the budget sizes of individual government agencies that contribute to military spending. If we knew the approximate portion of each agency budget that is devoted to military spending, we might be able to produce a credible total for the amount devoted to military activities, without actually finding the exact answer.

After exploring the resource, the data analyst learns the kinds of questions that can best be answered with the available data. With this insight, he or she can re-formulate the original set of questions.

- **Step 4. Determine the adequacy of your query's returned output**

Big Data resources can often produce an enormous output in response to a data query. When a data analyst receives a large amount of data, he or she is likely to assume that the query output is complete and valid. A query output is complete when it contains all of the data held in the Big Data resource that answers the query, and a query output is valid if the data in the query output yields a correct and repeatable answer.

A Google query is an example of an instance wherein query output is seldom seriously examined. When you enter a search term and receive millions of "hits", you may tend to assume that your query output is adequate. When you're looking for a particular Web page, or an answer to a specific question, the first output page on your initial Google query may meet all your needs. A thoughtful data analyst will want to submit many related queries to see which queries produce the best results. The analyst may want to combine the query outputs from multiple related queries, and will almost certainly want to filter the combined outputs to discard response items that are irrelevant. The process of query output examination is often arduous, requiring many aggregation and filtering steps.

After satisfying yourself that you've taken reasonable measures to collect a complete query output, you will still need to determine whether the output you have obtained is fully representative of the data domain you wish to analyze. For example, you may have a large query output file related to the topic of poisonous mushrooms. You've aggregated query outputs on phrases such as "mushroom poisoning", "mushroom poisons", "mushroom poison", "mushroom toxicity", and "fungal toxins". You pared down queries on "food poisoning," to include only mushroom-related entries. Now you want to test the output file to see if it has a comprehensive collection of information related to your topic of interest. You find a nomenclature of mushrooms, and you look for the occurrence of each nomenclature term in your aggregated and filtered output file. You find that there are no occurrences of many of the mushrooms found in the mushroom nomenclature, including mushrooms known to be toxic. In all likelihood, this means that the Big Data resource simply does not contain the level of detail you will need to support a thorough data analysis on topics related to poisonous mushrooms.

There is no standard way of measuring the adequacy of a query output; it depends on the questions you want to answer, and the analytic methods you will employ. In some cases, a query output will be inadequate because the Big Data resource simply does not contain the information you need; at least not in the detail you require. In other cases, the Big Data resource will contain the information you need, but does not provide a useful pathway by which your query can access the data. Queries cannot thoroughly access data that is not fully annotated, assigned to classes, and constructed as identified data objects.

Data analysts must be prepared to uncover major flaws in the organization, annotation, and content of Big Data resources. When a flaw is found, it should be promptly reported to the data manager for the resource. A good data manager will have a policy for accepting error reports, conducting investigations, instituting corrections as necessary, and documenting every step in the process.

– **Step 5. Describe your data**

Is the output data numeric or is it categorical? If it is numeric, is it quantitative? For example, telephone numbers are numeric, but not quantitative. If the data is numeric and quantitative, then your analytic options are many. If the data is categorical information (e.g., male or female, true or false), then the analytic options are limited. The analysis of categorical data is first and foremost an exercise in counting; comparisons and predictions are based on the occurrences of features.

Are all of your data objects comparable? Big Data collects data objects from many different sources, and the different data objects may not be directly comparable. The objects themselves may be annotated with incompatible class hierarchies (e.g., one data object described as a "chicken" may be classed as "Aves", while another "chicken" object may be classed as "Food." One data object described as "child" may have the "age" property divided into three-year increments up to age 21. Another "child" object may have "age" divided into 4-year increments up to age 16. The data analyst must be prepared to normalize assigned classes, ranges of data, subpopulations of wildly different sizes, different nomenclature codes, and so on.

After the data is normalized and corrected for missing data and false data, you will need to visualize data distributions. Be prepared to divide your data into many different groupings, and to plot and re-plot your data with many different techniques (e.g., histograms, smoothing convolutions, cumulative plots, etc.). Look for general features (e.g., linear curves, non-linear curves, Gaussian distributions, multi-modal curves, convergences, non-convergences, Zipf-like distributions). Visualizing your data with numerous alternate plotting methods may provide fresh insights and will reduce the likelihood that any one method will bias your objectivity.

– Step 6. Data reduction

An irony of Big Data analysis is that the data analyst must make every effort to gather all of the data related to a project, followed by an equally arduous phase during which the data analyst must cull the data down to its bare essentials.

There are very few situations wherein all of the data contained in a Big Data resource is subjected to analysis. Aside from the computational impracticalities of analyzing massive amounts of data, most real-life problems are focused on a relatively small set of local observations drawn from a large number of events. The process of extracting a small set of relevant data from a Big Data resource is referred to by a variety of names, including data reduction, data filtering, and data selection. The reduced data set that you will use in your project should obey the courtroom oath "the whole truth, and nothing but the truth."

Methods for reducing the dimensionality of data are described in Section 10.5 "Reducing Your Data". As a practical point, when the random and redundant variables have been expunged, the remaining data set may still be too large for a frontal computational attack using advanced methods. A good data analyst knows when to retreat and regroup. If something can be calculated to great precision on a large number of variables and data points, then it should be calculated with somewhat less precision with somewhat fewer variables and fewer data points. Why not try the small job first, and see what it tells you?

– Step 7. Select analytic algorithms (if absolutely necessary)

Algorithms are perfect machines. They work to produce consistent solutions; they never make mistakes; they need no fuel; they never wear down; they are spiritual, not physical.

Every computer scientist loves algorithms; if they could be re-assigned to their favorite position, most computer scientists would devote their careers to writing new algorithms.

If you peruse the titles of books in the Big Data field, you will find that most of these books emphasize data analysis. They focus on parallel processing, cloud computing, high-power predictive analytics, combinatorics methods, and the like. It is very easy to believe that the essential feature of Big Data, that separates it from small data, relates to analytic algorithms. [Glossary Cloud computing, Grid]

As algorithms become more and more clever, they become more and more enigmatic. Some of the most popular statistical methods can be used injudiciously, and these include p values and linear regression [24,25]. Journal editors know this. Consequently, when a scientist submits an article to a journal, he or she can expect the editor to insist that a statistician be included as a co-author. The editors have a valid point, but is it really helpful to forcibly insert a statistician into an unfamiliar project? Isn't there some risk that the inclusion of a statistician will provide a thin veneer of scientific credibility, without necessarily attaining a valid scientific conclusion? [Glossary *P* value, Linear regression]

The field of Big Data comes with a dazzling assortment of advanced analytic options. Who is really qualified to judge whether the correct method is chosen; whether the chosen method is implemented properly; and whether the results are interpreted correctly?

Analysts in search of analytic algorithms should consider these simple options:

– Stick with simple estimates.

If you have taken to heart the suggestion in Section 12.4, "Back-of-Envelope Analyses", to estimate your answers early in project development, then you have already found simple estimators for your data. Consider this option: keep the estimators, and forget about advanced algorithms. For many projects, estimators can be easily understood by project staff, and will provide a practical alternative to exact solutions that are difficult to calculate and impossible to comprehend.

– Pick better metrics, not better algorithms.

Sabermetrics is a sterling example of analysis using simple metrics that are chosen to correlate well with a specific outcome; a winning ball game. In the past several decades, baseball analysts have developed a wide variety of new performance measurements for baseball players. These include: base runs, batting average on balls in play, defense independent pitching statistics, defense-independent earned run average, fielding independent pitching, total player rating, or batter-fielder wins, total pitcher index, and ultimate zone rating. Most of these metrics were developed empirically, tested in the field, literally, and optimized as needed. They are all simple linear metrics that use combinations of weighted measures on data collected during ballgames. Though sabermetrics has its detractors, everyone would agree that it represents a fascinating and largely successful effort to bring objective numeric techniques to the field of baseball. Nothing in sabermetrics involves advanced algorithms. It is all based on using a deep understanding of the game of baseball to develop a set of simple metrics that can be easily calculated and validated.

- Micromanage your macrodata.

Much of the success of Big Data is attained by making incremental, frequent changes to your system in response to your metrics. An example of successful micromanagement for Big Data is the municipal CompStat model, used by Police Departments and other government agencies [26,27]. A promising metric is chosen, such as 911 response time, and a team closely monitors the data on a frequent basis, sometimes several times a day. Slow 911 response times are investigated, and the results of these investigations typically generate action items intended to correct systemic errors. When implemented successfully, the metric improves (e.g., the 911 response time is shortened), and a wide range of systemic problems are solved. Micromanaging a single metric can improve the overall performance of a department.

Departments with imagination can choose very clever metrics upon which to build an improvement model (e.g., time from license application to license issuance, number of full garbage cans sitting on curbs, length of toll booth lines, numbers of broken street lights, etc.) It is important to choose useful metrics, but the choice of the metric is not as important as the ability to effectively monitor and improve the metric.

As a personal aside, I have used this technique in the medical setting and found it immensely effective. During a period of about 5 years at the Baltimore VA Medical center, I had access to all the data generated in our Pathology Department. Using a variety of metrics such as case turn-around time, cases requiring notification of clinician, cases positive for malignancy, and diagnostic errors, our pathologists were able to improve the measured outcomes. More importantly, the process of closely monitoring for deficiencies, quickly addressing the problem, and reporting on the outcome of each correction produced a staff that was sensitized to the performance of the department. There was an overall performance improvement.

Like anything in the Big Data realm, the data micromanagement approach may not work for everyone, but it serves to show that great things may come when you carefully monitor your Big Data resource.

- Let someone else find an algorithm for you; crowd-source your project.

There is a lot of analytic talent in this world. Broadcasting your project via the Web may attract the attention of individuals or teams of researchers who have already solved a problem isomorphic to your own, or who can rapidly apply their expertise to your specific problem [28].

- Offer a reward.

Funding entities have recently discovered that they can solicit algorithmic solutions, offering cash awards as an incentive. For example, the InnoCentive organization issues challenges regularly, and various sponsors pay awards for successful implementations [29]. [Glossary Predictive modeling contests]

- Develop your own algorithm that you fully understand. You should know your data better than anyone else. With a little self-confidence and imagination, you can develop an analytic algorithm tailored to your own needs.

- **Step 8. Results are reviewed and conclusions are asserted**

When the weather forecaster discusses the projected path of a hurricane, he or she will typically show the different paths projected by different models. The forecaster might draw a cone-shaped swath bounded by the paths predicted by the several different forecasts. A central line in the cone might represent the composite path produced by averaging the forecasts from the different models. The point here is that Big Data analyses never produce a single, undisputed answer. There are many ways of analyzing Big Data, and they all produce different solutions.

A good data analyst should interpret results conservatively. Here are a few habits that will keep you honest and will reduce the chances that your results will be discredited.

- Never assert that your analysis is definitive.

If you have analyzed the Big Data with several models, include your results for each model. It is perfectly reasonable to express your preference for one model over another. It is not acceptable to selectively withhold results that could undermine your conclusions.

- Avoid indicating that your analysis provides a causal explanation of any physical process.

In most cases, Big Data conclusions are descriptive, and cannot establish physical causality. This situation may improve as we develop better methods to make reasonable assertions for causality based on analyses of large, retrospective data sets [30,31]. In the meantime, the primary purpose of Big Data analysis is to provide a hypothesis that can be subsequently tested, usually through experimentation, and validated.

- Disclose your biases.

It can be hard to resist choosing an analytic model that supports your pre-existing opinion. When your results advance your own agenda, it is important to explain that you have a personal stake in the outcome or hypothesis. It is wise to indicate that the data can be interpreted by other methods and that you would be willing to cooperate with colleagues who might prefer to conduct an independent analysis of your data. When you offer your data for re-analysis, be sure to include all of your data: the raw data, the processed data, and step-by-step protocols for filtering, transforming, and analyzing the data.

- Do not try to dazzle the public with the large number of records in your Big Data project.

Large studies are not necessarily good studies, and the honest data analyst will present the facts and the analysis, without using the number of data records as a substitute for analytic rigor.

– **Step 9. Conclusions are examined and subjected to validation**

"Sometimes you gotta lose 'til you win"

From "Little Miss" (song) by Sugarland

Validation involves demonstrating that the conclusions that come from data analyses are reliable. You validate conclusions by showing that you draw the same conclusion repeatedly in comparable data sets.

Real science can be validated, if true, and invalidated, if false. Pseudoscience is a pejorative term that applies to scientific conclusions that are consistent with some observations, but which cannot be confirmed or tested. For example, there is a large body of information that would suggest that earth has been visited by flying saucers. The evidence comes in the form of many eyewitness accounts, numerous photographs, and cover-up conspiracy theories. Without commenting on the validity of UFO claims, it is fair to say that these assertions fall into the realm of pseudoscience because they are untestable (i.e., there is no way to prove that flying saucers do not exist), and there is no definitive data to prove their existence (i.e., the "little green men" have not been forthcoming).

Big Data analysis always stands on the brink of becoming a pseudoscience. Our finest Big Data analyses are only valid to the extent that they have not been disproven. A good example of a tentative and clever conclusion drawn from data is the Titius-Bode Law. Titius and Bode developed a simple formula that predicted the locations of planets orbiting a star. It was based on data collected on all of the planets known to Johann Daniel Titius and Johann Elert Bode, two eighteenth century scientists. These planets included Mercury through Saturn. In 1781, Uranus was discovered. Its position fit almost perfectly into the Titius-Bode series, thus vindicating the predictive power of their formula. The law predicted a fifth planet, between Mars and Jupiter. Though no fifth planet was found, astronomers found a very large solar-orbiting asteroid, Ceres, at the location predicted by Titius and Bode. By this time, the Titius-Bode Law was beyond rational disputation. Then came the discoveries of Neptune and Pluto, neither of which remotely obeyed the Law. The data had finally caught up to the assertion. The Titius-Bode Law was purely descriptive; not based on any universal physical principles. It served well for the limited set of data to which it was fitted, but was ultimately discredited.

Let us look at a few counter-examples. Natural selection is an interesting theory, published by Charles Darwin, in 1859. It was just one among many interesting theories aimed at explaining evolution and the origin of species. The Lamarckian theory of evolution preceded Darwin's natural selection by nearly 60 years. The key difference between Darwin's theory and Lamarck's theory comes down to validation. Darwin's theory has withstood every test posed by scientists in the fields of geology, paleontology, bacteriology, mycology, zoology, botany, medicine, and genetics. Predictions based on Darwinian evolution dovetail perfectly with observations from diverse fields. The Lamarckian theory of evolution, proposed well before DNA was established as the genetic template for living organisms, held that animals passed experiences to succeeding generations through germ cells; thus strengthening intergenerational reliance on successful behaviors of the parent. This theory was groundbreaking in its day, but subsequent findings failed to validate the theory. Neither Darwin's theory nor Lamarck's theory could be accepted on their own merits. Darwin's theory is correct, as far as we can know, because it was validated by careful scientific observations collected over the ensuing 150 years. Lamarck's theory is incorrect, because it failed the validation process.

The value of big data is not so much to make predictions, but to test predictions on a vast number of data objects. Scientists should not be afraid to create and test their prediction models in a Big Data environment. Sometimes a prediction is invalidated, but an important conclusion can be drawn from the data anyway. Failed predictions often lead to new, more successful predictions.

Of course, every analysis project is unique, and the steps involved in a successful project will vary. Nonetheless, a manageable process, built on techniques introduced in preceding chapters, might be helpful. My hope is that as Big Data resources mature and the methods for creating meaningful, well annotated, and verified data become commonplace, some of the steps listed in this chapter can be eliminated. Realistically, though, it is best to assume that the opposite will occur; more steps will be added.

Section 15.5. After Failure

Programming can be fun, so can cryptography; however they should not be combined.

Charles B. Kreitzberg and Ben Shneiderman

In 2001, funds were appropriated to develop the National Biological Information Infrastructure. This Big Data project was a broad cooperative effort intended to produce a federated system containing biological data from many different sources, including Federal and State government agencies, universities, natural history museums, private organizations, and others. The data would be made available for studies in the field of resources management. On January 15, 2012, the resource was officially terminated, due to budget cuts. A Website announcement sits vigil, like a tombstone, marking the passage of an ambitious and noble life (Fig. 15.1).

Like the humans who create the resources, Big Data lives for a time, and then it dies. Humans often carry life insurance, or savings, to pay the costs of burial and to manage the distribution of an estate. In most cases, no preparations are made for the death of Big Data. One day it is there, and the next day it is gone.

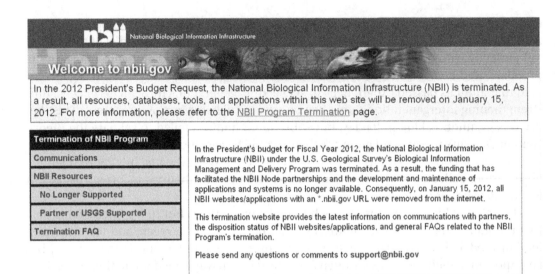

FIG. 15.1 A U.S. government website announcing the demise of the National Biological Information Infrastructure project, in 2012.

Abandonware is a term that is often applied to software developed under a grant. When the grant funding ceases, the software is almost always abandoned. Nobody assumes the responsibility of tying up loose ends, finding and removing bugs, distributing the software, or using the software in new projects. Similarly, when a Big Data project dies, the programmers and staff scrabble to find new jobs; nobody remains to clean up the leftovers. The software is abandoned. The identifier system and the data model for the resource are almost always lost to posterity. All those charts and all those registration numbers simply disappear. The data in the failed resource either slips into the bit-void or becomes legacy data; stored on disks, shelved, and forgotten. Legacy data can be resurrected, but it seldom happens.

Here are a two precautions that can salvage some of the pieces of a failed Big Data project.

– **Write utilities, not applications.**

Software applications can be envisioned as utilities with a graphic user interface. One software application may have the functionality of three or four utilities (e.g., spreadsheet plus graph plotter plus statistics suite, with the graphic user interface of a word processor). When a Big Data Resource is built, it will usually be accessed under an umbrella application. This application will support user queries, while shielding the user from the complexities of the data and the mechanics of passing the query over multiple, networked servers. When a Big Data resource is terminated, applications with a user interface to the resource retain no value. However, the small utility programs built into the application, may have enormous value to the builders of related Big Data resources. The best applications are modularized; built from working parts that can be extracted and used as modules or utilities. Utilities that apply generally to many data-related projects are valuable assets.

When you write code, employ responsible coding practices. There is a vast literature on coding etiquette. Programmers are always advised to write simple code, to comment their code (i.e., explain the software steps and algorithms within the programs, as well as any modifications to the software), and to avoid using variable names that have no obvious meaning. This book will not dwell on good programming practices, but suffice it to say that as programs become more and more complex, good software practices become essential.

– **Pay up-front to preserve your legacy data and your identifiers.**

Big Data resources are very expensive. It makes sense to put aside funds to preserve the data, in the event of failure. Data means nothing unless it is properly identified; hence, the data held in a Big Data resource must be preserved with its identifiers. If the data is identified and well annotated, it will be possible to re-integrate the data objects into a successor resource.

Preserving legacy data is particularly important in hospital information systems, which have a very high failure rate [3]. It is unacceptable to lose patient histories, just because the hospital bought bad software. Hospitals, medical centers and all Big Data systems that serve critical missions should set aside money, in an escrow fund, for this purpose.

Section 15.6. Case Study: Cancer Biomedical Informatics Grid, a Bridge Too Far

In a software project team of 10, there are probably 3 people who produce enough defects to make them net negative producers.

Gordon Schulmeyer

Some governments do an excellent job at analyzing their own failures. After the Ariane 5 exploded 37-post launch on its maiden space flight, the French government issued the results of their investigation [14]. When the Mars Climate Orbiter crashed on Mars, the U.S. government issued a report on their investigation [15]. The world might be a better place if we, as individuals, published our own investigative reports when we disappoint our friends and co-workers or fail to meet our personal goals; but self-accountability has never been a human strength.

In 2004, the National Cancer Institute launched an ambitious project, known as the Cancer Biomedical Informatics Grid (CaBig™), aimed at developing standards for annotating and sharing biomedical data, and tools for data analysis. An unusual aspect of this government project was that it had its own business model, "to create a front end that will make caBIG attractive for others to invest in and take on responsibility for downstream events" [32]. Further, it was "anticipated at the time that the caBIG effort would evolve into a self-sustaining community" [32].

When the effort began, its goals seemed relatively focused, but over the next 6 years, the scope of the project grew; it eventually covered software interoperability and Web-based services for sharing and analyzing data produced by virtually every biomedical specialty, from gene sequencing to radiology imaging to tissue specimens.

For a time, the project received generous support from academia and industry. In 2006, the Cancer Biomedical Informatics Grid was selected as a laureate in ComputerWorld's honors program [33]. ComputerWorld described the project as "Effectively forming a World Wide Web of cancer research," with "promises to speed progress in all aspects of cancer research and care." The laudable ambitions of the project came with a hefty price tag. By 2010, the National Cancer Institute had sunk at least 350 million dollars into the effort [32]. Though the project was ambitious, there were rumblings in the cancer informatics community that very little had been achieved. In view of past and projected costs, an ad hoc committee was assigned to review the program.

In the report issued to the public, in 2011, the committee found that the project had major deficiencies and suggested a yearlong funding moratorium [32]. Soon thereafter, the project leader left the National Cancer Institute, and the Cancer Bioinformatics Grid was terminated [34].

The ad hoc committee report went into considerable detail to explain the deficiencies of the program. Their observations serve as a general warning for overly ambitious Big Data efforts.

> The caBIG program has grown rapidly without adequate prioritization or a cost-effective business model, has taken on a continuously increasing and unsustainable portfolio of development and support activities, and has not gained sufficient traction in supporting critical cancer research community needs. The working group interviews indicate that the program has developed some extremely expensive software solutions that have not been adopted in a meaningful way by the NCI designated Cancer Centers, have competed unnecessarily with existing solutions produced by industry leaders that hold a 60% to 70% market share in the NCI-designated Cancer Centers, and ultimately have created an enormous long-term maintenance, administration, and deployment load for the NCI that is financially unsustainable [32].

Regarding analysis tools, the working group found that "the level of impact for most of the tools has not been commensurate with the level of investment. For example, many tools, such as caArray ($9.3M), have been developed at significant expense and without a clear justification, particularly since a number of similar commercial and open software tools already existed. It is indeed noteworthy and a lesson for the future that the more widely adopted Life Sciences tools have their roots in projects that were already fairly successfully developed by academic research institutions, whereas most of the caBIG™-initiated projects have been less successful and, ironically, much more expensive. Similarly, enormous effort was devoted to the development of caGRID ($9.8M), an environment for grid-based cloud computing, but the working group did not find evidence that it has empowered a new class of tools to 'accelerate the discovery of new approaches for the detection, diagnosis, treatment, and prevention of cancer' as envisioned" [32].

The working group found that the project participants were attracted to new technological innovations, even when those innovations provided no advantages. "In particular, the interviews suggest that the strategic goals of the program were determined by technological advances rather than by key, pre-determined scientific and clinical requirements. Thus, caBIG™ ended up developing powerful and far-reaching technology, such as caGRID, without clear applications to demonstrate what these technologies could and would do for cancer research." Regarding the value of new technology, the working group "struggled to find projects that could not have been implemented with alternative less expensive or existing technologies and software tools" [32].

I was a program director at the National Cancer Institute during the first two years of the project and had a front-row seat for the CaBig pageant. At the time, I thought that the project was too big, too complex, too ambitious, that it served too many interests (the intramural community at NIH, the extramural community of academic researchers, a profusion of industry contractor, and the Office of the Director at the National Cancer Institute), enlisted the assistance of too many people, had too much money, was lavished with too much hype, had initiated too many projects, and that the program directors operated with insufficient accountability. In a word, I was envious.

In the case of the Cancer Biomedical Informatics Grid, hindsight would suggest that the project may have benefited from the following:

1. Start small [35].

Projects should begin with a relatively small number of highly dedicated and energetic people. Studies have shown that as more and more money and personnel are added to a software project, the chances of success drop precipitously [36]. When you stop and think about it, most of the great programming languages, some of the most important operating systems, the design of the Internet, the language of the world wide Web, computer hardware advances, innovational corporations, and virtually all of the great algorithms in computer science, were created by one individual, or a small group of people (usually less than 5).

Some Big Data projects are necessarily large. The Human Genome project, touted as one of the largest and most successful Big Data projects in human history, involved hundreds of scientists. But when it came down to organizing the pieces of data into a coherent sequence, the job fell to one man. In a report published by the Center for Biomolecular Science and Engineering at UC Santa Cruz, Jim Kent "developed in just 4 weeks a 10,000-line computer program that assembled the working draft of the human genome" [37]. "Kent's assembly was celebrated at a White House ceremony on June 26, 2000" [37].

The Cancer Biomedical Informatics Grid began with teams of workers: contractors, government employees, academics, and advisors from various related informatics projects. The project sprang into existence too big to build.

2. Complete your initial goals.

Projects that are closely tied to a community (the cancer research community in the case of CaBIG™, tend to expand their goals based on the interests of enthusiastic community members. It is very common for project managers to be approached by individuals and small groups of people asking that their pet projects be added to the list of project goals. Requests that come from powerful and influential members of the community cannot be ignored. Somehow project managers must placate their constituents without losing focus on their original goals. It is not sufficient to show that the resource is managing to do a lot of different things; managers must be able to complete their critical goals. Failing to do so was a problem for CaBig™. As a counter-example, consider the amazing success of Google. Today, we enjoy the benefits of Google Earth, Google Maps, Google Books, and so one. But the Google people started with a simple, albeit large and all encompassing, search engine. Projects were added after the search engine had been created and popularized.

3. Do not try to do everything yourself.

There is a tendency today for large projects to seek representation in every ongoing effort that relates in any way to the project. For example, a standards effort in the cancer community might send representatives to every committee and standards project in the general area of health technology interoperability. Likewise, narrowly focused standards efforts (e.g., gene array specifications, microscopy specifications) often attract representatives from large corporations and from other standards committees. The push toward internecine activity is based on the belief that the ultimate goal of software interoperability and data integration can only be achieved by broadening the level of participation. In practice, these interchanges make it difficult for project managers to achieve any kind of focus. Each workday is diluted with de-briefings on other projects, conferences, and committee meetings.

Life is about balance. Managers need to have some sense of what is happening in the world outside of their own project, but they must not digress from their own goals. In many cases, careful attention to the Big Data fundamentals (e.g., specifying data fully, identifying and classifying data objects, achieving introspection), should suffice in creating a functional resource that can operate with other resources. Data is fungible; a well-specified data model can often to ported to other formats, other standards and other specifications, as needed.

4. Do not depend on things that you cannot depend on.

This tautology should never be forgotten or trivialized. It is quite common for Big Data resources to choose hardware, software, standards, and specifications that cannot possibly serve their needs. Oblivious to reality, they will choose a currently popular, but flawed methodology, hoping some miracle will save them. So strong is this belief that things will somehow "work out" that virtually every type of project proposal (e.g., grant application, business proposal, book deal) contains a hidden deus ex machina; an implied request that the reviewer suspend his disbelief. Allowing for human nature, some grant reviewers extend a "one-miracle per grant" policy. An applicant can ask for one miracle, but no more.

After CaBig™ was terminated, Barry Smith, a big thinker in the rather small field of ontology, wrote an editorial entitled "CaBIG™ has another fundamental problem: it relies on "incoherent" messaging standard" [38]. In his editorial Smith suggested that HL7, a data model specification used by CaBig™ could not possibly work, and that it had proven to be a failure by those people who actually tried to implement the specification and use it for its intended purposes [38]. At about the same time that CaBig was being terminated, the $17 billion interoperability project undertaken by the U.K.'s National Health Service was scuttled. This failed program had been called "the world's biggest civil information technology program" [8]. Back in 2001, a report published by the NHS Information Authority cited fundamental flaws in HL7 [39]. These flaws included intrinsic difficulties in establishing a workable identifier system. The report concluded that despite these problems, choosing HL7 was strategically correct, as it was the only data model with a process of continual review and update [39]. At the time, everyone was hoping for a miracle. The miracle did not come for CaBig™; nor had it come for the UK's interoperability project. No one can say to what degree, if any, HL7 flaws contributed to the downfall of these projects; but flaws in the data model could not have been very helpful.

5. Use existing, thoroughly tested open source technologies wherever possible.

There is a healthy appetite among information technologists for new and improved technologies. Though this preference may serve well for small and short-term projects, it is often counterproductive for Big Data efforts. All Big Data projects are complex, and there is seldom room to compound complexities by tossing innovative software into the mix. Unless there is a compelling reason to use new technology, it is usually much safer to use older technology that it adequate for the tasks at hand. Stable open source solutions, particularly when there is an active community that supports the software, is often the best choice. Aside from cost considerations (most open source software is free), there is the added benefit of longevity. Unlike commercial software, which can be discontinued or replaced by applications that are incompatible with installed versions, popular open source software tends to stay viable. In the case of CaBig™, the working group indicated that the project chose new technologies, when older technologies would suffice.

6. Avoid developing solutions for problems that nobody really cares about.

I have a colleague who works in the field of computer-aided diagnosis, an interesting field that has not yet garnered widespread acceptance in the medical community. Apparently, clinicians still prefer to reach their diagnoses through their own efforts, without the aid of computer software [40]. Perhaps, he thought, the aid of a computer program might be appreciated in areas where physicians were in short supply; such as developing countries.

He thought that it would be useful to find a market for computer aided diagnostic software somewhere in Africa. Of course, healthcare limitations in developing countries are often basic availability issues (e.g., access to hospitals, beds, antibiotics, sterile materials, equipment, and so forth). Access to diagnostic computer programs, developed on clinical data sets collected on U.S. patients, may not be "what the doctor ordered."

In the case of CaBig, the working group found that CaBig™ was developing powerful new technologies for which there was no apparent need. It is only human nature to want to make a difference in the world. For Big Data project managers, it is sometimes best to wait until the world asks you for your input.

7. A Big Data resource should not focus its attention on itself.

Big data projects often involve a great number of people. Those involved in such efforts may come to believe that the resource itself has an intrinsic value that exceeds the value of its rendered services [41]. Experience would indicate that most users, even avid users, are not interested in the details. The couch potato may sit transfixed in front of his television for hours on end, but he is not likely to pursue an interest in the technologies underlying signal transmissions, or high definition image construction. Likewise, the public has no interest in knowing any of the protocols and models that underlie a Big Data resource. As someone who was asked to review manuscripts published by the scientific contributors to CaBig™, it was my impression that publications emanating from this effort were largely self-congratulatory pieces describing various aspects of the project, in details that would have meaning only to the fellow members of the team [42].

Program managers should never forget that a Big Data resource is all about the data, and how the data is analyzed. The resource itself, rightly or wrongly, should not be the focus of public attention.

Section 15.7. Case Study: The Gaussian Copula Function

America is the only country where a significant proportion of the population believes that professional wrestling is real but the moon landing was faked.

David Letterman

It is important to remember that mathematics only describes observed relationships. It does not explain how those relationships are achieved. A clever physicist such as James Clerk Maxwell can write a set of equations that describe how electric charges and electric currents create electric and magnetic fields, but equations cannot tell us what the meaning is of magnetism. I would venture to say that there isn't a physicist who truly understands the meaning of magnetism, or electricity, or matter, or time.

What applies to the physicist applies to the economist. Recent experience with the Gaussian copula function provides a cautionary tale [43]. This formerly honored and currently vilified formula, developed for Wall Street, was used to calculate the risk of default correlation (i.e., the likelihood that two investment vehicles would default together) based

$$c_R^{\text{Gauss}}(u) = \frac{1}{\sqrt{\det R}} \exp\left(-\frac{1}{2} \begin{pmatrix} \Phi^{-1}(u_1) \\ \vdots \\ \Phi^{-1}(u_d) \end{pmatrix}^T \cdot (R^{-1} - I) \cdot \begin{pmatrix} \Phi^{-1}(u_1) \\ \vdots \\ \Phi^{-1}(u_d) \end{pmatrix} \right),$$

FIG. 15.2 For any readers who may be curious, this formula represents the density of the Gaussian copula function [45]. The Gaussian copula is a distribution over the unit multidimensional cube. R is the parameter matrix. I is the identity matrix. Theta is the inverse cumulative distribution function of a standard normal.

on the current market value of the vehicles, and without factoring in historical data (Fig. 15.2). The Gaussian copula function was easy to implement, and became a favorite model for predicting risk in the securitization market. Though the formula had its early detractors, it soon became the driving model on Wall Street. In about 2008, the function simply stopped working. At about the same time, stock markets collapsed everywhere on planet earth. In some circles, the Gaussian copula function is blamed for the global economic disaster [44]. [Glossary Correlation distance, Nongeneralizable predictor]

Glossary

Autocoding When nomenclature coding is done automatically, by a computer program, the process is known as "autocoding" or "autoencoding".

Autoencoding Synonym for autocoding.

BLOB A large assemblage of binary data (e.g., images, movies, multimedia files, even collections of executable binary code) that are associated that have a common group identifier and that can, in theory, be moved (from computer to computer) or searched as a single data object. Traditional databases do not easily handle BLOBs. BLOBs belong to Big Data.

Class-oriented programming A type of object-oriented programming for which all object instances and all object methods must belong to a class. Hence, in a class-oriented programming language, any new methods and instances that do not sensibly fall within an existing class must be accommodated with a newly created subclass. All invocations of methods, even those sent directly to a class instance, are automatically delivered to the class containing the instance. Class-oriented programming languages embody a specified representation of the real world in which all objects reside within defined classes. Important features such as method inheritance (through class lineage), and introspection (through object and class identifiers) can be very simply implemented in class-oriented programming languages. Powerful scripts can be written with just a few short lines of code, using class-oriented programming languages, by invoking the names of methods inherited by data objects assigned to classes. More importantly, class-oriented languages provide an easy way to discover and test relationships among objects.

Cloud computing According to the U.S. National Institute of Standards (NIST) and Technology cloud computing enables "ubiquitous, convenient, on-demand network access to a shared pool of configurable computing resources (e.g., networks, servers, storage, applications, and services) that can be rapidly provisioned and released with minimal management effort or service provider interaction" [46]. As the NIST definition would suggest, cloud computing is similar to Big Data, but there are several features that are expected in one and not the other. Cloud computing typically offers an interface and a collection of in-cloud computational services. Cloud data is typically contributed by a large community, and the contributed data is deposited often for no reason other than to provide convenient storage. These features are not expected in Big Data resources. Perhaps the most important distinction between a cloud computing and Big Data relates to mutability. Because cloud data is contributed by many different entities, for many purposes, nobody expects much constancy; data can be freely extracted from the cloud, or modified in place. In the cloud, the greatest emphasis is placed on controlling computational access to cloud data; with less emphasis on controlling the content of the cloud. In contrast, Big Data resources are designed to achieve a chosen set of goals using a constructed set of data. In most cases, the data held in a Big Data resource is immutable. Once it is entered into the resource, it cannot be modified or deleted without a very good reason.

Correlation distance Also known as correlation score. The correlation distance provides a measure of similarity between two variables. Two similar variables will rise and fall together [47,48]. The Pearson correlation score is popular, and can be easily implemented [49,50]. It produces a score that varies from −1 to 1. A score of 1 indicates perfect correlation; a score of -1 indicates perfect anti-correlation (i.e., one variable rises while the other falls). A Pearson score of 0 indicates lack of correlation. Other correlation measures can be applied to Big Data sets [47,48].

Fair use Copyright and Patent are legal constructs designed to provide intellectual property holders with the uninfringed power to profit from their creative labors, while still permitting the public to have full access to the holders' property. When Public use of copyrighted material does not limit its profitability to the copyright holder, then the "fair use" of the material is generally permitted, even when those uses exceed customary copyright limits. Most countries have some sort of "fair use" provisions for copyrighted material. In the United States, Fair Use is described in the Copyright Act of 1976, Title 17, U.S. Code, section 107, titled, Limitations on exclusive rights: Fair use. Here is an excerpt of the Act: "Notwithstanding the provisions of sections 106 and 106A, the fair use of a copyrighted work, including such use by reproduction in copies or phonorecords or by any other means specified by that section, for purposes such as criticism, comment, news reporting, teaching (including multiple copies for classroom use), scholarship, or research, is not an infringement of copyright. In determining whether the use made of a work in any particular case is a fair use the factors to be considered shall include (1) the purpose and character of the use, including whether such use is of a commercial nature or is for nonprofit educational purposes; (2) the nature of the copyrighted work; (3) the amount and substantiality of the portion used in relation to the copyrighted work as a whole; and (4) the effect of the use upon the potential market for or value of the copyrighted work. The fact that a work is unpublished shall not itself bar a finding of fair use if such finding is made upon consideration of all the above factors" [51].

Grid A collection of computers and computer resources (typically networked servers) that are coordinated to provide a desired functionality. In the most advanced Grid computing architecture, requests can be broken into computational tasks that are processed in parallel on multiple computers and transparently (from the client's perspective) assembled and returned. The Grid is the intellectual predecessor of Cloud computing. Cloud computing is less physically and administratively restricted than Grid computing.

Linear regression A method for obtaining a straight line through a two-dimensional scatter plot. It is not, as it is commonly a "best fit" technique, but it does minimize the sum of squared errors (in the y-axis values) under the assumption that the x-axis values are correct and exact. This means that you would get a different straight line if you regress x on y; rather than y on x. Linear regression is a popular method that has been extended, modified, and modeled for many different processes, including machine learning. Data analysts who use linear regression should be cautioned that it is a method, much like the venerable P-value, that is commonly misinterpreted [25].

Nongeneralizable predictor Sometimes data analysis can yield results that are true, but nongeneralizable (i.e., irrelevant to everything outside the set of data objects under study). The most useful scientific findings are generalizable (e.g., the laws of physics operate on the planet Jupiter or the star Alpha Centauri much as they do on earth). Many of the most popular analytic methods are not generalizable because they produce predictions that only apply to highly restricted sets of data; or the predictions are not explainable by any underlying theory that relates input data with the calculated predictions. Data analysis is incomplete until a comprehensible, generalizable and testable theory for the predictive method is developed.

P value The *P* value is the probability of getting a set of results that are as extreme or more extreme as the set of results you observed, assuming that the null hypothesis is true (that there is no statistical difference between the results). The *P*-value has come under great criticism over the decades, with a growing consensus that the *P*-value is often misinterpreted, used incorrectly, or used in situations wherein it does not apply [24]. In the realm of Big Data, repeated samplings of data from large data sets will produce small *P*-values that cannot be directly applied to determining statistical significance. It is best to think of the *P*-value as just another piece of information that tells you something about how sets of observations compare with one another; and not as a test of statistical significance.

Predictive modeling contests Everyone knows that science is competitive, but very few areas of science have been constructed as a competitive game. Predictive analytics is an exception. Kaggle is a Web site that runs predictive-modeling contests. Their motto is "We're making data science a sport." Competitors with the most successful predictive models win prizes. Prizes vary from thousands to millions of dollars, and hundreds of teams may enter the frays [52].

References

[1] Lohr S. Google to end health records service after it fails to attract users. The New York Times; June 24, 2011.

[2] Schwartz E. Shopping for health software, some doctors get buyer's remorse. The Huffington Post Investigative Fund; January 29, 2010.

[3] Heeks R, Mundy D, Salazar A. Why health care information systems succeed or fail. Institute for Development Policy and Management, University of Manchester; June 1999. Available from: http://www.sed.manchester.ac.uk/idpm/research/publications/wp/igovernment/igov_wp09.htm [viewed July 12, 2012].

[4] Littlejohns P, Wyatt JC, Garvican L. Evaluating computerised health information systems: hard lessons still to be learnt. Br Med J 2003;326:860–3.

[5] Linder JA, Ma J, Bates DW, Middleton B, Stafford RS. Electronic health record use and the quality of ambulatory care in the United States. Arch Intern Med 2007;167:1400–5.

[6] Patient Safety in American Hospitals. HealthGrades; July 2004. Available from: http://www.healthgrades.com/media/english/pdf/hg_patient_safety_study_final.pdf [viewed September 9, 2012].

[7] Gill JM, Mainous AG, Koopman RJ, Player MS, Everett CJ, Chen YX, et al. Impact of EHR-based clinical decision support on adherence to guidelines for patients on NSAIDs: a randomized controlled trial. Ann Fam Med 2011;9:22–30.

[8] Lohr S. Lessons from Britain's health information technology Fiasco. The New York Times; September 27, 2011.

[9] Dismantling the NHS national programme for IT. Department of Health Media Centre Press Release; September 22, 2011. Available from: http://mediacentre.dh.gov.uk/2011/09/22/dismantling-the-nhs-national-programme-for-it/ [viewed June 12, 2012].

[10] Whittaker Z. UK's delayed national health IT programme officially scrapped. ZDNet; September 22, 2011.

[11] Kappelman LA, McKeeman R, Lixuan ZL. Early warning signs of IT project failure: the dominant dozen. Inf Syst Manag 2006;23:31–6.

[12] Arquilla J. The Pentagon's biggest boondoggles. The New York Times (Opinion Pages); March 12, 2011.

[13] FIPS PUB 119-1. Supersedes FIPS PUB 119. November 8, 1985. Federal Information Processing Standards Publication 119-1 1995 March 13. Announcing the Standard for ADA; 1985. Available from: http://www.itl.nist.gov/fipspubs/fip119-1.htm [viewed August 26, 2012].

[14] Ariane 501 Inquiry Board Report. July 19, 1996. Available from: http://esamultimedia.esa.int/docs/esa-x-1819eng.pdf [viewed August 26, 2012].

[15] Mars Climate Orbiter, Mishap Investigation Board. Phase I Report; November 10, 1999. ftp://ftp.hq.nasa.gov/pub/pao/reports/1999/MCO_report.pdf.

[16] Sowers AE. Funding research with NIH grants: a losing battle in a flawed system. In: The Scientist. vol. 9. October 16, 1995.

[17] Pogson G. Controlled English: enlightenment through constraint. Lang Technol 1988;6:22–5.

[18] Schneier B. A plea for simplicity: you can't secure what you don't understand. Information Security; November 19, 1999. Available from: http://www.schneier.com/essay-018.html [viewed July 1, 2015].

[19] Vlasic B. Toyota's slow awakening to a deadly problem. The New York Times; February 1, 2010.

[20] Valdes-Dapena P. Pedals, drivers blamed for out of control Toyotas. CNN Money; February 8, 2011.

[21] Drew C. U-2 spy plane evades the day of retirement. The New York Times; March 21, 2010.

[22] Riley DL. Business models for cost effective use of health information technologies: lessons learned in the CHCS II project. Stud Health Technol Inform 2003;92:157–65.

[23] Ecker JR, Bickmore WA, Barroso I, Pritchard JK, Gilad Y, Segal E. Genomics: ENCODE explained. Nature 2012;489:52–5.

[24] Cohen J. The earth is round (p < .05). Am Psychol 1994;49:997–1003.

[25] Janert PK. Data Analysis with Open Source Tools. O'Reilly Media; 2010.

[26] Rosenberg T. Opinionator: armed with data, fighting more than crime. The New York Times; May 2, 2012.

[27] Hoover JN. Data, analysis drive Maryland government. Information Week; March 15, 2010.

[28] Howe J. The rise of crowdsourcing. Wired 2006;14:06.

[29] Cleveland Clinic: build an efficient pipeline to find the most powerful predictors, Innocentive; September 8, 1911. https://www.innocentive.com/ar/challenge/9932794 [viewed September 25, 2012].

[30] Robins JM. The control of confounding by intermediate variables. Stat Med 1989;8:679–701.

[31] Robins JM. Correcting for non-compliance in randomized trials using structural nested mean models. Commun Stat Theory Methods 1994;23:2379–412.

[32] An assessment of the impact of the NCI cancer Biomedical Informatics Grid (caBIG). Report of the Board of Scientific Advisors Ad Hoc Working Group, National Cancer Institute, March 2011.

[33] The ComputerWorld honors program case study. Available from: http://www.cwhonors.org/case_studies/NationalCancerInstitute.pdf [viewed August 31, 2012].

[34] Komatsoulis GA. Program announcement to the CaBIG community. National Cancer Institute. https://cabig.nci.nih.gov/program_announcement [viewed August 31, 2012].

[35] Olavsrud T. How to avoid Big Data spending pitfalls. CIO; May 08, 2012. Available from: http://www.cio.com/article/705922/How_to_Avoid_Big_Data_Spending_Pitfalls [viewed July 16, 2012].

[36] The Standish Group Report: Chaos. Available from: http://www.projectsmart.co.uk/docs/chaos-report.pdf; 1995 [viewed September 19, 2012].

[37] The human genome project race. UC Santa Cruz Center for Biomolecular Science and Engineering; Available from: http://www.cbse.ucsc.edu/research/hgp_race March 28, 2009.

[38] Smith B. caBIG has another fundamental problem: it relies on "incoherent" messaging standard. Cancer Lett 2011;37(16) April 22.

[39] Robinson D, Paul Frosdick P, Briscoe E. HL7 Version 3: an impact assessment. NHS Information Authority; March 23, 2001.

[40] Eccles M, McColl E, Steen N, Rousseau N, Grimshaw J, Parkin D, et al. Effect of computerised evidence based guidelines on management of asthma and angina in adults in primary care: cluster randomised controlled trial. BMJ 2002;325: October 26.

[41] Scheff TJ. Peer review: an iron law of disciplines. Self-Published Paper; May 27, 2002. Available from: http://www.soc.ucsb.edu/faculty/scheff/23.html [viewed September 1, 2012].

[42] Boyd LB, Hunicke-Smith SP, Stafford GA, Freund ET, Ehlman M, Chandran U, et al. The caBIG™ life science business architecture model. Bioinformatics 2011;27:1429–35.

[43] Berman JJ. Principles of big data: preparing, sharing, and analyzing complex information. Waltham, MA: Morgan Kaufmann; 2013.

[44] Salmon F. Recipe for disaster: the formula that killed wall street. Wired Magazine 17:03; February 23, 2009.

[45] Arbenz P. Bayesian copulae distributions, with application to operational risk management—some comments. Methodol Comput Appl Probab 2013;15:105–8.

[46] Mell P, Grance T. The NIST definition of cloud computing. Recommendations of the National Institute of Standards and Technology. NIST Publication 800-145NIST; September 2011.

[47] Reshef DN, Reshef YA, Finucane HK, Grossman SR, McVean G, Turnbaugh PJ, et al. Detecting novel associations in large data sets. Science 2011;334:1518–24.

[48] Szekely GJ, Rizzo ML. Brownian distance covariance. Ann Appl Stat 2009;3:1236–65.

[49] Berman JJ. Methods in medical informatics: fundamentals of healthcare programming in Perl, Python, and Ruby. Boca Raton: Chapman and Hall; 2010.

[50] Lewis PD. R for medicine and biology. Sudbury: Jones and Bartlett Publishers; 2009.

[51] Copyright Act, Section 107, Limitations on exclusive rights: fair use. Available from: http://www.copyright.gov/title17/92chap1.html [viewed May 18, 2017].

[52] Stross R. The algorithm didn't like my essay. The New York Times; June 9, 2012.

16

Data Reanalysis: Much More Important Than Analysis

Section 16.1. First Analysis (Nearly) Always Wrong

A new thing is just an old thing that hasn't had an opportunity to disappoint you.

Anon

As we were celebrating the many scientific breakthroughs attributed to Big Data analyses, a disquieting observation intruded upon the festivities. It seems that many of the fundamental studies in the field had yielded irreproducible results. Expensive studies, on massive volumes of data, were providing conclusions that were no more reliable than coin tosses. When we tried to understand why this was happening, we discovered that just about everything that could go wrong was indeed going wrong. Many of the faulty studies came from the field of medicine, but medical failures are easy to spot when patients do not respond to treatment as hoped. We really have no idea as to how many Big Data conclusions from the fields of business, politics, economics, and marketing have been leading us astray.

Consider these shocking headlines.

– "Unreliable research: Trouble at the lab" [1]. The Economist, in 2013 ran an article examining flawed biomedical research. The magazine article referred to an NIH official who indicated that "researchers would find it hard to reproduce at least three-quarters of all published biomedical findings, the public part of the process seems to have failed." The article described a study conducted at the pharmaceutical company Amgen, wherein 53 landmark studies were repeated. The Amgen scientists were successful at reproducing the results of only 6 of the 53 studies. Another group, at Bayer HealthCare, repeated 63 studies. The Bayer group succeeded in reproducing the results of only one-fourth of the original studies.

Principles and Practice of Big Data. https://doi.org/10.1016/B978-0-12-815609-4.00016-9

- "A decade of reversal: an analysis of 146 contradicted medical practices" [2]. The authors reviewed 363 journal articles, re-examining established standards of medical care. Among these articles were 146 manuscripts (40.2%) claiming that an existing standard of care had no clinical value.
- "Cancer fight: unclear tests for new drug" [3]. This New York Times article examined whether a common test performed on breast cancer tissue (Her2) was repeatable. It was shown that for patients who tested positive for Her2, a repeat test indicated that 20% of the original positive assays were actually negative (i.e., falsely positive on the initial test) [3].
- "Reproducibility crisis: Blame it on the antibodies" [4]. Biomarker developers are finding that they cannot rely on different batches of a reagent to react in a consistent manner, from test to test. Hence, laboratory analytic methods, developed using a controlled set of reagents, may not have any diagnostic value when applied by other laboratories, using different sets of the same analytes [4].
- "Why most published research findings are false" [5]. Modern scientists often search for small effect sizes, using a wide range of available analytic techniques, and a flexible interpretation of outcome results. The manuscript's author found that research conclusions are more likely to be false than true [5,6].
- "We found only one-third of published psychology research is reliable—now what?" [7]. The manuscript authors suggest that the results of a first study should be considered preliminary.
- "A reasonable doubt, the false promise of DNA testing" [8]. An account published in The Atlantic highlights the errors in the analysis of DNA mixtures.

It is now abundantly clear that many scientific findings, particularly those findings based on analyses of large and complex data sets, are yielding irreproducible results [9]. The economic cost attributed to irreproducibility in preclinical studies (i.e., preliminary medical research that occurs before testing on human subjects can begin), is estimated at $28 billion dollars per year, in the United States alone [10].

Anyone who attempts to stay current in the sciences soon learns that much of the published literature is irreproducible [9]; and that almost anything published today might be retracted tomorrow. This appalling truth applies to some of the most respected and trusted laboratories in the world [11–18]. Those of us who have been involved in assessing the rate of progress in disease research are painfully aware of the numerous reports indicating a general slowdown in medical progress [19–26]. For the optimists, it is tempting to assume that the problems that we may be experiencing today are par for the course, and temporary. It is the nature of science to stall for a while and lurch forwards in sudden fits. Errors and retractions will always be with us so long as humans are involved in the scientific process. For the pessimists among us, there seems to be something going on that is really new and different; a game changer. This game changer is the "complexity barrier", a term credited to Boris Beizer, who used it to describe the impossibility of managing increasingly complex software applications [27]. The complexity barrier applies to every modern Big Data project [28–30].

Some of the mistakes that lead to erroneous conclusions in complex, data-intensive research are well known, and include the following:

– Errors in sample selection, labeling, and measurement

Data errors and data documentation errors are common, and there are many examples, from the scientific literature, where such errors are documented [31–35]. Setting aside sampling errors, data analysts must contend with the problem of insufficient sampling [36]. Experimental acumen cannot compensate for a statistically inadequate number of sample specimens.

– Outright fraud

We do not know the frequency of scientific fraud, but there is no reason to believe that fraud is rare [18,37–39]. When a scientist's career is on the line, the temptations to fabricate or fudge data can be overwhelming. The literature is rich with examples of fraud committed in some of the most respected laboratories on earth [12,39–49]. Considering that Big Data is collected from a multitude of researchers, it should come as no great surprise to learn that some of the data included in Big Data resources is fabricated or fraudulent.

– Misinterpretation of the data

The most common source of scientific error is post-analytic, arising from misinterpretation of results [5,24,50–56]. Virtually every journal article contains, hidden in the introduction and discussion sections, some distortion of fact or misleading assertion. Scientists cannot be objective about their own work. As humans, we tend to interpret observations to reinforce our beliefs and prejudices and to advance our agendas [39].

Large, multi-institutional studies involving many human subjects, many specimens, and large collections of data, analyzed by teams of statisticians and computer scientists, carry the veneer of respectability. Manuscript reviewers may be reluctant to reject a work submitted by a group of 100 scientists. Nonetheless, large studies are just as susceptible to errors of data misinterpretation as are studies performed by a single investigator, on a small set of data. Today, some of the largest experimental studies and clinical trials produce results with very small differences between the experimental group and the control group (e.g., an extension of cancer survival time of two weeks, a 10% difference in biomarker levels). When the stakes are high, as is always the case with expensive multi-institutional studies, scientists will be inclined to exaggerate the benefits of a positive finding, no matter how insignificant the results may be. [Glossary Sponsor bias]

One of the most common strategies whereby scientists distort their own results, to advance a self-serving conclusion, is message framing [57]. In message framing, scientists omit from discussion any pertinent findings that might diminish or discredit their own conclusions. The common practice of message framing is often conducted on a subconscious, or at least a sub-rational, level. A scientist is not apt to read articles whose conclusions contradict his own hypotheses and will not cite disputatious works in his manuscripts. Furthermore, if a paradigm is held in high esteem by a majority of the scientists in a field, then works that contradict the paradigm are not likely to pass peer review. Hence, it is difficult for contrary articles to be published in scientific journals. In any case the message delivered in a journal article is almost always framed in a manner that promotes the author's interpretation.

It must be noted that throughout human history, no scientist has ever gotten into any serious trouble for misinterpreting results. Scientific misconduct comes, as a rule, from the purposeful production of bad data, either through falsification, fabrication, or through the refusal to remove and retract data that is known to be false, plagiarized, or otherwise invalid. In the United States, allegations of research misconduct are investigated by the Office of Research Integrity. Funding agencies in other countries have similar watchdog institutions. The Office of Research Integrity makes its findings a matter of public record [58]. Of 150 cases investigated between 1993 and 1997, all but one case had an alleged component of data falsification, fabrication or plagiarism [59]. In 2007, of the 28 investigated cases, 100% involved allegations of falsification, fabrication, or both [60]. No cases of misconduct based on data misinterpretation were prosecuted. Perhaps the Office of Research Integrity understands that the self-serving interpretation of data in entrenched deep within the human psyche.

Big Data analysis is difficult, and there is no way to work with complex information without making mistakes. This being the case, it seems prudent to assume that every scientific publication, based on the analysis of a single set of data, Big or small, should be considered tentative, until other laboratories confirm the findings.

Section 16.2. Why Reanalysis Is More Important Than Analysis

Prediction is very difficult, especially if it's about the future.

Niels Bohr

One of the nicest things about Big Data, aside from its ample size, is its permanence. We know that if we commit some error on the first analysis of a Big Data collection, we can always get a second bite at the apple; another chance to go back and correct our error, and to possibly produce a conclusion that is more important than anything we originally anticipated. Considering that many of the published studies coming out today are based on the analysis of massive sets of data, the importance of reanalysis cannot be overestimated.

Let us look at some of the specific roles filled by data reanalysis.

- **Verification of data and validation of conclusions**

Verification involves checking to determine whether the data was obtained properly (i.e., according to approved protocols), and that the data accurately measured what it was intended to measure, on the correct specimens, and that all steps in data processing were done using well-documented and approved protocols. Verification often requires a level of expertise that is at least as high as the expertise of the individuals who produced the data [61]. Data verification requires a full understanding of the many steps involved in data collection and can be a time-consuming task. In one celebrated case, in which two statisticians reviewed a microarray study performed at Duke University, the time devoted to the verification effort was reported to be 2000 hours [62]. To put this in perspective, the official work-year, according to the U.S. Office of Personnel Management, is 2087 hours. Because data verification requires deep knowledge of the data being studied, it stands to reason that data verification is greatly facilitated by preparing well-annotated data that supports introspection.

Validation is a different process than verification. Whereas verification checks the data upon which conclusions are drawn, validation checks to see if the conclusions drawn from the data are correct. Validation usually begins by repeating the same analysis of the same data, using the methods that were originally recommended. Obviously, if a different set of conclusions is reached, using the same data and methods, then the original conclusions failed validation. Validation may also involve applying a different set of analytic methods to the same data, to determine if the conclusions are consistent. Data can be legitimately analyzed by multiple different methods. The ability to draw the same conclusion from a data set, consistently, from multiple methods of analysis, is a type of validation.

Another type of validation involves testing new hypotheses, based on the assumed validity of the original conclusions. For example, if you were to accept Darwin's theory of evolution, then you would expect to find a chronologic archive of fossils in ascending strata of shale. This being the case, paleontologists provided independent validation to Darwin's conclusions.

Scientists rankle at the idea that their data must be inspected, reanalyzed, and sometimes repeated, by other scientists, including competitors. Scientists should understand that data validation requires a great deal of effort, and that the scientists who devote themselves to this task are often interrupting their own careers because they believe that the results under review are of sufficient importance to justify their sacrifice. **Also, the primary purpose of every validation effort is to legitimize the original work, not to discredit the work**. It is better to have a genuine scientific advancement than to have a huge waste of everyone's time and money. [Glossary Primary data, Secondary data]

There is an old saying that "God did not make us perfect, so he compensated by making us blind to our own faults." Verification and validation help us to compensate for our blindness, but we cannot attain definitive conclusions in every case. Realistically, the most we can expect is to verify that the data was obtained properly and to validate that the conclusions fit the results [52,62–66]. To help us, there is a rich literature containing guidelines for achieving validation, including a suite of helpful algorithms, software, devices, statistical methods, and mathematical models [61,67–70].

– **Clarifications and improvements upon earlier studies**

Original data sets can be reanalyzed using alternate or improved methods to attain outcomes of greater precision or reliability than the outcomes produced in the original analysis. Nowhere has this been more successful than in the field of forensics, where newer studies based on experience with large sets of data have led to fundamental changes in the way that forensic evidence is interpreted [8,71–74].

– **Performing additional analyses and updating results from earlier studies**

It is impossible to fully analyze a complex study on the first attempt. There will always be some analytic opportunity that was overlooked [75]. Furthermore, as new data arrives, the original data needs to be reanalyzed, with the newer data. In some cases, the newer data permits the data curator to fill in missing data points, to enter corrections in the original data, and to achieve a more accurate assessment of outlier data points in the original data set. It is a terrible waste to simply abandon an old project, when a reanalysis at some future time, would help tie loose ends and clarify unanswered questions remaining from the first analysis of the original data [76].

– **Extending the scope of the original study**

Sometimes, data collected for one project can be usefully merged with data collected in other projects. Such projects may pertain to previous, concurrent, or future works, just so long as they contain related data. In Chapter 17, we will be discussing data repurposing, which involves using Big Data to answer questions and to achieve results that were not anticipated by the designers of the Big Data resource.

Section 16.3. Case Study: Reanalysis of Old JADE Collider Data

Life can only be understood backwards; but it must be lived forwards.

Soren Kierkegaard

In the 1980s, the PETRA collider conducted a number of so-called atom smashing experiments designed to measure the force required to bind together quarks and gluons, the fundamental components of protons and neutrons. In 1986, the PETRA collider was decommissioned and replaced with colliders that operated at higher energy levels. Several decades passed, and advances in physics raised a new set of questions that could only be answered with observations on low-energy collisions; the kind of observations collected by PETRA and omitted by present-day colliders [77].

An effort to retrieve the 1980s data was spearheaded by Siegfried Bethke, one of the original scientists in PETRA's JADE project [78]. In the period following the decommissioning of PETRA, the original data had been dispersed to various laboratories. Some of the JADE data was simply lost, and none of the data was collected in a format or a medium that was directly accessible.

The project was divided into three missions, involving three teams of scientists. One team rescued the data for archived tapes and transferred the data into a modern medium and format. The second team improved the original JADE software, fitting it to modern computer platforms. By applying software that used updated Monte Carlo simulations, the second team generated a new set of data files. The third team reanalyzed the regenerated data using modern methods and improved calculations.

The project culminated in the production of numerous scientific contributions that could not have been achieved without the old JADE data. Success was credited, at least in part, to the participation of some of the same individuals who collected the original data.

Section 16.4. Case Study: Vindication Through Reanalysis

That which can be asserted without evidence, can be dismissed without evidence.
Christopher Hitchens

In 1978, Joseph Strauch published a phylogenetic taxonomy of Charadriiformes birds (i.e., a subclassification based on evolutionary descent), by studying their bones (i.e., via osteology) [79]. When he was finished his project, he left his osteologic data for others to reanalyze. As it happened, his conclusions stirred a controversy that persisted over several decades. Nearly 20 years later, Phillip Chu took a hard look at Strauch's measurements [80]. Chu re-coded Strauch's data to eliminate objectionable feature assignments. Chu conducted a parsimony analysis of the data rather than using Strauch's compatibility analysis; both being methods that establish phylogenetic order. In the end, Chu's study confirmed Strauch's findings.

It is not particularly easy to publish journal manuscripts that vindicate earlier works. Journal editors, traditionally, are interested in publishing new science; not in revisiting previously published studies. It is plausible that Chu's paper, reanalyzing Strauch's work, was worthy of publication only because Strauch's early work had been publicly challenged by his colleagues [81]. Journal editors should be receptive to reanalysis manuscripts as they often provide new insights that advance their fields [82]. In many cases, reanalysis is the most effect way by which scientific truth can be established.

Reanalysis can only be performed on studies for which data is available. Scientists can avoid having their studies reanalyzed by simply withholding their data from their colleagues.

Section 16.5. Case Study: Finding New Planets From Old Data

Many an object is not seen, though it falls within the range of our visual ray, because it does not come within the range of our intellectual ray, i.e. we are not looking for it. So, in the largest sense, we find only the world we look for.
Henry David Thoreau in Journal, 2 July 1857

Astronomers gather enormous amounts of information on stars. Such data includes direct photographic images of stars, using improved telescopes (e.g., Hubble Space Telescope), high-resolution spectroscopy data, X-ray data (e.g., from NASA's Chandra X-ray observatory). As it happens, if a star is orbited by planets, those planets will have some effect, over the course of time, on the measurements collected on the star [83].

Over the past decade, using preexisting star data, astronomers have found evidence for thousands of extrasolar planets (exoplanets). Some of the planet-hunting techniques include [83]:

- Transit method. Exoplanets dim the light received from a star during their transit.
- Radial velocity or Doppler method. Exoplanets can cause the star's speed to vary with respect to the speed at which the star moves toward or away from the earth, and this variation in speed causes a Doppler shift in the star's emitted spectral lines.
- Transit timing variation. If a star is orbited by multiple planets then the time when an exopolanet begins its transit across the star and the duration of its transit will vary depending on the other planets in the vicinity at the time of transit.
- Gravitational microlensing. Exoplanets orbiting a lensing star can produce perturbations in the measured magnification of the lensing phenomenon.
- Astrometry. Orbiting exoplanets can change the star's position in the sky.
- Pulsar timing. Orbiting exoplanets may cause small perturbations in the timing of radio wave pulsations. This method, which applies only to planets orbiting pulsars, was employed to find the first confirmed exoplanet in 1992.
- Direct imaging. When the exoplanets are large and the star is relatively close to the earth the exoplanets can be imaged directly by blocking the light produced by their star (Fig. 16.1).

Today, new methods of finding exoplanets are being developed. Existing data is being reanalyzed to accommodate new techniques as they arrive. Data that has already been

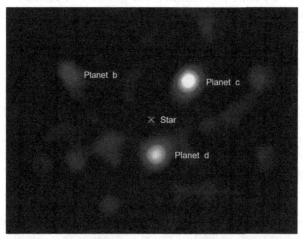

FIG. 16.1 Actual image of three exoplanets orbiting HR8799, 120 light years from earth, was obtained. The orbiting exoplanets were made visible in the image, by blocking out the image of their star. *From NASA, obtained with the Palomar Observatory's Hale Telescope, public domain.*

used to find exoplanets is being reanalyzed to validate the original conclusions, and to help find additional planets missed in the first analysis, and to uncover new information about exoplanets that have been discovered [84].

Glossary

Primary data The original set of data collected to serve a particular purpose or to answer a particular set of questions, and intended for use by the same individuals who collected the data.

Secondary data Data collected by someone else. Much of the data analyses performed today are done on secondary data [85]. Most verification and validation studies depend upon access to high-quality secondary data. Because secondary data is prepared by someone else, who cannot anticipate how you will use the data, it is important to provide secondary data that is simple and introspective.

Sponsor bias Are the results of big data analytics skewed in favor of the corporate sponsors of the resource? In a fascinating meta-analysis, Yank and coworkers asked whether the results of clinical trials, conducted with financial ties to a drug company, were biased to produce results favorable to the sponsors [86]. They reviewed the literature on clinical trials for anti-hypertensive agents, and found that ties to a drug company did not bias the results (i.e., the experimental data), but they did bias the conclusions (i.e., the interpretations drawn from the results). This suggests that regardless of the results of a trial, the conclusions published by the investigators were more likely to be favorable, if the trial were financed by a drug company. This should come as no surprise. Two scientists can look at the same results and draw entirely different conclusions.

References

[1] Unreliable research: trouble at the lab. The Economist; October 19, 2013.

[2] Prasad V, Vandross A, Toomey C, Cheung M, Rho J, Quinn S, et al. A decade of reversal: an analysis of 146 contradicted medical practices. Mayo Clin Proc 2013;88(8):790.

[3] Kolata G. Cancer fight: unclear tests for new drug. The New York Times; April 19, 2010.

[4] Baker M. Reproducibility crisis: blame it on the antibodies. Nature 2015;521:274–6.

[5] Ioannidis JP. Why most published research findings are false. PLoS Med 2005;2:e124.

[6] Labos C. It ain't necessarily so: why much of the medical literature is wrong. Medscape News and Perspectives; 2014 September 09.

[7] Gilbert E, Strohminger N. We found only one-third of published psychology research is reliable—now what? The Conversation; August 27, 2015. Available at: http://theconversation.com/we-found-only-one-third-of-published-psychology-research-is-reliable-now-what-46596 [viewed on August 27, 2015].

[8] Shaer M. A reasonable doubt, the false promise of DNA testing. The Atlantic; June 2016.

[9] Naik G. Scientists' elusive goal: reproducing study results. Wall Street J December 2, 2011.

[10] Freedman LP, Cockburn IM, Simcoe TS. The economics of reproducibility in preclinical research. PLoS Biol 2015;13:e1002165.

[11] Zimmer C. A sharp rise in retractions prompts calls for reform. The New York Times; April 16, 2012.

[12] Altman LK. Falsified data found in gene studies. The New York Times; October 30, 1996.

[13] Weaver D, Albanese C, Costantini F, Baltimore D. Retraction: altered repertoire of endogenous immunoglobulin gene expression in transgenic mice containing a rearranged mu heavy chain gene. Cell 1991;536(inclusive):65.

[14] Chang K. Nobel winner in physiology retracts two papers. The New York Times; September 23, 2010.

[15] Fourth paper retracted at Potti's request. The Chronicle; March 3, 2011.

[16] Whoriskey P. Doubts about Johns Hopkins research have gone unanswered, scientist says. The Washington Post; March 11, 2013.

[17] Lin YY, Kiihl S, Suhail Y, Liu SY, Chou YH, Kuang Z, et al. Retraction: functional dissection of lysine deacetylases reveals that HDAC1 and p300 regulate AMPK. Nature 2013;482:251–5. retracted November.

[18] Shafer SL. Letter: to our readers. Anesthesia and Analgesia; February 20, 2009.

[19] Innovation or stagnation: challenge and opportunity on the critical path to new medical products. U. S. Department of Health and Human Services, Food and Drug Administration; 2004.

[20] Hurley D. Why are so few blockbuster drugs invented today? The New York Times; November 13, 2014.

[21] Angell M. The truth about the drug companies. In: The New York review of books. vol 51. July 15, 2004.

[22] Quality of Health Care in America Committee, editor. Crossing the quality chasm: a new health system for the 21st century. Washington, DC: Institute of Medicine; 2001.

[23] Wurtman RJ, Bettiker RL. The slowing of treatment discovery, 1965–1995. Nat Med 1996;2:5–6.

[24] Ioannidis JP. Microarrays and molecular research: noise discovery? Lancet 2005;365:454–5.

[25] Weigelt B, Reis-Filho JS. Molecular profiling currently offers no more than tumour morphology and basic immunohistochemistry. Breast Cancer Res 2010;12:S5.

[26] Personalised medicines: hopes and realities. London: The Royal Society; 2005. Available from: https://royalsociety.org/~/media/Royal_Society_Content/policy/publications/2005/9631.pdf [viewed July 29, 2017].

[27] Beizer B. Software testing techniques. 2nd ed. Hoboken, NJ: Van Nostrand Reinhold; 1990.

[28] Vlasic B. Toyota's slow awakening to a deadly problem. The New York Times; February 1, 2010.

[29] Lanier J. The complexity ceiling. In: Brockman J, editor. The next fifty years: science in the first half of the twenty-first century. New York: Vintage; 2002. p. 216–29.

[30] Berman JJ. Data simplification: taming information with open source tools. Waltham, MA: Morgan Kaufmann; 2016.

[31] Sainani K. Error: What biomedical computing can learn from its mistakes. Biomed Comput Rev 2011;12–9 Fall.

[32] Bandelt H, Salas A. Contamination and sample mix-up can best explain some patterns of mtDNA instabilities in buccal cells and oral squamous cell carcinoma. BMC Cancer 2009;9:113.

[33] Knight J. Agony for researchers as mix-up forces retraction of ecstasy study. Nature September 11, 2003;425:109.

[34] Kuderer NM, Burton KA, Blau S, Rose AL, Parker S, Lyman GH, et al. Comparison of 2 commercially available next-generation sequencing platforms in oncology. JAMA Oncol December 15, 2016.

[35] Satter RG. UK investigates 800,000 organ donor list errors. Associated Press; April 10, 2010.

[36] Gerlinger M, Rowan AJ, Horswell S, Larkin J, Endesfelder D, Gronroos E, et al. Intratumor heterogeneity and branched evolution revealed by multiregion sequencing. N Engl J Med 2012;366: 883–92.

[37] Marshall E. How prevalent is fraud? That's a million dollar question. Science 2000;290:1662–3.

[38] Martin B. Scientific fraud and the power structure of science. Prometheus 1992;10:83–98.

[39] Berman JJ. Machiavelli's Laboratory. Amazon Digital Services, Inc.; 2010.

[40] Findings of scientific misconduct. NIH Guide 1997;26(23). July 18. Available from: http://grants.nih.gov/grants/guide/notice-files/not97-151.html.

[41] Findings of scientific misconduct. Department of Health and Human Services. Notice NOT-OD-01-048. July 10, 2001 Available from: http://grants.nih.gov/grants/guide/notice-files/NOT-OD-01-048.html [viewed October 8, 2009].

[42] Findings of scientific misconduct. NIH Guide 1997;26(15). May 9. Available from:http://grants.nih.gov/grants/guide/notice-files/not97-097.html.

[43] Findings of scientific misconduct. NOT-OD-02-020. December 13, 2001 Available from:http://grants.nih.gov/grants/guide/notice-files/NOT-OD-02-020.html.

[44] Findings of scientific misconduct. NIH Guide 1995;24(33). September 22. Available from:http://grants.nih.gov/grants/guide/notice-files/not95-208.html.

[45] Cyranoski D. Woo Suk Hwang convicted, but not of fraud. Cloning pioneer gets two years for embezzlement and bioethics breach. Nature 2009;461:1181. October 26.

[46] Fuyuno I, Cyranoski D. Doubts over biochemist's data expose holes in Japanese fraud laws. Nature February 1, 2006;439:514.

[47] Sontag D. In harm's way: research, fraud and the V.A.; abuses endangered veterans in cancer drug experiments. The New York Times; February 6, 2005.

[48] Oltermann P. 'Superstar doctor' fired from Swedish institute over research 'lies', The Guardian; March 24, 2016. Available at: https://www.theguardian.com/science/2016/mar/23/superstar-doctor-fired-from-swedish-institute-over-research-lies-allegations-windpipe-surgery [viewed March 24, 2016].

[49] Harris G. Diabetes drug maker hid test data, files indicate. The New York Times; July 12, 2010.

[50] Ioannidis JP. Is molecular profiling ready for use in clinical decision making? Oncologist 2007;12:301–11.

[51] Ioannidis JP. Some main problems eroding the credibility and relevance of randomized trials. Bull NYU Hosp Jt Dis 2008;66:135–9.

[52] Ioannidis JP, Panagiotou OA. Comparison of effect sizes associated with biomarkers reported in highly cited individual articles and in subsequent meta-analyses. JAMA 2011;305:2200–10.

[53] Berman JJ. Principles of big data: preparing, sharing, and analyzing complex information. Waltham, MA: Morgan Kaufmann; 2013.

[54] Ioannidis JP. Excess significance bias in the literature on brain volume abnormalities. Arch Gen Psychiatry 2011;68:773–80.

[55] Pocock SJ, Collier TJ, Dandreo KJ, deStavola BL, Goldman MB, Kalish LA, et al. Issues in the reporting of epidemiological studies: a survey of recent practice. BMJ 2004;329:883.

[56] McGauran N, Wieseler B, Kreis J, Schuler Y, Kolsch H, Kaiser T. Trials 2010;11:37.

[57] Wilson JR. Rhetorical strategies used in the reporting of implantable defibrillator primary prevention trials. Am J Cardiol 2011;107:1806–11.

[58] Office of Research Integrity. Available from: http://ori.dhhs.gov.

[59] Scientific misconduct investigations 1993–1997. Office of Research Integrity, Office of Public Health and Science, Department of Health and Human Services; December, 1998.

[60] Office of Research Integrity Annual Report 2007. Available from: http://ori.hhs.gov/images/ddblock/ori_annual_report_2007.pdf; June 2008 [viewed January 1, 2015].

[61] Committee on Mathematical Foundations of Verification, Validation, and Uncertainty Quantification; Board on Mathematical Sciences and Their Applications, Division on Engineering and Physical Sciences, National Research Council. Assessing the reliability of complex models: mathematical and statistical foundations of verification, validation, and uncertainty quantification. National Academy Press; 2012. Available from: http://www.nap.edu/catalog.php?record_id=13395 [viewed January 1, 2015].

[62] Misconduct in science: an array of errors. The Economist; September 10, 2011.

[63] Begley S. In cancer science, many 'discoveries' don't hold up. Reuters; March 28, 2012.

[64] Abu-Asab MS, Chaouchi M, Alesci S, Galli S, Laassri M, Cheema AK, et al. Biomarkers in the age of omics: time for a systems biology approach. OMICS 2011;15:105–12.

[65] Moyer VA, on behalf of the U.S. Preventive Services Task Force. Screening for prostate cancer: U.S. Preventive Services Task Force recommendation statement. Ann Intern Med May 21, 2011.

[66] How science goes wrong. The Economist; October 19, 2013.

[67] Oberkampf WL. Verification and validation in computational simulation. 2004 Transport Task Force Meeting. United States Department of Energy's National Nuclear Security Administration. Salt Lake City, Utah, April 29, 2004.

[68] General principles of software validation; final guidance for industry and FDA staff. U.S. Food and Drug Administration; January 11, 2002.

[69] Nuzzo R. P values, the gold standard of statistical validity, are not as reliable as many scientists assume. Nature 2014;506:150–2.

[70] Ransohoff DF. Rules of evidence for cancer molecular-marker discovery and validation. Nat Rev Cancer 2004;4:309–14.

[71] Tanner R. New science challenges arson conviction. Associated Press; December 31, 2006.

[72] Lentini JJ. Scientific protocols for fire investigation. CRC Press; 2006 [chapter 8].

[73] Grann D. Trial by fire: did Texas execute an innocent man? The New Yorker; September 7, 2009.

[74] Strengthening Forensic Science in the United States: A Path Forward. The Committee on Identifying the Needs of the Forensic Science Community and the Committee on Science, Technology, and Law Policy and Global Affairs and the Committee on Applied and Theoretical Statistics, Division on Engineering and Physical Sciences. The Natonal Academies Press, Washington, DC., 2009.

[75] Kangaspeska S, Hultsch S, Edgren H, Nicorici D, Murumagi A, Kallioniemi O. Reanalysis of RNA-sequencing data reveals several additional fusion genes with multiple isoforms. PLoS ONE 2012;7: e48745.

[76] Berman JJ. Repurposing legacy data: innovative case studies. Waltham, MA: Morgan Kaufmann; 2015.

[77] Curry A. Rescue of old data offers lesson for particle physicists. Science 2011;331:694–5.

[78] Biebel O, Movilla Fernandez PA, Bethke S. the JADE Collaboration. C-parameter and jet broadening at PETRA energies. Phys Lett 1999;B459:326–34.

[79] Strauch JG. The phylogeny of the Charadriiformes (Aves): a new estimate using the method of character compatibility analysis. Trans Zool Soc Lond 1978;34:263–345.

[80] Chu PC. Phylogenetic reanalysis of Strauch's osteological data set for the Charadriiformes. Condor 1995;97:174–96.

[81] Mickevich MF, Parenti LR. Review of "The phylogeny of the Charadriiformes (Aves):a new estimate using the method of character compatibility analysis." Syst Zool 1980;29:108–13.

[82] Bochdanovits Z, Verhage M, Smit AB, de Geus EJ, Posthuma D, Boomsma DI, et al. Joint reanalysis of 29 correlated SNPs supports the role of PCLO/Piccolo as a causal risk factor for major depressive disorder. Mol Psychiatry 2009;14:650–2.

[83] Bracewell RN, MacPhie RH. Searching for non solar planets. Icarus 1979;38:136–47.

[84] Khan A. Possible earth-like planets could hold water: scientists cautious. Los Angeles Times; November 7, 2012.

[85] Smith AK, Ayanian JZ, Covinsky KE, Landon BE, McCarthy EP, Wee CC, et al. Conducting high-value secondary dataset analysis: an introductory guide and resources. J Gen Intern Med 2011;26:920–9.

[86] Yank V, Rennie D, Bero LA. Financial ties and concordance between results and conclusions in meta-analyses: retrospective cohort study. BMJ 2007;335:1202–5.

17

Repurposing Big Data

Section 17.1. What Is Data Repurposing?

If you want to make an apple pie from scratch, you must first create the universe.

Carl Sagan

Big Data resources are so very difficult to create and maintain that they really should not be devoted to any single use. We might as well get the most for our investments, and this means that we should repurpose our data. Data repurposing involves taking pre-existing data and performing any of the following [1]:

– **Finding new uses for data**

Fingerprints have been used, since antiquity, as a method for establishing the identity of individuals. Fingerprints were pressed onto clay tablets, seals, and even pottery left by ancient civilizations that included Minoan, Greek, Japanese, and Chinese. As early as the second millennium BCE, fingerprints were used as a type of signature in Babylon, and ancient Babylonian policemen recorded the fingerprints of criminals.

Toward the close of the 19th century, Francis Galton repurposed fingerprint data to pursue his own particular interests. Galton was primarily interested in the heritability and racial characteristics of fingerprints, a field of study that can be described as a scientific dead-end. Nonetheless, in pursuit of his interests, he devised a way of classifying fingerprints by patterns (e.g., plain arch, tented arch, simple loop, central pocket loop, double loop, lateral pocket loop, and plain whorl). This classification launched the new science of fingerprint identification, an area of research that has been actively pursued and improved over the past 120 years (Fig. 17.1).

Principles and Practice of Big Data. https://doi.org/10.1016/B978-0-12-815609-4.00017-0

FIG. 17.1 U.S. Federal Bureau of Investigation Fingerprint Division, World War II. *From FBI, public domain.*

In addition to Galton's novel classification methods, two closely related technological enhancements vastly increased the importance of fingerprints. The first was the incredibly simple procedure of recording sets of fingerprints, on paper, with indelible ink. With the simple fingerprint card, the quality of fingerprints improved, and the process of sharing and comparing recorded fingerprints became more practical. The second enhancement was the decision to collect fingerprint cards in permanent population databases (literally, digital data). Fingerprint databases enabled forensic scientists to match fingerprints found at the scene of a crime, with fingerprints stored in the database. The task of fingerprint matching was greatly simplified by confining comparisons to prints that shared the same class-based profiles, as described by Galton.

Repurposing efforts have expanded the use of fingerprints to include authentication (i.e., proving you are who you claim to be), keying (e.g., opening locked devices based on an authenticated fingerprint or some other identifying biometric), tracking (e.g., establishing the path and whereabouts of an individual by following a trail of fingerprints or other identifiers), and body part identification (i.e., identifying the remains of individuals recovered from mass graves or from the sites of catastrophic events). In the past decade, flaws in the vaunted process of fingerprint identification have been documented, and the improvement of the science of identification is an active area of investigation [2].

Today, most of what we think of as the forensic sciences is based on object identification (e.g., biometrics, pollen identification, trace chemical investigation, tire mark investigation, and so on). When a data object is uniquely identified, its association with additional data can be collected, aggregated, and retrieved, as needed.

– **Performing original research that could not have been performed when the data was collected**

History is replete with examples of old data driving new discoveries. A recent headline story explains how century old tidal data plausibly explained the appearance of the iceberg that sank the titanic, on April 15, 1912 [3]. Records show that several months earlier, in 1912, the moon, Earth, and sun aligned to produce a strong tidal pull, and this happened when the moon was the closest to the earth in 1400 years. The resulting tidal surge was sufficient to break the January Labrador ice sheet, sending an unusual number of icebergs toward the open North Atlantic waters. The Labrador icebergs arrived in the commercial shipping lanes four months later, in time for a fateful rendezvous with the Titanic. Back in January 1912, when tidal measurements were being collected, nobody foresaw that the data would be examined a century later.

Clever scientists are finding that old data can be reanalyzed to answer questions that were not anticipated by the scientists who performed the original study. Getting new uses from old data is the most cost-effective means of conducting research, and should be encouraged [1]. [Glossary Data archeology]

– Creating novel data sets through data file linkages

Introspective data, data triples, and data schemas are concepts that had little resonance before the days of Big Data. Using techniques that link heterogeneous forms of data to Web Locations is the basis for the so-called Semantic Web, the largest Big Data resource available to everyone [4]. The Semantic Web can be imagined as one enormous data re-purposing project in which everyone pursues their own purposes. [Glossary Heterogeneous data]

For data professionals, repurposing will often involve one or more of the following efforts:

– Finding subsets in a population once thought to be homogeneous
– Seeking new relationships among data objects
– Creating new concepts or ways of thinking about old concepts based on a re-examination of data
– Fine-tuning existing data models
– Starting over and remodeling systems

Section 17.2. Dark Data, Abandoned Data, and Legacy Data

We need above all to know about changes; no one wants or needs to be reminded 16 hours a day that his shoes are on.

David Hubel

Every child believes, for a time, that the universe began with his own birth. Anything preceding his birth is unreal, and of no consequence. Many Big Data resources have a similar disregard for events that preceded their birth. If events occurred prior to the creation of the Big Data resource, then those events have no consequence and can be safely ignored. Of course, this is absurd. It is accurate to think of new data as the result of events that involved old data; nothing in the universe occurs in the absence of preceding events.

Today, a large part of data science is devoted to finding trends in data; determining the simple functions that model the variation of data over time, and predicting how data will change in the future. These analytic activities require prior data that is annotated with a time measurement. Analysis of such data often reveals long-term trends, short-term trends, and periodic trends, often with characteristic forms (e.g., linear, exponential, power series). Hence, new data has very little meaning when it is not interpreted along with old data.

It is a shame that legacy data gets such shabby treatment by Big Data creators. Old data often resides in obsolete formats, on obsolete media, without proper annotation, and is collected under dubious circumstances. The incorporation of legacy data into modern Big Data resources is a tall order, but we need to make an effort to save legacy data whenever possible. Managers of Big Data resources are often expected to absorb smaller, older data sets. We cannot just pretend that such data has a lesser role than new data.

The healthcare industry is a prime example of Big Data in search of a legacy. President Barack Obama had set a goal for every American to have a secure medical record. What might such records include? Let us consider the medical record for a hypothetical patient named Lily Livesy, age 92. Not only has Lily outlived her doctors; she has outlived most of her hospitals. Though she lived in one city all her life, several of the hospitals that administered her medical care have been closed, and the records destroyed. In the past thirty years, she has received medical care at various doctor's offices, and in various departments in various hospitals. Some of these hospitals kept paper records; some had electronic records. Only one of the hospitals had anything that might be likened to an integrated hospital information system that aggregated transaction records produced by the various hospital departments (pharmacy, pathology, radiology, surgery, medicine, and so on). This hospital initiated a new Electronic Health Record system in the year 2013. Unfortunately, the new system is not compatible with the same hospital's prior information system, and the old records did not transfer to the new system. Consequently, in the year 2019, Lily Livesy, age 92, has one Electronic Health Record, residing in one hospital's information system, with no contribution from any other medical facility, and this Electronic Health Record contains a secure identifier, but no actual medical information. Her 92 year-long medical history is virtually blank. The same data deficits would apply to millions of other Americans. This is why, despite our best intentions, complete medical records, extending from birth to death, for all American citizens, will not be attainable anytime this century.

Often, the utility of legacy data comes as an afterthought inspired by a preliminary analysis of contemporary data. If a cancer researcher notices that the incidence of a certain tumor is high, he or she would naturally want to know whether the incidence of the tumor has been increasing over the past five years, ten years, 15 years and so on. A forensic criminologist who collects a CODIS signature on a sample of DNA might desperately need to check his sample against CODIS signatures collected over the past five decades. The most useful Big Data resources reach back through time. [Glossary CODIS]

Legacy data plays a crucial role in correcting the current record. It is not unusual for people to rely on flawed data. If we knew the full history of the data, including how it was originally collected, and how it was modified over time, we might avoid reaching erroneous conclusions. Several years ago, newspaper headlines drew attention to a modern manual for prisoner interrogation, used by U.S. forces stationed in Guantanamo. It turned out that the manual was a republication of a discredited Chinese operations manual used during the Korean War. The chain of documentation linking the current manual back to the original source had been broken [5]. In another example of lost legacy data, a Supreme Court decision was based, in part, on flawed information; an earlier statute had been overlooked [6]. Had the legacy data been raised during deliberations, an alternate Supreme Court verdict may have prevailed. To know the history of a data source, we need access to the legacy data that documents the original sources of our data, and permits us to trace the modifications of the data, over time.

It is human nature to evaluate the world through direct observations. If we want to know the length of an object, we measure its length with a ruler. If we want to know the number of eggs in a basket, we count the eggs. There are times when direct observations are not the best way to understand our world. If we are clever, we can determine the height of an object by comparing the length of its shadow, with the length of the shadow of an object of known height. We can estimate the number of eggs in a basket by weighing the basket, with and without the eggs, and dividing the total weight of the eggs by the predetermined average weight of a single egg. When we have a wealth of descriptive data about many different objects in our environment, we can derive new meaning from old measurements. The remainder of this chapter is devoted to five cases in point.

Section 17.3. Case Study: From Postal Code to Demographic Keystone

When you get to a fork in the road, take it.

Yogi Berra

There are three ways to assign integers to objects: cardinals, ordinal, and nominals. Cardinals tell us the number of objects (e.g., 2, 5, or ten items). Ordinals give us a rank (e.g., 1st, or 5th, or 8th place in a list). Nominal means "in name only", and nominals are arbitrary numbers that help identify an object. Telephone numbers, social security numbers, and zip codes are nominals. Nominals can be added together or multiplied, and divided, but it would be pointless to do so. Despite its self-effacing definition and its limited mathematical utility, nominal data sets are among the most useful of legacy data resources [1].

Zip codes were contrived by the U.S. Postal service to speed the distribution of mail. The original 5-digit zip codes were introduced in the early 1960s, with each zip code representing a geographic area containing a roughly equivalent segment of the population. The first three digits of the zip code identify mail distribution centers, from which mail sorted by the remaining two digits is distributed to the proper post offices. In the 1980s, an additional 4 digits was appended to the zip code, identifying individual buildings within the boundary of the 5-digit code.

Because zip codes describe geographic and demographic areas, they can be assigned a longitude, latitude, and elevation, typically measured at the geographic center of its boundaries. All data to which a zip code is attached (e.g., addresses, charge card transactions, crime reports, occurrences of reportable diseases, deaths, electricity consumption, water resources, homes receiving cable television, broadband usage) can be organized with the zip code serving as its primary record key. The lowly zip code, intended as an aid to mailmen, has been repurposed to serve entrepreneurs, epidemiologists, resource managers, and many others.

Section 17.4. Case Study: Scientific Inferencing From a Database of Genetic Sequences

It [natural selection] is all about the survival of self-replicating instructions for self-replication.

Richard Dawkins [7]

With the exception of identical twins, parthenogenetic offspring, and clones, every organism on earth has a unique sequence of DNA-forming nucleotides that distinguishes its genetic material (i.e., it's genome) from the genome of every other organism. If we were to have a record of the complete sequence of nucleotides in an individual's genome, we could distinguish that individual from every other organism on earth, by comparing genome sequences. This would require a lot of digital storage for every organism. In the case of humans, the genome is 3 billion nucleotides in length. As luck would have it, because there is enormous variation in genome sequence, from individual to individual, the identity of human individuals can be established by sampling short segments of DNA [1].

CODIS (Combined DNA Index System) collects the unique nucleotide sequences of the equivalent 13 segments of DNA, for every individual included in the database [8]. Using CODIS, DNA sampled at a crime scene can be matched against DNA samples contained in the database. Hence, the identity of individuals whose DNA is found at a crime scene can often be established. In the absence of a match it is sometimes possible to establish the genetic relationship (i.e., paternal or maternal relatives) between crime scene samples and individuals included in the database.

CODIS serves as an example of a database with narrow scope (i.e., names of people and associated DNA sequences), and broad societal value. The basic design of the CODIS database can be extended to any organism. For example, a database of DNA samples collected from individual trees in a geographic location can establish the source of seeds or pollen grains sticking to an article of clothing, and this information might lead to the location where a criminal event transpired. A population database containing full genome DNA sequences could be used to determine the presence or absence of disease-causing genes in individuals or to predict the response of an individual to a particular drug [9–12].

Section 17.5. Case Study: Linking Global Warming to High-Intensity Hurricanes

You can observe a lot by watching.

Yogi Berra

The UK Hadley Centre maintains a database of sea surface temperatures, over a 5-degree latitude-longitude global grid, from the year 1850, to the present, and updated monthly [13]. This data tells us how the ocean temperature changes seasonally and geographically, over time. Kerry Emanuel found a new use for the Hadley data when he noticed an association between regionally increased ocean temperatures and particularly strong hurricanes spawned in these same regions. Reviewing 50 years of data, Emanuel confirmed that the intensity of hurricanes increased in step with the warming of the North Atlantic and Pacific oceans [14]. A data set, intended primarily for charting trends in water temperature and correlating those trends with the oceanic reach of sea ice, found a new use: forecasting the intensity of hurricanes [1].

Section 17.6. Case Study: Inferring Climate Trends With Geologic Data

We waste a lot of time waiting for spectacular new material. We haven't sat down and taken a very close look at the material we have.

Bettina Stangneth, historian and author of "Eichmann Before Jerusalem: The Unexamined Life of a Mass Murderer" [15]

Mountains are like icebergs made of rock. The bulk of a mountain is buried underground. When the top of the mountain is eroded, the weight of the mountain is reduced, and the mountain bobs upwards, a little bit. The natural process through which mountains are flattened, over eons, requires the erosion of its surface plus its ever-rising subsurface.

When water is sucked from a mountain, the mountain lightens and rises. Likewise, if the water is sucked out of a tectonic plate, the entire plate (i.e., the surface of the planet overlying the plate) will rise. The National Science Foundation's Plate Boundary Observatory provides precise measurements of ground positions from data generated by GPS (Global Positioning System) satellites. A group of scientists working at the Scripps Institution of Oceanography found that all of the ground stations in the western United States exhibited measurable uplift. In the period 2003–04, the western states rose an average of 0.15 inches, and the western mountains rose more than half an inch in the same period.

This wide rise coincides with a long drought in the west. It would seem that the only explanation for the uplift of the tectonic plate, and the greater uplift of the western mountains, is the loss of water, via evaporation, without replacement. So strong is the relationship between water loss and mountain rise that water resources in the west can now be tracked with GPS ground measurements [16,1].

Section 17.7. Case Study: Lunar Orbiter Image Recovery Project

The world is the totality of facts, not things. (Die Welt ist die Gesamtheit der Tatsachen, nicht der Dinge)

Ludwig Wittgenstein

Following the first Apollo mission to the moon (Apollo 11, July 20, 1969), the five subsequent Apollo missions left behind recording instruments on the lunar surface. The collective set of downlinked data received from these instruments in known as the Apollo Lunar Surface Experiments Package (ALSEP). More than 11,000 data tapes were recorded [1].

During the Apollo program, control and use of the tapes, as well as the responsibility to safely archive the tapes, was transferred among various agencies and institutions. When the Apollo mission ended, funds were low, and a portion of the data that had been distributed to various investigators and agencies was never sent to the official archives [17]. It should come as no surprise that, at the present time, about half of the ALSEP tapes are missing; their whereabouts uncertain. Of the available tapes, much of the data is difficult to access, due to the use of abandoned data media (i.e., 7 and 9 track tapes) and obsolete data formats [17].

Available ALSEP data, when converted into a modern data format, has proven to be a valuable asset, when reanalyzed with modern analytic tools (Figs. 17.2 and 17.3). For example, the first analyses of ALSEP's seismic data, conducted 35 years ago, indicated that about 1300 deep moonquakes had occurred during the period when the data was being downlinked. The field of seismic analysis has advanced in the interim. A reanalysis of the

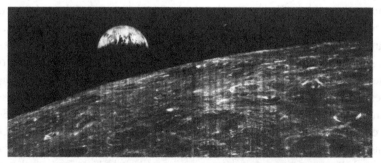

FIG. 17.2 Earth's first view of itself, from a location near the moon, by the United States Lunar Orbiter I, on August 23, 1966. *From U.S. National Aeronautics and Space Administration (NASA), public domain.*

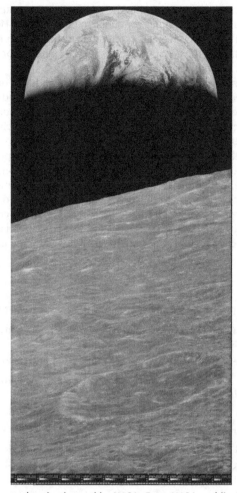

FIG. 17.3 Same image, but processed and enhanced by NASA. *From NASA, public domain.*

same data, using modern techniques, has produced an upward revision of the first estimate; to about 7000 deep moonquakes [17].

Today, there is a renewed push to find, collect and archive the missing ALSEP data. Why is there a sudden urgency to finish a chore that should have been completed decades ago? Simply put, the tapes must be restored before the last of the original investigators, who alone understand the scope and organization of the data, vanish into retirement or death.

Glossary

CODIS Abbreviation for Combined DNA Index System. CODIS is a collection of the unique nucleotide sequences of the equivalent 13 segments of DNA, for every individual included in the database [8]. The CODIS database is used by law enforcement personnel and contains identifying DNA sequences for individuals who have been processed within the criminal justice system. DNA obtained at a crime scene can be matched against DNA samples contained in the database. Hence, the identity of individuals whose DNA is found at a crime scene can often be established. In the absence of a match, it is sometimes possible to establish the genetic relationship (i.e., paternal or maternal relatives) between crime scene samples and individuals included in the database.

Data archeology The process of recovering information held in abandoned or unpopular physical storage devices, or packaged in formats that are no longer widely recognized, and hence unsupported by most software applications. The definition encompasses truly ancient data, such as cuneiform inscriptions stored on clay tablets circa 3300 BCE, and digital data stored on 5.25-inch floppy disks in Xyrite word-processor format, circa 1994.

Heterogeneous data Sets of data that are dissimilar with regard to content, purpose, format, organization, and annotations. One of the purposes of Big Data is to discover relationships among heterogeneous data sources. For example, epidemiologic data sets may be of service to molecular biologists who have gene sequence data on diverse human populations. The epidemiologic data is likely to contain different types of data values, annotated and formatted in a manner that is completely different from the data and annotations in a gene sequence database. The two types of related data, epidemiologic and genetic, have dissimilar content; hence they are heterogeneous to one another.

References

[1] Berman JJ. Repurposing legacy data: innovative case studies. Waltham, MA: Morgan Kaufmann; 2015.

[2] A review of the FBI's handling of the Brandon Mayfield case. U. S. Department of Justice, Office of the Inspector General, Oversight and Review Division; March 2006.

[3] Forsyth J. What sank the Titanic? Scientists point to the moon. Reuters; March 7, 2012.

[4] Berners-Lee T. Linked data—design issues. July 27, 2006. Available at: https://www.w3.org/DesignIssues/LinkedData.html [viewed December 20, 2017].

[5] Shane S. China inspired interrogations at Guantanamo. The New York Times; July 2, 2008.

[6] Greenhouse L. In court ruling on executions, a factual flaw. The New York Times; July 2, 2008.

[7] Dawkins R. The greatest show on earth: the evidence for evolution. New York: Free Press; 2009.

[8] Katsanis SH, Wagner JK. Characterization of the standard and recommended CODIS markers. J Forensic Sci 2012;58:S169–72.

[9] Guessous I, Gwinn M, Khoury MJ. Genome-wide association studies in pharmacogenomics: untapped potential for translation. Genome Med 2009;1:46.

[10] McCarthy JJ, Hilfiker R. The use of single-nucleotide polymorphism maps in pharmacogenomics. Nat Biotechnol 2000;18:505–8.

[11] Nebert DW, Zhang G, Vesell ES. From human genetics and genomics to pharmacogenetics and pharmacogenomics: past lessons, future directions. Drug Metab Rev 2008;40:187–224.

[12] Personalised medicines: hopes and realities. London: The Royal Society; 2005. Available from: https://royalsociety.org/~/media/Royal_Society_Content/policy/publications/2005/9631.pdf [viewed July 29, 2017].

[13] CISL Research Data Archive. http://rda.ucar.edu/datasets/ds277.3/ [viewed November 8, 2014].

[14] Roush W. The gulf coast: a victim of global warming? Technology Review; September 24, 2005.

[15] Schuessler J. Book portrays Eichmann as evil, but not banal. The New York Times; September 2, 2014.

[16] Borsa AA, Agnew DC, Cayan DR. Ongoing drought-induced uplift in the western United States. Science 2014;345:1587–90.

[17] Recovering the Missing ALSEP Data. Solar System Exploration Research Virtual Institute. NASA. Available from: http://sservi.nasa.gov/articles/recovering-the-missing-alsep-data/ [viewed October 13, 2014].

18

Data Sharing and Data Security

OUTLINE

Section 18.1. What Is Data Sharing, and Why Don't We Do More of It?

It's antithetical to the basic norms of science to make claims that cannot be validated because the necessary data are proprietary.

Michael Eisen [1]

Without data sharing, there can be very little progress in the field of Big Data. The reasons for this are simple:

− Research findings have limited value unless they are correlated with data contained in other databases.
− All findings, even those based on verified data, are tentative and must be validated against data contained in multiple datasets.
− Unless data is shared, scientists cannot build upon the work of others, and science devolves into a collection of research laboratories working in isolation from one another, leading to intellectual stagnation [1,2].
− Scientific conclusions have no credibility when the research community, oversight agencies, and the interested public cannot review the data upon which the findings were based and the details of how that data was measured.

Without data sharing, we do not have science. We just have people with their own agendas asking us to believe their conclusions. A long list of anguished position papers, urging researchers to share their data, has been published [3–8]. To be sure, there are technical obstacles to data sharing, but for every technical obstacle, there is a wealth of literature offering solutions [9–20].

Principles and Practice of Big Data. https://doi.org/10.1016/B978-0-12-815609-4.00018-2

Despite the imperatives of data sharing, scientists have been slow to adopt data sharing policies [6]. Because the issue of data sharing is so important to the field of Big Data, it is worth reviewing the impediments to its successful implementation.

Section 18.2. Common Complaints

Science advances funeral by funeral.

Folk wisdom

Here is a listing of the commonly heard reasons for withholding data from the public, along with suggested remedies.

– To protect scientists from "research parasites"

A recent opinion expressed by two editors of the New England Journal of Medicine, in an essay entitled "Data Sharing," expressed concern that a new brand of researcher uses data generated by others, for his or her own ends. The editors indicated that some front-line researchers characterize such individuals as "research parasites" [21]. The essayists suggested that researchers who want to use the data produced by others should do so by forming collaborative partnerships with the group that produced the original data [21].

The idea of collaborative partnerships may have been a reasonable strategy 30 years ago, before the emergence of enormous datasets, built from the work of hundreds of data contributors. Negotiating for data, with the promise of developing a mutually beneficial collaboration, is not a feasible option. Today, scientific projects may involve dozens or hundreds of scientists, with no single individual claiming ownership or responsibility for the aggregate data set. The individual contributors may have only the dimmest awareness of their own role in the effort. Under these circumstances, an outside investigator is unlikely to find an identifiable individual or group of individuals with the technical expertise, the scientific judgment, the legal standing, the ethical authority, and the strength of will, to negotiate a new collaboration and to surrender a large set of data.

Today, well-designed data sets can be merged with other sources of data, and repurposed for studies that were never contemplated by the original data contributors [22]. The goal of Big Data is to create data sets that can be utilized by the entire scientific community, with minimal barriers to access. **Characterizing data users as "research parasites," as witnessed in the New England Journal of Medicine article, misses the whole point of Big Data science** [21].

– To avoid data misinterpretation

Every scientist who releases data to the public must contend with the fear that a member of the public will misinterpret the data, reach an opposite conclusion, publish their false interpretation, and destroy the career of the trusting soul who was kind enough to provide the ammunition for his own execution. Teams of scientists developing a new drug or treatment protocol, may fear that if their data is released to the public, their competitors may seize the opportunity to unjustly critique their work and thus jeopardize their project.

Examples of such injustices have been sought, but not found [6]. There is no evidence that would lead anyone to believe that a misinterpretation of data has ever overshadowed a correct interpretation of data. Scientists have endured the withering criticisms of their colleagues since time immemorial. As they say, it comes with the territory. Hiding data for the purpose of avoiding criticism is unprofessional.

– **To limit access to responsible professionals**

Some researchers believe that data sharing must be a conditional process wherein investigators submit their data requests to a committee of scientists who decide whether the request is justified [23]. In some cases, the committee retains the right to review any results predicated on the shared data, with the intention of disallowing publication of results that they consider to be objectionable.

There are serious drawbacks to subjecting scientists to a committee approval process. The public needs unfettered access to the original data upon which published research results are based. Anything less makes it impossible to validate the conclusions drawn from the data, and invites all manner of scientific fraud. In the United States, this opinion is codified by law. The Data Quality Act of 2002 restrains government agencies from developing policies based on data that is unavailable for public review [24–27]. [Glossary Data Quality Act]

– **To sustain the traditional role of data protector**

Lawyers, bankers, healthcare workers, and civil servants are trained to preserve confidentiality; much like priests protect the confessional. It is understandable that many professionals are reluctant to share their data with the world-at-large. Nonetheless, it is unreasonable to hide data that has scientific value.

Before confidential information is released, data holders must be convinced that data sharing can be accomplished without breaching confidentiality, and that the effort spent in the process will yield some benefit to the individuals whose data is being appropriated. A large literature on the subject of safely sharing confidential data is readily available [28–33].

– **To await forthcoming universal data standards**

Trying to merge data sets that are disorganized is impossible, as is merging data sets wherein equivalent types of data are discordantly annotated. Because researchers and other data collectors use a variety of different types of software to collect and organize their data, obstacles raised by data incompatibility have been a major impediment to data sharing. The knee-jerk solution to the problem has always been to create new data standards.

In the past few decades, the standards-making process has evolved into a major industry [34]. There are standards for describing, organizing, and transmitting data. There are dozens, maybe hundreds, of standards, nomenclatures, classifications, and ontologies for the various domains of Big Data information, and all of these intellectual products are subject to multiple revisions. [Glossary Classification versus ontology]

The hunger for standards is insatiable [35]. The calls for new standards and new ontologies never seems to end. The many shortcomings of standards were discussed at length in Chapter 7, "Standards and Data Integration." It must be noted that the proliferation of standards, many of which are abandoned soon after they are created, has served to increase the complexity and decrease the permanence of Big Data resources [28,36].

Despite all the effort devoted to data standards, there is no widely adopted system for organizing and sharing all the different types of data encountered in Big Data resources. Perhaps the "standard" answer is not the correct answer, in this instance. Specifications, discussed in detail in Section 7.2, are a possible alternative to standards, and should be considered an option for those who are open to suggestion in this matter.

- **To protect legal ownership**

Ownership is a mercantile concept conferring the right to sell. If someone owns a cow, that means that they have the right to sell the cow. If you own a house, even a mortgaged house, then you have the right to sell the house. Let's focus on one particular type of confidential record that pertains to virtually everyone: the medical record. Who owns your confidential medical record? Is it owned by the patient? Is it owned by the medical center? Is it owned by anyone?

In law, there does not seem to be anyone who has the right to sell medical records; hence, it is likely that nobody can claim ownership. Still, medical institutions have a fiduciary responsibility to maintain medical records for their patients, and this entitles both patients and healthcare providers to use the records, as needed. Patients have the right to ask hospitals to send their medical records to other medical centers or to themselves. Hospitals are expected to archive tissues, medical reports and patient charts to serve the patient and society. In the United States, State health departments, the Centers for Disease Control and Prevention (CDC), and cancer registries all expect medical centers to deliver medical reports on request.

In all the aforementioned examples, data sharing is conducted without jeopardizing, or otherwise influencing, the data holder's claim of ownership.

- **To comply with rules issued from above**

It is not uncommon for researchers to claim that they would love to share their data, but they are forbidden from doing so by the lawyers and administrators at their institutions. Two issues tend to dissuade administrators from data sharing. The first is legal liability. Institutions have a responsibility to avoid punitive tort claims, such as those that may arise if human subjects complain that their privacy has been violated when their confidential information is shared. From the viewpoint of the institution, the simplest remedy is to forbid scientists from sharing their data. Secondly, institutions want to protect their own intellectual property, and this would include patents, drug discoveries, manufacturing processes, and even data generated by their staff scientists. Institutions may sometimes equate data sharing with poor stewardship of intellectual property. [Glossary Intellectual property]

When an institution forbids data sharing, as a matter of policy, scientists should argue that the institution cannot facilitate or promote its own research. Simply put, if Institution A does not share its data with Institution B, then Institution B will not share its data with Institution A. In addition, when Institution A publishes a scientific breakthrough, then Institution B will not find those claims credible, until their own researchers can review the primary data.

It is easy to forget that society, as a whole, benefits when scientific projects lead to discoveries. Without data sharing, those benefits will come at a glacial pace, and at great expense. Institutions have a societal obligation to advance science through data sharing.

– **To demand reimbursement**

Professionals are, by definition, people who are paid for their services. Professionals who go to the trouble of providing data to the public will want to be reimbursed. Data holders can be reassured that if they have created data at their own expense, for their own private use, then that data is theirs to keep. Nobody will take that data from them. But if the data holders have made public assertions based on their data, then they should understand that the scientific community will not give those assertions any credence, without having the data available for review.

The price that researchers pay for withholding their data (i.e., lack of validation of conclusions and professional obscurity) far exceeds the negligible costs of data sharing. Contrariwise, if the data is shared, and their results are validated, then their payment may come in the form of future grants, patents, prestige, and successful collaborations.

– **To avoid distributing flawed data**

Scientists are reluctant to release data that is full of errors. In particular, data curators may fear that if such data is released to the public, they will be inundated with complaints from angry data analysts, demanding that every error be corrected.

A 2011 study has shown that researchers with high quality data were, generally, willing to share their data [37]. Researchers who had weak data, that might support various interpretations and contrasting conclusions, were less willing to share their data. It is important to convince the scientists who create and hold data that the researchers who use their data, without asking permission and without forming collaborations, are not "research parasites"; they are the people who will validate good work, improve upon imperfect work, and create new work from otherwise abandoned data. The societal push toward data sharing should provide a strong impetus for scientists to improve the value of their data, so that they will something worth sharing.

Aside from corrections, all data sets need constant updating, and there are proper and improper ways of revising data (discussed in Section 8.1, "The Importance of Data that

Cannot Change"). Dealing with change, in the form of revised systems of annotations, and revised data elements, is part of the job of the data curator.

Institutions cannot refuse to share their data simply because their data contains errors or is awaiting revisions. Flawed data is common, and it's a safe bet that every large data set contains errors [38]. Institutions and scientific teams should hire professionals with the requisite skills to properly prepare, correct, and improve upon their data collections.

– **To protect against data hackers**

Properly deidentified, Big Data records may contain information that, when combined with data held in other databases, may uniquely identify patients [39]. As an obvious example, if a medical record contains an un-named patient's birth date, gender and zip-code, and a public database lists names of people in a zip-code, along with their birth dates and gender, it is a simple step to ascertain the identity of "deidentified" patients.

A specific instance, making national news headlines, may serve to clarify just how this may happen [40]. A 15-year-old boy was fathered using anonymously donated sperm. The boy wanted to know the identity of his biological father. A private company, had created a DNA Database from 45,000 DNA samples. The purpose of the database was to allow clients to discover kin by having their DNA compared with all the DNA samples in the database. The boy sent his DNA sample to the company, along with a fee. A comparison of the boy's Y chromosome DNA (inherited exclusively from the father) was compared with Y chromosome DNA in the database. The names of two men with close matches to the boy's Y chromosome were found.

The boy's mother had been provided (from the sperm bank) with the sperm donor's date of birth and birthplace. The boy used an online service to obtain the name of every man born in the sperm donor's place of birth on the sperm donor's date of birth. Among those names, one name matched one of the two Y-chromosome matches from the DNA database search. This name, according to newspaper reports, identified the child's father. [Glossary Y-chromosome]

In this case, the boy had access to his own uniquely identifying information (i.e., his DNA and specifically his Y chromosome DNA), and he was lucky to be provided with the date of birth and birthplace of his biological father. He was also extremely lucky that the biological father had registered his DNA in a database of 45,000 samples. And he was lucky that the DNA database revealed the names of its human subjects. The boy's success in identifying his father required a string of unlikely events, and a lax attitude toward subject privacy on the part of the personnel at the sperm bank and the personnel at the DNA database.

Regardless of theoretical security flaws, the criminal or malicious identification of human subjects included in deidentified research data sets is extremely rare. More commonly, confidential records (e.g., personnel records, credit records, fully identified medical records) are stolen wholesale, relieving thieves from the intellectually challenging task of finding obscure information that may link a deidentified record to the identity of its subject.

– **To preserve compartmentalization of data**

Most data created by modern laboratories has not been prepared in a manner that permits its meaningful use in other laboratories. In many cases the data has been compartmentalized, so that the data is disbursed in different laboratories. It is par for the course that a no single individual has taken the responsibility of collecting and reviewing all of the data that has been used to support the published conclusions of a multi-institutional project.

In the late 1990s, Dr. Wu Suk Hwang was a world-famous cloning researcher. The government of South Korea was so proud of Dr. Hwang that they issued a commemorative stamp to celebrate his laboratory's achievements. Dr. Hwang's status drastically changed when fabrications were discovered in a number of the manuscripts produced by his laboratory. Dr. Hwang had a habit of placing respected scientists as co-authors on his papers [41]. When the news broke, Hwang pointed a finger at several of his collaborators.

A remarkable aspect of Dr. Hwang's publications was his ability to deceive the coworkers in his own laboratory, and the co-authors located in laboratories around the world, for a very long time. Dr. Hwang used a technique known as compartmentalization; dividing his projects into tasks distributed to small groups of scientists who specialized in one step of the research process. By so doing, his coworkers never had access to the entire project's data. The data required to validate the final achievement of the research was not examined by his co-workers [42,41].

For several years, South Korean politicians defended the scientist, to the extent of questioning the patriotism of his critics. Over time, additional violations committed by Dr. Hwang were brought to light. In 2009, Hwang was sentenced in Seoul, S. Korea, to a two-year suspended prison sentence for embezzlement and bioethical violations; but he was never found guilty of fabrication [43].

Large data projects are almost always compartmentalized. When you have dozens or even hundreds of individuals contributing to a project, compartmentalization occurs quite naturally. In fact, what would you do without compartmentalization? Wait for every scientist involved in the project to review and approve one another's data? Today, large scientific projects may involve hundreds of scientists. Without compartmentalization, nothing would ever get published. The lesson here is that at the end of every research project, all of the data that contributed to the results must be gathered together as an organized and well-annotated dataset for public review.

– **To guard research protocols**

In every scientific study, the measurements included in the data must be linked to the study protocols (e.g., laboratory procedures) that produced the data. In some cases, the protocols are not well documented. In other cases, the protocols are well documented, but the researchers may have failed to follow the recommended protocols; thus rendering the data irreproducible. Occasionally, the protocols are the intellectual property of an entity other than the persons who created the data. In all these instances, the data holders may be reluctant to share their protocols with the public.

– **To conceal instances of missing data**

It is almost inevitable, when data sets are large and complex, that there will be some missing data points. In this case, data may be added "by imputation." This involves computing a statistical best bet on what the missing data element value might have been, and inserting the calculated number into the data set. A data manager may be reluctant to release to the public a database with "fudged" data.

It is perfectly legitimate to include imputed data points, on the condition that all the data is properly annotated, so that reviewers are aware of imputed values, and of the methods used to generate such values.

– **To avoid bureaucratic hurdles**

As discussed in Section 9.3, "Data that Comes with Conditions," institutions may resort to Kafkaesque measures to insure that only qualified and trusted individuals gain access to research data. It should come as no surprise that formal requests for data may take two years or longer to review and approve [44]. The approval process is so cumbersome, that it simply cannot be implemented without creating major inconveniences and delays, for everyone involved (i.e., data manager and data supplicant).

In the United States, federal agencies often seek to share data with one another. Such transactions require Memoranda of Understanding between agencies, and these memoranda can take months to negotiate and finalize [44]. In some cases, try as they might, data is not shared among federal agencies due to a lack of regulatory authorization that cannot be resolved to anyone's favor [44].

Hyper vigilance, on the part of U.S. Federal agencies, may stem from unfortunate incidents from the past that cannot be easily forgotten. One such incident, which attracted international attention, occurred when the United States accidentally released details of hundreds of the nuclear sites and programs, including the exact locations of nuclear stockpiles [45]. Despite their reluctance to share some forms of data, U.S. agencies have been remarkably generous with bioinformatics data, and the National Institutes of Health commonly attaches data sharing requirements to grants, and other awards.

In the U.S., Federal regulations impose strict controls on sharing identified medical data. Those same regulations specify that deidentified human subject data is exempted from those controls, and can be freely shared [32,33]. Data holders must learn the proper methods for deidentifying or anonymizing private and confidential medical data.

Whew! Where does this leave us? Data sharing is not easy. Nonetheless, published claims cannot be validated unless the data is made available to the public for review, and science cannot advance if scientists cannot build upon the data produced by their colleagues. Research institutions, both public and private, must find ways to deal with the problem, despite the difficulties. They might start by hiring scientists who are steeped in the craft of data sharing.

Section 18.3. Data Security and Cryptographic Protocols

No matter how cynical you become, it's never enough to keep up.

Lily Tomlin

Let us be practical. Nearly everyone has confidential information on their computers. Often, this information resides in a few very private files. If those files fell into the hands of the wrong people, the results would be calamitous. For myself, I encrypt my sensitive files. When I need to work with those files, I decrypt them. When I'm finished working with them, I encrypt them again. These files are important to me, so I keep copies of the encrypted files on thumb drives and on an external server. I don't care if my thumb drives are lost or stolen. I don't care if a hacker gets access to the server that stores my files. The files are encrypted, and only I know how to decrypt them.

Anyone in the data sciences will tell you that it is important to encrypt your data files, particularly when you are transferring files via the internet. Very few data scientists follow their own advice. Scientists, despite what you may believe, are not a particularly disciplined group of individuals. Few scientists get into the habit of encrypting their files. Perhaps they perceive the process as being too complex.

For serious encryption, you will want to step up to OpenSSL. OpenSSL is an open source collection of message digest protocols (i.e., protocols that yield one-way hashes) and encryption protocols. This useful set of utilities, with implementations for various operating systems, is available at no cost from:

https://www.openssl.org/related/binaries.html

Encryption algorithms and suites of cipher strategies available through OpenSLL include: RSA, DH (numerous protocols), DSS, ECDH, TLS, AES (including 128 and 256 bit keys), CAMELLIA, DES (including triple DES), RC4, IDEA, SEED, PSK, and numerous GOST protocols. In addition, implementations of popular one-way hash algorithms are provided (i.e., MD5 and SHA, including SHA384). OpenSSL comes with an Apache-style open source license. [Glossary AES]

For Windows users, the OpenSSL download contains three files that are necessary for file encryption: openssl.exe, ssleay32.dll, and libeay32.dll. If these three files are located in your current directory, you can encrypt any file, directly from the command prompt, as shown:

```
openssl aes128 -in public.txt -out secret.aes -pass pass:abcdefgh
```

The command line provides your chosen password, "abcdefgh" to the aes128 encryption algorithm, which takes the file public.txt and produces an AES-encrypted output file, secret.aes. Of course, once you've encrypted a file, you will need a decryption method. Here's a short command line that decrypts the encrypted file created by the preceding command line:

```
openssl aes128 -d -in secret.aes -out decrypted.txt -pass pass:
abcdefgh
```

We see that decryption involves inserting the "-d" option into the command line. AES is an example of a symmetric encryption algorithm, which means that the encryption password also serves as the decryption password.

Encrypting and decrypting individual strings, files, groups of files, and directory contents is extremely simple and can provide a level of security that is likely to be commensurate with your personal needs.

Here is a short Python script, aes.py, that encrypts all the files included in a list, and deposits the encrypted files in a thumb drive sitting in the "f:" drive.

```
import sys, os, re
filelist = ['diener.txt','simplify.txt','re-ana.txt', 'phenocop.
txt', 'mystery.txt','disaster.txt', 'factnote.txt', 'perlbig.txt',
'referen.txt', 'create.txt', 'exploreo.txt']
pattern = re.compile("txt")
for filename in filelist:
   out_filename = pattern.sub('enc', filename)
   out_filename = "f:\\" + out_filename
   print(out_filename)
   cmdstring = "openssl aes128 -in " + filename + " -out " + out_filename +
   " -pass pass:abcdefgh"
   os.system(cmdstring)
```

– **Public and private key cryptographic protocols**

Many cryptographic algorithms are symmetric; the password used to encrypt a file is the same as the password used to decrypt the file. In an asymmetric cryptographic algorithm, the password that is used to encrypt a file is different from the password that is used to decrypt the file. The encrypting password is referred to as the public key, by convention. The public key can be distributed to friends or posted on a public web site. The decrypting password is referred to as the private key, and it must never be shared.

How is a public/private key system used? If Alice were to encrypt a file with Bob's public key, only Bob's private key could decrypt the file. If Bob does not lose his private key, and if Bob does not allow his private key to be shared or stolen, then only Bob can decrypt files encrypted with his public key. Alice can send the encrypted file, without worrying that the encrypted file could be intercepted and opened by someone other than Bob.

As discussed, openssl can be run via command lines, from the system prompt (e.g., c:\> in Windows systems). Let's generate a public/private key pair that we'll use for RSA encryption.

```
openssl genpkey -algorithm RSA -out private_key.pem -pkeyopt
rsa_keygen_bits:2048
openssl rsa -pubout -in private_key.pem -out public_key.pem
```

These two commands produced two files, each containing a cryptographic key. The private key file is private_key.pm. The public key file is public_key.pem. Let's encrypt a file (sample.txt) using RSA encryption and the public key we just created (public_key.pem)

```
openssl rsautl -encrypt -inkey public_key.pem -pubin -in sample.txt -
out sample.ssl
```

This produces an encrypted file, sample.ssl. Let's decrypt the encrypted file (sample.ssl) using the private key that we have created (private_key.pem)

```
openssl rsautl -decrypt -inkey private_key.pem -in sample.ssl -out
decrypted.txt
```

In common usage, this protocol only transmits small message files, such as passwords. Alice could send a large file, strongly encrypted with AES. In a separate exchange, Alice and Bob would use public and private keys to transmit the password. First, Alice would encrypt the password with Bob's public key. The encrypted message would be sent to Bob. Bob would decrypt the message with his private key, thus producing the password. Bob would use the password to decrypt the large AES-encrypted file. A large variety of security protocols have been devised, utilizing public/private key pairs, suiting a variety of purposes.

The public/private keys can also be used to provide the so-called digital signature of the individual holding the private key.

To sign and authenticate a transferred data file (e.g., mydata.txt), the following three steps must be taken:

1. Alice creates a one-way hash of her data file, mydata.txt, and creates a new file, called hashfile, to hold the one-way hash value.
   ```
   openssl dgst -sha256 mydata.txt > hashfile
   ```
2. Alice signs the hashfile with her private key, to produce a digital signature in the file "signaturefile":
   ```
   openssl rsautl -sign -inkey private_key.pem -keyform PEM -in
   hashfile >signaturefile
   ```
3. Alice sends mydata.txt and the signature file("signature") to Bob
4. Bob verifies the signature, with his public key.
   ```
   openssl rsautl -verify -inkey public_key.pem -pubin -keyform PEM -in
   signaturefile
   ```
   The verified contents of the signature file is the original hash created by Alice, of mydata.txt
   ```
   SHA256(mydata.txt)=6e2a1dbf9ea8cbf2accb64f33ff83c7040413963e
   69c736accdf47de0bc16b1a
   ```
   This verifies Alice's signature and yields Alice's hash of her mydata.txt file
5. Bob conducts his own one-way hash on his received file, mydata.txt.
   ```
   openssl dgst -sha256 mydata.txt
   ```

This produces the following hash value, which is the same value that was decrypted from Alice's signature

```
SHA256(mydata.txt)=6e2a1dbf9ea8cbf2accb64f33ff83c7040413963e69c
736accdf47de0bc16b1a
```

Because the received mydata.txt file has the same one-way hash value as the sent mydata.txt file, and because Bob has verified that the sent mydata.txt file was signed by Alice, then Bob has taken all steps necessary to authenticate the file (i.e., to show that the received file is the file that Alice sent).

There are some limitations to this protocol. Anyone in possession of Alice's private key can "sign" for Alice. Hence the signature is not equivalent to a hand-written signature or to a biometric that uniquely identifies Alice (e.g., iris image, CODIS gene sequences, full set of fingerprints). Really, all the process tells us is that a document was sent by someone in possession of Alice's private key. We never really know who sent the document.

The signature does not attest to anything, other than that a person with Alice's key actually sent the document. There is nothing about the process that would indicate that she personally verifies that the content of the transmitted material is accurate or that the content was created by Alice or that she agrees with the contents.

Cryptography is fascinating, but experts who work in the field of data security will tell you that cryptographic algorithms and protocols can never substitute for a thoughtful data security plan that is implemented with the participation and cooperation of the staff of an organization [46,47]. In many instances, security breaches occur when individuals, often trusted employees, violate protocol and/or behave recklessly. Hence, data security is more often a "people thing" than a "computer thing". Nonetheless, if you are not a multi-million dollar institution, and simply want to keep some of your data private, here are a few tips that you might find helpful. If you have really important data, the kind that could hurt yourself or others if the data were to fall into the wrong hands, then you should totally disregard the advice that follows and seek the help of a professional security agent.

– Save yourself a lot of grief by settling for a level of security that is appropriate and reasonable for your own needs.

Don't use a bank vault when a padlock will suffice.

– Avail yourself of no-cost solutions.

Some of the finest encryption algorithms, and their implementations, are publicly available in OpenSSL and other sources.

– The likelihood that you will lose your passwords is much higher than the likelihood that someone will steal your passwords.

Develop a good system for passkey management that is suited to your own needs.

– The security of the system should not be based on hiding your encrypted files or keeping the encryption algorithm secret.

The greatest value of modern encryption protocols is that it makes no difference whether anyone steals or copies your encrypted files, or learns your encryption algorithm.

– File encryption and decryption should be computationally fast.

Fast, open source protocols are readily available.

– File encryption should be done automatically, as part of some computer routine (e.g., a backup routine), or as a chron job (i.e., a process that occurs at predetermined time).

You should be able to batch-encrypt and batch-decrypt any number of files all at once (i.e., from a command loop within a script), and you should be able to combine encryption with other file maintenance activities. For example, you should be able to implement a simple script that loops through every file in a directory, or a directory tree (i.e., all the files in all of the subdirectories under the directory), all at once, adding file header and metadata information into the file, scrubbing data as appropriate, calculating a one-way hash (i.e., message digest) of the finished file, and producing an encrypted file output.

– You should never implement an encryption system that is more complex than you can understand [48].

Your data may be important to you, and to a few of your colleagues, but the remainder of the world looks upon your output with glazed eyes. If you are the type of person who would protect your valuables with a padlock, rather than a safety deposit box, then you should probably be thinking less about encryption strength and more about encryption operability. Ask yourself whether the encryption protocols that you use today shall be widely available, platform-independent and vendor independent, 5, 10 or 50 years from now. Will you always be able to decrypt your own encrypted files?

– Don't depend on redundancy

At first blush, it would be hard to argue that redundancy, in the context of information systems, is a bad thing. With redundancy, when one server fails, another picks up the slack; if a software system crashes, its duplicate takes over; when one file is lost, it is replaced by its back-up copy. It all seems good.

The problem with redundancy is that it makes the system more complex. The operators of a Big Data resource with built-in redundancies must maintain the operability of the redundant systems in addition to the primary systems. More importantly, the introduction of redundancies introduces a new set of interdependencies (i.e., how the parts of the system interact), and the consequences of those interdependencies may be difficult to anticipate.

In recent memory, the most dramatic example of a failed redundant system involved the Japanese nuclear power plant at Fukushima. The plant was designed with redundant systems. If the power failed, a secondary power generator would kick in. On March 11, 2011, a powerful earthquake off the shore of Japan produced a tidal wave that cut the nuclear reactor's access to the electric power grid. The back-up generators were flooded by the same tidal wave. The nuclear facilities were cut off from emergency assistance; also due to the tidal wave. Subsequent meltdowns and radiation leaks produced the worst nuclear disaster since Chernobyl.

As discussed previously in this chapter, on June 4, 1996, the first flight of the Ariane 5 rocket self-destructed, 37 seconds after launch. There was a bug in the software, but the Ariane had been fitted with a back-up computer. The back-up was no help; the same bug that crippled the primary computer put the back-up computer out of business [49]. The lesson here, and from the Fukushima nuclear disaster, is that redundant systems are often ineffective if they are susceptible to the same destructive events that caused failure in the primary systems.

Computer software and hardware problems may occur due to unanticipated interactions among software and hardware components. Redundancy, by contributing to system complexity, and by providing an additional opportunity for components to interact in an unpredictable manner, may actually increase the likelihood of a system-wide crash. Cases have been documented wherein system-wide software problems arose due to bugs in the systems that controlled the redundant subsystems [49].

A common security measure involves backing up files and storing the back-up files off-site. If there is a fire, flood, or natural catastrophe at the computer center, or if the computer center is sabotaged, then the back-up files can be withdrawn from the external site and eventually restored. The drawback of this approach is that the back-up files create a security risk. In Portland Oregon, in 2006, 365,000 medical records were stolen from Providence Home Services, a division of Seattle-based Providence Health Systems [50]. The thief was an employee who was handed the back-up files and instructed to store them in his home, as a security measure. In this case, the theft of identified medical records was a command performance. The thief complied with the victim's request to be robbed, as a condition of his employment. At the very least, the employer should have encrypted the back-up files before handing them over to an employee.

Nature takes a middle-of-the-road approach on redundancy. Humans evolved to have two eyes, two arms, two legs, two kidneys, and so on. Not every organ comes in duplicate. We have one heart, one brain, one liver, one spleen. There are no organs that come in triplicate. Human redundancy is subject to some of the same vulnerabilities as computer redundancy. A systemic poison that causes toxicity in one organ will cause equivalent toxicity in its contra-lateral twin.

–   Save Time and Money; Don't Protect Data that Does not Need Protection

Big Data managers tend to be overprotective of the data held in their resources, a professional habit that can work in their favor. In many cases, though, when data is of a purely academic nature, containing no private information, and is generally accessible from alternate sources, there really is no reason to erect elaborate security barriers.

Security planning always depends on the perception of the value of the data held in the resource (i.e., Is the data in the Big Data resource worth anything?), and the risks that the data might be used to harm individuals (e.g., through identity theft). In many cases, the data held in Big Data resources has no intrinsic monetary value and poses no risks to individuals. The value of most Big Data resource is closely tied to its popularity. A resource used by millions of people provides opportunities for advertising and attracts funders and investors.

Regarding the release of potentially harmful data, it seems prudent to assess, from the outset, whether there is a simple method by which the data can be rendered harmless. In many cases, deidentification can be achieved through a combination of data scrubbing, and expunging data fields that might conceivably tie a record to an individual. If your data set contains no unique records (i.e., if every record in the system can be matched with another record, from another individual, for which every data field is identical), then it is impossible to link any given record to an individual, with certainty. In many cases, it is a simple matter to create an enormous data set wherein every record is matched by many other records that contain the same informational fields. This process is sometimes referred to as record ambiguation [51].

Sometimes a Big Data team is compelled to yield to the demands of their data contributors, even when those demands are unreasonable. An individual who contributes data to a resource may insist upon assurances that a portion of any profit resulting from the use of their contributed data will be returned as royalties, shares in the company, or some other form of remuneration. In this case, the onus of security shifts from protecting the data to protecting the financial interests of the data providers. When every piece of data is a source of profit, measures must be put into place to track how each piece of data is used, and by whom. Such measures are often impractical, and have great nuisance value for data managers and data users. The custom of capitalizing on every conceivable opportunity for profit is a cultural phenomenon, not a scientific imperative.

Section 18.4. Case Study: Life on Mars

You must accept one of two basic premises: Either we are alone in the universe, or we are not alone in the universe. And either way, the implications are staggering.

Wernher von Braun

On September 3, 1976, the Viking Lander 2 touched down upon the planet mars, where it remained operational for the next 3 years, 7 months and 8 days. Soon after landing, it performed an interesting remote-controlled experiment. Using samples of martian dirt, astrobiologists measured the conversion of radioactively-labeled precursors into more complex carbon-based molecules; the so-called Labeled-Release study. For this study, control samples of dirt were heated to a high temperature (i.e., sterilized), and likewise exposed to radioactively-labeled precursors, without producing complex carbon-containing molecules. The tentative conclusion, published soon thereafter, was that Martian organisms in the samples of dirt had built carbon-based molecules through a metabolic pathway [52]. As you might expect, the conclusion was immediately challenged, and remains controversial, to this day, nearly 32 years later [22].

In the years since 1976, long after the initial paper was published, the data from the Labeled-Release study has been available to scientists, for re-analysis. New analytic techniques have been applied to the data, and new interpretations have been published [52]. As additional missions have reached mars, more data has emerged (i.e., the detection of water and methane), also supporting the conclusion that there is life on mars. None of the data is conclusive; Martian organisms have not been isolated. The point made here is that the shared Labeled-Release data is accessible and permanent, and can be studied again and again, compared or combined with new data, and argued ad nauseum [22].

Section 18.5. Case Study: Personal Identifiers

Secret agent man, secret agent man
They've given you a number and taken away your name
Theme from the television show "Secret Agent", airing in the United States from 1964–66; song written by P. F. Sloan and Steve Barriby

We came to a conclusion in 2002. I don't think you can do it (create an electronic health record) without a national identifier.

Peter Drury [53]

An awful lot of the data collected by scientists concerns people (e.g., financial data, marketing data, medical data). Given everything discussed so far in this book regarding the importance of providing data object uniqueness, you would think that we would all be assigned our own personal identifiers by now.

Of course, nothing could be further from the truth. Each of us are associated with dozens, if not hundreds of irreconcilable identifiers intended to serve a particular need at a particular moment in time. These include bank accounts, credit cards, loan applications, brokerage and other investment accounts, library cards, and voter IDs. In the United States, a patient may be assigned separate identifiers for the various doctors' offices, clinics and hospitals that she visits over the course of her life. As mentioned, a single hospital may assign a patient many different "unique" identifiers, for each department visited, for each newly installed hospital information system, and whenever the admission clerk forgets to ask the patient if he or she had been previously registered. U.S. Hospitals try to reconcile the different identifiers for a patient under a so-called Enterprise Master Patient Index, but experience has shown the problems encountered are insurmountable. As one example, in Houston's patient index system, which includes 3.5 million patients, there are about 250,000 patients that have a first and last name in common with at least one other registrant, and there are 70,000 instances wherein two people share the same first name, last name, and birthdate [54]. There is a growing awareness that efforts at reconciling systems wherein individual patients are registered multiple times, are never entirely satisfactory [53].

The subject of data security cannot be closed without mention of the National Patient Identifier. Some countries employ a National Patient Identifier (NPI) system. In these cases, when a citizen receives treatment at any medical facility in the country, the transaction is recorded under the same permanent and unique identifier. Doing so enables the data collected on individuals, from multiple hospitals, to be merged. Hence, physicians can retrieve patient data that was collected anywhere in the nation. In countries with NPIs, data scientists have access to complete patient records and can perform healthcare studies that would be impossible to perform in countries that lack NPI systems. In the United States, where a system of NPIs has not been adopted, there is a perception that such a system would constitute an invasion of privacy. Readers from outside the United States are probably wondering why the United States is so insecure on this issue.

In the United States, the call for a national patient identification system is raised, from time to time. The benefits to patients and to society are many. Aside from its absolutely necessary role in creating data that can be sensibly aggregated and meaningfully analyzed, it also serves to protect the privacy of individuals by eliminating the need for less secure forms of identification (e.g., credit cards, drivers licenses).

Regardless, U.S. citizens are reluctant to have an identifying number that is associated with a federally controlled electronic record of their private medical information. To show its disdain for personal identifiers, the U.S. Congress passed Public Law 105-277, in 1999, prohibiting the Department of Health and Human Services from using its funds to develop personal health identifiers, without first obtaining congressional approval [55].

In part, this distrust results from the lack of any national insurance system in the United States. Most health insurance in the United States is private, and private insurers have wide discretion over the fees and services provided to enrollees. There is a fear that if there were a national patient identifier with centralized electronic medical records, insurers may withhold reimbursements or raise premiums or otherwise endanger the health of patients. Because the cost of U.S. medical care is the highest in the world, medical bills for uninsured patients can quickly mount, impoverishing individuals and families [56].

Realistically, no data is totally safe. Data breaches today may involve hundreds of millions of confidential records. The majority of Americans have had social security numbers, credit card information, and private identifiers (e.g., birth dates, city of birth, names of relatives) misappropriated or stolen. Medical records have been stolen in large number. Furthermore, governments demand and receive access to our confidential medical records, when they deem it necessary [57]. Forbidding National Patient Identifiers has not made us safe. [Glossary Social Security Number]

Maybe we should ask ourselves the following: "Is it rational to forfeit the very real opportunity of developing new safe and effective treatments for serious diseases, for the very small likelihood that someone will crack my deidentified research record and somehow leverage this information to my disadvantage?"

Suppose everyone in the United States were given a choice: you can be included in a national patient identifier system, or you can opt out. Most likely, there would be many millions of citizens who would opt out of the offer, seeing no particular advantage in having a national patient identifier, and sensing some potential harm. Now, suppose you were told that if you chose to opt out, you would not be permitted to enjoy any of the health benefits coming from studies performed with data collected through the national patient identifier system. New safe and effective drugs, warnings of emerging epidemics, information on side effects associated with your medications, biomarker tests for preventable illnesses, and so on, would be reserved for individuals with national patient identifiers. Those who made no effort to help the system would be barred from any of the benefits that the system provided. Would you reconsider your refusal to accept a national patient identifier, if you knew the consequences? Of course, this is a fanciful scenario, but it makes a point.

Glossary

AES The Advanced Encryption Standard (AES) is the cryptographic standard endorsed by the U.S. government as a replacement for the old government standard, DES (Data Encryption Standard). AES was chosen from among many different encryption protocols submitted in a cryptographic contest conducted by the U.S. National Institute of Standards and Technology, in 2001. AES is also known as Rijndael, after its developer. It is a symmetric encryption standard, meaning that the same password used for encryption is also used for decryption.

Classification versus ontology A classification is a system in which every object in a knowledge domain is assigned to a class within a hierarchy of classes. The properties of superclasses are inherited by the subclasses. Every class has one immediate superclass (i.e., parent class), although a parent class may have more than one immediate subclass (i.e., child class). Objects do not change their class assignment in a classification, unless there was a mistake in the assignment. For example, a rabbit is always a rabbit, and does not change into a tiger. Classifications can be thought of as the simplest and most restrictive type of ontology, and serve to reduce the complexity of a knowledge domain [58]. Classifications can be easily modeled in an object-oriented programming language and are non-chaotic (i.e., calculations performed on the members and classes of a classification should yield the same output, each time the calculation is performed). A classification should be distinguished from an ontology. In an ontology, a class may have more than one parent class and an object may be a member of more than one class. A classification can be considered a special type of ontology wherein each class is limited to a single parent class and each object has membership in one and only one class.

Data Quality Act In the United States the data upon which public policy is based must have quality and must be available for review by the public. Simply put, public policy must be based on verifiable data. The Data Quality Act of 2002, requires the Office of Management and Budget to develop government-wide standards for data quality [24].

Intellectual property Data, software, algorithms, and applications that are created by an entity capable of ownership (e.g., humans, corporations, universities). The entity holds rights over the manner in which the intellectual property can be used and distributed. Protections for intellectual property may come in the form of copyrights and patent. Copyright applies to published information. Patents apply to novel processes and inventions. Certain types of intellectual property can only be protected by being secretive. For example, magic tricks cannot be copyrighted or patented; this is why magicians guard their intellectual property so closely. Intellectual property can be sold outright, essentially transferring ownership to another entity; but this would be a rare event. In other cases, intellectual property is retained by the creator who permits its limited use to others via a legal contrivance (e.g., license, contract, transfer agreement, royalty, usage fee, and so on). In some cases, ownership of the intellectual property is retained, but the property is freely shared with the world (e.g., open source license, GNU license, FOSS license, Creative Commons license).

Social Security Number The common strategy, in the United States, of employing social security numbers as identifiers is often counterproductive, owing to entry error, mistaken memory, or the intention to deceive. Efforts to reduce errors by requiring individuals to produce their original social security cards puts an unreasonable burden on honest individuals, who rarely carry their cards, and provides an advantage to dishonest individuals, who can easily forge social security cards. Institutions that compel patients to provide a social security number have dubious legal standing. The social security number was originally intended as a device for validating a person's standing in the social security system. More recently, the purpose of the social security number has been expanded to track taxable transactions (i.e., bank accounts, salaries). Other uses of the social security number are not protected by law. The Social Security Act (Section 208 of Title 42 U.S. Code 408) prohibits most entities from compelling anyone to divulge his/her social security number. Legislation or judicial action may one day stop healthcare institutions from compelling patients to divulge their social security numbers as a condition for providing medical care. Prudent and forward-thinking institutions will limit their reliance on social security numbers as personal identifiers.

Y-chromosome A small chromosome present in males and inherited from the father. The normal complement of chromosomes in male cells has one Y chromosome and one X chromosome. The normal complement of chromosomes in female cells has two X chromosomes and no Y chromosomes. Analysis of the Y chromosome is useful for determining paternal lineage.

References

[1] Markoff J. Troves of personal data, forbidden to researchers. The New York Times, May 21, 2012.

[2] Markoff J. A deluge of data shapes a new era in computing. The New York Times, December 15, 2009.

[3] Guidance for sharing of data and resources generated by the molecular libraries screening centers network (mlscn)—addendum to rfa rm-04-017. NIH notice not-rm-04-014. July 22, 2004. Available from: http://grants.nih.gov/grants/guide/notice-files/NOT-RM-04-014.html [viewed September 19, 2012].

[4] Sharing publication-related data and materials: responsibilities of authorship in the life sciences. Washington, DC: The National Academies Press; 2003. Available from: http://www.nap.edu/openbook.php?isbn=0309088593 [viewed September 10, 2012].

[5] NIH policy on data sharing, 2003. Available from: http://grants.nih.gov/grants/guide/notice-files/NOT-OD-03-032.html [viewed September 13, 2015].

[6] Fienberg SE, Martin ME, Straf ML, editors. Sharing research data. Washington, DC: Committee on National Statistics, Commission on Behavioral and Social Sciences and Education, National Research Council. National Academy Press; 1985.

[7] Policy on Enhancing Public Access to Archived Publications Resulting from NIH-Funded Research. Notice Number: NOT-OD-05-022, 2005.

[8] Revised Policy on Enhancing Public Access to Archived Publications Resulting from NIH-Funded Research. Notice Number: NOT-OD-08-033. Release date: January 11, 2008. Effective date: April 7, 2008. Available from: http://grants.nih.gov/grants/guide/notice-files/not-od-08-033.html; 2008 [viewed December 28, 2009].

[9] Berman JJ. A tool for sharing annotated research data: the "Category 0" UMLS (unified medical language system) vocabularies. BMC Med Inform Decis Mak 2003;3:6.

[10] Berman JJ, Edgerton ME, Friedman B. The tissue microarray data exchange specification: a community-based, open source tool for sharing tissue microarray data. BMC Med Inform Decis Mak 2003;3:5.

[11] Berman JJ. De-identification. Washington, DC: U.S. Office of Civil Rights (HHS), Workshop on the HIPAA Privacy Rule's De-identification Standard; 2010 March 8–9.

[12] Berman JJ. Racing to share pathology data. Am J Clin Pathol 2004;121:169–71.

[13] Berman JJ. Principles of big data: preparing, sharing, and analyzing complex information. Waltham, MA: Morgan Kaufmann; 2013.

[14] Berman JJ. Data simplification: taming information with open source tools. Waltham, MA: Morgan Kaufmann; 2016.

[15] de Bruijn J. Using ontologies: enabling knowledge sharing and reuse on the Semantic Web. Digital Enterprise Research Institute Technical Report DERI-2003-10-29. Available from: http://www.deri.org/fileadmin/documents/DERI-TR-2003-10-29.pdf; October 2003 [viewed August 14, 2012].

[16] Drake TA, Braun J, Marchevsky A, Kohane IS, Fletcher C, Chueh H, et al. A system for sharing routine surgical pathology specimens across institutions: the Shared Pathology Informatics Network (SPIN). Hum Pathol 2007;38:1212–25.

[17] Sweeney L. In: Guaranteeing anonymity when sharing medical data, the Datafly system. Proc American Medical Informatics Association; 1997. p. 51–5.

[18] Sweeney L. Three computational systems for disclosing medical data in the year 1999. Medinfo 1998;9(Pt 2):1124–9.

[19] Malin B, Sweeney L. How (not) to protect genomic data privacy in a distributed network: using trail re-identification to evaluate and design anonymity protection systems. J Biomed Inform 2004;37:179–92.

[20] Neamatullah I, Douglass MM, Lehman LW, Reisner A, Villarroel M, Long WJ, et al. Automated de-identification of free-text medical records. BMC Med Inform Decis Mak 2008;8:32.

[21] Longo DL, Drazen JM. Data sharing. New Engl J Med 2016;374:276–7.

[22] Berman JJ. Repurposing legacy data: innovative case studies. Waltham, MA: Morgan Kaufmann; 2015.

[23] Frellick M. Models for sharing trial data abound, but with little consensus. www.medscape.com August 3, 2016.

[24] Data Quality Act. 67 Fed. Reg. 8,452, February 22, 2002, addition to FY 2001 Consolidated Appropriations Act (Pub. L. No. 106-554 codified at 44 U.S.C. 3516).

[25] Tozzi JJ, Kelly Jr WG, Slaughter S. Correspondence: data quality act: response from the Center for Regulatory Effectiveness. Environ Health Perspect 2004;112:A18–9.

[26] Guidelines for ensuring and maximizing the quality, objectivity, utility, and integrity of information disseminated by federal agencies. Fed Regist 2002;67(36) February 22.

[27] Sass JB, Devine Jr. JP. The Center for Regulatory Effectiveness invokes the Data Quality Act to reject published studies on atrazine toxicity. Environ Health Perspect 2004;112:A18.

[28] Berman JJ. Biomedical informatics. Sudbury, MA: Jones and Bartlett; 2007.

[29] Berman JJ. Threshold protocol for the exchange of confidential medical data. BMC Med Res Methodol 2002;2:12.

[30] Berman JJ. Confidentiality issues for Medical Data Miners. Artif Intell Med 2002;26:25–36.

[31] Quantin CH, Bouzelat FA, Allaert AM, Benhamiche J, Faivre J, Dusserre L. Automatic record hash coding and linkage for epidemiological followup data confidentiality. Methods Inf Med 1998;37:271–7.

[32] Department of Health and Human Services. 45 CFR (Code of Federal Regulations), Parts 160 through 164. Standards for Privacy of Individually Identifiable Health Information (Final Rule). Fed Regist 2000;65(250):82461–510 December 28.

[33] Department of Health and Human Services. 45 CFR (Code of Federal Regulations), 46. Protection of Human Subjects (Common Rule). Fed Regist 1991;56:28003–32 June 18.

[34] Kammer RG. The Role of Standards in Today's Society and in the Future. Statement of Raymond G. Kammer, Director, National Institute of Standards and Technology, Technology Administration, Department of Commerce, Before the House Committee on Science Subcommittee on Technology; September 13, 2000.

[35] Berman HM, Westbrook J. The need for dictionaries, ontologies and controlled vocabularies. OMICS 2003;7:9–10.

[36] Robinson D, Paul Frosdick P, Briscoe E. HL7 Version 3: an impact assessment. NHS Information Authority; March 23, 2001.

[37] Wicherts JM, Bakker M, Molenaar D. Willingness to share research data is related to the strength of the evidence and the quality of reporting of statistical results. PLoS ONE 2011;6:e26828.

[38] Goldberg SI, Niemierko A, Turchin A. Analysis of data errors in clinical research databases. AMIA Annu Symp Proc 2008;2008:242–6.

[39] Behlen FM, Johnson SB. Multicenter patient records research: security policies and tools. J Am Med Inform Assoc 1999;6:435–43.

[40] Stein R. Found on the Web, with DNA: a boy's father. The Washington Post; Sunday, November 13, 2005.

[41] Hwang WS, Roh SI, Lee BC, Kang SK, Kwon DK, Kim S, et al. Patient-specific embryonic stem cells derived from human SCNT blastocysts. Science 2005;308:1777–83.

[42] Wade N. Clone scientist relied on peers and Korean pride. The New York Times; December 25, 2005.

[43] Berman JJ. Machiavelli's Laboratory. Amazon Digital Services Inc.; 2010.

[44] National Academies of Sciences, Engineering, and Medicine. Innovations in federal statistics: combining data sources while protecting privacy. Washington, DC: The National Academies Press; 2017.

[45] Broad WJ. U.S. accidentally releases list of nuclear sites. The New York Times; June 3, 2009.

[46] Schneier B. Applied cryptography: protocols, algorithms and source code in C. New York: Wiley; 1994.

[47] Schneier B. Secrets and lies: digital security in a networked world. 1st ed. New York: Wiley; 2004.

[48] Schneier B. A plea for simplicity: you can't secure what you don't understand. Information Security; 1999. November 19. Available from: http://www.schneier.com/essay-018.html [viewed July 1, 2015].

[49] Leveson NG. System safety engineering: back to the future. Self-published ebook; 2002. Available at: http://sunnyday.mit.edu/book2.pdf [viewed September 22, 2016].

[50] Weiss TR. Thief nabs backup data on 365,000 patients. Computerworld; 2006. January 26. Available from: http://www.computerworld.com/s/article/108101/Update_Thief_nabs_backup_data_on_365_000_patients [viewed August 21, 2012].

[51] Noumeir R, Lemay A, Lina J. Pseudonymization of radiology data for research purposes. J Digit Imaging 2007;20:284–95.

[52] Bianciardi G, Miller JD, Straat PA, Levin GV. Complexity analysis of the viking labeled release experiments. Intl J Aeronaut Space Sci 2012;13:14–26.

[53] Beaudoin J. National experts at odds over patient identifiers, Healthcare IT News; 2004. October 18. Available at: http://www.healthcareitnews.com/news/national-experts-odds-over-patient-identifiers [viewed October 23, 2017].

[54] McCann E. The patient identifier debate: will a national patient ID system ever materialize? Should it? Healthcare IT News; 2013 February 18.

[55] Dimitropoulos LL. Privacy and security solutions for interoperable health information exchange perspectives on patient matching: approaches, findings, and challenges. RTI International, Indianapolis, June 30, 2009.

[56] Dalen JE. Only in America: bankruptcy due to health care costs. Am J Med 2009;122:699.

[57] Lewin T. Texas orders health clinics to turn over patient data. The New York Times; October 23, 2015.

[58] Patil N, Berno AJ, Hinds DA, Barrett WA, Doshi JM, Hacker CR, et al. Blocks of limited haplotype diversity revealed by high-resolution scanning of human chromosome 21. Science 2001;294:1719–23.

19

Legalities

Section 19.1. Responsibility for the Accuracy and Legitimacy of Data

At this very moment, there's an odds-on chance that someone in your organization is making a poor decision on the basis of information that was enormously expensive to collect.

Shvetank Shah, Andrew Horne, and Jaime Capella [1]

In 2031, lawyers will be commonly a part of most development teams.

Grady Booch

I am not a lawyer, and this chapter is not intended to provide legal advice to the readers. It is best to think of this chapter as an essay that covers the issues that responsible managers of Big Data resources worry about, all of the time. When I was a program director at the National Institutes of Health, I worked on resources that collected and analyzed medical data. My colleagues and I worked through the perceived legal risks that encumbered all of our projects. For the most part, our discussions focused on four issues: (1) responsibility for the accuracy of the contained data; (2) rights to create, use, and share the data held in the resource; (3) intellectual property encumbrances incurred from the use of standards required for data representation and data exchange; and (4) protections for individuals whose personal information is used in the resource. Big Data managers contend with a

Principles and Practice of Big Data. https://doi.org/10.1016/B978-0-12-815609-4.00019-4

wide assortment of legal issues, but these four problems, that never seem to go away, will be described in this chapter.

The contents of small data resources can be closely inspected and verified. This is not the case for Big Data. Because Big Data resources are constantly growing, and because the sources of the data are often numerous and not strictly controlled, it is a safe bet that some of the data is incorrect. The reflexive position taken by some data managers can be succinctly stated as: "It is not my problem!"

To a small extent, measures taken to improve the quality of data contained in a Big Data resource will depend on how the data will be used. Will the data be used for mission-critical endeavors? In the medical realm, will the data be used to make diagnostic or treatment decisions? These contingencies raise the stakes for Big Data resources, but the data manager's responsibility is largely the same, regardless of the intended use of the resource. Every Big Data resource must have in place a system whereby data quality is constantly checked, errors are documented, corrective actions are taken, and improvement is documented. Without a quality assurance plan, the resource puts itself in great legal jeopardy. In addition to retaining legal counsel, data managers would be wise to follow a few simple measures:

- **Make no unjustified claims.**

It is important that statements issuing from the Big Data resource, including claims made in advertisements and informational brochures, and verbal or written communications with clients, should never promise data accuracy. People who insist on accuracy should confine their attention to small data resources. If your Big Data resource has made no effort to ensure that the data is true and accurate, then you owe it to your users to indicate as much.

- **Require your sources to take necessary measures to provide accurate data.**

Sources that contribute to Big Data resources should have their own operation protocols, and these protocols must be made available to the manager of the Big Data resource. In addition, sources should certify that that their contributed data conforms, as best as they can ascertain, to their data policies.

- **Have procedures in place ensuring that the data provided by outside sources is accurately represented within the resource.**

Big Data managers should exercise reasonable diligence to ensure that the received data is legitimate, and to verify such data when it is received.

- **Warn your data users that their analytic results, based on the resource's data, must be validated against external data sources.**

It may seem obvious to you that conclusions drawn from the analyses of Big Data are always tentative, and must be validated against data from other sources. Sometimes data analysts need to be protected from their own naiveté, necessitating an explicit warning.

– **Open your verification procedures to review (preferably public review).**

Users find it unsettling to read exculpatory verbiage in user licenses, expressing that the data provider cannot guarantee the accuracy of the data and cannot be held liable for any negative consequences that might arise from the use of the data. At the very least, data managers should re-assure their users that reasonable measures have been taken to verify the data contained in the resource. Furthermore, those measures should be available for review by any and all potential data users.

– **Provide a method by which complainants can be heard.**

This may actually be one of those rare instances when the immutability of a Big Data resource is broken. If material is known to be illegal or if the material is a potential danger to individuals, then it may be necessary to expunge the data (i.e., violate data immutability).

– **Be prepared to defend your data and your procedures**

Big Data managers must understand their data. The conclusions drawn from their data may someday serve as evidence in legal proceedings, including all manner of arbitration and litigations, both civil and criminal. In the case of Daubert v Merrell Dow Pharmaceuticals, Inc., the U.S. Supreme Court ruled that trial judges must determine the relevance and adequacy of data-based evidence presented by expert witnesses. Judicial oversight is conducted through a pre-trial review that "entails a preliminary assessment of whether the reasoning or methodology underlying the testimony is scientifically valid and of whether that reasoning or methodology properly can be applied to the facts in issue" [2]. Hence, Big Data managers must constantly strive to assure that the data contained in their resources are fully described and linked to the protocols through which the data was obtained. Any verification processes, through which data is entered and checked into the resource, may be reviewed by government committees and courts.

When Big Data resources are used to influence the governmental process, special regulatory conditions may apply. The U.S. government passed the Data Quality Act in 2001, as part of the FY 2001 Consolidated Appropriations Act (Pub. L. No. 106-554) [3,4]. The Act requires Federal Agencies to base their policies on high quality data and to permit the public to challenge and correct inaccurate data [5]. The drawback to this legislation, is that science is a messy process, and data may not always attain a high quality. Data that fails to meet standards of quality may be rejected by government committees or may be seized upon by lobbyists to abrogate good policies that were based on the imperfect data [6–8]. [Glossary Data Quality Act]

Data managers chant a common lament: "I cannot be held responsible for everything!" They have a point, but their inability to control everything does not relieve them of their responsibility to exercise a high degree of data diligence.

Section 19.2. Rights to Create, Use, and Share the Resource

Free software is a matter of liberty, not price.

Richard Stallman

As mentioned earlier, ownership is a mercantile concept; the owner of an item is the person who can sell the item. If you own a cow, then you can sell the cow. Once the cow is sold, you no longer own the cow; the cow has a new owner. This simple ownership arrangement does not work well for Big Data. Data can be copied ad infinitum. In virtually all cases financial transactions that involve the transfer of data do not actually result in the loss of the data by the provider. The data provider continues to hold the data after the transaction has transpired. In the Big Data universe, Big Data is not "owned" in the usual sense of the word; data is intangible. This explains why the term "service" pops up so often in the information field (e.g., Internet Service Providers, Web Services, List Servers). Data is more often a service than an owned commodity. [Glossary Web service]

Because Big Data comes from many sources, different uses, and can be retrieved via federated queries across multiple resources (Big and small), the customary laws pertaining to property rights can be difficult to apply. Big Data managers need to know whether they have the right to acquire and distribute the data held in their resources. It may be easiest to think in terms of two separable issues: laws dealing with data acquisition, and laws dealing with data distribution.

Information produced through a creative effort (e.g., books, newspapers, journal articles) usually falls under copyright law. This means that you cannot freely obtain and distribute these materials. Exceptions would include books that fall into the public domain (e.g., books produced by the federal government, and books whose copyright term has expired). Other exceptions might include copyrighted material that fall under Fair Use provisions [9]. Fair Use provisions permit the distribution of copyrighted material if it is done solely for the public good, with no profit motive, and if it can be done in a way that does not financially harm the copyright holder (e.g., does not result in the loss of sales and royalties).

Most Big Data resources are primarily composed of raw data, along with annotations to the data. The data may consist of measurements of physical objects and events, and short informational attributes appended to abstract data objects. These types of data are generally not produced through a creative effort, and would not fall under copyright law. In the United States, the most cited precedent relevant to data acquisition is Feist Publishing, Inc. v. Rural Telephone Service Co. When Rural Telephone Co. refused to license their alphabetized listing of names and telephone numbers to Feist Publishing, Inc., Feist proceeded to copy and use the data. Rural Telephone Co. claimed copyright infringement. The court ruled that merely collecting data into a list does not constitute a creative work and was not protected by copyright.

European courts differ somewhat from American courts with regard to copyright protections. Like their American counterparts, Europeans interpret copyright to cover creative works, not data collections. However, the 1996 European Database Directive instructs courts to extend sui generis (i.e., one of a kind or exceptional) protection to databases. In Europe, databases created with a significant investment of time, effort and money cannot be freely copied for commercial use. The idea behind such a directive is to protect the investments made by database builders. By protecting the database owner the European law attempts to promote the creation of new Big Data resources along with the commercial activities that follow.

Insofar as Big Data resources have international audiences, differences in database laws across different countries can be very frustrating for data managers who strive for legal clarity. Consequently, providers and users often develop their own solutions, as needed. Acquisition of commercial data (i.e., data that does not belong to the public domain), much like access to commercial software, is often achieved through legal agreements (e.g., licenses or contracts) between the data providers and the data users.

Regarding laws dealing with holding and distributing data, the Digital Millennium Copyright Act of 1998 (DMCA) applies in the United States. This law deals primarily with anti-piracy security measures built into commercial digital products [10]. The law also contains a section (Title II) dealing with the obligations of online service providers who inadvertently distribute copyrighted material. Service providers may be protected from copyright infringement liability if they block access to the copyrighted material when the copyright holder or the holder's agent claims infringement. To qualify for liability protection, service providers must comply with various guidelines (i.e., the so-called safe harbor guidelines) included in the Act. In most instances, compliant service providers would also be protected from infringement claims when their sites link to other sites that contain infringing materials. [Glossary DMCA]

Whereas the DMCA provides some liability relief for inadvertent copyright infringers, the United States No Electronic Theft Act of 1997 (NET Act) makes possible the criminal prosecution of infringers who distribute copyrighted material for non-commercial purposes (i.e., for free) [11]. In the early days of the Internet, there was a commonly held, but unfounded, belief that copyrighted material could be held and distributed without fear of legal retribution, if no profit was involved. This belief, perhaps based on an overly liberal interpretation of the Fair Use provisions, came to an end with the NET Act.

Without delving into legal minutiae, here are a few general suggestions for data managers:

1. Require your sources to substantiate their claim that the data is theirs to contribute. Nobody should be submitting data that they do not own or that they do not have the right to distribute.
2. Require your sources to indicate that the data was collected in a manner that did not harm individuals and that the data can be distributed without harming individuals.
3. Use government data whenever feasible. Much of the best data available to Big Data resources comes absolutely free from the U.S. government and other governments that have a policy of contributing their official data to the public domain. Big Data resources

can freely copy and redistribute public domain government data. Links to the major sources of prepared U.S. government data are found at: http://www.data.gov/.

In addition, virtually all data collected by the government, including data collected through federal grants, and data used to determine public actions, policies, or regulations, can be requested through the Freedom of Information Act [12]. Many countries provide their citizens with the right to acquire data that was generated with government (i.e., taxpayer) funds.

4. Pay for legitimate data when feasible. It seldom makes good sense to copy a data set into a Big Data resource, if that data requires constant updating and curation. For example, a comprehensive list of restaurants, with their addresses and phone numbers, is always a work in progress. Restaurants open, close, move their locations, acquire new phones numbers, revise their menus, and modify their hours of operation. If there is a database resource that collects and updates this information, there may be little reason to replicate these activities within another data resource. It may make much more sense to license the database or to license access to the database. A federated data service, wherein queries to your Big Data resource are automatically outsourced to other databases, depending on the query subject, may be much more feasible than expanding your resource to include every type of information. In many circumstances the best and the safest method of using and distributing data may come from negotiating payments for external data.

Section 19.3. Copyright and Patent Infringements Incurred by Using Standards

She was incapable of saying please, incapable of saying thank you and incapable of saying sorry, all the while creating a surge in the demand for these expressions.
Edward St. Aubyn, in his book, "At Last"

As described in Chapter 7, the standards that you have been using in your Big Data resource may actually belong to somebody else. Strange as it may seem, standards are intellectual property and can be copyrighted, patented, or licensed. Not only may a standard be patented, but specific uses of the standard may also be patented, and the patents on uses of the copyright may be held by entities who were not at all involved in the creation of the standard.

If you choose to pay a license fee for the use of a proprietary standard, you might find that the costs exceed the sticker price [13]. The license agreement for the standard may impose unwanted restrictions on the use of the standard. For example, a standard may be distributed under a license that prohibits you from freely distributing the intellectual product of the standard (i.e., materials created through the use of the standard). This may mean that your users will not be able to extract and download data that has been formatted in conformance with the standard, or annotated with codes, numbers, terms or other information that could not have been created without the use of the standard. The same restrictions might apply to licensed software.

The building blocks of Big Data resources may hide intellectual property [14,13]. This is particularly true for software, which may inadvertently contain subroutines or lines of code that fall under a claim within an issued patent. One day, you might receive a letter from a lawyer who represents a patent holder, asserting that a fragment of code included in a piece of your software infringes his client's patent. The letter may assert the patent and demand that you cease using the patent holder's intellectual property. More commonly, the letter will simply indicate that a conflict has arisen and will suggest that both parties (your Big Data resource and the patent holder) should seek a negotiated remedy. In either case, most Big Data resources will keep a law firm on retainer for such occasions. Do not despair; the ultimate goal of the patent holder is to acquire royalty payments; not to initiate a lawsuit.

Big Data resources are complex and contain many different types of data objects that may have been transformed, annotated, or formatted by many different methods. The uses of these methods may be restricted under licenses, contracts and other legal contrivances. A few precautionary steps may help reduce your risks:

– Whenever possible, use free and open source standards, software, nomenclatures, and ontologies for all of your data annotations. Do not disparage free and open source products. In the world of Big Data, many of the best standards, data formats, nomenclatures, classifications, software, and programming languages are free and open source [15].
– Inventory your standards, software, nomenclatures, and ontologies. For each item, write a description of any restrictions that might apply to your resource.
– Investigate on the Web. See if there are any legal actions, active or settled, involving any of the materials you might use. Visit the U.S. Patent Office to determine whether there are patent claims on the uses of the standards, software, nomenclatures and ontologies held in your resource. Most likely, your Big Data resource will send and receive data beyond the U.S. Consult the World Intellectual Property Organization (WIPO). Do not restrict your search to proprietary materials. Free and open source materials may contain embedded intellectual property and other encumbrances.
– Talk to your legal staff before you commit to using any proprietary product. Your law firm will need to be involved in virtually every aspect of the design and operation of your Big Data resource.
– If you must use licensed materials, carefully read the "Terms of Use" in the agreement. Licenses are written by lawyers who are paid to represent their client (the Licensor). In most cases, the lawyer will be unaware of the special use requirements of Big Data resources. The Terms of Use may preclude the customary activities of a Big Data resource (e.g., sharing data across networks, responding to large numbers of queries with annotated data, storing data on multiple servers in widely distributed geographic locations). As noted previously, it is important to have a lawyer review license

agreements before they are signed, but the data manager is in the best position to anticipate provisions that might reduce the value of a Big Data resource.

Big Data would greatly benefit from a universal framework supporting resource interoperability [16]. At present, every data manager must fend for herself.

Section 19.4. Protections for Individuals

Everything is gone;
Your life's work has been destroyed.
Squeeze trigger (yes/no)?

Computer-inspired haiku by David Carlson

Data managers must be familiar with the concept of tort. Tort relates to acts that result in harm. Tort does not require an illegal act; it only requires a harm and a person or entity who contributes to the harm and who is liable for the damages. Tort works like this; if you are held liable for harm to another entity, then you must compensate the victim to an extent that makes the victim whole (i.e., brings the victim back to where he was before suffering harm). If the victim makes a case that the harm resulted from negligence or due to conditions that could have been corrected through customary caution, then punitive fees can be added to the victim's award. The punitive fees can greatly exceed the restorative fees. Consequently, it behooves every data manager to constantly ask themselves whether their Big Data resource can result in harm to individuals (i.e., the users of the data, or the subjects of the data). Needless to say, Big Data managers must seek specialized legal advice to minimize tort-related risks.

In the Big Data universe, tort often involves the harms that befall individuals when their confidential data files have been breached. I was raised in Baltimore, not far from the community of Catonsville. Catonsville was the site of a 1968 protest against United States involvement the Vietnam War. Nine anti-war activists stormed into a draft office, stole files, and publicly burned the files. The Catonsville 9 attained instant international notoriety. The number of files destroyed: 379. In the Big Data era the ante has been upped by many orders of magnitude. Today, when records are stolen or destroyed, you can expect the numbers to be in the millions, or even hundreds of millions [17].

In May, 2006, 26.5 million records on military veterans were stolen, including Social Security numbers and birth dates. The records had been taken home by a data analyst employed by the Department of Veterans Affairs. His laptop, containing all this information, was stolen. A class action lawsuit was brought on behalf of the 26.5 million aggrieved veterans. Three years later, the Department of Veterans Affairs paid $20 million to settle the matter [18]. In the United Kingdom, a copy of medical and banking records on 25 million Britons were lost in the mail [19]. The error led to the sudden resignation of the chairman of Her Majesty's Revenue and Customs [19].

There are occasions when security is broken, but no theft occurs. In these instances, resource managers may be unaware of the privacy breach for a surprisingly long period of time. Medical data collected on about 20,000 patients was posted on a public Web site in 2010. The data included patient names, diagnosis codes, and administrative information on admissions and discharges occurring in a six month period in 2009. The data stayed posted on the public Web site for about a year before a patient happened to see the data and reported the breach to the hospital [20]. Accidental breaches are common in many different fields [21].

Today, healthcare organizations must report data breaches that affect more than 500 people. Hundreds of such breaches have been reported. These breaches cost the health-care industry in excess of $6 billion annually, and the costs are increasing, not decreasing [17]. Other industries have data breaches but are not required to report incidents.

Industry costs do not reflect the personal costs in time, emotional distress, and money suffered by individuals coping with identity theft. In the Big Data field, everyone's deepest fear is identity theft. None of us wants to contemplate what may happen when another person has access to their financial accounts or gains the opportunity to create new accounts under the stolen identity.

Security issues are inseparable from issues related to privacy and confidentiality. We have dealt with some of the more technical issues of data security in Section 18.3, "Data Security and Cryptographic Protocols". In this chapter, we can review a few of the commonsense measures that will reduce the likelihood of identification theft.

1. Do not collect or provide information that will link an individual to his or her data record unless you really need the information. If you do not have information that links a record to a named individual, then you cannot inadvertently expose the information. Names, social security numbers, credit card numbers, and birth dates constitute the core information sought by identity thieves. Big Data resources should seriously consider whether such information needs to be stored within the resource. Does your resource really need to collect social security numbers and credit card numbers? Can the person's name be adequately replaced with an internal identifier? Do you need a birth date when a birth year might suffice? When these data items are necessary, do they need to be included in data records that are accessible to employees?
2. Work with deidentified records whenever possible. Deidentification may not be a perfect way to render records harmless; but it takes you very close to your goal. A thoughtfully deidentified data set has quite limited value to identity thieves.
3. All files should be encrypted whenever possible. Most breaches involve the theft of unencrypted records. Breaking an encrypted record is quite difficult and far beyond the technical expertise of most thieves.
4. Back-up data should be encrypted, inventoried, and closely monitored. Back-up data is a vulnerability. Thieves would be just as happy to steal your back-up data as your original data. Because theft of back-up data does not result in a system crash, such thefts can go undetected. It is very important to secure your back-up data and to deploy a system that monitors when back-up data is removed, copied, misplaced, destroyed, or otherwise modified.

Section 19.5. Consent

MRECs [Medical Research Ethics Committees] sometimes place extreme demands on researchers. These demands have included gaining consent for each step of the research and ensuring data are destroyed on completion of a project...

Louise Corti, Annette Day, and Gill Backhouse [22]

For data managers who deal with medical data, or with any data whose use puts human subjects at risk, consent issues will loom as a dominant legal issue. The reason why consent is a consuming issue for data managers has very little to do with its risks; the risks associated with obtaining improper consent are very small. Consent issues are important because consenting data can be incredibly expensive to implement. The consent process can easily consume the major portion of the data manager's time, and cost-effective implementations are difficult to achieve.

In the context of Big Data, informed consent occurs when a human agrees to accept the risk of harm resulting from the collection and use of their personal data. In principle, every consent transaction is simple. Someone involved with the Big Data resource approaches a person and indicates the data that he would like to collect for the data project. He indicates the potential harms that may occur if consent is granted. If relevant, he indicates the measures that will be taken to minimize the risk of harm. The human subject either signs, or does not sign, the consent form. If the subject signs the form, then his data can be included in the Big Data resource. [Glossary Informed consent, Bayh-Dole Act]

It is important that data managers understand the purpose of the consent form, so that it is not confused with other types of legal agreements between data owners and data contributors. The consent form is exclusively devoted to issues of risk to human subjects. It should not be confused with a commercial agreement (i.e., financial incentives for data use), or with an intellectual property agreement (i.e., specifying who controls the uses of the data); or with scientific descriptions of the project (i.e., determining how the data is to be used and for which specific purposes).

The term "informed consent" is often misinterpreted to mean that the patient must be fully informed of the details of the Big Data project with an exhaustive list of all the possible uses of their personal data. Not so. The "informed" in "informed consent" refers to knowledge of the risks involved in the study, not the details of the study itself. It is reasonable to stipulate that the data in Big Data resources is held permanently, and can be used by many individuals, for a wide variety of purposes that cannot be predetermined. Filling the consent form with detailed information about the uses of the resource is counterproductive, if it distracts from the primary purpose of the form; to explain the risks.

What are the risks to human subjects in a Big Data project? With few exceptions, Big Data risks are confined to two related consequences: loss of confidentiality and loss of privacy.

The concepts of confidentiality and of privacy are often confused, and it is useful to clarify their separate meanings. Confidentiality is the process of keeping a person's secret.

Privacy is the process of ensuring that the person will not be annoyed, betrayed, or harmed as a result of his decision to give you his secret. For example, if you give me your unlisted telephone number in confidence, then I am expected to protect this confidentiality by never revealing the number to other persons. I may also be expected to protect your privacy by never using the telephone number to call you unnecessarily, at all hours of the day and night (i.e., annoying you with your private information). In this case the same information object (i.e., your unlisted telephone number) is encumbered by confidentiality (i.e., keeping the unlisted number secret) and privacy (i.e., not using the unlisted number to annoy you).

To cover confidentiality risks the consent form could indicate that personal information will be collected, but that measures will be taken to ensure that the data will not be linked to your name. In many circumstances, that may be all that is needed. Few patients really care if anyone discovers that their gall bladder was removed in 1995. When the personal information is of a highly sensitive nature, the consent form may elaborate on the security measures that ensure confidentiality.

The risk of losing privacy is a somewhat more subtle risk than the loss of confidentiality. In practical terms, for Big Data projects, loss of privacy occurs when the members of the Big Data resource come back to the human subject with a request for additional information, or with information regarding the results of the study. The consent form should indicate any constraints that the Big Data resource has put into place to ensure that subjects are not annoyed with unwelcome future contacts by members of the project. In some cases the Big Data project will anticipate the need to recontact human subjects (i.e., to invade their privacy). In this case the consent form must contain language informing the subjects that privacy will not be fully protected. In many cases subjects do not particularly care, one way or the other. They are happy to participate in projects that will benefit society, and they do not mind answering a phone call at some future time. The problem for the Big Data resource will come if and when subjects have a change of heart, and they decide to withdraw consent.

Obtaining consent from human subjects carries its own administrative and computational challenges; many of which are unanticipated by Big Data managers. Consent-related tasks include the following:

1. Creating a legally valid consent form.

There are many ways to write a bad consent form. The most common mistake is inserting consent clauses among the fine-print verbiage of broader legal documents (e.g., contracts, agreements, licenses). This is a bad mistake for several reasons. The validity of informed consent can be challenged if an individual can claim that he or she was not adequately informed. The consent form should be devoted to a single topic, consent, and should not be inserted into other legal forms that require the subject's signature.

The consent form should be written in language that the average person can understand. In many cases, particularly in medical settings, informed consent should be read aloud by an individual who is capable of explaining difficult passages in the consent document.

Consent forms should not contain exculpatory clauses. For example, the consent form should not contain language expressing that the Big Data resource cannot be held liable for harm resulting from the use of the consenter's data. Neither should the form ask signers to waive any of their normal rights.

The consent form should have a signature section, indicating an affirmative consent. Certain types of informed consent may require the signature of a witness, and consent protocols should have provisions for surrogate signatures (e.g., of a parent or legal guardian). It is common for consent forms to provide an opportunity for subjects to respond in the negative (i.e., to sign a statement indicating that consent is denied). Doing so is seldom a good idea, for several reasons. First, the negative (non-affirmative) statement is not legally required and there are no circumstances for which a non-affirmative statement has any practical value. Secondly, individuals should not feel compelled to respond in any way to the consent form. If they freely choose to give consent, they can sign the form. If they do not wish to give consent, they should not be coerced to sign their names to a statement of denial. Thirdly, a non-affirmative statement can produce great confusion in the future, when an individual consents to having the same record used for another research project, or when the individual has a change of heart, and decides to provide consent for the same project.

The consent form should reveal circumstances that might influence a person's decision to provide consent. For example, if the investigators have a commercial interest in the outcome of the study, then that information should be included in the consent form. It is reasonable for individuals to fear that they might suffer harm if the investigators have something to gain by a particular outcome of an experiment or analysis.

Traditionally, consent is not open-ended. Consent generally applies to a particular project that is conducted over a specified period of time. Consent ends when the project ends. There has been a trend to lengthen the window of time to which consent applies, to accommodate projects that might reasonably be expected to extend over many years. For example, the Framingham study on heart disease has been in progress for more than 60 years [23]. If the Big Data project intends to use consented data for an indefinite period, as it almost always does, then the consent form must clarify this condition.

Most importantly, the consent form should carefully describe the risks of participation. In the case of Big Data analyses, the risks are typically confined to loss of confidentiality or loss of privacy.

2. Obtaining informed consent.

The U.S. Census is an established project that occurs every decade. The methods and the goals of the census have been developed over many decades. About 600,000 census workers are involved; their jobs are to obtain signed census forms from about 250 million individuals. The cost of each census is about $14 billion. Keeping these numbers in your mind, imagine that you are a Big Data manager. You maintain and operate a global Big Data resource, with data on over 2 billion individuals (8 times the population of the United States). You are informed by your supervisor that a new project for the resource will require you to obtain informed consent on the resource's catchment population. You are told that you will be assigned ten additional part-time workers to help you. You are given a budget of $100,000 for the project. When you complain that you need more help and a larger budget, you are told that you should use the computational power of the Big Data resource to facilitate the effort. You start looking for another job.

There are no easy ways to obtain informed consent. Popular marketing techniques that use automated or passive affirmations cannot be used to obtain informed consent. For example, opt out forms in which human subjects must take an action to be excluded from participating in a potentially harmful data collection effort are unacceptable. Informed consent must be affirmative. Forms should not be promissory (i.e., should not promise a reward for participation). Informed consent must be voluntary and uncompensated.

Consent must be obtained without coercion. Individuals cannot be denied customary treatment or access to goods and services if they refuse to grant consent. There are circumstances for which the choice of person who seeks informed consent may be considered coercive. A patient might feel threatened by a surgeon who waves a research-related consent form in their face minutes before a scheduled procedure. Big Data managers must be careful to obtain consent without intimidation.

The consent form must be signed if it is to have any legal value. This means that a Web page submission is unacceptable unless it can be reasonably determined that the person providing the consent is the same person who is listed in the submitted Web page. This would usually necessitate an authenticated password, at minimum. Issues of identity theft, password insecurity, and the general difficulty of managing electronic signatures make Web-based consent a difficult process.

The process of obtaining consent has never been easy. It cannot be fully automated because there will always be people whose contact information (e.g., email accounts) are invalid or who ignore all attempts at contact. To this date, nobody has found an inexpensive or labor-free method for obtaining informed consent from large numbers of individuals.

3. Preserving consent.

After consent has been obtained, it must be preserved. This means that the original paper document or a well-authenticated electronic document, with a verified signature, must be preserved. The consent form must be linked to the particular record for which it applies and to the protocol or protocols for which the consent applies. An individual may sign many different consent forms, for different data uses. The data manager must keep all of these forms safe and organized. If these documents are lost or stolen, then the entire resource can be jeopardized.

4. Ensuring that the consent status is kept confidential.

The consent forms themselves are potential sources of harm to patients. They contain information related to special studies or experiments or subsets of the population that include the individual. The consent form also contains the individual's name. If an unauthorized person comes into possession of consent forms, then the confidentiality of the individuality would be lost.

5. Determining whether biases are introduced by the consent process.

After all the consents have been collected, someone must determine whether the consented population introduces bias. The data analyst would ask: "Is the group of people who provide consent in any way different from the group of people who refuse to provide consent?" and, if so, "Will differences between the consenters and the non-consenters bias analytic outcomes?" A data analyst might look for specific differences among the consented and unconsented group in features that are relevant to the question under study. For example, for a medical disease study, are there differences in the incidence of the disease between the consenting group and the non-consenting group? Are there differences in the ages at which the disease occurs in consenters and non-consenters?

6. Creating a process whereby reversals and modifications of consent can be recorded and flagged.

In most cases, consent can be retracted. Retraction is particularly important in long or indefinite studies. The data manager must have a way of tracking consents and documenting a new consent status. For any future use of the data, occurring after the consent status has changed, the subject's data records must not be available to the data analyst.

7. Maintaining records of consent actions.

Tracking consent data is extremely difficult. Here are a few consent-related activities that Big Data managers must record and curate: "Does each consent form have an identifier?" "Does each consent form link to a document that describes the process by which the consent form was approved?" "If paper consent forms were used, can the data manager find and produce the physical consent document?", "Was the consent restricted, permitting certain uses of the data and forbidding other types of data uses?" "Is each consent restriction tagged for tracking?", "If the consent form was signed, is there a protocol in place by which the signature is checked to determine authenticity?", "Does the data manager have a recorded policy that covers situations wherein subjects cannot provide an informed consent (e.g., infants, patients with dementia)?", "Does the resource have protocols for using surrogate signatures for children and subjects who have guardians or assignees with power-of-attorney?", "Does the Big Data resource have policies that exclude classes of individuals from providing informed consent?", "Is there a protocol to deal with subjects who withdraw consent or modify their original consent?" "Does the resource track data related to consent withdrawals and modifications?"

8. Educating staff on the liberties and limitations of consented research.

Many Big Data managers neglect to train their staff on legal matters, including consent-related issues. Information technologists may erect strong mental barriers to exclude the kinds of legal issues that obfuscate the field of data law. Data managers have no choice but to persevere. It is unlikely that factors such as staff indifference and workplace incompetence will serve as mitigating factors when tort claims are adjudicated.

Section 19.6. Unconsented Data

> *The main point in our favor is that there is little or no case law, at least in the UK, which has unearthed any complaints by research participants about misuse of their contributions.*
>
> Louise Corti, Annette Day, and Gill Backhouse [22]

There are enormous technical difficulties and legal perils in the consent process. Is there some way of avoiding the whole mess?

I have worked for decades in an information-centric culture that has elevated the consent process to an ethical imperative. It is commonly held that the consent process protects individuals from harm, and data managers from liability. In the opinion of many of my colleagues, all confidential data on individuals should be consented into the database, unless there is a very good reason to the contrary.

After many years of dealing with the consent issue, I have reached a very different conclusion. To my way of thinking, consent should be avoided, if feasible; it should only be used as a last resort. In most circumstances, it is far preferable for all concerned to simply render data records harmless, and to use them without obtaining consent. As the dependence on consent has grown over the past few decades, several new issues, all having deleterious societal effects, have arisen:

1. Consent can be an unmerited revenue source for data managers.

When consent must be obtained on thousands or millions of individuals, the consenting costs can actually exceed the costs of preparing and using the data. When these costs are passed on to investors, or to taxpayers (in the case of public Big Data resources), it raises the perceived importance and the general cash flow for the resource. Though data managers are earnest and humble, as a rule, there are some managers who feel comfortable working on projects of dubious scientific value, and a low likelihood of success, if there is ample funding. Tasks related to the consent process cost money, without materially contributing to the research output. Because funding institutions must support consenting efforts, grant writers for Big Data projects can request and receive obscenely large awards, when consent is required.

2. The act of obtaining consent is itself a confidentiality risk.

The moment you ask for consent, you're creating a new security weakness, because the consent form contains sensitive information about the subject and the research project. The consent form must be stored, and retrieved as needed. As more and more people have access to copies of the consent forms, the risk of a confidentiality breach increases.

An irony of Big Data research is that the potential harm associated with soliciting consent may easily exceed the potential harms of participating as a subject in a Big Data project.

3. Consent issues may preoccupy data managers, diverting attention from other responsibilities.

There is a limit to the number of problems anyone can worry about. If half of your research effort is devoted to obtaining, storing, flagging, and retrieving consent forms, then you are less likely to pay attention to other aspects of the project. One of the chief lessons of this book is that, at the current time, most of our Big Data resources teeter on the brink of failure. The consent process can easily push a resource over the brink.

4. Consented research has been used for unintended purposes.

Once you have received permission to use personal data in a consented study, the data remains forever. Scientists can use this data freely, for any purpose, if they deidentify the data or if the original consent form indicates that the data might be used for future unspecified purposes. The latter option fueled the Havasupai lawsuit, to be discussed in the final section of this chapter.

As it happens, consent can be avoided altogether if the data in the resource has been rendered harmless through deidentification. Let's remember that the purpose of the consent form is to provide individuals with the choice to decline the risks associated with the use of their data in the Big Data resource. If there are no risks, there is no need to obtain consent. Data managers taking the unconsented path to data use need to ask themselves the following question. "Can I devise a way by which the data can be used, without risk to the individual?"

Exceptions exist. Regulations that restrict the use of data for designated groups of individuals may apply, even when no risk of harm is ascertained. Data confidentiality and privacy concerns are among the most difficult issues facing Big Data resources. Obtaining the advice of legal counsel is always wise.

The widespread use and public distribution of fully deidentified data records is a sort of holy grail for data miners. Medical records, financial transactions, collections of private electronic communications conducted over air and wire all contribute to the dark matter of the information universe. Everyone knows that this hidden data exists (we each contribute to these data collections), that this hidden data is much bigger than the data that we actually see, and that this data is the basic glue that binds the information universe. Nonetheless, most of the data created for the information universe is considered private. Private data is controlled by a small number of corporations who guard their data against prying eyes, while they use the data, to the extent allowed by law, to suit their own agendas. Why isn't Big Data routinely deidentified using methods discussed earlier, as discussed in Sections 3.6 and 3.7, and distributed for public review and analysis? Here are some of the reasons:

- Commercially available deidentification/scrubbing software is slow. It cannot cope with the exabytes of information being produced each year.

- None of the commercially available deidentification/scrubbing software does a perfect job. These software applications merely reduce the number of identifiers in records; they leave behind an irreducible amount of identifying information.
- Even if deidentification/scrubbing software actually were to perform as claimed, removing every identifier and every byte of unwanted data from electronic records, some records might be identified through the use of external database resources that establish identities through non-identifying details contained in records.
- Big Data managers are highly risk averse and would rather hoard their data than face the risk, no matter how unlikely, of a possible tort suit from an aggrieved individual.
- Big Data managers are comfortable with restricted data sharing, through legal instruments such as Data Use Agreements. Through such agreements, selected sets of data extracted from a Big Data resource are provided to one or a few entities who use the data for their own projects and who do not distribute the data to other entities. [Glossary Data sharing]
- Data deidentification methods, like many of the useful methods in the information field, can be patented. Some of the methods for deidentification have fallen under patent restriction, or have been incorporated into commercial software that is not freely available to data managers [24]. For some data managers, royalty and license costs are additional reasons for abandoning the deidentification process.
- Big Data managers are not fully convinced that deidentification is possible, even under ideal circumstances.

It may seem impossible, but information that is not considered identifying may actually be used to discover the name of the person linked to deidentified records. Basically, deidentification is easy to break when deidentified data can be linked to a name in an identified database containing fields that are included in both databases. This is the common trick underlying virtually every method designed to associate a name with a deidentified record.

Data managers who provide deidentified data sets to the public must worry whether there is, or ever will be, an available identified database that can be used to link fields, or combinations of fields, to their deidentified data, and thus link their records to the names of individuals. This worry weighs so heavily on data managers and on legal consultants for Big Data resources that there are very few examples of publicly available deidentified databases. Everyone in the field of Big Data is afraid of the legal repercussions that will follow when the confidentiality of their data records is broken.

Section 19.7. Privacy Policies

No keyboard present
Hit F1 to continue
Zen engineering?

Computer-inspired haiku by Jim Griffith

Discussions of privacy and confidentiality seem to always focus on the tension that results when the interests of the data holders conflict with the interests of the data subjects. These issues can be intractable when each side has a legitimate claim to their own preferences (businesses need to make profit, and individuals need some level of privacy).

At some point, every Big Data manager must create a Privacy Policy, and abide by their own rules. It has been my experience that legal problems arise when companies have no privacy policy, or have a privacy policy that is not well-documented, or have a privacy policy that is closed to scrutiny, or have a fragmented privacy policy, or fail to follow their own policy. If the company is open with its policy (i.e., permits the policy to be scrutinized by the public), and willing to change the policy if it fails to adequately protect individuals from harm, then the company is not likely to encounter any major problems.

Privacy protection protocols do not need to be perfect. They do, however, need to be followed. Companies are much more likely to get into trouble for ignoring their own policies than for following an imperfect policy. For a policy to be followed, the policy must be simple. Otherwise, the employees will be incapable of leaning the policies. Unknowable policies tend to be ignored by the unknowing staff.

Every Big Data project should make the effort to produce a thoughtful set of policies to protect the confidentiality of their records and the privacy of data subjects. These policies should be studied by every member of a Big Data project, and should be modified as needed, and reviewed at regular intervals. Every modification and review should be thoroughly documented. Every breach or failure of every policy must be investigated, promptly, and the results of the investigation, including any and all actions taken, must be documented. Competent data managers will make it their priority to see that the protocols are followed and that their review process is fully documented.

If you are a Big Data manager endowed with a overactive imagination, it is possible to envision all types of unlikely scenarios in which confidentiality can be breached. Nobody is perfect, and nobody expects perfection from any human endeavor. Much of law is based on a standard of "reasonableness." Humans are not held to an unreasonable standard. As an example, the privacy law that applies to hospitals and healthcare organizations contains 390 occurrences of the word "reasonable" [25]. A reasonable approach to confidentiality and privacy is all that can be expected from a complex human endeavor.

Section 19.8. Case Study: Timely Access to Big Data

Don't accept your dog's admiration as conclusive evidence that you are wonderful.
Ann Landers

In the clinical bioinformatics world, testing laboratories must have access to detailed population data, on millions of gene variants, with which to correlate their findings [26–31]. Specifically, genetics laboratories need to know whether a gene variant is present in the normal population that has no clinical significance; or whether variants are associated with disease. The lives of patients are put at risk when we are deprived of timely and open access to data relating genetic findings to clinical phenotypes.

In 2008, a 2-year-old child had a severe seizure, and died. In the prior year, the child had undergone genetic testing. The child's doctors were concerned that the patient might have Dravet's syndrome, a seizure disorder in which about 80% of patients have a mutation in the SCN1A gene. The laboratory discovered a mutation in the child's SCN1A gene, but remarked in their report that the mutation was a variant of unknown significance. That is to say that the reference database of sequence variants, used by the laboratory, did not contain information that specifically linked the child's SCN1A mutation to Dravet syndrome. In this circumstance, the laboratory report indicated that the gene test was "inconclusive"; they could neither rule in or rule out the possibility that the found mutation was diagnostic of Dravet syndrome.

Some time later, the child died.

In a wrongful death lawsuit filed by the child's mother, the complaint was made that two published reports, appearing in 2006 and 2007, had linked the specific SCN1A gene mutation, that was subsequently found in her child's DNA, with an epileptic encephalopathy [32]. According to the mother, the reporting laboratory should have known the significance of her child's mutation [32,33]. Regardless of the verdict rendered at this trial, the circumstances serve as fair warning. In the era of Big Data, testing laboratories need access to the most current data available, including the data generated by competing laboratories.

Section 19.9. Case Study: The Havasupai Story

Freeing yourself was one thing; claiming ownership of that freed self was another.
                                                                    Toni Morrison

For those who seek consent for research, the case of the Havasupai Tribe v. Arizona Board of Regents holds us in thrall. The facts of the case play out over a 21-year period, from 1989 to 2010. In 1989 Arizona University obtained genetic samples from several hundred members of the Havasupai Tribe, a community with a high prevalence of Type II diabetes. In addition to their use in diabetes research, the informed consent indicated the samples might be used for research on "behavioral and medical disorders," not otherwise specified. The researchers tried but failed to make headway linking genes sampled from the Havasupai tribe with cases of diabetes. The gene samples were subsequently used for ancillary studies that included schizophrenia and for studies on the demographic trends among the Havasupai. These ancillary studies were performed without the knowledge of the Havasupai. In 2003 a member of the Havasupai tribe happened to attend a lecture, at Arizona State University, on the various studies performed with the Havasupai DNA samples.

The Havasupai tribe was enraged. They were opposed to the use of their DNA samples for studies of schizophrenia or for the studies of demographic trends. In their opinions, these studies did not benefit the Havasupai and touched upon questions that were considered embarrassing and taboo, including the topic of consanguineous matings, and the prevalence rates of mental illnesses within the tribe.

In 2004, the Havasupai Tribe filed a lawsuit indicating lapses in the informed consent process, violation of civil rights, violation of confidentiality, and unapproved use of the samples. The case was dismissed on procedural grounds, but was reinstated by the Arizona Court of Appeals, in 2008 [34].

Reinstatement of the case led to lengthy and costly legal maneuvers. Eventually, the case was settled out of court. Arizona State University agreed to pay individuals in the Havasupai tribe a total of $700,000. This award is considerably less than the legal costs already incurred by the University. Arizona State University also agreed to return the disputed DNA samples to the Havasupai tribe.

If the Havasupai tribe had won anything in this dispute, it must have been a Pyrrhic victory. Because the case was settled out of court, no legal decision was rendered, and no clarifying precedent was established.

Though I am not qualified to comment on the legal fine-points, several of the principles related to the acquisition and use of data are relevant and can be discussed as topics of general interest.

First, the purpose of an informed consent document is to list the harm that might befall the individual who gives consent, as a consequence of his or her participation as a human subject. Consent relates only to harm; consent does not relate to approval for research. Laypersons should not be put into a situation wherein they must judge the value of research goals. By signing consent, the signator indicates that he or she is aware of the potential harm from the research, and agrees to accept the risk. In the case of samples or data records contributed to a Big Data resource, consenters must be warned, in writing, that the data will be used for purposes that cannot be specified in the consent form.

Secondly, most consent is obtained to achieve one primary purpose, and this purpose is customarily described briefly in the consent form. The person who consents often wants to know that the risks that he or she is accepting will be compensated by some potential benefit to society. In the case of the Havasupai Tribe v. Arizona State University, the tribe sought to exert control over how their DNA would be used [35]. It would seem that the Havasupai Tribe members believed that their DNA should be used exclusively for scientific efforts that would benefit the tribe. There is no ethical requirement that binds scientists to conduct their research for the sole benefit of one group of individuals. A good consent form will clearly state that research conducted cannot be expected to be of any direct value to the consenter.

Finally, the consent form should include all of the potential harms that might befall the consenter as a consequence of his or her participation. It may be impossible to anticipate every possible adverse consequence to a research participant. In this case, the scientists at Arizona State University did not anticipate that the members of the Havasuapai Tribe would be harmed if their gene data was used for ancillary research purposes. I would expect that the researchers at Arizona State University do not believe that their research produced any real harm. The Havasupai tribal members believe otherwise. It would seem that the Havasupai believed that their DNA samples were abused, and that their trust had been violated.

Had the original consent form listed all of the potential harms, as perceived by the Havasupai, then the incident could have been avoided. The Havasupai could have reached an informed decision weighing the potential benefits of diabetes research against the uncertain consequences of using their DNA samples for future research projects that might be considered taboo.

Why had the Havasupai signed their consent forms? Had any members of the Havasupai tribe voiced concerns over the unspecified medical and behavioral disorders mentioned in the consent form, then the incident could have been avoided.

In a sense, the Havasupai v. Arizona Board of Regents lawsuit hinged on a misunderstanding. The Havasupai did not understand how scientists use information to pursue new questions. The Board of Regents did not understand the harms that occur when data is used for legitimate scientific purposes. The take home lesson for data managers is the following: to the extent humanly possible, ensure that consent documents contain a complete listing of relevant adverse consequences. In some cases, this may involve writing the consent form with the assistance of members of the group whose consent is sought.

Glossary

Bayh-Dole Act The Patent and Trademark Amendments of 1980, P.L. 96-517. Adopted in 1980, the U.S. Bayh-Dole legislation and subsequent extensions gave universities and corporations the right to keep and control any intellectual property (including data sets) developed under federal grants. The Bayh-Dole Act has provided entrepreneurial opportunities for researchers who work under federal grants, but has created conflicts of interest that should be disclosed to human subjects during the informed consent process. It is within the realm of possibility that a researcher who stands to gain considerable wealth, depending on the outcome of the project, may behave recklessly or dishonestly to achieve his or her ends.

DMCA Digital Millennium Copyright Act, signed into law in 1998. This law deals with many different areas of copyright protection, most of which are only peripherally relevant to Big Data. In particular, the law focuses on copyright protections for recorded works, particularly works that have been theft-protected by the copyright holders [10]. The law also contains a section (Title II) dealing with the obligations of online service providers who inadvertently distribute copyrighted material. Service providers may be protected from copyright infringement liability if they block access to the copyrighted material when the copyright holder or the holder's agent claims infringement. To qualify for liability protection, service providers must comply with various guidelines (i.e., the so-called safe harbor guidelines) included in the Act.

Data Quality Act In the United States the data upon which public policy is based must have quality and must be available for review by the public. Simply put, public policy must be based on verifiable data. The Data Quality Act of 2002, requires the Office of Management and Budget to develop government-wide standards for data quality [3].

Data sharing Providing one's own data to another person or entity. This process may involve free or purchased data, and it may be done willingly, or under coercion, as in compliance with regulations, laws, or court orders.

Informed consent Human subjects who are put at risk must provide affirmative consent, if they are to be included in a government-sponsored study. This legally applies in the United States and most other nations, and ethically applies to any study that involves putting humans at risk. To this end, researchers provide prospective human subjects with an "informed consent" document that informs the subject of the risks of the study, and discloses foreseen financial conflicts among the researchers. The informed consent must be clear to laymen, must be revocable (i.e., subjects can change their mind and withdraw from the study, if feasible to do so), must not contain exculpatory language (e.g., no waivers of responsibility for the researchers), must not promise any benefit or monetary compensation as a reward for participation, and must not be coercive (i.e., must not suggest a negative consequence as a result of non-participation).

Web service Server-based collections of data, plus a collection of software routines operating on the data, that can be accessed by remote clients. One of the features of Web services is that they permit client users (e.g., humans or software agents) to discover the kinds of data and methods offered by the Web Service and the rules for submitting server requests. To access Web services, clients must compose their requests as messages conveyed in a language that the server is configured to accept, a so-called Web services language.

References

[1] Shah S, Horne A, Capella J. Good Data Won't Guarantee Good Decisions. Harvard Business Review; April 2012.

[2] Cranor C. Scientific Inferences in the Laboratory and the Law. Am J Public Health 2005;95:S121–8.

[3] Data Quality Act. 67 Fed. Reg. 8,452, February 22, 2002, addition to FY 2001 Consolidated Appropriations Act (Pub. L. No. 106-554 codified at 44 U.S.C. 3516).

[4] Bornstein D. The dawn of the evidence-based budget. The New York Times. May 30, 2012.

[5] Guidelines for ensuring and maximizing the quality, objectivity, utility, and integrity of information disseminated by federal agencies. Fed Regist February 22, 2002;67(36).

[6] Sass JB, Devine Jr. JP. The Center for Regulatory Effectiveness invokes the Data Quality Act to reject published studies on atrazine toxicity. Environ Health Perspect 2004;112:A18.

[7] Tozzi JJ, Kelly Jr. WG, Slaughter S. Correspondence: data quality act: response from the Center for Regulatory Effectiveness. Environ Health Perspect 2004;112:A18–9.

[8] Mooney C. Thanks to a little-known piece of legislation, scientists at the EPA and other agencies find their work questioned not only by industry, but by their own government, Interrogations, Boston Globe. August 28, 2005. http://archive.boston.com/news/globe/ideas/articles/2005/08/28/interrogations/?page=full [viewed November 7, 2017].

[9] Copyright Act, Section 107, Limitations on exclusive rights: fair use, Available from: http://www.copyright.gov/title17/92chap1.html [viewed May 18, 2017].

[10] The Digital Millennium Copyright Act of 1998 U.S. Copyright Office Summary, Available from: http://www.copyright.gov/legislation/dmca.pdf [viewed August 24, 2012].

[11] No Electronic Theft (NET) Act of 1997 (H.R. 2265). Statement of Marybeth Peters The Register of Copyrights before the Subcommittee on Courts and Intellectual Property Committee on the Judiciary. United States House of Representatives 105th Congress, 1st Session. September 11, 1997. Available from: http://www.copyright.gov/docs/2265_stat.html [viewed August 26, 2012].

[12] The Freedom of Information Act. 5 U.S.C. 552, Available from: http://www.nih.gov/icd/od/foia/5usc552.htm [viewed August 26, 2012].

[13] Gates S. Qualcomm v. Broadcom—The Federal Circuit Weighs in on "Patent Ambushes". December 5, 2008. Available from: http://www.mofo.com/qualcomm-v-broadcom—the-federal-circuit-weighs-in-on-patent-ambushes-12-05-2008 [viewed January 22, 2013].

[14] Cahr D, Kalina I. Of pacs and trolls: how the patent wars may be coming to a hospital near you. ABA Health Lawyer 2006;19:15–20.

[15] Berman JJ. Data simplification: taming information with open source tools. Waltham, MA: Morgan Kaufmann; 2016.

[16] Greenbaum D, Gerstein M. A universal legal framework as a prerequisite for database interoperability. Nat Biotechnol 2003;21:979–82.

[17] Perlroth N. Digital data on patients raises risk of breaches. The New York Times December 18, 2011.

[18] Frieden T. VA will pay $20 million to settle lawsuit over stolen laptop's data CNN. January 27, 2009

[19] Mathieson SA. UK government loses data on 25 million Britons: HMRC chairman resigns over lost CDs. ComputerWeekly.com 20 November 20, 2007.

[20] Sack K. Patient data posted online in major breach of privacy. The New York Times September 8, 2011.

[21] Broad WJ. U.S. accidentally releases list of nuclear sites. The New York Times June 3, 2009.

[22] Corti L, Day A, Backhouse G. Confidentiality and Informed Consent: Issues for Consideration in the Preservation of and Provision of Access to Qualitative Data Archives [46 paragraphs].

[23] Framingham Heart Study. NIH, U.S. National Library of Medicine. Clinical Trials.gov. Available from: http://www.clinicaltrials.gov/ct/show/NCT00005121 [viewed October 16, 2012].

[24] Berman JJ. Racing to share pathology data. Am J Clin Pathol 2004;121:169–71.

[25] Department of Health and Human Services. 45 CFR (Code of Federal Regulations), Parts 160 through 164. Standards for Privacy of Individually Identifiable Health Information (Final Rule). Fed Regist December 28 2000;65(250):82461–510.

[26] Gilissen C, Hoischen A, Brunner HG, Veltman JA. Disease gene identification strategies for exome sequencing. Eur J Hum Genet 2012;20:490–7.

[27] Bodmer W, Bonilla C. Common and rare variants in multifactorial susceptibility to common diseases. Nat Genet 2008;40:695–701.

[28] Wallis Y, Payne S, McAnulty C, Bodmer D, Sistermans E, Robertson K, Moore D, Abbs D, Deans Z, Devereau A. Practice guidelines for the evaluation of pathogenicity and the reporting of sequence variants in clinical molecular genetics, Association for Clinical Genetic Science; 2013. http://www.acgs.uk.com/media/774853/evaluation_and_reporting_of_sequence_variants_bpgs_june_2013_-_finalpdf.pdf [viewed May 26, 2017].

[29] Pritchard JK. Are rare variants responsible for susceptibility to complex diseases? Am J Hum Genet 2001;69:124–37.

[30] Pennisi E. Breakthrough of the year: human genetic variation. Science 2007;318:1842–3.

[31] MacArthur DG, Manolio TA, Dimmock DP, Rehm HL, Shendure J, Abecasis GR, et al. Guidelines for investigating causality of sequence variants in human disease. Nature 2014;508:469–76.

[32] Ray T. Mother's negligence suit against Quest's Athena could broadly impact genetic testing labs. GenomeWeb March 14, 2016.

[33] Ray T. Wrongful death suit awaits input from South Carolina supreme court. Genomeweb April 4, 2017.

[34] Appeal from the Superior Court in Maricopa County Cause No. CV2005-013190. Available from: http://www.azcourts.gov/Portals/89/opinionfiles/CV/CV070454.pdf [viewed August 21, 2012].

[35] Informed consent and the ethics of DNA research. The New York Times April 23, 2010.

20

Societal Issues

OUTLINE

Section 20.1. How Big Data Is Perceived by the Public

The greatest enemy of knowledge is not ignorance, it is the illusion of knowledge.
Stephen Hawking

Big Data, even the Big Data that we use in scientific pursuits, is a social endeavor. The future directions of Big Data will be strongly influenced by social, political, and economic forces. Will scientists archive their experimental data in publicly accessible Big Data resources? Will scientists adopt useful standards for their operational policies and their data? The answers depend on a host of issues related to funding source (e.g., private or public), cost, and perceived risks. How scientists use Big Data may provide the strongest argument for or against the public's support for Big Data resources.

The purposes of Big Data can be imagined as one of the following dramatic settings:

– **The Big Snoop (hoarding information about individuals, for investigative purposes)**

In this hypothesis, Big Data exists for private investigators, police departments, and snoopy individuals who want to screen, scrutinize, and invade the privacy of individuals, for their own purposes. There is basis in reality to support this hypothesis. Investigators, including the FBI, use Big Data resources, such as: fingerprint databases, DNA databases, legal records, air travel records, arrest and conviction records, school records, home ownership records, geneology trees, credit card transactions, financial transactions, tax records, census records, Facebook pages, tweets, emails, and sundry electronic residua. The modern private eye has profited from Big Data, as have law enforcement officers. It is unsettling that savvy individuals have used Big Data to harass, stalk, and breach the privacy of other individuals. These activities have left some individuals dreading future sanctioned or unsanctioned uses of big Data. On the up side, there is a real possibility that Big Data will serve to prevent crime, bring criminals to justice, and enhance the security of law-abiding citizens. The value of Big Data, as a method to reduce crime, has not fully engaged the public consciousness.

- **Big Brother (collecting information on individuals to control the general population)**

Modern governments obtain data from surveillance cameras and sophisticated eavesdropping techniques, and from a wide variety of information collected in the course of official operations. Much of the data collected by governments is mandated by law (e.g., census data, income tax data, birth certificates), and cannot be avoided. When a government sponsors Big Data collections, there will always be some anxiety that the Big Data resource will be used to control the public, reducing our freedoms of movement, expression, and thought. On the plus side, such population-wide studies may eventually reduce the incidence of terrorist attacks, confine the spread of epidemics and emerging diseases, increase highway safety, and improve the public welfare.

- **Borg invasion (collecting information to absorb information on a population)**

I assume that if you are reading this book on Big Data, you most likely are a Star Trek devotee, and understand fully that the Borg are a race of collectivist aliens who travel through galaxies, absorbing knowledge from civilizations encountered along the way. The conquered worlds are absorbed into the Borg "collective" while their scientific and cultural achievements are added to a Big Data resource. According to the Borg hypothesis, Big Data is the download of a civilization. Big Data analysts predict and control the activities of populations: how crowds move through an airport; when and where traffic jams are likely to occur; when political uprisings will occur; how many people will buy tickets for the next 3-D movie production. Resistance is futile.

- **Junkyard (a place to put our stuff)**

The late great comedian, George Carlin, famously chided us for wasting our time, money, and consciousness on one intractable problem: "Where do we put all our stuff?" Before the advent of Big Data, electronic information was ephemeral; created and then lost. With cloud computing, and with search engines that encompass the Web, and with depositories for our personal data, Big Data becomes an infinite storage attic. Your stuff is safe, forever, and it is available to you when you need it.

When the ancient Sumerians recorded buy-sell transactions, they used clay tablets. They used the same medium for recording novels, such as the Gilgamesh epic. These tablets have endured well over 4000 years. To this day, scholars of antiquity study and translate the Sumerian data sets. The safety, availability and permanence of electronic "cloud" data is a claim that will be tested, over time. When we are all dead and gone, will our data persist for even a fraction of the time that the Sumerian tablets have endured?

- **Scavenger hunt (searching for treasure)**

Big Data is a collection of everything, created for the purpose of searching for individual items and facts. According to the Scavenger hunt hypothesis, Big Data is everything you ever wanted to know about everything. A new class of professionals will emerge, trained to find any information their clients may need, by mining Big Data resources. It remains to be seen whether the most important things in life will ever be found in Big Data resources.

- **Egghead heaven (collecting information to draw generalized scientific conclusions)**

The National Science Foundation has issued a program solicitation entitled Core Techniques and Technologies for Advancing Big Data Science & Engineering [1]. This document encapsulates the Egghead hypothesis of Big Data.

"The Core Techniques and Technologies for Advancing Big Data Science and Engineering (BIGDATA) solicitation aims to advance the core scientific and technological means of managing, analyzing, visualizing, and extracting useful information from large, diverse, distributed, and heterogeneous data sets so as to: accelerate the progress of scientific discovery and innovation; lead to new fields of inquiry that would not otherwise be possible; encourage the development of new data analytic tools and algorithms; facilitate scalable, accessible, and sustainable data infrastructure; increase understanding of human and social processes and interactions; and promote economic growth, improved health, and quality of life. The new knowledge, tools, practices, and infrastructures produced will enable breakthrough discoveries and innovation in science, engineering, medicine, commerce, education, and national security-laying the foundations for US competitiveness for many decades to come [1]."

- **In your Facebook (a social archive that generates money)**

The underlying assumption here is that people want to use their computers to interact with other people (i.e., make friends and contacts, share thoughts, arrange social engagements, give and receive emotional support, and memorialize their lives). Some might dismiss social networks as a ruse whereby humans connect with their computers, while disconnecting themselves from committed human relationships that demand self-sacrifice and compassion. Still, a billion members cannot all be wrong, and the data collected by social networks must tell us something about what humans want, need, dislike, avoid, love, and, most importantly, buy. The Facebook hypothesis is the antithesis of the Egghead hypothesis in that participants purposefully add their most private thoughts and desires to the Big Data collection so that others will recognize them as unique individuals.

- **Much ado about nothing (Big Data does not qualify as anything new)**

According to some detractors, Big Data represents what we have always done, but with more data. This last statement, which is somewhat of an anti-hypothesis, is actually prevalent among quite a few computer scientists. They would like to think that everything they learned in the final decades of the 20th century will carry them smoothly through their careers in the 21st century. Many such computer scientists hold positions of leadership and responsibility in the realm of information management. They may be correct; time will either vindicate or condemn them.

Discourse on Big Data is hindered by the divergence of opinions on the nature of the subject. A proponent of the Nihilist hypothesis will not be interested in introspection, identifiers, semantics, or any of the Big Data issues that do not apply to traditional data sets. Proponents of the George Carlin hypothesis will not dwell on the fine points of Big Data analysis if their goal is limited to archiving files in the Big Data cloud. If you read blogs and magazine articles on Big Data, from diverse sources (e.g., science magazines, popular culture magazines, news syndicates, financial bulletins), you will find that the authors are all talking about fundamentally different subjects called by the same name; Big Data. [Glossary Semantics]

To paraphrase James Joyce, there are many sides to an issue; unfortunately, I am only able to occupy one of them. I closely follow the National Science Foundation's lead (vida supra). The focus for this chapter is the Egghead hypothesis; using Big Data to advance science.

Section 20.2. Reducing Costs and Increasing Productivity With Big Data

Every randomized clinical trial is an observational study on day two.

Ralph Horwitz

We tend to think of Big Data exclusively as an enormous source of data; for analysis and for fact-finding. Perhaps we should think of Big Data as a time-saver; something that helps us do our jobs more efficiently, and at reduced cost. It is easy to see how instant access to industry catalogs, inventory data, transaction logs, and communication records can improve the efficiency of businesses. It is less easy to see how Big Data can speed up scientific research, an endeavor customarily based on labor-intensive, and tedious experiments conducted by scientists and technicians in research laboratories. For many fields of science, the traditional approach to experimentation has reached its fiscal and temporal limits; the world lacks the money and the time to do research the old-fashioned way. Everyone is hoping for something to spark the next wave of scientific progress, and that spark may be Big Data.

Here is the problem facing scientists today. Scientific experiments have increased in scale, cost, and time, but the incremental progress resulting from each experiment is no greater today than it was fifty years ago. In the field of medicine, 50-year progress between 1910 and 1960 greatly outpaced progress between 1960 and 2010. Has society reached a state of diminishing returns on its investment in science?

By 1960, industrial science reached the level that we see today. In 1960, we had home television (1947), transistors (1948), commercial jets (1949), nuclear bombs (fission, fusion in 1952), solar cells (1954), fission reactors (1954), satellites orbiting the earth (Sputnik I, 1957), integrated circuits (1958), photocopying (1958), probes on the moon (Lunik II, 1959), practical business computers (1959), and lasers (1960). Nearly all the engineering and scientific advances that shape our world today were discovered prior to 1960.

These engineering and scientific advancements pale in comparison to the advances in medicine that occurred between 1920 and 1960. In 1921, we had insulin. Over the next four decades, we developed antibiotics effective against an enormous range of infectious diseases, including tuberculosis. Civil engineers prevented a wide range of common diseases using a clean water supply and improved waste management. Safe methods to preserve food, such as canning, refrigeration, and freezing saved countless lives. In 1941, Papanicolaou introduced the Pap smear technique to screen for precancerous cervical lesions, resulting in a 70% drop in the death rate from uterine cervical cancer, one of the leading causes of cancer deaths in women. By 1947, we had overwhelming epidemiologic evidence that cigarettes caused lung cancer. No subsequent advances in cancer research have yielded reductions in cancer death rates that are comparable to the benefits achieved with Pap smear screening and cigarette avoidance. The first polio vaccine and the invention of oral contraceptives came in 1954. By the mid 1950s, sterile surgical technique was widely practiced, bringing a precipitous drop in post-surgical and postpartum deaths. The great achievements in molecular biology, from Linus Pauling, James D. Watson, and Francis Crick, came in the 1950s.

If the rate of scientific accomplishment were dependent upon the number of scientists on the job, you would expect that progress would be accelerating, not decelerating. According to the National Science Foundation, 18,052 science and engineering doctoral degrees were awarded in the United States, in 1970. By 1997, that number had risen to 26,847, nearly a 50% increase in the annual production of the highest level scientists [2]. The growing work force of scientists failed to advance science at rates achieved in an earlier era; but not for lack of funding. In 1953, according to the National Science Foundation, the total United States expenditures on research and development was $5.16 billion, expressed in current dollar values. In 1998, that number has risen to $227 billion, greater than a 40-fold increase in research spending [2]. Most would agree that, over this same period, we have not seen a 40-fold increase in the rate of scientific progress.

Big Data provides a way to accelerate scientific progress by providing a large, permanent, and growing collection of data obtained from many different sources; thus sparing researchers the time and expense of collecting all of the data that they use, for very limited purposes, for a short span of time.

In the field of experimental medicine, Big Data provides researchers with an opportunity to bypass the expensive and time-consuming clinical trial process. With access to millions of medical records and billions of medical tests, researchers can find subpopulations of patients with a key set of clinical features that would qualify them for inclusion in narrowly focused, small trials [3]. The biological effects of drugs, and the long-term clinical outcomes, can sometimes be assessed retrospectively on medical records held in Big Data resources. The effects of drugs, at different doses, or in combination with other drugs, can be evaluated by analyzing large numbers of treated patients. Evaluations of drugs for optimal doses, and optical treatment schedule, in combination with other drugs, is something that simply cannot be answered by clinical trials (there are too many variables to control).

Perhaps the most important scientific application of Big Data will be as a validation tool for small data experiments. All experiments, including the most expensive prospective clinical trials, are human endeavors and are subject to all of the weaknesses and flaws that characterize the human behavior [4–6]. Like any human endeavor, experiments must be validated, and the validation of an experiment, if repeated in several labs, will cost more than the original study. Using Big Data, it may be feasible to confirm experimental findings based on a small, prospective studies, if the small-scale data is consistent with observations made on very large populations [7]. In some cases, confirmatory Big Data observations, though not conclusive in themselves, may enhance our ability to select the most promising experimental studies for further analysis. Moreover, in the case of drug trials, observations of potential side effects, non-responsive subpopulations, and serendipitous beneficial drug activities may be uncovered in a Big Data resource.

In the past, statisticians have criticized the use of retrospective data in drug evaluations. There are just too many biases and too many opportunities to reach valueless or misleading conclusions. Today, there is a growing feeling that we just do not have the luxury of abandoning Big Data. Using these large resources may be worth a try, if we are provided access to the best available data, and if our results are interpreted by competent analysts, and sensibly validated. Today, statisticians are finding opportunities afforded by retrospective studies for establishing causality, once considered the exclusive domain of prospective experiments [8–10]. One of the most promising areas of Big Data studies, over the next decade or longer, will be in the area of retrospective experimental design. The incentives are high. Funding agencies, and corporations should ask themselves, before financing any new and expensive research initiative, whether the study can be performed using existing data held in Big Data resources [11].

Section 20.3. Public Mistrust

Never attribute to malice that which is adequately explained by stupidity
Commonly attributed to Robert J. Hanlon, but echoed by countless others over the ages

Much of the reluctance to share data is based on mistrust. Corporations, medical centers, and other entities that collect data on individuals will argue, quite reasonably, that they have a fiduciary responsibility to the individuals whose data is held in their repositories. Sharing such data with the public would violate the privacy of their clients. Individuals agree. Few of us would choose to have our medical records, financial transactions, and the details of our personal lives examined by the public.

Recent campaigns have been launched against the "database state." One such example is NO2ID, a British campaign against ID cards and a National Identify Register. Other anti-database campaigns include TheBigOptOut.org which campaigns against involuntary participation in the United Kingdom medical record database and LeaveThemKidsAlone, protesting fingerprinting in schools.

When the identifying information that links a personal record to a named individual is removed, then the residual data becomes disembodied values and descriptors. Properly deidentified data poses little or no threat to humans, but it has great value for scientific research. The public receives the benefits of deidentified medical data every day. This data is used to monitor the incidence and the distribution of cancer, detect emerging infectious diseases, plan public health initiatives, rationally appropriate public assistance funds, manage public resources, and monitor industrial hazards. Deidentified data collected from individuals provides objective data that describes us to ourselves. Without this data, society is less safe, less healthy, less smart, and less civilized.

Those of us who value our privacy and our personal freedom have a legitimate interest in restraining Big Data. Yet, we must admit that nothing comes free in this world. Individuals who receive the benefits of Big Data, should expect to pay something back. In return for contributing private records to Big Data resources, the public should expect resources to apply the strictest privacy protocols to their data. Leaks should be monitored, and resources that leak private data should be disciplined and rehabilitated. Non-compliant resources should be closed.

There are about a billion people who have Facebook accounts wherein they describe the intimate details of their lives. This private information is hanging in the cloud, to be aggregated, analyzed and put to all manner of trivial, commercial purposes. Yet, many of these same Facebook users would not permit their deidentified medical records to be used to save lives. It would be ideal if there were no privacy or confidentiality risks associated with Big Data. Unfortunately, zero-risk is not obtainable. However, it is technically possible to reduce the imagined risks of Big Data to something far below the known risks that we take with every electronic monetary transaction, every transfer of information, every move we make in public places, every click on our keyboards, and every tap on our smartphones. Brave new world!

Section 20.4. Saving Us From Ourselves

Man needs more to be reminded than instructed.

Samuel Johnson

Ever since computers were invented, there has been a push toward developing decision-making algorithms. The idea has been that computers can calculate better and faster than humans and can process more data than humans. Given the proper data and algorithms, computers can make better decisions than humans. In some areas, this is true. Computers can beat us at chess, they can calculate missile trajectories, and they can crack encryption codes. They can do many things better and faster than humans. In general the things that computers do best are the things that humans cannot do at all.

If you look over the past half century of computer history, computers have not made much headway in the general area of decision-making. Humans continue to muddle through their days, making their own decisions. We do not appoint computers to sit in juries, doctors seldom ask a computer for their diagnostic opinions, computers do not decide which grant applications receive funding, and computers do not design our clothing. Despite billions of dollars spent on research on artificial intelligence, the field of computer-aided decision-making has fallen short of early expectations [12–15]. It seems we humans still prefer to make our own mistakes, unassisted. [Glossary Artificial intelligence, Machine learning]

Although computers play a minor role in helping us make correct decisions, they can play a crucial role in helping us avoid incorrect decisions. In the medical realm, medical errors account for about 100,000 deaths and about a million injuries each year, in the United States [16]. Can we use Big Data to avoid such errors? The same question applies to driving errors, manufacturing errors, construction errors, and any realm where human errors have awful consequences.

It really does not make much sense, at this early moment in the evolution of computational machines, to use computers to perform tasks that we humans can do very well. It makes much more sense to use computers to prevent the kinds of errors that humans commit because we lack the qualities found in computers.

Here are a few examples wherein Big Data resources may reduce human errors:

- **Identification errors.**

As discussed at length in Section 3.4, "Really Bad Identifier Methods," identification is a complex process that should involve highly trained staff, particularly during the registration process. Biometrics may help establish uniqueness (i.e., determining that an individual is not registered under another identifier) and authenticity (i.e., determining that an individual is who he claims to be). Computer evaluation of biometric data (e.g., fingerprints, iris imaging, retinal scan, signature, etc.) may serve as an added check against identification errors.

- **Data entry errors.**

Data entry error rates are exceedingly common and range from about 2% of entries up to about 30% of entries, depending on various factors including the data type (e.g., numeric or textual) and length of the entry [17,18].

As society becomes more and more complex, humans become less and less capable of avoiding errors. Errors that have been entered into a data resource can be very difficult to detect and correct. A warning from a computer may help humans avoid making some highly regrettable entry errors. Probably the simplest, but most successful, example of a computational method to find entry errors is the check-digit. The check-digit (which can actually be several digits in length) is a number that is computed from a sequence (e.g., charge card number) and appended to the end of the sequence. If the sequence is entered incorrectly, the computed check digit will be different from the check-digit that had been embedded as a part of the original (correct) sequence. The check-digit has proven to be a very effective method for reducing data entry errors for identifiers and other important short sequences; and a wide variety of check-digit algorithms are available. [Glossary Checksum]

Spell-checkers are another example of software that finds data errors at the moment of entry. In fact, there are many opportunities for data scientists to develop methods that check for inconsistencies in human-entered data. If we think of an inconsistency as anything that differs from what we would expect to see, based on past experience, than Big Data resources may be the proper repository of "consistent" values with which inconsistencies can be detected.

– **Medical errors.**

Computer systems can suspend prescriptions for which doses exceed expected values, or for which known drug interactions contraindicate use, or for which abuse is suspected (e.g., multiple orders of narcotics from multiple physicians for a single patient). With access to every patient's complete electronic medical record, computers can warn us of idiosyncratic reactions that may occur, and the limits of safe dosages, for any particular patient. With access to biometric identifying information, computers can warn us when a treatment is about to be provided to the wrong patient. As an example, in operating rooms, computers can check that the screened blood components are compatible with the screened blood of the patient; and that the decision to perform the transfusion meets standard guidelines established by the hospital.

– **Rocket launch errors.**

Computers can determine when all sensors report normally (e.g., no frozen o-rings), when all viewed subsystems appear normal (e.g., no torn heat shield tiles hanging from the hull), and when all systems are go (e.g., no abnormalities in the function of individual systems), and when the aggregate system is behaving normally (e.g., no conflicts between subsystems). Rockets are complex, and the job of monitoring every system for errors is a Big Data task.

– **Motor vehicle accidents.**

It is thrilling to know that computers may soon be driving our cars, but most motor vehicle collisions could be prevented if we would simply eliminate from our roads those human drivers who are impaired, distracted, reckless, or otherwise indisposed to obeying the legal rules of driving. With everything we know about electronic surveillance, geopositioning technology, traffic monitoring, vehicle identification, and drug testing, you might think that we would have a method to rid the roads of poor drivers and thereby reduce motor vehicle fatalities. More than 32,000 people die each year from traffic accidents in the United States, and many more individuals are permanently disabled. Big Data technology could collect, analyze, and instantly react to data collected from highways; and it is easy to see how this information could greatly reduce the rate of motor vehicle deaths. What are we waiting for?

Section 20.5. Who Is Big Data?

The horizons of physics, philosophy, and art have of late been too widely separated, and, as a consequence, the language, the methods, and the aims of any one of these studies present a certain amount of difficulty for the student of any other of them.
Hermann L. F. Helmholtz, 1885 [19]

To get the most value from Big Data resources, will we need armies of computer scientists trained with the most advanced techniques in supercomputing? According to an industry report prepared by McKinsey Global Institute, the United States faces a current shortage of 140,000–190,000 professionals adept in the analytic methods required for Big Data [20]. The same group estimates that the United States needs an additional 1.5 million data-savvy managers [20].

Analysis is important; it would be good to have an adequate workforce of professionals trained in a variety of computationally-intensive techniques that can be applied to Big Data resources. Nevertheless, there is little value in applying advanced computational methods to poorly designed resources that lack introspection and data identification. A high-powered computer operated by a highly trained analyst cannot compensate for opaque or corrupted data. Conversely, when the Big Data resource is well designed, the task of data analysis becomes relatively straightforward.

At this time, we have a great many analytic techniques at our disposal, and we have open source software programs that implement these techniques. Every university offers computer science courses and statistics courses that teach these techniques. We will soon reach a time when there will be an oversupply of analysts, and an under-supply of well-prepared data. When this time arrives, there will be a switch in emphasis from data analysis to data preparation. The greatest number of Big Data professionals will be those people who prepare data for analysis.

Will Big Data create new categories of data professionals, for which there are currently no training programs? In the near future, millions of people will devote large portions of their careers toward the design, construction, operation, and curation of Big Data resources. Who are the people best equipped for these tasks?

– **Resource Builders:**

– Big Data Designers

Big Data does not self-assemble into a useful form. It must be designed before any data is collected. The job of designing a Big Data resource cannot be held by any single person, but a data manager (i.e., the person who supervises the project team) is often saddled with the primary responsibility of proffering a design or model of the resource. Issues such as what data will be included, where the data comes from, how to verify the data, how to annotate and classify the data, how to store and retrieve data, how to access the system, and a thousand other important concerns must be anticipated by the designers. If the design is bad, the Big Data resource will likely fail.

– Big Data Indexers

As discussed in Section 2.4, indexes help us find the data we need, quickly. The Google search engine is, at its heart, an index, built by the PageRank algorithm. Without indexers and the algorithms that organize data in a way that facilitates the kinds of searches that users are likely to conduct, Big Data would have very little appeal. The science of indexing is vastly underrated in universities, and talented indexers are hard to find. Indexers should be actively recruited into most Big Data projects.

– Domain experts

It is impossible to sensibly collect and organize Big Data without having a deep understanding of the data domain. An effective data domain expert has an understanding of the kinds of problems that the data can help solve, and can communicate her knowledge to the other members of the Big Data project, without resorting to opaque jargon. The domain expert must stay current in her field and should regularly share information with other domain experts in her field and in fields that might be relatable to the project.

– Metadata experts

The most common mistake made by beginners to the metadata field is to create their own metadata tags to describe their data. Metadata experts understand that individualized metadata solutions produce Big Data than cannot be usefully merged with other data sets. Experts need to know the metadata resources that are available on the web, and they must choose metadata descriptors that are defined in permanent, accessible, and popular schemas. Such knowledge is an acquired skill that should be valued by Big Data managers.

– Ontologists and classification experts

As noted in Sections 5.6–5.8, it is almost impossible to create a good classification. Creating, maintaining, and improving a classification is a highly demanding skill, and classification errors can lead to disastrous results for a Big Data resource and its many users. Hence, highly skill taxonomists and ontologists are essential to any Big Data project.

– Software programmers

Software programmers are nice to have around, but they have a tendency to get carried away with large applications and complex graphic user interfaces. Numerous examples shown throughout this book would suggest that most of the useful algorithms in Big Data are actually quite simple. Programmers who have a good working knowledge of many different algorithms, and who can integrate short implementations of these algorithms, as needed, within the framework of a Big Data Resource, are highly useful.

In many cases, Big Data projects can operate quite well using free and open source database applications. The primary purpose of programmers, in this case, often falls to making incremental additions and adjustments to the bare-bones system. As a general rule, the fewer the additions, the better the results. Programmers who employ good practices (e.g., commenting code, documenting changes, avoiding catastrophic interactions between the different modules of an application) can be more effective than genious-grade programmers who make numerous inscrutable system modifications before moving to a higher-paying job with your competitor.

– Data curators, including legacy experts

Who is responsible for all this immutability that haunts every Big Data resource? Most of the burden falls upon the data curator. The word "curator" derives from the Latin, "curatus," the same root for "curative" and conveys that curators fix things. In a Big Data resource the curator must oversee the accrual of sufficiently annotated legacy and prospective data into the resource; must choose appropriate nomenclatures for annotating the data; must annotate the data; and must supplement records with updated annotations as appropriate, when new versions of nomenclatures, specifications, and standards are applied. The curator is saddled with the almost impossible task of keeping current the annotative scaffold of the Big Data resource, without actually changing any of the existing content of records. In the absence of curation, all resources will eventually fail.

It all seems so tedious! Is it really necessary? Sadly, yes. Over time, data tends to degenerate: records are lost or duplicated, links become defunct, unique identifiers lose their uniqueness, the number of missing values increases, nomenclature terms become obsolete, mappings between terms coded to different standards become decreasingly reliable. As personnel transfer, quit, retire, or die, the institutional memory for the Big Data resource weakens. As the number of contributors to a Big Data resource increases, controlling the quality of the incoming data becomes increasingly difficult. Data, like nature, regresses from order to disorder; unless energy is applied to reverse the process. There is no escape; every reliable Big Data resource is obsessed with self-surveillance and curation.

– Data managers

In most instances the data manager is also the data project manager in charge of a team of workers. Her biggest contribution will involve creating a collegial, productive, and supportive working environment for their team members. The database manager must understand why components of Big Data resources, that are not found in smaller data projects (e.g., metadata, namespaces, ontologies, identifier systems, timestamps) are vital, and why the professionals absorbed in these exclusively Big Data chores are integral to the success of the projects. Team training is crucial in Big Data efforts, and the data manager must help each member of her team understand the roles played by the other members.

The data manager must also understand the importance of old data and data permanence and data immutability.

– Network specialists

Big Data seldom exists in a silo. Throughout this book, and only as a convenience, data is described as something that is conveyed in a document. In reality, data is something that streams between clouds. Network specialists, not discussed in any detail in this book, are individuals who know how to access and link data, wherever it may reside.

– Security experts

Security issues were briefly discussed in Section 18.3, "Data Security and Cryptographic Protocols." Obviously, this important subject cannot be treated in any great depth in this book. Suffice it to say, Big Data requires the services of security experts whose knowledge is not confined to the type of encryption protocols described earlier. Data security is more often a personnel problem than a cryptographic puzzle. Breaches are likely to arise due to human carelessness (i.e., failure to comply with security protocols) or from misuse of confidential information (i.e., carrying around gigabytes of private information on a personal laptop, or storing classified documents at home). Security experts, skilled in the technical and social aspects of their work, fulfill an important role.

– **Resource Users**

– Data validators

What do we really know about the measurements contained in Big Data resources? How can we know what these measurements mean, and whether they are correct? Data managers approach these questions with three closely related activities: data verification, data reproducibility, and validation. As previously mentioned, verification is the process that ensures that data conforms to a set of specifications. As such, it is a pre-analytic process (i.e., done on the data, before it is analyzed). Data reproducibility involves getting the same measurement over and over when you perform the test properly. Validation involves showing that correct conclusions were obtained from a competent analysis of the data. The primary purpose of data validators is to show that the scientific conclusions drawn from a Big Data resource are trustworthy and can be used as the foundation for other studies. The secondary purpose of data validators is to determine when the conclusions drawn from the data are not trustworthy, and to make recommendations that might rectify the situation.

– Data analysts

Most data analysts carry a set of methods that they have used successfully, on small data problems. No doubt, they will apply the same methods to Big Data, with varying results. The data analysts will be the ones who learn, from trial and error, which methods are computationally impractical on large sets of data, which methods provide results that have no practical value, which methods are unrepeatable, and which methods cannot be validated. The data analysts will also be the ones who try new methods and report on their utility. Because so many of the analytic methods on Big Data are over-hyped, it will be very important to hire data analysts who are objective, honest, and resistant to bouts of hubris.

–   Generalist problem solvers

Arguably the most essential new professional is the "generalist problem solver," a term that describes people who have a genuine interest in many different fields, a naturally inquisitive personality, and who have a talent for seeing relationships where others do not. The data held in Big Data resources becomes much more valuable when information from different knowledge domains lead to unexpected associations that enlighten both fields (e.g., veterinary medicine and human medicine, bird migration and global weather patterns, ecologic catastrophes and epidemics of emerging diseases, political upheaval and economic cycles, social media activity and wages in African nations). For these kinds of efforts, someone needs to create a new set of cross-disciplinary questions that could not have been asked prior to the creation of Big Data resources.

Historically, academic training narrows the interests of students and professionals. Students begin their academic careers in college, where they are encouraged to select a major field of study as early as their freshman year. In graduate school, they labor in a sub-discipline, within a rigidly circumscribed department. As postdoctoral trainees, they narrow their interests even further. By the time they become tenured professors, their expertise is so limited that they cannot see how other fields relate to their own studies. The world will always need people who devote their professional careers to a single sub-discipline, to the exclusion of everything else, but the future will need fewer and fewer of these specialists [21]. My experience has been that cross-disciplinary approaches to scientific problems are very difficult to publish in scientific journals that are, with few exceptions, devoted to one exclusive area of research. When a journal editor receives a manuscript that employs methods from another discipline, the editor usually rejects the paper, indicating that it belongs in some other journal. Even when the editor recognizes that the study applies to a problem within the scope of the journal, the editor would have a very difficult time finding reviewers who can evaluate a set of methods from another field. To get the greatest value from Big Data resources, it is important to understand when a problem in one field has an equivalence to a problem from another field. The baseball analyst may have the same problem as the day trader; the astrophysicist may have the same problem as the chemist. We need to have general problem solvers who understand how data from one resource can be integrated with the data from other resources, and how problems from one field can be generalized to other fields, and can be answered with an approach that combines data and methods from several different disciplines. It is important that universities begin to train students as problem solvers, without forcing them into restrictive academic departments.

– Scientists with some minimal programming skills (not usually full-time programmers)

In the 1980s, as the cost of computers plummeted, and desktop units were suddenly affordable to individuals, it was largely assumed that all computer owners would become computer programmers. At the time, there was nothing much worth doing with a computer other than programming and word processing. By the mid-1990s, the Internet grabbed the attention of virtually every computer owner. Interest in programming languages waned, as our interest in social media and recreational uses of the computer grew. It is ironic that we find ourselves inundated with an avalanche of Big Data, just at the time that society, content with online services provided by commercial enterprises, have traded their computers for smartphones. Scientists and other data users will find that they cannot do truly creative work using proprietary software applications. They will always encounter situations wherein software applications fail to meet their exact needs. In these cases it is impractical to seek the services of a full-time programmer.

Today, programming is quite easy. Within a few hours, motivated students can pick up the rudiments of popular scripting languages such as Python, Perl, Ruby and R. With few exceptions, the scripts needed for Big Data analysis are simple and most can be written in under 10 lines of code [22–25]. It is not necessary for Big Data users to reach the level of programming proficiency held by professional programmers. For most scientists, programming is one of those subjects for which a small amount of training preparation will usually suffice. I would strongly urge scientists to return to their computational roots and to develop the requisite skills for analyzing Big Data.

– Data reduction specialists

There will be a need for professionals to develop strategies for reducing the computational requirements of Big Data and for simplifying the way that Big Data is examined. For example, the individuals who developed the CODIS DNA identification system (discussed in Section 17.4, "Case Study: Scientific Inferencing from a Database of Genetic Sequences") relieved forensic analysts from the prodigious task of comparing and storing, for each sampled individual, the 3 billion base pairs that span the length of the human genome. Instead, a selection of 13 short sequences can suffice to identify individual humans. Likewise, classification experts drive down the complexity of their analyses by focusing on data objects that belong to related classes with shared and inherited properties. Similarly, data modelers attempt to describe complex systems with mathematical expressions, with which the behavior of the system can be predicted when a set of parameters are obtained. Experts who can extract, reduce, and simplify Big Data will be in high demand and will be employed in academic centers, federal agencies and corporations.

– Data visualizers

Often, all that is needed to make an important observation is a visualized summary of data. Luckily, there are many data visualization tools that are readily available to today's scientists. Examples of simple data plots using matplotlib (a Python module) or Gnuplot (an open source application that can be called from the command line) have been shown. Of course, Excel afficionados have, at their disposal, a dazzling number of ways with which they can display their spreadsheet data. Regardless of your chosen tools, anyone working with data should become adept at transforming raw data into pictures.

– Free-lance Big Data consultants

Big Data freelancers are self-employed professionals who have the skills to unlock the secrets that lie within Big Data resources. When they work under contract for large institutions and corporations, they may be called consultants, or freelance analysts. When they sell their data discoveries on the open market, they may be called entrepreneurial analysts. They will be the masters of data introspection, capable of quickly determining whether the data in a resource can yield the answers sought by their clients. Some of these Big Data freelancers will have expertise limited to one or several Big Data resources; expertise that may have been acquired as a regular employee of an institution or corporation, in the years preceding his or her launch into self-employment. Freelancers will have dozens, perhaps hundreds, of small utilities for data visualization and data analysis. When they need assistance with a problem, the freelancer might enlist the help of fellow freelancers. Subcontracted alliances can be arranged quickly, through Internet-based services. The need for bricks-and-mortar facilities, or for institutional support, or for employers and supervisors, will diminish. The freelancer will need to understand the needs of his clients and will be prepared to help the client redefine their specific goals, within the practical constraints imposed by the available data. When the data within a resource is insufficient, the freelancer would be the best person to scout alternate resources. Basically, freelance analysts will live by their wits, exploiting the Big Data resources for the benefit of themselves and their clients.

– Everyone else

As public data becomes increasingly available, there will be an opportunity for everyone to participate in the bounty. See Section 20.7, "Case Study: The Citizen Scientists".

Section 20.6. Hubris and Hyperbole

Intellectuals can tell themselves anything, sell themselves any bill of goods, which is why they were so often patsies for the ruling classes in nineteenth-century France and England, or twentieth-century Russia and America.

Lillian Hellman

A Forbes magazine article, published in 2015, running under the title, "Big Data: 20 Mind-Boggling Facts Everyone Must Read," listed some very impressive "facts," including that claim that more data has been created in the past two years than in all of prior history [26]. Included in the article was the claim that by the year 2020, the accumulated data collected worldwide will be about 44 zettabytes (44 trillion gigabytes). The author wrote, as one of his favorite facts, "At the moment less than 0.5% of all data is ever analyzed and used, just imagine the potential here" [26].

Of course, it is impossible to either verify or to discredit such claims, but experience would suggest that only a small percentage of the data that is collected today is worth the serious attention of data analysts. It is quite rare to find data that has been annotated with even the most minimal information required to conduct credible scientific research. These minimal annotations, as discussed previously, would be the name of the data creator, the owner of the data, the legal restraints on the usage of the data, the date that the data was created, the protocols by which the data was measured and collected, identifiers for data objects, class information on data objects, and metadata describing data within data objects. If you have read the prior chapters, you know the drill.

Make no mistake, despite the obstacles and the risks, the potential value of Big Data is inestimable. A hint at future gains from Big Data comes from the National Science Foundation's (NSF) 2012 solicitation for grants in core techniques for Big Data. The NSF envisions a Big Data future with the following pay-offs [1]:

- "Responses to disaster recovery empower rescue workers and individuals to make timely and effective decisions and provide resources where they are most needed;
- Complete health/disease/genome/environmental knowledge bases enable biomedical discovery and patient-centered therapy; The full complement of health and medical information is available at the point of care for clinical decision-making;
- Accurate high-resolution models support forecasting and management of increasingly stressed watersheds and ecosystems;
- Access to data and software in an easy-to-use format are available to everyone around the globe;
- Consumers can purchase wearable products using materials with novel and unique properties that prevent injuries;
- The transition to use of sustainable chemistry and manufacturing materials has been accelerated to the point that the US leads in advanced manufacturing;
- Consumers have the information they need to make optimal energy consumption decisions in their homes and cars;
- Civil engineers can continuously monitor and identify at-risk man-made structures like bridges, moderate the impact of failures, and avoid disaster;
- Students and researchers have intuitive real-time tools to view, understand, and learn from publicly available large scientific data sets on everything from genome sequences to astronomical star surveys, from public health databases to particle accelerator simulations and their teachers and professors use student performance analytics to improve that learning; and
- Accurate predictions of natural disasters, such as earthquakes, hurricanes, and tornadoes, enable life-saving and cost-saving preventative actions."

Lovely. It would seem that there is nothing that cannot be accomplished with Big Data!

I know lots of scientists; the best of them lack self-confidence. They understand that their data may be flawed, their assumptions may be wrong, their methods might be inappropriate, their conclusions may be unrepeatable, and their most celebrated findings may one day be discredited. The worst scientists are just the opposite; confident of everything they do, say, or think [27].

The sad fact is that, among scientific disciplines, Big Data is probably the least reliable, providing major opportunities for blunders. Prior chapters covered limitations in measurement, data representation, and methodology. Some of the biases encountered in every Big Data analysis were covered in Chapter 14, "Special Considerations in Big Data Analysis." Apart from these limitations lies the ever-present dilemma that assertions based on Big Data analyses can sometimes be validated, but they can never be proven true. Confusing validation with proof is a frequently encountered manifestation of overconfidence. If you want to attain proof, you must confine your interests to pure mathematics. Mathematics is the branch of science devoted to truth. With math, you can prove that an assertion is true, you can prove that an assertion is false, you can prove that an assertion cannot be proven to be true or false. Mathematicians have the monopoly on proving things. None of the other sciences have the slightest idea what they're doing when it comes to proof.

In the non-mathematical sciences, such as chemistry, biology, medicine, and astronomy, assertions are sometimes demonstrably valid (true when tested), but assertions never attain the level of a mathematical truth (proven that it will always be true, and never false, forever). Nonetheless, we can do a little better than showing that an assertion is simply valid. We can sometimes explain why an assertion ought to be true for every test, now and forever. To do so, an assertion should have an underlying causal theory that is based on interactions of physical phenomena that are accepted as true. For example, $F = ma$ ought to be true, because we understand something about the concepts of mass and acceleration, and we can see why the product of mass and acceleration produce a force. Furthermore, everything about the assertion is testable in a wide variety of settings.

Big Data analysts develop models that are merely descriptive (e.g., predicting the behavior of variables in different settings), without providing explanations in terms of well-understood causal mechanisms. Trends, clusters, classes, and recommenders may appear to be valid over a limited range of observations; but may fail miserably in tests conducted over time, with a broader range of data. Big Data analysts must always be prepared to abandon beliefs that are not actually proven [21].

Finance has eagerly entered the Big Data realm, predicting economic swings, stock values, buyer preferences, the impact of new technologies, and a variety of market reactions, all based on Big Data analysis. For many financiers, accurate short-term predictions have been followed, in the long-run, with absolutely ruinous outcomes. In such cases, the mistake was overconfidence; the false belief that their analyses will always be correct [28].

In my own field of concentration, cancer research, there has been a major shift of effort away from small experimental studies toward large clinical trials and so-called high-throughput molecular methods that produce vast arrays of data. This new generation of cancer research costs a great deal in terms of manpower, funding, and the time to complete a study. The funding agencies and the researchers are confident that a Big Data approach will work where other approaches have failed. Such efforts may one day lead to the eradication of cancer; who is to say? In the interim, we have already seen a great deal of time and money wasted on huge, data-intensive efforts that have produced predictions that are not reproducible, with no more value than a random throw of dice [4,29–32].

Despite the limitations of Big Data, the creators of Big Data cannot restrain their enthusiasm. The following is an announcement from the National Human Genome Research Institute concerning their own achievements [33]:

"In April 2003, NHGRI celebrated the historic culmination of one of the most important scientific projects in history: the sequencing of the human genome. In addition, April 2003 marked the 50th anniversary of another momentous achievement in biology: James Watson and Francis Crick's Nobel Prize winning description of the DNA double helix" and "To mark these achievements in the history of science and medicine, the NHGRI, the NIH and the DOE held a month-long series of scientific, educational, cultural and celebratory events across the United States."

In the years following this 2003 announcement, it has become obvious that the genome is much more complex than previously thought, that common human diseases are genetically complex, that the genome operates through mechanisms that cannot be understood by examining DNA sequences, and that much of the medical progress expected from the Human Genome Project will not be forthcoming anytime soon [29,34,35]. In a 2011 article, Eric Lander, one of the luminaries of the Human Genome Project, was quoted as saying, "anybody who thought in the year 2000 that we'd see cures in 2010 was smoking something" [35]. Monica Gisler and co-workers have hypothesized that large-scale projects create their own "social bubble," inflating the project beyond any rational measure [36]. It is important that Big Data proselytizers, myself included, rein in their enthusiasm.

Section 20.7. Case Study: The Citizen Scientists

There is no reason why someone would want a computer in their home.
   Ken Olson, President and founder of Digital Equipment Corporation, in 1977.

In a sense, we have reached a post-information age. At this point, we have collected an awful lot of information, and we all have access too much more information than we can possibly analyze within our lifetimes. In fact, all the professional scientists and data analysts who are living today could not possibly exhaust the information available to anyone with Internet access. If we want to get the most out of the data that currently resides within our grasps, we will need to call upon everyone's talents, including amateurs. Lest we forget, every professional scientist enters the ranks of the amateur scientists on the day that he or she retires. Today, the baby boomer generation is amassing an army of well-trained scientists who are retiring into a world that provides them with unfettered access to limitless data. Hence, we can presume that the number of amateur scientists will soon exceed the number of professional scientists.

Historically, some of the greatest advancements in science have come from amateurs. For example, Antonie van Leeuwenhoek (1632–1723), one of the earliest developers of the compound microscope, who is sometimes credited as the father of microbiology, was a janitor. Augustin-Jean Fresnel (1788–1827) was a civil engineer who found time to make significant and fundamental contributions to the theory of wave optics. Johann Jakob Balmer (1825–1898) earned his living as a teacher in a school for girls while formulating the mathematical equation describing the spectral emission lines of hydrogen. His work, published in 1885, led others, over the next four decades, to develop the new field of quantum mechanics. Of course, Albert Einstein was a paid patent clerk and an amateur physicist who found time, in 1905, to publish three papers that forever changed the landscape of science.

In the past few decades a wealth of scientific resources has been made available to anyone with Internet access. Many of the today's most successful amateurs are autodidacts with access to Big Data [37–42]. Here are a few examples:

- Amateurs identifying the genes that cause human disease

In the field of medicine, some of the most impressive data mining feats have come from individuals affected by rare diseases who have used publicly available resources to research their own conditions.

Jill Viles is a middle-aged woman who, when she was a college undergraduate, correctly determined that she was suffering from Emery-Dreifuss muscular dystrophy. The diagnosis of this very rare form of muscular dystrophy was missed by her physicians. After her self-diagnosis was confirmed, she noticed that her father, who had never been told he had any muscular condition, had a distribution of his muscle mass that was suggestive of Emery-Dreifuss. Jill's suspicions initiated a clinical consultation indicating that her father indeed had a mild form of the same disorder and that his heart had been affected. Her father received a needed pacemaker, and Jill's shrewd observations were credited with saving her father's life. Jill pursued her interest in her own condition and soon became one of the early beneficiaries of genome sequencing. A mutation of the lamin gene was apparently responsible for her particular variant of Emery-Dreifuss muscular dystrophy. Jill later realized that in addition to Emery-Dreifuss muscular dystrophy, she also exhibited some of the same highly distinctive features of partial lipodystrophy, a disease characterized by a decrease in the fat around muscles. When the fat around muscles is decreased, the definition of the muscles (i.e., the surface outline of musculature) is enhanced. She reasoned, correctly, that the lamin gene, in her case, was responsible for both conditions (i.e., Emery-Dreifuss muscular dystrophy and partial lipodystrophy).

Jill's story does not end here. While looking at photographs of Priscilla Lopes-Schlief, an Olympic athlete known for her hypertrophied muscles, Jill noticed something very peculiar. The athlete had a pattern of fat-deficient muscle definition on her shoulders, arms, hips, and butt, that was identical to Jill's; the difference being that Priscilla's muscles were large, and Jill's muscles were small. Jill contacted the Olympian, and the discussions that followed eventually led to Priscilla's diagnosis of lipodystrophy due to a mutation on the lamin gene, at a locus different from Jill's. Lipodystrophy can produce a dangerous elevation in triglycerides, and Priscilla's new diagnosis prompted a blood screen for elevated lipids. Priscilla had high levels of triglyceride, requiring prompt treatment. Once again, Jill had made a diagnosis that was missed by physicians, linked the diagnosis to a particular gene and uncovered a treatable and overlooked secondary condition (i.e., hypertriglyceridemia). And it was all done by an amateur with Internet access [43]!

Jill Viles' story is not unique. Kim Goodsell, a patient with two rare diseases, Charcot-Marie-Tooth disease and arrhythmogenic right ventricular cardiomyopathy, searched the available gene datasets until she found a single gene that might account for both of her conditions. After much study, she determined that a point mutation in the LMNA gene was the most likely cause of her condition. Kim paid $3000 for gene sequencing of her own DNA, and a rare point mutation on LMNA was confirmed to be responsible for her dual afflictions [44]. In Kim's case, as in Jill's case, a persistent and motivated amateur can be credited with a significant advance in the genetics of human disease.

– 36-year-old satellite resurrected

The International Space/Earth Explorer 3 (ISEE-3) spacecraft was launched in 1978 and proceeded on a successful mission to monitor the interaction between the solar wind and the earth's magnetic field. In 1985, ISEE-3 visited the comet Giacobini-Zinner, and was thereupon given a new name, ICE (the International Cometary Explorer). In 1999 NASA, short of funds, decommissioned ICE. In 2008, NASA tried to contact ICE and found that all but one of its 13 observational experiments were still in operation, and that the spacecraft had not yet exhausted its propellant.

In April 2014, a citizens group of interested scientists and engineers announced their intention to reboot ICE [38]. In May, 2014, NASA entered into a Non-Reimbursable Space Act Agreement with the citizen group, which would provide the reboot team with NASA advisors, but no funding. Later in May, the team successfully commanded the probe to broadcast its telemetry (i.e., its recorded data). In September, the team lost contact with ICE. ICE will return to a near-earth position, in 17 years (Fig. 20.1). There is reason to hope that scientists will eventually recover ICE telemetry, and, with it, find new opportunities for data analysis [45].

In the past, the term "amateurish" was used to describe products that are unprofessional and substandard. In the realm of Big Data, where everyone has access to the same data, amateurs and professionals can now compete on a level playing field. To their credit, amateurs are unsullied by the kind of academic turf battles, departmental jealousies, and high-stakes grantsmanship ploys that produce fraudulent, misleading, or irreproducible results [46]. Because amateurs tend to work with free and publicly available data sets, their research tends to be low-cost or no-cost. Hence, on a cost-benefit analysis, amateur scientist may actually have more value, in terms of return on investment, than professional

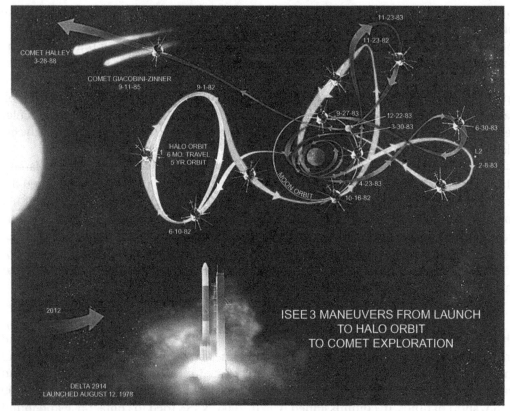

FIG. 20.1 Trajectory of the International Cometary Explorer. *From NASA, public domain.*

scientists. Moreover, there are soon to be many more amateur scientists than there are professionals, making it likely that these citizen scientists, who toil for love, not money, will achieve the bulk of the breakthroughs that come from Big Data science. Of course, none of these breakthroughs, from citizen scientists, would be possible without free and open access to Big Data resources.

Section 20.8. Case Study: 1984, by George Orwell

He who controls the past controls the future.

George Orwell

When you have access to Big Data, you feel liberated; when Big Data has access to you, you feel enslaved. Everyone is familiar with the iconic image, from Orwell's 1984, of a totalitarian government that spies on its citizens from telescreens [47]. The ominous phrase, "Big Brother is watching you," evokes an important thesis of Orwell's masterpiece; that an evil government can use an expansive surveillance system to crush its critics.

Lest anyone forget, Orwell's book had a second thesis, that was, in my opinion, more insidious and more disturbing than the threat of governmental surveillance. Orwell was concerned that governments could change the past and the present by inserting, deleting, and otherwise distorting the information available to citizens. In Orwell's 1984, previously published reports of military defeats, genocidal atrocities, ineffective policies, mass starvation, and any ideas that might foment unrest among the proletariat, were deleted and replaced with propaganda pieces. Such truth-altering activities were conducted undetected, routinely distorting everyone's perception of reality to suit a totalitarian agenda. Aside from understanding the dangers in a surveillance-centric society, Orwell was alerting us to the dangers inherent with mutable Big Data.

Today, our perception of reality can be altered by deleting or modifying electronic data distributed via the Internet. In 2009, Amazon was eagerly selling electronic editions of a popular book, much to the displeasure of the book's publisher. Amazon, to mollify the publisher, did something that seemed impossible. Amazon retracted the electronic books from the devices of readers who had already made their purchase. Where there was once a book on a personal eBook reader, there was now nothing. Amazon rectified their action by crediting customer accounts for the price of the book. So far as Amazon and the publisher were concerned, the equilibrium of the world was restored [48].

The public reaction to Amazon's vanishing act was a combination of bewilderment ("What just happened?"), shock ("How was it possible for Amazon to do this?"), outrage ("That book was mine!"), fear ("What else can they do to my eBook reader?"), and suspicion ("Can I ever buy another eBook?"). Amazon quickly apologized for any misunderstanding and promised never to do it again.

To add an element of irony to the episode, the book that was bought, then deleted, to suit the needs of a powerful entity, was George Orwell's 1984.

One of the purposes of this book is to describe the potential negative consequences of Big Data when data is not collected ethically, prepared thoughtfully, analyzed openly, or subjected to constant public review and correction. These lessons are important because the future reality of our Big Data universe will be determined by some of the people who are reading this book today.

Glossary

Artificial intelligence Artificial intelligence is the field of computer science that seeks to create machines and computer programs that seem to have human intelligence. The field of artificial intelligence sometimes includes the related fields of machine learning and computational intelligence. Over the past few decades, the term "artificial intelligence" has taken a battering from professionals inside and outside the field, for good reasons. First and foremost is that computers do not think in the way that humans think. Though powerful computers can now beat chess masters at their own game, the algorithms for doing so do not simulate human thought processes. Furthermore, most of the predicted benefits from artificial intelligence have not come to pass, despite decades of generous funding. The areas of neural networks, expert systems, and language translation have not met expectations. Detractors have suggested that artificial intelligence is not a well-defined sub discipline within computer science as it has encroached into areas unrelated to machine intelligence, and has appropriated techniques from other fields, including statistics and numerical analysis. Some of the goals of artificial intelligence have been achieved (e.g., speech-to-text translation), and the analytic methods employed in Big Data analysis should be counted among the enduring successes of the field.

Checksum An outdated term that is sometimes used synonymously with one-way hash or message digest. Checksums are performed on a string, block or file yielding a short alphanumeric string intended to be specific for the input data. Ideally, If a single bit were to change, anywhere within the input file, then the checksum for the input file would change drastically. Checksums, as the name implies, involve summing values (typically weighted character values), to produce a sequence that can be calculated on a file before and after transmission. Most of the errors that were commonly introduced by poor transmission could be detected with checksums. Today, the old checksum algorithms have been largely replaced with one-way hash algorithms. A checksum that produces a single digit as output is referred to as a check digit.

Machine learning Refers to computer systems and software applications that learn or improve as new data is acquired. Examples would include language translation software that improves in accuracy as additional language data is added to the system, and predictive software that improves as more examples are obtained. Machine learning can be applied to search engines, optical character recognition software, speech recognition software, vision software, neural networks. Machine learning systems are likely to use training data sets and test data sets.

Semantics The study of meaning. In the context of Big Data, semantics is the technique of creating meaningful assertions about data objects. A meaningful assertion, as used here, is a triple consisting of an identified data object, a data value, and a descriptor for the data value. In practical terms, semantics involves making assertions about data objects (i.e., making triples), combining assertions about data objects (i.e., merging triples), and assigning data objects to classes; hence relating triples to other triples. As a word of warning, few informaticians would define semantics in these terms, but I would suggest that most definitions for semantics would be functionally equivalent to the definition offered here.

References

[1] Core techniques and technologies for advancing Big Data science. National Science Foundation program solicitation NSF 12-499; June 13, 2012. Available from: http://www.nsf.gov/pubs/2012/nsf12499/nsf12499.txt [viewed September 23, 2012].

[2] National Science Board, Science & Engineering Indicators-2000. Arlington, VA: National Science Foundation; 2000 (NSB-00-1).

[3] Berman J. Precision medicine, and the reinvention of human disease. Cambridge, MA: Academic Press; 2018.

[4] Bossuyt PM, Reitsma JB, Bruns DE, Gatsonis CA, Glasziou PP, Irwig LM, et al. The STARD statement for reporting studies of diagnostic accuracy: explanation and elaboration. Clin Chem 2003;49:7–18.

[5] Ioannidis JP. Why most published research findings are false. PLoS Med 2005;2:e124.

[6] Ioannidis JP. Some main problems eroding the credibility and relevance of randomized trials. Bull NYU Hosp Jt Dis 2008;66:135–9.

[7] Pueschel M. National outcomes database in development. U.S. Medicine; December, 2000.

[8] Cook TD, Shadish WR, Wong VC. Three conditions under which experiments and observational studies produce comparable causal estimates: new findings from within-study comparisons. Journal of Policy Analysis and Management 2008;27:724–50.

[9] Robins JM. The control of confounding by intermediate variables. Statistics in Medicine 1989;8:679–701.

[10] Robins JM. Correcting for non-compliance in randomized trials using structural nested mean models. Communications in Statistics-Theory and Methods 1994;23:2379–412.

[11] Bornstein D. The dawn of the evidence-based budget. The New York Times; May 30, 2012.

[12] Ledley RS, Lusted LB. Reasoning foundations of medical diagnosis. Science 1959;130:9–21.

[13] Shortliffe EH. Medical expert systems—knowledge tools for physicians. West J Med 1986;145: 830–9.

[14] Heathfield H, Bose D, Kirkham N. Knowledge-based computer system to aid in the histopathological diagnosis of breast disease. J Clin Pathol 1991;44:502–8.

[15] Eccles M, McColl E, Steen N, Rousseau N, Grimshaw J, Parkin D, et al. Effect of computerised evidence based guidelines on management of asthma and angina in adults in primary care: cluster randomised controlled trial. BMJ 2002;325:941.

[16] Grady D. Study finds no progress in safety at hospitals. The New York Times; November 24, 2010.

[17] Goldberg SI, Niemierko A, Turchin A. Analysis of data errors in clinical research databases. AMIA Annu Symp Proc 2008;2008:242–6.

[18] Shelby-James TM, Abernethy AP, McAlindon A, Currow DC. Handheld computers for data entry: high tech has its problems too. Trials 2007;8:5.

[19] Helmholtz HLF. On the sensations of tone as a physiological basis for the theory of music. 2nd ed. London: Longmans, Green, and Co.; 1885

[20] Manyika J, Chui M, Brown B, Bughin J, Dobbs R, Roxburgh C, et al. Big data: the next frontier for innovation, competition, and productivity. McKinsey Global Institute; June 2011.

[21] Tetlock PE. Expert political judgment: How good is it? How can we know? Princeton: Princeton University Press; 2005.

[22] Berman JJ. Perl programming for medicine and biology. Sudbury, MA: Jones and Bartlett; 2007.

[23] Berman JJ. Ruby programming for medicine and biology. Sudbury, MA: Jones and Bartlett; 2008.

[24] Berman JJ. Methods in medical informatics: fundamentals of healthcare programming in Perl, Python, and Ruby. Boca Raton: Chapman and Hall; 2010.

[25] Berman JJ. Data simplification: taming information with open source tools. Waltham, MA: Morgan Kaufmann; 2016.

[26] Marr B. Big Data: 20 mind-boggling facts everyone must read. Forbes Magazine; September 30, 2015.

[27] Berner ES, Graber ML. Overconfidence as a cause of diagnostic error in medicine. American J Med 2008;121:S2–S23.

[28] Thaler RH. The overconfidence problem in forecasting. The New York Times; August 21, 2010.

[29] Cecile A, Janssens JW, vanDuijn CM. Genome-based prediction of common diseases: advances and prospects. Human Molecular Genetics 2008;17:166–73.

[30] Begley S. In cancer science, many 'discoveries' don't hold up. Reuters; March 28, 2012.

[31] Michiels S, Koscielny S, Hill C. Prediction of cancer outcome with microarrays: a multiple random validation strategy. Lancet 2005;365:488–92.

[32] Venet D, Dumont JE, Detours V. Most random gene expression signatures are significantly associated with breast cancer outcome. PLoS Comput Biol 2011;7:e1002240.

[33] Fifty years of DNA: from double helix to health, a celebration of the genome. National Human Genome Research Institute; April 2003. Available from: http://www.genome.gov/10005139 [viewed September 19, 2012].

[34] Wade N. Scientist at work: David B. Goldstein, a dissenting voice as the genome is sifted to fight disease. The New York Times; September 16, 2008.

[35] Cohen J. The human genome, a decade later. Technology Review; January–February, 2011.

[36] Gisler M, Sornette D, Woodard R. Exuberant innovation: the human genome project. Cornell University Library; March 15, 2010. Available from: http://arxiv.org/ftp/arxiv/papers/1003/1003.2882.pdf [viewed September 22, 2012].

[37] Brekke D. Quiet passing of unlikely hero. Wired; January 22, 2000.

[38] Meyer R. A long-lost spacecraft, now saved, faces its biggest test yet. The Atlantic; June 3, 2014.

[39] Chang K. A hobbyist challenges papers on growth of dinosaurs. The New York Times; December 16, 2013.

[40] Chang K. A spat over the search for killer asteroids. The New York Times; May 23, 2016.

[41] Chung E. Stardust citizen scientist finds first dust from outside solar system: dust collected by NASA spacecraft a decade ago finally identified and analyzed. CBC News; August 14, 2014.

[42] Mims FM. Amateur science—strong tradition, bright future. Science 1999;284:55–6.

[43] Epstein D. The DIY Scientist, the Olympian, and the Mutated Gene: how a woman whose muscles disappeared discovered she shared a disease with a muscle-bound Olympic medalist. ProPublica; January 15, 2016.

[44] Fikes BJ. The patient from the future, here today. The San Diego Union-Tribune; March 5, 2014.

[45] Berman JJ. Repurposing legacy data: innovative case studies. Waltham, MA: Morgan Kaufmann; 2015.

[46] Berman JJ. Machiavelli's Laboratory. Amazon Digital Services, Inc.; 2010

[47] Orwell G. 1984. Signet, Tiptree, U.K., 1950.

[48] Pogue D. Amazon.com plays big brother with famous e-books. The New York Times; July 17, 2009.

Index

Note: Page numbers followed by *f* indicate figures.

PICTURING MACHINES
1400–1700

TRANSFORMATIONS
STUDIES IN THE HISTORY OF SCIENCE AND TECHNOLOGY
JED BUCHWALD
GENERAL EDITOR

Mordechai Feingold, editor, *Jesuit Science and the Republic of Letters*

Sungook Hong, *Wireless: From Marconi's Black-Box to the Audion*

Myles Jackson, *Spectrum of Belief: Joseph von Fraunhofer and the Craft of Precision Optics*

Mi Gyung Kim, *Affinity, That Elusive Dream: A Genealogy of the Chemical Revolution*

Janis Langins, *Conserving the Enlightenment: French Military Engineering from Vauban to the Revolution*

Wolfgang Lefèvre, editor, *Picturing Machines 1400–1700*

William R. Newman and Anthony Grafton, editors, *Secrets of Nature: Astrology and Alchemy in Early Modern Europe*

Alan J. Rocke, *Nationalizing Science: Adolphe Wurtz and the Battle for French Chemistry*

PICTURING MACHINES
1400–1700

EDITED BY WOLFGANG LEFÈVRE

THE MIT PRESS

CAMBRIDGE, MASSACHUSETTS

LONDON, ENGLAND

This book was set in Times by Heinz Reddner, MPI for the History of Science, Berlin.

Library of Congress Cataloging-in-Publication Data

Picturing machines 1400–1700 / edited by Wolfgang Lefèvre.
    p. cm. — (Transformations)
   ISBN 978-0-262-12269-6 (hc), 978-0-262-55088-8 (pb)
   1. Mechanical drawing. 2. Engineering graphics. I. Lefèvre, Wolfgang, 1941–
  II. Transformations (M.I.T. Press)

T353.P55 2004
604.2—dc22

                                      2003070620

CONTENTS

PART V: PRACTICE MEETS THEORY

APPENDIX

INTRODUCTION

WOLFGANG LEFÈVRE

The engineers and architects of the Renaissance are renowned not only for the universality of their genius and the audacity of their creations but also for their drawings. Leonardo da Vinci's famous drawings of technical devices, although unparalleled in many respects, are just one instance of a practice of drawing in the realm of early modern engineering that came into being at the end of the Middle Ages and eventually addressed a broad audience through the *Theatres of Machines* in the last third of the sixteenth century. The new types and methods of graphic representation developed and used by Renaissance engineers have long attracted the attention of historians of art, architecture, science, and technology. Apart from their often fascinating aesthetic qualities, these drawings have been particularly appreciated as historical documents that testify to the development of technology, the spread of perspective, the psychological roots of technological creativity, and the beginnings of modern scientific attitudes.

As an unintentional consequence of this appreciation, however, little attention has focused on the significance that these drawings had, not for present historians, psychologists, and philosophers, but for the historical actors themselves, that is, for the mechanicians, engineers, and architects of the age. Why did they produce drawings? For whom and for what purposes? What were the prerequisites for this drawing practice, what were the contexts, and what were the consequences? In short, how did drawing shape the practice and the notions of early modern engineers? Those were the questions from which the idea of this volume arose.

Among the many shared views of this volume's authors (who also differ with respect to several aspects of its topic), there is the conviction that these questions can be successfully approached only by studies that dispense with large generalizations as regards the cultural, technological, intellectual, and aesthetical significance of early modern engineering drawings—generalizations that hampered rather than promoted an adequate recognition of them in the past. They believe firmly that what is needed instead is studies that actually go into the specific details and properties of these drawings—studies that, in addition, focus thoroughly on different aspects of such properties, regardless of whether or not the pursuit of these different aspects leads to a new large synthesis.

The authors' emphasis thus lay on analysis, fine-grained studies, and close attention to details, and all of them refrained from premature synthesis. The extent to which their studies nevertheless form a connected and consistent whole was surprising even for them. However, the authors being experts in this field of historical investigation, the chapters' connectedness and coherence may be more obvious to them than to a broader readership. This introduction therefore cannot dispense with giving some indications of the context for these studies. The following outline of the historical setting of these drawings as well as of their crucial aspects constitutes neither shared starting points nor jointly achieved results of the authors, but the attempt of the editor to provide a sketchy topography in which each chapter's choice of issues and perspectives can be located.

TECHNICAL DRAWINGS: SYMPTOM AND INTEGRAL PART OF THE TECHNOLOGICAL TRANSFORMATIONS IN THE EARLY MODERN AGE

In the culture of the West, technical drawings, that is, drawings traced by technicians for professional purposes or those derived from them, appeared only at the end of the Middle Ages and flourished for the first time during the Renaissance. The emergence and development of such a drawing practice is in itself a remarkable fact that indicates profound changes in the social labor process of the West during this age. Traditional production did not employ technical drawings. This holds for agriculture, then and for a long time to come the most important domain of social labor in terms of both the amount of people employed and the wealth produced. For farming, breeding, and growing according to the standards given at the time, and even for manufacturing intricate tools like ploughs, no technical drawings were needed. Surveying, which may come to mind in this context, was not yet a normal part of the agricultural practice. As regards the realm of ordinary crafts, the second important domain of social labor, one encounters by and large the same picture. Almost all of the crafts performed their professional tasks without drawings. An important exception was, naturally, the decorative arts, which passed on established figurative and ornamental patterns through *exempla*, that is, pattern books *(Musterbücher)*. But among the drawings of these *exempla*, only a few could properly be called technical drawings. It was not developments within these established fields of production to which technical drawings owed their birth. Rather, they owed their emergence and development to new sectors of production that transgressed the limits of the traditional labor processes still prevailing, in terms of both the depth of division of labor and the technical procedures applied. And the employment of such drawings is indicative of just these transgressions.

Technical drawings appeared first at the construction sites of Gothic cathedrals. The oldest extant architectural plans date from the thirteenth century. Warfare was the next sector where technical drawings were utilized. Beginning with a few instances in the second half of the fourteenth century and blooming in the fifteenth, drawings of all sorts of assault devices, and increasingly of guns and gun-mountings, heralded the era of early modern machine drawings. Since the middle of the fifteenth century, they were complemented and gradually even outnumbered by drawings of civil devices such as mills of all kinds, cranes and other hoisting devices, and different kinds of pumps and further water-lifting machines for mining and irrigation. Along with this, though apparently following somewhat secluded paths, technical drawings of ships testify to a developmental stage of the ship-building craft where traditional crafts methods, though still indispensable, no longer sufficed.

The emergence and spread of technical drawings thus was part and parcel of new developments in certain exceptional fields in the realm of production that could be called the high-tech sectors of the age. Moreover, these drawings were connected with those very features of all or some of these fields of advanced production by which they distinguished themselves from the traditional ways of production in agriculture and ordinary crafts. The following five features deserve particular attention in this respect, although not all of them occur in each of the sectors of advanced technology.

First, technical drawings were connected with new forms of division of labor characteristic of some of the new high-tech production sectors—forms that developed first at the construction sites of Gothic cathedrals and later, outside architecture, above all along with shipbuilding and mining on a large scale. In these sectors, the flat hierarchy of the typical craft workshop was replaced by complex structures of cooperation, responsibilities, and command. The chief engineer or architect, himself subordinated to clerical or secular commissioners or boards of commissioners, had to instruct and coordinate masters and subcontractors from different crafts who, for their part, directed their teams. The task of coordination often comprised the harmonization of work carried out at different times and different places. Such intricate forms of cooperation among the different parties involved in a project of advanced technology necessitated not only new forms of communication but also new means of communication. Technical drawings are perhaps the most striking of those new means.

Second, technical drawings were connected with new forms of knowledge propagation brought about by the new production sectors of advanced technology. In these sectors, experience with and knowledge gained through new technologies could not be exchanged and circulated effectively by means of the undeveloped and slow communication mechanisms of the medieval crafts. Rather, for these purposes, too, new means of communication were needed, technical drawings included.

Third, and in close relation to this, technical drawings were connected with new forms of learning and instruction that developed along with the new high-tech production. The traditional form in which craftsmen passed on knowledge and skills to the next generation, that is, the system of apprenticeship and journeymen's traveling that rests essentially on learning by doing, proved to be insufficient to acquire all of the capabilities required by the advanced technologies. First forms of schooling technical knowledge developed. In fact, the art of producing and reading technical drawings constituted a central part of the curriculum of the first technical schools, as is indicated in the very name of the first of these schools, namely the *Accademia del disegno* in Florence founded in 1563.

Fourth, technical drawings also were closely related to fundamental changes in the body of practical knowledge induced by the advanced production sectors. This body no longer comprised only the traditional experiences and skills of practitioners, but combined them with elements of knowledge that originated in sciences. The expanding employment of geometrical constructions and theorems in several practical contexts is particularly characteristic of this development. The broad range of competence required by the new technologies is best represented by a social figure who came into being with the new fields of advanced production, namely the engineer, who was in charge not only of designing and planning ambitious projects but also of their actual realization through the coordinated cooperation of different kinds of crafts. For these engineers, experience in several crafts was no less important than competence in design and planning, which, for its part, included learned knowledge as well as drawing capabilities.

Fifth, technical drawings were connected with the establishment of technology as a matter of public interest. Pictorial representations of machines and devices of all sorts were a chief means by which the protagonists of the new advanced production sectors managed to attract the attention of the educated public for technological issues. With this, technology became, for the first time in the culture of the West, appreciated as a valuable sphere of culture. The technological transformations of the early modern period entailed a cultural transformation in which drawings played a significant role.

TECHNICAL DRAWINGS IN EARLY MODERN ENGINEERING

The practical role of technical drawings in early modern fields of advanced production was not fixed once and for all. Naturally, for which tasks technical drawings could be employed depended partly on the needs of the engineering process in its concrete social embedment and partly on the capacity of drawings to meet such needs. However, both those needs and that capacity underwent changes—changes that often were intertwined. By using drawings and, above all, by developing particular pictorial languages, engineers discovered the possibilities given by the medium of drawings and broadened the spectrum of tasks for which drawings proved to be useful. Thus, an investigation of the actual functions of engineering drawings requires attention to many different aspects.

Mediating Parties

The most obvious function of technical drawings was a social one, that is, their function as a means of communication between different individuals or, more to the point, different kinds of individuals. They served as a means of communication between practitioners, either those of equal rank or those in a hierarchical relation; between contracting parties, commissioners, and responsible practitioners, as well as the latter and subcontracting practitioners; and between practitioners and a broader public interested in new technology. Mere contemplation of the different constellations of interacting parties makes immediately obvious that "communication" is much too vague a term to characterize the variety of roles technical drawings played in such interactions. In one case, they may have conveyed proposals; in another, they may have documented an agreement; in yet others, they may have fixed decisions, given instructions, served as a basis for consultations and negotiations, exchanged experiences, imposed and secured control, advertised inventions, services, and projects, taught a community of readers, illustrated arguments, and so on. Each of these different cases of social interaction in which technical drawings are employed comprises not only a specific constellation of differing interests but also a specific constellation of differing experiences, competence, and knowledge. Thus, technical drawings could only function successfully as a means of communication and mediation if they allowed for the different and differently informed views of the parties involved. Drawings that served understanding between practitioners of equal rank jointly engaged in a certain project highlighted, or for that matter omitted, other aspects and details of a technical object than drawings that tried to convey an idea of the object to a commissioner. Each case required a different emphasis on completeness, precision, and neatness, and different distinctness as regards information about, for instance, the function of the device in question or its construction.

Studying technical drawings thus can reveal an entire world of social relations between different groups of individuals in which technological projects were embedded. And, conversely failing to take the concrete mediating role of a certain technical drawing into account may seriously impair the understanding of it. Giving testimony as much to the social form of early modern high-tech production as to the technology employed, the technical drawings of the age must be read with social-historical expertise as well as with technological competence. However, the ways in which the extant drawings were passed on often destroyed their original relations to other materials and thus important hints on which a reconstruction of this social context could have rested. As some of this volume's chapters show, it often is impossible to determine the original purpose and the circumstances of a technical drawing. In some cases, however, an analysis of the drawing style may prove helpful.

Between Pictures and Plans

The variety of ways in which technical drawings were employed as means of communication in the social world around the most advanced technologies of the age involuntarily draws attention to the astonishingly flexible capability of drawings as a medium. How did this medium succeed in serving so many different communication tasks, each of which had different conditions, demands, and purposes? A partial answer to this question may be found in the fact that, on closer inspection, the medium of drawings proves to comprise a whole of graphic languages—that of pictures, of diagrams, of plans, and so on. From a semiotic perspective, each of these different graphic languages follows particular rules and grammars. Accordingly, each of them demands particular expertise for rendering objects in the framework of its rules and also for reading and understanding such renderings. Architectural plans of some complexity, for instance, are only partly understandable to nonexperts. Obviously, different graphical languages presuppose different knowledge and are therefore involved in the discriminations that govern the social distribution of knowledge in a culture. Furthermore, each of these graphic languages has its specific advantages and disadvantages. The graphic language that rules the construction of orthographic plans, for instance, may be unequaled as regards precise information about angles, distances, proportions, and so on. But it cannot compete with the language of perspective rendering when the purpose is to convey an impression of the object as a solid body situated in a surrounding space. Thus there is not only a social aspect of the employment of a specific graphic language but also a material one, that is, the aspect of the possibilities and limits of rendering characteristic of each of these languages. These differences between the material capacities of the various graphic languages by their very nature suggest which specific language is suitable in a certain case. However, the practitioner must additionally consider which of these languages is best understood by the audience in question. Compromises between these two concerns may be unavoidable, and such compromises are, indeed, a characteristic feature of the technical drawings of this age.

Architects and engineers of the early modern period did not only make use of multiple different graphical languages—and often in a virtuose manner—in a certain sense, they also must be regarded as the inventors of these languages. True, from the literary estate of Classic Antiquity, they inherited some clues to the graphic languages employed by their ancient predecessors; and equally true, they could build on some drawing conventions and geometrical techniques used by medieval architects and technicians. They did not have to start from scratch. Yet, as regards the different projection techniques developed in Antiquity—orthographic projection, perspective, geographic projection—each of them was reinvented rather than rediscovered in the Renaissance, and subsequently further developed and refined in an absolutely autonomous manner. Furthermore, against the background of the spread of perspective rendering in the fine arts, the schematic style characteristic of representing machines in the Middle Ages was replaced by a specific style of rendering machines on a single sheet, which was furthermore supplemented by the elaboration of an arsenal of artificial views such as cutaways, exploded views, and so on.

Surveying the styles and techniques of picturing developed from the modest beginnings in the late Middle Ages up to the famous drawings of Leonardo and to the splendid *Theatres of Machines* at the end of the sixteenth century, one encounters an admirably rich world of pictorial languages that was engendered along with the advanced technologies of the age. The range of these languages—from sketches, to perspective views from a deliberately chosen fictitious viewpoint, and to thoroughly constructed projections—testifies to the wealth of graphic abilities the community of engineers and architects commanded. At the same time it mirrors the broad spectrum of competences that had become characteristic of this community—ranging from practical skill and artisanal experience to learned knowledge.

Shaping Engineering

Among the many striking features of early modern engineering drawings, there is one that deserves particular attention. In contrast to architecture, scaled orthographic plans—ground plans, elevations, sections—were the exception rather than the rule in the realm of machine engineering. In this realm, pictorial representations in a quasi-perspective style prevailed. This finding suggests that, in this age, drawings were not as indispensable a means for designing and manufacturing machines as plans and blueprints are today. In the inceptive stage of the designing process, sketches may have been employed by Renaissance engineers in by and large the same manner as by contemporary engineers—I will come back to this in a moment. But, as far as the extant engineer drawings of the age show, the subsequent manufacturing process was not guided by exact plans as it is today. Even the few known instances of employment of orthographic plans in the manufacturing process suggest that these plans served as a means of orientation rather than as a blueprint. The Renaissance engineers could apparently confine themselves to telling the craftsmen in charge of execution some decisive details and leaving the concrete shaping of the machine parts to them. However, this reliance on craftsmen was not reliance on personal experience and tacit knowledge alone. Rather, they were relying on craftsmen who were well equipped with a rich arsenal of geometrical aids developed over centuries—several drawing instruments, templates of all sorts, practitioners' techniques of creating nontrivial geometrical shapes and developing one geometric figure from another. If one takes this arsenal of practical geometrical aids into account, the machine drawings of Renaissance engineers may appear less inappropriate for the manufacturing process than at first glance. Taken together with the geometrical means at the construction site, early modern machine drawings become recognizable as means of the real construction process.

The employment of drawings in design processes deserves further attention. In these processes, drawings are not simply visualizations of ideas. Rather they function as material means that shape ideas. Their role is that of models that simultaneously provide a fictitious and a real opportunity to test possible arrangements of machine parts, try out new combinations and alternative shapes of these parts, and so on. As material creations in space, drawings are subjected to the laws of space and thus represent real conditions as regards possible spatial relations of rendered objects. (The famous drawings by M. C. Escher, which seemingly transgress these laws, confirm this impressively.) The advantages of such flat models on paper over solid ones made of wood or clay are obvious. They can be created and changed almost instantly. Presupposing some drawing skill on the part of the engineer, their restraint to two dimensions can be compensated for almost completely. They are incomparably flexible and allow unfettered experimentation. They furthermore allow unparalleled concentration on issues of interest thanks to the possibility of omitting all interfering or distracting parts of the device in question and reducing it to its essentials. However, the limits of these models on paper are obvious as well. Being two-dimensional creations, their representational potency ends when the focus is no longer on shapes of machine parts, their spatial relations, and the kinematic significance of such relations—when physical dimensions come to the fore, encompassing the mass and force of the designed object. Nevertheless, within their limits, drawings became an indispensable means of the design process and in this way shaped the very element of the engineering practice on which its fame as an outstandingly innovative activity essentially rests.

THE VOLUME'S FOCUS AND ARRANGEMENT

As stated above, this volume focuses on the functions and significance technical drawings actually had for the professional practice of early modern engineers and architects. It will not address other interesting aspects such as the aesthetic quality of these drawings, the manifold levels of meanings that these drawings had beyond the context of engineering, their place in the visual culture of the Renaissance, and so on. But, as may be obvious after this short introductory outline, despite this concentration on the engineering context, the subject matter is so rich in aspects, dimensions, and structures that one single book cannot aspire to cover it all. Furthermore, as also stated at the beginning, the authors of this volume do not try to present a synthetic view of early modern engineering drawings in the practical context of engineering. Instead they confine themselves to detailed investigations of such drawings under a variety of viewpoints that pertain to this context. If they nevertheless claim to offer more than just a collection of articles that deal with some aspects of this theme, this confidence rests on the conviction that they have concentrated their efforts on such aspects of the topic that are essential for an adequate understanding of technical drawings as means of early modern engineering. The five parts into which the book is divided represent these aspects.

Without anticipating the short introductions that precede each of these parts, their respective focus can be indicated as follows. Part I, entitled *Why Pictures of Machines?* and containing chapter 1 by Marcus Popplow, is about basic categories for ordering the huge and extraordinarily diverse store of extant technical drawings of the age with respect to their origin, purposes, functions, and contexts, thereby providing a first survey of this material. Part II, with the title *Pictorial Languages and Social Characters*, which contains chapters 2 and 3 by David McGee and Rainer Leng, respectively, is occupied with the development of the specific style of early modern machine rendering and the question of whether and how this style can be considered a response to the different social functions of these drawings. Part III, *Seeing and Knowing*, with chapter 4 by Pamela O. Long and chapter 5 by Mary Henninger-Voss, addresses knowledge linked with technical drawings, be it the knowledge presupposed and/or conveyed by them, or knowledge that is presupposed but cannot be conveyed. Part IV, *Producing Shapes*, with chapters 6 through 8 by Filippo Camerota, Wolfgang Lefèvre, and Jeanne Peiffer, respectively, focuses on the development of drawing techniques that required, either from the beginning or in the course of their evolution, familiarity with learned knowledge such as geometry or geometric optics. Part V, finally, *Practice Meets Theory*, with chapter 9 by Michael S. Mahoney, is dedicated to technical drawings at the interface between practical and theoretical mechanics.

To a certain extent, particularly as regards the fifteenth century, the arrangement of the volume also reflects main stages in the historical development of early modern engineering drawings. After chapter 1, which provides an analytical survey of the technical drawings of this age, chapters 2 and 3 discuss the emergence and the first stage of the specific early modern type of machine rendering in fifteenth-century Italy (Taccola) and Germany *(Büchsenmeistertraktate)*, respectively. Whereas chapter 7 contains an outline of the development of architectural plans up to the first decades of the sixteenth century, chapter 4 presents the mature stage that machine drawings achieved with Giorgio Martini and Leonardo da Vinci in the second half of the fifteenth century, a maturity that was not surpassed by the succeeding development of early modern engineering drawing. This further development is not traced coherently in this volume, with the conspicuous consequence that the famous *Theatres of Machines* from the last decades of the sixteenth and the beginning of the seventeenth centuries are not discussed as a special issue. This omission resulted mainly from the volume's focus on the functions and significance technical drawings had in the engineering practice. Although partly deriving from those of immediate practical context, and although constituting in no way a separate or particular genus of engineering drawings, the drawings of the *Theatres* served the propagation rather than the real practice of engineering. The same holds for the rightly renowned woodcuts in Georgius Agricola's *De re metallica* from 1556. The fact that this treatise and those *Theatres*, the majority of which is accessible through exemplary modern editions, have enjoyed thorough scholarly attention in the last decades made the decision easier to refrain from addressing them in favor of much less investigated issues such as decision making by means of plans (chapter 5), drawing techniques of a learned character (chapters 6 through 8), and the relation between engineering drawings and theoretical mechanics (chapter 9).

ACKNOWLEDGMENTS

The idea of this volume arose after a small conference on engineering drawings of the Renaissance, which was organized by the editor together with David McGee and Marcus Popplow and held at the Max Planck Institute for the History of Science, Berlin, in summer 2001. The authors who eventually joined in this book project circulated drafts of their chapters several times and met in winter 2002, again at the MPI in Berlin, for a thorough discussion of each contribution as well as of the focus and arrangement of the volume as a whole. My first thanks go to them. Seriously engaged in the volume's topic and cooperating in an unusual spirit of good fellowship, they never failed in promptly and patiently reacting to all of the greater and smaller demands such a book entails. Our joint work was an exciting and rewarding experience!

Warm thanks also go to James Bennett, Raz Chen, Judith V. Field, Bert S. Hall, Volker Hoffmann, Alex G. Keller, Antoinette Roesler–Friedenthal, and Thomas B. Settle, who contributed to the 2001 conference and discussed earlier versions of a number of the volume's chapters both critically and constructively.

The planning, organization, and realization of such an extensive and costly production process was only possible with the backing of the Max Planck Institute for the History of Science, Berlin. Jürgen Renn, who runs Department I of the institute, where preclassical mechanics constitutes the subject of one of the current long-term projects, provided the requisite support for the enterprise and accompanied it through its course with unceasing sympathy and encouragement. Many thanks!

Furthermore I would like to thank Jed Z. Buchwald for the decision to include the book in his Transformations series. Dealing with a topic that relates to the history of both science and technology, the volume could not have found better company.

Like machines, to actually come into being, books need to be not only designed (written) but also manufactured. My thanks go therefore to Angelika Irmscher, Heinz Reddner, and Susan Richter, who assisted in the editorial work and helped to bring the manuscript with all the figures to its final camera-ready form. Finally, I want to thank Sara Meirowitz, responsible for the science, technology, and society books at the MIT Press, who discreetly and reliably provided every support needed.

PART I
WHY PICTURES OF MACHINES?

INTRODUCTION TO PART I

In the pictorial world of the Middle Ages and the Renaissance, depictions of technical objects occur everywhere, in paintings and frescoes, on stained-glass windows, in reliefs carved in wood, chiselled in stone, cast in bronze, in illuminated Bibles, prayer-books, books of hours, on single-sheet woodcuts and engravings, in manuscripts and printed treatises with illustrations, on sketches, in notebooks, on plans—everywhere. Only a small fraction of them are technical drawings and thus subjects for this book. Presupposing a rather pragmatic definition, the authors of this volume consider those drawings to be technical that were either traced (or commissioned to be traced) and used by technicians in the pursuit of their professional life or derived from such practitioners' drawings. Of these technical drawings, only machine drawings and a few architectural drawings are dealt with in this volume. However, despite this concentration, the number of drawings addressed here is still huge. According to some experts' estimation, for the period 1400–1700 alone, one has to reckon with five to ten thousand drawings of machines and machine parts.

The unsatisfactory vagueness of such guesses results from the fact that nobody knows how many such drawings might be buried in such locations as the archives of states, cities, dioceses, monasteries, and princely families, and in the manuscript departments of libraries and museums. The expectation that hitherto unknown materials will surface there in the future rests primarily on a suspicious feature of the known material. The bulk of this material consists in presentational drawings that were published in booklets and books—manuscript books *(Bilderhandschriften)* as well as printed ones—in the early modern period. Only a small part consists of workshop drawings pertaining either to the documents of commissioners of machinery and buildings or to the private store of engineers and architects themselves. This fact can certainly be explained by the assumption that drawings of the design and construction process of technical artifacts were not kept but thrown away after a certain span of time. However, the unbalanced ratio of the presentational and the workshop shares of the known machine drawings of this period may also result, at least to some extent, from the storage and display policy of archives and libraries in past eras, when the cultural divide between the realm of technology and the realm of fine arts and literature was still prevalent.

For a first orientation, table I.1 showing the most important sources of early modern engineering drawings, to which the chapters of this volume frequently refer, may be convenient.

The prevalence of the presentational over the workshop material among the extant engineering drawings poses a serious problem to our enterprise. For a picture of the actual role technical drawings played in the practice of engineers and architects, workshop drawings are naturally of far more significance than the presentational ones. The latter may be telling in this respect as well. But only on the basis of thorough investigations of the drawings actually used in the design and construction processes will one be able to determine to what extent and through which of their features certain presentational drawings, too, give testimony to the use of drawings in early modern engineering.

Table I.1. Prominent Sources of Early Modern Machine Drawings		
Time	Workshop Drawings	Presentational Manuscripts or Books
1250	Villard de Honnecourt (A, B)	
1300		
1350		Vigevano (C2)
1400		Kyeser (C2)
		Master Gun-makers' Booklets (C1)
1450	Taccola (B)	Taccola (C2)
		Valturio (1472) (D1)
		Anonymous of the Hussite Wars (C1)
		Giorgio Martini (D1)
1500	Leonardo da Vinci (B)	
	A. da Sangallo (B)	Vitruvius (1521) (D3)
		Tartaglia (1537) (D2)
1550		Agricola (1556) (D1)
		Ceredi (1567) (D1)
		Besson (1569) (C3)
		Monte (1577) (D2)
	Holzschuher (A, B)	Errard (1584) (C3)
	Schickhardt (A, B)	Ramelli (1588) (C3)
		Pappus (1588) (D3)
		Heron (1589) (D3)
		Lorini (1597) (D1)
1600		Zonca (1607) (C3)
		Zeising (1607) (C3)
		Caus (1615) (C3)
		Strada (1617) C3)
		Branca (1629) (C3)

A: Design and construction drawings
B: Sketch-books and notebooks
C: Collections of drawings with or without explanatory text
  C1: Practitioner booklets
  C2: Representational manuscripts
  C3: Theatres of machines
D: Drawings in treatises and editions
  D1: In technological treatises
  D2: In treatises on mechanics
  D3: In editions of classical sources

Likely the most crucial problem for an interpretation of technical drawings of the early modern period is the determination of the probable purpose, or purposes, a certain drawing served. Even in the case of drawings in books with an explicit introduction, this purpose can be dubious when, for instance, experts and nonexperts are equally addressed by the book. On the other hand, as regards workshop drawings, it is sometimes impossible to determine the purpose. "Why pictures of machines?"—this seemingly simple question proves to be an intricate one. Familiarity with a considerable amount of the extant material is presupposed to outline some characteristic features of these engineering drawings and to propose convincing and useful categories according to which this material may be ordered. Presenting such an outline and proposing those categories, chapter 1 by Marcus Popplow not only provides an analytic survey of the subject of this volume but also prepares the ground for the subsequent chapters.

WHY DRAW PICTURES OF MACHINES?
THE SOCIAL CONTEXTS OF
EARLY MODERN MACHINE DRAWINGS

MARCUS POPPLOW

INTRODUCTION

Early modern machine drawings long have been studied with the purpose of recon-structing details of the machine technology employed in their age of origin.[1] In this context, two distinct groups of sources traditionally have received broad attention and by now, for the most part, have been edited: the numerous manuscripts by Leonardo da Vinci and the representational machine books. From the late Middle Ages the lat-ter served to present spectacular engineering designs to a broader public, first in manuscript form and then in print. Regarding the reconstruction of early modern machine technology, the investigation of both Leonardo's machine drawings and the designs of the machine books always has been confronted with one central problem: It is often difficult to determine clearly the realizability of the designs presented. Thus, research on these sources long has focused on efforts to differentiate more clearly which of their designs represented machines actually in use in the early mod-ern period and which of them should rather be regarded as products of the contempo-raries' imagination.

Which role was assigned to the medium of drawing by early modern engineers themselves? And what effects did the employment of drawings have on the communi-cation of existing knowledge and the production of new knowledge on contemporary machine technology? Such questions about the practical as well as cognitive func-tions of the means of representation used by contemporary engineers have been focused on more closely only recently. This is true for the drawings considered here as well as for three-dimensional models of machines.[2] This delay corresponds to the fact that even today, sixteenth-century engineering drawings with more practical functions, preserved as single sheets or personal sketch-books, are to a great extent unpublished and thus less accessible.[3] This chapter places special emphasis on such

---

1   The topic of this chapter has been presented to a workshop at the Max Planck Institute for the History of Science (Berlin) and a *Journée des Études* at the Centre Koyré (Paris). I am grateful to the participants for their commentaries and suggestions, in particular for the extensive discussion by Pamela O. Long. An earlier version of this chapter has been published as "Maschinenzeichnungen der 'Ingenieure der Renaissance'" in *Frühneuzeit-Info* 13(2002), 1–21.
2   See Ferguson 1992, Hall 1996, Lefèvre 2003. For the related topic of the visual representation of the trajectories of projectiles, see Büttner et al. 2003. For the early modern employment of scaled-down models of machines, see Popplow (in press).
3   In the pioneering study by Ferguson (1992) on visual thinking in the history of engineering, Leonardo's manuscripts and presentational machine books are the only sources from the fifteenth and sixteenth centuries.

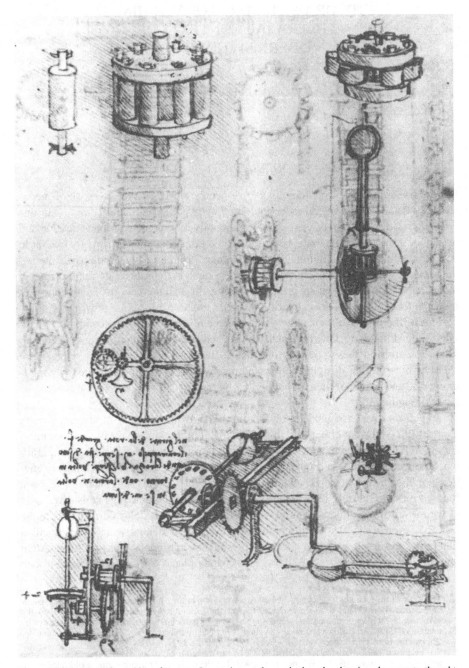

Figure 1.1. Studies of machine elements. It remains unclear whether the drawing documents thought experiments or objects actually assembled in the workshop. Drawing by Leonardo da Vinci. (Photo: Biblioteca Nacional Madrid, Codex Madrid, fol. 10ᵛ.)

less formal drawings not addressed to a broader public. It is such sources that document how the medium of "drawing" was indispensable for planning, realizing, and maintaining large-scale technological projects in the early modern period.

The aim of this chapter is to work out a classification of the contexts in which early modern machine drawings were employed.[4] This task confronts a number of difficulties regarding the interpretation of corresponding source material. For Leonardo's machine drawings, it long has proved difficult to determine the purposes they originally served (figure 1.1). Some have been identified as proposals for innovations of specific mechanical devices, for example, his series of drawings on textile machines (figure 1.2). Others have been interpreted as didactic means of conveying his tremendous knowledge on the behavior of machine elements to others, and some obviously served theoretical functions.[5] It furthermore has been suggested that Leonardo used drawings for recording trials with three-dimensional objects made in his workshop.[6] As regards the wealth of machine drawings preserved from early modern authors other than Leonardo, it has been argued convincingly that these must be differentiated according to their functions of documentation, communication, or design.[7] However, confronted with the source material considered below, which early modern engineers employed in the

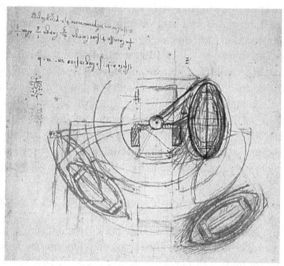

Figure 1.2. Detail of machine for weaving braids. Drawing by Leonardo da Vinci. (Codex Atlanticus, fol. 884[r].)

process of realizing mechanical devices, it has proved extremely difficult to assign precisely these functions to such drawings. With regard to these difficulties of interpretation, a different approach is taken here. As a first step, sixteenth-century machine drawings are differentiated according to four main contexts of employment: First, they served to present devices to a broader public; second, they could take on a role in the concrete manufacturing process; and third, they could constitute part of an engineer's personal archives. Fourth, and this final group to some extent amounts to a special case, engineering drawings could merge into or be connected with considerations of a more theoretical nature. Of course, such a classification does not exclude

---

4 For a classification of medieval technical drawings, see Knobloch 1997.
5 See Truesdell 1982, Maschat 1989, and Long (this volume).
6 See Pedretti 1982 and Long (this volume).
7 See Lefèvre 2003.

the possibility that one and the same drawing could be employed in more than one of these four contexts over the course of time.

The following remarks are limited to describing the situation in the sixteenth century without investigating the question of the origins of the employment of machine drawings for more practical purposes in the Middle Ages. With the exception of the numerous illustrated gunners' manuals,[8] early engineering drawings from the fourteenth and fifteenth centuries have been preserved almost exclusively in the context of the production of presentational manuscripts. However, it is hard to imagine that the numerous fifteenth-century manuscript machine books could have been produced without any foundation in some less formalized practice. Scattered textual evidence that still awaits closer investigation indeed testifies to a more informal employment of machine drawings as early as the beginning of the fifteenth century.[9] Furthermore, it must be noted that any attempt to explain when and why drawings came to be employed in mechanical engineering in the Middle Ages must take into consideration the tradition of late medieval architectural drawings.[10] As the different roles of machine builder, architect, and fortification engineer emerged more clearly only in the course of the sixteenth century, it can be assumed that the employment of such a crucial medium as drawing in earlier periods still showed similar characteristics in all three of these fields.[11] As the focus of this chapter lies on the social contexts of employing machine drawings in the sixteenth century, neither will the development and use of different graphic techniques—most prominently, changes induced by the invention of perspective in the fifteenth century—be investigated here.[12]

1. PRESENTING DEVICES TO A BROADER PUBLIC

The above-mentioned machine books in manuscript[13] and in print[14] served to present machines to a broader public, formally continuing a manuscript tradition dating back to Antiquity and the Arab Middle Ages.[15] This public initially consisted of courtly audiences before expanding ever more to learned laymen and fellow technical experts during the sixteenth century. Drawings and later woodcuts and engravings allowed

---

8   See Leng (this volume).
9   Documents from the *fabbrica* of Milan cathedral mention in passing that proposals for a mechanically driven stone-saw were to be submitted first in the form of a drawing before the most promising designs were required to be presented as scaled-down models. See Dohrn–van Rossum 1990, 204–208.
10   See Lefèvre (this volume).
11   Contexts of employing architectural and fortification drawings in the early modern period have received only scarce attention to date. See Schofield 1991 and Frommel 1994a. Architectural treatises from the fifteenth century onwards often contain explicit discussions of the role of the medium of drawing in the design process. See Thoenes 1993.
12   See Ferguson 1992, Lamberini (in press), and Camerota (this volume). For the density of information conveyed and the broad variety of graphic techniques employed in Leonardo's machine drawings, see Hall 1976b; Heydenreich et al. 1980; and Galluzzi 1982. The argument brought forth by Samuel Edgerton, according to which geometrically constructed perspective drawings in sixteenth-century machine books paved the way for the "geometrization of nature" in the "Scientific Revolution" has, by now, been refuted convincingly. See Mahoney 1985 and Hall 1996, 21–28.
13   See Hall 1982a, Hall 1982b, Galluzzi 1993, Galluzzi 1996a, Friedrich 1996, Leng 2002, Long 2001, 102–142, and Leng (this volume).
14   See Keller 1978, Knoespel 1992, and Dolza and Vérin 2001.
15   See Lefèvre 2002 and Hill 1996.

these audiences to study siege engines, mills, water-lifting devices, and other examples of early modern machine technology. In the fourteenth and fifteenth centuries, these manuscripts contained mostly military devices. Well-known examples are the manuscripts assembled by Guido da Vigevano (c. 1335), Konrad Kyeser (1405), Mariano Taccola (1449), Roberto Valturio (1455, printed in 1472), and the author known as "Anonymous of the Hussite Wars" (c. 1470/1480). The first pioneering manuscripts showing engines for civil purposes were composed in Italy, again, by Mariano Taccola (c. 1430/1440) and Francesco di Giorgio Martini (c. 1470/1480). While large devices for military and civil purposes had existed only rather vaguely in the visual memory of medieval contemporaries, this situation now changed, at least for those among whom these manuscripts circulated. With the regard to the works of Konrad Kyeser, Mariano Taccola, and Francesco di Giorgio Martini, some dozens or even hundreds of copies of the original manuscripts have been discovered. They still await closer investigation concerning the questions of who commissioned them and who was responsible for the artistic process of producing the manuscript copies.[16] Towards the end of the sixteenth century, with the printed machine books of Jacques Besson (1578), Jean Errard (1584), Agostino Ramelli (1588), Vittorio Zonca (1607), Heinrich Zeising (1612ff), Salomon de Caus (1615), Jacopo Strada (1618), and Giovanni Branca (1629), machines for civil purposes became a subject of learned knowledge as well. In addition to the printed *Theatres of Machines,* a number of manuscripts have been preserved, which very likely document a preparatory stage of publication. One example is a manuscript version of Jacques Besson's "Theatrum instrumentorum et machinarum" (1571/72), which was later published posthumously.[17] The intention of publication also can be presumed in the case of a manuscript of the Florentine scholar Cosimo Bartoli (c. 1560/70),[18] which already shows typical traits of machine books: complete views of devices as well as additional detailed views, carefully ordered text sections and labels of reference (figure 1.3).

Late medieval and early modern machine books all have a comparable structure: Full-page images of technical devices are each accompanied by a more or less detailed text explaining their general features. While these explanations often consist only of a few lines in the early manuscripts—in some cases, there are no textual explanations at all—in sixteenth-century works, the length of the explanatory texts grew considerably. This is especially true for the printed machine books. Yet to be investigated is the question as to whether this growth of textual information corresponded to a shift in the contexts of employment of these works. It is well possible that some authors of the earlier manuscripts assumed that the inspection of their manuscripts would be accompanied by oral explanations. Authors of the printed sixteenth-century works, in contrast, from the start had to presuppose a "silent reader" who had to be provided with more detailed explanations of the functioning of the devices presented.

---

16 See Leng 2002 and Scaglia, 1992. For the manuscripts of Konrad Kyeser, see Friedrich 1996.
17 See Keller 1976.
18 See Galluzzi 1991, 223.

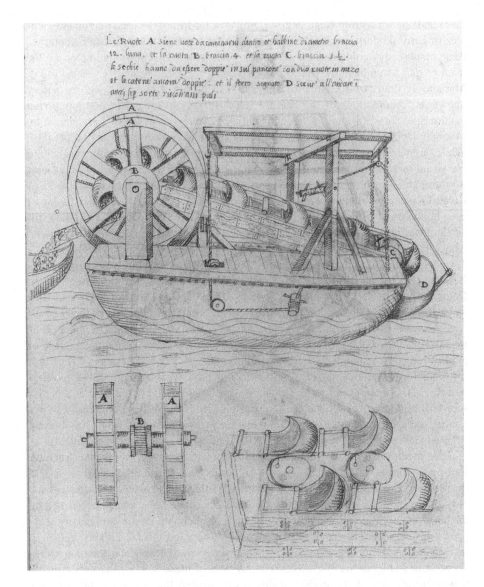

Le Ruote A. sieno uote da caminarui dentro et habbino diametro braccia
12. luna, et la ruota B. braccia 4. et la ruota C. braccia 3½.
le secchie hanno da essere doppie in sul pancone con duo ruote in mezo
et le catene ancora doppie : et il ferro segnato D secur all'ancore i
anzi sep sorte riscotrassi pali

Figure 1.3. High-quality drawing of a dredger with didactic implications. Note the details of the mechanism drawn separately below. Drawing by Cosimo Bartoli, c. 1565. (Florence, Biblioteca Nazionale Centrale, Palatino E.B. 16.5 (II), fol. 60[r], courtesy Ministero per i Beni e le Attività Culturali, all rights reserved.)

Regarding techniques of graphic representation, the perspective illustrations in printed books on machines did not differ much from fifteenth-century presentational manuscripts: An elevated viewpoint enabled the spectator to discern machine elements that would remain hidden if the device were represented from the front on

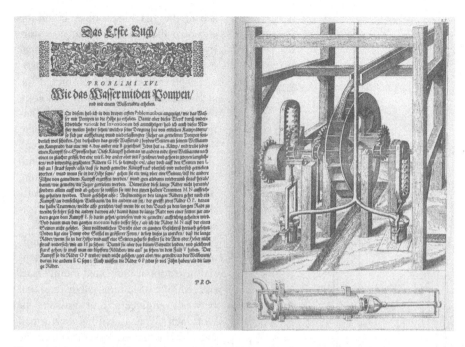

Figure 1.4. Presentation of pumps in a printed machine book. Below a separate drawing to emphasize technical details of the arrangement to drive the pumps' pistons. (Caus 1615, fol. 23^r.)

ground level. However, the perspective techniques now had become more refined. In continuation of fifteenth-century manuscripts, total views of a device were accompanied by separate drawings of technical details usually also rendered in perspective (figure 1.4). Other graphic techniques like horizontal or vertical sections or ground plans, more difficult to read for the lay spectator, are found only rarely in printed machine books (for a late exception, see figure 1.5).

It has often been stressed that representational manuscripts and printed machine books mirrored technical reality only to a very limited extent. Over time, however, this interpretation has changed considerably. In earlier research it was sometimes argued that the authors of these books did not yet dispose of modern exactness in their technological descriptions. More recently, such "playfulness" has been interpreted as a response to specific expectations placed on technical experts, especially in the context of court culture. From this perspective, early modern machine books appear as a distinctive genre characterized by a carefully selected information: "Unrealistic" designs in general might well be interpreted as expressions of anticipated future achievements. Usually, such designs represented combinations of machine elements, which themselves were already employed in practice. Indeed, authors of the printed machine books often stressed that the designs they presented also were to inspire their colleagues to try out ever new combinations of machine elements to improve traditional machine technology and to extend its fields of application. In this

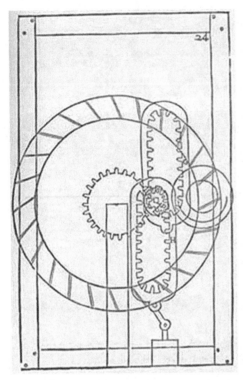

Figure 1.5. Vertical section of machine elements of the water-lifting device shown in figure 4. (Caus 1615, fol. 24ʳ.)

sense, the machine books could be ahead of their time without necessarily losing their relation to technical practice. A number of European territorial powers explicitly promoted the application of mechanical technology by granting privileges for the invention of newly designed mechanical devices.[19] This practice, which spread all throughout Europe in the sixteenth century, provides a very concrete background for the sense of experimentation conveyed by the broad variety of designs in the machine books.

Machine books served the communication between engineers and potential investors or interested members of the republic of letters, rather than that between engineers and artisans or workmen. The functions performed even by the late medieval manuscripts were manifold: They could serve to entertain or to present factual information, or to prove the erudition of princely commissioners and promote the self-advertisement of technical experts. Such implications of the practice of authorship in early modern engineering have only recently began to be investigated more closely.[20] Assembling textual and visual information on machine technology with the aim of presentation to others created a new kind of reflecting knowledge. This new kind of knowledge distinguished the engineer from the ordinary artisan and thus underlined the legitimacy of claims to higher social status of these technical experts.

Even though machine books served as a kind of visual inventory of contemporary technical ideas even among the engineers themselves, they played only a marginal role for engineers' everyday practice. The materiality of technology was often ignored in these presentational treatises. Machines in these books should be understood as a product of the engineer's brain, his *ingenium*; their material realization was not the topic of these books. The organizational activities of the engineer on the building site were mentioned as scarcely as materials, measurements or gear ratios— it was considered self-evident that such factors had to be established at a later point

---

19 See Popplow 1998b.
20 See Long 2001.

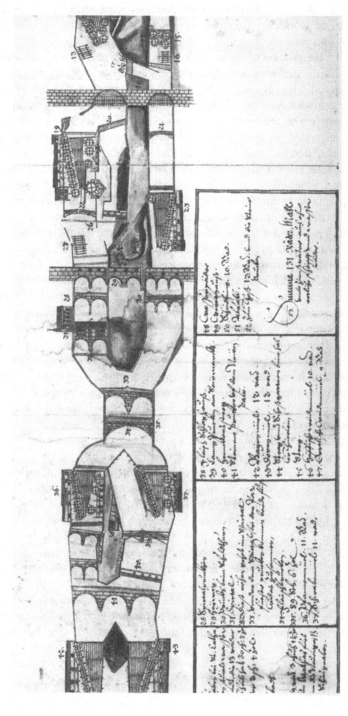

Figure 1.6. Survey plan by the town's master builder Wolf Jacob Stromer of the Nuremberg mills alongside the river Pegnitz (detail), 1601. (Photo: Germanisches Nationalmuseum Nürnberg, HB 3089, Kaps 1055h.)

of time at the site. However, illustrations of machines were often incorporated in lively landscapes or workshop scenarios to suggest the possibility of immediate employment.

Information on the process of creating illustrations for presentational treatises is scarce. For most of the earlier manuscripts, the persons known as "authors," such as Guido da Vigevano, Konrad Kyeser, or Roberto Valturio, were responsible only for the texts and the composition of the treatises and commissioned the production of the illustrations to artists who remained anonymous. The selection of these artists and how they were instructed about which devices they had to illustrate, and how and with which technical details, remains unclear. This is also true for the woodcuts and engravings in the later printed works. That artists visited each machine at its original site is documented, as an exception, in the case of Georgius Agricola's preparations for his encyclopaedia on mining, *De re metallica,* published in 1556. His letters show that he had to send different artists to the mining centres of Saxony several times until he found one who produced drawings of the machines employed there in a quality sufficient to serve as templates for the woodcuts.[21] It is difficult to imagine that this was a standard practice, however; artists sometimes might have drawn from three-dimensional models of machines or from some sort of sketch. In any case, the production process of such books on machines presupposes some tradition of more informal machine drawing, of which only faint traces have been preserved from the period preceding Leonardo da Vinci's notebooks.[22]

In addition to their incorporation in machine books, machine drawings with representative functions have also been preserved as single leaves. A special case is provided by plans kept in communal archives representing, for example, a town's waterways. From early fifteenth-century Basle, such a plan has been preserved with a coloured scheme of the different waterworks crossing the town. It is assembled of pieces of parchment and is, in total, nearly ten meters long.[23] In other cases, such plans also depicted water-mills alongside the town's waterways in symbolized form, that is, as mill-wheels turned ninety degrees laterally (figure 1.6).

A different example of a carefully composed machine drawing with representative functions has been preserved among the documents of Württemberg master builder Heinrich Schickhardt (1558–1635). It shows a device that had been built a short time earlier by the carpenter Johannes Kretzmaier, probably under supervision of Schickhardt himself, to provide the castle of Hellenstein near Heidenheim with water (figure 1.7).[24] Between the source and the castle, a height difference of a total of ninety meters had to be overcome. The drawing emphasized only the core element of the transmission machinery: a combination of a lantern and an oval rack. It served to transfer the rotary motion provided by the water-wheel to the reciprocating motion of the horizontal beam driving the piston rods of the pumps. Indispensable construction details of the device, like the guide rails of the rack, are missing. The lower part

---

21 See Kessler–Slotta 1994.
22 See McGee (this volume).
23 See Schnitter 2000.
24 See Müller 2000.

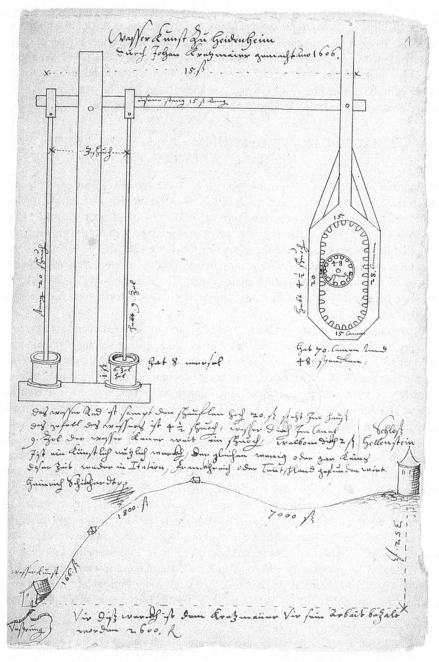

Figure 1.7. Commented presentational drawing of the pumps supplying Hellenstein castle. Drawing by Heinrich Schickhardt, 1606 or after. (Stuttgart, HStA, N220 T149, all rights reserved.)

of the drawing serves to visualize the concrete setting of employment: at the bottom left, the building that housed the device; above right, Hellenstein castle. In addition to specifications of measurements and the performance of the device, a written commentary signed by Schickhardt testifies to the documentary character of the sheet: "This is an artificial and useful device of which only few or even none are to be found neither in Italy, nor in France or Germany in these times."[25] The exact purpose for which this drawing had been produced nevertheless remains unclear.

2. MACHINE DRAWINGS IN THE PROCESS OF REALIZING MECHANICAL DEVICES

Starting in the late Middle Ages, the separation of the social roles of engineer and artisan became more clearly discernible. While the former was responsible for the design and the organization of a given project, the latter carried out the actual work. This development is discussed as one of the central prerequisites for the growing relevance of drawings to mechanical engineering since the late Middle Ages.[26] In the sixteenth century, in any case, drawings in the process of realizing mechanical devices served the engineer to communicate with the investor on the one hand and (although presumably to a lesser extent) with the artisans carrying out the work on the other.

Communication with the investor especially concerned the preparatory stage of realizing mechanical devices. With regard to the competition among early modern engineering experts, drawings could serve to present engineers' abilities at a foreign court, even though for such purposes the demonstration of scaled-down models presumably was preferred, because of the more immediate impression it created. Both of these media played a central role in the above-mentioned practice of granting privileges for inventions.[27] Applicants for such a privilege in a certain territory often submitted drawings or models to underline the credibility of their inventions. Such presentations, however, were not necessarily required, as in any case the inventor had to prove the realizability of his invention after the privilege had been granted by constructing a test specimen in full size in the course of the subsequent six to twelve months. In some cases, applicants presented a whole set of inventions by means of illustrated manuscripts, which ultimately strongly resembled manuscript machine books. This was true in the case of the above-mentioned manuscript by Jacques Besson, the designs of which were protected by a privilege in 1569 when the compilation was presented to King Charles IX.[28] A quite similar manuscript was composed in 1606 by the Spanish engineer Jerónimo de Ayanz for King Philipp III.[29] To obtain a privilege for inventions, Ayanz presented drawings of forty-eight of his inventions,

---

25   "Ist ein künstlich nutzlich werckh, der gleichen wenig oder gar keins dieser Zeit weder in Italien, Franckhreich oder Teütschland gefunden wirt." Stuttgart, HStA, N220 T149.
26   See McGee (this volume).
27   See Popplow 1998b.
28   See Keller 1976, 76.
29   See Tapia 1991, 53–256.

most of which were presented as sketches in perspective and accompanied by extensive descriptive texts.

A manuscript composed with quite different intentions, under the supervision of Duke Julius the Younger of Brunswick-Wolfenbüttel around 1573, represents an attempt to employ catalogues of designs of mechanical devices for concrete regional innovations.[30] The manuscript assembled illustrations of devices and instruments that were to facilitate and speed up labor in the quarries and on the building sites of the duchy. Moreover, it contained a list of persons active in the duchy's administration and declared that they were obliged to consult the volume accordingly to improve the technical equipment available at the sites for which they were responsible. Some of the illustrations had been copied from earlier manuscripts, others show instruments reportedly already employed elsewhere in the duchy, and a few represent inventions allegedly made by Duke Julius himself (figure 1.8). While a number of formally similar manuscripts mentioned above seem to have been composed rather for reasons of prestige or entertainment at court, this manuscript was thus composed with the aim of practical employment. In this context, the designs of the Wolfenbüttel manuscript clearly refer to local circumstances in the Wolfenbüttel duchy, a trait that is not discernible in other cases. Attempts to turn designs encountered in such manuscripts into practice are, of course, quite conceivable in other cases as well, but they have not yet been documented. With the deliberate intention of realization after his death, the unique sketches of mills left behind by Nuremberg patrician Berthold Holzschuher were similarly meant to serve as a guideline to construction, although in a purely private context.[31]

In the concrete process of realizing mechanical devices, drawings helped the engineer to bridge the different locations of decision processes and the actual realization of a project—the court or the town hall and the building site. These contexts are especially easy to discern with regard to examples from the broad collection of some two to three hundred loose leaves containing drawings of all sorts of mills and water-lifting devices by the above-mentioned Heinrich Schickhardt. Schickhardt, who served the dukes of Württemberg for decades as master builder and engineer,[32] in general does not appear as an ingenious inventor of new devices, but rather as somebody trying to provide the duchy with up-to-date technology that had already proven its efficacy elsewhere. In contrast to machine drawings in Leonardo da Vinci's manuscripts, composed roughly one hundred years earlier, Schickhardt's collection contains no theoretical reflections at all, whereas the relationship of his drawings to actual technical projects is extensively documented. Schickhardt's ability to employ all kinds of graphic techniques for drawing mechanical devices might have been above average. Nevertheless, it seems that his drawings represent an extraordinary case of preservation rather than an extraordinary way of using the medium. Thus they most likely testify to standardized practices in early modern engineering.

---

30 See Spies 1992.
31 See Leng (this volume).
32 See Schickhardt 1902, Popplow 1999, Bouvard 2000.

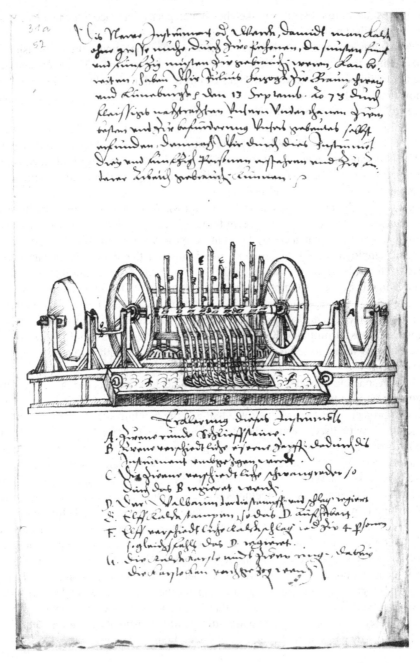

Figure 1.8. Presentational drawing of a device for stamping and mixing lime allegedly invented by Duke Julius of Brunswick-Wolfenbüttel, c. 1573. (Photo: Niedersächsisches Staatsarchiv Wolfenbüttel, Instrumentenbuch I, fol. 31^r.)

Around 1600, Heinrich Schickhardt supervised the building of several mills in Montbéliard, a project extensively documented in his papers.[33] One of the devices realized was a paper-mill. A survey drawing of this mill shows, in an idealized way, the most important parts of its inventory—two of the basins where soaked rags were reduced to pulp as the raw material for the production of paper have been carefully omitted to leave space to show such components as the mill's press (figure 1.9). At

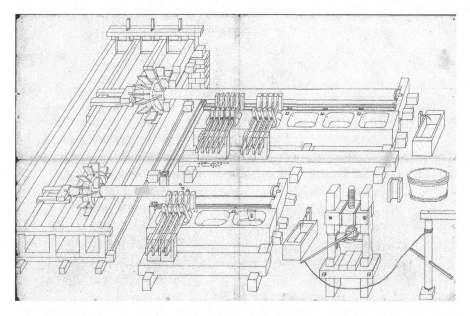

Figure 1.9. Inventory of a paper-mill in Montbéliard. Drawing by Heinrich Schickhardt, c. 1597. (Stuttgart, HStA, N220 T193, all rights reserved.)

which point of time the drawing was made is not definitively clear, but it is very likely that it was composed before the mill was actually built. A tiny comment written below the upper left basin says: "There shall only be four stamps in one hole" (instead of five as shown here). And indeed, the stamp at the extreme left is marked as obsolete by several diagonal lines. Most probably, a drawing like this was presented to the Duke of Württemberg for his formal approval or to keep him informed about such a costly project. Why the changes to details of the design were later documented in the way seen here remains unclear, however. Among several more detailed drawings of parts of this paper-mill is one showing a vertical section of one of the stamps, complete with measurements and, again, disclosing several corrected features (figure 1.10). Others concern the press, for example. Plans of the different storeys of the building have been preserved as well.

---

33   See Bouvard 2000, 63–77.

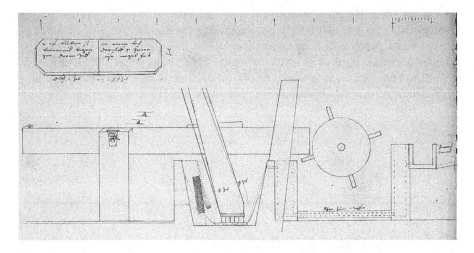

Figure 1.10. Vertical section of the cam-shaft and one of the stamps of a paper-mill in Montbéliard. Drawing by Heinrich Schickhardt, c. 1597. (Stuttgart, HStA, N220 T186, all rights reserved.)

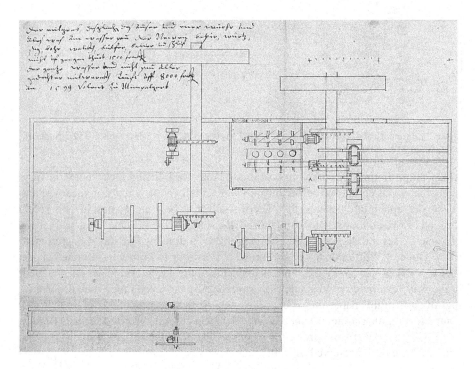

Figure 1.11. Ground plan of a fulling-/stamp-/grinding-/polishing-/drilling-/sawmill in Montbéliard, earlier version. Drawing by Heinrich Schickhardt, c. 1597. (Stuttgart, HStA, N220 T182, all rights reserved.)

Documents pertaining to another of the devices realized by Heinrich Schickhardt in Montbéliard document well the role of drawings in the relationship between engineer and artisan. In the course of constructing a combined fulling-/stamp-/grinding-/polishing-/drilling- and sawmill, Schickhardt again used different kinds of graphic representations, among them two ground plans. The first of these plans is a preliminary study of the disposition of the different mechanisms most probably rendered before the mill was actually built: A closer look reveals one set of stamps crossed out and a small note says that the water-wheels have to be set a greater distance from each other (figure 1.11). The latter addition shows that Schickhardt produced such plans to scale, and a roughly drawn scale is indeed to be found on the plan near the water-wheels. The second plan of the same mill shows that these changes had been carried out (figure 1.12).

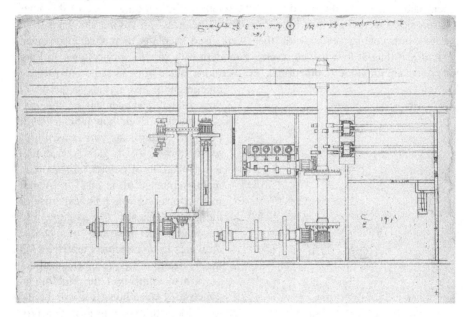

Figure 1.12. Ground plan of a fulling-/stamp-/grinding-/polishing-/drilling-/sawmill in Montbéliard, later version. Drawing by Heinrich Schickhardt, c. 1597. (Stuttgart, HStA, N220 T182, all rights reserved.)

The interesting thing about these two plans is that the second plan, at least, not only served to record what Schickhardt had planned, but was also part of the contract between Schickhardt and the carpenter who actually built the mill—as becomes clear from copies of documents preserved together with these drawings. Schickhardt, like most of his colleagues, was always engaged in several projects in different places at any given time. As he himself once wrote, he was in most cases responsible only for the design of a building or machine. He left plans and other information for the artisans to use, coming back weeks or months later to check on the realization of the project. In the case of the mill discussed here, Schickhardt composed a document on behalf of the Duke of Württemberg on 24 October, 1597, which specified how the mill was to be built by the carpenter. The text included the remark that "everything concerning the mechanisms and the rooms should be made properly and diligently according to the drawing."[34] The drawing mentioned is the second plan, which indeed corresponds in detail to Schickhardt's written description. At a later point in time, Schickhardt again noted the change of one detail, both on the plan and on the margin of the written document: The carpenter had provided the axle of the spice mill with three cams. This, however, resulted in the mill working "too fast." Two cams, Schickhardt remarked, sufficed in this case. The importance thus placed on this detail is somewhat puzzling, however, because the document for the carpenter had mentioned only the gearing of the mills without specifying such details as the number of teeth on the toothed wheels. This is also true for other aspects of the project. The document laid down only the breadth and the width of the building; none of the other measurements were fixed in written form. This "openness" proves that drawings from the sixteenth century, even when they were used as plans to realize mechanical devices and thus at first glance resemble modern orthographic projections, still are not equivalent to modern blueprints. Furthermore, such drawings did not provide unambiguous instructions on the three-dimensional arrangement of the machine parts.[35] Even though Schickhardt provided the artisans with a wealth of information, a lot of "gaps" concerning the realization of certain machine elements remained to be filled in by oral instructions or through the expertise of the artisans. Finally, this example also documents the proximity of machine drawings to architectural drawings. As the realization of large mechanical devices also comprised the building in which they were housed, it can be assumed that drawings used in that process adhered to standards similar to that of plans used in the construction process of buildings, for example, larger residential houses. Such reciprocal dependencies of machine drawings and architectural drawings remain open to future investigation.

Figure 1.13. Documentation of the size of a leather disc for sealing pistons in a pump cylinder. Drawing by Heinrich Schickhardt, 1603. (Stuttgart, HStA, N220 T150, all rights reserved.)

---

34 "alles an mülwercken und gemecher dem abriß gemeß sauber und fleißig gemacht." Stuttgart, HStA, N220 T182.

35 See Lefèvre 2003.

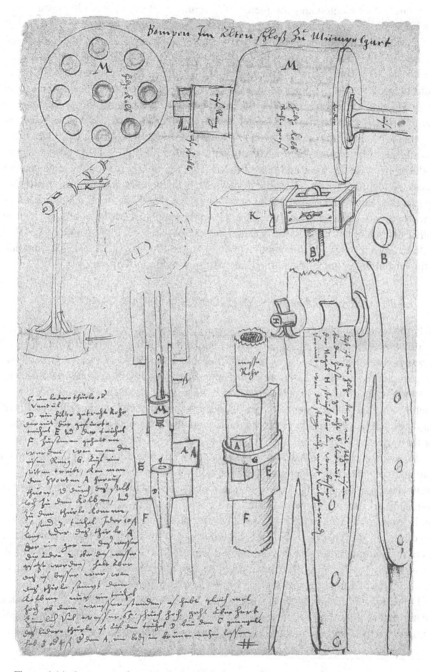

Figure 1.14. Inventory of parts of pumps for Montbéliard castle. Drawing by Heinrich Schickhardt, 1603. (Stuttgart, HStA, N220 T150, all rights reserved.)

A rarely documented and completely different type of drawing could play a minor role in the process of realizing mechanical devices: drawings determining the size of workpieces. Such dimensioning was, of course, a procedure that had long been required in any kind of building project and was solved by different means such as moulds and templates. Given its increasing availability in the sixteenth century, paper could be used for such a purpose as well. Such procedures seem especially likely in the production of the numerous toothed wheels for clockwork and automata. An example of this kind of drawing, again from Schickhardt's legacy, concerns a leather ring that served to seal up pistons moving up and down in pumping cylinders. This drawing with the remark "leather disc for the pumps"[36] (figure 1.13) probably was produced because the wear and tear of these discs frequently made their replacement necessary such that it was advisable to always have new discs at hand. Another drawing makes clearer the context of the employment of this disc: Here Schickhardt was concerned with restoring the pumps for the water supply of Montbéliard castle. The leather disc is to be found on the upper right part of the page, represented by the thin circle fixed to the right of the wooden piston marked "M" (figure 1.14). The function of this visual inventory of the parts of the pump is, again, not discernible.

3. MACHINE DRAWINGS AS ENGINEERS' PRIVATE ARCHIVES

Early modern engineers in many cases assembled personal archives with drawings of their own projects and drawings of devices realized by others. To be sure, the sorts of drawings discussed so far also could find their place in such collections. The following paragraphs, however, after briefly discussing drawings that served to illustrate engineers' own thought experiments and to document their own experiences with machines or machine elements in their workshop, will concentrate on different sets of drawings that helped engineers record the design of mechanical devices they saw during their travels.

As has been remarked above, it is still open to what extent Leonardo da Vinci's drawings of machine elements represented not only thought experiments, but arrangements of objects that had been tested in his workshop (figure 1.15).[37] In other engineers' documents known to date, hardly any drawings with these two functions can be discerned. This makes it extremely difficult to judge the role they might have played in the design practice of the fifteenth and sixteenth centuries in relation to three-dimensional arrangements or scaled-down models of machines. An early example of drawings that might be interpreted as thought experiments are a number of small studies of war ships in a manuscript by Mariano Taccola.[38]

Drawings produced by engineers while traveling are documented to a much greater extent than the thought experiments mentioned in the preceding paragraph. Parallel to artists' and architects' practices of keeping model books for reproducing

---

36 "lederne scheiblein zu den pompen." Stuttgart, HStA, N220 T150.
37 See Long (this volume).
38 See McGee (this volume).

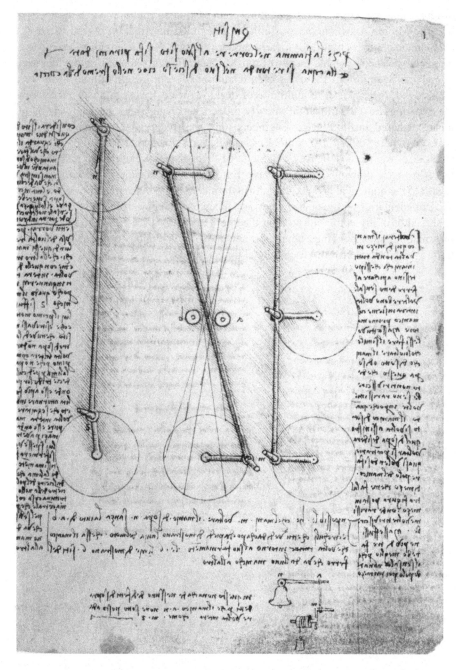

Figure 1.15. Arrangement of linkages to transfer circular motion. Drawing by Leonardo da Vinci. (Photo: Biblioteca Nacional Madrid, Codex Madrid, fol. 1^r.)

different kinds of objects at some future point of time, a tradition reaching far back
into the Middle Ages, engineers similarly assembled information on the variety of
mechanical engines employed in early modern Europe. For the engineer, such drawings were an indispensable means of quickly recording information on devices seen
elsewhere. Even if, for example, standard solutions for the design of flour-mills were
widespread, there existed a multitude of designs for devices employing more complex gearing, as standardization in early modern mechanical engineering was by no
means fostered institutionally. Engineers thus kept records of remarkable devices
seen elsewhere, either in notebooks—usually, for practical reasons, of relatively
small size—or on loose leaves.[39] In spite of their diversity, both north and south of
the Alps such drawings seem to have followed some standard conventions with
regard to the numerical and textual information conveyed. Measurements, gear ratios,
and commentaries on the device's performance appear regularly, as either personally
observed or orally communicated on the site. Especially with regard to this body of
information, such drawings obviously adhered to conventions quite different from
those which characterized presentational treatises.

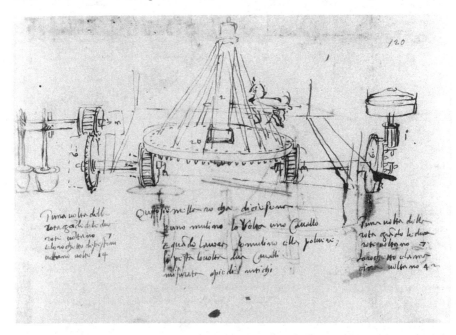

Figure 1.16. Sketch of a combined flour- and stamp-mill in Cesena. Drawing by Antonio da Sangallo the
Younger. (Florence, Gabinetto Disegni e Stampe, U1442A^r.)

---

39   See, for example, the diary of two journeys to Italy by Heinrich Schickhardt, illustrated with numerous
     drawings of buildings and machines. Schickhardt 1902, 7–301.

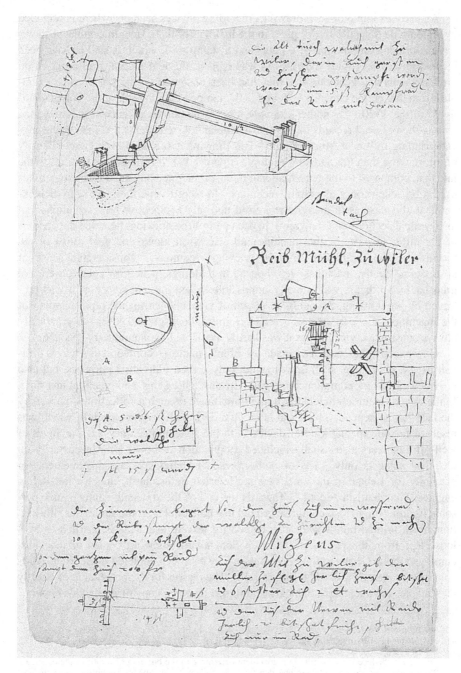

Figure 1.17. Sketches of a combined fulling- and grinding-mill at Wiler near Montbéliard. Drawing by Heinrich Schickhardt, c. 1610. (Stuttgart, HStA, N220 T241, all rights reserved.)

An early example of such drawings recording devices seen elsewhere is provided by Antonio da Sangallo the Younger in the first half of the sixteenth century. It shows a horse-driven combination of a flour- and a stamp-mill, which is said to have been located in Cesena (figure 1.16).[40] As was typical, the leaf includes information on measurements and gear ratios as they had been observed on the site. Similar examples from that period can also be found in the famous sketch-book of the Volpaia family from the 1520s.[41] Another example, one of numerous of such leaves included among the personal records of Heinrich Schickhardt, represents a combined fulling- and grinding-mill near Montbéliard.[42] This drawing exhibits an even greater density of information than the Sangallo example (figure 1.17). It furthermore testifies to the fact that, compared to the presentational treatises, engineers in such cases sometimes used a broader variety of graphic techniques to record what they had seen. Schickhardt, in this case as in many others, used not only perspective representations, but also vertical sections and top views. In each part of the drawing, he noted dimensions of the different parts of the machine and also wrote down the gear ratios of the machinery. The production of such drawings required a considerable sense of abstraction. Firstly, machines were housed in buildings, which, of course, were not transparent, so that it was actually impossible to see the machinery as it was portrayed by the drawing. Secondly, the point of view chosen for the representation of the machine—at a certain distance and slightly above ground level—is practically always constructed virtually, as it was hardly available to contemporary spectators.

How exactly engineers later made use of information recorded in this way is difficult to say. Of course, such drawings not only served individual purposes, but also provided a basis for communication with artisans, colleagues, and potential investors. In the case of Heinrich Schickhardt, the importance of such a collection is proven by the fact that he kept such leaves at home in a desk with drawers, each of which was reserved for one special kind of device.[43] In the sixteenth century, such archives of well-off engineers also usually included a collection of books—not necessarily on technical subjects only. Lists of books owned by engineers are documented, for example, for Leonardo da Vinci; again, Heinrich Schickhardt; and for the Italian engineer Giambattista Aleotti.[44] Towards the end of the sixteenth century, such personal libraries might also comprise printed machine books. At the same time, illustrations from printed books were also copied for private use. This is testified to by a loose leaf from the papers of Heinrich Schickhardt showing copies of machines employed in the German mining regions as they had been depicted in Georgius Agricola's *De re metallica* of 1556 (figure 1.18). The reason for the production of exactly these copies are unclear, as Schickhardt himself owned a copy of Agricola's book.

40 See Frommel 1994c, 418.
41 For this manuscript in general, see Brusa 1994, 657–658.
42 See Bouvard 2000, 60–63.
43 This can be deduced from a note in a document pertaining to the building of a mill in Pleidelsheim: "Mühlwehr wie das gemacht; ist beü den wasser gebeüen in der oberen Schubladen zuo finden." Stuttgart, HStA, N220 T212.
44 For Leonardo, see Leonardo 1974, II fol. 2^v–3^r and Leonardo 1987, 239–257; for Schickhardt, see Schickhardt 1902, 331–342; for Aleotti, see Fiocca 1995.

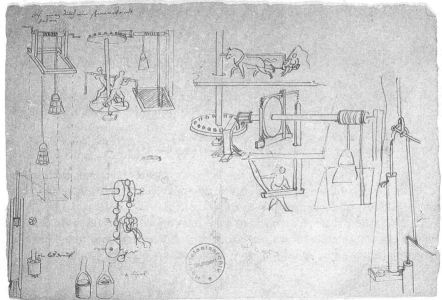

Figure 1.18. Top: Lifting device. Woodcut from Agricola 1556, 167. Bottom: Copies from Agricola's *De re metallica* by Heinrich Schickhardt, c. 1605. (Stuttgart, HStA, N220 T151, all rights reserved.)

Finally, another drawing from the legacy of Antonio da Sangallo the Younger might serve to illustrate the difficulties of unambiguously assigning early modern machine drawings to the three contexts of representation, realization, and documentation discussed so far. The leaf shows different views of a pump (figure 1.19).[45] As

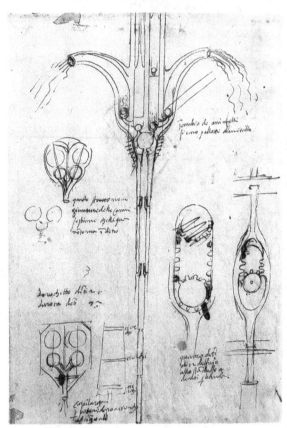

the brief explications make clear, its peculiarity was that the valves usually employed in pumps had been replaced by metal balls. From this innovative feature it could be assumed that the leaf represented a study that was not connected to a particular project. From the information given on the numbers of teeth of the toothed wheels, it could also be assumed that the drawing shows a device that was actually in use. Even if this was indeed the case, it would still remain unclear whether the device had been designed by Antonio da Sangallo or whether it represented a device made by others, which had been investigated during his travels. Ultimately, it remains open which purpose such a documentation actually served. The analysis of early modern machine drawings is often confronted with such problems of interpretation. To narrow down the possibilities of interpretation, descriptive texts, text fragments on the drawing and textual documents preserved with the drawings again and again prove most useful. Where such additional material is missing—which is often the case due to the frequent separation of pictorial and textual sources practised in a number of European archives some decades ago—interpretation is often confronted with considerable difficulties.

Figure 1.19. Studies of pumps with the usual valves replaced by metal balls. Drawing by Antonio da Sangallo the Younger. (Florence, Gabinetto Disegni e Stampe, U847A^r.)

---

45   See Frommel 1994c, 335.

4. DRAWINGS SERVING THEORETICAL CONSIDERATIONS OF MACHINES

This fourth category of machine drawings represents a special case in the classification proposed here. Up to this point, machine drawings have been classified according to the social context of their employment: presentation to a broader public, realization of concrete projects, storing information for the engineer's own use. The theoretical analysis of machines by means of drawings appears to be orthogonal to these categories, as such an approach might be found in each of these three categories. To be sure, the definition of "theoretical" in the context of early modern engineering drawings is still an open question. In general, the sixteenth-century theory of mechanics is understood as consisting in the analysis of the simple machines based on the lever and the balance. However, there also are engineering drawings that testify to general reasoning on machines without any reference to preclassical mechanics and its visual language of geometrical diagrams. Drawings of standardized types of mills in a treatise by Francesco di Giorgio Martini could be adduced as an early example,[46] a drawing by Antonio da Sangallo that will be discussed below as a later one. Engineers' considerations of working principles of machines in drawings like these might be labelled "theoretical" as well, but the establishment of corresponding definitions lies beyond the scope of the present contribution. The following paragraphs thus mainly concern the appearance of the visual language of preclassical mechanics in sixteenth-century engineering drawings.

In the sixteenth century, mechanics gradually emerged as an independent discipline. This process was inseparably connected to the reception of ancient sources. Pseudo-Aristotle's "Mechanical Problems" were now edited and commented upon as well as the works of Archimedes, Hero of Alexandria and later those of the Alexandrine mathematician Pappus. Additional sources comprised medieval treatises in the tradition of the *scientia de ponderibus,* most prominently those of Jordanus Nemorarius. All of these approaches were founded on the theoretical analysis of the balance with unequal arms and the lever by means of geometrical proofs. This common basis facilitated attempts in the sixteenth century to unify all of these different strains from the Greek and Hellenistic eras and the Arab and European Middle Ages. In this context, special attention was devoted to the classification of the five simple machines (lever, wedge, winch, screw, and pulley) dating back to Hero of Alexandria, which, for example, guided Guidobaldo del Monte in structuring his influential "Mechanicorum liber" (1577). As early as the late fifteenth century, as soon as the work on "rediscovered" texts on mechanics began, engineers strove to use this body of theory to investigate more closely the properties of the sixteenth-century machinery with which they were dealing. As the analysis of the simple machines proceeded by means of geometrical proofs to determine relationships of distance, force, weight and velocity, graphical representations played an important role. The corresponding visual language is documented, for example, in the illustrations of Guidobaldo del Monte's treatise and was presented concisely on the title page of the German translation of

---

46 See Long (this volume).

Figure 1.20. Geometrical analysis of the simple machines as the foundation of mechanics, and their practical application. Frontispiece of Daniel Mögling's *Mechanischer Kunst-Kammer Erster Theil* (Frankfurt 1629).

Guidobaldo's work in 1629 (figure 1.20). The frontispiece presented an overview of the simple machines and their geometrical analysis, alluding to their practical application as well.

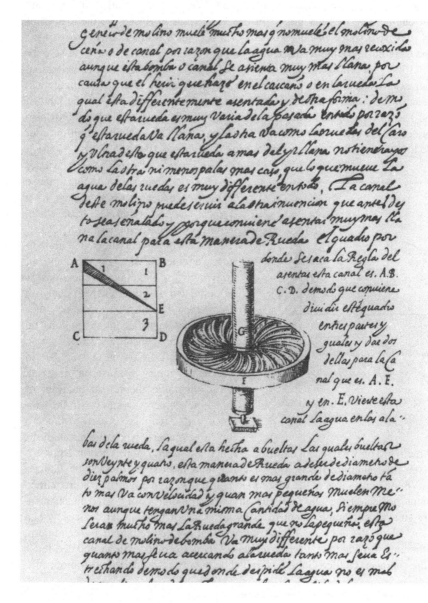

Figure 1.21. Diagram to determine the inclination of a conduit to drive a horizontal water-wheel. (Photo: Biblioteca Nacional Madrid, Los veintiún libros …, Mss. 3372–3376, fol. 290^r.)

The few examples known to date that combined engineering drawings with geometrical analysis in terms of the simple machines are found in engineering treatises,
in which they aimed to underline the author's acquaintance with the foundations of
contemporary science. The author of the Spanish engineering treatise "Twenty-one
books of engineering and machines" in the 1570s thus analysed the inclination of
conduits to drive a water-wheel by means of a geometrical diagram (figure 1.21).[47]
Giuseppe Ceredi, physician to the Dukes of Parma and Piacenza, in a treatise published in 1567 concerning the application of the Archimedean screw to irrigation,
also dealt extensively with the theory of the balance and the lever as the theoretical
foundation of the analysis of machines.[48] Arguing for the superiority of the kind of
crank he had chosen to drive his Archimedean screws by manpower, Ceredi also
incorporated geometrical abstractions in the illustration of his own solution in order
to allude to the scientific reasoning underlying his choice (figure 1.22).

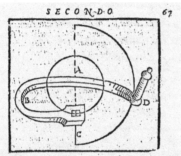

*Tutto cio che è dal punto. A'. fino al punto.'B'. è fouerchio, quanto
fia per la forza, che uiene dalla natura del nette : perche la parte,
che è dal punto. B: al punto. C. fe gli oppone con uguale potenza.
Voglio dire, che facendofi il cerchio nel punto. B. farà uinto in diametro da quello, che farà fatto nel punto.'D. & dal grado di tal
uittoria nafce tutta la forza. Ma perche quando il motore ha
mandato l'eftremità del nette uerfo il centro del mondo., effendo il*

Figure 1.22. Geometrical analysis in terms of the
lever of a crank to drive an Archimedean screw.
(Ceredi 1567, 67.)

A similar kind of explanation was later given by Simon Stevin in his treatise "De Weeghdaet" with reference to a typical crane operated by a tread-wheel employed on early modern riversides.[49] In private documents of sixteenth-century engineers, the employment of drawings for theoretical reflections is documented more extensively only in Leonardo da Vinci's notebooks. His analysis of such factors as friction and the strength of materials, in particular, still appears to have been singular. In sixteenth-century manuscript material, no comparable theoretical analysis of machine elements is known. In the process of realizing early modern machines, such theoretical analyses
seemed hardly to play a role. An exception, which, however, again concerns a preliminary stage of evaluating the design of a machine, was later reported by Galileo
Galilei. While Galileo was at the Florentine court, a foreign engineer who remained
anonymous presented the Duke of Tuscany with a model of a geared mechanism
allegedly suitable for employment in different kinds of mechanical devices.[50] The
crucial fact about the engineer's proposal was that his device entailed a pendulum,
which, the engineer claimed, greatly increased its performance. In an undated letter
sent to the engineer, Galileo Galilei, who had been present at the demonstration, sub-

---

47   See Turriano 1996, fol. 290r.
48   See Ceredi 1567.
49   See Stevin 1955, 344.
50   See Galilei 1968b.

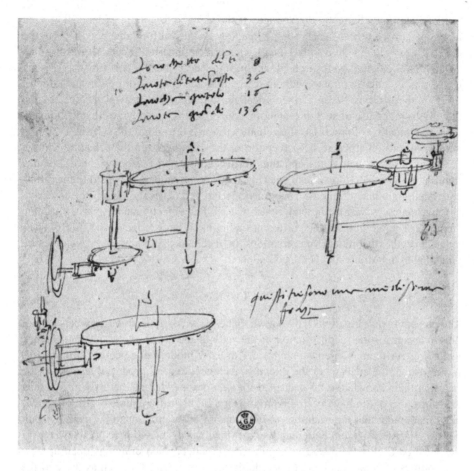

Figure 1.23. Study on different combinations of the same gears for a horse-driven flour-mill. Drawing by Antonio da Sangallo the Younger. (Florence, Gabinetto Disegni e Stampe, U1487A^{r}.)

stantiated to the engineer why he considered this hope to be unfounded. In part of this letter, Galileo reduced the major features of the model presented to a geometric drawing in order to enable its study according to the principles of the balance. The original drawing, however, has not been preserved. This example shows that such theoretical analyses in the context of discussing the design of particular machines can be expected above all in court contexts. In this framework, the "scientific" foundation of personal judgements gained increasing importance over the course of the sixteenth century.

In engineers' personal accounts, it was primarily measurements and gear ratios that were reported extensively. However, it does not seem that they generally used this information as a starting point for further reasoning of a more general nature. The only drawing known so far that points in such a direction is, again, part of the collection of Antonio da Sangallo the Younger. It shows ways to combine identical gearing assembled differently in space, stating that they are all "of the same power"[51] (figure 1.23). This comment shows that in early modern engineering practice, concepts like "force" or "velocity" were used by engineers to describe the performance of such devices as a matter of course, without referring to the contemporary scientific definitions of these terms. Even if theoretical reasoning in mechanics did not emerge directly from the use of such prescientific concepts, it seems obvious that the intensified dissemination of preclassical mechanics towards the end of the sixteenth century, at least in Italy, sharpened the perception of the gaps between the scientific and the colloquial use of such terms and thus further stimulated reasoning among figures familiar with both cultures. Such gaps also became obvious with regard to the different visual grammars of engineering drawings and geometrical diagrams of preclassical mechanics, which in the end concerned similar objects, namely basic machine elements. The merging of such different traditions of knowledge raised fruitful challenges for the theoretical investigations in mechanics pursued, for example, by Guidobaldo del Monte and Galileo Galilei.

CONCLUSION

The preceding investigations have shown that early modern machine drawings are not only relevant for the reconstruction of the state of the art of contemporary machine technology. Although an epistemic history of early modern engineering still remains to be written, the analysis of the drawings technical experts produced testifies to the fact that their knowledge far exceeded the *tacit knowledge* of the artisan: Machine drawings turn out to have been the product of a highly differentiated form of knowledge that could take on a number of functions in different contexts of employment. This is especially evident with regard to the drawings discussed here, which were, for the most part, closely related to engineering practice. Such drawings open up new possibilities of contextualizing Leonardo da Vinci's drawings as well as those of the more representative machine books.

---

51   "[...] sono una medesima forza." Frommel 1994c, 448.

PART II
PICTORIAL LANGUAGES AND SOCIAL CHARACTERS

At first glance, almost all of the machine drawings of the fifteenth and sixteenth centuries may appear unprofessional to a modern beholder. They in no way resemble the orthographical plans and schematics that engineers trace and employ today. Usually omitting crucial details, often representing instead superfluous particulars of apparently rhetoric nature, and rarely giving measurements, they may even evoke the suspicion that they are creations of interested laymen with limited competence as regards technological matters. Taking modern blueprints as standard, it is, indeed, hard to imagine that drawings of this kind were of any use for the practice of early modern engineers.

It is particularly the seemingly naiveté of these drawings that is misleading. As the chapters by David McGee and Rainer Leng show, these drawings are anything but naïve. Even the most awkwardly drawn ones do not testify to unskilled attempts of practitioners groping for any manner of depicting intricate technical objects. Rather, such drawings testify to first experimental steps toward depicting them in a specific style of rendering that distinguishes early modern machine drawings from the machine drawings of the Middle Ages as well as those of the modern age.

The few extant machine drawings from the Middle Ages, which David McGee discusses in the first part of his chapter, show a preceding style of technical drawings reminiscent of modern schematics. Focusing exclusively on some of the essential parts of a certain device and their arrangement, no effort is made to represent its appearance. Because of this, they are often as unintelligible for a nonexpert as modern engineering drawings. However, medieval technical drawings share one feature with the seemingly naïve engineering drawings of the early modern period that distinguishes them from modern ones. They, too, omit crucial details, do not give measurements, and leave the question of the device's dimensions unanswered.

Against the background of this preceding style of machine rendering, two points become immediately clear. First, the early modern style of representing machines was developed in a rejection of an earlier professional style of machine rendering and, thus, cannot be regarded as naïve and unprofessional itself. Second, the fact that it shares with the preceding style a practice of neglecting crucial details, measurements, and dimensions, calls for a reconstruction of the engineering practice in these ages that accounts for the apparent usefulness of such incomplete drawings. Both points together warn us not to judge the early modern machine drawings from the viewpoint of present engineering drawings. Rather, one has to try to interpret them in their actual setting, that is, in the actual practice of early modern engineering.

The two chapters of part II do exactly this. They trace the origin and first developmental stages of the specific language of early modern machine drawings. The chapter by McGee explores the emergence of the new style with the Sienese engineer Mariano di Jacopo, called Taccola, whose drawings constituted the starting point for the Italian tradition of early modern engineering drawings, which became influential for the entire West. The chapter by Leng focuses on the less-known pictorial catalogues of German master gun-makers from the fifteenth century. They are of particular interest with respect to the development of the language of early modern machine drawings for two reasons. First, the bulk of them addresses exclusively fellow experts; their pictorial language thus must have fit exactly the practical needs of these practitioners. Second, in these drawings, one can study how artificial views such as cutaways and exploded views were elaborated step by step.

Investigating carefully and in detail the pictorial means developed and employed, the two chapters inevitably lead to the social conditions that shaped the designing and manufacturing of machines in the fifteenth century. And insights into these conditions lead, in turn, to a better understanding of the peculiarities of the new drawing style, which looked so strange at first glance. This style now appears as a highly artificial compromise between the different ends these drawings had to fulfill at the same time, namely to present the device in question to nonexperts such as (potential) commissioners, on the one hand, and, on the other, to give the masters in charge of the (possible) construction of the device all of the technical information they needed. In reconstructing this double-faced nature of the early modern machine drawings, these chapters restore to these drawings their genuine technical character, which is easily eclipsed by their more obviously representational function for a contemporary beholder.

THE ORIGINS OF EARLY MODERN
MACHINE DESIGN

DAVID MCGEE

From approximately 1450 to 1750, machine designers made use of a singular kind of graphic representation. Where craftsmen used partial representations or none at all, and architects used plans and elevations to show different views of a building, machine designers stuck to a tradition of one drawing and one view for one machine in both sketches and presentations. Such a long and stable association between machine design and one kind of drawing calls for explanation. This chapter attempts to lay some of the groundwork for such an explanation through an investigation of the origins of the early modern tradition of machine drawing in the period before 1450.

Any attempt to delve into the origins of early modern machine design is beset by methodological problems. One set of these problems arises from the nature of the evidence. There are only a handful of extant manuscripts and they are spread across three succeeding centuries. Such a distribution compels the historian of machine design to develop an account of long-term historical trends, with all the problems that implies, knowing that any such account can only rest on a slender foundation. Such a predicament generates a real need to get all we can out of the available evidence. But doing so is not easy. Early technical drawings are extraordinarily difficult to interpret. Confronted with a drawing from the fifteenth-century notebook of artist-engineer Mariano Taccola (figure 2.1), for example, it is not at all clear what is important, why these particular devices are found together, why they are drawn at different sizes, or even what the drawings are for. Indeed, one rather quickly realizes that we simply do not have the conceptual tools we need to get at this kind of evidence.

A second set of problems is of our own making, beginning with our Platonic approach to design, where design is taken to be a more or less entirely cognitive affair that is concerned exclusively with the *conception* of form.[1] Our tendency is to take drawings as expressions of ideas, then move backwards from the drawings into the minds of the artists, either to describe what they were thinking, or, since knowing what they were thinking is almost impossible, to characterize their mentality, or to opine about cognition. This is to leap from the methodological frying pan into the fire, jumping from evidence we do not understand very well into the even more treacherous realms of psychology—perhaps even to celebrate "tacit knowledge," or "nonverbal thinking"—long before we have really come to grips with the evidence at hand.[2]

---

1   A good example is Ferguson 1992. But the emphasis on conception and psychology is rampant in the design professions. See, for example, Lawson 1980; Akin 1986; and Rowe 1987.

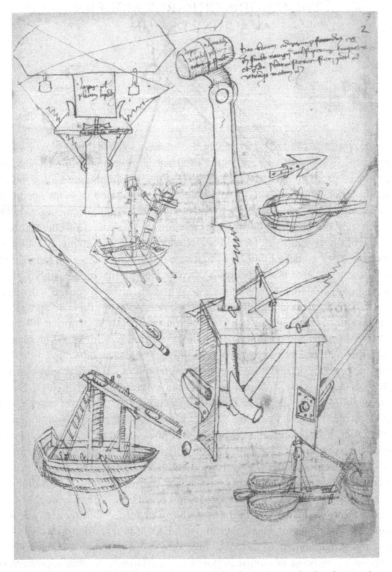

Figure 2.1. A typically difficult to interpret folio from the notebook of an early Renaissance engineer. Drawing by Mariano Taccola. (Munich, Bayerische Staatsbibliothek, Codex Latinus Monacensis 28800, f. 2^r; reproduced in Taccola 1971.)

---

2 My concern here is twofold. On the one hand it seems to me that historians do a tremendous disservice to their discipline by attempting to devolve historical explanation off onto the "truth" of psychological theories (which are, of course, constantly changing), thereby abandoning the effort to build up historical explanation itself. On the other, I very much worry about theories that locate the essential evidence in the mind, which is to base a theory on evidence which (like tacit knowledge) is unknowable by definition.

A related problem stems from modern familiarity with the conventions of linear perspective and our almost unconscious acceptance of two assumptions that go with it. One assumption is that, when people look at things, they really do see something very much like a perspective drawing—and especially that they see things from a distinct, single viewpoint. The second is that making pictures is about rendering the appearance of things as they are seen either in reality or in the mind. These assumptions are problematic in themselves, and extremely problematic when it comes to pre-perspective drawings, which do not render the overall appearance of things accurately, and frequently do not have viewpoints. Nevertheless, on the assumption that the artists had perspective images in their minds, we proceed to interpret their drawings in terms of perspective, using all its language of viewpoints and picture planes. Apart from anachronism, the usual result is an interpretation of the actual drawing as simply a poor rendering of the much clearer mental image, due either to inadequate drawing methods or simple incompetence. We have all read of the childlike "naiveté" of early machine drawings and of the "mistakes" their artists are supposed to have made.[3] This line of reasoning leads to an account of the development of drawing over time as a process of closer and closer approximation to the more accurately realistic representational methods of linear perspective, which is not only teleological, but takes for granted that people should want such a thing, thereby relieving us of the need to actually offer an explanation. More important, discounting the *actual* properties of early drawings as "error," this approach dismisses the *actual* evidence in favor of what is *assumed* to be in the mind, when it is closer attention to the evidence that we need.

One last set of difficulties may be referred to as the Princes problem. It stems from the observation that none of the authors of early machine drawings were actually machine makers, and that most of the early manuscripts up to 1550 were created for princely patrons. Much fine work has been done by Paolo Galluzzi and Pamela Long using the written content of these manuscripts to stress the importance of courtly humanism on the development of Renaissance machine design, and especially on the emergence of "engineers" as a distinct social group.[4] Yet, an unspoken assumption is that the fact of a princely audience explains something about the contents and properties of the drawings. For example, the more fanciful drawings or impossible machines can be regarded as "dreams," or a form of play, composed by amateurs as entertainment.[5] The point here is that on this basis we tend to push the drawings away as objects of serious attention, worthy of investigation for what they can tell us.

To help combat these problems, this chapter proposes that early machine drawings should be analyzed in terms of design, where design is fundamentally regarded as a form of *doing* rather than thinking, and as *the process by which artifacts get the*

---

3   Perhaps the worst example is Edgerton 1991.
4   See particularly Galluzzi 1993 and Long 1997.
5   Thus even Gille 1966 suggested that the authors of early manuscripts were not the "true agents" of technological progress, and that invention was mostly a pastime, a form of amusement, an intellectual game. See also Hall 1979b.

dimensions they actually have. A moment's reflection reveals that many artifacts get their final dimension only when they are actually made, something that was particularly true of early machines. That is to say, the definition of design employed here can include both conception *and* construction, things we habitually regard as separate on the basis of our understanding of modern design techniques. One methodological principle to arise out of this definition is that, unless we have good evidence to think otherwise, it should be assumed that the drawings we see were considered by their authors to be perfectly adequate to the process of completing a design. The second is that, in order to get the most out of premodern machine drawings, it will often be much more useful to consider what came *after* them, rather than what went before them. That is to say, it will be crucial to consider the context of construction—even though we can be reasonably sure that none of the machines seen in this chapter was ever built.

The claim here is that by paying attention to *doing* we can get a better understanding of the properties of early modern drawings, and through a better interpretation of the evidence we can arrive at a better tale of development over time. To support this claim, the chapter examines four episodes in the development of early machine drawings, beginning with the lodge books of Villard de Honnecourt, followed by the manuscripts of Guido da Vigevano and Konrad Kyeser. These brief studies set the stage for an examination of the drawings of Mariano Taccola—a man who may justly be regarded as the father of the style of machine drawing that would remain intact for three hundred years, and continues to dominate even now in the twenty-first century.

A word of warning to the reader. This is a methodological chapter. The normal historical apparatus, particularly the discussion of historical context, has been kept to a minimum. The intention is to outline an argument as clearly as possible, rather than provide a history, so that the approach may be judged, so to speak, by its good works.

1. THE CONTEXT OF CONSTRUCTION

The value of looking at early machine drawings in terms of the context of construction can be illustrated with a famous drawing of a self-powered sawmill by Villard de Honnecourt, the French architect active circa 1250 whose sketch-books provide us with some of the earliest machine drawings in the West (figure 2.2).[6]

Villard's depiction of the sawmill already displays two basic characteristics of machine drawings that would remain stable for the next 500 years. The first is that Villard has adopted what may be called the principle of one machine, one drawing. The second is that there are no measurements, to which we may add that parts of the arrangement are clearly missing. In short, final dimensions have not been determined.

However, perhaps the most striking thing about this drawing is the visual confusion it presents to the modern viewer, a confusion that stems from our expectation, honed by years of experience with perspective drawings, that one drawing of one

---

6   The standard version of Villard's sketch-book is Hahnloser 1972.

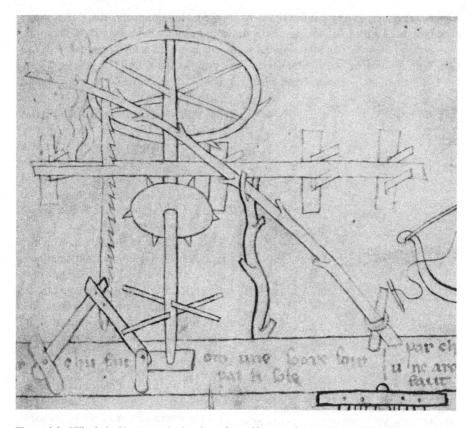

Figure 2.2. Villard de Honnecourt's drawing of a self-powered sawmill, c. 1225. (Detail of Paris, Bibliothèque Nationale de France, MS Fr 19093, f. 44; reproduced in Hahnloser 1972, plate 44.)

machine should also have one viewpoint. To us, Villard appears to have rotated different parts of the sawmill away from their "true" position, relative to any viewpoint and, moreover, has done so without any system. The timber seems to be shown from the top, the water-wheel from a three-quarters view, the saw blade from the side, the cams that bring the saw blade down from a different side, the tree-branch that brings the blade up from the side at one end but the top at the other, while the point at which the saw blade meets the timber seems to be shown from the top, side and three-quarters view all at once. We read the drawing as having multiple viewpoints at the same time, some of them mutually exclusive. Hence the difficulty we have understanding the picture.

Given our confusion, it is essential to point out that Villard can and does draw in different and (at least to us) "better" ways. In his architectural drawings, for example, Villard uses plans and elevations to show two sides of the same building in different drawings. Moreover, it turns out that Villard is also able to draw machines in a "better" way, as can be seen in his depiction of a new arrangement for fixing the hub of a wagon without having to cut the shaft (figure 2.3). Here it appears that Villard may even have used a compass to construct a drawing that simultaneously shows both the true shape of the wheel and the way the arrangement would actually appear from one side. With its apparent viewpoint, we read this drawing quite easily.[7]

The main point, however, is not whether one of Villard's drawing styles is better

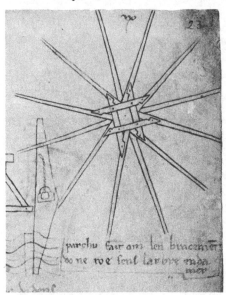

Figure 2.3. Villard de Honnecourt's drawing of a wagon wheel. (Detail of Paris, Bibliothèque Nationale de France, MS Fr 19093, f. 45; reproduced in Hahnloser, 1972, plate 45.)

than another, but recognition of the fact that Villard has *different* drawing styles available to him. This immediately rules out interpretations of the style used in the drawing of the sawmill as "naïve" or "childlike," or indicative of some odd medieval way of "seeing" the world (since if styles are equated with world-views, then there are at least three world-views in Villard, which is ridiculous).[8] More important, acknowledging that Villard had different drawing styles available to him leads to the realization that he *could* have drawn his sawmill differently, and thus that he *chose* to depict the sawmill in the way that he did.[9] Evidently, he did so because he thought this style was best for his purpose.[10] What purpose could that be?

This is where consideration of the context of construction comes in, beginning with the fact that a machine like Villard's, like any machine of its kind at the time, would have to be adapted to the site on which it was built. At different locations, the water-wheel would have to be of a different diameter, the drive shafts would have to be longer or shorter, the support

---

7   I say "apparent viewpoint" because the fact that the eye seems to be fixed at the center of the wheel may simply have been the result of using a compass, and thus imply nothing about Villard having any notion of viewpoint, even though he was trying to depict the hub device as it would appear.

8   See for example, Booker 1963, and also Edgerton 1991.

9   The importance of the availability of different styles of mechanical drawing for practitioners is also noted by Marcus Popplow (this volume) in his discussion of the technical practice of Heinrich Schickhardt.

10   I forego what could be a lengthy discussion of why Villard, who may be taken to be indicative of premodern technical practice in general, uses different styles of drawings for different kinds of objects: the flat style for machines, plans and elevations and stonecutting diagrams for buildings, etc. I suggest, however, that part of the answer is to be found in the context of construction. There was a much greater need for precision in the construction of a building, whose parts must fit together precisely, than there was in the construction of the wooden machinery of the Middle Ages. The need for precision implies a greater need to control the worker's hands. Plans represent the first step in excercising that control, stonecutting drawings the next. As I suggest below, no such control over the hands of the machine maker is wanted, but rather precisely the opposite.

structures adapted to the river bank, and so on. In other words, the design of the machine would not be completed until it was actually built. A second point is that the necessity of adapting the machine to the site implies the existence of a *person* who can do just that. And not just any person, but an *expert* capable of completing the arrangement and establishing the final dimensions of the machine for himself. The question then becomes—what needs to be conveyed to such a person? Not precise dimensions or complete arrangement. That would be pointless, since each version of the sawmill must be different from every other.[11] Rather, all one really needs to provide is an indication of the main parts and their general relationship to each other. Knowing what the parts are, the expert can *deduce* how they must go together, as well as whatever other parts are needed while building the machine on site. The same point is made by Rainer Leng in his chapter, concerning the graphic representations of fifteenth-century gunmakers.[12]

Given the existence of an expert, it is not necessary for either the overall machine or its parts to be rendered "realistically," as they would actually appear.[13] All that is needed is to represent the characteristics of the parts in such a way that the expert can understand what they are. Mere icons will do, and the iconic nature of Villard's drawing is nicely illustrated by the buckets or paddles attached to his water-wheel on the lower right.[14] They are on backwards. If interpreted realistically, the water-wheel would drive the timber *away* from the saw blade rather than toward it. However, if we eschew the idea that Villard made a "mistake," in favor of the assumption that the buckets did exactly what Villard intended them to do in this drawing, we realize that the buckets are no more than signs that says to the expert "water-wheel," rather than, say, "grist-wheel." On this account it becomes superfluous to assume that Villard had any notion of rotating parts to show them as they would appear from *different viewpoints*. In fact, the evidence does not suggest any idea of a viewpoint whatsoever. Rather, the drawing takes on the character of a logical schema built of icons, intended for the eyes of an expert builder who does not need to know exactly how either the whole or the parts would appear. All he needs to know is the main parts and their general arrangement.[15]

This interpretation of Villard's sawmill rests on the proposition that early machine designers had the context of construction by experts in mind and made their drawings accordingly. Alas, it must be admitted that there is not enough evidence in Villard's

---

11 In other words, the dimensions are not missing because the drawing represents only an idea, as might be argued. They are not included because they are unnecessary.
12 The same idea is suggested by Hall 1996, 9, who observes that missing or distorted elements in drawings almost automatically suggest the ability to supply the missing details on the part of those who have to use them.
13 As Leng also points out in his chapter.
14 The word "icon" points to one of the typical language difficulties encountered in the discussion of drawings. I use the word to mean a graphic element whose purpose is to show essential or characteristic attributes of a machine part so that it can be *named* in the virtual context of the drawing, not to show how a machine part actually appears so that it could be *recognized* when seen in the real world. In this sense a machine icon is a symbol, like an Egyptian hieroglyph. But like a hieroglyph, the symbol is not an arbitrary choice like a letter. It is an abstraction of the visual appearance of the real thing, making it difficult to say precisely that one drawing is supposed to show a machine as it appears, while another does not.

notebooks to confirm or refute this contention. Striking confirmation, however, can be found in the work of Guido da Vigevano.

2. PARTS, PICTURES, AND PATRONS

Guido da Vigevano was an Italian medical doctor from Pavia who found service with various kings and queens of France in the fourteenth century. He is of interest here for his *Texaurus Regis Francae*, an illustrated manuscript, which he wrote in 1335 to advise Phillip VI of France on the conduct of a proposed crusade to the Holy Land.[16] The first part of the manuscript concerns such things as proper diet and how to avoid being poisoned. The second part consists of 14 folios of text and drawings about the kind of military equipment the king would need to take with him.

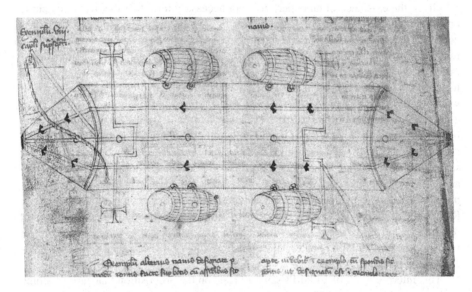

Figure 2.4. Guido da Vigevano's modular paddle boat, c. 1335. (Paris, Bibliothèque Nationale de France, Manuscrit latin 11015, *Texaurus regis Francie aquisitionis terre sancte de ultra mare,* f. 49[r]; reproduced in Vigevano 1993, plate IX.)

---

15 The argument that drawings like Villard's were understandable to experts naturally implies the existence of a shared set of visual icons among those using the flat style in any given locale. The existence of such a set raises the question of what would happen if an item had to be indicated for which there was no icon, perhaps because it was entirely new. Without arguing the issue at length, it might be suggested that Villard's spoke and hub represent one attempt to show a new item more realistically, as does the branch in his sawmill. Both, though in very different ways, might be seen as attempts to show unknown items as they would appear—perhaps the only alternative.

16 This expedition never took place. For the original text and marvelous reproductions of the drawings, see Vigevano 1993.

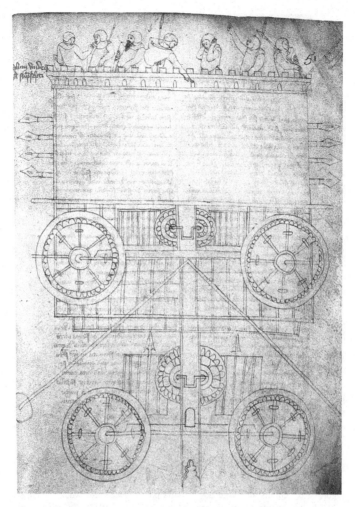

Figure 2.5. Guido da Vigevano's cranked assault wagon. (Paris, Bibliothèque Nationale de France, Manuscrit latin 11015, *Texaurus regis Francie aquisitionis terre sancte de ultra mare*, f. 51ʳ; reproduced in Vigevano 1993, plate XIII.)

Figure 2.4 shows Guido's proposal for a modular crank-and-paddle powered paddle-boat.[17] From it we see that Guido (and/or his illustrator) was working in the same "flat" style as Villard.[18] We have one machine and one drawing. There are no dimensions and little concern for overall appearance.[19] Guido is more rigorous than Villard in epicting all of his parts flat on the paper, so that none are shown in a three-quar-

---

17 One of Guido's guiding ideas was that very little timber was to be found in the Holy Land so that the equipment needed for the campaign would have to be prefabricated in France in pieces small enough to be carried on horseback. For an English translation, see Rupert Hall 1976a.

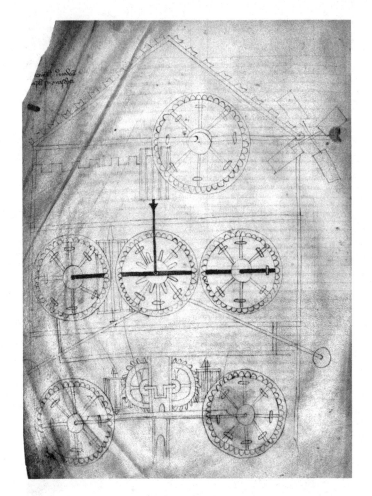

Figure 2.6. Guido da Vigevano's wind-powered assault wagon. (Paris, Bibliothèque Nationale de France, Manuscrit latin 11015, *Texaurus regis Francie aquisitionis terre sancte de ultra mare*, f. 52ᵛ; reproduced in Vigevano 1993, plate IV.)

ters view. But the emphasis is again on the parts, which are again shown as icons. The flattened ends of the barrels, for example, are not shown to indicate volume, but to show that these items are casks.

18 Of the three extant copies of Guido's manuscript, none are the original, so that none of the drawings we have can be said to be by Guido's own hand. It is not known for certain whether Guido himself made the original drawings or had them done by an illustrator, although there is a passage in the text where he suggests that one of his assault devices could be covered with a protective quilt, and that this should be done "as the maker sees fit, for I could not sketch all this in the figure," Hall 1976a, 29. For the sake of convenience, however, I will still refer to the drawings in the manuscript as "Guido's drawings."

19 Although indications of size are given in the accompanying text, they are extremely general.

More important is what Guido says about the text associated with this drawing, which is that "a skilled man will easily understand this because I cannot write it more clearly."[20] A few pages later, after describing the front wheels of his famous crank-powered assault wagon (figure 2.5), Guido observes that: "the rear wheels should be similarly prepared in this way, and all these things shall be arranged by the master millwright who knows how to match these wheels together."[21] Similarly, referring to his famous drawing of a wind-powered assault wagon (figure 2.6), Guido comments on his description of the gearing by saying that "all these matters are the concern of the master millwright and especially the master windmill-wright."[22] These passages clearly show that, while writing his text and using the same style of drawing as Villard, Guido was indeed thinking about the context of construction and particularly about construction by skilled experts. The last two passages make it explicit that Guido believed his drawings would be fully comprehensible to these experts, and that he fully expected them to be able to figure out from the drawings how to complete the design, deducing the necessary details from the arrangement of well known icons of lantern gears, cranks, and so on.[23]

It is worth stressing how different this design relationship is from that of today, where the designer is expected to provide complete drawings and complete specifications of dimensions, parts, and arrangement, and the worker is expected to make the parts (and thus the machine) exactly as shown. But Guido not only expects the participation of expert machine builders, he *needs* them to make positive, creative contributions. In fact, his text is full of comments to the effect that "whoever is to do this will provide for the best," and that various parts will be "made as those doing the job see fit."[24] In other words we see that medieval machine design was *not* conceived to be a matter of individual conception, of making a drawing and that was that. It was conceived as a process in which at least two people were needed to give a machine the final dimensions it would actually have when it was actually built. As in Villard, it was not necessary, nor of any use, to provide dimensions or a realistic depiction of the overall machine to the experts who would build the machines and were quite capable of determining final dimensions and arrangements for themselves. Flat, iconic representations would do.

However, in the very passages where Guido says his drawings will be perfectly adequate for experts, he reveals that the flat, iconic style of the machine builders was *not* understandable to everyone—and particularly not to the ideal reader to whom the

---

20 Hall 1976a, 26.
21 *Ibid.*, 27.
22 *Ibid.*, 29.
23 Having confirmed the idea of a style of drawings intended for expert makers, the question arises as to the existence of a tradition of technical drawing in this style, stretching back in time. Lefèvre 2002 has shown that whether or not one regards the illustrations as in any way original or authentic, the drawings found in military manuscript treatises on machines and mechanics prove that a flat style tradition of machine drawing existed in both the Byzantine and Arabic worlds from at least the early Middle Ages onward. His examples, as well as those in this chapter, show that there was considerable flexibility in the application of this style, over and above its shared conventions.
24 To give a typical example, Guido writes that the platform of one of his assault towers is to be supported by iron struts: "as the makers shall see fit, for it is not possible to set out every detail. But when someone undertakes this work he will himself make provision for this platform." Hall 1976a, 22.

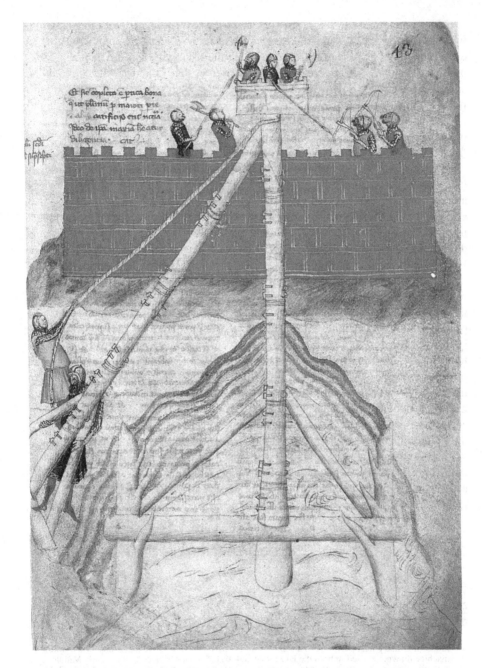

Figure 2.7. One of Guido da Vigevano modular assault towers, where shading is used to show three dimensions. (Paris, Bibliothèque Nationale de France, Manuscrit latin 11015, *Texaurus regis Francie aquisitionis terre sancte de ultra mare,* f. 43^r; reproduced in Vigevano 1993, plate IV.

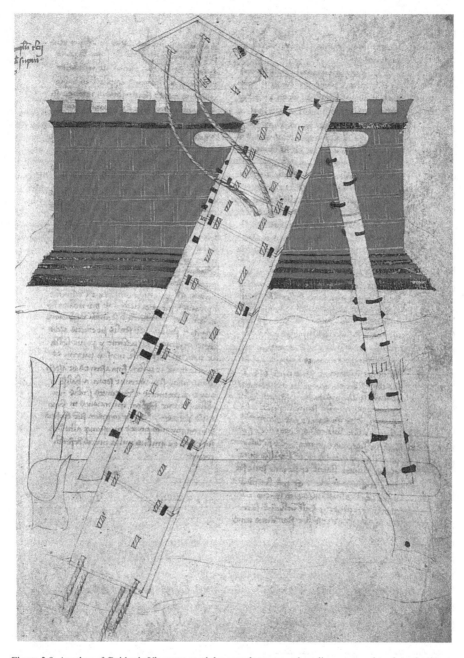

Figure 2.8. Another of Guido da Vigevano modular assault towers, where lines are used to show the three-dimensional volume of the timbers. (Paris, Bibliothèque Nationale de France, Manuscrit latin 11015, *Texaurus regis Francie aquisitionis terre sancte de ultra mare,* f. 44ʳ; reproduced in Vigevano 1993, plate IV.)

manuscript is dedicated, namely, his patron the King. This is significant because it points to Guido's recognition of the participation of a *third* person in the design process: namely, the patron who may not determine any dimensions himself, but whose role is nevertheless crucial, since without his approval, no money will be spent and no finalization of the dimensions will ever take place.

In light of Guido's awareness of the fact that the flat style created problems of communication with patrons, two of his drawings may be singled out as particularly important. The first shows an assault tower resting on three big legs, composed of short wooden modules held together by iron pins (figure 2.7).

The poles in this drawing are not represented by mere lines as the flat style would demand. Instead, shading has been added to give the poles three-dimensional volume, and thereby indicate that they are round. The second drawing (figure 2.8) shows a modular assault ramp, up which men are supposed to run until they reach the topmost section, which they flip over the ramparts of the enemy fortress. Here, the ramp has been drawn to show two sides of the modules at the same time, and lines have been drawn in to indicate the volume of the modules without the use of shadowing.[25] In both drawings we also notice that the machines have been set in scenes, with the fortress being assaulted in the rear.

There would appear to be two possible explanations for this combination of setting and appearance, depending on the audience. It may be argued, for example, that Guido's goal was to show the purpose of his machines to the patron and that he set them in their context of use in order to show what they were for. Since it would have been pointless to set the normal flat icons in such a setting, he therefore added the third dimension to the parts in order to show everything as it would appear. Alternately, it may be argued that Guido's modular ramps were quite new. Hence, there were no icons for them in the established repertoire that could be used to communicate with the makers. Since there was no other choice, it was necessary to attempt to show the parts as they would appear.[26] The setting was added to help makers to identify the parts by deduction from the purpose.

It seems most likely that both factors were in play, and that Guido was attempting to find a means of communicating to both audiences at once. If so, these drawings may be taken to indicate the horns of the early machine designer's dilemma. In trying to show the appearance of his assault towers, Guido may have succeeded in giving a better idea of their purpose, but his ability to show the parts to the expert/maker has suffered. Without referring to the text, for example, it is extremely difficult to recognize the spikes or the modules for what they are.[27]

How to devise a style of representation that could answer the needs of both machine makers and patrons at once? This problem may be taken as one of the major difficulties facing machine designers in the fourteenth and fifteenth centuries. A

---

25 Several more of Guido's drawings also show three dimensions without shadowing.
26 As noted above, the same may be seen in Villard's drawing of a sawmill, where the tree branch used as the return spring seems to be shown as it would appear, there being no standard mechanical icon for such a thing.

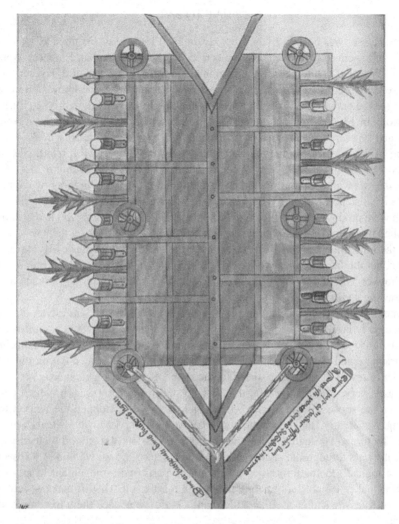

Figure 2.9. A protected assault chariot in the flat style from Konrad Kyeser's *Bellifortis*. (Göttingen, Niedersächsischen Staats- und Universitätsbibliothek, Cod. Ms. philos. 63, fol. 24^r; reproduced in Kyeser 1967.)

sequence of drawings in Konrad Kyeser's *Bellifortis* provides fascinating evidence about the nature of the solution to the problem.

---

27 There is a wonderful confirmation of the difficulty of interpreting the parts correctly in the Yale manuscript version of this drawing, where the illustrator has not only changed the spikes into dowels, but spread them out along the length of the poles so that they have nothing to do with pinning the modular sections of the poles together. Rather, to judge by what the Yale illustrator does with the pegs in other drawings, he seems to think that they were supposed to be steps, by which one could climb up the poles to the assault platforms. See the images of the New York manuscript included in Vigevano 1993.

3. FLAT TO FAT

Konrad Kyeser was born in Eichstadt, Bavaria, in 1366. He was trained as a medical doctor and spent his later years in the service of the Holy Roman Emperor Wenceslaus of Prague. Despite being a medical man, or perhaps because of it, Kyeser seems to have gained considerable military experience. In any case, we know that he was present at the embarrassing defeat of Sigismund of Hungary by the Ottoman Turks at Nicopolis in 1396. Four years later, Sigismund imprisoned his half-brother Wenceslaus and Kyeser was banished to Eichstadt. There he composed his *Bellifortis*, or Strong in War, dedicating the manuscript to the weak Emperor Ruprecht, who succeeded Wenceslaus in 1400, and who was in turn ousted by Sigismund in 1410.[28]

Bellifortis was a popular work among the princes of its time, judging by the many copies in existence. Much of its popularity must have been due to its lavish drawings, which were produced by illuminators from the Prague scriptorium who had also been banished by Sigismund, and apparently passed through Eichstadt on their way home. How the final drawings in *Bellifortis* relate to any original drawings of Kyeser is not known, but the fact that they were made by various hands means that Kyeser's work, like those of Villard and Vigevano, once again reflects the availability of many different drawing styles.

One of the styles found in many of Kyeser's drawings is the "flat" style, of which a particularly lavish version is found in figure 2.9. It shows a covered assault wagon, armed with small guns. Here we see the continuation of the principle of one machine, one drawing, and no dimensions. Every part of the device is shown flat on the paper in order to reveal its characteristic or iconic shape. Realistic appearance is not a concern, as is very clearly shown by the chain at the front of the wagon, which passes between the spokes of the wagon wheels and would have stopped the wheels from turning. Apparently, this did not bother the illustrator, who knew the wheels would in reality be turned 90 degrees from the position shown and thus that the chain would not really pass through the spokes at all. As this picture shows, we find in Kyeser the kind of logical schematic used to communicate parts and arrangement to an expert machine maker, capable completing the design for himself. That is, makers were also a part of Kyeser's audience. Like Villard and Guido, he needed them to complete his designs.

Kyeser, however, may also be assumed to have faced Guido's problem. The "flat" style might be fine for communicating with wheelwrights, but an unwarlike emperor could not be expected to readily understand them. Thus, in Kyeser we would expect to find a further development of techniques for drawing machines as they would appear. One sequence of drawings is particularly telling about the nature of this development.

The sequence consists of several protective shields, which were to be rolled forward in battle (figure 2.10). All of them appear to be drawn by the same hand. The

---

28   For biographical information on Kyeser's life, see Quarg in Kyeser 1967 and Long 1997.

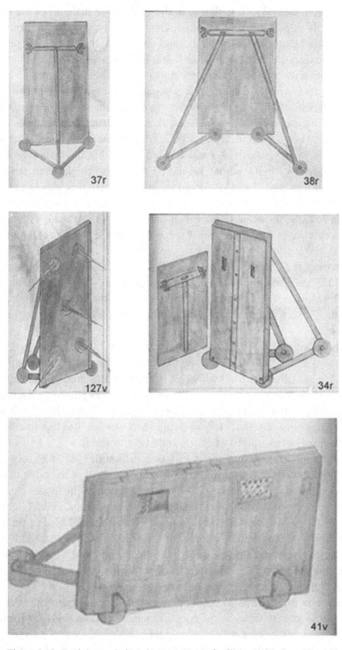

Figure 2.10. Evolving technique in a sequence of rolling shields from Konrad Kyeser's *Bellifortis*. (Details from Göttingen, Niedersächsischen Staats- und Universitätsbibliothek, Cod. Ms. philos. 63, fol. 37^r, 38^r, 127^v, 34^r, and 41^v; reproduced in Kyeser 1967.)

first shield is from folio 37^r. It is drawn in the "flat" style and shows the shield as a rectangle, and its three wheels as iconic circles. The next drawing is from folio 38^r and also shows a rectangle for the shield and circles for the wheels. Here, the artist seems to have realized that the triangular frame for the three wheels of his first shield would have got in the way of anyone trying to shelter behind it, and so his second shield rolls on four wheels, each pair on a separate assembly. These assemblies are splayed out, the closest wheels being wider apart than the farther wheels, which are also higher on the page. This may indicate that the artist was already beginning to think about issues of appearance, but a definite turn towards appearance is seen in folio 127^v. Here we have the same four-wheeled shield, but the front wheels are partially obscured by the shield as they should be, while the shield itself is shown as a rhomboid rather than a square, as if turned in space. Turned this way, the artist seems to have realized that three sides of the shield could be seen at once. Accordingly, he has drawn in lines to show the thickness of the shield. Unfortunately, the attempt to show the appearance of the shield seems to have presented many problems to our artist. Over and above the odd treatment of the spikes on the front, he has shown the thickness on the top and right of the shield when it should have been indicated on the top and left. He has also continued to draw the wheels and their assemblies in the flat style. The closest wheel strut is drawn at right angles to the shield, when it should tilt up. The farther wheel assembly should have been completely obscured, but is still partly visible.

Given these difficulties, the shield drawn in folio 34^r represents a true breakthrough. The artist has now indicated the thickness on the proper sides of his rhomboid shield. He has tilted the front wheel assembly upwards so that it recedes into the picture space. He has once again shown the front wheels as partially obscured. These are major developments despite the fact that he has still drawn the wheels as circles and, the advantages of the "flat" iconic dying hard, despite the fact that he has still shown the back wheel assembly when it should be obscured.

It is fascinating to see that these last two difficulties are the very problems attacked in folio 41^v. Here we sense the growing confidence of the artist in rendering the thickness of the shield, which is now more or less correct in appearance and even indicates the jointing of the timbers. The rearmost wheel assembly has rightly disappeared. The front-most wheel assembly is tilted up as it should be. For the first time, the artist attempts to render the thickness of the wheel assembly struts. For the first time, he tries to show the thickness of the wheels, and even how the wheels go through the shield, although he has been let down by his continuing practice of starting his wheels as circles when they should be oblongs.

Our artist, however, seems to be quite aware of this, as can be seen in the final drawing of our sequence, which is the colored image found on folio 73^v (figure 2.11). This drawing now shows men behind a shield, which they are rolling along the ground, and such is our artist's confidence that he has even tried to show the thickness of a V-shaped shield. The back wheel assembly is properly obscured. The thickness of the front wheel assembly is shown. He also appears to have recognized that, if shown from the side, the wheels would really appear as circles and he would not have to deal with the problem of drawing ellipses. Unfortunately, it seems as if he was again unable to resist the temptation to provide further information, and so has drawn in the thickness of the wheels as they would appear from below.

What is truly fascinating about this sequence is that it moves from the "flat" style to what we might call a "fat" style. But the shift clearly did not take place all at once. It did not take place as a result of the application of some established drawing scheme, or involve any rigorous sense of visual geometry (even though the artist uses a compass and straightedge). Rather, it appears the artist adopted the idea that his task was to show the *parts* as they would appear and then proceeded piecewise, replacing the original icons one at a time, perfecting the appearance of individual parts one after another. That is to say, he did not start with an idea of overall appearance or with a set of top-down rules that only needed to be applied. On the contrary, the overall appearance of the shield in the final drawing is built *up*, one part at a time. On this basis, it may be suggested that the goal was not only to adapt the flat style to the needs of patrons, but to *preserve* its original ability to convey enough information about parts and arrangement to builders that they could finish the design for themselves. The author has succeeded. In the later drawings of the sequence, we can quite clearly tell which parts are which, and yet we easily understand the purpose of the machine.

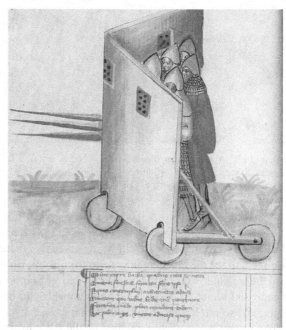

Figure 2.11. A perfected drawing of a rolling shield from Konrad Kyeser's *Bellifortis*. (Details of Göttingen, Niedersächsischen Staats- und Universitätsbibliothek, Cod. Ms. philos. 63, fol. 73ᵛ; reproduced in Kyeser 1967.)

Following the sequence of shield drawings provides the key to interpreting what is undoubtedly the masterpiece of technical representation in Kyeser's manuscript. This is the magnificent picture of a trebuchet seen in figure 2.12. The background in the original is richly figured in red, the ground a stately gray, the timbers bright gold. Every timber and every part is given a three-dimensional shape. The overall impres-

Figure 2.12. A magnificent drawing of a trebuchet from Konrad Kyser's *Bellifortis*. (Göttingen, Niedersächsischen Staats- und Universitätsbibliothek, Cod. Ms. philos. 63, fol. 30ʳ; reproduced in Kyeser 1967.)

sion is of a style so completely different from that of Villard de Honnecourt that it is difficult not to think in terms of some sort of "proto-perspective." This, however, would be to interpret the drawing in terms of something that had not yet happened. It also assumes the artist had a "better" image in his mind, which included a distinct point of view. But we should recall that this was not the case with the final picture of the rolling shields. With the lessons of the sequence of shields in mind, the picture of the trebuchet may be interpreted quite differently.

Passing over the fact that we once again have one drawing, one machine and no detailed measurements, it should be remarked that there is no overall point of view.[29] We seem to see the machine in general from above right. Yet we see the throwing arm and the weight bucket from above left. The right support of the triangular frame in the foreground seems to be pictured from above right, but the left support from below right, and so on. Rather than think the artist *had* a point of view but implemented it badly, we are better off to think that he did the same thing as the author of the shield sequence. The artist drew the machine one part at a time, giving each part its three dimensionality in an effort to depict its appearance. The overall appearance is only an artifact of this piecemeal procedure.

We may acknowledge that the picture contains perfectly adequate information for the expert builder. We may also acknowledge that the result was more than good enough to indicate the use of the machine to the patron. Both goals are accomplished through the depiction of appearance, but it seems clear that even though designers had adopted the idea of depicting the appearance of individual machine parts, they had not yet adopted the idea of a viewpoint as a means of regulating the appearance of machines as wholes.[30]

This, however, is the very step taken in the work of Mariano Taccola.

---

29 Interestingly, there are four numbers written on the drawings, seeming to indicate the proper length of some parts. They would not, however, relieve the maker of the necessity of determining the final dimensions of each part for himself. There is, for example, no indication of the relative thickness of any timbers.

30 In the case of the trebuchet, we have a drawing of a well-known machine. Leng (this volume) examines the application of this drawing style to gunpowder weapons as a whole new kind of artifact.

4. INVENTION OF THE SKETCH

Mariano di Jacobi detto Taccola was born in Siena in 1381. We know that he was paid for wooden sculptures in the Duomo of Siena in 1408, which suggests he trained as an artist. In 1424, he became secretary of a quasi-religious housing and hospital organization, which suggests he may have trained as a notary. In 1434 Taccola lost his secretary's job, becoming a *stimatore* or quantities estimator. In 1441 he became an inspector of Siena's roads, bridges and, possibly, water supply. He died shortly after 1453. This is almost all we know for certain about Taccola's life as it relates to his actual technical experience.[31]

During his life, Taccola prepared two technical manuscripts. The first is generally referred to as *De Ingeneis*, and was to have consisted of four books on civil and military technology. Books 3 and 4 were finished for presentation to the Emperor Sigismund in 1433. Books 1 and 2 were never completed. Fifteen years later, however, Taccola set to work on a second manuscript, variously known as *De Rebus Militaribus* or *De Machinis*, completed in 1449. What is significant for our purposes is that Taccola worked out many of the ideas he presented in *De Machinis* by filling the unfinished pages of Books 1 and 2 of *De Ingeneis* with hundreds of rough sketches, turning them into a sort of notebook. Examining these sketches and comparing them to the drawings in *De Machinis* we are able to follow a person actually working out technical ideas for the first time in history.[32]

An example of one of Taccola's final drawings from *De Machinis* is shown in figure 2.13. It shows two ships, one with a rotating stone dropper, the other with a sort of swinging-boom "fire dropper." In both examples we note Taccola's adherence to the principle of one machine, one drawing, and the absence of measurements, clearly locating these drawings in the mechanical tradition of design we have been examining. That is to say, Taccola does not expect to make these machines himself. He expects they will be made by experts who can complete the design and determine final dimensions for themselves. To these experts it is necessary only to indicate the parts and their arrangement.

Yet one has to acknowledge that Taccola accomplishes his goal in a strikingly different way compared to what had been done before. Gone is the flat style of the Middle Ages. Looking down from above right, Taccola is able to show us three sides of his ships at once so that they have volume and take up space. He adds to this sense of three-dimensional mass and volume through the use of shadowing. Furthermore, in place of the "flat" style, we finally have the adoption of the idea of a single viewpoint to go with the idea of a single drawing of a single machine, showing the overall appearance of the machine as well as the appearance of the parts at the same time.

---

31 For Taccola's life and works see Taccola 1972. For more on the context of the Sienese engineers and further details of Taccola's life, see Galluzzi 1993.

32 The first two books of *De Ingeneis,* also comprising the "notebook," are found in Taccola 1984a. The third book is found in Taccola 1969. Two different manuscript versions of *De Machinis* are found in Taccola 1984b and Taccola 1971.

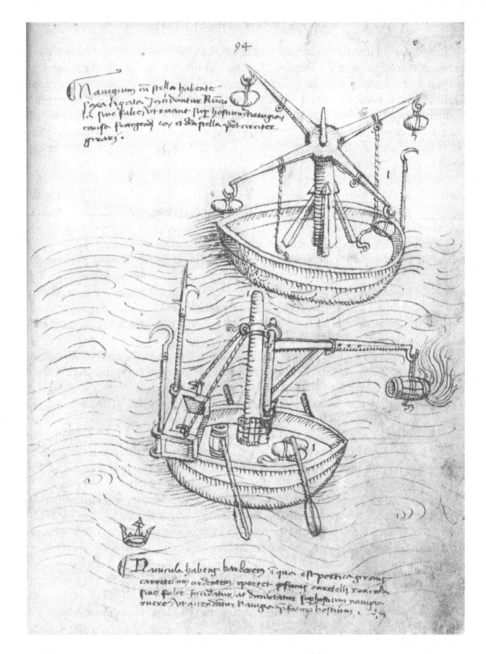

Figure 2.13. A page from Mariano Taccola's *De Machinis* of 1449. (Munich, Bayerische Staatsbibliothek, Codex Latinus Monacensis 28800, fol. 94^r; reproduced in Taccola 1971.)

What brought about this change in style? It has been said that Taccola's *De Machinis* is not really an engineer's book but a prince's book, the implication of this statement being that Taccola's drawings are presentation drawings, intended to entertain as a form of technical fantasy, and that is why they have the properties they do.[33] To a certain extent, this would be in keeping with the trend of development seen in Vigevano and Kyeser, where the attempt to satisfy the needs of the patrons resulted in a shift away from icons to the depiction of machines as they would appear. However, this cannot be the whole story. Figure 2.14 is a page from Taccola's notebook. It contains preliminary sketches for the droppers presented in *De Machinis*. These drawings have the *same* properties as the finished drawings—one machine, one drawing, no measurement, single viewpoint, etc. That is to say, Taccola used the same style to work out his ideas as he did to present them, and hence the story about a presentation style is at best incomplete. Further consideration of Taccola's preliminary drawing style is called for.

Figure 2.14. Sketches of "droppers" from Taccola's notebook. (Detail of Munich, Bayerische Staatsbibliothek, Codex Latinus Monacensis 197 Part 2, fol. 3ʳ; reproduced in Taccola 1984a.)

---

33 Long 1997.

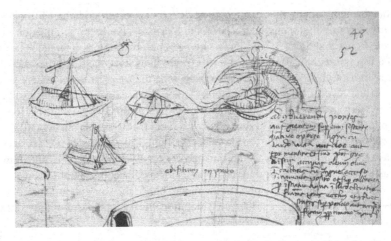

Figure 2.15. Details of folio 52ʳ from Taccola's notebook, showing several sketches of ships with protective shields, and one ship with a "grappler." (Detail of Munich, Bayerische Staatsbibliothek, Codex Latinus Monacensis 197 Part 2, fol. 52ʳ; reproduced in Taccola 1984a.)

Assuming the explanation for Taccola's new style must lie in what it allowed him to *do*, we can pursue the issue by turning to another page of the notebook (figure 2.15). Here we see that Taccola has sketched three different kinds of protected attack boats: one with a stone dropper, one with a ram, and one with a large hook or "grappler" on the side. We immediately see that his technique has enabled him to quickly generate three alternatives. Using paper, he is able to store them. Stored, they can be *compared*. In short, Taccola's style provided him with a graphic means of technical exploration.

In this particular case we can attain a closer understanding of the nature of this exploration by turning to another folio (figure 2.16), where the "grappler" boat shows up again in the lower left. This sketch still has the grappler on the side, which is to be dropped by a pulley. In the drawing immediately above it, the grappler is moved to the bow, attached to the boat by a stoutly constructed hinge, and dropped by a weighted, overhead boom. In the sketch above that, Taccola keeps the bow grappler, adds one to the stern, but removes the protective shielding while he considers a system of dropping the grapplers from double pulleys in the mast. The double-pulley mast is then retained in the sketch at the lower right, where one grappler is replaced with an assault ladder and the protection replaced. What we see here is that Taccola's technique does more than allow the rapid generation of alternatives. It allows a systematic holding of some elements steady, while others are explored in successive iterations of the design.[34] More than exploration, Taccola's style provides him with a method of systematic investigation through variation.

----

34 The importance of doing this is stressed in Jones 1970, 22.

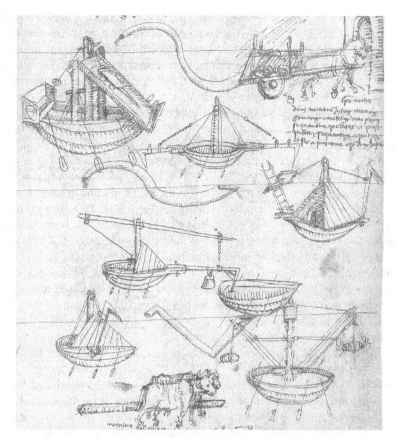

Figure 2.16. Detail of folio 60ʳ from Taccola's notebook, showing several sketches of ships with "grapplers." (Detail of Munich, Bayerische Staatsbibliothek, Codex Latinus Monacensis 197 Part 2, fol. 60ʳ; reproduced in Taccola 1984a.)

But systematic investigation of what? To what considerations are Taccola's variations a response? To answer such questions, it is helpful once again to consider the context of construction—this time the seemingly simple fact that by drawing ships on paper, Taccola does *not* have to go to the trouble of actually making them. Using paper, he does not have to deal with the real costs of labor and materials. Indeed, he does not have to deal with real materials at all, and he can also work in the absence of real dimensions.[35] Thus, we see the masts of his grapplers rise out of the various boats in figure 2.16, but Taccola does not have to be concerned with how big the masts, the boat, or the grapplers should really be. More important, working in the absence of real materials and real dimensions, Taccola is able to *work in the absence of real physics.*

---

35 Jones 1970, 22 emphasizes the importance of being able to work in the absence of real cost and real material constraints.

In the absence of materials, dimensions, and physics, Taccola is able to iterate freely, in both senses of the word. There is no cost of construction and no cost of failure. He is not constrained as the expert maker would be. However, the absence of physics makes it clear that Taccola's variations in design did not result from the consideration of *physical* problems related to issues in mechanics. The information necessary for any such considerations is just what is missing. For instance, given the absence of dimensions, he could not know either factually or intuitively whether his boats would have the necessary displacement to float themselves, let alone to carry the equipment and protection he proposes, with adequate stability. Indeed, in the absence of dimensions, he could not even calculate the problem of the levers around which all his designs are organized.[36]

Instead, it appears that Taccola's drawing technique allows him to pass *beyond* the context of construction to investigate the future context of *use*. Thus, there is nothing physically impossible about placing the grappler on the side of a boat, as Taccola does in his initial sketches (figure 2.16). The pulley arrangement would work. But it is not very likely to be effective in battle, particularly if the enemy ship approaches from the other side. Seemingly aware of this, in his next sketch Taccola considers dropping the grappler from the bow, which could be more effective, although the stern would be unprotected. Hence, the double-grappler of his next drawing. Similarly, there is nothing physically better about dropping grappling hooks from overhead beams, as opposed to pulleys, or double pulleys. Again, the cause of variation appears to be Taccola's awareness of awkwardness or ineffectiveness for the humans who actually have to use them.[37]

It may be significant that this process of systematic variation, iteration, and exploration of the context of use continues into the final drawings of *De Machinis*. Thus, in figure 2.17 we have what almost looks like the original design, but with a grappler at the front using his perfected hinge arrangement and the pulley dropper. In figure 2.18 we have a sort of grappler and assault ladder combined, but a more sensible form of protection than the original pyramid. In figure 2.19 we have a ladder and grappler combination, with a flat deck for protection, while double pulleys on one mast have been abandoned for double masts with a single pulley each. Such a sequence of drawings flowing from the notebook to *De Machinis* strongly suggests that Taccola's "presentation" drawings are actually an extension of his investigative technique, rather than the reverse. If so, it would be possible to argue that Taccola developed his inno-

---

36 One might claim that Taccola varied his design in response to his *intuitive* understanding of mechanical problems. My response is not to say that this is impossible, but rather that is an argument for which there can be no proof, because the whole advantage of the only evidence available—the drawings—is that they *eliminate* the need for physical considerations on the part of the artist.

37 It may be argued here that I have violated my own injunction and started to discuss what Taccola was thinking. There is a difference, however. Following the approach presented in this paper, one is able to describe Taccola's thinking as sequence of reasoning, and moreover to point to the evidence for assertions about that reasoning. This is quite different from opining about "cognition," or "tacit" knowledge, particularly in that one can argue about the actual evidence at hand, and not about human nature, whether one particular psychological theory is valid and so on. With more space, I would further argue that, while doing may not always be regarded as thinking, thinking may always be regarded as a form of doing.

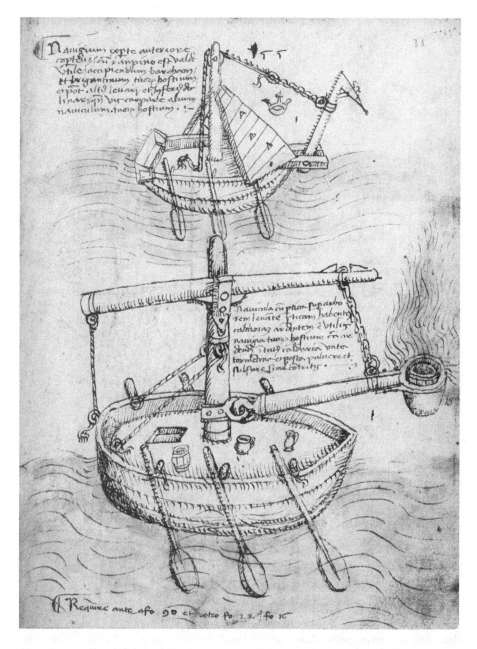

Figure 2.17. A presentation drawing of a "grappler" ship from Taccola's *De Machinis*. (Detail of Munich, Bayerische Staatsbibliothek, Codex Latinus Monacensis 28800, fol. 55ʳ; reproduced in Taccola 1971.)

Figure 2.18. A second presentation drawing of a "grappler" ship from Taccola's *De Machinis*. (Detail from Munich, Bayerische Staatsbibliothek, Codex Latinus Monacensis 28800, fol. 81ᵛ; reproduced in Taccola 1971.)

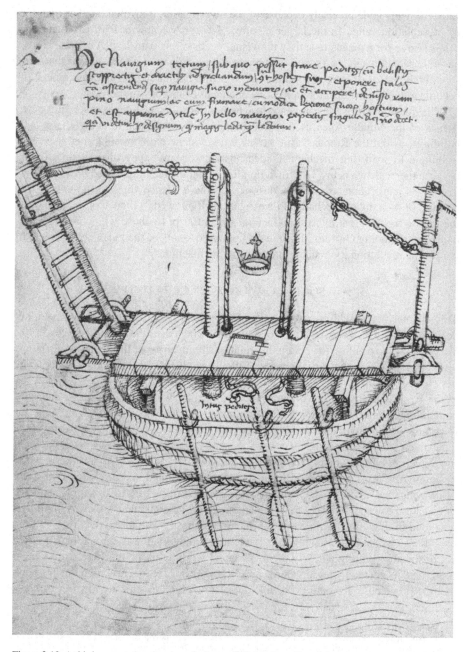

Figure 2.19. A third presentation drawing of a "grappler" ship from Taccola's *De Machinis*. Unfortunately, part of the hook has been cut off the page. (Detail of Munich, Bayerische Staatsbibliothek, Codex Latinus Monacensis 28800, fol. 90ᵛ; reproduced in Taccola 1971.)

vative drawing style precisely because of the investigation it involved, rather than for its presentation value. In the larger picture, however, we may at least say that Taccola's search for a means to satisfy the visual needs of his patrons had a momentous, but unexpected result—a graphic technique for investigation and invention that was soon, in the hands of Francesco di Giorgio Martini, Leonardo da Vinci, and others, to become crucial to the development of mechanical technology in the West.

The question arises, of course, as to whether Taccola's technique was really new. Artists of the Middle Ages used representations of appearance to show chairs and buildings, as did the Romans. But we have no evidence to show that they used the technique to work out mechanical ideas. Is this because they did not do so? Or because the evidence is lost? My own opinion is that the technique used by Taccola *is* new in its application to machine design, and the evidence suggests he adopted it because of the kind of mechanical investigations it allowed.[38] Indeed, it may even be the reason for the great interest in Taccola's manuscript by those who consulted and copied it, including Giorgio Martini and Leonardo—not for the machinery depicted, but for the revolutionary manner of depicting the machinery.

5. THE SOCIAL GROUNDS OF CREATIVITY

This chapter began with a question about the stable relationship between machine design and the use of one machine, one view drawings over the period 1400 to 1750. We are now in a position to explain this relationship in terms of the creation of a powerful "paper tool" in the early decades of the fifteenth century, as used by Taccola, if not created by him.[39]

To be sure, the use of one drawing for one machine was already part of a tradition stretching back into the Arabic and Byzantine past, when the flat style could be used to communicate to workers, and perhaps to the learned readers of ancient treatises on mechanics.[40] But in the fifteenth century, a switch away from logical schema towards the depiction of appearance, followed by the adoption of the viewpoint (before the invention or at least diffusion of linear perspective), resulted in the creation of a style of drawing that could do three things at once. First, it continued to supply enough information to expert makers that a machine could be constructed, even though the new kind of drawing was no more complete than those made in the flat style.[41] Second, the new kind of drawing provided information that was comprehensible to patrons, demonstrating not only the mechanical working of the machine, but its purpose, and its method of use. Third, as is clear from Taccola's notebook, the new method provided a new and powerful tool of investigation. Not a tool of invention,

---

38 Although its origins may lie in the painterly techniques of Giotto and Duccio. If so, then we might look to explain Taccola's use of his technique in terms of the artists' timely possession of a new kind of tool in a changing context of patronage, where both machine building in the service of the state and humanistic investigations of ancient technologies were extremely serious business.

39 See Goody 1977, and Klein 2003.

40 Again see Lefèvre 2002. But also see Hill 1996, and Jazari 1974.

41 As Popplow points out in his chapter, even Schickhardt's highly detailed drawings intended for his workers left out critical information about gears.

one should be careful to say, since Vigevano was clearly capable of inventing in the flat style, but certainly a tool of more *rapid* investigation of combinations and their use in future contexts. The somewhat halting development of rolling shields in Kyeser's manuscript, as opposed to Tacola's rapid development of several kinds of assault ships, is illustrative.

Another issue raised in the introduction was the "Princes problem," arising from the fact that many early manuscripts were intended for a princely audience. What was the effect of this audience on mechanical drawings? The explanation offered here is that the need to communicate more clearly with patrons drove a change from the depiction of logical, iconic schemas to the depiction of appearance. This development was under way before the invention of linear perspective. However, this was not the only factor in play. The evidence of Kyeser's *Bellifortis* suggests that the method of depiction was not simply borrowed from existing courtly styles of painting and illu-mination. Rather, as the sequence of rolling shields suggests, the method of depiction had to be worked out, step after step. But as the drawings in Kyeser's manuscript also show, one of the most important constraints on the development of the new style was the need to preserve the ability to communicate parts and arrangement to makers. It is not that there was only one audience for these drawings. There were always two.[42]

This chapter also started with a plea for coming to grips analytically with the evi-dence provided by the properties of early machine drawing. I've tried to show that there can be a benefit in doing so, in the form of three perhaps surprising results.

One surprise is that, although we tend to regard drawings as evidence of thinking, as extensions of cognitive processes of design in the mind's eye, the properties of early machine drawings actually point more immediately to what might be referred to as a social process of design in which the second person is the real expert, the person who actually determines the final dimensions of the machine that is made. This is a fundamental context of machine design that would remain the same for several centu-ries after Taccola. Indeed, mechanical drawing would only begin to change in the late eighteenth century when the shift from wood to metal, and the holy grail of replace-able parts, set off a determined attempt to gain control over the hands of machine makers.[43]

The second surprise is the methodological one, in which consideration of the con-text of construction proves to be crucial to understanding the properties of the machine drawings we see—even though it may well be that the machines were never made. This is important, because we truly need to remind ourselves that design draw-ings are really about what is to happen after them, not what went before them. It is in this context of what *will* happen that they are made. Surely it ought to be important in trying to interpret them.

The third surprise follows from an examination of Taccola's use of drawings to investigate new kinds of machines, investigations that seem much closer to the cogni-

---

42 This only makes sense, since a prince might at any time order a machine from a manuscript to be built, and the authors of drawings were, in most cases, not capable of actually building machines themselves.
43 See the discussions in Alder 1997; Baynes and Pugh 1981; Noble 1984; Smith 1977.

tive processes of design as we normally conceive them. But even here, consideration of construction proves important, leading to the realization that Taccola used these drawings precisely because they meant he did not have to build experimental machines, but was free to innovate as he pleased precisely because he was able to work in the absence of the real constraints of dimensions, materials, labor and physics.

Now on the one hand, it seems to me that the absence of physical considerations in the drawings creates big problems for anyone hoping to conscript Taccola and other early artist-engineers as conceptual builders of the scientific revolution. Neither formal physics nor even intuitive physics are there in the evidence. On the other hand, it seems to me that the lack of physics and the evident impossibility of many early machine drawings has often misled us into thinking that they are merely forms of creative play, fantasy—in other words, just fooling around.

There is, however, another way to look at it. If we bring our expert maker back into the design process, we realize that real constraints are not actually gone, so much as displaced into the future—and, particularly, that they are displaced onto the shoulders of another person. It is the job of the expert builder not only to make the machine, but to make it work, meaning that dealing with the physics was the problem of the builder, and not of someone like Taccola. If so, and if it is the absence of constraints that is the key to innovation on paper, it is possible to suggest that the social situation was the very ground of the individual creativity we see in early machine drawings, and that it was the recognized existence of expert makers (so often disparaged as mere craftsmen) that created the very space in which the Renaissance design of machines on paper was possible.

SOCIAL CHARACTER, PICTORIAL STYLE, AND THE GRAMMAR OF TECHNICAL ILLUSTRATION IN CRAFTSMEN'S MANUSCRIPTS IN THE LATE MIDDLE AGES

RAINER LENG

INTRODUCTION

In the late Middle Ages a number of well-known authors created illustrated technical manuscripts that are by now acknowledged not only by technical historians. The sketch-book of Villard de Honnecourt,[1] the *Texaurus Regis Franciae* by Guido da Vigevano,[2] Konrad Kyeser's *Bellifortis*,[3] and the writings of Mariano Taccola[4] are available through editions and widely quoted in scholarly literature.[5] Villard's sketch-book and a part of Taccola's works are recordings with a more or less private character. The other manuscripts are characterized by dedications to high-ranking aristocrats. Guido dedicated his *Texaurus* to the French king Philipp VI (1328–1350), Konrad Kyeser bestowed his *Bellifortis* on two German kings, Wenzel (1376–1400, † 1419) and Ruprecht I (1400–1410), and Mariano Taccola prepared a luxurious copy of *De ingeneis* for King Sigismund (1410–1437).[6]

Hence the best-known pictorial manuscripts can be counted among the courtly literature. As David McGee remarks,[7] the special relationship between the author and the addressee has consequences for the way in which technical items are presented. Being "not really an engineers' but a princes' book," these pictorial manuscripts must ensure understanding. They therefore usually show the whole device in action depicted in scenery or settings. They focus on illustrative graphic presentation, not on physical function or conveying machine building know-how precisely. Provided that there are explicit technical contents to be conveyed, there must have been a further recipient besides the addressee, a real expert who knew about the practical dimensions neglected in the drawings.

---

1 For a modern facsimile edition, see Hahnloser 1972; the most important literature is discussed in Binding 1993, 207–224, Barnes 1982 (bibliography), and Bechmann 1993.
2 Vigevano 1993; for author and work see also Hall (A.R.) 1976a, Hall (B.S.) 1978, Hall (B.S.) 1982c, Alertz 2001.
3 Kyeser 1967 from Ms. Göttingen Philos. 63 and Kyeser 1995 from Ms. Göttingen Philos. 64 and 64a; see Berg and Friedrich 1994, Friedrich 1996 and Leng 2002, I 109–149.
4 *De rebus Militaribus (De Machinis):* Taccola 1984b and 1971, *De ingeneis* and the notebook: Taccola 1984a. For the complete work and the most important literature, see Degenhart and Schmitt 1980–82, vol. 4.
5 For instance Hall (B.S.) 1976b, Hall (B.S.) 1982a, see also Popplow 2001, 251–263 and the chapter by David McGee in this volume, both with further references.
6 For the background see Knobloch in Taccola 1984b, 11 and Degenhart and Schmitt 1980–82, IV 21–27.
7 See the chapter by David McGee in this volume.

This specific tradition of explaining technology for high-ranking laymen by means of drawings can be found throughout the entire fifteenth century. Besides the well-known books mentioned, this is also apparent in two unpublished manuscripts dealing with the art of warfare, which were produced for the emperors Sigismund (1410–1437) and Frederick III (1440–1495).[8] Even the first printed technical book— Roberto Valturio's *De re militari*, dedicated to Sigismundo Pandulfo Malatesta (1417–1468) (publ. Verona 1472)—is adapted to the simple rules of explaining machinery in settings.[9] In a large series, not even a step-ladder is drawn without the wall it leans on, nor a pump shown without a stretch of water.

Less known, however, is that a number of pictorial catalogues appeared soon after the year 1400,[10] written by those very skilled expert makers onto whose shoulders authors like Kyeser or Taccola displaced the practical realization of their technical suggestions "without physics."[11] On the "social scale"[12] of authors of manuscripts conveying technical knowledge, these experts are ranked at least two levels lower than Kyeser and the other celebrities mentioned above. And this holds for their audience as well. They are mere craftsmen, but rare specialists in their métier, who put pen to paper for the first time to write down their knowledge in words and sketches. This fact raises a number of questions. How did disparaged tradesman come to writing, given the fact that proficiency in craftsmanship had obviously coped without written manuals for centuries? Did Vigevano or Kyeser supply them with examples, or did they start drawing on their own? Are there significant differences in the pictorial means to those of the technical literature for the courtly context? What were the social circumstances or backgrounds of these catalogues, what function did they serve, which readers did they address? And finally: what kind of expressiveness did their graphic techniques display and how can their development be explained? If technical drawings are part of a universal graphical language to transmit technical knowledge, what is the grammar of this language?

1. THE BEGINNINGS

The beginnings of craftsmen's analysis of technology through writing and picturing are marked by a sample of four illustrated manuscripts, which can be dated to the first

---

8   Zurich, Zentralbibliothek, Ms. Rh. Hist. 33b, c. 1420–1440, dedicated to emperor Sigismund, and Vienna, Kunsthistorisches Museum, P 5014, c. 1440–1450; for the Zurich manuscript see Grassi 1996, for both directly related manuscripts see Leng 2002, I 221–230 and for a description of the manuscripts, II 315–318 and 417–422.
9   Valturio 1472, mainly a tract about antique Roman warfare, but also with some contemporary devices, was dedicated to Sigismundo Pandulfo Malatesta between 1447 and 1455 and spread across Europe in the form of roughly 20 illuminated manuscripts before printing; see Rodakiewicz 1940 81f., Ricossa and Bassignana 1988, 172, Popplow 2001, 263–266. Almost all of the woodcuttings from the printed edition were used to illustrate the German translation of Vegetius by Ludwig Hohenwang in 1475 (dedicated to the landgrave of Stühlingen), see Fürbeth and Leng 2002, 36–53.
10 Widely ignored for decades and only seldom mentioned, for example in Hall (B.S.) 1979b 21ff. and 127–130; some of these were investigated by Knobloch 1996 in the context of technical drawing. The outstanding figure of Leonardo usually seems to reduce all other authors of pictorial catalogues to the rank of minor figures; for example in Gille 1964 or Parsons 1968.
11 See David McGee in this volume.
12 Hall (B.S.) 1979a, 54.

quarter of the fifteenth century. There are, firstly, two closely related manuscripts, one from Vienna and one from Munich, that contain images of warfare devices. In both cases only the first sheets contain marginal notes or annotations. The Viennese manuscript dates to 1411.[13] The Munich *Anleitung Schießpulver zu bereiten, Büchsen zu laden und zu beschießen*, created in the first quarter of the fifteenth century, presents on only 22 raw single leaves a selection of the Viennese manuscript.[14] The missing drawings may have been lost before binding, such that it also might have been a pattern of the Viennese codex. Characters and style are, on first sight, more primitive. In comparison, a third manuscript, also from Vienna and today kept in the Museum for the History of Art, contains extensive text passages in which both single images and series of plain pictures are integrated.[15] It was compiled between 1410 and 1430. The last is a manuscript from Nuremberg, a fragment of a formerly much larger volume, written between 1420 and 1425.[16] It contains a series of pictures of military devices with explanatory rhymes.

The authors remain anonymous. Neither their names nor the names of their addressees are known. In all four cases the composition is very simple: standard letters, wash drawings, paper instead of vellum, ordinary flexible bindings. Only the Viennese cod. 3069 has a vellum-bound wooden cover. It also contains several pictures from the *Bellifortis*. Therefore we may assume that parts of it were written for an addressee at court;[17] at the very least, the author must have had access to a courtly library. The main part of the codex is based on early craftsmen's material, and even the *Bellifortis* pictures are selected according to technical criteria. All scenes containing mythological or literary items have been sorted out. However, the other three manuscripts of this sample were issued by anonymous craftsmen, i.e. by master gunmakers, who were experts on modern military technology in the fifteenth century. These few early manuscripts constitute the roots of an independent literary species of illuminated manuscripts about military technology that were compiled by experts, away from the courts. This tradition remains vital throughout the whole fifteenth and early sixteenth century. About 50 of them are still preserved today.

The spread of these early pictorial catalogues and of all subsequent master gunmakers' books of the fifteenth century was concentrated in the South of Germany. Although gunpowder technology had been common knowledge in Europe since the first decades of the fourteenth century, and many German master gun-makers served in other countries, no comparable manuscripts have been found in Italy, France or England.

The tradition of master gun-makers' drawings was unique and engendered independently of the older machine-drawing tradition. Only single copies of the Latin writings of Villard and Vigevano were distributed in Italy and France—inaccessible for the uneducated practitioners. Taccola, on the other hand, started with his machine

---

13 Vienna, Österreichische Nationalbibliothek, cod. 3069; see Leng 2002, II 334–336 and I 172–180.
14 Munich, Bayerische Staatsbibliothek, cgm 600, ed. Leng 2000.
15 Vienna, Kunsthistorisches Museum, P 5135, see Leng 2002, II 319–323 and I 180–195.
16 Nuremberg, Germanisches Nationalmuseum, Hs. 25.801; see Leng 2002, II 266f and I 191–194.
17 See Leng 2000, 25f and Leng 2002, I 179f.

drawings at least two decades after the first master gun-makers' books. The only one from whom they possibly could have benefited was Konrad Kyeser, who wrote his *Bellifortis* six years before the first precisely dated master gun-makers' book in the Viennese cod. 3069 of 1411. But among the four eldest manuscripts, only one has a distant relationship to the *Bellifortis* as mentioned above. But Kyeser's drawings did not inspire the anonymous author: the first part of the manuscript is completely independent. All pictures derived from the *Bellifortis* are later additions to the second part of the codex, and none of them is listed in the opening table of contents on the first two sheets.

Although the master gun-makers' tradition was unique, these craftsmen did not, of course, "invent" drawing in general for their purposes. Some of the characteristic features of the older courtly tradition, such as the use of different viewpoints or the depiction of integral devices on one sheet, strikingly similar conventions or rules, can be found in Villard's sketch-book as well as in Vigevano's and Kyeser's illustrated manuscripts on war.[18] But considering the fact that no master gun-maker had ever seen them (during their first attempts) it is plausible to assume that they used commonly known pictorial techniques of the Middle Ages, which could be seen in any illustrated book, in church glass and murals. These common elements of a pictorial language, together with their very special everyday experiences with single rough drawings for new constructions, constituted the foundation for the first pictorial collections by craftsmen in complete, unique and—as opposed to single sheet files—lasting manuscripts.

2. THE SOCIAL FUNCTION OF CRAFTSMEN'S MANUSCRIPTS

Immediately at the beginning of each manuscript we find drawings or texts with technical content. None of the authors seems ever to have thought it necessary to address his readers or to explain the purpose of his writing. Thus, the social function of these writings has to be inferred by analyzing the social status of the author and by interpreting the text and illustrations. We learn more about the master gun-makers by studying their contracts.[19] There is an increasing number of such working contracts from the 1370s onwards. Most of the master gun-makers had been working as blacksmiths or metal-workers in medieval urban settlements. They left their traditional workplaces in order to specialize and thus to climb the social ladder. They often changed employers, they were very well paid and relatively rare. Lords and cities depended on their detailed know-how of chemistry and weapon technology. Their high mobility, their specific and valuable knowledge were the social basis of the first master gun-makers' codices.

For centuries, technical know-how had been passed on orally in a settled urban tradition. The solitary traveling gun-maker could not rely on such oral tradition. His precious knowledge was always in danger of being lost in perilous military engage-

---

18 See the chapter by David McGee in this volume.
19 See for the following Leng 1996.

ments. Only by setting down it on paper could master gun-makers ensure that their knowledge be preserved enduringly for journeymen, apprentices and lords. Knowledge, as well as times, were changing: Military technology was developing quickly, instructions for the production of powder and additional substances became increasingly complicated, and their number was increasing. Master gun-makers were no longer able to memorize those instructions. The *Fireworkbook of 1420* had already demanded that "Der Meister sol auch kennen schreiben und lesen."[20] Obviously there had been frauds among the traveling masters. Those who were able to prove their knowledge by producing written evidence were more likely to find employment.[21] For these experts, the way out of the "sub-literate groups"[22] into literacy was not too far and, in any case, of big advantage.

At the beginning there may have been only single sheets, notes about gunpowder recipes, single sketches of technical devices. These early documents of master gun-makers' technology have been lost. Today we can trace them back only by studying the codices that were compiled later from this early material. Looking at the oldest codices one can discover their relationship to those single-file collections.[23]

Apart from the task of passing on innovative knowledge, and from a partial presentational function, these writings had one main social intention: the exchange of technical know-how. Communication among master gun-makers took place by way of books: Master gun-makers mention their colleagues' books, and they find themselves mentioned, too: "… ze einer bedutnws eins andern puochs …";[24] "… diesen sinn den suoch / in dem andern buoch …"[25] Wherever single characteristic drawings can be found in more than one manuscript, conclusions can be drawn about the relationship between these codices and, thus, the exchange of drawings can be assumed.

Certainly the relationships among images of commonly known devices is hard to prove. But the Viennese manuscript cod. 3064, which can be dated to around 1440, is a nice example of parallel intertextual communication: in the inserted gunpowder recipes, the writer names his colleagues who had invented them.[26] It seems that the anonymous author was in close contact to his colleagues in the south of Germany, and they seem to have exchanged recipes quite frequently. The high mobility of master gun-makers and their regular meetings at the major sites of military conflict, for which large numbers of this trade were called together, favored such a literary communication. It is likely that on such occasions sketches were exchanged as well. This

---

20 Hassenstein 1941, 16: "... wann der stuck souil sind die darzu gehoerendt / die ein yetlicher guetter püchsen meister künden soll / vnd die ein mayster on die geschrift in seinem sinne nit gedencken kann ..." See also Hall (B.S.) 1979a, 47–58.
21 See Schmidtchen 1990 30, Leng 2002, I 105f.
22 Hall (B.S.) 1979a, 48.
23 See Leng 2002, I 151–154.
24 Vienna, Österreichische Nationalbibliothek, cod. 3069, fol. I^r;—see Leng 2000, 25.
25 Nuremberg, Germanisches Nationalmuseum, Hs. 25.801, f. 15^r—see Leng 2002, I 193.
26 Vienna, Österreichische Nationalbibliothek, cod. 3064, f. 9^v "magistrum conradum," f. 10^r "ulreichs mawrers puluer," f. 10^v "mayster johannes von Österreich," f. 12^v "also zugelt der marckgraf von rotel Salpeter" (who is also mentioned in Cologne, Archiv der Stadt Köln, W* 232 f. 84^r and Frankfurt, Bibliothek des Instituts für Stadtgeschichte, Reichssachen Nachträge Nr. 741, 29), f. 16^v "magistrum iohannem," again f. 25^v "magistrum iohannem de austria" and f. 26^v "secundum iohannem." For a description of the manuscript see Leng 2002, II 331 and I 238.

explains why some characteristic drawings can be found in a number of pictorial catalogues all of which are not mere copies, but individual sketch-books.

The openness with which the master gun-makers apparently communicated their know-how and even inventions is remarkable.[27] Of course, issues of secrecy arose on occasion. The patrons of these experts were interested in spreading their knowledge. In the case of the master gun-maker's death or absence, other servants or citizens should be able to serve the ordnance, too. Therefore some contracts obliged the master gun-makers to teach few persons about shooting and the production of gunpowder.[28] But sometimes this produced undesirable competitors. In 1468 the master gun-maker of Lucerne claimed damages from the council, because some citizens, to whom he was obliged to convey his knowledge, had started producing gunpowder on their own. He complained about the reduction of his income.[29] For such reasons master gun-makers could be interested in maintaining secrecy as regarded their monopoly of know-how within their closer sphere. On the other hand, master gun-makers were cleared for classified military information. No one knew better the attack or defence potential of their patrons. Because of this, the employers themselves were vitally interested in forbidding the conveyance of knowledge, and included strict clauses in their employment contracts. Problems of secrecy occurred for economic or military reasons, but they never seem to have prevented the transfer of technological know-how in drawings among the master gun-makers. The group of skilled experts was too small and too scattered—normally one master gun-maker per city!—for serious competition. It seems they felt themselves to be colleagues obliged to each other by their dangerous and exclusive profession more than as distrustful competitors.

The early production of illuminated technical manuscripts by craftsmen, who were not writing for the courts but for themselves and for their colleagues, is autonomous. It is not influenced by the presentational manuscripts such as those produced by Kyeser and Taccola. No master gun-maker would have understood the ambitious, literary Latin hexameters of the *Bellifortis*! Besides, the audience these craftsmen address is a different one. The drawings serve as a memory aid in the first place, but they also pass on knowledge to journeymen, apprentices, and colleagues. For the first time we encounter texts and pictures that have been produced for communication between skilled experts.

3. PICTORIAL STYLE: PRIMITIVE OR FUNCTIONAL?

The pictorial style, especially that of the early master gun-maker books, often has been contemptuously called "crude" or "primitive."[30] Of course, the style of the sketches is rough. The simple feather drawings couldn't compete with the magnificent book illumination of Taccola or Kyeser. But as they were not meant for courtly

---

27 For the problem in general, see Long 2001, especially 117–122 about openness and secrets in writings on gunpowder artillery and machines, including some of the above-mentioned manuscripts.
28 For some examples see Leng 1996, 314f.
29 See Hess 1920, 22.
30 Gille 1964, 49f. and even Hall (B.S.) 1979b, 21f.

recipients or addressees, they didn't have to compete with such illustrations. However, if they served the communication between specialists, they had to employ graphic techniques that permitted the transmission of knowledge by means of drawings. In some aspects these techniques did not work smoothly, of course, and cannot be compared with modern visualizations of technical devices. However, this must be considered in light of the fact that these uneducated authors had just stepped out of an oral tradition in the direction of technical literacy.

Even in a first perusal of the craftsmen's manuscripts, one is struck by a significant difference from the presentational manuscripts: only the single device is depicted, and this—in contrast to the courtly manuscripts—without any context, scenery or settings. This didn't mean reducing the pictorial means by dispensing with unnecessary information. While the presentational manuscripts needed an illustrative context to demonstrate the function and the purpose of the apparatus to the courtly audience—all mechanical laymen—this contextual information was not necessary for communication among technicians.

An example taken from one of the oldest master gun-makers' sketch-books illustrates this fact (figure 3.1): The author outlines a piece of artillery with multiple tubes that is meant to improve the slow rate of fire. A scene to illustrate the employment—targets, walls, enemies, as we would expect in the presentational manuscripts—is not necessary. The skilled expert was able to imagine the advantage at first sight. Instead, the raw sketch conveys additional information: A massive block mount is necessary to bear the weight of the tubes. The four tubes must be installed in radial order on a wooden disc and reinforced with two cross-shaped beams. The figure of a gunner revolving the disc explains its function. The tubes are attached to the beams by clamp straps. The disc is fastened on the mount with an iron screw. Thus the horizontal adjustment is variable by means of a spin-

Figure 3.1. Sketch of a multiple gun using different viewpoints. Drawing from the first quarter of the fifteenth century. (Munich, Bayerische Staatsbibliothek, cgm 600, fol. 13^r; Leng 2000, 116.)

dle. At the bottom of the mount, a device for additional vertical adjustment is indicated, showing that the disc could also be inclined. These indications all give crucial elements of the construction, enough for another skilled craftsman to figure out and determine the final dimensions of such a weapon on the basis of his individual experiences in machine making. Detailed information about materials, measurements or any special mechanisms could be omitted. In any case, in the age before early industrial mass-production, every craftsman had to develop made-to-measure solutions for each single device to make it work.

By restricting the raw sketch to the essential functional elements of the construction, it was quite possible to convey the knowledge necessary to build a working multiple-tube gun. This functionality of the drawing must be kept in mind by the modern observer, who might experience "visual confusion" because of the drawing's use of multiple viewpoints instead of correct perspective rendering.[31] The mould is shown from a front view, while the disc with the tubes—which is horizontal in reality—is turned about 90 degrees to show it in plain view within the same sketch. Before the introduction of more efficient elevated viewpoints[32] or the spreading of real perspective drawings,[33] the use of multiple viewpoints cannot be judged as pictorial primitivity. On the contrary, it allows a highly effective compression of the data needed to convey technical information by means of one drawing. In contrast to a depiction by modern orthogonal plans, which would demand at least two plans, a late medieval craftsman transmitted the necessary information on only one sheet. And his recipient, used to decoding the multiple viewpoints, had no difficulties extracting all relevant information he needed.

A last comparison demonstrates the difference between craftsmen's and presentational manuscripts (figure 3.2): The early master gun-makers' book in the Viennese cod. 3069 shows a lathe on f. 90[r]. The drive, tools, workpiece, even the slide in a variable guide and a storage space for other tools can be clearly recognized. On f. 21[v] a more simple device is shown: caltrops, with detailed information about the wooden stakes with iron barbs, and a trick for rendering them harmless. Once more we are confronted with a concentration on the single device, renunciation of superfluous settings and accentuation of functional elements. Based on the material presented in this manuscript, a professional illustrator of about 1450 compiled a codex on warfare technologies for Emperor Frederick III.[34] Here the translation of technical sketches back into a courtly context required a reduction of technical precision in favor of the addition of settings. The lathe is less detailed and was embedded in a workshop scene, and the caltrops had to be integrated in a scenery that shows how they were used to put obstacles in the infantry's way. These drawings appear sophisticated, but they are only useful to a courtly observer. On the other hand, the craftsman, who did not need such contextual information, would miss a detailed depiction of the devices. From his point of view the sophisticated presentation was primitive. He preferred the "crude" but more informative sketches.

---

31 See the chapter by McGee in this volume for the viewpoints in Villard and by others.
32 For the development of elevated viewpoints or perspective drawing, see the chapters by McGee and Popplow in this volume.
33 See the chapters by Camerota, Lefèvre, and Peiffer in this volume.
34 Vienna, Kunsthistorisches Museum, P 5014, see above note 8.

Figure 3.2. Top: Devices without scenery. From the *master gun-makers' book* from 1411 (Vienna, Österreichische Nationalbibliothek, cod. 3069, fol. 90^r (left) and 21^v (right).) Bottom: Illuminated codex for courtly audience with additional scenery. Pictures based on the material of the *master gun-makers' book* from 1411, but dedicated to Emperor Frederick III. (Vienna, Kunsthistorisches Museum, P 5014, fol. 110^r (left) and 49^r (right); Leng 2002, I plate 12.)

4. DEVELOPING A GRAMMAR OF TECHNICAL DRAWING

With the depiction of complete devices from different angles, the craftsmen and master gun-makers of the early fifteenth century invented a pictorial language that permitted the exchange of technical information. The grammar of this language was simple but functional, reduced to two rules: Show the whole device without any distracting settings, and use as many viewpoints as necessary to convey precise information. The amount and the quality of the transported know-how, however, remained limited. The reader had to take a lot of trouble to translate or even decipher information. Without his own experience as a reference system in the background, these first attempts at a pictorial language had to fall on deaf ears, or rather on blind eyes. There was the serious danger of misunderstanding single elements. With the increasing complexity of technical devices, a sophistication of the rules of the pictorial grammar was inevitable. Analysis, variation, segmentation, and new combinations of single elements were promising means to improve the performance of a pictorial language.

Of course the master gun-makers of the late Middle Ages never expressed in words their primitive rules of communication by means of drawings; only some Renaissance artists attained this degree of reflection.[35] But in an intuitive way they discerned that the once successful process of pictorial communication could be developed further by varying devices, by separating single elements from the integral device, and by extending the catalogue of pictorial rules, in part playfully, in part with concrete intentions. We can find an early example of a prolific attempt to expand the first two simple rules in the already mentioned Viennese manuscript P 5135, compiled between 1410 and 1430. Although the traditional rule of showing only complete devices is predominant, the author in some cases tried out the use of sophisticated pictorial rules. Especially the sketch of the multiple tube we know from the Munich cgm 600 (figure 3.1) apparently was considered to be unsatisfactory. The draft conveyed an approximate idea of the function, but the two viewpoints chosen did not allow the decisive interior mechanisms to be shown in a precise way. Only a master who had built comparable machines before could gain real benefit from this "workshop drawing." But in order to convey the necessary information to a less experienced master or to a journeyman, the precision of pictorial means had to be improved. Distinctly discernible is the attempt to depict separately such components of the whole device (figure 3.3, top left), which until then had been partly concealed or merely indicated. The sketches on f. 93^r–95^r (figure 3.3, top left, bottom left and right) demonstrate such an attempt at pictorial segmentation and variation.

Hidden components or larger parts blocking the observer's view were given their own single sketches for a more precise depiction. In a mental process, the everyday experience of building complete machinery from single components was transferred to improved didactic presentation and the conveyance of technical knowledge among the group of skilled experts. The capacity of pictorial means was increased by adding a further rule of communication: show even the hidden parts in single sketches.

---

35   See the chapters by Camerota, Lefèvre, and Peiffer in this volume.

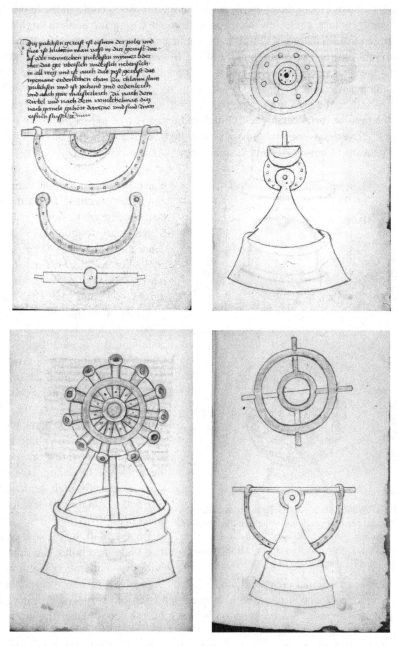

Figure 3.3. Pictorial segmentation and variation, rare examples for depicting single components. From Anonymous, *Pixen, Kriegsrüstung, Sturmzeug vnd Fewrwerckh,* between 1410–30. (Vienna, Kunsthistorisches Museum, P 5135, fol. 55[r] (top left), fol. 93[r] (top right) and 94[v] / 95[r] (bottom); Leng 2002, I plate 10.)

Although a rare exception among the master gun-makers' books in the first half of the fifteenth century, these few sketches were trend-setting for modern engineering drawings and their custom of dissolving complicated devices into single components and assemblies. Soon after the middle of the fifteenth century a noticeable push in pictorial means is perceptible. Several developments may have stimulated the step to more efficient designs. We can detect technological improvements as well as increased self-confidence among the writing craftsmen, who now began to sign their names on their pictorial catalogues. Gun foundries produced lighter and more powerful tubes.[36] Even conventional arms, the so-called "Antwerk," became more sophisticated by means of improved smithcraft and new woodworking technologies.[37] This resulted in a multiplication of ordnance parks and a great variety of gun types, brought forward especially by the adaptation of gun carriages and foundations for all conceivable purposes. The mounts no longer had to be carved from massive trunks, as we noticed in figures 3.1 and 3.3, but could promptly be composed of prefabricated boards and beams from the wide-spread sawmills.[38] The sophisticated elements of construction are soon reflected in the improved pictorial means of the master gun-makers' books. The increasing number of parts and the more complicated interaction between them obviously reinforced the tendency to enlarge the catalogue of rules for their depiction.

Figure 3.4. Sketch of a multiple gun, using elevated viewpoint and perspective. From Johannes Formschneider, *master gun-makers' book,* Bavaria, third quarter of the fifteenth century. (Munich, Bayerische Staatsbibliothek, cgm 734, fol. 67^v; Leng 2000, 52.)

While the sketches of the *Anonymous of the Hussite Wars*,[39] dated to around 1440, hardly knew more than undressed boles, the drawings that the Nuremberg gun-maker Johannes Formschneider created around 1460 show precise wooden constructions in nearly correct perspective view, with detailed depiction of every single beam, pivot hole, nail, and spindle.[40] A comparison between two sketches makes the difference

36 See Schmidtchen 1997, 312–392 and Schmidtchen 1990, 192–210 with further references.
37 Schmidtchen 1990, 210–220 and Schmidtchen 1982, with a large number of figures from master gun-makers' codices.
38 See Lindgren 2000, also with notes about the consequences for pictorial catalogues.
39 Hall (B.S.) 1979b, for the dating of the manuscript and for further literature see Leng 2002, I 231–233.

obvious. Figure 3.1 represents the oldest type: raw construction in multiple view-points. The same type of gun (figure 3.4), but half a century later in Formschneiders manuscript, leaves us with a completely different impression. Shape and function of the foundation are clearly perceptible. Even without experience of this special type, but based only on the sketch, a skilled master gun-maker should have been able to build a copy of this multiple-tube gun. Short marginal notes on the purpose of the piece as a turret gun and on the maximum weight of the gun barrels provide further information. But besides more precise technical information, the main development is a new pictorial style. Perspective had changed radically. The alert eye no longer had to rebuild the construction from different viewpoints in a mental process. The new and single viewpoint moved to one point lying on the middle of the axis between the two angles from which the gun had been shown before. That migration enabled the engineering draftsman to maintain a general view of the device while, at the same time, presenting formerly hidden parts.

From then on, the elevated view-point and, for the most part, the

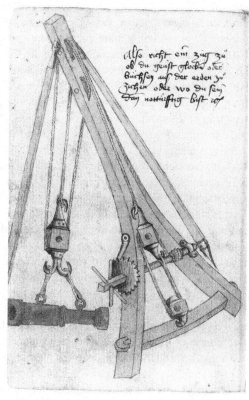

Figure 3.5. Hoisting device with over-dimensional depiction of mechanisms in comparison to the weight. From Johannes Formschneider, *master gun-makers' book,* Bavaria, third quarter of the fifteenth century. (Munich, Bayerische Staatsbibliothek, cgm 734, fol. 61ᵛ.)

---

40 The first Formschneider manuscript appears to be Munich, Bayerische Staatsbibliothek, cgm 734 (c. 1460), see Leng 2002, II 206–209. On f. 60ᵛ the author gave some autobiographical information. He compiled the book by order of the Nuremberg council for his successor, Master Wagmeister. Previously he had served in the Nuremberg armoury for 30 years as a master gun-maker and "adventurer": *Item lieber her wagmeister diese stück hab ich eüch gemacht mer auff fürdrung ewer gnedigen herren dan von des geltz wegen dar vmb bitt ich eüch freuntlichen vnd fleissiglichen mit gantzem ernst Ir wolt euch diese stück enpfolhen lassen sein vnd in rechter guoter huot halten als ich sy dan gehalten hab in meiner huot wol xxx iar in nürnberg ... Johannes formschneider büchsen meister vnd guoter aben teürer.* A more complete catalogue of Formschneider's drawings exists in the later codex Munich, Bayerische Staatsbibliothek, cgm 356 (c. 1480–90), see Leng 2002, II 198–201. Other manuscripts compiled between 1460 and 1500 based on Formschneider's materials can be found in Frankfurt, Stadt- und Universitätsbibliothek Frankfurt am Main, Ms. germ. qu. 14, Gotha, Universitäts- und Forschungsbibliothek Erfurt/Gotha, Chart B 1032, Munich, Archiv des Deutschen Museums, Hs. 1949–258, Stuttgart, Württembergische Landesbibliothek, Cod. milit. 4° 31 and Weimar, Stiftung Weimarer Klassik/Anna-Amalia-Bibliothek, Q 342. For this group of manuscripts see Leng 2002, I 239–249.

optional laterally relocated viewpoint, predominated in nearly all sketches. This came closer to the real perspective already known in Italy. But the conveying of functional principles remained the primary concern, and this sometimes caused violations of the correct use of perspective and scale according to the traditional medieval attitude, so that what was more important could be enlarged. We notice for example in figure 3.5 that Formschneider put special value on the order of pulleys and the correct route of the ropes to make clear the mode of operation of the lifting device. Marginal notes explained the advantage of easily lifting or pulling heavy gun barrels or bells.[41] But more important than the indication of its (evident) use was the over-dimensional depiction of the central elements of pulleys, spindle, gear wheel, and locking spring. Once more, and for good reasons, portrayal of functionality ruled over correct per- spective rendering.

With the increasing complexity of mechanical devices, even this pictorial process came up against the limits of feasibility. Again new pictorial means had to be con- ceived to make the catalogue of drawing rules fit for more intricate tasks. Hoisting devices, for example, were built for ever heavier weights with more and more ropes and pulleys. This could no longer be elucidated properly by means of drawing. In a manuscript from 1524 we find an example for an almost touching but extremely cre- ative attempt to extend and to vary pictorial rules in order to solve this problem.[42] The author was the founder and master gun-maker Christoph Seßelschreiber, who is still widely known for his bronze casting for the mausoleum of Emperor Maximilian I († 1519) in Innsbruck.[43] His book *Von Glocken- und Stückgießerei* con- tains a set of complicated hoisting devices on f. 69–72. In accordance with the tan- gled course of the ropes, Seßelschreiber punched the sheets and pulled a number of fine threads through the paper holes representing the real ropes over the portrayed beams and pulleys (figure 3.6).

Of course this was more a didactic peculiarity born out of necessity. But it shows the often playful and inventive method of varying elements of technical communica- tion by means of drawing. However, it proved to be more promising to follow the course of segmentation of machinery we have mentioned as manifested in some sketches from the beginning of the fifteenth century. Even in the comparatively early Formschneider manuscripts we have noticed that the general view of objects appears more and more dissolved. Single parts and mechanisms of central importance for the function are often drawn separately and are sometimes enlarged beside the general view. Figure 3.7 (top) from a Munich manuscript, derived from the Formschneider drawings in the last quarter of the fifteenth century, shows the change in this manner of proceeding.[44] The elevation mechanism hidden in the interior of the gun mould is repeated, dissolved in its single parts, next to the position of its invisible effect. Some

---

41 Munich, Bayerische Staatsbibliothek, cgm 734, f. 60ᵛ: "Also richt ein zug zu, ob du geust glocken oder büchsen aus der erden zuo zichen oder wo du seiner dann nottürfftig bist. Etc."
42 Munich, Bayerische Staatsbibliothek, cgm 973, see Hartwig 1927, for description and literature see Leng 2002, II 151–154 and I 225–227 with a biography of Seßelschreiber.
43 See Egg 1969, 122 and Gürtler 1996, 88f.
44 Munich, Bayerische Staatsbibliothek, cgm 356, 50f.; see above, note 40.

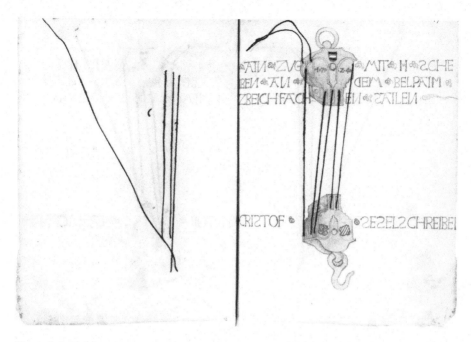

Figure 3.6. Hoisting device with fine threads representing the course of ropes. From Christoph Seßelschreiber, *Von Glocken- und Stuckgießerei,* Munich 1524. (Munich, Bayerische Staatsbibliothek, cgm 973, fol. 70^v / 71^r.)

marginal notes seemed essential to explain the correct location: "inn das gerüst durch—ausser die stangen." As soon as this manner of depicting separately important but in reality invisible parts was introduced into the syntax of drawing, similar depictions managed to do with reduced forms. The piece of artillery in figure 3.7 (bottom) uses the same gear rack to adjust the height of the tube. Here mere pictorial hints and textual references were sufficient—a process we can call the "introduction of relative pronouns in pictorial grammar."

Later technicians grasped this rule and perfected its handling. The *Buch der stryt*,[45] composed by the Palatinate master gun-maker Philipp Mönch in 1496, who had a particular liking for gear rack and worm-gear drive, shows the separated and enlarged elements of central drive mechanisms on nearly every sheet (figure 3.8).

This tried and tested idea of separating single mechanisms soon seemed to be promoted to a common principle, which again deserved variation and consequent sophistication. Further steps to an enlarged catalogue of rules are apparent in some drawings of the famous *Medieval Housebook*.[46] The identities of both the draftsman of the planets, tournaments and other genre scenes, who decisively influenced

---

45 Heidelberg, Universitätsbibliothek, cpg 126, for a description see Leng 2002, II 151–154 and I 225–227; Berg and Friedrich 1994, 175, 178.
46 Facsimile and commentary volume ed. by Waldburg 1997; the *Housebook* is private property of the Counts Waldburg–Wolfegg in the castle Wolfegg, Southern Germany.

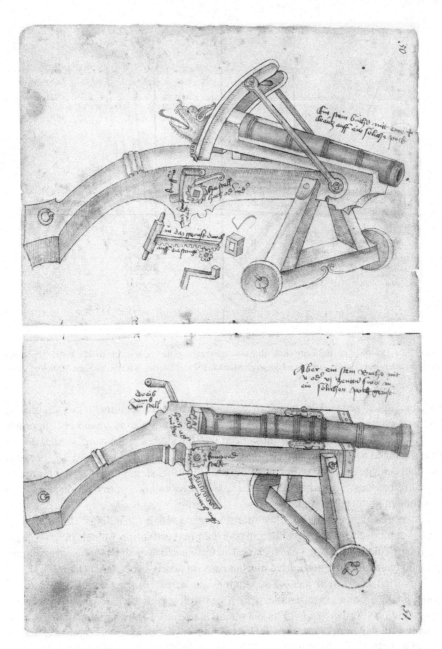

Figure 3.7. Guns and foundations with separately depicted elevation mechanisms and marginal notes. From a later copy of Johannes Formschneider, *master gun-makers' book,* Bavaria, fourth quarter of the fifteenth century. Munich, Bayerische Staatsbibliothek, cgm 356, p. 50f.)

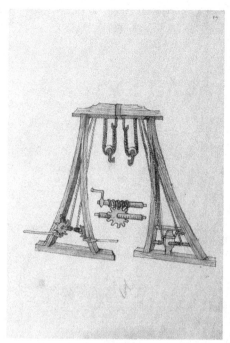
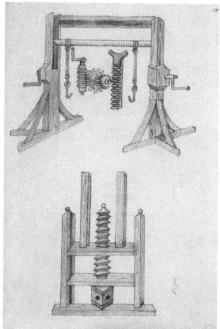

Figure 3.8. Hoisting devices with separately depicted hidden worm-gear driving. From Philipp Mönch, *büch der stryt vnd buochßen,* Heidelberg 1496. (Heidelberg, Universitätsbibliothek, cpg 126, fol. 19[r] and 20[r]; Leng 2002, I plate 22.)

the style of Dürer, and that of the patron of the manuscript are among to the great unsolved mysteries of the history of art. Notwithstanding that the courtly contexts put the codex in the milieu of the presentational manuscripts, large parts of it deal with technological questions, produced perhaps by a second, probably engineering drafts-man. These mining[47] and warfare materials[48] on the fifth through ninth gatherings are clearly related to the Formschneider drawings. In some sketches the author went beyond the segmentation of mechanisms to something like a modern exploded view, dismantling the machines into all single parts. A battle wagon pushed by six horses, for example, is shown with every detail of the chassis separated (f. 52[v]).[49] Another battle wagon with a supporting framework on wheels, movable sides and roof is also depicted in all its single component parts (f. 51[r]). And the sketches of a hoisting device (f. 54[v]), a thread-cutting lathe with a single slide rest, a tool-holder and a feed in a dovetailed guide, even with a replaceable cutting die (f. 54[r]) and the components of a traveling screen, equally testify to a true enjoyment of pictorial dismantling (figure 3.9).

---

47 See Ludwig 1997.
48 See Leng 1997.
49 For the following, see the description of all pictures of the *Housebook* in Waldburg 1997, 42, 48–51.

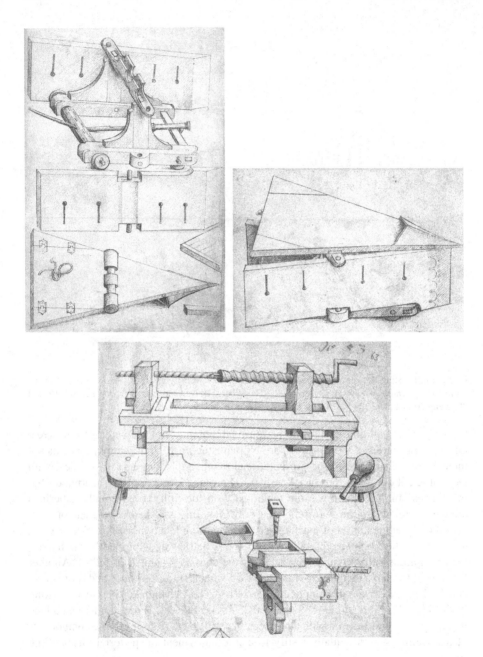

Figure 3.9. Component parts of a battle wagon and a thread-cutting lathe. From the *Medieval Housebook,*
Southern Germany, c. 1480.

Other drawing methods became more sophisticated, too. Two sketches of mills, one indirectly driven via a horizontal dropped shaft and pin gears, the other direct via a vertically dropped shaft with bar and counterbalance weight (f. 48^{r+v}) are shown from a perfect elevated and vertically shifted viewpoint. Every single part remains visible to a large extent. Almost nothing is hidden.

Revealing, too, is the pictorial implementation of a large mine with smelting hut, crushing machine and interior for the furrow drain process, which had just been invented at the time (f. 36^{r}–38^{r})[50]. At first an introductory general view shows the complete plan. Then a fictitious round through the fitting-out of the mine follows. The order of the drawings from f. 36^{v} onwards is absolutely remarkable. Beginning with the source of energy, an undershot water-wheel, the draftsman reveals the whole course of the system of forces following the shafts. Along the main shaft, which drives several side shafts via pin wheels, he leads the observer through the hut. On the way we meet a crushing machine, shaft-driven bellows for the furnaces and a pumping plant (figure 3.10).

Here the pictorial segmentation serves the depiction of various combined technical processes in a large smelting hut, which could no longer be described instructively in one drawing. With the *Medieval Housebook* the segmentation of a pictorial "sentence" about a single device was enlarged and transformed into the segmentation of a complete plant, out of which a pictorial "text" arises.

Figure 3.10. Interior of a smelting hut; water-wheel driving a crushing machine with pinwheel transmission of a camshaft indicated, continued on the next sheet with the shaft-driven bellows for the furnaces. From the *Medieval Housebook,* Southern Germany, c. 1480.

---

50 Described by Ludwig 1997, 127–135.

5. USING THE GRAMMAR

With the introduction of these pictorial elements, the language of technical drawings necessary for communication among outstanding craftsmen in the later Middle Ages was complete. The catalogue of rules as well as the resulting possibilities had evolved continuously since the beginning of the fifteenth century. At first they were restricted to three instructions we can summarize as follows: 1. Show the whole device; don't annoy your beholder by using settings—he is as skilled as you and will recognize the purpose. 2. Use as many viewpoints as necessary to show all important parts and functions. 3. (Seldom used): Only in case of unbearable overlapping, leave out some parts and show them separately.

But at the end of the fifteenth century the grammar of technical illustration had become more extensive and elaborate by trial and error, by creativity and by the sheer necessity dictated by more complicated machines. The drawings following these conventions became more expressive and instructive, even for craftsmen without respective experience. Confined to the essential rules, the new catalogue of sophisticated rules can be paraphrased as follows:

1. Draw simply but in a concentrated manner. Try hard to manage with one sheet of paper for each device, but avoid mere hints that might possibly confuse a less skilled beholder. Therefore draw all particular details distinctly; if necessary, enlarge items that might help to understand the function. Spare your distant colleague's having to work out his own solutions for a result you have already achieved.
2. Do without multiple viewpoints, but pay extreme attention to the one you ultimately choose. Move laterally and upward until the maximum is visible and a minimum concealed.
3. If elements decisive for the function cannot be shown in a general view, then separate them from the context. Draw them enlarged next to their correct position. Form a pictorial "subordinate clause" and join it with a "conjunction" in the form of notes or lines to the "main clause." If such a separately rendered part or mechanism is applied in several sketches, a brief indication will suffice in the following pages.
4. If the machine is so complicated that too many parts will remain hidden and too many necessary pictorial "subordinate clauses" will obscure the whole context, then restructure the "text." Make a series of new "main clauses." Show all parts separately and connect them again by adding a general view in order to enable the beholder to discern their relationship.
5. If you are dealing with such an extended ensemble of mechanisms and machines that all single elements would result in a unfathomable puzzle of sketches, then exploit all your pictorial resources of analysis, segmentation, and structuring. Form a "text." Tell a story. Present a general view as introduction and orientation. Then hold on to sequences of shafts, forces, and functions. There you may fix

other "subordinate clauses" until the whole pictorial "text" fits together seamlessly in your beholder's mind.

Naturally, no craftsman ever was capable of recording in words the grammar of drawing outlined above. But the rules existed in the minds of craftsmen as a common convention for the conveying of technical information by the mere combination of ink lines—it was no coincidence that in the sixteenth century Italian engineers emphasized that geometrical elements serve as an "alphabet of design!"[51] Finally, every system of grammatical signs has to reach a high degree of complexity before it is ripe for a descriptive grammar. But, of course, the language of drawing also worked very well without scientific analysis.

How well it worked and which creative possibilities were opened up by the handful of sophisticated rules is apparent in some manuscripts compiled at the transition from the fifteenth to the sixteenth century. At this time the wealth of drawings had become so great that some extensive collections had amassed in voluminous codices. The council of Frankfurt asked an unknown master gun-maker to compile a collection of nearly all known texts and drawings about military devices, especially those derived from the various Formschneider manuscripts.[52] For this compilation on 178 paper sheets, the anonymous craftsman made full use of all known pictorial means. In particular, the complete spectrum of gun types is represented in all imaginable variations and with all graphical tricks.

But the most amazing consequences of the use of the pictorial grammar are found in two thick and large-format manuscripts, both created in the first decade of the sixteenth century.[53] The Erlangen *Kriegsbuch* and the Weimar *Ingenieurkunst- und Wunderbuch* each contain over 600 sketches on paper, or rather, parchment. The Erlangen codex belonged to the Franconian nobleman Louis von Eyb, the Younger († 1521), administrator of the Upper Palatine territories and counsellor of his Palatine Earl during a period that spanned the War of Bavarian Succession (1504).[54] The Weimar codex, depending on the identification of the patron's coat of arms, was compiled either for the Upper Palatine noblemen von Wolfstein or the Franconian counts von Hohenlohe.[55] Both manuscripts are closely related, as is evident in the duplication of whole series of pictures. Although a partial presentational function can be suspected, the material of nearly all sketches derives directly from craftsmen, especially master gun-makers and mill experts.

51 For the quotations of Girolamo Cataneo and his pupil Giacomo Lanteri, see the chapter by Marry Henninger-Voss in this volume.
52 Frankfurt, Stadt- und Universitätsbibliothek Frankfurt am Main, Ms. germ. qu. 14, on the front cover *supralibros* with the coat of arms of the city and inscription *DIS BVCH GEHORT DEM RADE ZV FRANCFORT;* for a description see Leng 2002, II 107–110 and I 267–269, plate 23 and 24.
53 For the tendency towards voluminous collections of pictorial catalogues at the end of the fifteenth century, see Leng 2002, I 269–279.
54 Erlangen, Universitätsbibliothek Erlangen-Nürnberg, Ms. B. 26; see for contents and also for the patron's biography, Keunecke 1992–93, and Leng 2002, II 97–100.
55 Weimar, Stiftung Weimarer Klassik / Anna-Amalia-Bibliothek, fol. 328; for contents see Kratzsch 1979, Kratzsch 1981, Hall (B.S.) 1979b, 40f; on the discussion of the coat of arms, see Leng 2002, II 292–296 and Metzger 2001, 253–264. An edition on CD-ROM is being prepared by Christoph Waldburg.

In particular, an extensive series of mills contained in both manuscripts evinces a new desire to combine pictorial elements.[56] Small and large foundations with one or two millstones, main shafts and side shafts, direct and indirect driving, gear wheels and crankshafts, driven by bars, ropes, by hand, by water-wheels or tread-wheels for men or horses, were combined at will. Once craftsmen had learned to dismantle a depicted machine in separate parts and mechanisms as well as to analyse the structure of a "sentence," the task of reassembling them into various combinations was child's play.

Thus, with a restricted variety of forms they were able, as in every language, to produce a nearly unlimited number of pictorial expressions. In return, these multiplying possibilities of combination in depiction encouraged technical experimentation with different drives, measurements and energy sources.

This process of segmentation and the subsequent pleasure in varied reconstruction is best apparent in a large series with several sketches on each of the 15 sheets in the Weimar *Ingenieurkunst- und Wunderbuch*. They present the possibilities of a modular system for the construction of battering instruments or ladders with claws (figure 3.11).[57] Composed of a handful of iron parts such as bars, screws, nuts, claws or rollers, more and more different devices are generated step by step in a nearly endless series of permutations. On these sheets the playful combination of morphological and syntactical elements of the grammar of drawing has indeed produced a complete pictorial text.

6. DESCRIBING THE GRAMMAR

At the beginning of the sixteenth century, the handful of rules for conveying technical information in drawings had produced a large range of pictorial expressions among a small group of specialized craftsmen. But at this point we must note that this process grinds to a halt. Several reasons can be traced. The conveying of technical information over long distances by drawings developed by single innovative craftsmen was supported mainly by traveling master gun-makers. But this special job profile changed at the beginning of the sixteenth century. The construction of guns and foundations was no longer an individual process mastered by only a few skilled experts. Large foundries with preindustrial mass production run by the early modern sovereigns took their place. But with these foundries, the conveying and passing on of knowledge regained a stable sphere, comparable to that in medieval urban settlements before the master gun-makers had left their workplaces as blacksmiths and metalworkers. Hardly any drawings were produced in the foundries. Surely they used rough sketches, too, but there was no need to communicate them or to put them down on lasting media. Moreover, issues of competition and secrecy now became more decisive. Direct oral exchange was desirable and again possible. The production of

---

56 Weimar, Stiftung Weimarer Klassik / Anna-Amalia-Bibliothek, fol. 328, spread over f. 1ᵛ–38ʳ and Erlangen, Universitätsbibliothek Erlangen-Nürnberg, Ms. B. 26, f. 124ʳ–143ʳ.
57 Weimar, Stiftung Weimarer Klassik / Anna-Amalia-Bibliothek, fol. 328, in two parts on f. 282ʳ–287ᵛ and 279ʳ–305ᵛ.

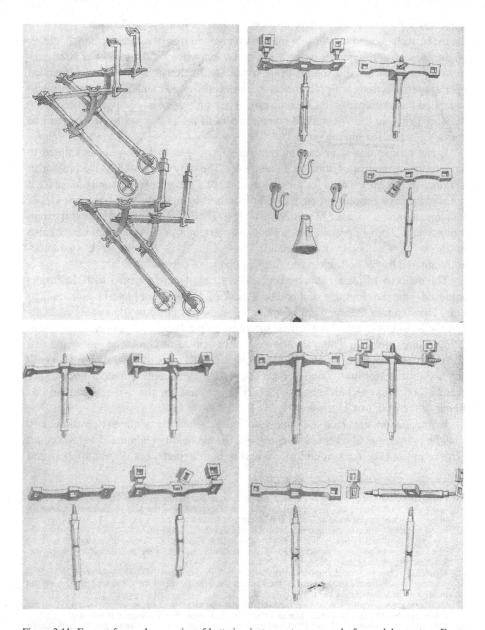

Figure 3.11. Excerpt from a large series of battering instruments composed of a modular system. From Anonymous, *Ingenieurkunst- und Wunderbuch,* Southern Germany, c. 1500. (Weimar, Stiftung Weimarer Klassik, Anna-Amalia-Bibliothek, fol. 328, fol. 282^v (top left), fol. 283^v (top right), fol. 284^r (bottom left) and fol. 285^r (bottom right); Leng 2002, I plate 28.)

guns was separated from their professional use. The formerly independent status of
the extensively skilled master gun-makers changed to that of a mere piece-laborer.

In the sixteenth century we still notice an extensive production of master gun-
makers' books. They no longer dealt with construction, but contain long-winded
operating instructions with pictures as mere decorative elements.[58] Illustrations with
separated parts or assemblies became rare. They only appear, equipped with precise
measurements, in the attempts of some masters in the armouries to standardize the
various gun and foundation types.[59]

It can be mentioned in passing that the master gun-makers never attempted to
analyse their grammar of drawing theoretically. The only one who dared to present an
illuminated full explanation for the construction of a gunner's quadrant by means of a
ruler and pair of compasses, an exception admired by his colleagues, was the Upper
Palatine master gun-maker Martin Merz († 1501).[60] The inscription on his still extant
tombstone in Amberg praises him not only as outstanding master gun-maker, but also
as an excellent authority in mathematics and geometry.[61] But his work was apparently ignored by his colleagues; no further copies are known.[62]

The majority of master gun-makers and craftsmen were satisfied with the results
achieved—the development of a practical working language of pictorial communication. The step into theory, the description of rules was up to the educated architects
and the artist engineers.[63] After their inception in the Italian Renaissance, the circulating printed tracts by Dürer contributed to the spread of combined views and real
perspective drawings even among technicians.[64] Not until the end of the sixteenth
century did the results of real perspective drawings become apparent both in the
sketches of some technicians like Schickhardt[65] and in the wide-spread printed
Renaissance Books of Machines.[66]

In conclusion, one pictorial catalogue from the middle of the sixteenth century
shall be pointed out as a remarkable instance for the quality technical drawings could
achieve against the background of conveying both experiences of practitioners and

---

58 For example the *Buch von den probierten Künsten*, compiled by the Cologne blacksmith and master
    gun-maker Franz Helm in 1535, spread over Germany, especially in the new sovereigns' libraries, in
    more than 75 extant manuscripts, each with about 60 illustrations; ed. by Leng 2001; for an overview
    of the numerous other master gun-maker books see Leng 2002, I 330–367.
59 For example, Dresden, Staats- und Universitätsbibliothek Dresden, C 363: *hantbiechlin vnd ausszug
    von meinen erfindungen*, compiled 1570–80 by the Danzig master of the armouries Veitt Wolff von
    Senftenberg, see Leng 2002, II 88–91.
60 Munich, Bayerische Staatsbibliothek, cgm 599, f. 66[r]–101[v], see Leng 2002, II 88–91 and I 250–255.
61 Amberg parish church, outside next to the entrance; see the figure of the tombstone along with the
    inscription in Schmidtchen 1977, 182.
62 One more copy possibly can be identified in a manuscript mentioned by Schneider 1868 129f. in the
    Library of the Princes of Liechtenstein in Vaduz, although today this manuscript is missing.
63 For their contributions, see Long 1985, and several other chapters in this volume. For Brunelleschi or
    Leon Battista Alberti see Fanelli 1980, 3–41; Gärtner 1998; on the development of perspective, recently
    Bätschmann and Giaufreda 2002, 12–18; for Leonardo, see Parsons 1968, 15–93.
64 Dürer 1983 (1525) and 1969 (1525); for the extensive literature, see Mende 1971, 479–483. See also
    the chapters by Lefèvre and Peiffer in this volume.
65 See Popplow in this volume with further references.
66 For example Besson 1569; Errard 1584; Ramelli 1588; Zonca 1985 (1607); Zeising 1607; Branca 1629;
    Böckler 1661. For this genre, see Popplow 1998c, 103–124 and Popplow 1998a, 47, 54f.; Bacher 2000
    with several figures.

theoretical analysis. In the year 1558 the Nuremberg patrician Berthold Holzschuher
(† 1582),[67] finance broker,[68] owner of several mines in Tyrol[69] and passionate technician and inventor, drew up his holographic will. He left his eldest son nothing but a
bundle of sketches with (as he thought) revolutionary driving methods for carriages
and mills.[70] The book was to be kept secret and handed over by the executor to his
son after Berthold's death. The heir was asked to call on cities, sovereigns, and kings
with these drawings and thus to get rich on his well-deserved user fees.

Here communication by means of drawings had to surmount temporal distances,
and Holzschuher by necessity had to ensure that the sketches were perfectly comprehensible without his help. Therefore he displayed the full range of pictorial means
available in the sixteenth century. Of course, the complicated gearing of the carriages
was influenced by the curious vehicles in Emperor Maximilian's 1526 posthumously
printed *Triumpf*,[71] and the style of technical drawing by Dürer's architectural tracts.
But the modernity of Holzschuher's drawings is fascinating when compared with
actual blueprints.

For all devices we find logical sequences of general views, segmentation into
mechanisms and into every single component, precise depiction of gear wheels,
axles, and transmissions, everything designed exactly with a ruler and a pair of compasses. Dozens of erased help lines and hundreds of pricks on each sheet bear witness
to his detailed efforts. The last sheet of his will, a folded two-page perspective drawing of a large mill moving 16 millstones, constitutes the highlight of the book (figure
3.12). Beforehand every single component was introduced. Additionally, three horizontal sections on the ground level, the shaft system level, and the top view made
clear the interaction of framework, components, and the system of forces.

The most amazing fact is that all drawings are completely true to scale. Next to
each sketch explanatory notes point out whether a quarter, a half or a complete inch
on the paper corresponds to one Roman foot (c. 29,8 cm). Sometimes extra scales are
added at the margins of the plans. Character legends join up each component to written explanations. Not only the purpose of the devices, their construction and combination is exhaustively explained step by step, but also the method of depiction. The
last will of Berthold Holzschuher therefore can be considered as one of the first modern blueprints containing both a theory of technical drawing and an expressive example of their application.

Holzschuher's sketches remained a never-realized and soon-forgotten curiosity.
None of the devices would have worked because he had misjudged gear ratio step-up
and frictional losses. But their creation once more shows the social significance of

---

67 For his biography, see Koch 1972, 579f.
68 For his financial activities, amongst them a revolutionary dowry insurance, see Ehrenberg 1890 and
    Krieg 1916.
69 See Kunnert 1965, 246–249.
70 Nuremberg, Germanisches Nationalmuseum, Hs. 28.893; for some short notes and a few figures see
    Neuhaus 1940 and Rauck 1964; for a description of the manuscript see Leng 2002, II 269–271. The
    author is preparing an chapter to be printed in the "Mitteilungen des Vereins für die Geschichte der Stadt
    Nürnberg."
71 Facsimile ed. by Appelbaum 1964, on the carriages, see 90–99.

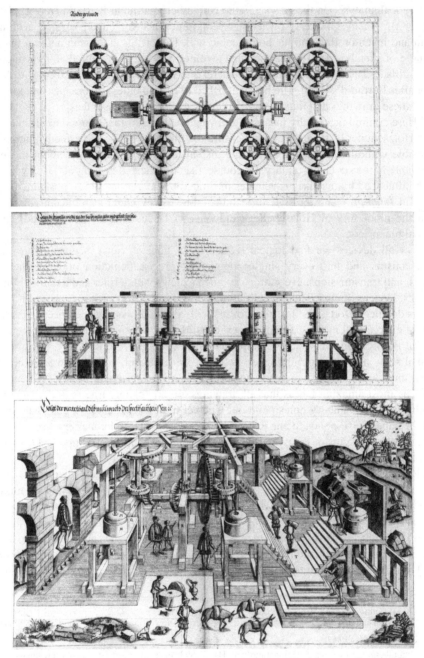

Figure 3.12. Scaled section plans and perspective view of a large mill with a complicated shaft and gearing system. From Berthold Holzschuher, *Last will and book of his inventions,* Nuremberg 1558. (Nuremberg, Germanisches Nationalmuseum, Hs. 28.893, fol. 31^v / 32^r, fol. 39^v / 40^r, and fol. 41^v / 42^r.)

craftsmen's manuscripts. The more skilled experts had to convey increasingly complicated technical know-how over spatial and temporal distances, the more sophisticated and efficient became the basis of communication, with a catalogue of drawing rules serving as a pictorial grammar. Their tentative development and practical testing from the beginning of the fifteenth century onwards led to the creation of some conventions in the following century, which are obeyed in technical drawing to this very day. Thus some fundamental blueprint techniques may be rooted not only in educated early modern engineering, but also in later medieval craftsmen's practice.

PART III
SEEING AND KNOWING

INTRODUCTION TO PART III

The topic of this section is the intricate relationship between technical drawings and knowledge—knowledge those drawings presuppose and/or allude to, invoke, transmit explicitly or implicitly, and/or knowledge, finally, that they help to gain. Besides "literacy" with regard to the different pictorial languages employed, the reading of technical drawings presupposes a spectrum of knowledge of different provenance—technological knowledge regarding both the construction and the working of the depicted device, expertise as regards the practical context of its intended employment, mechanical assumptions or other convictions of how natural things work, and sometimes even mathematical and geometrical knowledge. Accordingly, what these drawings actually convey depends to no little extent on the beholder's familiarity with the represented object and its whereabouts. They tell different things to experts and nonexperts, to the artisans who construct the device, to other experts who employ it professionally, and to lay people who are interested in it for any reason.

Technical drawings thus constitute a sort of focus where different kinds of knowledge come together and intersect. As such a focus, they radiate, in turn, into different domains of knowledge whose relations to the realm of engineering thus became recognizable. They are therefore mirrors of the intellectual world of early modern engineering if one succeeds in reconstructing the ways in which knowledge was connected with them.

The fact that technical drawings presuppose so much knowledge of all sorts is of course due in part to the limited representational capacity given by their restriction to two dimensions, and to the special semiotic shortcomings of each of the different pictorial languages. Technical drawings thus may appear to imply rather than display and convey knowledge. However, such a view not only would be at odds with the general fact that these drawings were an indispensable means of communicating ideas, proposals, solutions, agreements, and the like. It also would miss a dimension of early modern engineering drawings that is of particular significance in this context. Rather than depending mainly on presupposed knowledge, these drawings contributed essentially to the transformation of implicit, tacit knowledge of the crafts into explicit knowledge. They proved to be a chief means of rendering implicit knowledge explicit.

The chapter by Pamela Long studies the ways in which Francesco di Giorgio Martini and Leonardo da Vinci, two outstanding masters of engineering drawing, promoted the articulation of experiences and knowledge of engineering by means of both drawings and texts. Struggling for clarity, both were innovative in coining new terminology as well as in introducing new pictorial formulas, thereby creating an intertwined whole of pictorial and verbal elements that represented practitioners' knowledge with combined forces.

The relationship between image and text in illustrated manuscripts and books on engineering issues, which deserves much more attention than could be paid in this volume, is of particular interest with respect to the advantages and shortcomings of the two media as regards the articulation of technological knowledge. Both Giorgio

Martini's treatises and Leonardo's notebooks demonstrate impressively that such an articulation and explication cannot be achieved by either of the two media alone. Or, to put it more precisely, they show that the capabilities to spell out technological knowledge each of them possesses can only be exploited fully when each elucidates the other. In this field of knowledge, images explain words no less than words images. Furthermore, the combination of verbal and pictorial means of representation opened up new possibilities for connecting the knowledge of this realm of practice with a variety of branches of learned culture. Pamela Long's chapter shows that this opportunity was immediately realized and seized by Giorgio Martini and Leonardo, albeit in quite different manners.

The notebooks of Leonardo also testify to a further aspect of the relationship between drawings and knowledge, namely to the employment of drawings for the obtainment of insights—insights of a more theoretical nature as well as more practical ones. When employed in this way as a means of thinking, technical drawings and knowledge exhibit an interplay that is challenging to both the historical actors and the historian who tries to determine which knowledge was tacitly presupposed, which actually represented, and which gained by drawings used for this purpose.

This interplay is the topic of the chapter by Mary Henninger-Voss, who discusses the employment of plans by civilian and military experts in the service of Venice for consulting and deciding on crucial questions of fortification at the turn of the fifteenth century. When applied for these purposes, a fortification plan resembles an iceberg. For what can actually be seen on the plan is only the visible peak, whereas the bulk of information it contains for an expert is as invisible to a nonexpert as the main body of an iceberg under water. Moreover, since specialists of different expertise interact in such decision-making, military architects and military commanders in this case, the whole range of knowledge presupposed by, connected with, and derivable from a fortification plan is not possessed by any of the specialists, but only by an ideal team of experts representing the state of the arts involved. What is obvious in this case holds for the relationship between technical drawings and knowledge in general: it is mediated by the social relations among the actors involved in advanced technological projects.

4

PICTURING THE MACHINE:
FRANCESCO DI GIORGIO AND LEONARDO DA VINCI
IN THE 1490s

PAMELA O. LONG

In memoriam
Carolyn Kolb (1940–1994)

At the start of his notebook on machine elements and mechanics, *Madrid Codex I*, which he created in the 1490s, Leonardo da Vinci, at the time a client of Ludovico Sforza of Milan, suggests that his treatise would provide a respite from fruitless labors, namely the attempt to discover perpetual motion.[1] "I have found among the other superfluous and impossible delusions of man," he writes, "the search for continuous motion, called by some perpetual wheel." He continues that for many centuries, almost all those who worked on hydraulic and other kinds of machines devoted "long search and experimentation with great expense" to this problem. "And always in the end, it happened to them as to the alchemists, that through a small part the whole is lost." Leonardo offers to give such investigators "as much respite in such a search as this my small work lasts." An additional advantage is that readers will be able to fulfill their promises to others so that they will "not have to always be in flight as a result of the impossible things that were promised to princes and rulers of the people." He recalls many who went to Venice "with great expectation of gain, and to make mills in dead water. But after much expense and effort, unable to move the machine, they were obliged to move themselves away with great haste from such a place."[2]

Leonardo may well be referring to attempts by his older contemporary, the architect/engineer Francesco di Giorgio, to construct water-powered mills in "dead water" (*aqua morte*) by a system powered by continuously circulating water. Yet evidence from several notebooks attests that Leonardo himself on several occasions investi-

---

1   I would like to thank Wolfgang Lefèvre for organizing the workshop at the Max Planck Institute for the History of Science that prompted me to write this chapter. Versions of the chapter were presented at the Joint Colloquium, Princeton University Program in History of Science and University of Pennsylvania Department of History and Sociology of Science; and at the Department of History and Philosophy of Science, Indiana University at Bloomington. I thank the participants of these colloquia as well as the members of the Max Planck workshop for their extremely helpful comments which helped me to revise the chapter. I also thank Guido Giglioni for detailed discussions of Francesco di Giorgio's Italian, Steven Walton for discussions concerning Leonardo's cranks, and Bob Korn for assistance with models and "photoshop" enhancement of images.
2   Leonardo 1974, I f. 0^r, and IV 2, "Io ho trovato infra l'altre superchie e impossibile credulità degli omini la cierca del moto continuo, la quale per alcuno è decta rota perpetua"; "co'llunga cierca e sspe[ri]mentatione e grande spesa"; "E ssenpre nel fine intervenne a lloro come alli archimisti, che per una picola parte si perdea il tutto"; "tanto di quiete in tale cierca, quanto durerà questa mia picola opera"; "none aranno senpre a stare in fughe, per le cose impossibile promesse ai principi e regitori di popoli."; "con grande speranze di provisioni, e fare mulina in acqua morta Che non potendo dopo molta spesa movere tal machina, eran costretti a movere con gran fu[ri]a sè medesimi di tale aer."

gated the ways in which such a machine might function. Moreover, the issues of mechanical power and friction that were the focus of such attempts were also essential components in the design and construction of all machines. Leonardo's comment points to common areas of concern among engineers in the late fifteenth century, as well as to differences of opinion about what might work.[3]

The comment on perpetual motion is one of several unrelated introductory paragraphs that Leonardo tried out on the first sheet of *Madrid I*. While he condemns attempts at perpetual motion, he also alludes to the practices of engineers, their construction of mills, and their attempts to find efficacious ways to power them. Yet his treatise is not a practical manual for working engineers, but rather a broad-ranging investigation of machine elements, such as cranks and gears, and their motions, and it includes study of statics and mechanics. Leonardo lived in a culture in which writing books about the mechanical arts and machines had become a significant practice. Clearly a notable influence on his own writing were the treatises on architecture and engineering written by the Sienese architect/engineer, Francesco di Giorgio.

For both Francesco and Leonardo, machines were physical entities that performed certain designated tasks, but they also were significant within larger intellectual and cultural arenas as objects of study within investigations of motion and mechanics. Both men depicted intricate mechanisms pictorially and described their motions in detail. Machines became modalities for understanding certain problems in the natural world, such as motion. The fifteenth- and sixteenth-century development of the genre of "machine books," often created for presentation to patrons, suggests that machines had acquired symbolic value as well, perhaps representing both technological efficacy and the power and authority of princes. Pictorial images of machines within codices on architecture and engineering became increasingly commonplace in the fifteenth century. Yet further study is needed to investigate the *ways in which* machines were pictured in this period, and the uses to which those pictorial representations were put.

In this chapter I examine and compare several of the mechanical drawings of Francesco di Giorgio Martini and Leonardo da Vinci. The two men met, possibly a few years before they were both called to Pavia for consultation on the construction of the Pavian cathedral in 1490s. They shared similar backgrounds. Both were trained initially as painters, both expanded their opportunities for work through the astute use of patronage, both worked at a variety of kinds of tasks—engineering, military engineering, sculpture, architecture, and painting. Both struggled to learn Latin as adults, and both created illustrated treatises and notebooks on a variety of topics, including machines. Leonardo studied at least one of Francesco's treatises as is evidenced by his hand-written glosses on the manuscript copy of *Trattato I* in Florence (Ashburnham 361 in the Biblioteca Mediceo-Laurenziana). Leonardo's list of the books he owned includes a book by the Sienese engineer, although he did not specify which one. Clearly the two men benefited from their meeting and influenced one another.[4]

---

3   Francesco di Giorgio Martini 1967, I 148 and f. 35^v (tav. 66); and see Reti 1969.
4   See Pedretti 1985; and Marani 1991.

Francesco and Leonardo both concerned themselves with real machines in the context of their engineering and architectural work carried out for their patrons. In addition, in their writings and illustrations, they connected machines to learned knowledge traditions in obvious and not so obvious ways. Nevertheless, the ways in which these two engineer/authors approached machines pictorially, and the ways in which they attempted to join their practical knowledge of machines to traditions of learning were quite diverse. A comparison of their dissimilar approaches allows a better understanding of how mechanical knowledge in the practical sense came to be connected to learned traditions in the late fifteenth century. Both men made these connections by means of illustrated writings, or what in some cases might be better called collections of images inscribed with small amounts of text. As a shorthand way of codifying their differences, I suggest that Francesco di Giorgio created a humanist book in which his interest in engineering finds a place alongside his architectural interests, whereas Leonardo da Vinci created a treatise more directly connected to the scholastic traditions of statics and kinematics.

Each in his own way was carrying on what was a well-established and rich tradition of visualizing machines by the late fifteenth century. There were a few ancient books that contained illustrations of machines—most notably the *De architectura* by the Roman architect Vitruvius, the original drawings of which had been lost, and Greek manuals on siege machines, the so-called poliorcetic treatises, containing beautiful, hand-colored illustrations of siege ladders, rams, and other military equipment.[5] Occasionally machinery is depicted in notebooks and treatises in the medieval centuries, such as a sawmill on one of the sheets created by Villard de Honnecourt in the thirteenth century, and plows and mills depicted in miniature in books of days and in almanacs. Nevertheless, there is an unmistakable expansion of drawings depicting machines and instruments from the late fourteenth century. Examples include the treatise and drawings on the astrarium by Giovanni da Dondi; the treatise on military weapons and war machines by Guido da Vigevano, the machines depicted in Conrad Kyeser's *Bellifortis*; the numerous drawings of guns, gun carriages, and other machines in fifteenth-century German manuscripts; machine drawings in a manuscript treatise *Liber instrumentorum bellicorum* by Giovanni Fontana; the well-known notebooks and treatises of the Sienese notary, Mariano Taccola, which were known to both Francesco di Giorgio and Leonardo; and the military machines depicted in Roberto Valturio's *De re militari libri XII*, by far the most popular military book of the fifteenth century, copied many times and printed in 1472.[6]

Writings about the mechanical arts—painting, sculpture, architecture, fortification, artillery, pottery, mining, and metallurgy, including machines and visual repre-

---

5 For a spectacular eleventh/twelfth-century copy of the poliorcetic writings with numerous drawings of military apparatus and machines, see Biblioteca Apostolica Vaticana, Cod. Vat. gr. 1164, esp. ff. 88^v– 135^r. And see Schneider 1912, esp. 4–6 and plates; Galluzzi 1996a, 22; Rowland 1998, 35–37; and Lefèvre 2002, 112f.

6 For an introduction to the scholarly literature and editions of the treatises, see esp. Aiken 1994; Alertz 2001; Barnes 1982; Hall 1978; Hall 1976a; Leng 2002; Long 2001, 102–142; and Popplow 1998a. See also the chapter by David McGee in the present volume. For a lucid essay on attitudes toward the mechanical arts from antiquity to the Renaissance, see Summers 1987, 235–265.

sentations of machines—proliferated in the fifteenth and sixteenth centuries. The complex reasons for this expansion of authorship in the mechanical arts include what historians of technology have called technological enthusiasm, a delight in the technology of machines in itself, regardless of economic or practical implications.[7] Machines were utilized to solve technological problems and to consider more theoretical issues involving statics and kinematics. As they became the focus of pictorial and textual authorship, they became incorporated into literate discourse in a variety of ways. Machines accompanied by visual images of machines and machine parts, including multiple variations of machine elements such as gearing, and of entire machines such as mills and guns, became a significant focus of authorship. While such books had distinct relationships with the actual practice of making and operating machinery and inventing new configurations of actual machinery, it is also true that they entailed a kind of practice separate from material fabrication and operation. Machine drawings do not necessarily provide an undistorted lens into material practices. The relationships of images and textual discussions to material practice is a matter for investigation rather than a given. Moreover, as will be seen in the discussion below, the relationship of "machine books" to the cultures of learning and of practice were never uniform, but rather were complex and diverse from one text and author to another.

1. FRANCESCO DI GORGIO: ENGINEER AND ASPIRING HUMANIST

Francesco di Giorgio (1439–1501) was the son of a poulterer who had left this occupation in the countryside to work for the commune of Siena. Francesco thereby grew up in Siena at the time when the notary Mariano Taccola was still alive; he may well have known the older man and was well acquainted with the latter's engineering notebooks and treatises. Taccola had completed *De ingeneis* in the 1420s and *De machinis* in the 1430s and 1440s. Both were filled with drawings of machines and instruments and often included short commentaries. Taccola, who called himself the "Sienese Archimedes," influenced Francesco in the latter's production of notebooks and treatises. For example, Francesco's earliest notebook, the so-called *codicetto* in the Vatican Library, contains a substantial number of copies from the notebooks of Taccola. Francesco combined his interest in machines and engineering with architecture, including the design and construction of buildings and the study of the Roman architect Vitruvius's *De architectura*. Francesco, who initially worked as a painter and sculptor, went on to become one of the best known architects and engineers of the late *quattrocento*. He worked in Siena, in Urbino for Federico da Montefeltro, in Naples for king Alfonso II, and elsewhere.[8]

Close investigation of Francesco's notebooks and writings have made clear his struggles to learn Latin, an effort that he seems to have made primarily to understand

---

7   For technical authorship, see Long 2001, and for technological enthusiasm, Post 2001, esp. xviii–xx and 285–308.
8   For Francesco, see esp. Betts 1977; Fiore and Tafuri, eds. 1993; Scaglia 1992; and Toldano 1987. For Taccola, see Taccola 1971; Taccola 1972; Taccola 1984a; and Taccola 1984b.

the *De architectura*. He eventually was able to produce an Italian translation of the text in the 1480s. An early attempt at writing an architectural treatise and translating Vitruvius can be seen in a manuscript known as the *Zichy Codex* in Budapest, which, Carolyn Kolb cogently argued, includes a copy of an early treatise by Francesco. Francesco's better known writings include two major treatises on architecture, military engineering, and machines. He wrote the first, known as *Trattato I*, a later version of *Zichy*, during his first major sojourn to Urbino from 1477 to 1480 as a result of the interest (and perhaps urging) of his patron, Federico da Montefeltro. He wrote the second treatise, known as *Trattato II*, later, perhaps in the years 1487 to 1489, or as has recently been suggested, even later after his return from Naples to his native Siena in 1496. The precise chronology of Francesco's notebooks and treatises is a complex issue that is the focus of ongoing investigation and scholarly disagreement. Yet all agree that his two major treatises are quite different from each other and that one is earlier than the other. A comparison of the two works as they concern machines underscores these differences. *Trattato I* is the detailed treatment that reflects the concerns of a practising engineer. *Trattato II* sets out the topics according to more general principles, and follows some of the practices of humanist authorship.[9]

Trattato I exists in two copies created by professional scribes. The first, L, created around 1480 to 1482, is in the Biblioteca Mediceo-Laurenziana (Codice Ashburnham 361) in Florence, and the second, T, containing some of Francesco's amendments and additions, created around 1482–1486, is in the Biblioteca Reale (Codice Saluzziano 148) in Turin. The two manuscripts form the basis for the twentieth century edition, edited and transcribed by Corrado Maltese and Livia Maltese Degrassi and published in 1967. As Massimo Mussini suggests, the specificity of the technical details of the drawings of *Trattato I* make it certain that Francesco himself was closely involved in their execution. The most complete version (T) is divided into eighteen sections, or topics, which are not labelled in the original manuscript. Many of the sections represent certain categories of problems or projects confronted by engineers. There is a section on forts, but also one on the bridges of forts and on other types of defense. The chapter on theatres is followed by a separate section on columns. Other sections include practical measurement, lifting machines and mills; ways of conducting water, metals (not present in L), military arts and machines, convents, instruments, bells, towers, and gardens. The abundant drawings are found within the columns of text and in the margins of the sheets and were probably executed by Francesco himself, as was established by Mario Salmi in 1947 by comparison with other of Francesco's known drawings. Although the exact circumstances of the origins of the two manuscripts cannot be reconstructed fully, Mussini's suggestion that they may have been created in Francesco's workshop in Siena is an attractive one.[10]

---

9   Francesco di Giorgio Martini 1967. For the *Zichy codex*, see Kolb 1988; and see Mussini 1991a, 82–88, 108–109, 121–124, note 75–79; and Mussini 1991b. For Francesco's translation of Vitruvius, see Fiori 1985; and Scaglia 1985.

Trattato I contains several sections that focus specifically on machines—military machines, cranes and lifting devices, and mills of various kinds. In the following I look at one of these sections, called by the editors, "levers of wheels and mills." This section takes up seven folios or fourteen sheets. The text is written in carefully blocked-out columns, two columns per sheet, by a professional scribe. Interspersed with the columns are box-shaped drawings of mills with their wheels and gears carefully depicted. There are a total of 58 such drawings which, taken as a whole, explain many variations of mills. There are water powered mills including those with horizontal, overshot, and undershot wheels, each in a variety of configurations. There are dry mills powered by animals, mills turned by cranks, and windmills. Taken as a whole, the images reveal two characteristics. First, most of the mills are set into neat, near-square boxes that are aligned with the columns of text. The drawings are made to conform to scribal guidelines, namely the width of the columns of text, square blocks of text and block-like images of mills reflecting one another. Second, it is clear that Francesco is interested in the gears and their interconnections, the axles or shafts, and the ways in which they are powered. For most mills, he provides details of the actual size of the desirable dimensions, such as the diameter of the wheels, gears, and axles, and their thickness.[11]

Francesco begins his section on mills with a discussion of the lifting capacity of a wheel—a discussion that pertains to wheels as to levers in general. He provides a formula for calculating this capacity based on the size of the wheel and of its axle. Take half the diameter of the wheel, he says, and half the diameter of the axle and calculate how many times the second goes into the first. If a diameter of the wheel is 10 braccia and the diameter of the axle is one braccia, you take half of each. Then calculate how many times one-half (half the diameter of the axle) goes into five (half the diameter of the wheel). The result is 10. So, he concludes, every pound of weight on the circumference of the wheel will lift 10 pounds on the axle. A schematic drawing of half a wheel and axle accompanies the discussion (figure 4.1).[12]

Immediately following, Francesco asserts that the lift of wheels for mills and many other kinds of machines are guided by these figures. Nevertheless it is difficult to demonstrate everything that is written, because there is such a great variety of things, one the opposite of the next. Although he has shown what he calls the "Ragione dela lieva," or "Reasoning of the lever," he recognizes that it is insufficient for the great variety of particular kinds and varieties of machines that engineers and others might devise. Therefore, he concludes that it is necessary to make a model. "Many things," he cautions, "to the mind of the architect seem easy and that necessarily succeed, that putting them into effect many lacks are found, which are repaired

---

10 Francesco di Giorgio Martini 1967, vol. 1, which includes a facsimile of the sheets containing drawings of T. For a brief but masterful description of the manuscripts and the scholarship surrounding their origins and dating, see Mussini 1993. See also Salmi 1947; and Scaglia 1992, 154–159 (no. 62, for Manuscript L) and 189–191 (no. 80, for Manuscript T), although it should be noted that Scaglia's claim that the two manuscripts were created at the monastery of Monte Oliveto Maggiore is without foundation in evidence and has been cogently disputed by Mussini.
11 Francesco di Giorgio Martini 1967, I ff. 33^v–40^r of facsimile pages, tav. 62–75. See also Marchis 1991.
12 Francesco di Giorgio Martini 1967, f. 33^v of facsimile pages, tav. 62.

Figure 4.1. Explanation of lifting capacity of a wheel and mills. From Francesco di Giorgio Martini, *Trattato I*. (Turin, Biblioteca Reale, Codice Saluzziano 148, fol. 33^v; courtesy Ministero per i Beni e le Attività Culturali della Repubblica Italiana.)

124

PAMELA O. LONG

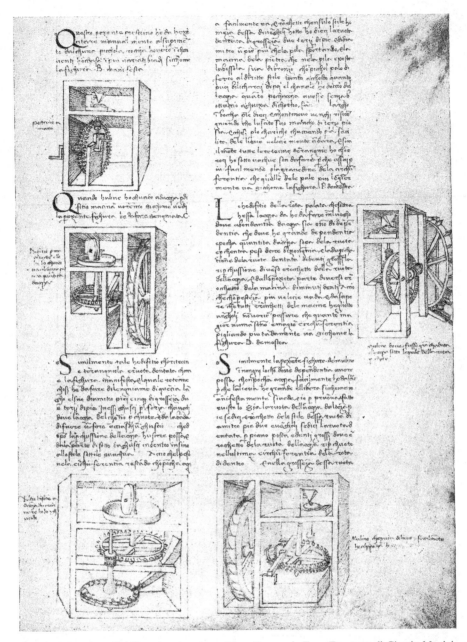

Figure 4.2. Drawings of mills and explanations of how they work. From Francesco di Giorgio Martini, *Trattato I*. (Turin, Biblioteca Reale, Codice Saluzziano 148, fol. 34^r; courtesy Ministero per i Beni e le Attività Culturali della Repubblica Italiana.)

[only] with difficulty. I for myself have seen the experience of a sufficient number of the inventions that here will be demonstrated, not relying on myself [i.e., my ideas alone]."[13] Francesco here shows the calculations meant to determine a mathematical way of measuring the lifting power of wheels, while at the same time underscoring the necessity of practical experience and of making models to avoid costly mistakes.

Francesco then turns to the construction of individual mills (figures 4.2 and 4.3). His treatment consists of a succession of descriptions and drawings of special kinds of mills. He describes and illustrates 58 different varieties. While most are for grinding grain, several serve other purposes. For example, he tells us that mills C and P (the two lower mills in the left column of figure 4.2) are for grinding olives and/or *guado*, the leaves of the plant *crocifer*, which were crushed in mills to produce a blue dye used in the textile industry. They are different from grain mills, especially because they lack a grain chute and feature an arrangement of the mill stones in which one stone sits ver-

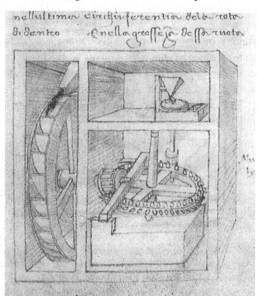

Figure 4.3. Enlargement of overshot mill E (lower right) of figure 4.2.

tically on top of the horizontal stone, an arrangement that differs from that of grain mills in which two horizontally placed stones turn against each other. Francesco also describes suitable locations for particular mills. For example, mill B (the top drawing in the left column of figure 4.2) is a small mill powered by a crank designed for a small fort, a convent, or a house. He tells readers that it is designed to grind a variety of different kinds of grains.[14]

In his descriptions of wheels, Francesco carefully explains to what type of mill it belongs and then goes on to describe the wheels and gearing, giving what he considers to be the appropriate measurements, and other details such as the best number of rods on lantern gears. In the case of water-wheels, he specifies which kind of wheel is appropriate for specific situations given the availability of water. If the water is some-

---

13 *Ibid.,* 141–142, and f. 33ᵛ of facsimile pages (tav. 62), "molte cose all'animo dell'architetto paia facile, e che riuscir il debba, che mettendolo in effetto gran mancamenti in essi truova, in ne' quali con difficoltà reparar vi pub. Io per me delle invenzioni che qui demostrate seranno, d'assai buona parte, in me non confidando, spirienza ho veduta." The phrase "raggione dela lieva" is written above the diagram of the half wheel and axle on f. 33ᵛ of the manuscript.

14 *Ibid.,* 143–145, f. 34 of facsimile (tav. 63).

what sporadic, overshot wheels are better, whereas undershot wheels are okay for situations where a continuous, strong flow of water is available.[15]

Take, for example, mill E, the one on the bottom right of figure 4.2 (enlarged in figure 4.3). The mill is powered by an overshot water-wheel. The water spills from a funnel on top of the wheel, turning the wheel. On the shaft of the wheel is a lantern gear, which rolls over the vertical teeth placed around the top circumference of the horizontal crown gear wheel. In the back, another lantern gear is attached to the shaft, which turns the mill stone. This lantern gear rolls on the horizontal teeth of the same wheel. Francesco has also drawn the mill's wooden frame and the grain funnel into which the grain to be ground into flour is to be poured, and the millstones. The latter two components are drawn in disproportionately small sizes.[16]

Here is what Francesco tells his readers about this mill powered by an overshot water-wheel:

> Similarly the present figure of the mill in every place where it can have contingency with little water, it is easy to make, because the capacity to lift is great and free, as is apparent, and I through my own trial have made and seen. Let the water-wheel be 16 feet high. The lantern gear of the axle of this wheel, a diameter of two feet and [with] 16 rods. The toothed horizontal wheel places its large teeth where the lantern gear of the water-wheel revolves over the large teeth on the top circumference of the toothed wheel. And in the width [rim] of this wheel, [are] the small teeth, and because the lantern gear of the millstone makes them turn, given their thickness and their size, and the outer circumference of the wheel where the teeth are, the millstone goes quickly and with facility. And let the diameter of this wheel be 8 feet, its axle short in the manner of an acorn, made [thus] because it goes more strongly. The lantern gear of the water-wheel, let its rods be [made] in the manner of rollers, that its motion light passing lifts strongly enough, thus as figure E demonstrates.[17]

Francesco thus informs his readers that the mill is good for situations with an uncertain or uneven water supply because of its great power. (He recognizes that overshot water-wheels are more efficient than undershot.) He describes the actual workings of the machine and provides some of its dimensions. His writing as a whole is important for its extensive development of technical vocabulary and this passage illustrates that. Taken as a whole, his writing presents by far the most extensive technical vocabulary of any previous author, something that makes his texts unique and important sources for the study of mills and other contemporary technologies.[18]

---

15 *Ibid.*
16 *Ibid.*
17 *Ibid.*, 144–145, f. 34 of facsimile (tav. 63), "similmente la presente figura del mulino in ogni loco dove dependenzia avere possa con poca acqua facilmente è da fare perché la lieva è grande e libera, siccome manifestamente si vede, e io per pruova fatto e visto l'ho. Sia la ruota dell'acqua d'altezza piè sedici. Er rocchetto dello stile d'essa ruota diamitro piè due e vergoli sedici. La ruota dentata per piano posta i denti grossi dove er rocchetto della ruota dell'acqua ripercuote nell'ultima circunferenzia della rota di dentro. E nella grossezza d'essa ruota i minuti denti, i quali er rocchetto della macina girando, per la spessezza loro e la grossezza e ultima circunferenzia della ruota dove i denti sono, veloce mente e con prestezza la macina va. E sia el diamitro d'essa ruota piei otto, el bilico suo corto a guisa di ghianda fatto perché più saldamente e con fermezza va. Er rocchetto della rota dell'acqua, a guisa di rulli i suo' vergoli sia, ch'el moto suo leggermente passando assai più forte leva, siccome la figura E demostra."
18   Calchini 1991.

In *Trattato I* Francesco provides a detailed written account, illustrated by numerous drawings, of his architectural and engineering knowledge. Mills were essential elements of many building projects from castles to forts, and they were also essential to any town or city. Taken as a whole, Francesco provides numerous examples of variations. Knowledge of such variations was essential to any practising architect/engineer of his time. Presented with variations of geography, power supply and potential use, the master must devise or decide upon a mill with particular characteristics for specific sites. Yet Francesco was not writing a manual for other practitioners. Rather, he was detailing his knowledge of practice in a codex at the instigation of his patron, Federico da Montefeltre, for the benefit of a more elite readership. The professionally copied text and carefully drawn images, on vellum bound in leather, consisted of a display of engineering knowledge for patrons, not a manual for use by other practitioners. Nevertheless, Francesco's careful detailing of numerous variations points to his close identification with the concerns of practicing engineers, a group that included himself.

Francesco's later treatise, Trattato II, is extant in two copies, one in Siena (S.IV.4), known as S (the earlier version) and another, Magliabecchiano (II.I.141) in Florence, known as M. Trattato II is organized differently from Trattato I, with fewer chapters, the addition of learned introductions, and more frequent references to ancient texts such as the *Naturalis historia* of Pliny and the *De architectura* of Vitruvius. Some of the more general topics of Trattato I survive as chapters in the later treatise, for example, the city, temples, fortresses, and harbours. Corrado Maltese, the editor of the 1967 edition of the treatise, supposed that it was the work of a humanist hired to revise Francesco's earlier treatise. The more recent investigations of Mussini, however, underscore that the treatise is the work of Francesco himself, with only short introductory passages added by another. This later treatise reflects Francesco's greater knowledge of Vitruvius. In addition, he provides more general treatments. He has moved from detailing many particular mills in Trattato I to more general accounts of far fewer mills, an approach more suitable to a readership of patrons and university-educated men.[19]

Trattato II is divided into seven books or *trattati*, as Francesco himself called them. There is a preamble, a book of common principles and norms, and then treatments of houses, ways to find water, castles and cities, temples, forts, ports, and a section on machines, "Machines to pull weights and conduct water, and mills," and then a conclusion. The material on machines that was scattered among various sections of *Trattato I* has received a new treatment in this treatise in one section only, section seven, which concerns lifting machines of various kinds and mills. In all, eighteen machines are displayed in drawings and described, a radical reduction from the earlier work, which contains drawings and descriptions of 58 mills alone, and

---

19 Francesco di Giorgio Martini 1967, I xi–xlviii; Mussini 1993, 358–359. The Magliabecchiano manuscripts contain several other works by Francesco as well—the *Trattato* is on ff. 1–102, Francesco's translation of Vitruvius on ff.103–192, and on ff. 193–244ᵛ a collection of drawings of military machines and fortification designs.

treats many other machines as well in separate sections, including lifting cranes and military machines.[20]

Offering a rationale for this new approach, Francesco explains: "Thus indeed I will provide a drawing *(la figura)* of some mills, so that through those, other similar ones may be able to be discovered by readers."[21] He thereby suggests that readers will be able to read about one kind of mill, seemingly understand its principles, and then discover other types. At the same time, it appears that his simplifying strategy involves more than simply wanting to present a more general and rationalized account of mills. This becomes apparent when he immediately plunges into a lengthy biographical lament: "More and more times have I thought about not wishing to reveal any of my machines," he says, "for the reason that I have acquired knowledge of that with my great cost of experience and grave inconvenience, leaving in part the things necessary to life." Nevertheless, experience has shown him "the effect of ingratitude." People do not realize, he continues, that "experience cannot be acquired truly without long time and expense, and independent of other useful cares." When they desire a machine or instrument, they see the design, and "seeming to them a brief thing, they scorn the fatigue of invention." Worse, Francesco concludes, they claim these inventions as their own.[22]

Francesco immediately turns from his bitter complaints to a story from antiquity, which he draws from the tenth book of the *De architectura*. Vitruvius tells the story of two engineers, Diognetus and Callias. The Roman architect relates that the people of Rhodes decided to withdraw support from their tried and true engineer, Diognetus, and to give it instead to a newcomer named Callias. Callias had arrived and presented a lecture in which he displayed a drawing or model *(exempla)* of a spectacular defensive machine, a crane that could sit at the top of the wall of the city and pick off siege machines by lifting them up and rotating them around and then placing them inside the walls, rendering them ineffective. However, soon King Demetrius the Besieger arrived at the island with a fleet of ships and a huge siege machine, which he set up immediately. As the assault continued, the people became increasingly frightened of their prospective defeat and enslavement, while Callias demurred that the machine for which he had shown a drawing in his lecture could not be made big enough to deal with Demetrius's huge siege machine. After much begging, a lucrative contract, and the supplication of the city's virgins, Diognetus was persuaded to return to his position of city engineer. He saved the city by instructing the Rhodians to breach the wall in the area of the siege machine. He had them throw large quantities of water and sewage on the ground around the machine, so that it got stuck in the mire. Finding his siege machine inoperative, Demetrius the Beseiger decided that the Rhodians were unbeatable, and promptly sailed away.[23]

---

20 Francesco di Giorgio Martini 1967, vol. 2.
21 *Ibid.*, II 492, "Si ancora di alquanti pistrini metterò la figura acciò che, per quelli, delli latri simili da li lettori possino essere trovati."
22 *Ibid.*, II 492–493.
23 *Ibid.*; and Vitruvius, 10.16.3–8.

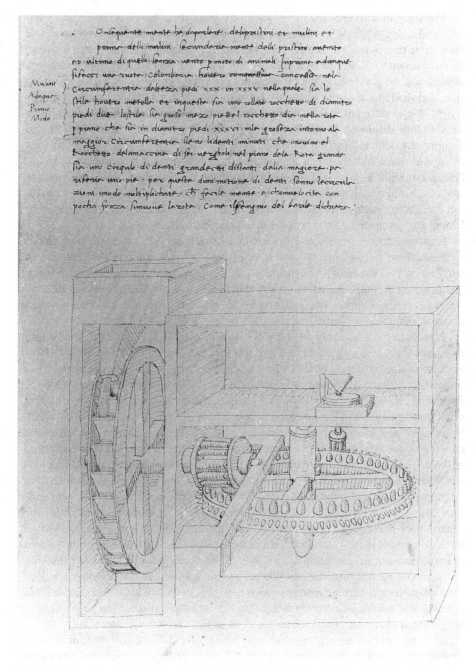

Figure 4.4. Overshot mill. From Francesco di Giorgio Martini, *Trattato II*. (Florence, Biblioteca Nazionale Centrale, Codice Magliabechiano II.I.141, fol. 95ᵀ; courtesy Ministero per i Beni e le Attività Culturali della Repubblica Italiana.)

The story underscores the importance of loyalty to a trusted architect and engineer, and it also suggests that actual practice is far more important than the theoretical knowledge presented in a lecture. Its appeal to Francesco undoubtedly also rested in its vivid account of the role of a drawing. By means of this ancient story, Francesco underscores the point that neither models nor drawings can be easily transformed into an actual working machine, primarily because they do not forewarn about problems arising due to different scales. While Vitruvius used one word, *exempla*, to explain what Callias displayed to the Rhodians, Francesco uses two—"some models and designs of machines" *(alcuni modelli e disegni di macchine)*. It is best not to trust those who have merely taken a model or drawing from someone else, such as, Francesco might as well add, those who had stolen drawings from him. Drawings cannot substitute for actual machines, nor can an "ignorant and presumptuous" architect like Callias stand in for the experience of an "ingenious and expert architect" like Diognetes—or indeed, like Francesco di Giorgio himself.

Francesco then proceeds to a discussion of machines. Of the eighteen machines he treats, there are six kinds of *argani*, or lifting machines comprised of capstans or windlasses, four kinds of *viti*, machines for lifting or pulling using screws, a water pump, and eight mills, including six dry mills or *pistrini* (including a windmill), and two water-mills or *mulini*, one powered by a horizontal wheel and the other powered by an overshot water-wheel.[24]

Francesco discusses only one type of overshot water-wheel mill, which he draws in a large size on a single page underneath the text describing it (figure 4.4). This machine is similar to the overshot wheel discussed above, one of five (E) on the sheet in *Trattato I* (figure 4.2, lower right and figure 4.3). His description in the later treatise is similar, but not the same as that for the analogous machine in *Trattato I*. He gives the dimensions of various parts such as the water-wheel and the crown-wheel. He explains that the wheel turns easily because the vertical teeth on top rim of the crown-gear are larger than the horizontal teeth on the lateral rim, and there are fewer of them. The smaller teeth on the lateral rim of the crown-wheel turn the smaller lantern wheel attached to the axle that turns the millstone. The drawing clearly shows the difference in size and number of the two rows of teeth on the crown-gear wheel. It also shows the different sizes of the two lantern gears, the one on the axle of the water-wheel being larger than that on the millstone axle. Francesco explains that these differences in the size of the teeth in the crown-wheel and the size of the lantern gears are what make the machine turn easily. It is notable that in his earlier drawing of the analogous machine such differences are not apparent in the drawing, nor does the explanation concerning ease of motion appear in the text. This is an example then, of using one machine to explain a more general principle (that of the gear differential), instead of enumerating varieties of mills suitable for diverse situations.[25]

The changes that are apparent between *Trattato I* and *Trattato II* are indicative of Francesco's efforts to write a more learned treatise, efforts that were influenced by

---

24 Francesco di Giorgio Martini 1967, II 495–504 and tav. 317–331.
25 *Ibid.*, II 500–501; and f. 95 (tav. 325). Cf. *ibid.*, I 492–493 and fol. 34 (tav. 63).

humanist practices of authorship. Rather than showing a large variety of particular machines, he discusses a smaller number chosen on the basis of types (for example, as we have seen, he discusses only one overshot water-wheel mill). He devotes more space and larger drawings to each type. He also begins each chapter with a general introduction, and refers often to ancient passages relevant to the topics that he treats. In Francesco's treatises, it is notable that the engineer becomes visible as an individual who suffers the hardships of invention, and works as a practitioner. Francesco decries the thefts of invention and the patrons' reliance on drawings alone rather than on the engineer's extensive practice. In *Trattato II* especially, the human engineer emerges as a striking figure. This is in part a result of Francesco's engagement with the *De architectura* and other ancient texts, and in part the result of his engagement with contemporary humanist writings such as the treatises of Alberti, Valturio, and Filarete.[26]

2. LEONARDO DA VINCI AND THE ELEMENTS OF MACHINES

In the early 1490s when Leonardo da Vinci (1452–1519) was working on *Madrid I*, he had extensive contact with Francesco di Giorgio when they were both consultants on the cathedral of Pavia. Possibly they had met earlier, in the late 1480s. Leonardo's project of a treatise on machines and mechanics clearly came about at least in part because of the influence of Francesco di Giorgio. In turn, Francesco was influenced by the earlier Sienese author of machine writings, Mariano Taccola. Francesco copied some of Taccola's text into his own notebooks, especially the small, early notebook referred to as the *codicetto*. Similarly, Leonardo copied some of Francesco's text into *Madrid Codex I*. And as mentioned previously, Leonardo put his own notations on several sheets of the manuscript copy of *Trattato I*, Ashburnham 361, now in Florence. Clearly Francesco di Giorgio and Leonardo influenced each other; their work contains common elements, but can also be distinguished in fundamental ways.[27]

Madrid I (Ms. 8937 of the Biblioteca Nacional in Madrid) consists of two notebooks that, it has been argued cogently, were bound together by Leonardo himself. In the first half, up to folio 92, the folios progress forward in normal fashion from left to right, written of course in Leonardo's characteristic mirror script. It consists of a treatise on practical mechanics and elements of machines. Leonardo numbered this half of the codex on the rectos of each folio through the first six quires. The second half of the work progresses backwards from the back of the treatise toward the centre—from right to left. It consists of the more theoretical part of the work—a treatise on the science of weights *(scientia de ponderibus)*. Leonardo numbered this half in mirror numbers on the verso starting at the back and going towards the front (folios 95–191 in the current foliation). All the sheets of both halves are the same size, all bear the same recurring watermark, and the whole contains only two protective flyleaves.[28]

---

26 See Grafton 2000; Long 2001, 122–133; Filarete (Antonio Averlino) 1972.
27 Galluzzi 1991; Johnston 2000, 24–105; Pedretti 1985; Tocci 1962; and Zwijnenberg 1999, esp. 35–46.

Madrid I contains many illustrations of machines and devices, which in part come out of Leonardo's studies of motion, but which are also grounded in his practice as an engineer. In the 1490s Leonardo was a client of the Sforza family in Milan, patronage that he acquired primarily on the basis of his abilities as an engineer and productive artist. Sforza patronage was bestowed on a group that included both learned humanists and artisan practitioners including Leonardo. The Sforza created a court that produced painting, sculpture, and buildings to their greater glory. They also strenuously involved themselves in the armaments industry and in canal building undertaken to improve transportation and agriculture primarily for commercial purposes. Court patronage, material production, and commerce were closely interrelated endeavors.[29]

The first half of *Madrid I* is filled with beautiful drawings of machines and devices, and machine parts surrounded by detailed textual explanations concerning how they work and how they demonstrate the ways in which the natural world works. It treats the motions of toothed wheels, springs, clock escapement mechanisms (what Leonardo calls *tempi*), continuous and discontinuous motions of various sorts—counterweights, chain gears, screws, pinions and wheels, endless screws, lifting devices, mills, keys and locks, and crossbows, among others. There is evidence that Leonardo carefully planned his treatise. In an examination of the manuscript in Madrid, Marina Johnston has recently discovered that the drawings were made first by creating preparatory drawings with metal or bone point before ink was applied. Other evidence of careful preparation includes sheets in the *Codex Atlanticus* in Milan that contain drawings that are repeated in *Madrid I* as well as blocks of text, crossed out, that are carefully written into *Madrid I*. It is clear in many cases that Leonardo drew the images first and then wrote around them. Detailed investigation of the specific pictorial and textual content of single sheets in the codex is particularly fruitful since many of the sheets function more or less as self-contained units on a single topic or on two or more related topics.[30]

For example, on folio 0 verso are two lines stating the subjects at hand. A single line on the right reads "On the Placing of the Figure" *(Del posare della figura)* (figure 4.5). On the left is the two line heading: "Book entitled of quantity and Power" *(Libro titolato de quantità e potentia)*. Below these two headings is a paragraph, which (I surmise) Leonardo added at a time later than when he created the rest of the page, inserting a comment into an available space. This comment concerns a nine year old boy named Taddeo who played the lute in Milan on Sept. 28, 1497 and who is considered the best lute player in Italy. Below the passage about Taddeo is drawn the figure of a crank mechanism. The crank itself can be seen in the back of an appa-

28   Leonardo 1974, vol. 3 (Reti's introduction to the codices); and Johnston 2000, 24–105, who provides a cogent analysis of the structure of the manuscript that adds significantly to Reti's foundational discussion; and see Maschat 1989.

29   For the Sforza court as it developed, see Catalano 1983; Chittolini 1989; Ianziti 1988; Lubkin 1994; and Welch 1996. For Milanese canal building, see Parsons 1968, 367–419.

30   Leonardo 1974, I f. 9[r], IV 24–25 (springs, clock escapements, continuous and discontinuous motions); I f. 10[r], IV 27–28 (chain gears); I f. 15[r], IV 39–40 (screws); I f. 19[r], IV 47–48 (endless screws); and I f. 46[v], IV 89–90 (mills). For a discussion of the organization of the work as a whole and of individual pages, see esp. Johnston 2000, 24–105; and see Zwijnenberg 1999, 100.

Figure 4.5. The "globulus" and other mechanisms with grooves. Drawing by Leonardo da Vinci. (*Codex Madrid I*, fol. 0ᵛ; courtesy Biblioteca Nacional, Madrid.)

ratus in which a rectangular frame is moved in and out from the centre of a quasi-cir-
cular form with irregularly curved edges. Leonardo writes that

> you will be able to perform all unequal motions as shown by this drawing with the art of
> the imperfect circle, which could also be called globulous or flexuous. This instrument
> will, in its revolutions, drive the front of member f to varying distances from its center.[31]

The heading underneath this passage labels the topic of discussion "Of instruments that perform poorly because of a little something that is at fault or in excess."
In the ensuing paragraph, Leonardo notes that the performance of an instrument is
often hindered "only through wishing to put the line of a length of a part where it cannot enter." It is as if one would pretend to put 8 into 7 which is impossible. From this
situation, he continues, "the constructors of said instruments remain dismayed, stupe-
fied, that they seem spellbound over their work when it involves movement." He
notes that this especially occurs when the movement is to take place in fittings or
channels. At the bottom of the page, Leonardo sketches two examples of apparatus in
which a weight hung over a pulley is dragging a board through grooves or channels.
He notes that if the board does not touch the channel at every point through the entire
groove, "it grows and would be able to enter then in said channel with its diameter."
Hence "its line comes to grow in a way that the space is not capable of such growth."
The sketch shows two figures of boards in channels. The figure on the left displays a
longer board, which contains writing that informs that the facility of movement will
improve as the length of the board increases. A second sketch on the right shows a
shorter board further back on the apparatus. The channels or grooves are labelled.
The board is placed at the start of its course along the channels, and because of its
shortness, will, according to Leonardo's textual note, have more trouble being pulled
along.[32]

Viewing the page as a whole, all of its elements seem at least in part related, with
the exception of the comment about Taddeo, which was probably added later. One
general theme concerns apparatuses that contain elements, which move along in
grooves or channels. Such mechanisms are powered either by a crank as in the globulous circle, or by descending weights as in the two figures at the bottom of the page.
Like Francesco di Giorgio, Leonardo is concerned about slight discrepancies or small
glitches in the apparatus, which cause it to function with less than optimum efficiency. Leonardo's interest in efficient function is also evident in his observation that
the length of the board being pulled in grooves affects the way in which it works.
This conclusion may or may not be intuitively obvious—it seems likely that
Leonardo had constructed this and similar mechanisms with slight variations to test
the ways in which they worked.

Leonardo's textual descriptions of the mechanical elements on this page strike me
as being notably awkward. This conclusion can be evaluated only by reading the Italian; Reti's English translations (which I have sometimes modified in my own English

31   Leonardo 1974, I f. 0ᵛ, IV 2–3.
32   *Ibid.*

citations) at times present more coherent and often more abstract restatements than the Italian seems to warrant. Leonardo seems continually challenged to find appropriate words to describe the workings of machine elements, as well he might given the lack of a developed technical vocabulary for machines in the written Italian of his day. Yet Leonardo's own part in developing such a vocabulary seems much less extensive than Francesco di Giorgio's. Leonardo's textual descriptions are essential to his illustrations and his illustrations are crucially necessary to understanding his writing. Particularly taking into consideration that Leonardo attempts to discuss the ways in which his devices work, text and drawings, respectively, would be relatively incomprehensible one without the other.

Leonardo is interested in the actual motions of the apparatus and his statements appear to be based on actually constructing many of the elements and causing them to move. I draw this conclusion from the fact that throughout *Madrid I* he describes hundreds of minute local motions of numerous machine elements and their variations. For instance, he refers to the difficulties of cranking the rectangle back and forth in the globulous circle, and he notes that a longer board in the grooves moves more easily than a shorter one. Although it is presumptuous to assume to know the extent or limits of Leonardo's own intuition as it concerned the motions of elements of machines, it seems improbable that he or anyone else would understand intuitively the direction and nature of all the small variations of motion, and the difficulties of all the various elements of machines that he discusses without trying at least some of them out. And we know that Leonardo was extensively involved with actual engineering projects such as canal building and large-scale casting and that he hired German artisans to help him construct apparatus. Moreover, he frequently refers to his own experience.[33]

At the same time, Leonardo here often carefully separates his investigations of local motions and particular elements of machines from actual whole working machines, despite the fact that his ability to invent and draw whole machines is spectacularly evident on some folios of this codex and elsewhere, especially in the *Codex Atlanticus*. Yet he does not seem interested in presenting a brief for engineers, but rather in pursuing an investigation of local motion and the operation of particular machine elements. His is a contribution to a learned investigation, which may well also have practical implications. He often fails to mention the actual possible uses of machine elements—and in fact, many of them (such as the globular crank with moving rectangle described above) may have been invented solely as an experimental apparatus to explore the local motion and operation of particular elements.

On the following folio Leonardo treats cranks joined by connecting rods, and their motions (figure 4.6). The heading for the sheet comprises a single word, "Questions" *(Quisiti)*, reminiscent of scholastic practices. The question he has in mind follows: "Why does the flame rushing to its [natural] place take the shape of a pyramid while water becomes round in its descent, that is, at the end of the drop?" After pos-

---

33 *Ibid.,* III 40–41 for Leonardo's references to the various German artisans that he hired; and for explicit references to experiments, I ff. 77^r, 78^r, and 122^v, and IV 179–180, 183–184, and 325–366.

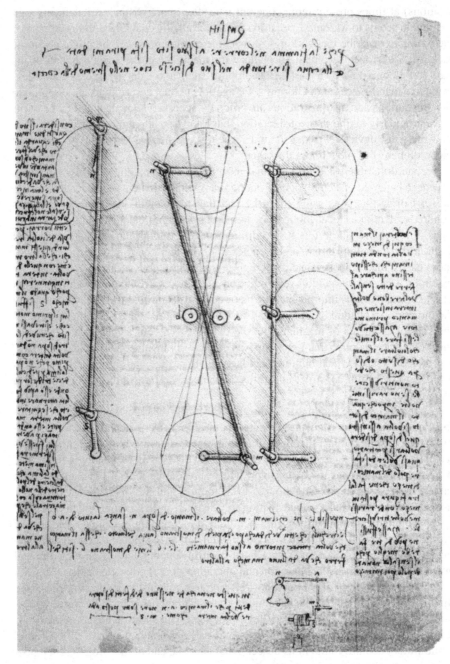

Figure 4.6. Crank Mechanisms. Drawing by Leonardo da Vinci. (*Codex Madrid I*, fol. 1ʳ; courtesy Biblioteca Nacional, Madrid.)

ing this question involving the differing shapes of two different elements as they seek their natural place, Leonardo launches into a discussion of what might be called the shape of motion of a device consisting of two cranks joined by a connecting rod. He illustrates the same device twice within a chronological time sequence, first on the left, and then in the middle after the crank has been moved. The depiction on the far right shows a variation with three cranks.

Concerning the apparatus on the left, Leonardo writes:

> Consider the position of these two cranks and observe the motion made by the lower crank in its movement to the left. You can see that it is moving down; the upper crank would be raised, but the length of the *sensale* [connecting rod] does not permit it, and as a consequence the crank returns. Should you wish to turn the other in a complete revolution by turning one of the cranks with the aid of such an instrument you would be deceiving yourself. Inasmuch as crank S would be the prime mover, being in the same position as can be seen, the upper crank would not be able to make a complete revolution like the first one because it can not overcome the perpendicular line that unites the centres of the axles. Therefore this crank will have more facility to turn back than to make a complete revolution. But if by chance the crank advances, it will stop violently when the line that unites the centre of the axle with the centre of the ring takes the same position as the central line that goes through the *sensale*, uniting both cranks.[34]

He thus describes the exact movement of the cranks and the connecting rod, assuming that the motor is the lower crank marked S. The central figure shows where the two cranks would normally end up, assuming that the lower crank in the image on the right is moved to the left, or in a counter clock-wise direction. Normally the top crank moves to the 6 o'clock position and then reverses itself, ending in the position shown in the middle image. Occasionally the top crank goes all the way around. This is what Leonardo says in his text, this is what he shows in the two images, and indeed, in a model that I made of this apparatus, this is what happened.

The central image shows the same apparatus that is shown on the left, with the cranks in different positions—the positions where they would normally end up when the bottom crank is moved in a counter clock-wise direction. Leonardo uses this central crank apparatus to make further statements about possible motions. In the paragraph directly underneath the three images, he notes that it would be impossible for the lower crank m to turn the upper crank n without the aid of a and b—he is referring to two pegs or rollers in the centre. He tells us that the circles show the movement of the cranks as they turn on their bearings and that the six lines show the six positions of the iron rod that connects the cranks. On the right side of the page, Leonardo discusses the three cranks attached by connecting rods. He says that if you turn the cen-

---

34   *Ibid.*, I f. 1ʳ, IV 5–6, "Considera il sito di questi due manichi, e guarda il moto che vol fare il manico di sotto, andando in verso man sinistra. Vedi che va declinando, e 'l manico di sopra si vorebe alzare e lla lungheza del sensale nol concede, onde torna indirieto. Se ttu vorrai, per causa del voltare l'un di questi maniche, che ll'altro insieme con quello dia volta intera, tu ti ingannerai. Inperochè qua[n]do il manico S si ffussi lui il primo motore, e che e' si trovassi nel sito che ttu vedi, l'altro di sopra non dara volta intera come 'l primo, perchè non passa la linia perpendiculare de' cientri de' lor poli. Onde esso ha più facilità a ttornare indirieto che seguitare la volta intera. Ma sse pure esso andassi inanzi per aventura, lui spesse volte si fermerà fortissimamente, quando la linia che va dal cientro del polo al cientro dell'anello, sia tutt' una cosa colla linia centrale che passa pel sensale che va dall' uno manico all' altro."

tral crank around, all the cranks that might be connected with the *sensale* would make a complete turn, going along with the central crank, which is the prime mover. However, if you attempt to turn all the cranks by turning the upper or lower cranks, "you would succeed only very seldom." He explains that when the lower crank turns to the left, the upper crank wants to turn in the opposite direction, almost as if the middle crank were an axle as in the centre figure, the rollers or pegs function as an axle. It would work better, Leonardo adds, if you would use more than three cranks, because then the *sensale* would have no axle in the middle.[35]

Here Leonardo is thinking about different arrangements of cranks and connecting rods, the ways that they would move, and whether they would function well or poorly, depending upon small variations. At the bottom of the page he provides a drawing of a very different apparatus with a crank that swings back and forth. Here none of the problems mentioned above would occur, he says, because the crank a n is not supposed to make a complete revolution as is m s.[36]

These two folios, albeit representing only a very small number of the hundreds of devices and elements of machines treated in *Madrid I*, provide a window into Leonardo's interests and methods. First, he explores specific kinds of mechanisms and devices. On the folios discussed here, he treats two kinds—first, mechanisms that move in channels, and second, cranks. Leonardo uses textual explanations and visual images in his investigations. Both are essential and I believe that they are on par with one another. In most cases here and throughout the codex, neither the text nor the image is fully understandable without the other. Leonardo experiments on paper with many specific variations. Certainly in at least some instances he also constructs actual devices. But even on paper, he creates structural variations and describes in detail the ways in which direction of motion and facility of functioning are affected by variations in structure. It is evident that he places natural elements (such as the drop of water and the flame) and mechanical elements within the same conceptual world, and believes that investigating the one can help illuminate the other. Yet Leonardo's interest in issues of friction and facility of motion are indications that, while philosophical issues such as the nature of motion fascinate him, he has one foot firmly planted within the arena of engineering practice.

In significant ways, *Madrid I* in itself constitutes a kind of investigation. Rather than presenting the *results* of "research"—it *is* the research. This is the case because for Leonardo, drawing and writing taken together constitute not only the tools of investigation, but the investigation itself. He devises a number of different variations as images on paper, and thinks about the ways that these variations influence direction and ease of motion. He is interested in both structural variations of different kinds of devices, and in the local motions that result from those variations. Such variables have profound practical manifestations. Leonardo is also interested in them as part of his general study of motion, a study, which is local and observational, which

---

35 *Ibid.*
36 *Ibid.*

sometimes must have involved making models, and which uses the creation of visual images as a way of both thinking about and explicating variations.

Leonardo's devices function on some level as observational tools. He constructs various devices and then observes and describes how they move, in what direction and how easily. It seems probable that he actually constructed or had constructed by others at least some of the devices that he draws. We know that he frequently hired German artisans to make machines and devices for him. He also mentions from time to time his own experimentation.[37] I believe that some of the motions that he describes probably could only be learned by making the device itself and observing how it could be made to move.

3. FRANCESCO, LEONARDO, AND THE VISUAL CULTURE OF MACHINES

Francesco di Giorgio and Leonardo shared similar interests that include machines and mechanisms. Although Francesco treated whole machines, while Leonardo often treats machine elements, the contrast is not as great as might appear at first sight. Francesco often focuses on only certain parts of the machines he discusses, such as the water-wheel and gears. He provides drawings of whole machines but is primarily concerned with certain working parts. Both men focus on how well the machine works—the factors that encourage or inhibit easy functioning. For both men, such concerns come out of their engineering practice. Both present variations, although Leonardo goes much further in his investigation of variations as a general focus, rather than as part of utilitarian machines.

By considering their respective treatises as a whole, the larger context within which each frames his treatment of machines becomes clear. From this point of view, there are fairly striking differences. In his *Trattato II*, Francesco's chapter on machines is part of a treatise on architecture and fortification. The Sienese architect worked on his treatises while he also learned Latin and worked to understand and translate Vitruvius's *De architectura*. He was familiar with the earlier treatises on architecture by the humanist Alberti and the architect Antonio Averlino called Filarete. His view of architecture includes a framework in which human proportions determine the mathematical symmetries of an ideal architecture, which reflect the ideal proportions of cosmos. His anthropomorphic ideal architecture was informed by both ancient and near-contemporary writings as well as by contemporary architectural practice.[38] Francesco attempted to write a humanist treatise on architecture and fortification in which machines played an essential part.

Considered as a whole, Leonardo's treatise has been described as a bifurcated work in which the two halves mirror each other. The second half, starting from back and moving to front, that concerns statics and kinematics, mirrors the first half that treats elements of machines, including numerous variations, and their motions. The

---

37 See footnote 33.
38 Millon 1958; Saalman 1959; and Tigler 1963. For the larger tradition of architectural authorship to which Francesco contributed, see Long 1985.

second half includes discussions of passages from the writings of Jordanus of Nemore and other late mediaeval writings on statics. Leonardo connects his discussions of machines with the an interest in statics and kinamatics, placing his work within this earlier tradition but also departing from it in his fundamental treatment of machines and mechanisms and their many variations, and in his investigation by means of both writing and drawing. On many sheets, including the one concerning cranks discussed above, Leonardo introduces statements concerning the motions of natural entities—here they are drops of water and flames. He seems to suggest an analogy between mechanical and natural motions; at the very least, he is introducing the two within the same framework and on the same page. Elsewhere in the treatise, such analogies are far more explicit. In general *Madrid I* investigates connections between the natural and the artifactual worlds with implied or explicit comparisons of natural and mechanical movements.[39] Such small-scale comparisons and analogies are quite different from Francesco's large-scale microcosm/macrocosm framework connected by mathematical proportions. Both authors compare the natural to the mechanical worlds, but in very different ways.

Francesco di Giorgio and Leonardo da Vinci both undertook projects of self-conscious authorship on topics of mechanics, mechanical arts, and machines. For both, visual images played a fundamental role in their authorial practices. In different ways, their writings were connected to engineering practice on the one hand, and to learned culture on the other. Their treatises reflect the fifteenth-century transformation of the status of drawing and painting from mechanical to liberal art, a transformation far from complete in the 1490s. The fact that visual images could play such a fundamental role in treatises that aimed to contribute to learned culture, and the fact that machines themselves were embedded in these learned treatises, signals a broad cultural transformation in which the mechanical arts and visual images have become newly significant within the larger culture, including the culture of learning. Both Francesco and Leonardo contributed to this trend in part because their images of machines and machine parts in themselves constituted virtuoso displays of technical facility. Those images also functioned as explanatory tools within detailed discussions of machines, written discussions relatively unfamiliar to any of the diverse cultures of learning in the fifteenth century.

Within their project of authorship Francesco and Leonardo articulated, both visually and in writing, the tacit knowledge of machines that they had acquired within the compass of engineering practice. The process of authorship itself clearly required each author to make visible and to articulate what he knew tacitly—to rationalize that tacit knowledge to make it understandable to practitioner and non-practitioner alike. We can suppose that the process of articulation encouraged the authors themselves to think about both the principles and the working of the machines that they described. This process is evident in Francesco's explanation of the gear differential, and is evident in Leonardo's drawings as well. Leonardo's experimentation by means of drawing machines and writing about them is grounded in his tacit knowledge of how

39 Kemp 1991.

machines are made and how they work. What is not always clear is the location of the line between what he knows tacitly and what he learns from the process of drawing and writing about mechanisms. Both authors make the tacit knowledge of their own practice available to readers, including learned readers who would not have possessed such knowledge.

In this chapter I have compared Francesco di Giorgio and Leonardo's treatments of machines, and pointed to their shared culture of engineering practice, and their common experience of functioning within client/patronage relationships. I have pointed to their substantial intellectual similarities and their significant differences. What I have not done is suggest that Leonardo's work represents an "advance" over that of Francesco di Giorgio's. Rather, each author identified with and developed his writing within the context of diverse strands of contemporary learned culture— Francesco within the context of humanist writing, Leonardo within the context of scholastic traditions. I would suggest that their two diverse ways of relating to and advancing learned culture point significantly to transformations within that culture.

Neither the writings of Francesco or Leonardo represent a point within a linear development to "classical" mechanics. At least in some cases, as Michael Mahoney has emphasized, classical mechanics as it developed in the seventeenth century moved toward increasing use of mathematics and greater abstraction.[40] Quite in contrast, both Francesco and Leonardo were interested in local motion, specific force and power, in overcoming small, real glitches, and in overcoming friction. Their conceptualizations of machines remained firmly grounded in engineering practice and in detailed, complicated images that were ready referents to actual or at least imaginable machines.

While their writings did not lead in a direct line to classical mechanics, they did make a contribution to the development of empiricism and experimentalism, which were central to the emergence of the new sciences in the subsequent centuries. Francesco's and Leonardo's practice of authorship in itself is significant in this regard. Both men utilized their knowledge of engineering practice and their skill in the visual arts to explicate machines and machine parts within treatises inspired by noble patrons and connected to traditions of learning. Their writings contributed to a culture of knowledge in which instrumentation, tools, machines, and indeed, drawings, came to play a crucial role in legitimating knowledge claims about the world, including its natural and its mechanical components.

---

40 See the chapter by Mahoney in this volume; see also Mahoney 1985 for his initial discussion of this idea.

MEASURES OF SUCCESS: MILITARY ENGINEERING AND THE ARCHITECTONIC UNDERSTANDING OF DESIGN

MARY HENNINGER-VOSS

> The office of the architect is to make the fortress secure from the four prime offenses: because in the ordering of our fortifications we will always have before our eyes, as the principle goal, that it should be secured from battery, from scaling [the walls], from shoveling [trenches], and from mines. And as the major part of the offense comes from artillery, so the principle defense will be from the same.

> Galileo Galilei, Trattato di Fortificazione

As a teacher of military engineering, Galileo began with compass constructions of basic geometric figures for the purpose of design. Immediately, however, he reminded his students that the design was to meet the particular goal of defense. For that "we will always have before our eyes" the offense that would never have been depicted directly in the design.[1] While historians of technology have made much of the engineer's visual knowledge, the "eye of the engineer" in sixteenth-century military engineering was not necessarily confined to what was present on the surface of the design itself. How then could a design be judged? What sorts of knowledge were required to visualize not only the building itself, but its probable success?

There is a difference between seeing an engineer's drawings and reading those drawings. It is perhaps because of this essential "reading" metaphor that we so often talk about a "lexicon" of design elements and the evolution of different "grammars" of drawing. I wonder however if our metaphors are not misleading us, re-encoding the act of interpretation of engineering drawings with a self-enclosed act of language recognition, rather than pointing us toward the difficulties in higher interpretation of meaning. In sixteenth-century formal treatises, engineers themselves often compared good design favorably to the thick description of words. A good model made transparent the window between the project design and the project's success. The actual ways in which designs of large public projects were interpreted and judged, however, belie the transparency of "good design." Engineers' designs interest me precisely because they are representations not of linear discourse, but of complex relationships between things—some of them depicted and measurable, and some of them unseen, though often reduced to a measure of something that is depicted. In the largest, most public engineering projects of sixteenth-century Italy—those concerning fortifications and hydraulics—it was the meaning of those measures and the nature of the

---

1   Galilei 1968a, 84.

relationships between design factors that became most hotly contested in the evaluation of designs.

The whole point of Eugene Ferguson's work, *The Eye of the Engineer,* was simply to root engineering practice in the design process.[2] Ferguson delved neither into the cognitive act of design, nor into the social act of design interpretation and realization. His message was simply to emphasize the nonverbal design function over the development of theoretical constructs in engineering, and to catalog the engineer's use of design over the past five centuries. He roots the forms and shapes of modern engineering practices, for good or ill, in the military engineering of sixteenth-century Italy and seventeenth-century France. It is the "fortress mentality" that remains for Ferguson the (perhaps unfortunate) heritage of engineers, the zealous accounting of "inputs" and "outputs."[3]

The milieu of Italian military engineers will also be my focus here. I am not interested in finding there a root of modern engineering, however. Or if I am, I am more concerned with the attributes of design that the twentieth-century theorist J. Christopher Jones outlined.[4] Like many scholars, Jones traced modern technique to a Renaissance design-by-drawing method that replaced a craft approach. Jones emphasized that the design allowed an engineering style that could overcome the limitations of the single craftsman's practices. The design enabled large, complex engineering tasks to be divided and coordinated; it stored tentative decisions for one part of the design while other parts were developed; it served a predictive function for what would work. I am less interested here in the particular assertions Jones has made or not made, but want to emphasize that these features of design all relate to a context in which technological tasks *are* large, evolve over years, and require a minimization of risk. As David McGee has shown, Jones's depiction of the "craft approach" may not be importable to early modern technical work;[5] however, Jones does offer two insights that may illuminate the large-scale engineering projects of that period. First, when designs are made by teams—as they certainly were for cinquecento fortifications—we must change the way in which we view design production. Second, the control and judgment process is set up to predict the success of a thing that does not yet exist, based on the performance of things that do. Jones shifts the discussion, appropriately, to what a design *does* at particular stages of the process.

Engineers' drawings do things; they do things in particular social contexts, and to be useful rely on an implicit human organization and economy of knowledge. As so many authors in this volume have shown, what a drawing means depends both on the techniques that convey that meaning and the place that the drawing has within the design and construction process. This essay draws on work in both these avenues, but is an investigation of how meaning was constructed around predictive operation for the planning of cinquecento fortifications. In their models' scaled dimensions, military engineers encoded meanings that depended on an "architectonic" understanding.

---

2   Ferguson 1992.
3   *Ibid.,* 70.
4   Jones 1970.
5   McGee 1999, 209–214.

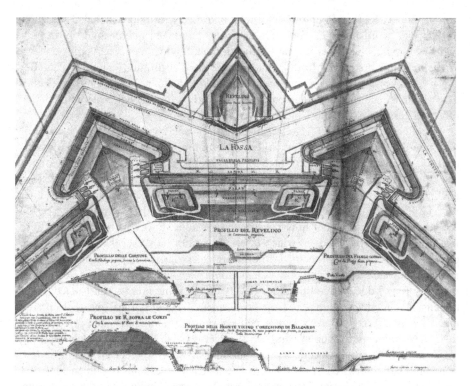

Figure 5.1. Plan and profile for two bastions at the fortifications of Palma, 1595. This is among the most finely rendered plans, in fine wash, neatly labeled. The plan should give the width and length of walls and thus proportions between firing positions; the profile should provide the width and height of the walls, as well as its composition (dirt, rock fill, etc.) and the proportions of wall, scarps (embankments that fortified walls), and trench. From Filippo Besset di Verneda, Parte di pianta e corrispondenti profile di tutta la fortezza di Palma, 1595. (Venice, Archivio di Stato, Raccolta Terkuz, n. 42.)

This architectonic understanding was teleological in that it tied form strictly to function; it was an understanding that placed the forms depicted into the larger context of the terrain to be defended, and sought to determine structure within the parameters of gunpowder warfare. While engineers challenged each other to "see" on the surface of the design the relationship between projected defense and probable offense, the "reading" of the design required not only eyes, but work—intellectual, technical, and sometimes physical—both on and off the plane of the design. The predictive meaning of a design was seldom transparent. Rather, it was the site of negotiation, controversy, and investigation. As we examine how fortifications models took on meaning, easy divisions between theory and practice dissolve.

What sort of drawings are we talking about? While measured drawings may be rare for sixteenth-century machine drawings, fortifications plans are littered with numbers and calculations, from the preparatory to the final drawings. In large part we will be looking at the construction and evaluation of models in the context of realization, to import Marcus Popplow's category, and these invariably included measurements.[6] The words "model" and "design" were employed interchangeably in both published architectural treatises on fortification and in documents pertaining to government administration of fortification works. The substitution of one for another cannot always be counted on since models could be made in different media—either a two-dimensional representation on paper, or a three-dimensional construction made of wood, gesso, or some other material.[7]

Architectural historians have assumed that presentation models were usually of the three-dimensional sort, although they also point out that Pietro Cataneo's mid-century architectural treatise, in which military architecture is prominent, encouraged students to develop perspective techniques so that perspective drawings might serve as a cheaper, more portable substitute.[8] The historiographical assumption in either case is that those unaccustomed to looking at architectural designs would see the project better in a small but fully constructed version. However, the letters we have that accompanied the models (which often we do not have) seem to refer to paper designs; and yet very few perspective drawings of projected fortifications are found in state archives. Further, both princes and military leaders had a high degree of fortifications knowledge. While I think it is likely that at some point a constructed three-dimensional model was required by governmental commissions, most discussion over models appears to have taken place over simple ground plans and profiles, and their measurements. Although perspective drawings also were included in teaching manuals for military engineers, the ground plan and profile received greatest attention, and Galileo insisted in his own treatise that with these all the dimensions of a fortification could be known[9] (figure 5.1). Another reason that most of our documents appear to refer to ground plans may have been that even fine ground plans could easily be copied and sent to the various decision-makers by pricking the outlines over another piece of paper (figure 5.2). Here I will only be concerned with these measured, scaled, drawn plans.

Models were not simply maps, nor were they simply sketched-out ideas. Drawing may be the nonverbal representation of information or ingenuity, but the design model is comparable (but not reducible) to mathematical solution-finding. In this way, we can see that engineers worked with a certain number of "knowns" and an array of "variables" in order to project a design that would work under all foreseeable conditions of attack. The process of identifying these knowns and variables was divided among the men who created the design and those involved in the control and evaluation of the design. The military design model was constantly moving between

6 Popplow in this volume.
7 Scheller 1995; Goldthwaite 1980; Rosenfeld 1989.
8 Cataneo 1554, book 1, f. 1; Adams and Pepper 1986, 190; Adams and Pepper 1994, 64; Ackerman 1954.
9 Severini 1994; Pellicanò 2000, 113–115. Filippo Camerota's studies of projection techniques in this volume, testify to perspective techniques that would have been measurable and useful in the context of fortifications planning, and yet I simply do not know if these techniques were used by military engineers in the planning process during the sixteenth century. In general, perspective drawings as part of presentations of fortifications plans (as opposed to reconnaissance, commemorative, or mnemonic uses) only become conspicuous in the seventeenth century.

spheres of knowledge. The design had to be seen now with the eyes and tools of proportion, now with the eyes and experience of military strategy, and now with the eyes and analyses of mechanical physics.

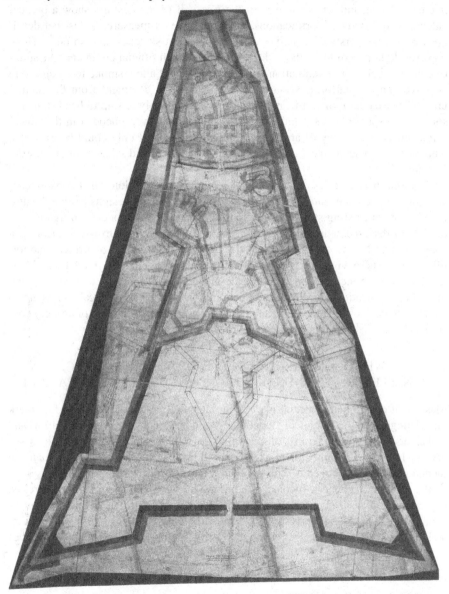

Figure 5.2. Plan of the lower fortress, called San Cataldo, Perugia, Rocca Paolina, 1540. The plan is color-keyed and has been extensively pricked at angles, an indication that it served as a basis for copies From the workshop of Antonio da Sangallo. (Florence, Gabinetto dei Disegni e Stampe, 272 Architettura r.)

The hierarchies around fortification models also meant that many design features that had less to do with the gross necessities of strategy were handled at other levels of the process. The sketch-books of men like Antonio da Sangallo or Michele Sanmicheli, premier military architects of the first half of the century, show a broader palette for the study of fortifications, including layered measured plans that developed over time, perspective studies, mapping studies, sketches, and so on.[10] These however do not seem to be those drawings that elicited official comment. Questions of embrasure sizes and ventilation chimneys that could allow smoke to escape from enclosed cannon platforms seem to have elicited little comment from the higher-ups.[11] The execution of fortification interiors was probably left up to head stonemasons.[12] The spare plans submitted to governments, however, encoded in their measurements a broad array of architectonic relationships that only could be evaluated with compass in hand, and with a view to the experiences and sciences of gunpowder warfare.

I present here a series of vignettes that address the question: what constituted a good model in cinquecento military engineering? How could designs predict the utility of a not yet existing structure? I have examined textbook models and training, an episode of disagreement between engineers, and the directions to which theoretical considerations behind designs could lead (i.e., to a "new science" of artillery and fortifications). These vignettes show different levels on which the models could be inspected. These include adherence to the mathematical practices that were fundamental to fortifications designs; the reduction of vast arrays of political and technical factors into debates over measure, and the philosophical investigation of artillery concerns relating to design decisions.

1. WHAT GOOD ARE MODELS? FROM VASARI'S "LIFE OF BRUNELLESCHI" TO BUONAIUTO LORINI'S "DELLE FORTIFICATIONI"

Mastery of design had a special place in architectural practice. It was the first mark that distinguished the "architect" functionally and hierarchically from the masons and carpenters.[13] To use the terms of fifteenth- and sixteenth-century Italian art theorists, *disegno* had to be an embodiment of *ingegno*; to judge design was also to evaluate the mental tools that fashioned it. This *ingegno* referred both to imaginative fantasy and to the rules and sciences (perspective, knowledge of nature, etc.) that controlled expression in *disegno*. For figurative arts and civil architecture, there remained a tension between the imaginative and the scientific aspects of *ingegno*.[14] In military architecture, expression of fantasy was obviously limited, and the array of sciences

---

10 See especially Adams and Frommel 1994.
11 E.g. see the correspondence of engineer Francesco Malacrede and generals Pallavicino and Savorgnano, in Venice, Archivio di Stato di Venezia, *Archivio proprio Contarini*, B. 8, ff. vii^r to xviii^r.
12 Adams and Pepper 1994, 62.
13 Pevsner 1942; Ackerman 1954.
14 Kemp 1977; Kemp 1990; Puppi 1981.

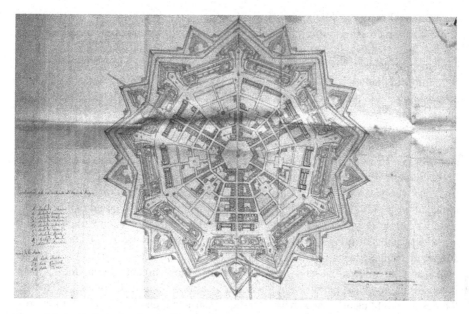

Figure 5.3. An ideal circuit of walls. The bastioned curtain wall is surrounded by a star-shaped "covered way," and in between the rays of the stars are rivelin outworks, additions popular in the seventeenth century. Pianta di Palmanova. (Venice, Biblioteca Museo Correr, Cod. Cic. 3486/IV.)

and practices in which the architect was to be conversant ranged from geometric construction to tactical savvy to artillery aim.[15]

There is a reason why sixteenth-century military architects began to identify themselves more exclusively as "engineers," and distance themselves from civil architects who dealt in "the secrets of the statue-makers" guild."[16] Military architecture was completely reconfigured in the first decades of the sixteenth century. Due to the demands of mounting and withstanding the new gunpowder weapons, high crenellated walls and round towers gave way to a polygonal circuit of walls and spade-like bastions (the enceinte known as the *trace italienne*), and a variety of platforms, trenches, outworks, earthworks, countermines, and scarping techniques (figure 5.3). The design of fortifications became more closely associated with the design of machines, and engineers began even referring to the fortifications system as "la macchina." The new fortifications offered little leeway for aesthetic choice or even traditional knowledge, as the innovative new architecture was to meet the requirements of gunpowder weapons.[17] At the same time, military engineers mediated between stonemasons (and legions of unskilled laborers), military leaders, and

---

15 Hale 1977; de la Croix 1963, 30–50; Lamberini 1987 and 1990; Manno 1987; Fiocca, 1998, *passim*.
16 Lorini 1597. See also Hale 1977. I have noticed the same phenomenon in the roles of the Florentine court where Bernardo Buontalenti and Ammanati are both called "architetto" or "ingegnero" in alternate years.
17 Wilkinson 1988.

governmental patrons while their training became ever more associated with mathematical arts. Increasingly through the century, fortifications engineers became members of large teams of professionals, beyond the cooperation and contributions of workshop assistants.[18] While workshop practices still thrived around figures such as Bernardo Buontalenti in Florence and the Genga family in Urbino, military engineers increasingly made their primary alliances among military leaders who had charge of engineering.

It was perhaps for these very reasons that the great sixteenth-century proponent of the dignity of arts of *disegno*, Giorgio Vasari, paid relatively little attention to military architecture. In his *Life* of Francesco di Giorgio Martini, he even lamented that Francesco had become too carried away with military concerns. While Vasari wanted to found design in a "science" that adhered to Aristotelian notions of theoretical knowledge, he also characterized design as a manifestation of inscrutable genius. Vasari's *Life of Filippo Brunelleschi* unwittingly lays out the tension between scientific clarity and skillful genius in models made for large public works. The tale is paradigmatic of many tales told in the history of this field.

In any large-scale building project in which government administration was involved, the model appears as an object that is at once clear and contested. That is, the design might clearly show what is to be constructed, and even how. But the predictive quality, whether the actual structure will "work," was in question when innovation was introduced. This is demonstrated in Vasari's *Life of Brunelleschi*. The story illustrates the way in which models served as manifestations of the inventiveness and knowledge of the architect, even as, paradoxically, judgment of them could be a problematic issue. According to Vasari, Brunelleschi secretly worked up models for the building of the duomo on the Florentine Cathedral, including detailed studies of lifting machines and scaffolding and a complete scaled wood model. These he produced at opportune moments. The wardens of the Cathedral, however, remained ever hesitant to award complete authority to Filippo, and the models produced by rival architects continually confused them, even as the cupola rose daily under Brunelleschi's direction.[19]

Vasari's *Life of Brunelleschi* is a story of heroic *ingegno* tied to meticulous *disegno*. Brunelleschi's designs and models are in fact testaments to his *ingegno*. Without them, we would not be convinced how much of the construction of the duomo lay in the direct control of the architect, and how much was contributed by the master stonemasons, carpenters, and metal workers on the job. Vasari makes it clear that every detail of the construction owes its genesis to a sketch or model from Brunelleschi's hand. Or rather from his head, since Vasari emphasizes that the workers were "poor men who worked by their hands," and that work could not proceed without Brunelleschi's direction. Rather, Brunelleschi taught the artisans everything—he constantly made models for the stonecutters and ironworkers, we are told, "from wood, wax, or even turnips." Control over the models meant control over the project from

---

18 Lamberini 1986.
19 Vasari 1963.

top to bottom, and if Vasari could link the design function to high intellectual activity, so much the better.[20]

Models in Vasari's story have a transparency of meaning on one level. For example, Vasari recounts the story in which Brunelleschi refuses to show his designs for putting the cupola on the Cathedral without pillars or other supports during the wardens' convocation of architects, since then the other masters too would know how to do it. Within the time frame of the *Life*, however, models as a class of object appear to have had a murkier aspect. The model of Brunelleschi's rival, Lorenzo Ghiberti, was rewarded with the incredible sum of 300 ducats, but, in Vasari, exhibits little or nothing of use. The models of the other masters are imperfect, and only confuse those whose "fickleness, ignorance and lack of understanding" led them to favor other contenders. But how does one escape such ignorance? How does a warden or patron *judge* a model?

Vasari's story only vaguely accounts for even Brunelleschi's knowledge—his years of study in Rome, where he "drew every sort of building." However, as Vasari himself would tell us (and fifteenth-century contemporaries such as Leon Battista Alberti), Brunelleschi's architecture of the Florentine Duomo is a signal chapter in the innovative approaches fifteenth-century architects took in creating a renaissance world competitive with the ancient one. It was a feat not only in the management of geometrical form, but of the weighty stones that went into its construction.[21] Whether such knowledge could have been "seen" in Brunelleschi's models in and of themselves, as Vasari seems to suggest, is another matter. The Florentine architect's own advice to the "Sienese Archimedes," the civil and military machine designer Mariano Taccola, suggested only that Taccola should not "squander his talent" explaining his inventions to ignorant persons who would only keep them from being heard "in high places and by the right authority." Reveal them only to an "appropriate gathering of men of science, philosophers, and masters in the mechanical art," Brunelleschi exhorted.[22] This is hardly an exhortation to "craft secrecy" as it has sometimes been interpreted, but how such a gathering might be constituted in the fifteenth century, except through the informal coming together of friends, is less clear. An institutional framework for men learned in design and its sciences was in fact to be part of Vasari's vision in his establishment of Florence's *Accademia del Disegno* in 1563.[23]

---

20 Obviously the sort of models that Brunelleschi might have made with wood scraps and turnips are not related to the sort of scaled drawings we address here. However, it is instructive that Vasari employs these stories to build up a general profile of Brunelleschi's design abilities. While Vasari had his own professional motives for presenting Brunelleschi in this light, even in the generation following Brunelleschi's death, Tuccio Manetti also heroized Brunelleschi's models, and his story of Brunelleschi's turnips are the source for Vasari, as are many other elements of the account. Howard Saalman's conservative reading of the documents related to work on the Duomo, however, brings a healthy skepticism to the view of Brunelleschi's enormous superiority over other craftsmen builders. See text and notes of Manetti 1970. Both Manetti and Vasari referred to models as a way of underscoring Brunelleschi's authorship of the building and differentiating him from other building professionals.
21 See Settle 1978.
22 Battisti 1981, 20–21.
23 Barzman 1989; Wazbinski 1987; Ward 1972; Kemp 1990, 132–135.

The *Accademia del Disegno* was in fact a kind of quasi-state institution for the training and organization of painters, sculptors, and architects—men who sought release from the confines of crafts guilds, and whose distinction was precisely their knowledge of design. Patronized by the Medici Dukes, the *Accademia* served as the primary consulting agency on all engineering projects and public works.[24] Officers were appointed directly by the Duke, and the academy fostered military as well as civil architects. Vasari's own nephew, the academician Giorgio Vasari the Younger, collected numerous designs pertaining to military engineering, from various measuring instruments to artillery pieces.[25] Galileo's friend and academician, Ludovico Cigoli, wrote a treatise on perspective that featured military application in its iconography, and the illegitimate Medici sons, Don Giovanni and Don Antonio, were associated with the *Accademia del Disegno* in part because they were expected to be skilled in military architecture. In 1595, when Don Giovanni was serving Emperor Rudolf II in the Hungarian wars, he sent home to Florence for "good engineers and men of design."[26] Buonaiuto Lorini, a Florentine who later became a senior engineer in the employ of Venice, was also encouraged by the Dukes of Florence to master design and to pursue a military career.

Lorini in fact published a treatise, *Delle fortificationi*, which outlined the expectations of a good working design. The purpose of the treatise is emphasized through the inclusion of a dialogue between a count delayed at Zara (in Dalmatia) and Lorini himself. It centers around the problem of determining which plans will succeed, and what qualifications of an engineer are requisite. Design, Lorini suggests, is necessary to all arts, but especially to command.

> On *disegno* depends the true understanding of all things: it enables one to show that great perfection which the *ingengo* of a man may have, who both imitates the wonderful works made by nature and by art, and also shows everyone, and makes them understand each of his concepts. And therefore *disegno* is of such value that whoever masters it can truly say that it is very easy to express perfectly any work that he wants to put forth. This is because through this [*disegno*] are shown all inventions and their elements—approving the good and emending the poor. But also the sites of the country, that is the land, the sea, and whatever natural and artificial features are at work there, are all represented on a simple piece of paper, whether you make your demonstration as things really are, or should be. Also, being able to see the design of things would not only be useful, but necessary—particularly with regard to explicating and making understood our concepts, as it would be. For example, if you wanted to present and make understood the construction in a city that has been done, or needs to be done, simply with words, it would turn out impossible not only to be able to judge its perfections and imperfections, but also to know its true form. On the other hand in *disegno* made with measure, one can make this demonstrable.[27]

This lengthy passage exposes much about expectations on models and the work they were to do. Lorini conceives of *disegno* as a process, which allows one to represent

---

24 Ward 1972, 111–116.
25 The younger Vasari's manuscripts are found in the Biblioteca Riccardiana, Florence.
26 Florence, Archivio di Stato Firenze, Carte Alessandri: Archivio Privato di D. Giov. De Medici, Filza 8, f. 47–49.
27 Lorini 1597, 32. On Lorini at Zara, see Promis 1874, 640.

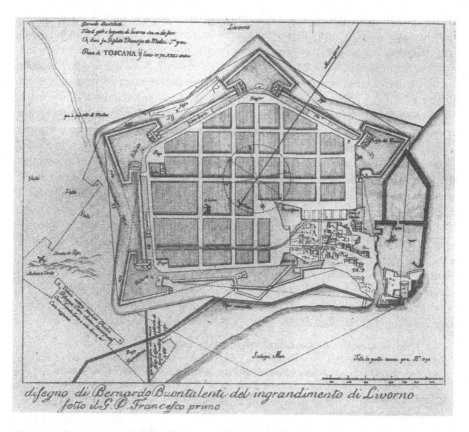

Figure 5.4. This printed plan of Bernardo Buontalenti shows the topography surrounding the city of Livorno, the existing fortifications, and (in dotted line), Buontalenti's suggestion for the enlargement of the city walls. (Anonymous 1796.)

both concepts and things as they really are. *Disegno* here is especially valuable because concepts and things, invention and nature, can *interact together* on the simple piece of paper (figure 5.4). While this may be a specific interpretation of the role of the architect, Lorini highlights here the importance of clarifying one's invention through design. The design can say what words cannot. It can synthesize details so that proper proportions and relations can be seen. Alternatively, a designer might separate details into elements that can be more clearly delineated. Here design takes on the characteristic of transparency. One can *see* true forms; perfections and imperfections are manifest. And yet, Lorini will maintain, such knowledge of design is founded not only on mathematical relations but on practice as well. Design is a kind of touchstone of theory and practice, and should demonstrate the inventor's depth of knowledge in each.

But how would all of this answer the count's initial dilemma: there always seem to be different opinions, and much opposition whenever a fortress is built. How can there be any rules for fortification beyond the various opinions of the different inventors? Lorini is quick to reply that the engineer, like the physician, relies on both theory—("having its foundation and every perfection from mathematics")—and practice.[28] But how can evidence of "practice" also be read off the design? Indeed, this is the unvoiced problem that peeks around the confident statements of a great number of military engineering treatises. How is the local, and individual, knowledge of an engineer's practical knowledge to be assessed, and how should it enter the design process? One answer implicit in Lorini's treatise is that opinions on fortifications are expressed as design, and there are rules for design, expressed in measure. Measure captures not only the theoretical, mathematical feature Lorini constantly alludes to, but it often appears in engineering treatises as a kind of boiled-down experience of practice itself. That is, measurement often appears to stand not only for a particular act of measure, but also for repetitive experience itself.

Lorini's patron, the fortifications governor and artillery general Giulio Savorgnano, claimed to have spent his fortune in the service of the Venetian State so that "I leave nothing but the great quantity of designs and honorable writings that are to my merit."[29] Indeed one of the most copied of Savorgnano's writings is a list of rules by which design could be judged. When Savorgnano wanted to outline rules for the princes and counselors who would decide among designs put forth by soldiers, engineers, and captains, he warned,

> To be able to have a solid judgment on this science [of fortification] is not so easy, nor is it enough to know how to design on paper *cortinas* and bastions in different ways, since there could be proposed many forms and different materials to the Prince who must make the decision. For this reason, the counselors reason well when they hear one [plan], but then when they have to compare those that disagree, the counselors that have never been to war remain confused and irresolute.

Savorgnano followed up with 25 rules for the judgment of fortifications. As it turns out, the guidelines this artillery general and superintendent of fortresses provides to ameliorate a lack of battle experience serves itself as a kind of measuring stick to be held to each proposed design. Almost every one of Savorgnano's rules is a parameter of measure that must be met. For example, the first rule is that the distance between the flank of one bastion to the outer angle of the next (or the protruding point of the bastion) should be 200 passi so that enemy artillery will not be able to shoot at the inside of the bastion from the point of the counterscarp without facing defensive fire. The second rule defines the flank of a bastion at 30 passi since that is how much room would be needed to operate two cannons.[30]

Experience, like the landscape itself, could be represented in measure. It is in this way that designs offered to place before one's eyes what did not yet exist, in relation to both the landscape that did exist and the dictates of experience. Through design, as

28 *Ibid.*, 54
29 Venice, Archivio di Stato di Venezia, Mat. Mist. Nob., B. 5, letter dated 15 May, 1592.
30 Giulio Savorgnano, "Regule delle fortificatione Moderna di Ill.mo et Ecc.mo Sig.or Giulio Savorgnano," Venice, Archivio di Stato di Venezia, Mat. Mist. Nob., B. 13, f. 3^r–3^v. See also Manno 1987.

Lorini suggested, the experienced engineer could determine the best proportions of the parts of a fortification plan, and how it would perform in given situations. But of course the final determination of what plan was "best" did not fall to the military architect. Models were the centerpiece to a circuit of opinions in which the engineer had only one voice, and *qua* engineer (as opposed to as a military officer), not a terribly significant voice.[31]

Models were the first reports sent to senior commanders or engineers, or sometimes the only thing they would see before making their own emendations. So Francesco Maria della Rovere I, the Duke of Urbino and master architect of the Venetian fortification system before his death in 1538, obtained the "models and measures" for fortification at Vicenza from a m[aestro] Agostino so he could prepare to go over the details when he reached Vicenza himself.[32] Another fortifications expert in the employ of Venice, Sforza Pallavicino, apparently made a number of suggestions on models sent to him for various projects in progress within Venetian territories, sometimes without ever having seen the site at all. He had only the designs to rely on.[33] This was of course routinely the case for the final decision-makers, especially in the case of Venice. An entire department of government, the *provveditori alle fortezze*, was instituted primarily to look after and track the flow of these models, and records relating to their progress.[34] Over 40 senators were expected to review the models of one of Michele Sanmicheli's contested fortification projects, and presentations were conducted regularly in the Senate.[35] Hence Savorgnano's "rules."

2. EPISTEMOLOGICAL ANCHORS: CARTOGRAPHIC SURVEY AND GEOMETRIC DESIGN

Measured design was the bond between the experience of the engineer and his pretensions to science. It was also the chain of good faith that bound the lieutenants, commanders, junior and senior engineers, and the counsels of government.[36] It was the means by which decisions that could affect the lives of hundreds of men, or thousands of people, could be made at a distance of hundreds of miles. It is not coincidental that military engineers tried to impress on their patrons the epistemological foundation of their practice—not merely for rhetorical effect, but often in conscientious earnestness. Fortifications designs were as much a product of human imagination as any other building, but they had to fit both a real present physical reality (the terrain of the place to be fortified) and a future possible reality (the eventuality of siege). Military engineers had one oar in cartographic methods, which promised to reduce to scale exactly physical landscapes and standing walls, and another in the geometric methods, which allowed the design of "perfect" forms of desired propor-

---

31 Morachiello 1988, 45–47.
32 Promis 1874, 109.
33 Venice, Archivio di Stato di Venezia, Mat. Mist. Nob., B. 7, ff. 10^r to 20^r.
34 J. R. Hale 1983b.
35 Venice, Archivio di Stato di Venezia, Sen. Terr, Reg. 33, ff. 70^v–71^r and 135^v–136^r. See also Hale and Mallet 1984, 250; see Lamberini 1987.
36 Lamberini 1987, 17.

tionalities. Textbooks extolled the virtues of these practices, and could also serve as a
kind of book of models of various structures and techniques that could be adapted to
particular sites.

The method of Sanmicheli is illuminating. A report on work he did for Duke
Francesco Sforza II related,

> On arriving, Maestro Michele several times went round the site measuring it, then he got
> down to drawing, first in the form it has now, then, after making a separate sketch show-
> ing the additions he proposed for making it as strong as possible, he made a further sur-
> vey to check the levels of the ground with a view to an effectual defense against mines.

Only then did Sanmicheli supply "some finished drawings" to the Duke.[37] It appears
often to have been the case that military engineers drew designs directly on maps of
surveyed sites. One such production in the Marciana Library shows the outline of the
old city walls in pink, the ones recently constructed in green, and the architect's proposed amendments of new bastions and a star-shaped rivelin in yellow.[38] The tendency could be universalized. "Where steps the military engineer steps the
chorographer," wrote the military engineer Lelio Brancaccio.[39] Indeed, survey and
mapping techniques had long been important to the architect's tool box, and was one
of the largest areas of overlap between architects and mathematicians. The practical
mathematical abacco tradition was common training for early modern Italian artifi-
cers, and both mathematicians and architects knew survey techniques.[40]

Measuring instruments, most of which relied on simple triangulation, made measuring by sight easily accessible. Most were adaptations of the astrolabe or quadrant,
but tracts on measuring instruments multiplied, and these were mostly written by military engineers, or authors catering to a military audience. There was the *verghe
astronomiche* of the engineer Antonio Lupicini, the mathematician Niccolo Tartaglia's instrument in *Nova Scientia,* Mutio Oddi's *squadra* and *polimetro,* and numerous others. Lorini suggested readers turn to one of the compendia of measuring
instruments, Cosimo Bartoli's Modo di Misurar (figure 5.5).[41] Measuring by this
choice of sighting instruments must have been quite a boon to professionals for
whom survey was a vital part of any project, and reconnaissance of enemy strong-
holds was a high priority.[42]

The seriousness with which this task was taken is testified by an episode in the
tutelage of Buonaiuto Lorini. Lorini had been sent to Bergamo by his patron Giulio
Savorgnano to make some designs of the city. He proved headstrong and "impatient,"
as Savorgnano discovered when he reached the city. Lorini had mis-measured a curtain wall by 30 *passi*, and was off on some of his angles by five or six degrees.
Employing a particular quadrant, Savorgnano had "busted his brains" to show the

---

37 Hale 1977, 18.
38 Venice, Biblioteca Nazionale Marciana, Ms. Cl. 7, no. 2453 (10493). This appears to be a late
    seventeenth-century manuscript detailing information on the fortifications of Venice.
39 Lelio Brancacio 1590.
40 Rossi 1998; Veltman 1979; Adams 1985.
41 Lupicini 1582; Tartaglia 1537; Oddi 1625 and 1633; Bartoli 1564.
42 See for example Adams et al. 1988.

Figure 5.5. Instrument for measuring angles. From Bartoli 1564. Bartoli offered a compendium of mathematical techniques of measurement.

younger engineer his errors. He had Lorini re-do his model, and Buonaiuto produced some excellent designs. Full of fury, and horrified that some of the faulty designs had reached the patrician Giacomo Contarini, Savorgnano wrote the patrician, "If I were an absolute ruler like the Duke of Florence, I would have [Lorini] castrated, as they do to unmanageable horses who become good once they are castrated."[43] But it was probably Lorini's hands that Savorgnano wanted to rein in. Despite Savorgnano's rage over Lorini's faulty measuring, he probably needed the Florentine's facility in drawing in order to make fine, detailed plans. Lorini was one of the most talented of apparently numerous men of design Savorgnano kept as "familiars."

43   Venice, Archivio di Stato di Venezia, Archivio Proprio Contarini, B. 19, dated 6 August 1586.

If the cartographic survey formed a basis of models, the geometrical construction of the architect's projected fortification on top of it was the most prominent feature. In teaching design to young architects and *cavalieri*, this aspect of the practice also became freighted with theoretical references. Girolamo Cataneo emphasized that this "most divine, most certain, most useful science of numbers and measures" was the basis of all arts, but especially military discipline, in which it was needed to order battle, design fortifications, and understand artillery.[44] Cataneo's repeated association of his design methods with the great "Platonic philosopher" Euclid was also featured in the engineering treatise of Cataneo's pupil, Giacomo Lanteri.[45] We can see the geometrical progress of Antonio de Medici, the natural son of Francesco I of Florence, under the tutelage of the court mathematician, by don Antonio's copy-book. The compass marks etch the geometrical construction of regular polygons, from a triangle to sequentially greater-number-sided shapes. The final exercise is a pentagon on which regular bastions at each angle have been constructed. Ostillio Ricci, better known as Galileo's teacher, also taught his pupils, pages at the Florentine court, from Leon Battista Alberti's measuring tract, and is associated with a work on the instrument known as the *archimetro* (figure 5.6). And so it came to pass that mathematicians—like Cataneo, and Galileo's teacher, Ostillio Ricci, and Galileo himself—became teachers of aspiring young architects and well-born soldiers.

The emphasis on geometry and measurement—discussion of which was invariably studded with references to certainty and to the philosophical heritage of this knowledge, sometimes going back to the Chaldeans—accomplished two things. First, it served to put architecture on a par with liberal arts; this marked architects above common workmen, and made the duties of architecture more consistent with the education of noblemen and leaders. Second, knowledge of geometrical elements served, as Lanteri has Cataneo say, as an "alphabet" for learning design.[46]

Knowledge of measuring instruments and the Euclidean basis of the geometric design of fortifications was the knowledge shared between engineers, their military patrons and their sovereigns, regardless of their own war experiences. At top levels, engineering was often seen as a principal aspect of the role of a captain general or of an artillery general.[47] Fortification design was a topic of avid interest for princes such as Francesco Maria della Rovere and Charles V, who corresponded and disputed about the topics of where to plant artillery, where to put the gate in a curtain wall, or how to cover flanks with fire from bastions.[48] Numerous other princes were also fascinated by the new fortifications methods, as often indicated by the prefatory remarks of engineering treatise writers. Some had a personal stake in the matter, as was the case for Cosimo I of Florence whose control over fortifications in his own territories was a hard-won political right granted by the Hapsburg emperor who had supported

---

44 Girolamo Cataneo 1584, III f. 2^r.
45 Lanteri 1557. Euclid the geometer was confused with the Platonic philosopher Euclid Megarensis throughout the sixteenth century.
46 Lanteri, 7.
47 For example, see in Savorgnano's collection Venice, Archivio di Stato di Venezia, Mat. Miste Nob., B. 14, f. 1^v; Collado 1606, first dialogue.
48 Promis 1874, 114

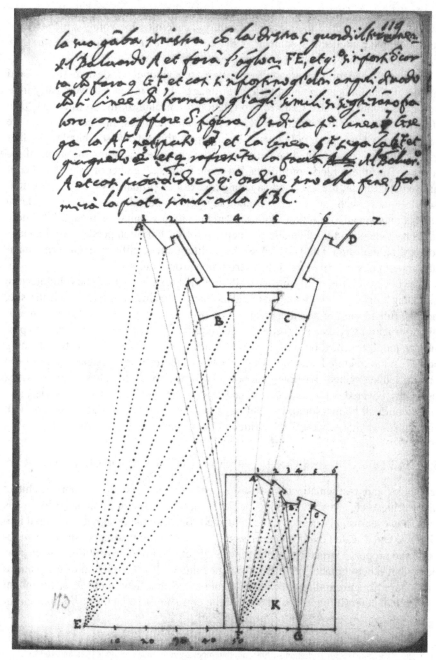

Figure 5.6. The fortification is reduced exactly to scale. From L'Uso Dell'Archimetro, attributed to Ostilio Ricci. (Florence, Biblioteca Nazionale Centrale Firenze, MS Magliabechiana VII 380, f. 118 ᵛ; courtesy Ministero per i Beni e le Attività Culturali, all rights reserved.)

the Medici rule.[49] Others were themselves military figures, as were della Rovere and Sforza. Near the end of the sixteenth century, academies for young *cavalieri* also multiplied. The intellectual content of these schools centered around the first six books of Euclid, and a plethora of measuring tracts.[50]

Geometry gave military architects clarity if not soundness of design. They shared a mode of communication and a lexicon of images. Even in the relatively early treatise of the architect Pietro Catanco (no relation to Girolamo), the fortifications take on their "orders" from the number of sides of their enceintes. In Cataneo, these are always depicted as regular polygons.[51] In large part, this is because a number of other design choices are predicated on the sides of the enceinte, including the shape of the bastions. Thus the "orders" of fortification came from the number of sides. Even Giovan Battista Zanchi, the seasoned captain who extolled primarily practice in his treatise, followed such an ordering, and suggested that those wanting to learn fortifications spend some time with an "architectural manual of models."[52] These would provide not only schematic geometrical forms, but often perspective drawings "to show all that should be built." With the proliferation of military engineering manuals, copy-books incorporated designs from numerous sources.[53]

However, even the most geometrically-inclined textbook of military architecture (including Lorini's) would warn its readers that all fortification had to suit the site, and take full account of possible means of offense and defense. Survey and geometrical construction were the propaedeutic arts required by anyone who expected to produce or judge fortification designs. They were, of course, insufficient for planning models. What mattered was how well adapted to the task the designer had utilized fortification elements—bastions, cannon platforms, exits, trenches, curtain walls, casemates, covered ways, and more. The heights, wall thickness, bastion angles, distances could all be measured, but had to be measured by knowledge outside design. Nothing shows this so well as the written opinions about these designs.

3. WHAT DESIGNS ARE GOOD ENOUGH? CONTOURS OF CONTROVERSY

Textbooks and mathematicians offered instruction on how the military architect might achieve clear, measured models. What made the design good and why it was good, however, usually needed to be explained. Because the design itself should ideally have had a scaled one-to-one correspondence with the landscape, standing structures, and proposed structures, the model could act as a kind of substitute object. This meant that other engineers and commanders could assess the relation of the planned fortification to surrounding landscape features, and that they could in effect mount assaults and defenses on it. That the measurements should stand the test of this sort of

49 Diaz 1976, chap. 2.
50 Hale 1983c and 1983d; Torlontano 1998; for contemporary reference see Oddi 1625 and 1633.
51 Pietro Cataneo 1567.
52 Zanchi 1554, 56.
53 Oxford, Bodleian Library, Canon Ital. 289, "Regole di Fortificatione" is one such copy-book. It is comprised almost entirely of designs copied from Lorini, Tensini, Brancaccio, and others.

analysis shows us that the knowledge embodied in a design was much more extensive than a simple lexicon of forms. But that knowledge was almost never of a certain nature, despite its expression in the "certain science" of geometry, and therefore designs also embodied opinion.

Fortifications textbooks consistently regaled engineers with precepts on how to derive (*cavare*, literally "to dig out') the form of a fortification. The form is derived from the landscape; the defense is derived from the offense, the length of wall between bastions is derived from the distance artillery shoots; the angle of bastions is derived from the number of sides. That is, nearly every dimension is designed in reaction to conditions or factors that may or may not be represented. One had to see not just a model, but the way walls might hold up to the expected barrage of artillery, and the maneuvers the design would make possible given various tactics of the enemy. And of course, costs were measured, too. As a whole, the fortifications design had to be inspected as a machine design might, since the fortress increasingly became machine-like in that all elements had to relate to each other for mutual defense.[54] Unlike most machine drawings of the sixteenth century, however, fortifications were drawn to scale, and usually included actual measurements.[55] Measurement became the terms in which debate could be conducted. All measures—the length of curtains, area of bastions and platforms, depth and width of moats, among many others—also had to be related both to environmental factors and to each other, and that is where "seeing" and controversy crossed paths. Here I will look at just one exchange in a debate over the goodness of a design that was a single salvo in a long battle between rival fortifications experts.[56]

After the Venetians lost Cyprus, they were faced with dwindling control over their Adriatic possessions. The fortification of Corfu and Zara became incendiary topics in the late 1570s. Corfu was routinely figured as "the principal frontier of our empire as well as of all Christendom," and Zara was the Venetians' chief trading and naval base on the Dalmatian coast.[57] Venice had its top fortifications experts and engineers working on these projects, and also hired, as many other states often did, an outside inspector and consultant. Ferrante Vitelli was given leave from his post as Governor-General for the Duke of Savoy to be able to inspect the strategic Venetian sites where stronger defenses were to be erected, and to review the models for construction. Savorgnano was also involved in the Corfu and Zara projects and sided with the Governor-General, Sforza Pallavicino, against Vitelli's opinions. Pallavicino had been involved with the Corfu fortifications early on; he had sent his opinions on a first phase of fortification there before the fall of Famagosta (1571), and had by 1578 made plans for expanded defenses at Zara.[58] The two experts' criticism of each other would become bitter and very public. Already in October of 1578, Pallavicino had

---

54 Hale 1983a; Manno 1987; Wilkinson 1988.
55 See McGee 1999; Poni 1985; Marchis and Dolza 1998.
56 See debate over Palmanova in La Penna 1997.
57 Hale and Mallet 1984, 443–444.
58 Venice, Archivio di Stato di Venezia, Mat. Miste Nob., B. 7, ff. 18^{v}–20^{r}. For Pallavicino's career, see Promis 1874, 447–463; Mallet and Hale 1984, 302–308.

expressed disgust with Vitelli's own designs for Corfu, and the Governor-General was further irritated when he received from the Venetian Signoria Vitelli's criticism of his model for Zara along with a re-worked design. What designs were good enough? According to Vitelli, Pallavicino's was not.

Vitelli had "corrected" a number of elements, Pallavicino wrote back to the Signoria, in a way that would only end up being less effective, and much more costly. Vitelli had not "seen" that most of the problems he addressed had already been accounted for in the design. For example, Vitelli objected that one could enter a bastion through the opening in the trench where the gate was, "but did not see that the cannon platform near the lobe of the bastion (*orrechie*), and the platform on the bastion, could not be entered," since they were covered from the point of the opposite bastion—"a feature he didn't have on his fortress in Corfu." Moreover, according to the plans sent by the Signoria, the lobe that Vitelli had re-designed for the Grimano bastion did not give direct [access] to the inside of the structure, "as your Serenity can measure with a line, and will be able to know whether I'm telling the truth or not." Also, Vitelli had written that the shots from the bastion Santa Marcella would not be longer than 230 passi, and so would not cover the entire length of the wall. First of all, Pallavicino pointed out, the distance [between Santa Marcella and the next bastion], even if it was a little out of range for the archebuses [to strike] "in full force," was still within range for the archebuses to do some damage. More importantly, "the defect of this distance" was already corrected for in his designs by the *dente* (jagged protrusions) in the curtain, which would only be 160 passi away. "I don't know if he did not see it, or if it slipped his mind," Pallavicino noted. Also, the Signoria will note that in Pallavicino's original design, there is a *piazza* for at least three artillery pieces, which will also be well-protected by the *dente*, being higher. Vitelli had criticized Santa Marcella for being too low, too short, and the front angle too acute. However, the shortest side was 30 passi—the same measure as a side of the Citadella bastion in Vitelli's Corfu design; Santa Marcella's longest side measured 60 passi, and the angle was obtuse.[59]

Pallavicino's reply to his Venetian employers was an attempt to convince the Venetian patricians to see as he did by drawing their attention to the significance of the measures in his model, and how they relate. We can see here that measures are not so clear-cut, even apparently in comparison to guidelines like Savorgnano's. Several things are of note. First, Pallavicino and Vitelli continually refer to actions that take place on the design as if it were a real structure. The exception to this is at the one point when Pallavicino wants measurement to be a truth-telling exercise, and invites the senators to measure for themselves on the plane of the design whether there would be the necessary access to a bastion. Second, some measurements sometimes appear to have some obvious meaning that all the actors must have recognized. In particular, the length of the short side of the Santa Marcella bastion (30 passi) had appeared in Savorgnano's rules as the minimum room necessary to mount two cannons on a bastion. The fact that the angle of the bastion is obtuse also relates to what

---

59   Venice, Archivio di Stato di Venezia, Mat. Mist. Nob., B. 9, ff. 44^r–52^r.

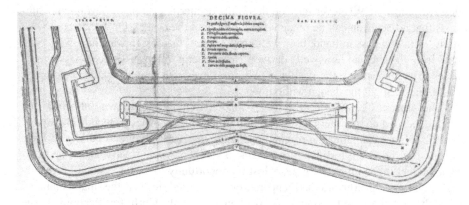

Figure 5.7. Lines of fire on a fortification plan. While these lines were regularly drawn on models in textbooks, private collections, and battle representations, they are seldom found drawn directly on a design submitted for a commission. From Girolamo Cataneo 1584.

was by 1578 a precept. A wide angle on a bastion gave it much less vulnerability at the tip where there could be a spot blind to defensive fire. On the other hand, when the difficulty of the length above 230 passi came up, Pallavicino first responded with a consideration of relative force with which an archebus shot could be made at that length. As Antonio Manno has pointed out, artillery shot was the unit in which a fortification was measured. However, the variety of positions and kinds of artillery, combined with their variable impact for various reasons, meant that this unit could not be a single standard. Nevertheless, in the language of design, measurements were the terms by which questions of tactics could be addressed (figure 5.7).

Vitelli, as a veteran inspector, had his own list of considerations for planning fortifications. It ran over thirty pages long, and was an attempt to enumerate everything that should be looked at during a site visit, and how to imagine every possible means of offense and defense in every possible landscape, under every possible administrative condition.[60] What is interesting is that what could often come out at the other end of this process—the examination of an exhaustive array of variables and considerations—were rules (measures) for judgment found in the writings of Savorgnano, Pallavicino, and Orsini.[61] That is, the "rulebook" measures could only be an approximate control on the infinite uncertainties of tactics and artillery performance.

Direct measure in and of itself did not decide anything; it only provided the basis on which one could form an opinion. Savorgnano commented on one of Vitelli's presentations to the Venetian Signoria, "These engineers of yours, whom I like to call "soffistici," want to make everything sound black-and-white, and do not understand the principles of things that make designs and not words, so that with a compass in hand, one can make them see their errors, or if they design something well, they

---

60 Vitelli's "Istruzioni" are transcribed in Promis 1874, 606–638.
61 Orsini's rules are collected in Venice, Archivio di Stato di Venezia, Mat. Mist. Nob. Pallavicino's military precepts are included in Ruscelli 1568, fol. 55–56, and Promis 1874, 447–462.

deserve praise."[62] Whatever may be the faults or merits of a plan, Savorgnano suggests that no decision can be rendered at all without the design. Only then could one study, with compass in hand, how well proportioned the design was, how well "dug out" from the site and how well "derived" from enemy offense. The compass was as necessary to "reading" the design as were the eyes, but neither eyes nor compass could reason about the design without words.

The process by which a design was finalized could be a drawn-out affair, and often depended on the internal organization of the state *vis a vis* its technical corps.[63] By 1582, four years after the original exchange between Vitelli and Pallavicino, the question of the Corfu fortifications became an embarrassment. Savorgnano and Pallavicino had in the intervening years lost no opportunity to criticize Vitelli, privately and publicly. Savorgnano had apparently even tried to get the mathematician Guidobaldo del Monte to publish a critical analysis of the Corfu fortifications.[64] The Senate declared that two provveditors-general should go to Corfu with the engineers Lorini and Bonhomo "because for many years now it has been known to all rulers and in all places that various objections have been raised to the fortifications in Corfu." Venice wanted to have these differences of opinion resolved "for the benefit of Christendom and ourselves."[65] Both Lorini and Bonhomo were clients of Savorgnano, and it would be their job to help the provveditors see in stone what could not be decided in design.

4. WHAT IS GOOD DESIGN? PRINCIPLES

The first person to claim in print that one might be able to judge the design of a fortress based on "principles of things that make designs" was the practical mathematics teacher, Niccolò Tartaglia. Tartaglia attempted to turn the arts of artillery into a mathematical science in his *Nova Scientia* of 1537.[66] He extended his researches into military arts, and published a series of didactic dialogues in *Quesiti et Inventioni* nine years later.[67] Tartaglia expressed the dream of judging designs merely by looking at them in the first dialogue of the sixth book of *Quesiti*. The first question revolves around the fortifications of Turin. The military engineer and artillery general Gabriel Tadino asked Tartaglia, was this not proof that "the *ingegno* of man has reached at present the most sublime level that it is possible to reach?" Although admitting that he has no experience of fortification at all—indeed has only seen a few—Tartaglia replies that the *ingengo* of a design is known only by the form, not the material. Having only his mathematical wits to rely on, Tartaglia had to insist that whatever the strength of the material construction, the flaws of the design itself were manifest in

62 Opinion on Vitelli (and other engineers employed by Venice) in Venice, Archivio di Stato di Venezia, Mat. Mist. N., B. 11, f. 210, quoted in Manno 1987, 228.
63 For Florence, see Lamberini 1986, 1987 and 1990; Casali and Diana 1983, Fara 1988.
64 Henninger-Voss 2000, 248–250.
65 Venice, Archivio di Stato di Venezia, SS. Reg. 83, 58–58^{v} (16 Feb), quoted from Hale and Mallet 1984, 444.
66 Tartaglia 1537.
67 Tartaglia 1546.

form only. Form by itself could add to the "strength" of the walls. To Tadino's insistence that Turin is as well fortified as any city could be, Tartaglia answers, "In this your reverence is greatly deceived."[68]

Tartaglia's analysis of the Turin fortifications is primarily based on its quadratic form and the proportionate smallness of its bastions. In order to make the heaviest strikes against a fortification wall, the cannonballs must strike perpendicular to the wall. With a quadratic design, there would be ample opportunity to do so. The small, acute bastions could only offer inadequate covering fire to the curtain walls. On the one hand, it is difficult to believe this was such disappointing news to Tadino as Tartaglia represented it to be. The quadratic form was already out of favor among fortifications designers, and Tadino had a long distinguished career as a fortifications superintendent and artillery general for Venice, Malta, and the Holy Roman Empire.[69] On the other hand, the Turin fortifications purportedly gave the city the aspect of an austere stronghold, and were something of a conversation piece.[70] In the 1540s new four-sided fortifications were still being built, or at least refit, such as those at Pistoia. However, by the next generation of engineers, four-sided fortifications would be entirely condemned.[71]

It is not so clear whether this rule of fortification was spread more by participation in the profession or by books such as Tartaglia's. In 1554, Giovan Battista Zanchi authored the first published architectural treatise devoted entirely to fortification.[72] He drew heavily on Tartaglia, so that even the small "map" of the Turin walls, represented as a perfect square with tiny bastions, seems to have been drawn from the mathematician's earlier book (figure 5.8).

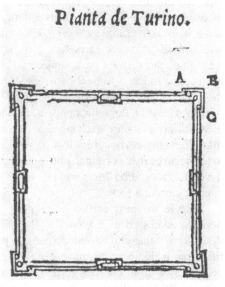

Figure 5.8. The plan of Turin from Tartaglia 1546.

An interesting aspect of Tartaglia's exchange with Tadino in *Quesiti et inventioni* is that Tartaglia constantly refers the general back to his drawing board. In the questions following the discussion of Turin, Tartaglia suggests that there are six "conditions" for an unassailable fortification. Tadino challenged Tartaglia "who can neither design or plan nor build a model" to explain them. Yet after naming each (for example, that no enemy be allowed to shoot perpendicularly to any walls), Tartaglia invari-

68 *Ibid.*, 64[r]–65[r].
69 See Promis 1874, 41ff.
70 Pollak 1991, chap. 1.
71 E.g. Savorgnano in Venice, Archivio di Stato di Venezia, Mat. Mist. Nob., f. 13, see 2[r]; Alghisi 1570.
72 Zanchi 1554.

ably refers the solution (and the explanation itself) to what could be seen on a model. Tadino protests that he has never seen a fortification that could not be battered perpendicularly; Tartaglia replies, "You could see in a design how this could be achieved."[73] Tartaglia claims that the bastions and *cortina* could be designed in such a way that the enemy would be forced to plant their battery only in a disadvantageous spot near the bastion. Tadino promises to return in a few days with "a plan designed by my own hand" that would fulfill these conditions. Tartaglia expresses his confidence in the general: anyone advised by Tartaglia could make such a design, much less Tadino "who is the height of *ingegno*."[74] As it turns out, however, the mathematician "who can neither plan nor build a model" has one, and slowly he begins to entice Tadino with it. After Tartaglia states his third and fourth conditions (that ordered battle arrays be vulnerable to defensive fire in four directions and that if the enemy does batter down a wall, it would be too dangerous for them to enter), Tadino "begins to imagine" how this could be done. "But I want to consider this better, and make a little model, because in making models a thing is better clarified, and then we will see if my opinion is the same as yours." Tartaglia answers again that in Tadino's design he will probably discover Tartaglia's own models. The general adds that "the practice, reasoning, and dispute on a subject makes one discover many things ... and one becomes introduced to new particulars that afterward he can think over and easily discover."[75] The process of design itself reveals relationships, but that process is itself—at least in Tartaglia's hopeful depiction of it—tied up with a larger discourse in which novelties are exchanged and examined. By the end of this series of questions, Tartaglia coyly agrees that he will only give up his models to Tadino himself. What he gives in his original 1546 edition, however, is primarily a list of dimensions: the cortina should be 7 feet wide up to 10 feet high, then tapered to 2 feet, but with a counterscarp of 8 feet ...

Only in the next edition of *Quesiti* (1554) did Tartaglia provide some actual designs, addressed as a *gionta* to the philosopher Marc Antonio Morosini. These designs are rather clumsy, but do put into publication some of the newest ideas in fortification, including a bastion with a lowered cannon platform that appears to have been derived from one designed by Antonio da Sangallo for a Roman fortification (figure 5.9). Tartaglia, who apparently counted a number of architects among his disciples, admitted that this design was in fact supplied to him by a student. A later sixteenth-century annotator of the *Quesiti* noted that the very same bastion was to be found in the treatise of Carlo Theti, and that Theti had probably stolen it from someone else.[76] The annotator bemoaned the state of such "novelties." I think, however, such comments misunderstand the nature of Tartaglia's "inventions."

---

73 Tartaglia 1546, f. 65^r.
74 *Ibid.*, f. 66^r.
75 *Ibid.*, f. 66^v.
76 Exemplar in Museo della storia della scienza in Florence, once believed to have belonged to Galileo.

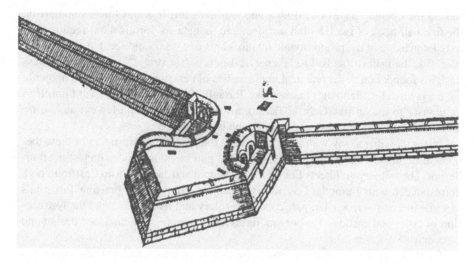

Figure 5.9. A plan of a bastion reproduced in the 1554 edition of Tartaglia 1546. The bastion appears to involve the innovations of Antonio da Sangallo from a few years before.

Tartaglia prized not his own design ability, but his ability to reason over designs, and put them in a discourse that could analyze and mathematize the design problem. His own work on ballistics with which the major portion of *Quesiti* is concerned could have had real design effects. For example, in the first book, he attempted to prove that a cannon placed in a valley an oblique 100 passi bellow a fortification would make more damage shooting upwards than a cannon placed on a hill directly across from the wall and only 60 passi away. This of course would hardly be an expected result. Tartaglia analyzed the common gunner perception that perpendicular hits caused the most damage in mathematical terms of positional gravity. By referring readers to weights on a balance, Tartaglia showed that the angle of impact mattered less than the speed and trajectory of the cannonball. Thus the mathematician claimed to correct faulty extensions of practical knowledge as in the case above.[77] Evidence that such studies *did* become considerations of designers is indicated by some of Tartaglia's plagiarists. When Walter Ryff published large portions of Tartaglia's books translated into German, it was as part of a large quarto volume that was meant to serve the total geometrical architect, and was published as one section in larger treatments of perspective and architecture.[78]

This study of artillery impact through philosophical traditions of analysis was one that would be developed for the remainder of the century by mathematicians up to and including Galileo.[79] Cataneo's Euclidean "alphabet" was the basis of design—

---

77 Henninger-Voss 2002.
78 Ryff 1547 and 1558. This was the second and third part of a large volume that also included other architectural treatises that drew on mathematical sources. A facsimile edition exists: Hildesheim/New York, 1981. I owe this reference to Rainer Leng.
79 Laird 1991.

indeed the opening pages of Lorini's and Galileo's fortifications tracts summarized the first half book of Euclid—but architectonic thought for fortifications required not only knowledge of the proportionate relationships that one could see, but also begged ones (like ballistics) that had to be analyzed in separate study.[80] It was nevertheless design's foundation in the practical mathematics of design and measure that no doubt gave a practical mathematics teacher like Tartaglia the foothold to present himself as an advisor to a man like Gabriel Tadino, a man so distant in profession and so far Tartaglia's superior in status, experience, and education.

It is impossible to say with any surety what the precise manifestation of these theoretical excursions on the impact of moving weights could have been on fortifications design. The military architects Francesco de Marchi and Jacopo Fusto Castriotto both corresponded with Tartaglia. Castriotto opened his letter floridly praising Tartaglia's *ingegno* in every science, but particularly "artillery and its effect ... and the fortification of cities and castles." He begged Tartaglia's opinion on an enclosed design and discourse.[81]

On the other hand, a less impressed Giulio Savorgnano, who had been featured as one of the interlocutors in *Quesiti,* had his dwarf send Tartaglia some further questions, "to make him figure out some delightful things."[82] The questions are almost entirely questions of what the cause of some effect is, like why the compass does not point to true north, or puzzles, like whether a mortar will have a greater effect if it is shot from a height down, or from a low position upwards. They are questions that either no one knew, or that Savorgnano seems already to have known either from his own battle experience, or from the vast number of experiments he performed with artillery at his own castle, Osopo. They are challenges to Tartaglia's claims to discover causes of what is already known by experience, and to be able to know by mathematical analysis what other men find out only through direct experience.

As a way of "ratiocination" that predicted the outcome of experience, Tartaglia's books were useful as teaching aids for readers of little experience, and even as a guide for those who had much. They became standard references for artillery manuals for another century. Even in eighteenth-century France, Tartaglia's relatively crude conceptualization of projectile motion appears to have been employed by common gunners, while Academy-trained artillery officers and designers were trained in the new rational mechanics that had developed out of the Galilean and Newtonian traditions. And the merits of this mathematical education were still unclear.[83] Tartaglia himself would have thought of himself as helping soldiers "understand the principles of things that make designs." But it is also clear that Tartaglia wanted to affect the understanding of practice not only of men like Gabriel Tadino and Giulio Savorg-

80 For identity of the Lorini–Galileo tracts, see Lamberini 1990.
81 Urbino, Urbino Università, Fondo del Commune, B. 77, fasc. VI, dated 27 Dec. 1549.
82 The questions are copied in Venice, Archivio di Stato di Venezia, Mat. Mist. Nob., B. 13, ff. 55ᵛ–57ᵛ. They are apparently misdated, however, as 1542, since question 26 (57ᵛ) begins "I saw what you said about granulated [gun]powder," a discussion that could not have been "seen" until the *Quesiti* of 1546. For serious views of the relation between Savorgnano and Tartaglia, see Manno 1987; Ventrice 1998; Keller 1976a.
83 Alder 1997, 93 and *passim.*

nano, but also of decision-makers—such as patricians and secretaries to princes—who were constantly called upon to judge design, but who often had no other lens with which to view them other than the teachings of experts and books.[84]

***

Recent studies of engineering history have focused on the visual basis of the "knowledge" of the engineer, and have been skeptical of the contributions "theory" has made to engineering practices. Eugene Ferguson, for example, specifically relates specialized training and theoretical argumentation to the engineer's need to legitimize his services before patrons.[85] It seems to me, however, that the intervention of practical mathematicians like Tartaglia was a natural extension of the design practice of architects, which was itself a mathematical practice. Certainly Tartaglia saw this as one point of entry for his services. Even if men like Giorgio Vasari had consciously attempted to heighten the status of design professionals by uniting them and claiming a theoretical basis for their practice, that practice had for a long time been sophisticated geometrically, as Wolfgang Lefèvre and Filippo Camerota have so amply shown, right down to the art of the stonemason. What changed in sixteenth-century military architecture was that the design process was shared by a far greater variety of men; models were evaluated within a more developed hierarchy (indeed were central to new bureaucratic agencies); and that the nature of its goals were tied to the concurrent, but exterior, developments in strategy and artillery. The spheres of knowledge broadened which, in the terminology of Michael Mahoney, had to be "bootstrapped" in the construction of fortifications. While engineers continued to employ a language of "seeing" in their opinions on designs, their drawings referred in their measure, execution, and planning to an ever more complex "architectonic" synthesis of geometrical techniques, strategic imagination, and newly emerging sciences of weights in motion.

---

84 [Henninger-]Voss 1995; Cuomo 1997.
85 Ferguson 1992, 65–73.

PART IV
PRODUCING SHAPES

INTRODUCTION TO PART IV

The topic of this section is the origins and development of the geometric techniques used for drawings and geometrical constructions employed in the process of planning and producing devices or buildings. Thus, it also discusses the origins and development of such techniques as an integral part and epitome of the professional knowledge and skill of engineers and architects.

Early modern depictions of machines, on which most of this volume's chapters focus, do not immediately make accessible the wealth of drawing techniques employed by the engineers and architects of the age. Apart from a few orthographic plans of machine parts, since the days of Taccola machines were usually depicted in a style that looks like a kind of naïve perspective rendering at first glance. No geometry seems to be needed for such creations. On closer inspection, however, it becomes obvious that these depictions not only use deliberately advantages of rendering in perspective, in particular as regards the choice of the viewpoint, but often employ different projection techniques such as military perspective. When one starts to pay attention to these technical details, these seemingly naïve depictions of machines appear embedded in the whole range of geometrical projection techniques that were partly rediscovered, partly reinvented, and in any case autonomously developed and elaborated in the Renaissance—such as perspective projection, orthographic projection, combined views, oblique projection, special projections in the realm of cartography, and stereometric constructions for special architectural purposes. Taking into account as well that engineering, architecture, and, as figures like Leonardo, Dürer, Raphael, and Michelangelo prove, even fine arts were trades not as secluded in this age as later, it becomes clear that the different drawing techniques developed in close connection. They form a whole of techniques and procedures for the production of technical drawings developed by and available to the broad community of engineers, architects, and artists.

The chapter by Filippo Camerota traces in detail the development of the most important methods employed for technical drawings—linear perspective, military perspective, shadow projection, double projection, and stereometry. The chapter's emphasis lies on the codification of each of these methods, that is, the codification of both a method's theoretical principles and the procedures and techniques that can legitimately be applied in its pursuit.

The distinction between drawing methods or styles such as perspective or orthographic projection, on the one hand, and of techniques and procedures through which they were realized, on the other, deserves attention. The whole of drawing methods available to engineers, architects, and artists consisted not only of a variety of styles, each of which had different advantages and shortcomings. It also was a many-faceted whole as regards the techniques and procedures used to achieve a rendering in each of these styles. To give an example: to trace a true perspective rendering of an object, one can construct it according to the rules of geometry and geometrical optics—if one has sufficient command of these rules and of the construction techniques required by them; if not, one can use mechanical or optical aids and achieve the same result.

Techniques and procedures of drawing thus can be distinguished according to the knowledge that they require on the part of the draftsman. It makes sense to speak of learned drawing techniques in contrast to more practical ones. To a large extent, this distinction between learned and practical drawing procedures coincides with that between learned geometry and practitioners' geometry. In our context, this distinction is of particular significance in two respects—with regard to the roots of the different drawing styles and techniques, on the one hand, and, on the other, with regard to the actual employment of these different techniques by engineers, architects, and artists in their professional practice.

The chapter by Wolfgang Lefèvre treats the question of the roots. It studies Antonio da Sangallo's and Albrecht Dürer's invention of the combined views technique, a method of plan construction on which modern plan constructions essentially rest, and traces its learned background as well as its roots in medieval practitioners' geometry. The chapter by Jeanne Peiffer discusses the practical employment of drawing techniques of a learned character. In pursuing the fate of highly developed drawing techniques elaborated by Dürer among artists of sixteenth-century Germany, and further up to their mathematical codification by Desargues at the beginning of the seventeenth century, it traces a history of a gradual divorce of learned and practical drawing procedures.

6

RENAISSANCE DESCRIPTIVE GEOMETRY: THE CODIFICATION OF DRAWING METHODS

FILIPPO CAMEROTA

The need for drawing as the visual expression of ideas forcefully emerges in the first treatises of the Renaissance, particularly in the field of architecture. As Leon Battista Alberti wrote, "the drawing will be a precise, uniform representation, conceived in the mind ... and brought to completion by a person endowed with ingenuity and culture."[1] It appeared obvious, and not only to a humanist such as Alberti, that drawing was considered a cultivated language, accessible only to those who knew its basic rules. According to Filarete, "knowing how to read a drawing is more difficult than drawing," so that, while there were many "good masters of drawing ... if you ask one according to what rule he has drawn a certain building ... he will be unable to tell you."[2] A good architect, then, had to be both "practical and knowledgeable," in the words of Francesco di Giorgio;[3] that is, he had to possess an adequate *dottrina*, or knowledge of the disciplines essential to his profession; he had to have *ingegno*, or inventive talent and the ability to solve technical problems; and he had to know *disegno*, or drawing, in order to correctly represent the product of his ingenuity and to convey his thought immediately and unequivocally to others. These are the basic premises underlying the extraordinary efforts expended by Renaissance theoreticians to codify the rules of graphic language that laid the foundations for what, near the end of the eighteenth century, Gaspard Monge was to call "descriptive geometry."

This work of codifying continued over a rather long span of time and was of course favored by the spread of printing and the rise of the academies. In this chapter the various stages of codification in writing will be examined in chronological order, grouped under such general headings as painters' drawings, military engineers' drawings, the projection of shadows, architectural drawings, and those of stonemasons.

In the *corpus* of Renaissance treatises, the intention of instructing an architect, an engineer, or a "practical and knowledgeable artist" emerges from the structure conferred on the development of the themes, which range from elements of practical geometry and workshop expedients to geometrical demonstrations based on Euclid's *Elements*. Considering the great abundance of *ingegno* in the construction sites and workshops—as brilliantly demonstrated by works of art and architecture—the theoreticians' objective was that of forming a *doctrine* of drawing that not only would serve to refine the techniques of the profession but also, and more importantly, to raise the intellectual status of those whose art was grounded in geometry. This need

---

1   Alberti 1975, I.1, transl. from Alberti 1988.
2   Filarete 1972, 157–158.
3   Francesco di Giorgio Martini 1967, VI 489–490.

was felt not only on the artistic but also on the technological level, where verbal descriptions of the complicated mechanical devices that drove worksite and war machines were now intentionally corroborated by the great descriptive value of drawings. This significant change with respect to medieval tradition is well documented in the manuscripts of Mariano di Jacopo, known as Taccola, and Francesco di Giorgio, where a progressive perfecting of techniques of representation may be noted, undoubtedly favored by the progress then being achieved in the field of pictorial perspective.[4] Francesco di Giorgio's appeal to the need for *drawing* as the visual expression of *ingenuity* is echoed by the even stronger admonition of Leonardo, who in one of the Windsor folios points out the inadequacy of words to describe the works produced by man: "But let me remind you that you should not stammer with words … since you will be greatly surpassed by the work of the painter."[5]

1. PAINTER'S GEOMETRY: THE DEVELOPMENT OF LINEAR PERSPECTIVE

The first attempt to establish rules and principles for geometric drawing was made by Leon Battista Alberti, who between 1435 and 1436 composed two works, closely linked and addressed primarily to painters: *Elementa picturae* and *De pictura*.[6] The title of the first book explicitly recalls the important treatise by Euclid, which had been the chief reference point for the mathematical sciences since the Middle Ages. Euclid's *Elementa* was a fundamentally important text in both the universities, where it furnished the geometric basis for studies on optics, astronomy and the other arts of the *quadrivio*, and in the abacus schools, where practical mathematics was taught to youths who were to go on to commerce and the crafts. During the fourteenth century the abacus schools underwent notable development, especially in Florence, and among the pupils were youths destined to become architects, painters, goldsmiths or sculptors. Filippo Brunelleschi, for instance, was one of them.[7] In composing his "Euclid for painters," Alberti thus addressed himself to a public already familiar to some extent with the principles of practical geometry. His indications serve to construct, by relatively simple steps, plane figures both in their true shape and diminished through perspective. It was only the humanist presumption of being able to explain the problems of drawing without drawings that kept these instructions from being immediately comprehensible. In this sense Alberti could be one of those "most worthy authors" mentioned by Francesco di Giorgio who, having chosen to explain their concepts "with characters and letters … and not by figured drawing," had made the content of their works obscure to many, being "rare those readers who, not seeing the drawing, can understand it."[8] In spite of its lack of drawings, *Elementa picturae*

---

4   See McGee and Long in this volume. On problems of representation in the treatises of Taccola and Francesco di Giorgio, see also Galluzzi 1996b, 24ff.; and Lamberini 2001, 3–10.
5   See Francesco di Giorgio Martini 1967, VI 489–490; and Leonardo, Windsor, Royal Library, K/P 162[r] (the passage is quoted in Galluzzi 1996b, 84).
6   *Elementa picturae*: see Alberti 1973, III 112–129; and Gambuti 1972, 131–172. *De pictura*: see Alberti 1973, III. For an English-language edition, see Alberti 1972.
7   See Manetti 1976, 52. On abacus schools, see Van Egmond 1980; and Franci and Toti Rigatelli 1989.
8   Francesco di Giorgio Martini 1967, VI 489–490.

was designed to furnish the rudiments necessary for an understanding of *De pictura*, where the themes dealt with, explains Alberti, "will be easily understood by the geometer. But he who is ignorant of geometry will understand neither these nor any other discussion on painting."[9]

The *De pictura* has no drawings either, but in this case the work is more conceptual than practical. Apart from his explanations of the method for constructing a floor in perspective and for using an instrument to draw from life, Alberti is interested mainly in defining the optical-geometric principles that govern the plane representation of three-dimensional space. The practical procedures were in fact already familiar to painters. Alberti had been stimulated to write on this subject expressly by the perspective skill of artists such as Brunelleschi, Masaccio, Ghiberti, and Donatello. On the plaster underlying the *Trinità* in Santa Maria Novella, for instance, Masaccio had left incised a practical but impeccable method for drawing the perspective of a foreshortened circle within a square; while in the *Nativity* of San Martino alla Scala, Paolo Uccello had left a fine example of what is today termed "distance point construction," which was at the time a practical way of representing the progressive decrease in size of a perspective floor directly on the painting, employing three vanishing points, one central and two lateral.[10] What was lacking was a text allowing artists to explain the reasons for their mode of operation, an "entirely mathematical" book—stated Alberti—which would explain the abstract concepts of geometry through the language of painters. Since the painter "studies only how to counterfeit what he sees" (I.2), Alberti is concerned with explaining the manner in which the eye perceives the reality around it, how the lines of sight measure sizes "almost like a pair of compasses" (I.5-6), how the Euclidean theory of similar triangles explains the relationship between real and apparent size (I.14), how in order to construct similar triangles it is necessary to cut the lines of sight (I.13), and how, in final analysis, "painting is none other than the intersection of the visual pyramid" (I.12). Comprehension of this last concept is entrusted to a fitting metaphor, which describes the painting "as an open window through which I look at what will be painted there" (I.19).

The way in which these concepts could be usefully transformed into a method of representation is explained by Alberti through a procedure that critics have always interpreted as an abbreviated form of the more laborious construction attributed to Filippo Brunelleschi.[11] Basically, Alberti seems to have developed a painters' method through which perspective could be drawn directly on the painting, thus avoiding the long architects' procedure that called for the use of preliminary draw-

---

9   Alberti 1973, III 53.
10 The method used by Masaccio consists of foreshortening the square inscribed in the circle, dividing each side into the same number of parts, and joining each point to the matching one on the contiguous side. This traces a series of tangents to the inscribed circle, which immediately appears in elliptical form. The traces of this construction method are clearly visible at the base of the capitals on the front columns. The construction method adopted by Paolo Uccello to foreshorten the perspective floor consists instead of tracing three points on the horizon line, one central and two at the ends, of joining the divisions of the ground line to those of these points, and tracing the parallel straight lines identified by the intersections of the three bands of converging lines. This construction method can be seen in the sinopia of the fresco now in the storage deposit of the Uffizi; see Camerota 2001, 93.
11 See Grayson 1964, 14–27.

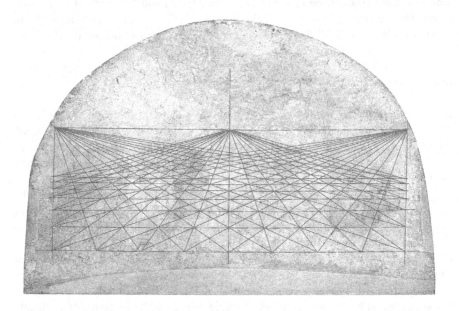

Figure 6.1. Perspective scheme constituting the sinopia of Paolo Uccello's *Natività*. (The fresco and its sinopia are now in the Uffizi Gallery, Florence.)

ings in plan and elevation. With Alberti's method, the principle of intersection of the visual pyramid could be applied to the construction of the flooring alone, which, being conceived as a modular grid, furnished the measurements for all of the objects appearing within the depth of the perspective field; the sinopia of Paolo Uccello's *Natività* shows how this was a real pictorial requisite (figure 6.1). Piero della Francesca developed this construction method still further, teaching artists to transpose points from the plan to perspective utilizing only the central vanishing point and a diagonal line.[12] The three points used by Paolo Uccello were to find a sort of codification in the work of Jean Pelerin le Viator, who in the first illustrated printed book on perspective (1505) described the methodical use of the so-called "tiers points," two lateral points similar to those used in architecture to draw the curve of a pointed arch.[13] The "third point" was not to be more appropriately termed the "distance point" until the writings of the great codifiers of the sixteenth century: Albrecht

---

12 Piero della Francesca 1984, I.28–29.
13 Viator 1505, in Ivins 1973. In the work of Viator "tiers points" are often used to construct buildings or interiors in an angular view. The term "tiers point" also appears in the notebook of Villard de Honnecourt (see Hahnloser 1972, 115–116) and is used to indicate the center of the compass in drawing a pointed arch. To trace this type of arch, the architect first draws the springing line, then points the compass at each end of the line, with the opening equal to the length of the line, and traces two arcs of a circle which intersect on the central axis. The "third points," i.e. the centers of the compass at the ends of the line, are also the points of convergence of the joint lines between the stone or brick ashlars of which the arch is built.

Dürer, who used it to teach how to operate according to the "the shortest way;" Sebastiano Serlio, who explained, not without some uncertainty, its relationship with "Albertian" construction and Giacomo Barozzi da Vignola, who was to make it the basis of his "second rule of perspective" (figure 6.2).[14]

The "first rule"—the "longest way" in Dürer's words—was instead represented by the method attributed to Brunelleschi by Giorgio Vasari, namely the methodical use of the plan and elevation views as graphic documents necessary to measure the intersection of the visual pyramid with the picture plane (figure 6.3).[15] Since this was

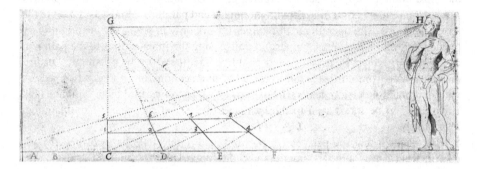

Figure 6.2. Second rule of perspective (G, vanishing point; H, distance point). Woodcut by Giacomo Barozzi da Vignola. (Vignola 1583.)

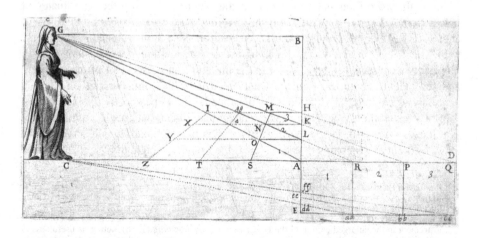

Figure 6.3. First rule of perspective (AE, picture plane in plan; AB, picture plane in elevation). Woodcut by Giacomo Barozzi da Vignola. (Vignola 1583.)

14 See Dürer 1983, IV figure 59–61; Serlio 1545, II 19; Vignola 1583, 98ff.
15 Vasari, *Vita di Filippo Brunelleschi*, in Vasari 1906, II 332: "*egli trovò da sè un modo ... che fu il levarla* [la prospettiva] *con la pianta e profilo e per via della intersecazione.*"

a problem of solid geometry—the intersection of a cone by a plane—it was necessary to adopt a procedure up to then used mainly by architects, the only one that allowed a three-dimensional object to be visualized in a two-dimensional drawing. The first clear description of this procedure was given by Piero della Francesca who, while acknowledging the greater operational difficulties involved, considered it "easier to demonstrate and to understand ..."[16] The two projections in plan and elevation, in fact, made the basic geometric model—the section of the visual cone—increasingly evident. Only this procedure could, moreover, guarantee a correct representation of the most complex objects such as vaults, capitals, *mazzocchi* and even the human head, which Piero geometricized with meridians and parallels, almost as if it were the terrestrial globe. The operational difficulties lay not only in preparing the preliminary plan and elevation drawings, but also in transposing the intersections to the painting, an operation, which Piero suggested could be facilitated by using simple "instruments': a hair from a horse's tail to trace the intersections in orthogonal projection with maximum precision, and various strips of paper to measure their "latitude" and "longitude" to be transferred to the base and to the vertical edges of the painting, respectively.

The concept that the perspective position of a point could be measured by two coordinates, just as cartographers defined the topographic position of a place, was very clearly expressed in the invention of the first instruments for perspective drawing. If the grid of Leon Battista Alberti derives from the method of *quadrettatura* employed by painters to enlarge their drawings, the so-called "sportello," or window, of Albrecht Dürer seems to show a more direct link with the cartographers' grid.[17] A frame with two orthogonal wires used to transfer to a drawing the coordinates of places from the Ptolemaic tables was described in the *Cosmographia* by Petrus Apianus (1524). It shows features similar to those of Dürer's instrument.[18] The "sportello," described in the fourth book of the *Underweysung der Messung* (1525), represents a constant reference point in the literature on perspective of the sixteenth and seventeenth centuries. Egnazio Danti cited it in his commentaries on Vignola expressly to explain "in what consists the foundation of perspective," since with it "we will see distinctly both the visual cone and the plane that cuts it" (figure 6.4).[19] The line of sight is in fact visualized by a thread tied to a nail, which serves as the eye. The point at which this thread traverses the frame, which marks the pictorial sur-

---

16 Piero della Francesca 1984, III 128ff.
17 In *De pictura* (Alberti 1973, III), II.31, Alberti describes the use of a transparent "veil" woven of thicker threads, which formed a reference grid used for drawing from life. An object seen through this grid could be easily traced on a drawing sheet with a *quadrettatura* of the same size. Dürer describes this instrument in the enlarged edition of the *Underweysung der Messung* (1538), noting its usefulness in enlarging drawings. Each square in the grid could in fact correspond to a sheet of the desired size. In the first edition of the treatise (1525), the German painter had proposed the use of two different instruments: a simple pane of glass on which the artist could draw directly the scene appearing through it, and a more refined frame fitted with a shutter, which allowed an object to be drawn by measuring the lines of sight "by means of three threads." One of the threads traversed the frame simulating the line of sight, while the other two defined with their intersection the picture plane and contemporaneously fixed the point of intersection between the plane and the line of sight.
18 See Apianus 1524, I 60.
19 Vignola 1583, I.3 55–56.

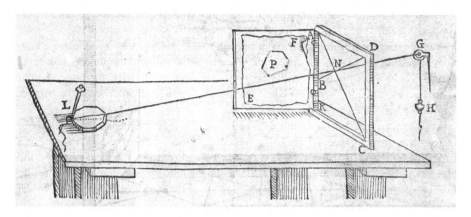

Figure 6.4. Albrecht Dürer's "sportello." Woodcut by Egnazio Danti. (Vignola 1583.)

face, is identified by the two orthogonal threads, which measure its "longitude" and "latitude." The intersection of the threads is then simply transferred onto the drawing sheet once the "sportello" has been closed.

The idea of materially representing the line of sight with a thread or cord had already been proposed by Francesco di Giorgio Martini in his treatise on practical geometry.[20] The context and the evidence of the drawing, which illustrates the perspective instrument, testify to the close relationship between the geometry of painters and that of surveyors (figure 6.5). This relationship also transpires in writings by Leonardo and Luca Pacioli, making it highly probable that Brunelleschi's invention sprang from the elaboration of a perspective procedure well known to surveyors and already taught in the abacus schools.[21] The concept of intersection of the visual pyramid was in fact basic to the methods used to measure distances by sight. An instrument described

Figure 6.5. Practical demonstration of linear perspective. Drawing by Francesco di Giorgio Martini. (*La praticha di gieometria*, Florence, Biblioteca Medicea Laurenziana, Ashb. ms 361, c. 32ᵛ.)

in the late fourteenth century by the abacist Grazia de' Castellani to measure where "the eye went with the line of sight," is entirely similar to the one illustrated by Francesco di Giorgio, and the geometric laws that could be applied to measure the

20 Francesco di Giorgio Martini 1967, I 139–140 (c. 32ᵛ).

distance of a point were the same as the ones that could be used to represent its perspective image.[22]

On the basis of this common appurtenance to the principles of Euclidean geometry, perspective underwent significant development precisely in the field of measurements. The process of codification having been launched with the fundamental contributions of Dürer, Serlio, Daniele Barbaro, Vignola, and Guidobaldo del Monte, the use of instruments for the practical, fast execution of any type of perspective, from that of a single object to an urban or landscape veduta, spread rapidly.[23] These instruments proved especially useful in the military field, where the need to represent correctly fortresses and territories could not be satisfied by directly surveying the place to be drawn. This military function of linear perspective emerges from a first testimonial by Egnazio Danti, who dedicates his edition of Vignola's treatise to the captain of the papal army Giacomo Boncompagni, explaining that perspective

> also offers great advantages in attacking and defending fortresses, since it is possible with the instruments of this Art to draw any site without approaching it, and to have not only the plan, but also the elevation with every detail, and the measurement of its parts in proportion to the distance lying between our eye and the thing we wish to draw.[24]

Many were the instruments devised for this perspective-topographic scope: the *distanziometro* of Baldassarre Lanci, for example, a refined invention by one of the most active military engineers of Cosimo I; the *gnomone* of Bernardo Puccini, another Medicean engineer; the *proteo militare* of Bartolomeo Romano; and even the *sportello* of Dürer, utilized by Pietro Accolti to explain how it is possible "from a certain perspective drawing, carried out with the said Instrument, to investigate, and represent its Geometrical Plan, and the quantity of each of its parts."[25] Accolti explains that, once the perspective drawing of a fortress has been executed, and the positions of the vanishing point and distance point established, it is possible to reconstruct the layout of the building by applying the perspective procedure inversely (figure 6.6). A point, which, in the perspective view, is identified by the intersection of two straight

21   In the *Codex Atlanticus* (Leonardo 1973–1980) are various sheets containing sketches and notes dedicated to perspective and methods of measuring by sight: 103^{br} (36vb), 119^r (42^{rc}), 339^r (122vb), 361^r, 400^v (148vb), 672^v (248va), (265rb). On sheet 103^{br}, for example, Leonardo writes: "Quella linia dov'è segnata in testa a, si chiama l'occhio. Quella dov'è segnato b, si chiama parete, cioè dove si tagliano tutte le linie che vengano all'occhio," and notes at the center of the drawing "Da misurare ogni distanza o altezza che tu vuoi." Luca Pacioli deals with the problem of perspective in his *Summa de aritmetica* (see Pacioli 1494): *Trattato geometrico, Distinctio octava, Capitulum secundum*, cc. 296^{r-v}. On the connection between measuring methods and Brunelleschi's perspective, see Krautheimer 1956, 238; Kemp 1978; Baxandall 1972, 86–108.
22   See Arrighi 1967, last paragraph of the "*quarta distinzione che tratta del modo di misurare chol'ochio, cioè chon strumenti*" (cc. 407^v–412^v).
23   In addition to the previously mentioned treatises by Dürer, Serlio, and Vignola, see Barbaro 1568 and Del Monte 1600.
24   Vignola 1583, dedication.
25   Baldassarre Lanci's *distanziometro*, now in the Florence Museum of the History of Science, inv. n° 152, 3165, is described by both Daniele Barbaro (Barbaro 1568, IX, IV) and Egnazio Danti (Vignola 1583, I.3 61). The *gnomone* is illustrated in detail by Bernardo Puccini in a treatise entitled *Modo di misurar con la vista* (1570–71), ms., Florence, Biblioteca Nazionale, Fondo Nazionale II–282, file. 15 (see Lamberini 1990, 351–403). Bartolomeo Romano also dedicates an entire treatise to his singular measuring "dagger" (Romano 1595). For the topographic operation of Dürer's *sportello*, see Accolti 1625, XVI.

lines, converging at the vanishing point and at the distance point respectively, will be identified in the plan by the intersection of a straight line perpendicular to the picture plane and one oriented at 45°. This procedure is a precursor of today's photogrammetry, conferring on perspective a leading role even among the most refined measurement methods of the Renaissance.

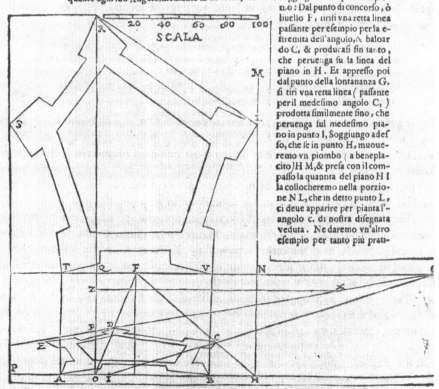

Figure 6.6. Method for working out the plan of a fortress from its perspective drawing. Woodcut by Pietro Accolti. (Accolti 1625.)

2. PERSPECTIVE FOR SOLDIERS AND INSTRUMENT MAKERS: FIRST THOUGHTS ON ORTHOGRAPHIC PROJECTION

The inverse procedure that could be used to determine the layout of a building by means of perspective testifies to the efforts exerted in the military field to render measurable even that which traditionally was considered false and deceptive. Throughout the sixteenth century the need to have a drawing that was both "pictorial" and measurable had favored the dissemination of a practical, rapid system of "perspective" through which the geometric characteristics of the object represented could be retained unaltered. Girolamo Maggi, who published the treatise *Della fortificatione della Città* in 1564, incorporating in it the writings of Jacopo Castriotto, explicitly mentions the use of this type of representation as specific to military architecture.

> Let no one think to see in these drawings of mine either methods or rules for perspective, firstly because, not being a soldier's technique, I would not know how to do it, and then because, due to the foreshortening, too much of the plan would be lost. These drawings serve instead expressly to show the plans, and this is called *soldierly perspective* (figure 6.7).[26]

The drawing proposed by Maggi is what is now called "military axonometry," retaining a term, which indicates its origin, or at least its predominant usage, in fortified architecture. Bonaiuto Lorini calls it "the most common perspective," Bartolomeo Romano "spherical perspective," Giovanni Battista Bellucci "perspective that serves for practical uses," and again in the nineteenth century, the first codifier of this method, William Farish, was to call it "isometric perspective."[27]

This kind of perspective without foreshortening, widespread since antiquity, had been throughout the Middle Ages the *prospettiva* of mathematicians. In treatises on practical geometry, as in the earlier land-surveying codes, geometric solids were represented in precisely this manner, according to a practice that was to remain in vogue throughout the Renaissance, even among the greatest theoreticians of linear perspective. Piero della Francesca, for instance, utilized it for the drawings in the *Trattato d'Abaco* and the *Libellus de quinque corporibus regularibus*, to lay emphasis on the geometric characteristics of bodies rather than on their apparent image.[28] Leonardo employed it superbly to represent machines and mechanical elements (figure 6.8), and the greatest mathematicians of the sixteenth century continued to use it, at least up to Federico Commandino and Guidobaldo del Monte, who were the first to introduce the perspective drawing into strictly scientific contexts. This change was hailed with enthusiasm by Bernardino Baldi, who praised the work of Commandino

> for the clarity of the language and the diligence of the figures, in which the employment of the art of perspective avoids that crudeness in which those who follow and have followed depraved usage and barbarous custom incur and have incurred in the past.[29]

---

26 Maggi and Castriotto 1564, II.3 40 (my italics). On this subject, see Scolari 1984.
27 See Lorini 1597, 32–33; Romano 1595, last plate. Sanmarino 1598, 1–6. The isometric drawing was to be definitively codified starting with the work of Farish 1822.
28 Piero della Francesca 1995.
29 Baldi 1998, 518.

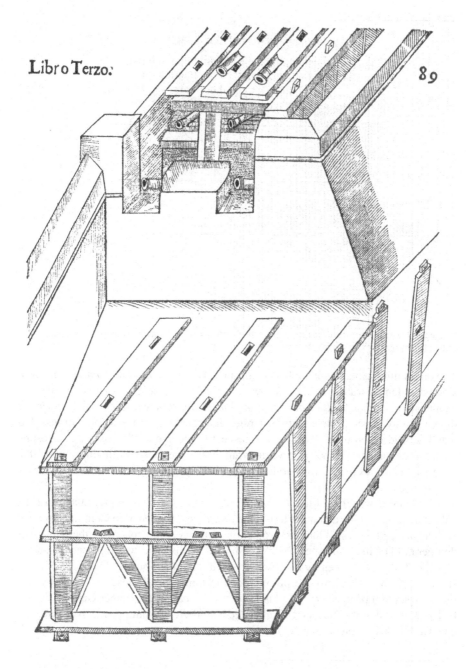

Figure 6.7. A bastion in "soldierly perspective." Woodcut by Girolamo Maggi and Jacopo Castriotto. (Maggi and Castriotto 1564.)

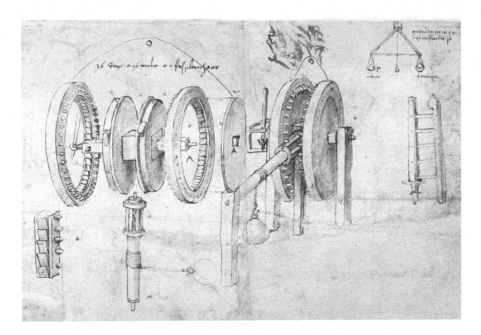

Figure 6.8. Isometric view of a winch with two wheels. Drawing by Leonardo da Vinci. (*Codex Atlanticus,* Milan, Biblioteca Ambrosiana, c. 30^v.)

The plates illustrating Commandino's edition of Ptolemy's *De analemmate* were considered a demonstration of this. So were those, at a later date, of Guidobaldo del Monte's *Planisphaeriorum universalium theoricae*, where the celestial sphere is drawn strictly in perspective with the circles deformed into ellipses, more or less flattened depending on their orientation in respect to the eye of the observer.[30] Despite such a systematic use of perspective drawing, it was precisely within this context that the foundations were laid for parallel projection, typical of the so-called "soldierly perspective."

The context was that of theoretical discussions of the plane representation of the celestial sphere. In his comment to Ptolemy's *Planisfero*, Federico Commandino had sufficiently demonstrated that the projection of the celestial sphere on the plane was an operation of linear perspective.[31] The boreal hemisphere was represented on the equatorial plane as if seen by an observer standing at the South Pole and, as demonstrated by Daniele Barbaro, the problem had become one of those discussed in treatises on pictorial perspective.[32] In his *De astrolabo catholico*, Reiner Gemma Frisius had sustained that the universal planisphere also derived from perspective, since the eye was situated at one of the equinoxes and the circles were projected onto the sol-

30 See Commandino 1562, 80; and Del Monte 1579.
31 See Commandino 1558.
32 See Barbaro, 1568, VI.2.

stitial colure, according to the projection procedure of Ptolemy, which Aguilonius was later to call "stereographic projection."[33] The universal planisphere described by Gemma Frisius had been widely known since the Middle Ages under the name of "saphaea azarchielis" (Azarchiel's plate) and, unlike the Ptolemaic planisphere, it allowed the utilization of only one planispheric plate for all latitudes. The circles in the sphere were reduced to arcs of a circle, and the entire planisphere could be easily drawn with a compass.

From the mid-sixteenth century on, a new universal planisphere based on the assumption that the distance of the eye could be infinite found increasing use (figure 6.9). It had been invented by Juan de Rojas Sarmiento, a pupil of Frisius, who maintained that this procedure, too, derived from perspective, on the condition, as Gemma Frisius was to explain, that one could imagine placing the eye at an infinite distance along the straight line passing through the equinoxes.[34] The points of the sphere were in fact transferred onto the plane according to what Aguilonius was to call "orthographic projection," i.e., by means of an array of parallel rectilinear lines used to represent both the parallels and the ecliptic by straight lines. Although perspective and orthographic projection belonged conceptually to the same projection principle, it was still difficult to imagine the eye as placed at an infinite distance. For Guidobaldo del Monte, for instance, placing the eye at an infinite distance meant "putting it in no place," a concept that "is abhorrent to perspective itself."[35] Nonetheless, apart from whether or not the parallel lines should converge at a point—a crucial problem in Postulate V of Euclid's *Elements*—Guidobaldo points out an obvious drawing error in Rojas' construction. He notes that the Spanish astronomer sustained that the meridians should be drawn as arcs of a circle, and that Gemma Frisius, while acknowledging that these were curves of another kind, did not know their geometric nature, so that he suggested drawing them by points. Guidobaldo explains that the curves of the meridians were none other than ellipses, adopting a demonstration already given by Federico Commandino in his edition of Ptolemy's *De analemmate*

---

33 Gemma Frisius 1556, I 4: "Astrolabum nostrum [universale] Spahera item plana est, ex visu defluxu similiter ut praecedens descripta. Verum eo solum differì, quod oculus non in polo, sed in Aequinoctiali constitutur, atque ita oppositum oculo hemisphaerium in planum perpendiculum obiectum visu describitur ..." See also Aguilonius 1613, 503: In dealing with question of perspective representation, Gemma Frisius suggests that the planisphere could also be drawn mechanically, by depicting an armillary sphere with an instrument such as Dürer's window (I 2–3). This method had in fact already been applied by Leonardo, as suggested by a drawing in the *Codex Atlanticus* (f. 1ᶠ⁻ᵃ) which illustrates a painter using a pane of glass to draw an armillary sphere. Presumably Dürer himself used a pane of glass for this purpose. The invention of the instrument dates from the years in which the German painter was working as a cartographer (1514–15), i.e., when he drew the splendid terrestrial globe in perspective for Johann Stabius and the representation of the Earth according to the third Ptolomaic method, which his friend Willibald Pirckheimer was to publish in his edition of the *Geografia* (Ptolemy 1525, VII).

34 Rojas Sarmiento 1550. On developments in orthographic projection, see the chapter by Lefèvre in this volume; see also Dupré 2001, 10–20.

35 Del Monte 1579, 141: "Da ciò appare quanto riduttive siano le loro parole per spiegare la sua origine. Giovanni De Rojas infatti omise del tutto dove bisognava collocare l'occhio. Gemma Frisio invece stabilisce che l'occhio (ove possibile) venga collocato a distanza infinita, cosa che senz'altro corrisponde a non collocarlo in nessun luogo. A quale condizione è infatti possibile che qualcosa nasca dalla prospettiva se l'occhio si allontana a distanza infinita? Senza dubbio ciò ripugna alla stessa prospettiva ..." See Guipaud 1998, 224–232.

188   FILIPPO CAMEROTA

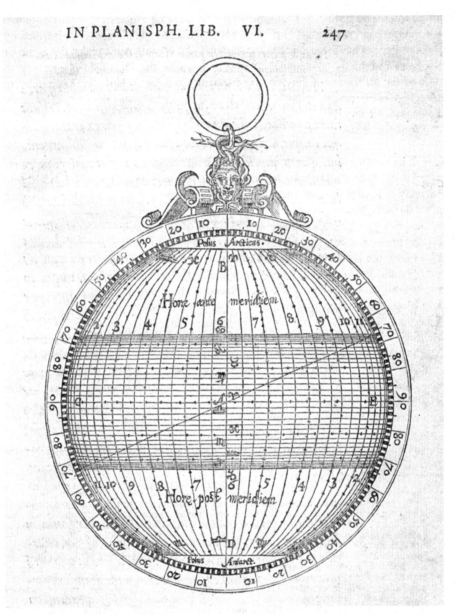

Figure 6.9. Orthographic projection of the celestial sphere (the "point of view" is at infinity). Woodcut by Juan de Rojas Sarmiento. (Rojas Sarmiento 1550.)

of 1563.[36] Considering the difficulty of tracing as many ellipses as there are visible meridians, all different from one another, Guidobaldo proposed the use of an ellipsograph of his invention, which he presumably also used to draw the extraordinary perspective representations of the celestial sphere, which illustrate his text.[37] Within this debate on orthographic projection may also be included the only perspective instrument of the Renaissance designed to obtain a perfect "isometric perspective," the one described in Hans Lencker's *Perspectiva* (1571), which materializes the direction parallel to the lines of projection by means of a metal stylus, always orthogonal to the drawing sheet (figure 6.10).[38]

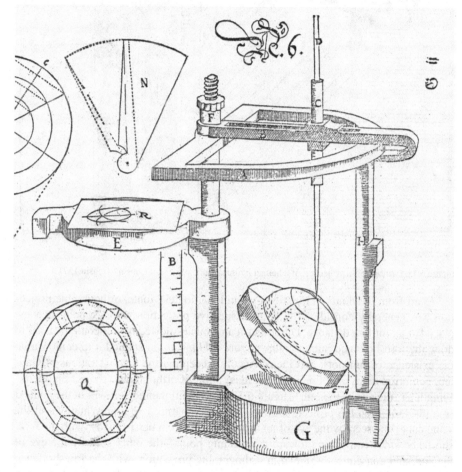

Figure 6.10. Instrument for orthographic projections. Drawing by Hans Lencker. (Lencker 1571.)

---

36 Del Monte 1579, 167: "Egli [il Rojas] credeva, non bisogna dimenticarlo tuttavia, che noi dovessimo disegnare archi di cerchio nel tracciare sul planisfero i meridiani." Then he quotes Gemma Frisius's words: "gli stessi meridiani, poiché non si conosce bene [quale sia] il loro tracciato, vengono ricavati per punti secondo un percorso disuguale; e non essendo questa operazione alla portata di un artefice qualsiasi, avviene che spesso si giunge ad un discorso privo di senso sia nel disegno sia anche nell'uso" (186). See Gemma Frisius 1556, I 4: "Meridiani verò lineis cursus anomalis, quae neque circoli sunt, tantum per puncta adsignata manu diligenti traductae."
37 Del Monte 1579, II 105.
38 See Lencker 1571, 21.

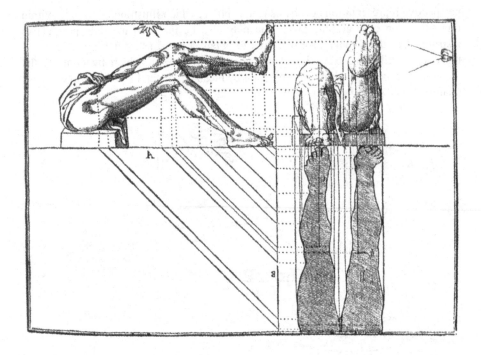

Figure 6.11. Orthographic projection of a human figure. Woodcut by Jean Cousin. (Cousin 1571.)

Apart from the doubts expressed by Guidobaldo del Monte, orthographic projection was generally considered to be a perspective projection with the point of view placed at an infinite distance. Dürer and Jean Cousin showed how it could be used to draw the foreshortened human figure (figure 6.11). Galileo applied it to demonstrate the existence of sunspots; and Pietro Accolti, accepting the demonstrations of Galileo, conferred a physical image on the ideal point of orthographic projection, identifying it as "the eye of the Sun" (figure 6.12).[39] Accolti mentions it within the context of a discussion on the projection of shadows, comparing two representations of the same cube, one seen by the eye of man and thus drawn in perspective—with the foreshortened sides converging toward a vanishing point—the other seen by the eye of the sun, and consequently drawn in orthographic projection, with the foreshortened sides parallel. "It is thus we consider that the aforesaid drawing should be, for repre-

---

39 See Dürer 1996, and Cousin 1571. Galileo employed parallel projection to explain the appearance of sun spots in the telescope (*Istoria e dimostrazioni intorno alle macchie solari e loro accidenti*, Roma 1613, 34; in Galilei 1890–1909, V 213–215). See Accolti 1625, III XXVIII, *Altra invenzione per conseguire la naturale incidenza de lumi, et dell'ombre sopra diversi piani ove vanno a cadere*: "Et insegnandoci il testimonio del senso visivo … che il Sole … manda l'ombre sue, parallele sul piano …; così restiamo capaci potersi all'occhio nostro, et in disegno far rappresentazione di quella precisa veduta di qualsivoglia dato corpo, esposto all'occhio (per così dire) del Sole, quale ad esso Sole gli si rappresenta in veduta: … così intendiamo dover essere il suddetto disegno, per rappresentazione di veduta del Sole terminato con linee, et lati paralleli, et non occorrenti a punto alcuno in Prospettiva."

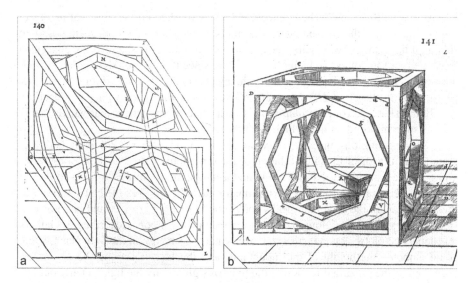

Figure 6.12. Two cubes in perspective: one viewed by a hypothetical eye situated in the sun (left), the other viewed by a human eye. Woodcuts by Pietro Accolti (Accolti 1625).

sentation of the viewpoint from the Sun, terminated with lines, and parallel sides, not concurring at any point in Perspective."[40] Although the author's objective was that of describing an alternative mode of representing shadows in perspective, employing a preliminary drawing, which he called *ombrifero*, or generator of shadows, this preliminary drawing contained all of the elements, which, with Girard Desargues, were to lead to definition of the ideal point in projective geometry. The cube appears to us as if we were looking at it from the position of the sun, that is, from a viewpoint at infinite distance. Among its many other merits, the dissemination of Galileo's telescope should probably also be credited with having favored the acceptance of this concept, which was still for Guidobaldo del Monte foreign to the physical reality of perception.

3. THE GEOMETRY OF LIGHT:
TOWARDS A THEORY OF SHADOW PROJECTION

Pietro Accolti's *ombrifero* drawing is a "singular," though hardly practical, attempt to solve a problem of crucial importance to painters but not yet defined on the level of projection: that of *sbattimenti*, or shadow projection. According to Accolti, up to then the authors concerned with perspective had gone "adventuring," concentrating mainly on the case of a punctiform light source, which produces "rays that are pyramidal, and concurrent," and ignoring the case, which would instead have been more useful to painters, that of the exposure of objects to daylight, which produces "paral-

---

40   Accolti 1625, 134.

lel rays, as the Sun does." The isometric cube was for Accolti the image of an object seen by the eye of the sun; but since the sun is at the same time both seeing eye and source of light, the cube appears free of shadows, or rather the shadows are concealed by the parts of the object, which produce them, since the lines of sight coincide with the rays of light. In order to see them, the eye would have to be shifted to another position, as in the cube drawn in perspective, where the shadows are produced by simply transferring from the *ombrifero* drawing the points where that which lies in front is superimposed on that which lies behind.

The considerations made by Accolti are an immediate reflection of Galileo's arguments in favor of his interpretation of the orographic nature of the moon. As we know, Galileo sustained that the moon had a mountainous surface similar to that of the Earth. The craggy profile of the terminator, he believed, was the clearest proof of this, even though the mountain peaks could not be seen where it would be most reasonable to expect them, along the circumference of the moon. According to Galileo, this depended on two main factors: the overly narrow breadth of the optical angle that could embrace the height of the mountains, and the fact that along the circumference—given the grazing direction of the lines of sight—only the lighted part could be seen. As any painter well knew, in the absence of shadow the perception of relief is lost.[41] If in fact we imagine an irregular surface lit up by the sun, explains Galileo,

> the sun, or one who might stand in its place, absolutely would not see any of the shadowy parts, but only those that are lit up; because in this case the lines of sight and of illumination proceed along the same straight paths, nor could there be shadow there where the ray of light arrives, so that none of the dark parts could be seen.[42]

To see the shadows it would be necessary for "the line of sight to rise above the aforesaid surface more than the solar ray," i.e., for the eye to look at the object from another point of view, just as Accolti was to propose.

The definition of two types of *sbattimenti*, or shadow projection, those produced by the light of a torch and those produced by sunlight, had appeared in treatises on pictorial perspective since Alberti's *De pictura*. Alberti, however, did not venture beyond the simple statement that "the light of the stars casts a shadow the same size as the body, but firelight makes them larger," although he suggested that a practical, rapid mode of drawing the perspective of a circle with precision could be that of projecting the shadow after having "positioned it, with reason, in its place."[43] Piero della

---

41 See for example, Cigoli 1992, f. 82ᵛ: "Gl'oggetti veduti dalla parte luminosa, per la scarsità dell'ombra non hanno rilievo." The question had been brought up by Galileo himself in a letter to Cigoli on the superiority of the arts in which, claiming the supremacy of painting over sculpture, he concluded: "Et avvertasi, per prova di ciò, che delle tre dimensioni, due sole sono sottoposte all'occhio, cioè lunghezza e larghezza ... Conosciamo dunque la profondità, non come oggetto della vista per sé et assolutamente, ma per accidente e rispetto al chiaro et allo scuro ..." (letter dated 26 June 1612, in Galilei 1890–1909, XI 340–343).

42 Letter dated 1 September 1611 to Christopher Grienberger, in Galilei 1890–1909, XI 178–203, specifically, 185.

43 See *De pictura* (Alberti 1973, III), I.11. The manner of drawing an ellipse as the projection of the shadow of a circle is mentioned in II.34; Alberti does not mention the type of light source but it is probable that he had in mind a torch collocated in place of the viewpoint. For a history of the theory of shadows, see Da Costa Kauffmann 1975, 258–287; *idem* 1993; see also De Rosa 1997.

Francesca deliberately chose not to deal with color, that third part of painting, which was supposed to comprise what Alberti called the "receiving of lights." But near the end of his treatise Piero describes a perspective procedure based not so much on the concept of *intersection* as on that of *projection*. An object placed between the eye and the painting is in fact projected by the lines of sight exactly as would be done by the rays of light from a punctiform light source.[44] The coincidence between the shadow of the object and its perspective image is well visualized in a drawing by Leonardo in which the effects of the projection of a sphere produced by an eye are compared with those produced by a candle.[45] To Leonardo we owe the studies that delve deepest into the question of shadow projection. The idea of dedicating an entire treatise to the problem of light and shadow had led him to investigate the phenomena of lighting effects with the great acuity characteristic of his theoretical reflections. In his writings we find important considerations on the difference between shadow proper (*ombra congiunta*) and cast shadow (*ombra separata*); on the separating line, or rather on that "means," which separates the shadowy part from the lighted one; on the *lume universale*, or diffused lighting, which produces half-light and sfumato; and on secondary light, the light reflected from any opaque body, which lightens the shadows of nearby objects.[46] Alberti too had theorized on *raggi flessi*, reflected rays, noting that they bring with them "the color which they find on the surface" of the object that reflects them: "Notice that one who walks in the meadows under the sun has a greenish tinge to his face."[47]

In spite of these keen observations, no geometric rule for the correct representation of shadows was established. Leonardo seems convinced, in fact, that shadows cannot be described geometrically, while

> the features can be made to gleam with veils, or panes of glass interposed between the eye and the thing that you wish to make shine, shadows are not included under this rule, due to the inconsistency of their terms, which are for the most part confused, as is demonstrated in the book on shadows and lights.[48]

The innovative practice of representing shadows geometrically was introduced by Albrecht Dürer, who in the four famous plates of the *Underweysung der Messung* demonstrated their principles of projection with crystalline clarity.[49] In the first panel the artist drew the shadow of a cube in plan and elevation, precisely localizing the source of punctiform light (figure 6.13a). In the second plate the cube and its shadow are drawn in perspective through the so-called "laborious procedure," i.e., Brunelleschi's legitimate construction (figure 6.13b). The third panel illustrates the method of creating shadow through the so-called "shorter way," Alberti's method, which is confused here with construction using the distance point (figure 6.13c). Lastly, the fourth

---

44 Piero della Francesca 1984, 210–215, LXXVIII–LXXX.
45 Leonardo da Vinci 1986, *Manoscritto C*, fol. 9.
46 Leonardo's reflections on the projection of shadows, the subject of a lost book "of shadows and lights," are found scattered through various manuscripts. They were collected by his pupil Fancesco Melzi in the fifth part of the *Libro di pittura*; see Leonardo da Vinci 1995.
47 See *De pictura* (Alberti 1973, III), I.11.
48 Leonardo da Vinci 1995, II 304, § 413.
49 Dürer 1983, IV: text to figure 52–58.

plate shows the finished drawing, without construction lines and with a pictorial touch in the gradation of the cast shadow and the representation of the sun as source of light (figure 6.13d). On the one hand, this last plate reveals Dürer's sensitivity to the phenomenon of diffraction that causes lightening of the shadow in the parts nearest to the light. On the other hand there is, as has been noted by critics, an obvious incompatibility between this drawing and the previous plates, where the light source is clearly punctiform and at a finite distance.[50] In that position, in fact, the sun would have cast a longer shadow, since its horizontal projection would necessarily have been on the line of the horizon. This "oversight" seems to proclaim a basic problem in understanding the difference between punctiform light source and light source at

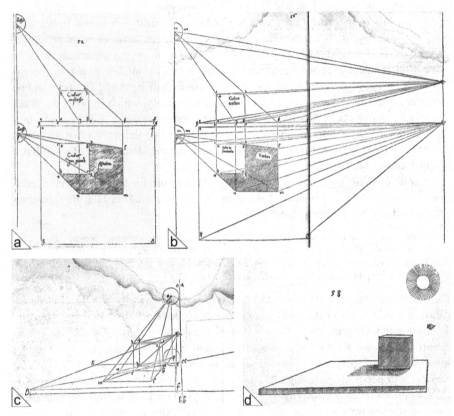

Figure 6.13. The perspective of shadows: a) plan and elevation of a cube with a light source and cast shadow; b) the cube and its shadow drawn in perspective according to the "laborious procedure" (intersection of the visual pyramid in plan and elevation); c) the cube and its shadow drawn in perspective according to the "shorter way" (use of the distance point); d) final drawing of the cube and its shadow in perspective. Woodcuts by Albrecht Dürer. (Dürer 1983.)

---

50 See Da Costa Kauffmann 1993, 49–78.

an infinite distance. The problem, however, was to be ignored throughout the century. As noted by Pietro Accolti, the treatise authors had focused exclusively on the representation of shadows projected by a torch or a candle. Dürer's scheme was to be proposed again, without substantial modifications, at least up to François d'Aguillon's *Opticorum libri sex* (1613), where we find it specified that

> those things lighted by lamps ... should be expressed *scaenographicae*, that is in perspective, that are exposed to the direct rays of the sun should be presented *orthograpicae*, that is in orthographic projection.[51]

Aguilonius, however, does not indicate how this orthographic projection should be represented in a perspective drawing, and critics have often noted that no solution was to be given before the crucially important studies of Girard Desargues.[52]

A correct solution for representing shadows projected by the sun was, in reality, given by Ludovico Cigoli, one of the painters closest to Galileo. The fact that his manuscript on *Prospettiva pratica* remained unpublished certainly prevented this rule from undergoing further development, and it even seems to have escaped the notice of those theoreticians who presumably had access to the text—Pietro Accolti and Matteo Zaccolini.[53] In the section entitled *Degli sbattimenti*, Cigoli describes three types of shadows: those larger than the illuminated body, produced by a punctiform light source, for which he adopts Dürer's scheme; those smaller than the illuminated body, as with an object placed before a window, where he gives instructions for producing the gleam (*barlume*); and those the same size as the illuminated body, which are the so-called *sbattimenti del Sole*, or "shadows projected by the Sun." Since *sbattimenti* of this kind are "contained by parallel lines" lying on the horizontal plane, Cigoli was concerned with finding their convergence point on the horizon, adopting the theory of the so-called "points of convergence" elaborated by Guidobaldo del Monte in his *Perspectivae libri sex*.[54] Guidobaldo had explained that, beyond the vanishing point and the distance point, on the horizon there could be determined as many vanishing points, or points "of convergence," as were the directions of the straight lines of the object to be represented. After having drawn the object, Cigoli then establishes the direction of the solar rays, their representation by means of a straight line drawn in plan and elevation, and finds the point of convergence of that straight line "and its parallels."

Cigoli notes that this rule was necessary since "the Sun being too big and too far away, we cannot draw its plan and profile." In the case of a punctiform light source, in dealing with a point in space at a close distance, it was both possible and necessary to determine its exact position by means of the plan and elevation views. Cigoli continues to use the terms "plan" and "height" even as concerns perspective construction with a distance point, indicating that these two spatial references—the horizontal

---

51 Aguilonius 1613, 683.
52 Bosse 1647, 177–178.
53 See Cigoli 1992, 79^v–82^v. Left unpublished, as it still is, was also the important manuscript by Matteo Zaccolini, *Della Descritione dell'ombre prodotte de' corpi opachi rettilinei* (see Florence; Biblioteca Medicea Laurenziana, Ashburnham ms. 1212).
54 Del Monte 1600, II VIff.

plane and the vertical one—should still be considered measuring loci even in their apparent deformation. As is clearly apparent in the plates by Guidobaldo del Monte as well, the representation of shadow was determined by the contemporary presence of these two planes: the horizontal one that received the shadow of the object, and the vertical one that visualized the projection of the cone of light. This favored, or rather demanded, that the method of representation commonly used in architecture—projection in plan and elevation—should become the necessary condition for any problem relevant to the geometric description of space: "Since the reason for the appearance of any object whatsoever derives from proceeding with strict reason in imitating it," wrote Cigoli with exemplary clarity, "it is necessary to know its shape and position, something which is obtained by means of two figures known as plan and profile, which reveal to us the three dimensions: length, width and depth."[55]

4. THE SPECIFICITY OF ARCHITECTURAL DRAWING: THE DOUBLE PROJECTION

Theoretical reflections on orthographic projection and the dissemination of rules for perspective had contributed to codifying a method of representation that had for centuries been specific to architecture. The need to define the shape and size of the building to be constructed called for a linear, measurable drawing that clearly expressed the architect's idea with no concessions to the pleasure of the eye. "Between the graphic work of the painter and that of the architect," wrote Leon Battista Alberti in *De re aedificatoria*,

> there is this difference: the former endeavors to portray objects in relief on the panel through shading and the shortening of lines and angles; the architect instead, avoiding shading, represents reliefs through a plane drawing, and represents in other drawings the form and extension of each facade and each side utilizing real angles and non-variable lines, as one who desires that his work not be judged on the basis of illusory appearance, but evaluated exactly on the basis of controllable measurements.[56]

Although the numerous drawings of Renaissance architecture show that forms of mixed representation survived for years, with shaded perspectives or combinations of perspective and orthogonal projections, the general intention on the theoretical level was that of conferring on the architectural drawing a specific nature of its own, able to describe "the acuteness of the conception, not the accuracy of the execution."[57]

On the difference between the painter's drawing and that of the architect, Raphael too was highly explicit when in his famous letter to Leon X he described precisely the characteristics of the architectural drawing:

> The drawing of buildings pertaining to the architect is divided into three parts, of which the first is the layout, that is, the plan, the second is the outer wall with its ornaments, and the third is the inner wall, also with its ornaments …[58]

---

55 Cigoli 1992, 17^r.
56 Alberti 1975, II.1. On the subject, see Thoenes 1998 and Di Teodoro 2002.
57 Alberti 1975, II.1. The latter quotation refers to the use of the models, which Alberti preferred simple and linear rather than too elaborate and colorful.

The three parts clearly recall the three "species" of Vitruvius, with the difference that to *ichnographia* and *orthographia*, Raphael adds "the inner wall," i.e., the section. The need for this third part was to lead Daniele Barbaro, alone among Vitruvius' commentators, to interpret as "profile" the third Vitruvian species, that of *scaenographia,* which was instead traditionally synonymous with perspective.[59] Raphael himself interprets the third Vitruvian species as *prospectiva,*[60] but in the letter to Leo X this kind of drawing is considered a pictorial method to be utilized only during the design stage.

> And although this mode of drawing in perspective is proper to the painter, it is nonetheless convenient to the architect as well because, just as it is good for the painter to know architecture in order to draw the ornaments well measured and with their proportions, so the architect is required to know perspective, through which exercise he can better imagine the building with its ornaments.

The architectural concept is instead to be conveyed entirely by orthogonal projections.

> And in such drawings there is no diminishing ... Because the architect cannot take any correct measurement from a diminished line, as is necessary to that artifice which requires all of the measurements to be perfectly made and drawn with parallel lines, not with those that seem so but are not ...

Drawing in orthogonal projection had been applied in architecture since antiquity. The extraordinary working drawings in the temple of Apollo at Didime, for example, show a refined ability to visualize three-dimensional shape through the two projections in plan and elevation.[61] Good examples of the correct use of these projections appear again in the drawings on parchment from the Gothic period, but in these, as in the ancient drawings, it is by no means taken for granted that the two projections should be aligned, nor that they should be drawn to the same scale. The famous drawing of the dome on St. Maria del Fiore, done by Giovanni di Gherardo da Prato as an illustration for the public discussion of a building problem, shows how even in the early fifteenth century, and within a context anything but marginal, the custom of representing plan and elevation views on the same sheet in arbitrary positions and scales was still widespread.[62] This situation began to change with the correct use of orthogonal projections, thanks to the dissemination of linear perspective. To represent the perspective image of a point it was in fact necessary to define its exact position in space, and this could be done only by first drawing the projection in plan and elevation according to a determined reduction ratio.

The drawings of Piero della Francesca are the first and finest examples of this mode of representation. In Theorem XIII of *De prospectiva pingendi* the three kinds of drawing indicated by Raphael, plan, elevation, and section, appear not only drawn

---

58 On the letter to Leo X, see Di Teodoro 1994.
59 See Vitruvio 1567, I.2.2 30.
60 See Vitruvio 1975, I.2.2.
61 See Haselberger 1980 and 1983.
62 The drawing is now in the Archivio di Stato di Firenze, Exhibition Inv. n° 158.

to the same scale but also superimposed one above the other in the same drawing (figure 6.14). Undoubtedly, Piero thus intended to explicitly show the coincidence

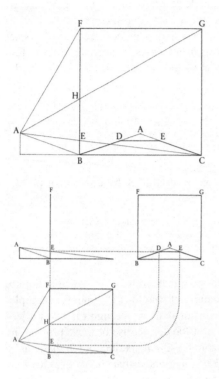

between viewpoint and vanishing point, demonstrating that he had perfectly mastered the problem of overturning planes. And this is not an isolated example. Francesco di Giorgio Martini demonstrates in the same way the coincidence between "center," i.e., "point and end of all lines" (the vanishing point) and "counter-center," i.e., "the eye which looks at the point," by overturning on the picture plane the orthogonal plane passing through the eye of the observer, and thus offering the first clear representation of the distance point perspective method.[63] A decisive contribution to the dissemination of double representation (plan and elevation) and triple orthogonal projection (plan, elevation and section, or lateral elevation) was made by Albrecht Dürer.[64] The first of his treatises, *Underweysung der Messung* (1525), is designed to furnish the necessary geometrical instructions not only to painters but also to "goldsmiths, sculptors, stonemasons, woodworkers, and others who base their art on the correctness of the drawing."[65] The themes

Figure 6.14. Perspective construction of a square horizontal plane. Drawing by the author, based on Piero della Francesca *De prospectiva pingendi*, c. 1475, I, XIII.

dealt with range from conic sections to the drawing of letters, from the proportioning of architectural elements to the design of solar clocks, from measurement techniques to perspective drawing. Each subject is illustrated with drawings of exemplary clarity, whose potential value to architectural drawing is fully demonstrated by the refined illustrations to the subsequent treatise on fortifications, *Etliche underricht, zu befestigung der Stett, Schlosz, und flecken* (1527).

But the most extraordinary applications of orthogonal projection are found in the treatise, published posthumously, on the proportions of the human body: *Hieirin sind begriffen vier Bücher von menschlicher Proportion* (1528). Here Dürer's refinement is expressed through a series of multiple projections designed to provide geometric

---

63 Francesco di Giorgio Martini 1967, I 139–140 (c. 32ᵛ). On Piero della Francesca, see Di Teodoro 2001.
64 See Lefèvre in this volume.
65 Dürer 1983, preface.

control even over the various positions of the human body (figure 6.15). Piero della Francesca's attempt to demonstrate that it was possible to geometricize even a geometrically undefined body such as the human head was developed by Dürer into a methodical system, subjecting the infinite variations of nature to the control of geometry. In these drawings appear orthogonal projections of bodies not only in frontal but

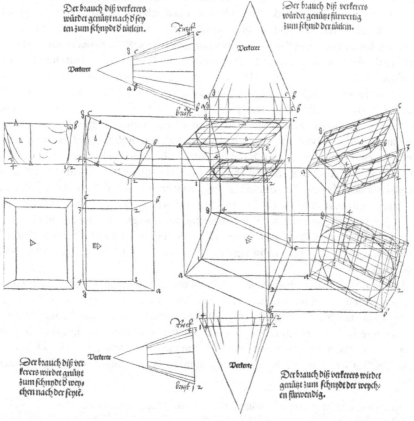

Figure 6.15. Multiple projections of a human torso. Woodcut by Albrecht Dürer. (Dürer 1996.)

also in oblique view, seen from below or above: real stereometries of human torsos, limbs and heads. Educated in a family of goldsmiths, Dürer possessed the expertise to dominate almost entirely the field of geometric constructions, but he himself stated that this type of drawing originated in a precise professional context, that of the stonemasons. In his dedication to Willibald Pirckheimer, he in fact writes that whoever wishes to confront the study of proportions must have

> well assimilated the manner of measuring and must have understood how all things must
> be extracted from the plan and elevation, according to the method daily practiced by the
> stonemasons.[66]

5. STONEMASONS' GEOMETRY:
ROTATION, OVERTURNING, AND DEVELOPMENT

According to Dürer, construction sites were the places where drawing was practiced at its highest level. On construction sites, the transition from the project drawing to actual construction was made through two basic elements: the wooden model—which visualized the building in three dimensions, furnishing all of the information necessary for carrying out the work, from construction system to decoration—and the working drawing on the scale of 1:1, which furnished the exact measurements of each individual architectural element. The working drawings were carried out on the site progressively as construction proceeded, so that each new element to be worked, in wood or in stone, could be perfectly adapted to the part already constructed. These drawings were made, under the supervision of the architect, by the carpenters or stonemasons themselves, who then had to shape each architectural element from a formless beam or a rough-hewn block of stone. The previously mentioned drawings of the Temple of Apollo at Didime show exactly this process of detailed definition of dimensions and shapes. The drawings, engraved in stone, would have furnished the necessary information to all of the stonemasons who succeeded one another during the long period of construction of the building. Similar graphic documents are found in some Gothic cathedrals, as for example on the terrace of the choir in the Cathedral of Clermont-Ferrand, where a drawing on the scale of 1:1 of one of the portals on the facade has survived.[67]

The working drawing had to be extremely precise, detailed but also essential, free from any sign that had no structural function. The essential nature of such drawings is clearly indicated by the manuals of Matthäus Roriczer and Hans Schmuttermayer, who divulge for the first time the "secrets" of the worksite drawing, revealing how knowledge of the method applied to extract the elevation from the plan had favored a stratagem that today would be called ergonomic. By utilizing a particular geometric

---

66 Dürer 1996, letter of dedication to Willibald Pirckheimer: "Darumb thut einem yglichen der sich diser kunst vndersteen will not / das er zuuor der messung wol vnderricht sey / vnd einen verstand vberkome / wie alle ding in grund gelegt / vnd auffgezogen sollen werden / wie dann die kunstlichen Steinmetzen in teglichem geprauch habenn."
67 The drawing is an elevation of the northern portal of the Cathedral of Clermont-Ferrand; see Sakarovitch 1998, 126.

figure—a square inscribed in another square rotated 45°—it was possible to deduct from the plan drawing alone the widths of all of the other points in the piece to be worked.[68] A particularly vivid picture of the techniques employed in medieval work-sites appears in the famous notebook of Villard de Honnecourt, which describes various graphic procedures for proportioning architectural elements and the human figure, enlarging the size of the design drawing to that of the working drawing, measuring heights and distances, tracing pointed arches and, above all, for designing complex structural features such as a stone arch over an oblique passageway or in a circular wall.[69]

Building this type of arch called for great technical skill that implicated an uncommon knowledge of static and geometric factors. The oblique vault was generated from a cylinder that traversed a straight parallelepiped, i.e., the wall thickness, at an angle. For the arch to fulfill its static function, the joints between the ashlars had to lie on planes perpendicular to the wall sector. Due to the oblique nature of the passage, the intrados surfaces of the various ashlars were not uniformly curved but twisted to varying degrees, and the joint lines appeared as elliptical segments. In all likelihood the problem was solved through a method now termed "by squaring," which consisted of obtaining the oblique piece from the cubic block by the orthogonal projection of its shape onto the faces of the cube. The ensemble of the six orthogonal projections, which make up a cube is precisely the system adopted by Dürer in representing the stereometry of the human torso, and it is presumably this method to which the German painter was referring in the passage quoted above.

In his drawings Villard illustrates how these problems of stereotomy could be solved on the worksite by the appropriate use of a square or ruler. The use of these instruments, however, seems to implicate the knowledge of a more refined system of geometric projections. In addition to the square and the ruler, Villard traces signs, which indicate the number of ashlars of which the arch is composed and the angles of the different cuts necessary to transform the cubic block of a single ashlar into an oblique parallelepiped. The problem was probably solved with the aid of scale models, but the signs left by Villard hint at the application of a graphic procedure not unlike the one codified later in subsequent treatises on stereotomy. The first description of the drawing methods employed by stonecutters was given by Philibert De L'Orme, whose *Premier tome de l'architecture* (1567) constitutes today, under this aspect, an exception in the panorama of Renaissance architectural treatises.[70] Books III and IV of this important treatise are entirely dedicated to "the art of stone cutting," which is presented here essentially as the "art of geometric drawing," indicating that the problem of cutting stones lay above all in the correct execution of the drawing. The importance attributed by De L'Orme to this aspect of architecture emerges already in the title-page, where eight "stereotomic icons" illustrate the graphic resolution of some of the most significant cases of stereotomy: the hemispheric vault, the

---

68 Roritzer 1486; Schmuttermayer 1489.
69 See Bechman 1993.
70 De L'Orme 1567. On the subject of stereotomy in the work of De L'Orme, see Pérouse de Montclos 2000; Potié 1996; Sakarovitch 1998.

Figure 6.16. Title-page with stereotomic diagrams. Woodcut by Philibert De L'Orme (*Le premier tome de l'architecture*, Paris 1567).

three-dimensional arch over a cylindrical wall, the so-called "vis de Saint-Gilles," i.e., the barrel vault over a helicoid staircase, the "trompe de Montpellier," i.e., the conical vault supporting a cylindrical volume, and the so-called "trompe quarré," the conical vault supporting a square volume (figure 6.16). As proven by the names still used to define some of these vaults, they are architectural types coming from the Romanesque and Gothic tradition of Southern France. The so-called "appareilleurs" (stonemasons) were those who guaranteed the correct execution of these works by superintending the cutting and laying of the stones, an extremely delicate operation that required technical expertise and a thorough understanding of the geometric development of bodies.

On the level of geometry, the problem to be solved was that of the intersection between two or more volumes. The so-called "trompe quarré," for instance, originated from the intersection of a horizontal half-cone and a vertical square volume. The ashlars of which it is composed have faces belonging to both of these figures and could be worked only after their exact geometric development had been determined. The stonemason first had to define the "slope line" in real size, a line that varied for each ashlar, and then to develop the intrados surface in order to design the so-called "panels," the templates used to cut the blocks of stone. The "par panneaux" construction method, described by De L'Orme, differs from the one called "par équarrissement" in that it could be used to design beforehand the exact shape of each individual face of the oblique block, thanks to a refined system of projections, rotations and overturning (figure 6.17). De L'Orme notes the innovative aspect of this method and claims credit for its invention, well aware of the fact that he was the first to write on

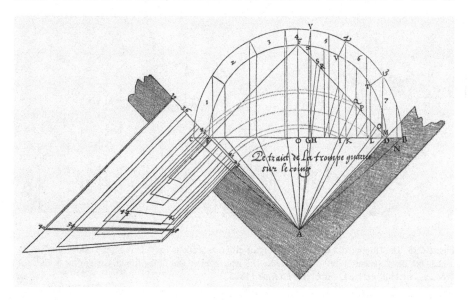

Figure 6.17. Constructive drawing for the "trompe quarrée." Woodcut by Philibert De L'Orme. (De L'Orme 1567.)

the subject. The French architect proclaims his ambition to raise "the art of geometric drawing" from its status of worksite practice to that of architectural theory, proposing to "review Euclid and accommodate his theory to the practice of our architecture."[71] This ambition, however, was to find no adequate realization. De L'Orme avoids facing the problem on the mathematical level and offers no definition of his refined procedures, which he explains instead in a practical and verbose manner.

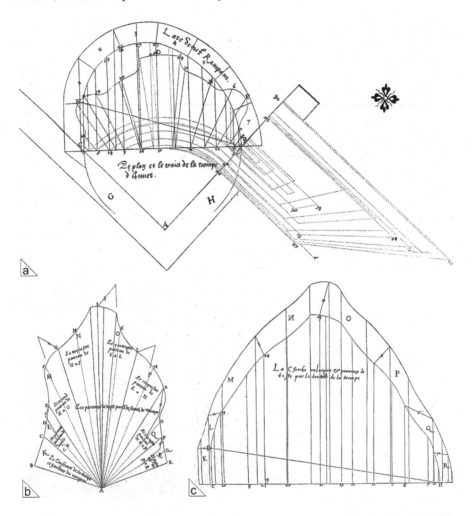

Figure 6.18. The "trompe d'Anet": overturning of different sections for determing the "slope line" of the ashlars (a), development of the vault's intrados (b), and development of the frontal edge of the ashlars (c). Woodcuts by Philibert De L'Orme. (De L'Orme 1567.)

---

71   De L'Orme 1567, III f. 62[r].

The design of the "panels" can be compared to the development of the faces of the polyhedrons and the sphere illustrated by Albrecht Dürer, in his treatise on geometry designed to teach how to construct those geometric solids materially.[72] But while the faces of the polyhedrons are all regular geometric figures—triangles, squares, pentagons, hexagons, and lunes in the case of the sphere—the faces of an oblique ashlar are irregular figures, also delimited by curved lines. In the absence of earlier graphic testimony, we must believe that De L'Orme's method consisted exclusively of total mastery of drawing; that is, of the application of a general projection principle, which can be used to resolve not merely a particular shape but any invention the architect is capable of imagining. Not by chance a large portion of the two books on stonecutting is dedicated to the detailed explanation and illustration of a new "trompe," of which De L'Orme writes, "I am certain that no worker in this kingdom has ever heard tell of it before."[73] This was the "trompe surbaissée, bias et rampante," a variation of the "trompe de Montpellier" distinguished by an elliptical shape, a rampant arch, of the generatrix of the cone. The invention of this "trompe" dates from 1536, when De L'Orme, upon returning from his voyage to Rome, built the first example of it in the Hôtel Bullioud at Lyon. About ten years later he created an even more refined variation in the Castle of Anet, which represents his masterpiece of stereotomy (figure 6.18). While the "trompe" of Lyon supported a simple cylindrical volume, that of Anet bore up a mixtilinear cylinder, which conferred on the line of intersection with the elliptical cone an undulating profile of bold audacity, both technically and formally: "which I wanted to make of a form so strange as to render the trompe of the vault more difficult, and beautiful to behold."[74]

De L'Orme evidently pursued a stereotomic virtuosity that emerged as a stylistic sign, and sought a new aesthetics based on geometry rather than on ornamentation. And it was in the name of this principle, in fact, that he criticized the father of Renaissance classicism, Donato Bramante, whose helicoid staircase in the Vatican Belvedere he considered an exemplary case of geometric incongruity. "If the architect who built it had known the geometric method of which I speak," wrote De L'Orme, he would have operated so that each element followed the geometric nature of helicoid development, even the bases and the capitals of the columns, which instead, in keeping with Vitruvius' rules, Bramante had drawn as if they supported "a straight, horizontal portico."[75] This geometric incongruity is also criticized by Juan Caramuel de Lobkowitz who in the next century was to base the principles of "oblique architecture" expressly on stereotomic geometry.[76] The solution suggested by De L'Orme for Bramante's staircase is the one employed in the tribune of the Cathedral of Saint-Étienne-du-Mont in Paris, which also appears in a drawing in the Louvre traditionally attributed to the French architect.[77] Of the helicoid staircase built in the Castle of

---

72 See Dürer 1983, IV figure 29–34.
73 De L'Orme 1567, IV 91^r.
74 *Idem*, IV.2 89.
75 *Idem*, IV.19 124.
76 Caramuel de Lobkowitz 1678, tratado VI, articulo XII.
77 Paris, Cabinet des Dessins du Louvre, n. 11114; see Blunt 1958, figure 139.

Fontainbleau, De L'Orme notes the "great artifice and the amazing difficulty" presented by the intersection of the three cylinders, which define, respectively, the perimeter of the staircase, the horizontal vaults of the flying buttresses and the oblique vault of the helicoid staircase.[78] His *esprit de géometrie* reaches perhaps its greatest aesthetic heights in the chapel in the Castle of Anet, which is a true architectural expression of the art of "extracting the elevation from the plan" (figure 6.19). The de-

Figure 6.19. Plan of the chapel in the Castle of Anet. Engraving by Androuet du Cerceau. (du Cerceau *Les plus excellents bastiments de France*, Paris 1579.)

78   De L'Orme 1567, IV.19 125.

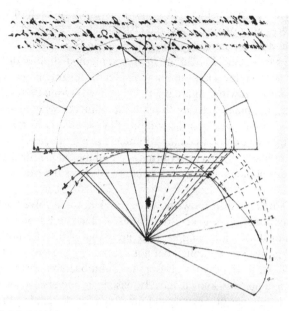

Figure 6.20. Constructive drawing for the "trompa de Monpeller."
Drawing by Alonso de Vandelvira. (Vandelvira *Libro de traças de
piedras,* ms., c. 1575.)

sign of the floor is in fact a perfect orthogonal projection of the three-dimensional
arches opening onto the cylindrical perimeter and of the amazing coffered hemi-
spherical vault. It is as if the real-size drawing made for executing the "panels" for
cutting the stones had been transformed from a worksite document into a decorative
motif.

The natural ease with which the stonemasons mastered drawing for stonecutting
is also demonstrated by the only other treatise on stereotomy from the sixteenth century: the *Libro de traças de cortes de piedras* by Alonso de Valdelvira.[79] The Spanish
architect composed this work between 1575 and 1590; although presumably familiar
with De L'Orme's writings, he developed the theme in an entirely independent manner. The *Libro de traças* is in fact the first specialized treatise on stereotomy: a refined
compendium of worksite techniques, which is even more eloquent than the work of
De L'Orme as regards the stonemasons' knowledge of geometry. The introduction
echoes that of many treatises on practical geometry and perspective:

> The beginning of the drawing is the point, and from it proceeds the line, and from lines
> the area and the surface, and from the surfaces the bodies on which all operations are car-
> ried out, and thus the species of the basis of this art are three: the lines which are those
> that surround the surfaces; the surfaces which are those that delimit the bodies, and the
> body which is the corporeal substance of the construction; (the species of the forms of
> these bodies are infinite).[80]

---

79 Valdelvira 1977.

As to the individual stone ashlar a description is given of each of its six faces, the method for determining its shape, and the instrument used to work it and to verify its installation: a mobile square called *baivel* (*biveau* in French). In the construction of the "trompa de Monpeller," Vandelvira applies a method slightly different from that of De L'Orme, which allows less superimposition of the graphic indications and thus easier reading of the drawing (figure 6.20). While De L'Orme draws the generatrix of the cone on a secant plane, passing through the points at which the cylindrical volume meets the rectilinear walls of the building, Vandelvira plots it on a tangent plane, displaying the intersection between the volumes much more clearly. Furthermore, the Spanish architect does not effect overturning to find the "slope line," but immediately constructs the development of the cone, which visualizes the true shape of the intrados "panels."

Vandelvira's method also shows how the problem to be solved was that of determining the transformation of geometric figures: a plane arch that is transformed into a three-dimensional arch for example, or a circle that is transformed into an ellipse. The concept of projection is implicit but not necessarily present in all of these operations. Only with the work of Girard Desargues did it become certain that this geometry of transformations was knowledgeably based on universally valid principles of projection. "This manner of carrying out drawings for the cutting of stones," writes Desargues in the *Pratique du trait* (1643), "is the same as the method of practicing perspective without employing any third point of distance, or of any other type, that lies outside of the painting ..." And further on he reiterates:

> ... so that there is no difference between the manner of depicting, reducing or representing anything in perspective, and the manner of depicting, reducing or representing in a geometrical drawing, so that geometrical drawing and perspective are no other than two species of the same gender, which can be enunciated or demonstrated together, using the same words ...[81]

Viewed in the light of hindsight, this truth emerged throughout the Renaissance. Albrecht Dürer, in particular, was the artist who came closest to a general formulation of "descriptive geometry." He put together the technical and geometric knowledge of many professional categories—painters, architects, cartographers, stonemasons, astronomers and instrument makers—demonstrating how the *dottrina* of drawing lodged in the workshops of craftsmen more than in the minds of geometers. He himself invented new graphic and mechanical procedures for geometric drawing: multiple orthogonal projections, the projection of shadows, the famous perspective instruments, and the extraordinary compasses for drawing ellipses, spirals, and new curves not yet defined on the geometric level. His numerous technical and geometric inventions, often developed and experimented within the workshops, and subsequently disseminated through the press, are a striking example of how ingenuity progressively favored the evolution of drawing toward a doctrine that was later to be reinforced by the geometric concepts of infinity, projection, and homology.

---

80   *Idem*, introduction.
81   See Desargues 1640; in Desargues 1864, I 305–362 (in particular, 305ff.).

7

THE EMERGENCE OF COMBINED ORTHOGRAPHIC PROJECTIONS

WOLFGANG LEFÈVRE

INTRODUCTION

Modern engineering employs all kinds of drawings—from sketches, over several orthographic plans and combinations of such plans, through a variety of schematics and diagrams, up to all sorts of illustration drawings in perspective.[1] Plans, and in particular combined plans, constitute the core of this abundance of engineering drawings, at least in the process of designing and manufacturing.[2] It is therefore surprising that plans are rather the exception than the rule among the drawings traced and used by engineers of the Renaissance.[3] From Taccola's well-known illustrated manuscripts of the first half of the fifteenth century up to the famous *Theatres of Machines* of the late sixteenth century, perspective drawings dominate the picture. And of the few plans known from this age, hardly any could be called a blueprint in the present understanding of the term.

Destruction or interrupted transmission can be ruled out as the main causes of this remarkable absence of plans,[4] which thus indicates a practice of engineering that, as a rule, designed and manufactured intricate devices without plans. A reconstruction of this praxis must focus on issues like the organization and division of labor in projects where engineers were involved, the prevalence of adaptations of traditional solutions to local circumstances, the almost complete lack of manufacturing the different parts of a mill, a water lifting device or a machine at different locations, the role of models and templates, the education of the craftsmen, and so on.

However, such a reconstruction of a practice of engineering without plans can easily be mistaken as an explanation of the absence of plans that argues: Since these engineers could do without plans, they didn't take pains to construct ones. But this conclusion is only valid if one can presuppose that Renaissance engineers would have been able to construct plans if they had wanted to. Are there reasons to assume that they had sufficient command of the techniques necessary for drawing orthographic plans? Now the fact that almost all of them also worked as architects appears to be a

---

1 See, for instance, Rising and Amfeldt 1964.
2 I disregard here the sweeping changes that the computer and specific electronic applications for engineering drawing brought about in the design process during the last two decades.
3 See Lefèvre 2003.
4 By their very nature, plans that actually served as a means of designing and manufacturing can be expected only either in the private files of the engineers themselves and of private customers, or in the archives of public commissioners. And it is rather probable that many of these documents are indeed lost. However, the few cases where such materials survived and are known confirm the overall picture of the marginality of plans in the realm of Renaissance engineering. The Württemberg engineer and architect Heinrich Schickhardt (1558–1635), whose private files are preserved and held in the Hauptstaatsarchiv Stuttgart, is probably a good case in point—see Popplow 1999.

very strong indication. Rather, what seems to require explanation is the strange fact that people who were familiar with orthographic plans when constructing buildings very rarely took advantage of these means when functioning as engineers. But again, this involves an unproven presupposition. This time, one would assume that all basic techniques of constructing orthographic plans were known and employed in the realm of Renaissance architecture. But is this true?

In the literature on Renaissance engineering, architecture, or perspective painting, it often is taken for granted that the engineers and architects of this age had basically the same techniques of constructing plans at their disposal as their modern colleagues. Since the drawing of orthogonal plans is considered a rather trivial task in contrast to perspective depiction, the technique of constructing plans is usually supposed to be given where plans occur: for instance, in ancient Mesopotamia or ancient Egypt. The technique of constructing plans, thus, has the appearance of a cultural achievement without a really historic development: its origin is hidden in the dawn of the earliest cultural developments, and its history seems to be that of a mere technical refinement of a procedure, the essence of which was given with its first invention and remained the same up to the present day.

The period under scrutiny here proves that this picture is wrong. For the Renaissance was the time of a decisive turn in the history of plan construction. The double-faced character of such periods of change is clearly discernible. On the one hand, one encounters techniques and conventions of producing plans that differ distinctively from present ones, not only in some secondary aspects, but in substance. (And one may ask whether the plans from the Renaissance convey an impression of how the technique of constructing plans might have looked in earlier ages, not only in the Gothic period, which is essentially the same with respect to plans, but also in classic Antiquity.) On the other hand, the modern techniques of constructing plans were actually invented—albeit rarely employed—in just this period. This chapter, therefore, has two goals. It tries to show that Renaissance (and Gothic) plan construction was not based on a technique that is essential for present-day plan construction, namely the combined views technique. Furthermore, it tries to show that this technique was invented at the turn of the fifteenth century and introduced to architecture by Antonio da Sangallo the Younger and Albrecht Dürer in the second decade of the sixteenth century, and that it took the entire sixteenth century before it became a standard technique in architecture.

Before coming to how the chapter proceeds, a brief clarification of terms like plan, elevation, and combined views may be useful. In this chapter, the term "plan" denotes any kind of orthographic projection, that is, a projection where each line that connects a point of the depicted object with its representation on the drawing meets the drawing plane at right angles.[5] In this way, it is warranted that plans depict true angles and true distances, or, in the case of scaled plans, true proportions among distances. It is this fidelity with regard to angles and distances that gives plans their spe-

---

5   More precisely, I use the term "plan" to denote all kinds of multiplanar orthographic projections. For a contemporary classification of projections, see, for instance, French 1947, 91.

cial value for architecture and engineering and distinguishes them from perspective pictures. The latter systematically distort certain angles and certain distances in order to depict the object as it appears to the eye. Perspective drawings, therefore, have the advantage of conveying an impression of the object as a solid and of its being located in three-dimensional space. In contrast, plans can only offer flat views of their objects. Thus, in order to depict a three-dimensional object by orthographic projection, at least two plans from different points of view have to be constructed.[6]

This is the principal reason why in architecture and engineering different kinds of plans are employed: ground plans, elevations, and sections, to name just the most common. Usually, the drawing planes of different plans that depict the same object from different points of view are perpendicular to each other, elevations or sections to the ground plan, and different elevations to each other (with the exception of the parallel drawing planes of elevations from exactly opposite points of view). Presupposed, furthermore, that these plans are drawn to the same scale, they can be unified in an unambiguous manner. The technical term "combined views" denotes a particular kind of such combinations of plans, namely the combination of a ground plan with one or two elevations, the planes of which cut one another at right angles (figure 7.1).[7] This combination layout is not just unambiguous but furthermore allows, in the case of three views, the position of a point in one of the plans to be deduced if this point is represented on the two others. Thus combined views are not only a means of repre-

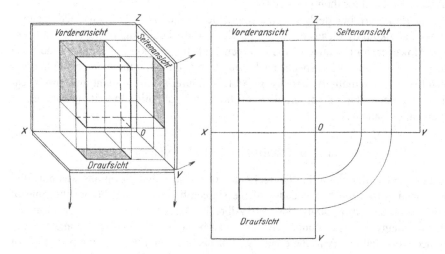

Figure 7.1. Combined views. (Bachmann and Forberg 1959, 168.)

6   For the purposes of this chapter, it is not necessary to discuss the different kinds of axonometric projections—isometric projection, dimetric projection, trimetric projection—which try to find a compromise between the advantages of both perspective drawings and plans. It is also not necessary to go into the different kinds of oblique projections, the best known of which is probably the "cavalier projection." For the problems of such projections, see French 1947, 459ff. and 468f.
7   Such combinations of a ground plan and one elevation are usually called "two-view drawings" or "double views" and those of a ground plan with two elevations "three-view drawings" or "triple views."

senting an object, but at the same time a means of constructing orthographic plans. To my knowledge, triple views—a ground plan with two elevations—became the standard form probably not before the nineteenth century. What is usually found before this century, beginning with a few exceptional cases around 1520, are double views— a ground plan and only one elevation—which are, nevertheless, suitable as a means for the construction of plans.

There is one kind of plan in particular that needs to be constructed by means of combined plans, namely, elevations of objects with surfaces that are not parallel to the drawing plane. Elevations, thus, are indicative of the presence or absence of the combined views technique. They can be read as symptoms. That is the reason why elevations from the period before 1500 are in the focus of the *first part* of this chapter, which comes to the conclusion that combined views were not in use before the sixteenth century. This means that, before this century, architects had command of a technique of constructing plans that differs significantly from our technique, not only in range but also in its inner structure.

The *second part* of the chapter gives some context of this technique by providing a concise survey of the role plans played in building practice up to the sixteenth century. The main result of this survey is that no urgent needs of the construction practice stimulated the invention of a general method of deriving elevations from ground plans. Actually, there was no such need. The architects of the age had at their disposal all means needed for their purposes.

The *third part* then deals with the emergence of the method of constructing elevations of any kind by means of combined views at the beginning of the sixteenth century. However, no narrative of the emergence of this method is attempted, merely a discussion of some of its aspects, background, and context without any claim to completeness or systematicity. I focus on Albrecht Dürer's achievement and only occasionally point to that of Antonio da Sangallo to draw attention to striking parallels or interesting contrasts.

1. THE PROBLEM OF ELEVATIONS

At first glance, the claim that the technique of constructing elevations was not fully developed before the first decades of the sixteenth century, not before the time of Antonio da Sangallo (1484–1546) and Albrecht Dürer (1471–1528), seems untenable. Too many pieces of counter-evidence can be adduced. Texts may come to mind immediately that testify to the contrary.[8] For instance, in a well known paragraph of his *De Re Aedificatoria*, Leon Battista Alberti (1404–1472) not only mentioned explicitly elevations of front and side façades of a building, but also stressed the difference between the perspective drawings of painters and the orthographic plans of architects.[9] The famous *Lettera a Leone X* from the beginning of the sixteenth century, which was formerly attributed to Raphael and later to Castiglione, is another

---

8   For these texts, see also part 4 of the chapter by Filippo Camerota.
9   See Alberti 1966 book II, chapter 1.

example that could be quoted.[10] This text, too, stressed the difference between pictorial drawings and plans and admonished architects to refrain from shortening in the manner of perspective drawings when constructing elevations. The circumstantial and sometimes even awkward way in which the letter tries to make this point is, however, rather indicative of how little ingrained the construction of orthographic elevations still was at that time.[11]

However, written texts alone cannot settle the question, as the case of Vitruvius' distinction between *Ichnographia* (ground plan), *Orthographia* (elevation), and *Scaenographia* (a kind of perspective) shows:[12] Because hardly any examples of such drawings are actually extant, one can only speculate how these different kinds of projections may have looked and how they were constructed and applied by Roman architects. With regard to the Renaissance and even to the Middle Ages, we are in a better situation. Even among the architectural drawings of the thirteenth and fourteenth century, that is, among the oldest known architectural plans *(Baurisse)* of the Occident, not only ground plans can be found but also elevations. For the mediaeval countries north of the Alps, one can point to the Reims palimpsest from the beginning of the thirteenth century[13] or the elevations of the façade of the Strasbourg cathedral from the same century,[14] and for Italy, to the elevations for the façade of the Orvieto *Duomo* from the beginning of the fourteenth century.[15] These elevations already display such a high standard of draughtmanship that it was taken as a extremely strong indication, if not almost proof, that the beginnings of architectural plans in the Occident must be dated to a time that long preceded this early evidence.[16]

Obviously, our claim needs qualification. For this, a closer look has to be taken at the architectural elevations from before the sixteenth century. There are sometimes elevations of complete façades of cathedrals, in particular the splendid west façades, or of church towers, mostly, however, elevations of parts of such façades, of walls with windows or arcades, of rose windows, and the like. In other words, what these elevations depict are those sides of buildings or of certain parts of a building whose planes are exactly parallel to the drawing plane. The same holds for elevations of the period, which belonged to the realm of carpenters and goldsmiths, even though only conditionally with respect to the latter. It goes without saying that it is much easier to construct the elevation of such a parallel surface than that of an object with oblique sides. It would, therefore, be premature to conclude from such façade elevations

---

10 Text in Bruschi et al. (ed.) 1978, 469ff.
11 See Germann 1993, 106; see also Thoenes 1993, 566f.
12 See Vitruvius I.2 (Rose 11:23ff.).
13 See Branner 1958.
14 See Anonymous 1912.
15 See Degenhart et al. (ed.) 1968 vol. I.3, table 25 and 27. As the rendering of the portals shows, these elevations are not purely orthogonal, but also employ techniques of shortening in the manner of perspective drawings. This combination of different drawing techniques is characteristic of Italian architectural elevations and sections before the sixteenth century, and even during this century as will be shown.
16 See Booz 1956, 67. Whether or not one is ready to accept Booz's conclusion, there is in any case indisputable evidence for a highly developed architectural draughtmanship that precedes the first known plans, namely the *"Ritzzeichnungen,"* that is, architectural drawings scratched in stone on the floor or on walls of mediaeval cathedrals—see Schoeller 1980 and 1989.

whether or not their creators had techniques at their disposal that enabled them to construct elevations of objects regardless of their position to the drawing plane. In other words, it is not clear whether these architects had command of a general technique of constructing elevations, or only of a partial one that was restricted to the case of parallel surfaces. This is the decisive question in our context. What I am claiming exactly is that such a general technique was not part of the professional skills of architects before the sixteenth century.

The preserved elevations themselves offer the strongest testimony in favor of this claim. Due to the very nature of the depicted objects, it was not always possible to avoid oblique sides when constructing elevations of the parallel surfaces of a building. And in such cases the construction of these oblique parts is always "wrong" according to our present standards. The "errors" that Villard de Honnecourt (fl. 1225–

Figure 7.2. Villard de Honnecourt's elevation of Laon Cathedral. (Paris, Bibliothèque nationale de France, ms.fr 19093, fol. 19; courtesy Bibliothèque nationale de France.)

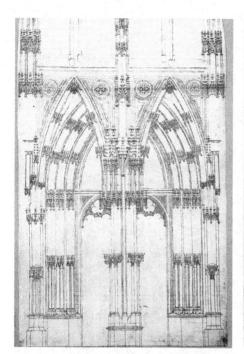 

Figure 7.3. West portal of Regensburg Cathedral, c. 1500. (Vienna, Kupferstichkabinett der Akademie der Künste, Inv. Nr. 16.871; courtesy Kupferstichkabinett der Akademie der Künste.)

1250) committed in this respect are famous (figure 7.2). But, if they are considered as plans at all, they may prove nothing more than that Villard was not "an accurate draftsman."[17] If, however, such "errors" also occur in elevations of outstanding mastery, which, furthermore, were drawn centuries after Villard (figure 7.3), one can no longer resort to individual faults but must acknowledge a general pattern. Focusing on this pattern, Peter Pause found out that the "errors" on elevations from the countries north of the Alps follow certain rules such that one can make out a small set of conventions, which have nothing to do with orthographic projection or with linear perspective.[18] One of these conventions, for instance, consisted in a simple cut-off rule: By cutting off a vertical strip of a window or arcade, the draftsman indicated that they belonged to an oblique wall (figure 7.4). From the age of Villard[19] until the sixteenth century, architects followed these conventions when rendering the shortening of oblique sides. This practice would be absolutely unintelligible if these architects had been able to construct such oblique surfaces in a correct orthographic manner.

---

17 See Branner 1963, 137.
18 See Pause 1973, 57ff.
19 For Villard, this cut-off rule is described in Hahnloser 1972, 32.

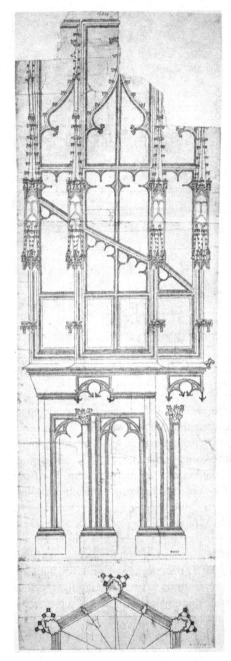

Figure 7.4. "Schnegg" of Konstanz Cathedral. (Vienna, Kupferstichkabinett der Akademie der Künste, Inv. Nr. 17.028; courtesy Kupferstichkabinett der Akademie der Künste.)

The absence of a general technique for constructing elevations of objects, whatever their position to the drawing plane, could also be shown by pointing out a different practice that was then common in Italy. Italian architects did not follow these conventions of their northern colleagues, but used linear perspective when rendering such oblique parts of buildings. As Wolfgang Lotz could show for sections, one encounters no purely orthographic sections in late fifteenth- and early sixteenth-century Italy. Rather, up to the time of Antonio da Sangallo, architectural sections always combined orthographic with perspective rendering.[20] This is all the more remarkable since these sections prove that a highly developed skill in constructing perspective drawings obviously was already remarkably widespread among Italian architects at the turn of the fifteenth century (figure 7.5). Thus it has the appearance that, at least in Italy, architects were able to construct perspective renderings in an unrestricted manner earlier than orthographic elevations. (I will come back to the relations of these two techniques in the last part of this chapter.)

It seems, perhaps, quite unbelievable that men capable of constructing correct perspective drawings[21] would not be able to construct all kinds of orthographic elevations. A possible reason why this appears astonishing might be that one underestimates the difficulties associated with the task of constructing elevations. And it is indeed probable that this underestimation is a result of the cultural fact that, among all of the modern techniques of geometric projection that emerged in the Renaissance, only linear perspective has enjoyed the undivided and continuing attention of a broad audience ranging from art historians over historians of culture, mathematics, and science up to semiotic experts, philosophers, and psychologists. The modern projection techniques of cartographers, astronomers, engineers, and architects, on the other hand, the foundation of which was also laid in this age, are taken to be either trivial or only of technical interest for certain experts.

In order to make the difficulties accompanying the construction of an elevation a bit more discernible, I should stress that drawing an elevation does not mean portraying something, but constructing a representation—a flat model, as Nelson Goodman might say.[22] This is obvious in the case of elevations for a planned building drawn in the design process. But even when the task is to record an already existing building by means of elevations, such elevations have to be constructed point by point in conformity with distances and angles taken from the real building using appropriate instruments. This architectural recording (*Bauaufnahme*) much more closely resembles the mapping of surveyors than the depicting of painters. Vision is suspended in

---

20 See Lotz 1977. As mentioned above, this also holds for the earliest known Italian elevations, that is, the elevations of the façade of the *duomo* in Orvieto from the beginning of the fourteenth century, and the famous elevation of the *campanile* of *Santa Maria del Fiore* in Florence from the middle of this century, which is thought to be a copy of an original drawing by Giotto. See for the latter Degenhart et al. (ed.) 1968, vol. I.3, table 66 and vol. I.1 89ff. See also Evans 1997, 166f.

21 Apart from a few rare exceptions, we do not actually know how perspective drawings were produced in the fifteenth and in the early sixteenth century. It thus seems well advised in our context to recall an easily neglected difference between the drawing of perspective pictures and elevations: Whereas the former could be produced without construction by means of a variety of mechanical-optical devices then available, there was (and is) no escape from constructing in the case of the latter.

22 See Goodman 1976, 172f.

Figure 7.5. San Pietro in Montorio. Section from 16th-century Italy. (London, Sir John Soane's Museum, Codex Coner, fol. 34; courtesy Sir John Soane's Museum.

favor of construction, that is, in favor of a complex process of measuring, recording measurements taken, processing these data, and, based on them, tracing the graphic representation according to geometrical rules.

As a means of the designing process, the tracing of elevations seems to be less complex, except for the difficulties of non-trivial geometrical constructions and accurate tracing in general, which I ignore here. The desired measurements—for instance, the height and width of the windows or the breadth of the wall between them—can be translated directly into appropriate geometrical lines and shapes on the plan. But this is true if, and only if, the plane of the wall or window represented is thought to be parallel to the drawing plane. In all other cases, one has to derive the correct translation from the ground plan. And this derivation is exactly the point where the combined views come into play. From a ground plan drawn to the same scale as the elevation and attached to it in the right way, one can "deduce" the correct positions of these problematic parts by means of ruler and compass. Moreover, given a list with the intended heights of all parts in question, one can develop the entire elevation from the ground plan. On the other hand, without this technique of combined plans or any substitute for it, it is simply impossible to determine the exact position and accurate size of the representations of those parts that are thought to lie in an oblique plane. An octagon like the Florentine baptistery is a case in point (figure 7.6). There is no other way to determine the distances between the vertical edges of such a simple and regular geometric body on the elevation than that of deriving them from the ground plan.

The extant elevations from before 1500 show, thus, that, apart from two or three exceptions,[23] which confirm rather than challenge the rule and will be discussed below in the third part, combined views were not in use before the sixteenth century. This is also confirmed by the telling fact that the famous treatises on architecture of the fifteenth century—besides the treatise by Alberti and the letter to Pope Leo X mentioned above, the *Trattato di Architettura* by Antonio Averlino, known as Filarete (1400–1469);[24] the treatises *Architettura ingegneria e arte militare*[25] and *Architettura civile e militare*[26] by Francesco di Giorgio Martini (1439–1501); and the drafts on architecture of Leonardo da Vinci (1452–1519)[27]—are absolutely silent on this technique.[28] Thus one can assume that combined views were actually unknown in this age. Not even Cesariano's Vitruvius edition of 1521 gives any indication of a change in this respect.[29] But if combined views had not yet been discovered as a decisive means of constructing plans, it is no longer strange that one does not encounter correct elevations of objects with oblique sides before Dürer and Antonio da Sangallo.

---

23 See Sakarovitch 1998, chap. 1.
24 Filarete 1972, English: Filarete 1965.
25 Francesco di Giorgio Martini 1967, vol. I.
26 Francesco di Giorgio Martini 1967, vol. II.
27 Edited in Bruschi et al. (ed.) 1978, 277ff.
28 See Sakarovitch 1998, 55ff. With respect to Alberti's treatise on architecture, see also Thoenes 1993, 569.
29 Note that, in Cesariano's edition, the plate to book I, chapter 2, which displays example drawings for Vitruvius' *orthographia*, also includes drawings that are not orthographic elevations but rather perspective drawings.

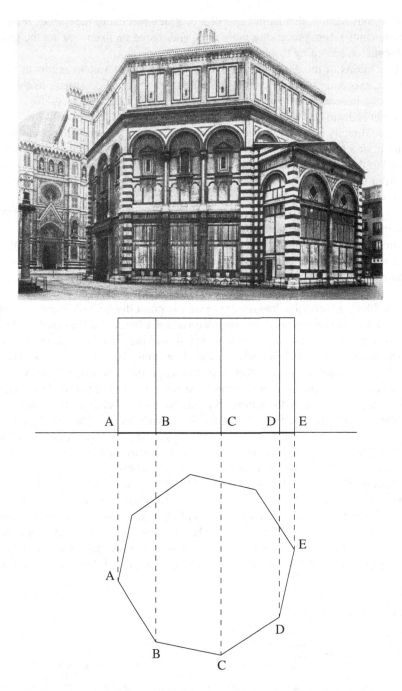

Figure 7.6. Deriving the elevation of an octagon from its ground plan. Drawing by the author.

On the contrary, it would be a miracle if the architects of the age had been able to construct correct elevations despite the lack of this technique.

Summing up this first part of this chapter, it can be stated that, in the realm of architecture, the technique of combined views does not occur in fully developed form until Antonio da Sangallo and Albrecht Dürer, that is, not before the first decades of the sixteenth century. Before going into these achievements and their context and prerequisites, it might be useful to take a quick glance at the role that plans and drawings played in the process of designing and constructing buildings until the end of the fifteenth century.

2. PLANS IN ARCHITECTURE UNTIL 1500

Regarding the role of plans for the architectural practice of the Middle Ages and the early Renaissance, the most striking feature of this practice is probably what can be called its improvisatorial character. As a rule,[30] the architectural features of a planned building were not fixed in all their aspects and details in advance. Commissioner and architect confined themselves to appoint only main features when contracting. Above all two reasons seem to be responsible for this practice. First was the custom of postponing decisions on certain questions to a time when they could be made in light of the growing building. Second, and probably more important, was the fact that many features needed no explicit agreement because they were obvious within the given tradition of construction. All that had to be done was to adapt the canonical rules and conventions of this tradition to local circumstances. And this adaptation was tacitly included in the overall design of the building upon which commissioner and architect had agreed.

The famous controversial conferences of the Milan cathedral council between 1389 and 1401, which are remarkable in so many respects,[31] provide a good example of the custom of postponing decisions. On the one hand, these conferences testify to ongoing revisions of former decisions, that is, to a thorough planning process. On the other hand, however, they testify to a lack of planning in advance that must be inconceivable for a contemporary architect. The elementary features of the foundations of walls and piles had not only been designed, but actually laid and constructed when the question of the height of the naves ignited the controversy. The essential question of the height of the cathedral was obviously not fixed along with the ground plan. *Santa Maria del Fiore*, the Florence cathedral, is another case in point. After the original foundations had been dismissed and replaced by significantly larger ones, an endless process of decisions and revisions of decisions took place with regard to the overall shape of the cathedral, which the *Operai di duomo* finally cut off with a somewhat iconoclastic act, namely by destroying the competing models.[32] However, what remained unsettled by this finality and engendered the subsequent "drama"[33] of the

---

30 See for the following Müller 1990, 14ff. where the most important literature is discussed.
31 See Ackerman 1949.
32 See Lepik 1994, 35.
33 Saalman 1980, 11.

cupola—not just a decorative spire on top of the crossing, but a weighty, important part of the cathedral, structurally as well as aesthetically—was the fundamental question of whether or not the huge octagonal crossing, which was already built could be vaulted with a cupola of the appointed shape. A similar drama was played out a century later in Rome when the new *San Pietro* church was built.[34] These three prominent examples may represent the innumerable cases up to the beginning of the sixteenth century where the definite design of essential parts of an building was not appointed in advance along with the ground plan, but left to later decisions.

In the light of this improvisatorial building practice, it becomes intelligible why one encounters only ground plans and elevations or sections of certain parts of a building among the preserved architectural plans of the age but, to my knowledge, never a set of plans that determines all main features of a building. The basis of this astonishingly flexible practice was a traditional framework of rules and conventions for how to build a sacral or profane building—a basis, however, which always became perilous when, as in the prominent cases mentioned, the designed buildings went beyond the limits of this tradition. The fact that these rules and conventions[35] had guiding functions for architects and craftsmen of the age equivalent to those plans have today explains why so many features of a building were not fixed in advance by means of plans. In Lorenz Lechler's *Unterweisungen* of 1516,[36] for instance, one of the few preserved texts that document such rules and conventions,[37] one finds several rules, including alternative ones that connect the measures of the essential parts of a church, such as the width of the main nave with its height, the width of the arcades, the thickness of piles, walls, buttresses, the measures for the adjacent chorus, and so on. The actual constructed churches were, of course, not mere executions of a set of rigid rules and conventions. Rather, these rules formed so flexible a system that decisions were necessary in any case as to which of the countless possibilities was to be realized in the situation at hand. And the architect often furthermore had to accommodate wishes of his commissioner that were at odds with the rules of the art.[38] On the basis of this system of rules, however, such decisions and accommodations could be appointed without a comprehensive set of plans. A ground plan and elevations of some critical parts usually sufficed for this purpose.

It is known that, in the countries north of the Alps, plans were part of the contracts between architects and clerical or secular commissioners. Their function was the same as that of the wooden models in Italy, namely to document the design the con-

---

34 See Bredekamp 2000.
35 See for the following Müller 1990, 78ff.
36 A diplomatic copy of the three preserved manuscripts of the booklet can be found in Coenen 1990, 174–266.
37 Apart from Lechler's booklet, there are five more such practitioners' texts written by architects for architects: 1)*Von des Chores Maß und Gerechtigkeit* of about 1500, of which only an incomplete nineteenth-century edition exists but neither the original nor contemporary copies (see Coenen 1990, 25); 2) the so-called *Wiener Werkmeisterbuch* from the fifteenth century; 3) Matthäus Roriczer's *Büchlein von der Fialen Gerechtigkeit* of 1486, and 4) his *Geometria Deutsch*, written slightly later; 5) Hans Schmuttermayer's *Fialenbüchlein* of the 1480s. Coenen 1990 provides diplomatic copies of all of these texts. Shelby 1977 provides an modernized edition and English translation of the two booklets by Roriczer and the one by Schmuttermayer.
38 See, for instance, Lechler's complaints: Coenen 1990, pp. 179, 233, 262.

tracting party had agreed upon.[39] But it is not known how sketchy or sophisticated these plans were. Some scholars regard the bulk of the preserved plans, and in particular the splendid elevations, as illustration drawings addressed to commissioners.[40] The reason for this ascription is mainly that these plans look too elaborated for drafts and too inaccurate for working drawings *(Werkrisse)*, that is, plans that were used in the construction process.[41] This leads to the question of exactly what functions plans had in the process of designing on the one hand, and for the construction process on the other, given the flexible character of the architectural practice outlined above.

Beginning with the function of plans in the architectural design process of the Middle Ages and the early Renaissance, there is little accord among the experts. Since no single plan is known that without any doubt served the purpose of design, and the same holds for sketches before Leonardo, one can only speculate about the graphic means that the architects of the age used when designing a sacral or profane building. It thus comes as no surprise that one meets the whole spectrum of possible assumptions in the literature, ranging from designing without any kind of plans[42] over sophisticated geometrical design procedures[43] up to designing with all kinds of orthographic plans.[44] One point, however, seems to be clear: Plans and other graphic representations were no means of reflection or thought experiments regarding the static of the planned building. Leonardo's sketches in the *Codex Madrid* are absolute exceptions. Another point seems to be at least very likely, namely the use of grids for designing ground plans. The famous plan of the Abbey of St. Gall from the ninth century obviously uses this grid technique, which can be traced back to ancient Egypt. There is no reason to assume that this suitable graphic technique went lost in the course of the Middle Ages.

With respect to the function of plans in the construction process, one should distinguish between reduced plans that are scaled down and full-size plans that are constructed to full scale (1:1). It goes without saying that, in contrast to full-size plans, reduced plans do not yield any information that can be picked up mechanically by means of compasses or comparable instruments. Rather, one first has to process the information of these plans through both arithmetical and geometrical procedures before measures and shapes can be transferred to the building materials. According to the literature, it is still an open question whether or not the few reduced plans we have can be taken to be accurately scaled plans, and therefore to be real working drawings.[45] But even if one regards some or all of them as working drawings, it has the appearance that their use was not widespread,[46] probably above all because the

---

39 See, for instance, Kletzl 1939, 4.
40 See, for instance, Saalman 1959, 103; see, on the other hand, Lepik 1994, 15.
41 For the term "working drawing," see French 1947, 307, for *Werkrisse*, see Kletzl 1939, 4ff.
42 See, for instance, Booz 1956, 68f.
43 Speculations about geometrical design procedures for the entire building, called *Proportionierungs-systeme*, go back to the *Neugothik* of the nineteenth century; for a principal critique see Hecht 1997; see also Müller 1990, 39ff.
44 See, for instance, Lepik 1994, 15.
45 See Booz 1956, 75ff., Hecht 1997, 361ff., Lepik 1994, 15f. I do not raise here the related question of reduced plans with dimensions and measures marked on them.
46 See for the following Müller 1990, 15ff.

craftsmen were not trained to use them. Furthermore, ground plans usually also did not function as real working drawings. The standard procedure for setting out the lines of foundations was not to transfer step by step the lines of a ground plan to the building ground, but to construct the design immediately on this ground with pegs and strings, whereby a plan, if one was used at all, served as a means of orientation rather than as a blueprint.

In contrast to reduced plans, full-size plans seem to have been an indispensable means of the construction process since the appearance of the Gothic style. As already mentioned, the oldest of such 1:1 constructions, that is, the *Ritzzeichnungen* scratched on walls or stone floors, precede the oldest known plans traced on parchment or wooden drawing tables. These plans usually concerned parts of the building with complicated shapes and curvature—e.g., arcades, windows, rose windows, and

Figure 7.7. Full-size drawing scratched in the terrace floor of Clermont-Ferrand Cathedral. (Photo: Robert Berger, Clermont-Ferrand.)

so on.[47] They often more resemble geometric constructions from which the form of a certain part of, say, a window can be taken mechanically, than they do plans of this window (figure 7.7). These constructions are thus closely related to other means employed in building practice for the creation of complicated shapes, ranging from special kinds of diagrams for the derivation of the rib systems of complicated vaults *(Bogenaustragung)*[48] to templates, which came into use in mediaeval architecture along with a standardization of stone-dressing at the beginning of the thirteenth century.[49] These geometric aids of the construction process, most of the full-scale *Ritzze-ichnungen* included, probably do not immediately come to mind when plans are mentioned and, indeed, can hardly be called architectural plans in the usual sense of the term. They are, however, reminiscent of the fact that such plans, if used in the construction process, are just one of many geometric means that support the creation of the desired shapes of a building. They point furthermore to geometrical practices of craftsmen that may be of interest with respect to the roots of the combined views technique as developed by Dürer.

Yet, before coming to the emergence of this technique, this concise survey of the use of plans in architecture before 1500 should be summed up by stressing that the architects and craftsmen of the age could obviously do very well without a general method of deriving elevations from ground plans. As Schofield put it for Italy:

> There can be little doubt that the orthogonal was used in the fifteenth century in the same way for the same type of architectural features, i.e., flat exteriors and flat interiors and details, but would not have been used for general views of interiors or exteriors where there were large-scale features that projected or receded from the front plane. Indeed, in the case of large-scale architectural units, it is to be doubted that a reasonably accurate, or even moderately inaccurate elevation drawing in perspective accompanied by dimensions (either drawn-on or on list) was any less useful than a fully marked-up orthogonal. The reason for thinking this is simply that a careful perspective drawing, especially one with dimensions marked on it, was better able than an orthogonal to provide precise information about details, large or small, that receded or projected from the frontal plane. This is confirmed by the practice of architects drawing in the early sixteenth century.[50]

Whether actually "better able" or only equally well, the architects of the age had in any case all means at their disposal needed to fulfill their tasks. Thus, in this investigation of the emergence of the combined view technique, one has to start from the fact that, obviously, no urgent needs of the construction practice stimulated Dürer or da Sangallo to create this new technique. Given this situation, it also comes as no surprise that this new method was only hesitantly adopted by sixteenth-century architects and became a standard technique of the art not before the seventeenth century.

---

47 For the full-scale *Ritzzeichnungen*, see Schoeller 1989 and 1980, as well as Davis 2002.
48 For such plans, see Müller 1990, 152ff.
49 See Müller 1990, 126ff. and Kimpel 1983. See also Shelby 1971.
50 Schofield 1991, 129.

3. THE GENERAL METHOD OF CONSTRUCTING ELEVATIONS

From the perspective of the history of mathematics, the new method in question is nothing but descriptive geometry in an early—actually in its first—but nevertheless manifest stage of development. However, in this first stage, this branch of mathematics looks not at all like what one expects from a scientific field. First, Dürer and da Sangallo were no mathematicians, notwithstanding the fact that they had obviously some mathematical knowledge and were in touch with men who can be regarded as mathematicians.[51] Second and more importantly, they established no theory. Da Sangallo did not publish a single treatise,[52] and Dürer's treatises do not argue for the new method of constructing plans but give, among many other things, practical step-by-step instructions on how to derive certain elevations from ground plans. Particularly his *Underweysung der Messung* from 1525,[53] the most important text in this context, looks like a book of recipes rather than a theoretical text. It is, thus, not sufficient to state that the beginnings of descriptive geometry were imbedded in the practice of artists. Rather, one has to realize that descriptive geometry was one of the techniques employed in this practice at the time and nothing else. Such practitioners' knowledge took on a scientific character only later, beginning at the turn of the sixteenth century, when men with scientific interests discovered its importance and reshaped it in theoretical frameworks which, at the same time, detached it from its origins.[54] The subject matter here is not descriptive geometry as it emerged through this appropriation and transformation by savants, but descriptive geometry as it was in its beginnings, that is, as practitioners' knowledge.[55]

In Dürer's treatises, one encounters for the first time a construction of an elevation by combined views in the first book of the *Underweysung der Messung*. The topic of this first book is techniques of constructing several shapes and curves by means of ruler and compasses. After a few pages with more or less trivial drawing tasks leading to the construction of several spirals (*Schnecken*—snails), Dürer almost casually goes on to elevations of helices, that is, of three-dimensional spirals. It is important to note that he himself does not use terms like projection or elevation in this passage as

---

51 For Dürer see, for instance, Steck 1948, 6 and 85ff. (note 8); for da Sangallo, see, for instance, Frommel 1994b, 3.
52 It is known, however, that da Sangallo considered publishing a commented edition of Vitruvius—see Frommel 1994b, 36.
53 Albrecht Dürer, *Underweysung der Messung / mit dem zirckel undrichtscheyt / in Linien ebnen unnd gantzen corporen / durch Albrecht Dürer zusamen getzogen / und zu nutz allen kunstlieb habenden mit zu gehörigen figuren / in truck gebracht / im jar. MDXXV*, published in Nuremberg. For a modern facsimile edition see, for instance, Dürer 1983. There exists no modern German edition of this treatise that would comply with present standards of critical editions. A facsimile edition with an English translation was edited by Walter L. Strauss—see Dürer 1977. Jeanne Peiffer edited a French edition along with an excellent introduction and valuable appendixes—see Dürer 1995.
54 For Dürer, see Peiffer 1995, 123.
55 This is the reason why Dürer and da Sangallo are not considered mathematicians in this chapter, which may appear strange in several respects. This categorization rests neither on a disdain of their mathematical abilities nor on their social standing—which contemporary mathematician enjoyed a higher standing than da Sangallo? The decisive point in our context is that these men developed and practised the new projective geometry as a technique of craftsmen and not as a doctrine. Furthermore, this technique cannot be taken to be an applied science. Rather, the later science of projective geometry emerged from this technique.

Strauss' English translation suggests.[56] Rather, he employs a practitioners' language in the fashion of the booklets of Roriczer or Schmuttermayer when describing the procedure: "to draw it [sc. the snail] up from below over itself" ("sie von unden uber-sich ziehen") or "the snail drawn up out of the ground" ("der schneck auß dem grund auf gezogen"). This plastic language emphasizes the essential feature of the technique, namely that the helix has to be developed out of the ground plan. He also stresses explicitly that there is no other technique of constructing the elevation of the helix than this development out of the ground plan. This said, he goes on to teach the technique by giving minute instructions on how to proceed step by step.

The first helix in book I has the particular feature that it becomes steeper and steeper as it winds up (figure 7.8). Dürer mentions an architectural application of this helix that is not entirely clear[57] and assures generally that this helix is of manifold use. Next he displays a second helix that winds up in equal degrees (figure 7.9). Again and more intelligibly than in the first case, he stresses the architectural benefit of this helix, namely for masons in charge of constructing a spiral staircase. His wording[58] actually indicates that such a helix was already known and used by masons. No matter what masons' plans Dürer had in mind,[59] it is interesting to see that Dürer was fully aware that the practical character of such designs had two dimensions: They were set down for the benefit of several crafts, on the one hand, and they were rooted in practitioners' knowledge, on the other. In what follows, I will deal with these two dimensions of the new method of constructing elevations—with its applications as an important context and with its prerequisite knowledge as an important background.

Beginning with the applications of the new method, architecture, on which this chapter focuses and which, surprisingly, was mentioned as the first such field of application by Dürer, naturally comes to mind with respect to da Sangallo. The latter is the first architect known who systematically constructed elevations by combined views when designing buildings (figure 7.10). He also used this technique for recording existing buildings *(Bauaufnahmen)*, in particular, edifices or ruins of Antiquity, which were zealously investigated by the members of the Bramante and Sangallo circle in Rome (figure 7.11).[60] For Dürer, too, the benefit of the new method for architecture was obviously more than a mere side effect of a painter's achievement. He dedicated the first half of the third book of the *Underweysung* to architectural issues and employed the new method there where appropriate.[61] Moreover, Dürer, himself not active as an architect but ambitious in engineering like so many of the great artists of

---

56 See Dürer 1977, 61. Unfortunately, Strauss' translation is not always reliable and often fails at exactly those passages that are decisive for understanding.
57 "[...] schnecken steyg / in ein durn dach [...]": spiral staircase in a spire? See Dürer 1977, 61 and Dürer 1995, 153.
58 "[...] schneckenlini [...] die auch die steyn metzen zu den stygen gebrauchen [...]"—snail line [...] which is also used by masons for stairs. See Dürer 1977, 67 and Dürer 1995, 155.
59 Ground plans of staircases can be found, for example, in the sketch-book of master WG (fl. about 1500) which is in the collection of the Städelsche Kunstinstitut Frankfurt am Main—see Bucher 1979, plates WG 19, 26, and 35.
60 See Frommel 1994b, 9f.
61 See book III, figures 1, 2, 5, 6, 8, and 12;—in Dürer 1977, pp. 186, 188, 200, 202, 208, 222; in Dürer 1995, pp. 232, 234, 246f., 249, 258.

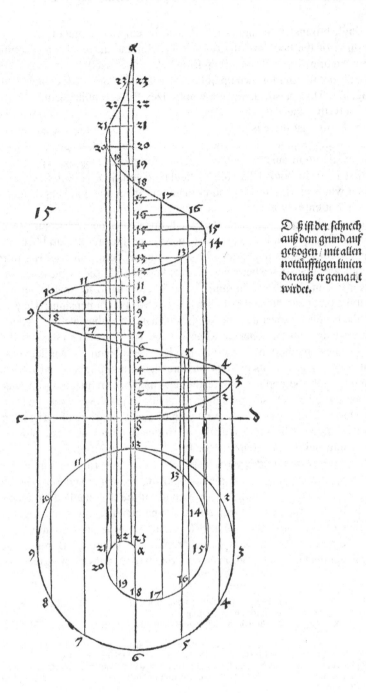

Figure 7.8. Elevation of a helix. (Dürer 1983, figure I.15.)

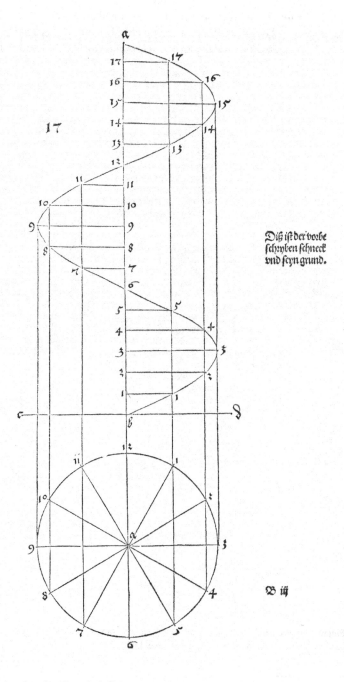

Figure 7.9. Another elevation of a helix. (Dürer 1983, figure I.17.)

the age, published a treatise on fortification that contained architectural plans including combined views.[62] These are the first architectural plans of this kind ever published in print. However, as contended above, sixteenth-century architects adopted this

Figure 7.10. Project for a chapel. Drawing by Antonio da Sangallo the Younger,
c. 1535. (Florence, Gabinetto Disegni e Stampe, U172A^r.)

62 Albrecht Dürer, *Etliche underricht / zu befestigung der Stett / Schloß / flecken*, published 1527 in
    Nuremberg. For a modern facsimile edition, see Dürer 1969.

Figure 7.11. Tomb of Theodoric recorded by Antonio da Sangallo. (Florence, Gabinetto Disegni e Stampe, U1536A[r] and 1406A[r].)

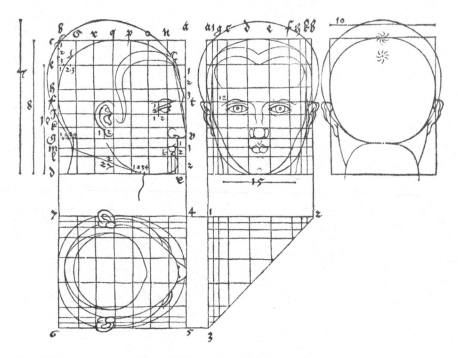

Figure 7.12. Combined plans. (Dürer 1996.)

new method only very reluctantly since they could still do very well without this innovation. Only in exceptionally complex cases, like the new *San Pietro* in Rome, did the familiar methods of designing demonstrate their limitations. This state of affairs did not change until the emergence of the baroque style in architecture.[63]

The field of application of the new method that Dürer doubtless had in mind primarily is that of art, and in particular that of painting—the term "painting" being used here to comprise drawing, engraving, etching, calligraphy, and so on. The most developed examples of the technique of combined views one can find in Dürer's writings occur in his treatise on the proportions of the human body,[64] one of the famous Renaissance treatises on painting. Here one even encounters the arrangement of three-view drawings (figure 7.12). Of particular interest is the invention of the triangle, which allows the connection of those lines in the ground plan and the front view that cannot be connected directly by straight lines. Dürer called this triangle "der ubertrag," the transfer, and gave step-by-step instructions on how to construct it. But,

---

63 As in the case of the Gothic style four hundred years previously, baroque architecture required new drawing methods, probably less because of its constructional problems and dimensions than because of its ornamental features, which included shapes of unheard-of complexity.

64 Albrecht Dürer, *Hierinn sind begriffen vier bücher von menschlicher Proportion*, published 1528 in Nuremberg. For a modern facsimile edition, see Dürer 1996.

as usual, he stuck to the manner of practitioners' booklets and refrained from giving geometrical explanations of such issues as, say, why the distances among the connecting lines are preserved by the "ubertrag."

The importance of painting as a context of the new method is indicated not only by the fact that, in Dürer's work, its applications in connection with painting prevail over those in any other field, but above all by the intimate relation of the new method to the perhaps most important technical achievement of Renaissance painting, namely to linear perspective. Actually, the new method was first developed in the context of linear perspective.

When the development of perspective depiction entered its first manifest stage in the fifteenth century, it was accompanied by treatises, with or without diagrams, that tried to establish rules on how to construct perspective pictures.[65] Furthermore, the first of those treatises, Alberti's *De pictura* (c. 1435), already offered reflections on such rules. These essays implicitly employed combinations of plans in the manner and in continuation of those diagrammatic representations of the visual cone that had been used in optics since antiquity, and also in mediaeval treatises on surveying. These diagrams are implicit in the arguments about how to construct perspective pictures even when, as in the case of Alberti's treatise, no figures are attached.[66] Being introduced and handled in this new context, it seems to have been only a question of time until these diagrams, which are essentially ground plans and elevations, were taken to be such plans, and until the potential of combined plans for constructing perspective pictures was realized by one of the artists.

Indeed, only about forty or fifty years after Alberti's treatise, this potential of plans was discovered, developed, and set down by Piero della Francesca (c. 1420–1492). In his treatise *De prospectiva pingendi,* written about 1480, Piero employed the technique of combined plans not only for the construction of the points where the visual rays cut the drawing plane,[67] but also for the establishment of the so-called *construzione legittima* of perspective depiction.[68] Furthermore, he apparently was fully aware of the fact that the correctness of rules for perspective constructions can only be assessed and demonstrated by means of combined plans.[69] What is more important in the context of this chapter, Piero also used this technique to construct orthogonal projections of the objects that were to be depicted in perspective. Aiming, for the sake of generality, at demonstrations that hold for every object, regardless of how its surfaces might be situated with respect to the drawing plane, he demonstrated how to transform the orthogonal depiction of an object in such a way that it appears

---

65 Note that I do not deal here with the more or less unknown rules that painters of the Renaissance actually followed when producing a perspective picture, but only with stated rules documented in texts.

66 See §§ 14–20 of Alberti 2000. For this background of Alberti's treatment of perspective depicting, see part 1 of the chapter by Filippo Camerota.

67 See, for instance, figure XLVI in Piero della Francesca 1984.

68 See Panofsky 1955, 249f.; see also Kemp 1990, 27ff. The term *construzione legittima* was first coined at the beginning of the seventeenth century by Pietro da Fabbrizio Accolti who ascribed this method to Alberti—see Field 1997, 30.

69 See Andersen 1992, 15.

to be turned around, tilted, shifted, and so on. In doing so, he unfolded in a masterly manner the possibilities of deriving plans from plans.[70]

One thus encounters the first mature instance of the combined views technique not in the domain of architecture, but in that of painting.[71] Although this achievement appears almost natural in the context of fifteenth-century perspective depiction and contemporary essays to establish its rules, it is perhaps not by chance that it was eventually attained by a man of outstanding mathematical education.[72] In the form developed by Piero, this technique went far beyond both the needs and the knowledge of an artist trained to the usual standards of the age. Thus it can be doubted that this method would have spread among artists even if Piero's treatise had been published.[73] It seems rather likely that his treatise would have been eclipsed by treatises such as Viator's *De artificiali perspectiva* (1505),[74] which avoided complex constructions and provided, instead, handy rules and examples appropriate to the needs and the grasp of the artists. This conjecture is also backed by the fate of Dürer's exposition of perspective construction in his *Underweysung*, which is almost as complex as Piero's.[75] This specific exposition's career in the sixteenth century rested on the Latin translation of the book,[76] that is, on its separation from the sphere of common artists.[77] The famous question of whether or not Dürer had knowledge of some of Piero's achievements[78] can, therefore, be put in this way: How was Dürer able to capitalize on Piero, if he actually did so,[79] or, if he did not, to invent the technique of combined views independently?

As already mentioned above, Dürer was interested in mathematical questions[80] and had command of mathematical knowledge that, though certainly not comparable with that of Piero, was at least uncommon among his colleagues.[81] In the fourth book

70 See, for instance, figures LII–LIV in Piero della Francesca 1984. In not yet published notes of lectures taught at the University of Lüneburg, Diethelm Stoller reconstructed by means of diagrams the transformation steps that Piero only described (Piero della Francesca 1984, 145ff.) but did not depict.

71 In Piero's treatise, one finds the technique of combined views also employed for rendering buildings or parts of them—see Piero della Francesca 1984, figures XLI, XLII, or LIX and LX. But his purpose was, of course, the depiction of such buildings and not their construction.

72 Piero wrote two books on mathematical topics (*Trattato d'abaco* and *De corporibus regolaribus*) and was the teacher of Luca Pacioli. Several of the achievements that were formerly ascribed to the latter are owed to Piero—see, for instance, Mancini 1916.

73 *De prospectiva pingendi* was not published until 1899, his other writings subsequently in the twentieth century. Before his rediscovery in the last century, Piero was almost forgotten even as painter.

74 A facsimile edition of this booklet can be found in Ivins 1973.

75 This exposition can be found at the end of the fourth book—see Dürer 1977, 364–389 and Dürer 1995, 335–351.

76 According to Steck 1948, 106f., two different Latin editions were published in 1532 and, apparently, a third one in 1534, all of them in Paris.

77 For the rather ambiguous impact of Dürer's exposition on contemporary artists, see Peiffer (in this volume), Kemp 1990, 61f., Panofsky 1955, 253 and 257. In his usual attention to practical applicability, Dürer added two mechanical devices for perspective drawing (see Dürer 1977, 387ff. and Dürer 1995, 351ff.) which can be considered derivatives of Alberti's *velum*.

78 See, for instance, Panofsky 1955, 251f. Despite some striking parallels, it seems very unlikely that Dürer ever had access to Piero's perspective treatise itself. Even if he actually obtained knowledge of the *construzione legittima* through an unknown intermediary, it still remains an open question whether he received the "original" Piero or only something derived from him.

79 See Peiffer 1995, 100.

80 Fortunately, among the few preserved letters of Dürer, there is one from 1522 addressed to the imperial architect Johann Tscherte, which testifies that the two men exchanged thoughts about the solution of geometric problems—see letter no. 40 in Rupprich 1956–69, I 94f.

of his *Underweysung*, for instance, he is dealing with the Delian problem,[82] that is, a highly theoretical problem of classic geometry, notwithstanding the fact that Dürer, as always, saw possible practical applications of its solution. Another example of his mathematical interests is closer to the topic of this chapter. In the first book of the *Underweysung*, he dealt with conic sections or, more precisely, with the problem of how to draw the curves of these sections.[83] And here one encounters a further field where the technique of combined views was applied, for Dürer constructed the ellipse, the parabola, and the hyperbola by means of combined plans (figure 7.13).

Strikingly, ellipse constructions by means of combined plans also occur in the papers of da Sangallo (figure 7.14).[84] It is sheets like these that confirm unambiguously that da Sangallo used combined orthogonal plans as a general technique and not only in the context of architectural plans. Scrutinizing the sheet, however, it is not entirely clear whether da Sangallo actually tried to produce the ellipse in the same way as Dürer, or whether he considered the ellipse to be a distorted image of a circle seen from an oblique point of view, and therefore constructed, by means of combined views, the perspective setting in which a circle would appear as an ellipse.[85] In any case, the amazing resemblance of da Sangallo's arrangement of the plans in this construction to Piero's usual arrangement raises the question of whether the former had knowledge of the latter's treatise on perspective.[86] Given the fact that the members of the Bramante and Sangallo circle obviously had command of advanced methods for constructing perspective drawings, it might well be the case that even the architect da Sangallo first acquired the combined views technique in the context of linear perspective.

Linear perspective thus appears not only as an important field of application for the combined views technique, but also as one of its roots. However, not the handy perspective technique accessible for and used by the common artist of this age, but only a geometrically modelled perspective could be exploited as such a source by men of some learning. Therefore it is not only the practical, but at the same time the learned background of the new method that one encounters in linear perspective. Did Dürer's technique of combined views also originate from genuine practitioners' knowledge?

As discussed above, the technique of combined views is essentially a general graphic technique of deriving plans from plans, and, in particular, elevations from

---

81   The classic work on Dürer and mathematics is Staigmüller 1891. Steck 1948 provides a still very useful collection of all kinds of biographical materials relating to Dürer and mathematics.

82   Geometrically phrased, the problem of finding the side of a cube with double the volume of a given cube—see Dürer 1977, 347–363 and Dürer 1995, 324–334.

83   On contemporary endeavours to find convenient techniques or tools for drawing these curves, see Rose 1970.

84   See U 830A recto in Frommel 1994a, 332.

85   In this chapter, I cannot go into the obvious problems of this construction. There are traces that the transformation of a circle into a ellipse through oblique projection was already discussed in Antiquity—see theorems 34 and 35 in Euclid's *Optics* and proposition 53 in book VI of Pappus' *Mathematical Collections*.

86   This question is also suggested by da Sangallo's two constructions of the famous *mazzocchio*—U 831 A recto and U 832 A recto (Frommel 1994c, 333)—which strongly resemble those of Piero—see Piero della Francesca 1984, figure L. For the two constructions of da Sangallo, see Frommel 1994c, 150.

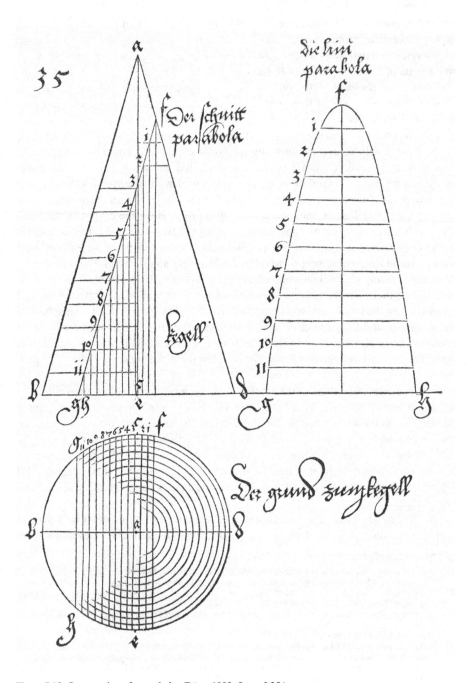

Figure 7.13. Construction of a parabola. (Dürer 1983, figure I.35.)

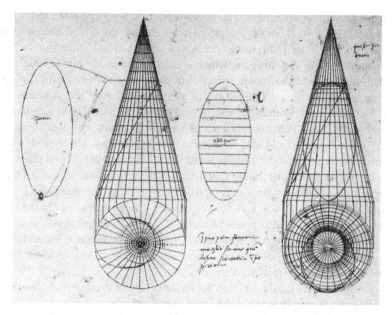

Figure 7.14. Construction of an ellipse. Drawing by Antonio da Sangallo the Younger. (Florence, Gabinetto Disegni e Stampe, U830A[r].)

ground plans. If one abstracts from the learned background of geometric construction and representation that shaped the development of this technique in the context of linear perspective, an elementary practical feature of this technique seems to be decisive, namely the treatment of two different plans, the ground plan and the elevation, as a unit. For the graphic derivation of the elevation from the ground plan in a way suggests itself if, but only if, both plans are treated as a unity. A systematic coordination of plans that allows their treatment as a unity seems, thus, to be the practical key for the combined views technique. Does one encounter such a coordination of plans, presupposing plans of the same scale and with perpendicular planes, in the realm of the practitioners, and if so, in which craft?

Considering the conclusions of our survey of the use of plans in the building trade of the age, the contemporary architectural practice of plan drawing seems to be a very unlikely candidate. Instances of ground plans and elevations, or sections for that matter, of one and the same part of a building are generally very rare before the sixteenth century. And in the few plan sets of this kind preserved, ground plan and elevation (section) appear almost never to be drawn to the same scale, least of all unambiguously coordinated (if at all), and are in the most cases even traced on different sheets.[87] The few combined plans mentioned but not discussed above appear all the more interesting: Among the preserved architectural drawings produced prior to Dürer and da Sangallo, there are indeed two or three plans that are, or might be considered to be, instances of such a coordination.[88]

The most beautiful example and at the same time the only one that can unambiguously be taken to be such a coordination of plans is the ground plan and elevation of the tower of the *Münster* in Freiburg (Breisgau), preserved in Vienna (figure 7.15).[89] It goes without saying that this plan, which is dated to sometime after 1400,[90] cannot be taken as testimony to the drawing practice of common architects of the age. On the contrary, it is one of a very small number of exceptional plans[91] that pose exactly the same questions regarding their roots and background as does Dürer's technique of combined views. One particular feature of the set of plans, however, hints at the direction in which one should seek the origins of this technique. On closer inspection, the ground plan turns out to be a whole set of several superposed ground plans, which represent horizontal sections through different levels of the tower (figure 7.16). Such a superposition of plans, which may appear strange from the perspective of later architectural plan conventions, was not only a wide-spread feature of ground plans of towers at that time,[92] but also points to a chief means of designing common among stonemasons and stonecutters.[93] Rather than fixing one's attention on architectural plans when looking for roots of the combined views technique, one should follow this hint and look at the practice of stonemasons.

Interestingly, Dürer himself related the coordination of ground plan and elevation not to architectural drawings but to the art of stonemasons:

> It is, therefore, necessary for anybody who wants to venture this art to be well acquainted in advance with measurement and to obtain an understanding of how ground plan and elevation of all things need to be drawn [literally: how all things need to be laid into ground and drawn up] as practised by the skilled stonemasons every day.[94]

It seems noteworthy that Dürer does not address the combination of ground plan and elevation as a new technique just developed by the stonemasons of his day. On the contrary, he actually seems to express the view that this technique had always pertained to the art of stone-dressing. Indeed, when stonemasons carve outlines on two

---

87   The same holds for the drawings of goldsmiths with which the young Dürer had become familiar in the workshop of his father. The famous collection of such design drawings *(Goldschmiederisse)* held by the Basel Kupferstichkabinett displays those patterns that were characteristic for architectural plans: plenty of "wrong" elevations, though also a remarkable amount of "right" ones, cases of resorting to perspective rendering, separation of ground plans and elevations, no single instance of combined views, and so on. See Falk 1979, figures 371–702.

88   See, for instance, Sakarovitch 1998, 43. Sakarovitch's second example, i.e., Antonio di Vicenzo's famous plan for the Milan cathedral from 1389, which inserts sections of three naves into the ground plan, is a rather problematic instance—see the discussion of this plan in Hecht 1997, 158ff.

89   Kupferstichkabinett der Akademie der bildenden Künste Wien, Inv.-no. 16874; reproduced, for instance, in Koepf 1969, figure 78 and in Recht 1989, 413. Sakarovitch 1998, 42 reproduces a different plan of this tower held in the Germanische Nationalmuseum Nürnberg. For this plan, see Broda 1996, 34ff. Another drawing that can, to some extent, be regarded as such a combined view, is the plan of the east façade of Clermont cathedral held in Clermont-Ferrand—see Recht 1989, 420.

90   See Koepf 1969, 16.

91   Some of which are only copies or derivations of this plan—see Koepf, *ibid.*

92   See, for instance, Koepf 1969, figure 12, 15, 16, 65, etc.

93   See Shelby 1977, 74ff.

94   "Darumb thut einem yglichen der sich diser kunst understeen will not / das er zuvor der messung wol underricht sey / und einen verstand uberkome / wie alle ding in grund gelegt / und auffgezogen sollen werden / wie dann die kunstlichen Steinmetzen in teglichem geprauch habenn […]." Dürer 1996, dedication to Pirckheimer, my translation.

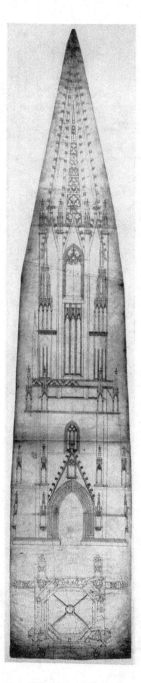

Figure 7.15. Ground plan and elevation of Freiburg Münster tower. (Vienna, Kupferstichkabinett der Akademie der Künste, Inv. Nr. 16.874; courtesy Kupferstichkabinett der Akademie der Künste.)

Figure 7.16. Ground plan with superposed levels. Detail from figure 7.15.

or three adjacent sides of a square hewn stone, these sides appear—at least in hind-sight—as different views of the intended finished stone, which are attached to one another in such a way that simply folding them into the same plane would suffice to make them veritable combined views. It seems therefore very likely that techniques equivalent to that of interconnected plans were used by stonemasons from time immemorial. Indications of such a practice can be found in ancient Egypt.[95] Unfortunately, for want of documents, one does not know very much about the details of this practice for the epochs prior to the sixteenth century, when it became the highly developed art of stereometry; even later, it was transformed into a branch of theoretical geometry.[96] No wonder that the one and only document of this art from the Middle Ages, which can be found in the aforementioned sketch-book of Villard de Honnecourt from the first half of the thirteenth century,[97] attracted the attention of many scholars. Although this document poses still insurmountable obstacles for our understanding,[98] some general features of this technique became comprehensible. Drawing a general conclusion of his investigations into Villard's sketch-book, Lon Shelby characterized the geometry of mediaeval masons as

> constructive geometry, by means of which technical problems of design and building were solved through the construction and physical manipulation of simple geometrical forms: triangles, squares, polygons, and circles.[99]

This telling characterization is confirmed by two documents from the late fifteenth century, which give insight into a technique of developing geometrical forms in the designing process of special parts of Gothic churches. This technique, which is closely related to the construction of ground plans and elevations, is taught in two of the practitioners' booklets mentioned above, namely the booklets on designing pinnacles of Matthäus Roriczer (fl. 1460–1490) and of Hans Schmuttermayer (fl. 1480–1520).[100] This stonemasons' design technique is not the invention of these two men but goes back, as Roriczer himself assumes, to the "iungkeh(er)rn von prage,"[101] that is, the Parlers in the second half of the fourteenth century, and may be even older as regards its principal features.

What is taught in the booklets consists mainly in sophisticated rules for developing a pinnacle's horizontal forms and dimensions at different levels by turning and transforming a square in certain ways, rules which fit Shelby's characterization perfectly. The result of these manipulations of a square is a diagram (figure 7.17) that provides the measures of essential parts of the pinnacle and can at the same time be

---

95 See, for instance, the elevations of capitals from Abu Fodah (Heisel 1993, figure Ä30 and Ä31) in connection with the famous elevations of a shrine from Ghorab (*ibid.*, figure Ä19 and Ä20); see also Sakarovitch 1998, 24ff.
96 See chap. 2 and 3 of Sakarovitch 1998. See also part 4 of the chapter by Filippo Camerota.
97 See Hahnloser 1972, tables 39, 40, and 41.
98 See *ibid.*, 104ff.; see also Shelby 1971 and 1972; see furthermore Bechmann 1993, chap. 5 and 6.
99 Shelby 1972, 409.
100 Coenen 1990 offers diplomatic copies of these booklets, Shelby 1977 an modernized edition and English translation. See for the following, for instance, Müller 1990, 59ff. and the introduction of Shelby 1977.
101 Coenen 1990, 312.

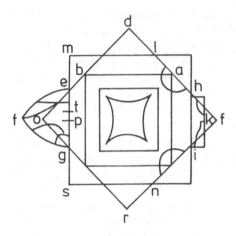

Figure 7.17. Scheme of horizontal sections of a pinnacle. From Matthäus Roriczer *Büchlein von der Fialen Gerechtigkeit*, 1486. (Coenen 1990, 337.)

taken as a ground plan of the pinnacle that consists of superposed horizontal sections at different levels. Taking the presumably old age of this stonemasons' design technique into account, it seems very likely that one encounters here the roots of the striking convention of superposed ground plans observed in the case of the plan set of the Freiburg *Münster* tower.

In Shelby's "constructive geometry" of the masons, one thus apparently encounters a practical geometry that constituted the background of the architectural plan construction of the age. True, this geometry does not aim firstly at the construction of plans but at that of diagrams, templates, and the like as means of determining and manufacturing the shapes of stones or composed parts of a building. But it makes little sense to assume that the master masons did not rely on the resources given by this constructive geometry when they had to construct plans. Moreover, as remarked above, plans that were used in the construction process—e.g., the full-scale plans scratched on the floor or walls of the building *(Ritzzeichnungen)*—were just one of many geometric means that supported the creation of the desired shapes of a building. They are products of the same practical geometry by which the masons constructed templates or diagrams like that in the booklets of Roriczer and Schmuttermayer. At least with respect to the countries north of the Alps, it may be well advised, therefore, to look at architectural plans of this age from the perspective of this constructive geometry rather than of our projective geometry.

The booklets also teach how to derive the "body" ("leib") of the pinnacle, that is, its vertical shapes, from these superposed horizontal sections.[102] The method taught is not that of a graphic derivation of an elevation by treating the two drawings as a unit, but that of determining the measures of the "body" by processing those of the ground plan in such a way that certain proportions between breadth and heights were met.[103] This is in accordance with the purpose of the two booklets, which do not aim at constructing plans, but at determining measures and forms. Nevertheless, if one abstracts from the development of the vertical measures of the "body" and focus on the horizontal ones, one encounters procedures of how to take distances from the ground plan using a compass and how to transfer them to the diagram of the "body" to construct its elevation.[104] In other words, the booklets give testimony to geometri-

---

102 See, for Roriczer, Coenen 1990, 316ff. and Shelby 1977, 90ff., for Schmuttermayer, Coenen 1990, 355ff. and Shelby 1977, 129ff.
103 See Shelby, 70f.

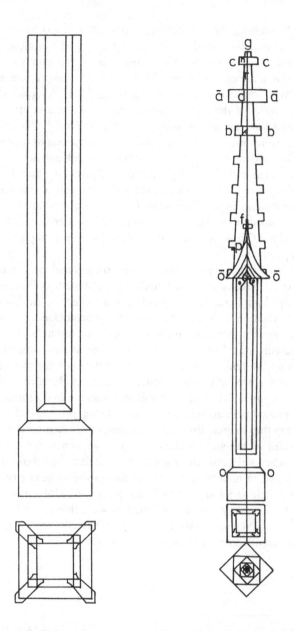

Figure 7.18. Ground plan and elevation coordinated. From Matthäus Roriczer *Büchlein von der Fialen Gerechtigkeit,* 1486 (left), and Hans Schmuttermeyer *Fialenbüchlein,* c. 1485 (right). (Coenen 1990, 337 and 365.)

---

104 Additionally, the advantage of superposed ground plans becomes visible if it is, as in the case of pinnacles, about the determination of the vertical shapes of building parts that narrows towards the top.

cal methods of determining the vertical shapes of building parts based on the ground plan, which are of immediate bearing for the construction of elevations.

With respect to Dürer's method of combined views, two points of the booklets' elevation construction are of special interest. The one is of linguistic nature. Roriczer uses the words "den grunt auszczihen" and "ausczug"[105] to designate the action and the result, respectively, of the construction of an elevation, and also indicates that these denominations are the established terms of the craft.[106] These are, however, exactly the terms Dürer uses for the construction of elevations as quoted above: "sie [die Schnecke] von unden ubersich ziehen" or "der schneck auß dem grund auf gezogen."[107] The language that Dürer uses in the *Underweysung* is obviously closely related to, if not dependent on, that connected with the practical geometry of masons as documented in the contemporary masons' booklets. This is one of the reasons why Dürer scholars regard it an established fact since Staigmüller's essay from 1891 that the knowledge of the Gothic masons and goldsmiths form a important background for his practical geometry.[108]

The second interesting point concerns the arrangement of the figures in the masons' booklets. As already said, the booklets teach a method of deriving the measures of the "body" from those of the ground plan that is geometric, but does not consist in a graphic derivation of the elevation from the ground plan by arranging the two plans in such a way that they can treated as a unit. Accordingly, the booklets' diagrams that represent ground plans and elevations are usually separated from each other and drawn to different scales. But, in each of the booklets, there also is one or two drawings where a ground plan is coordinated with an elevation in the manner of combined views (figure 7.18). It is, of course, not possible to determine whether this arrangement of ground plan and elevation can be taken as a trace of a more graphic method of deriving the one plan from the other and, thus, whether such a method was in use in the realm of the stonemasons' practical geometry, but is not dealt with explicitly by Roriczer and Schmuttermayer. But these drawings show in any case that it was obviously not an unfamiliar feature of this practical geometry to coordinate plans of the same scale and with perpendicular planes, a coordination, which can be regarded as the practical key for the combined views technique. It therefore has the appearance that this geometry constitutes the practitioners' knowledge sought as a root and background of Dürer's technique of combined views.

---

105 Shelby (*ibid.*, 90) translates with "extrapolate the base plan" and "extrapolation," respectively. It is essential to be aware of the close relation between the German "ziehen," from which the verb "ausziehen" and the noun "Auszug" are derivatives, and "zeichnen:" "eine Linie ziehen" means "to draw a line;" thus, "draw out of the ground" may be a more adequate translation since "to draw" means both "to pull" and "to trace."

106 "... merck der wirt gehayssen ..." (... note, it is called ...)—Coenen 1990, 316 and Shelby 1977, 90.

107 "[...] to draw it [sc. the snail] up from below over itself"—"the snail drawn up out of the ground." See text and caption to figure 15 of the first book of Dürer 1983.

108 See Günther 1887, 358, Staigmüller 1891, 50ff., Steck 1948, 83f. (note 6), and Peiffer 1995, 51ff.

8

PROJECTIONS EMBODIED IN TECHNICAL DRAWINGS: DÜRER AND HIS FOLLOWERS

JEANNE PEIFFER

INTRODUCTION

The aim of this chapter is to focus on technical drawings as mediators between practical and theoretical knowledge. Albrecht Dürer's *Underweysung der Messung* (Nuremberg 1525) offers some interesting examples of drawings embodying workshop techniques and aspects of theoretical knowledge. My reading of these drawings will make apparent a hitherto unnoticed aspect of Dürer's notion of *Messung*, namely construction in a visual space. In the second section, I look at the changes that occur in the understanding of that notion in its use by the immediate followers of Dürer, mainly in Nuremberg. Special attention is paid to the use of plans and instruments in the representation practices developed from Dürer's methods. The last section offers an outlook into the early seventeenth century, where on the one hand one encounters investigations of a more mathematical character with men like Egnazio Danti, Guidobaldo del Monte, and especially Girard Desargues, and on the other, some of their methods are discussed, criticized, and rejected by practitioners like the Parisian stonecutters. The chapter closes with a short depiction of the debate between Abraham Bosse and the French Academy, a debate in which the notion of representing things as they are perceived, and not as they are, is violently rejected.

1. CONSTRUCTIVE GEOMETRY IN A VISUAL SPACE: ALBRECHT DÜRER

When Dürer calls the geometry, the instructions of which he is assembling, *Messung* (from *messen*, to measure), he is referring to a geometry that owes a lot to the Euclidean *Elements*, but is far from demonstrative geometry. It is a constructive, concrete and material geometry that Dürer has in mind, one which has to do with real objects and artifacts that the artist can place before him ("vornemen"). He only once uses the Latin "geometria," which clearly refers to Euclid. Thus *gemessen linien* are curves constructed pointwise by ruler and compass. The term *gemessen* may also evoke three-dimensionality as in *gemessen leng* (II.30),[1] or the materiality of a line, as opposed to the abstract lines of Euclidean geometry. In his *Underweysung der Messung*, Dürer aims to offer his companions a range of geometrical forms, regularly constructed by ruler and compass, be the construction exact or approximate. I will concentrate in what follows on another aspect of his *Messung*, an aspect which has to do with vision. Some of Dürer's drawings take into account the fact that the objects

---

1 See Dürer 1983, II.30. I adopt this condensed form to refer to Dürer's *Underweysung der Messung*: book II, figure 30 or the text preceding the figure.

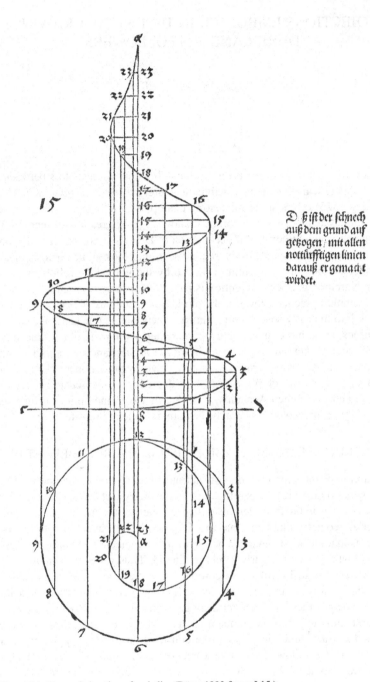

Figure 8.1. Plan and elevation of an helix. (Dürer 1983 figure I.15.)

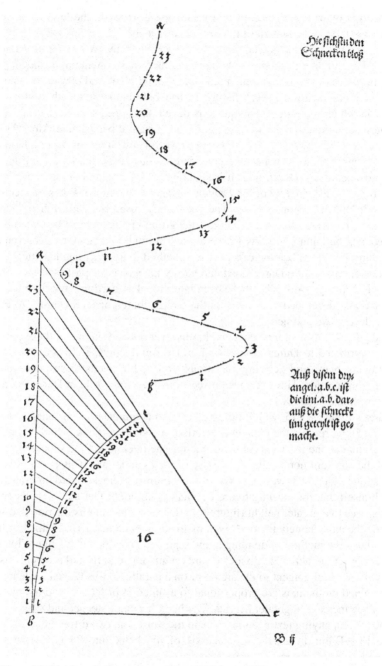

Figure 8.2. Division of a line segment. (Dürer 1983 figure I.16.)

he conceives of will be displayed in a visual space. How do the laws of perception intervene in the representation of the perceived objects?

In Dürer's geometry, objects are often represented by two of its projections, a ground plan and an elevation. As Wolfgang Lefèvre has convincingly shown in section II of his chapter, orthographic plans like ground plans and elevations, and especially their combinations, were hardly in use before the sixteenth century.[2] The context in which the latter technique was developed appears to be that of painters, like Piero della Francesca, who were interested in central perspective. Albrecht Dürer gives in his *Underweysung der Messung* powerful testimony of an excellent command of these methods, which are considered today to be the first occurrences of descriptive geometry. Dürer himself is silent on his method and his sources. We only know from the introduction of his *Proportionslehre*[3] that he considers this technique to be well known to stonemasons. The vocabulary[4] used by Dürer mirrors material procedures, like extracting a curve out of the ground plan, and can be traced back to the medieval building crafts. As I have argued in my book,[5] Dürer's achievement is outstanding for its application of a practical method to abstract mathematical objects like skew curves. While Dürer's text can hardly have been inspiring—he was hardly read—his drawings were. They have been interpreted as starting point for a branch of geometry that developed only towards the end of the eighteenth century in France, namely descriptive geometry.

And yet, there is a strange element, hardly ever noticed, in one of these drawings, which contributed to Dürer's fame. As Lefèvre rightly says, architectural recording like drawing plans and deriving elevations from them, needs to suspend vision in favor of construction. However, vision seems to play some role in Dürer's construction of the helix (figure 8.1).

Dürer's drawing I.15, which Lefèvre has also commented in his chapter (see his figure 7.15), represents the plan and elevation of a skew curve, a cylindrical helix on which a conical one is seated. In the elevation, the intervals of the subdivision of the axis of the cone are not equal, while those of the cylinder are. This is not obvious in Dürer's figure I.15, as it was engraved by the "Formschneider" based on drawings by Dürer himself, but the ensuing figure I.16 leaves no doubt that the intervals grow as the axis rises to a greater height (figure 8.2). Dürer even explains the rule that served to divide the axis, namely the one given by his figures I.8 and I.16. In this last figure, Dürer adapts the method of dividing a line segment, described in his figure I.8, to the construction of the helix. He divides an arc *be* into equal parts and projects this division on the vertical tangent to the arc *ba* or on a parallel to this tangent. The intervals thus obtained on the axis are proportional to a function of $tg(\alpha/n)$ and are no longer equal. They increase at greater heights. The helix becomes steeper and steeper.

The accompanying text is ambiguous in the sense that two different constructions collide. First Dürer divides the vertical axis of the helix into 24 equal parts. This

---

2   See also Sakarovitch 1998, chap. 1, for a depiction of the combined plan and elevation technique.
3   Dürer 1996.
4   See for instance Peiffer 1995, 55 and Lefèvre in this volume.
5   Peiffer 1995.

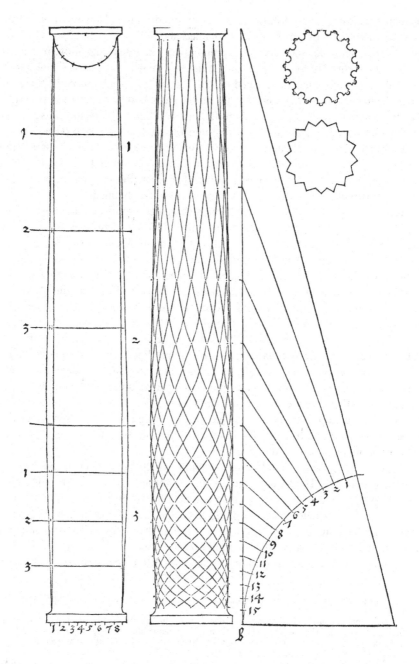

Figure 8.3. Construction of a convex column. (Dürer 1983 figure III.7.)

instruction is immediately followed by a contradictory one: "However I want here to have the intervals grow in the order explained above"[6] (i.e. in his figure I.8). First, Dürer explains how to construct a skew curve by extracting its elevation from its plan. Second, if the curve is applied to the construction of a spiral staircase in the spire of a tower, then the steps must be reduced in length in orderly progression the higher they rise in the tower, i.e. the helix is conical, and the steps will grow progressively steeper such that the axis is divided as in figure I.16 above. In the case of the construction of a mathematical object, the intervals dividing the axis of its elevation are of equal length. In an architectural context, however, when building a circular staircase in the spire of a tower, the subdivisions have to be altered in order to conform with the experience of steps becoming smaller and steeper when climbing up the tower. Dürer offers a construction that allows the desired effect to be obtained.

Book III of the *Underweysung*, which is dedicated entirely to architectural studies, offers some other applications of the same dividing technique. In the opening section, after some definitions of volumes, cones and pyramids, Dürer designs primarily columns, monuments and towers. Elements of decoration, like sun-dials and lettering, close the book. Dürer explains how to generate a column or a pointed solid, a pyramid or a cone, from a given ground plan. For instance, if the ground plan is a circle, he produces a column by raising this circle up to the desired height[7] and a cone by raising it towards a point.[8] More generally, the "kunstreychen bauleut," i.e. the artful builders in Strauss' translation,[9] know how to give the right volume, "in rechter maß beleyben," to a ground plan drawn with simple lines. They also know, from the given sections, how to decorate the columns and their different elements.[10] When constructing convex columns, Dürer applies the procedure described above (figures 8.1 and 8.2) to decorate the shaft with spiraling lines (figure 8.3). He wastes not a single word on that application, which apparently is straightforward to him, except that he insists on using the triangle *abc* in order to expand the intervals as they ascend, and to contract them as they descend.

The construction of a round twisted column, which follows, yields an even nicer example (figure 8.4). Here Dürer starts with a description of the dimensions and proportions of the column he wants to construct, i.e. effectively, the description of a profile. He uses the diameter above its plinth as a module. The proportions he gives are those of a ionic column in Vitruvius' *Architecture*. Then he goes on with the ground plan "from which to extract the column while twisting it."[11] As the figure clearly shows, this plan is composed of three circles with different diameters touching each

---

6   "Ich will aber hie die felt ubersich in einer ordnung erlengen / wie voren angetzeigt" (Dürer 1983, fol. B^r).
7   "Ich nym zum ersten ein cirkelrund felt … und far eben mit ubersich so hoch ich will / so wirdt ein runde seulen darauß" (Dürer 1983, fol. G^r). From this quotation, the role played by plans and elevations in the design of columns seems straightforward: the elevation fixes the different levels to which the forms contained in the ground plan will rise.
8   "Ich far aber auß allen forgemelten gründen ubersich / so hoch ich wil in ein spitz / so werden kegel darauß" (Dürer 1983, fol. G^v).
9   Dürer 1977, 191.
10   I am paraphrasing fol. G$_{ij}$^{v} of Dürer's *Underweysung*.
11   "leg ein grund darauß du dise seulen winden must" (Dürer 1983, fol. H^v).

other. Their peripheries are numbered with sixty subdivisions, beginning with point *a*, the center of the large circle. In the smallest circle, point *c* has the number 6. Then, continuing with the middle-sized circle, along half his periphery, Dürer counts up to 18 and starts with 19 on the large outer circle. Its whole circumference is subdivided into points 19 to 42, then the division continues on the second half of the middle-sized circle until the point *c* is reached with the number 54. The second half of the small circle is divided into points 55 to 60, which coincides with the center *a*. These points in the ground plan show how the axis of the column must be twisted.

Next, the axis of the column in profile is divided into 60 irregular intervals. The procedure used by Dürer is by now familiar. He projects an equally divided arc on the vertical axis. Thus the intervals increase as they ascend. They all fit along a compass opened to a constant angle, or, in other words, in the triangles defined by the branches of the compass and the intervals, the latter are the sides opposing the equal angles. According to the fourth postulate of Euclid's *Optics*,[12] these intervals would appear equal to an eye located at point *f*. Thus, let's presume that Dürer's diagram refers to a visual space in which the laws of Euclidean optics apply, i.e. that the visual rays proceed in straight lines from the eye, and that the collection of such rays constitutes a cone, of which the vertex is at the eye and the base at the surface of the objects seen. The apparent size of a visible object is determined by the visual angle that encloses the extremities of the object and the vertex of which is located in the observer's eye. In such a visual space, the division process used by Dürer induces that the growing vertical intervals of that subdivision are perceived as equal by an eye *f*.

In other words, in his diagram, Dürer takes into account the fact that a magnitude seen from underneath appears smaller than it is. To correct this perception, he increases the intervals as they ascend in accordance with the order of Euclidean optics. I will come back to the presence of an optical theory in these sophisticated practitioner's drawings of architectural and mathematical objects. But for the moment let me stress the highly innovative state of the construction as it is further described by Dürer. We have seen that he conceives of the ground plan, with its three circles divided into the numbers 1 through 60, as a horizontal projection of a helix, which is to represent the twisted axis of a column. That's why Dürer divides the elevation into 60 horizontal planes, numbered from 1 at the bottom up to 60. Each horizontal section of the column is represented by a horizontal line segment whose length is equal to the diameter of the column, the axis of the column being an axis of symmetry. In order to twist the axis, Dürer applies the technique described in his figure I.15 (figure 8.1), and applies to each horizontal line segment the displacement, taken from the ground plan, which the corresponding point of the axis undergoes when twisted. He then imagines that each point of the twisted axis of the column can be the center of a sphere whose diameter is precisely that of the column. If all their circumferences are connected by a line, the contour of the column appears. The surface is thus defined (in our modern terms) as the envelope of a family of spheres with constant radius and

---

12 Let it be assumed that things seen under a larger angle appear larger, those under a smaller angle appear smaller, and those under equal angles appear equal. Quoted from Lindberg 1976, 12.

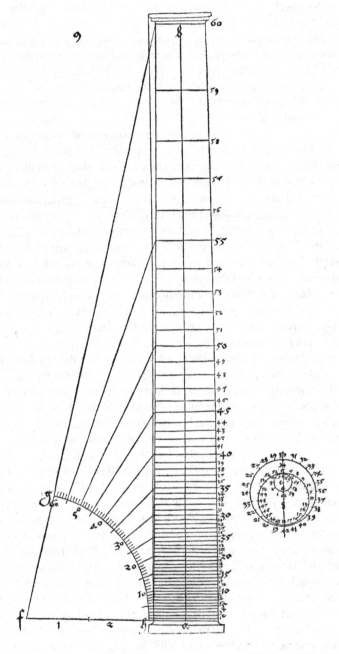

Figure 8.4. Construction of a round twisted column. (Dürer 1983 figure III.9–10.)

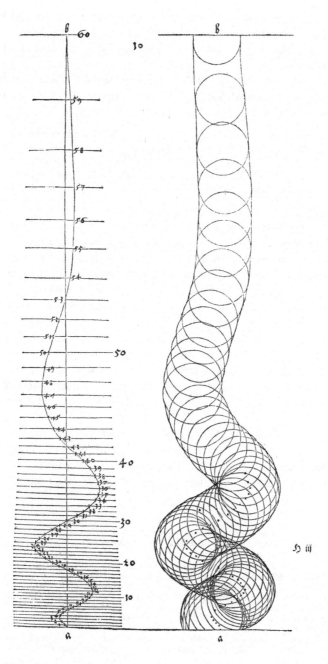

(Figure 8.4 continued)

with their centers on a curve. Dürer's drawing certainly inspired the founder of descriptive geometry, Gaspard Monge,[13] when in his lessons held centuries later at the newly founded *Ecole de l'an III* in revolutionary Paris, he taught on the so-called "surfaces de canaux."

In my opinion, it is in this part of his *Underweysung* that Dürer's excellent command of methods later called descriptive is most obvious. His last sentence of this section shows a good understanding of the nature of sections:

> Whereas in a round column the horizontal lines represent only round surfaces and the lines are placed evenly one on top of the other, in the case of twisted columns these lines no longer lie one flat on top of the next; instead, they are displaced, or moved to and fro above or below the next horizontal line, or to one side of it. They are transformed into oblique or curved shapes. This is shown in its simplest form in the following diagrams.[14]

But let's come back to the optical theory possibly underlying the drawings we have commented upon. While Dürer clearly describes how to divide the vertical lines in the drawings discussed above, he is silent on a possible connection to vision. He writes in a context where Euclidian optics[15] is known to some extent, and we know that he himself is well-read on this subject. When in Venice, Dürer bought, in 1507, a copy of Euclid's *Elementa*, edited and commented upon by Bartolomeo Zamberti[16]. This book contains a Latin translation of Theon's recension of Euclid's *Optica*. In Dürer's estate in London[17] we find a manuscript in Dürer's hand containing an extract, in German, of the axioms and propositions of that optics. Except for theorem 16, which is incomplete, the translation is correct. Hans Rupprich has shown that theorem 17 is not in Dürer's, but in Willibald Pirckheimer's hand. That was for him an opportunity to illustrate the kind of collaboration that the painter had established with his humanist friend: while Pirckheimer translated Euclid's text into German, Dürer wrote it down.

Moreover, Dürer bought in 1523 "ten books useful to painters" from the Senate of Nuremberg, among which a copy of a manuscript of Witelo's "De perspectiva," which had belonged to Regiomontanus' library and was inherited by the astronomer Bernhard Walther. The manuscript, today in Basel,[18] doesn't show any testimony of a

---

13 Gaspard Monge, *Application de l'analyse à la génération des surfaces courbes* (1807), fol. XIV. See Bruno Belhoste and René Taton, Leçons de Monge in Dhombres 1992, 267–459, 301 for the "surface de canaux."

14 See Dürer 1977, 213. In Dürer's original German version: "Aber so in der geraden seulen die zwerchlini all rund ebnen bedeuten und gerad auf einander stend / so beleyben doch die selbenn linien in der windung der krumen seulenn nicht mer blat auf einander / sunder schieben / hencken / und keren sich hin und her / ubersich undersich und nach der seyten / unnd werden schlemet ablang rundlecht linien darauß. Dise hab jch nach dem schlechtesten nachfolget aufgeryssen" (Dürer 1983, fol. $H_{ij}{}^r$).

15 On the spread of Euclidean optics in medieval and Renaissance Italy, see for instance Federici Vescovini 1983, and Cecchini 1998a.

16 *Euclidis megarensis philosophi platonici mathematicarum disciplinarum janitoris : habent [...] elementorum libros XIII [...] Bartolomeo Zamberto Veneto interprete*, Venice 1505. Dürer's copy with his monogram is kept today at the Herzog August Bibliothek Wolfenbüttel. It bears the following inscription: "daz puch hab ich zw Venedich vm ein Dugatn kawft im 1507 jor. Albrecht Dürer" (This book I bought it in Venice for one ducat in the year 1507).

17 London, British Library, Add. Ms. cod. 5228, fol. 202 and fol. 211–219; 5229, fol. 77.

18 See Steinmann 1979, who identified Dürer's copy.

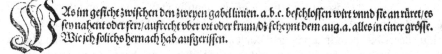

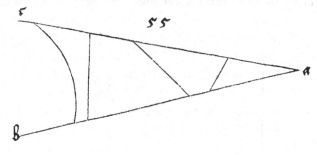

Figure 8.5. Diagram by Albrecht Dürer to the angle axiom of Euclidean optics. The axiom in Dürer's terms: "Whatever is seen [in the field] enclosed by the two forked lines ab and ac and touches them, be it near or far, vertical, oblique, or curved, it appears to the eye in the same size." (Dürer 1983 figure IV.55; translation from Dürer 1977, 373, modified by the author.)

reading by Dürer. Last but not least, Dürer includes the famous axiom of angles in Book IV of his *Underweysung* (figure 8.5). Like his Italian counterparts and sources, Dürer begins his exposé on the rules of perspective with some elements on vision that are clearly Euclidean.

Thus there can be no doubt that Dürer was aware of the existence of the Euclidean laws of optics where the apparent magnitudes of the perceived objects depend on the angle at which they are seen. And yet he doesn't explicitly relate this postulate to the technique detailed above. How are we to explain this silence? One possible explanation is that he learned the technique in a workshop environment and that he applies it tacitly. His knowledge of Euclid probably has its roots in quite a different context, namely in his encounter with Italian perspectivists. This knowledge was then later developed with the active help of Pirckheimer. A last example, taken from Book III of Dürer's *Underweysung*, seems to point to a practical workshop context. This time, the technique is applied to proportioning the letters of inscriptions located at a great height on columns, towers or walls. Dürer finishes his instructions by stating:

> This division can be applied to all sorts of things and not only to letters, especially to decorating towers by paintings located at different levels, in such a way that images located higher seem equal to those which are located lower.[19]

This points to a well-known practice of painters and architects. On the other hand, this example, as the diagram (figure 8.6) already shows, clearly establishes the link

---

19 My translation, see Dürer 1977 for a better one, if accurate. In Dürer's original words: "soliche teylung hat nit allein stat in den pustaben / sonder in allen anderen dingen / und in sonders so man einen hohen thuren in allen gaden mit bildwercken ziren will / also daß die oberen bild gleych den underen scheynen kan durch disen weg geschehen" (Dürer 1983, fol. K^v).

we were missing between vision and the observer's eye, "dein gesicht," *c*. The vertex of the triangle *abc*, used to subdivide *ab*, is identified with an eye, and the arc, which is divided in equal arcs, is an angle of vision. Moreover, Dürer is aware that the technique described here in a practical context is identical to that which he has applied to the construction of an abstract mathematical object: the helix, to which he refers explicitly.[20]

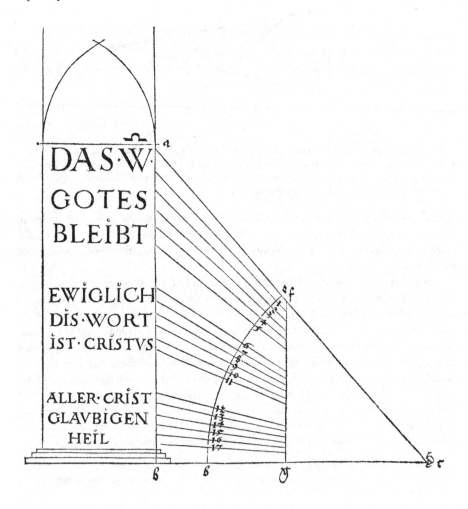

Figure 8.6. Proportioning the letters of inscriptions. (Dürer 1983 figure III.[28].)

---

20 "welicher an ein thuren schreyben will das man die oberst zeyl der bustaben als wol gesech zu lesen als die underst / der mach sie oben gröser dann unden / durch ein solichen weg / stell dein gesicht so weyt von dem thurn / und in der höch wie du wild / diß sey ein punct *c* und nym für dich den weg des dryangels *abc* der 16. figur des lini büchleins/ ..." (Dürer 1983, fol. K^v).

Of course, the technique of correcting optical errors—equal magnitudes appear smaller to the observer's eye when situated higher—has a long history,[21] going all the way back to Greek classical art. Plato[22] already mentions artists who lend their sculptures proportions that appear harmonious without really being so. Recall the story reported by Pliny, or by Iohannes Tzetzes (in the twelfth century), of a contest between Pheidias and Alcamenes for a Athena to be placed on top of a column. While Alcamenes' statue was beautifully proportioned, Pheidias offered a statue with distorted members and he was barely saved from lapidation. The situation reversed completely when the statues were put on top of the column. Vitruvius also gives advice to correct "falsa iudicia oculorum." The procedure used to do this is the one based on Euclidean optics, known in the Western world since its Latin translation "De visu" and circulated in manuscript form in the second half of the twelfth century.[23] It was probably part of the tradition of practical optics[24] as established by Simi and Camerota.

All of the examples we have discussed in Dürer point to an architectural context, even the helix, which Dürer thinks of as at a staircase in the spire of a circular tower. Thus, the division technique, the aim of which is to make things located at different heights appear equal, may well be a workshop technique distinct from, even if linked with, the Euclidean theory of vision.[25] But on the other hand, his examples don't fit the conditions of effective practice. They seem to be a kind of school exercise as shown by the location of the eye at the bottom, and not at man's eye level. The procedure may have been taught in practical abacus schools.

To sum up, an optical theory underlies Dürer's architectural drawings, and even his diagram of a helix, when this mathematical object is used in an architectural context. The stonemason's practice of representing things by means of a ground plan and a vertical section, an *Auszug*, enables Dürer to visualize the object in three dimensions and to display it in a visual space. In the construction process itself, he takes into account the laws according to which vision proceeds. His vertical *Auszug* is an elevation only if we look at it in a space ordered by Euclid's postulate.

Dürer's diagrams are thus located at the intersection of practice and theory. This interpretation is confirmed by at least one Italian follower of Dürer: Daniele Barbaro. In *La pratica della perspettiva* (1568), he comments on Dürers "artificio," "instrumento,"[26] and on

> the rule and form of the quadrant ..., which serves to fix the proportions of the letters or figures which are located high up in some column or wall. With the help of it, painters and architects know to divide the heights in proportional parts[27]

---

21 See Frangenberg 1993 and Cecchini 1998b.
22 Plato *Sophistes*, 235–236.
23 See for instance Cecchini 1998a.
24 See Simi 1996 and Camerota 1998.
25 In a paper entitled "L'optique euclidienne dans les pratiques artisanales de la Renaissance," to appear in *Oriens-Occidens. Sciences, mathématiques et philosophie de l'Antiquité à l'Âge classique*, I try to find traces of this procedure closely linked to the practical tradition of "measuring by eye."
26 See Barbaro 1568, 23.

and establishes a double link with the painters' and architects' knowledge on one hand, and with Euclid's *Optics* on the other, whose angle axiom he quotes.[28]

The technique described by Dürer, developed considerably in the baroque world, where it became important in creating artificial effects such as accelerating perspectives, but that's beyond the scope of the present study.

2. FROM EUCLIDEAN OPTICS TO PERSPECTIVE: THE CASE OF SIXTEENTH-CENTURY *KUNSTBÜCHER*. WITH PARTICULAR EMPHASIS ON HANS LENCKER'S *PERSPECTIVA* (1571)

In this section, the emphasis will be on the use Dürer's immediate followers in Nuremberg made of orthographic plans. As it is well known, Dürer's instructions were not immediately understood by the practical men to whom his book was addressed. In their eyes, these instructions needed some mediations, simplifications and further explanations to be useful and applicable to the crafts. Thus during the sixteenth century a host of booklets, the so-called *Kunstbücher*, was published mainly, but not exclusively, in Nuremberg, the aim of which was explicitly pedagogical. For instance, Augustin Hirschvogel,[29] painter on glass in a Nuremberg workshop, author of a *Geometria* (1543) and later "Mathematicus" of the city of Vienna, intends to offer beginners a better understanding of the basic principles and the practical applications of the art of measuring, "die edle und nützliche kunst des messens (Perspectiva in Latein genant)," unlike previous books, which are often obscure and hide the most necessary and noble aspects.[30] To a certain extent these books may thus grant us a glimpse of what was transmitted to apprentices in sixteenth-century Nuremberg, especially in the domain of constructive geometry, plan and elevation techniques, as well as in perspective.

The first point I want to make is that the understanding of Dürer's notion of "Messung" seems to have undergone some change. It is now used as a German translation of the Latin "perspectiva." Count Johann II of Simmern in his *Underweysung der Kunst des Messens* (1531) had already transcribed the Latin *Perspectiva* in German as "kunst des Augenmeß" (art of measuring by eye).[31] The notion of representation, namely in the form of perspective painting, is thus added to Dürer's constructive geometry. To be more precise, the visual appearance of things displayed in a three-

---

27   "Dalle dette cose si comprende la regula, & la forma del quadranto di Alberto Durero, col quale egli proportiona le lettere, overo le figure, che sono nell'altezza di qualche colonna o parete. Dalche sono avertiti i Pittori, overo gli Architetti a partire le altezze in parti proportionate" (Barbaro 1568, 9).

28   "Le cose, che si vedeno sotto anguli eguali pareno eguali" (Barbaro 1568, 10, § G).

29   On Hirschvogel, see Kühne 2002.

30   See the dedicatory letter of Hirschvogel 1543, fol. a2^r: "Nach dem bißher / durch unsere vorfordern / ein langezeit dise edle und Nützliche kunst des messens (Perspectiva in Latein genant) in Deutscher sprach verborgen gehalten / und den gemeinen man / zulernen schwerlich zu bekommen / Auch den meren Thail in Griekischer und Lateinischer sprach verfast / Was derselben bücher in Truck sein kommen / etwas tunckel / und das notigist und fürnembst / gewönlich dahinten behalten worden / Das dann manchen kunstbegirigen / der solchs gelesen / den buchstaben verstanden / aber zu keinen handtbrauch gelangt hat / abschewlich / und verdrießlich ist gewesen."

31   Johann II von Simmern, in the title of his 1546 *Perspective*, addressed his instructions to "all those who want to use the art of measuring by eye (called Perspectiva in Latin)."

dimensional space and depending on the visual angle are replaced by perspective representations of those same things, the foreshortening of which depends on their distance from the eye. The means to obtain those images remain in part the same, namely orthographic plans, which can be conceived of as belonging to architecture,[32] as Walter Ryff states in the perspective book of his *Der fürnembsten / notwendigsten / der gantzen Architectur angehörigen Mathematischen und Mechanischen künst / eygentlicher bericht* (1547), where he translates passages by Alberti, Serlio, and Vitruvius into German. Ryff (or Rivius), a physician who spent some years in Nuremberg and published a German translation of Vitruvius' architecture, is a quite prolific writer on perspective. He is conscious that teaching perspective by the written word is a difficult task namely because "of the solids on the plane raised out of the ground plan."[33] And yet without this technique of drawing geometric plans ("Geometrischer grund"), nobody will be able to master the art of perspective, which Ryff identifies with Vitruvius' *Scenographia*.

By including perspective in their understanding of "Messung," these men—like Dürer—see it as part of a practical geometry in which the observer's eye plays a role. But instead of constructing three-dimensional objects, they represent them on a picture plane, as Dürer already did in the very last section of his *Underweysung der Messung*. But while Dürer clearly referred the picture plane to the intersection of the visual pyramid, this Albertian model is completely absent from the *Kunstbücher*. These seem related rather to that older tradition of practical (applied) optics.Going back to the Arabs, it was taught in the medieval abacus schools, and its methods were used by astronomers, geographers, and architects in order to deduce from the appearances of distant or inaccessible magnitudes, their real measures. According to the Arab philosopher al-Farabi (tenth century),[34] who lists this practical optics in his classification of the mathematical disciplines, it also includes optical corrections of the sort studied in my first section. So, at least in Nuremberg, although the Albertian rules are available in Ryff's writings, very rough perspective workshop methods develop as part of the art of measuring by eye. Note that for Francesco di Giorgio Martini, the painter's perspective also belonged to that tradition of "misurazione con la vista."

"Messung," as understood by the authors of the sixteenth-century Nuremberg *Kunstbücher*, thus aims to construct perspectival representations of more or less complex solids, most of them polyhedral. Some applications to architectural settings are discussed in most of these treatises. As far as the methods are concerned, they essentially consist in drawing plans and their perspectives in already foreshortened planes. Hirschvogel, for instance, teaches in his *Geometria* "how to obtain the ground plans of regular and irregular solids, to throw them into perspective, and to give an eleva-

---

32 Ryff 1547, in the lengthy summary of Book III, speaks of: "der Architectur angehörigen Kunst von der Geometrischen Messung."
33 See Ryff 1547, fol. cc$_{iij}$: "Obgleich wol die trefflich kunst der Perspectiva vast schwer und müsam / schrifftlich zuhandlen und tractieren / fürnemlichen aber der Corper halben / auff der ebne aus dem grund auffgezogen."
34 See Jolivet 1997, 260.

tion."[35] He starts from a ground plan, called "Stainmetzenfierung," or the stonemasons' square (figure 8.7). Next he constructs a foreshortened square, called "Geometria der perspectiva." This same square, foreshortened once for all solids, is used through the whole book as a kind of template. The bulk of the book is dedicated to explaining in detail how to divide the square or "geometria," i.e. how to draw in the geometrical square the ground plan of the solid , which is to be represented in perspective. This

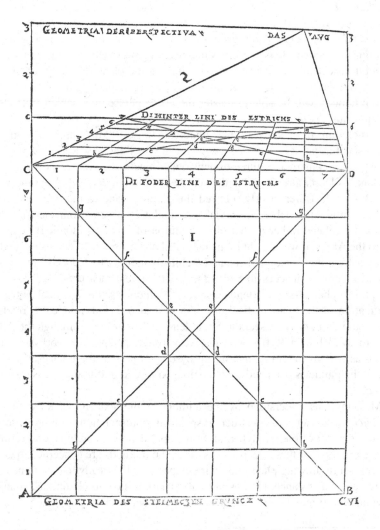

Figure 8.7. Stonemason's plan and *Perspectiva*. (Hirschvogel 1543 figure 2.)

---

35   That's what he announces in the title of his *aigentliche und grundtliche anweysung / in die Geometria / sonderlich aber / wie alle Regulierte / und Unregulierte Corpora / in den grundt gelegt / und in das Perspecktiff gebracht / auch mit jren Linien auffzogen sollen werden.*

appears to be the most important step in the construction process, at least the one, which deserves ample explanation and description. To throw this ground plan into perspective, Hirschvogel uses the method known in the literature as the diagonal method (which I prefer to call the template method). To keep from losing his way in transferring the division of the ground-plan to the foreshortened one, he makes use of colours and small symbols like + and ↓, thus allowing a glimpse into the workshop projection techniques. Finally he uses an elevation of the object to complete the perspective view of it.

Heinrich Lautensack, a goldsmith and painter from Frankfurt, who follows Hirschvogel's instructions (without always completely understanding them) in his *Underweysung deß rechten gebrauchs deß Circkelß und Richtscheyts, auch der Perspectiva ...* (1563), comments on the basic use of the ground plan in those perspective drawings:

> Those plans which are drawn in this way are called plans in the Geometria or stonemasons' plans, because here you may see how all things are in the ground plan. If you are to throw them into perspective you thus know where to put things while foreshortening.[36]

These so-called stonemasons' plans are also central in the practical method described by the goldsmith Hans Lencker in his *Perspectiva*, which was to appear seven years later and which earned him a position at the Saxon court in Dresden.

Lencker's *Perspectiva* (1571) is addressed to craftsmen who build in wood or stone, like carpenters and stonemasons. Before they began working so, Lencker entreated them to visualize the planned construction in small models[37] obtained with the help of perspective, i.e. plane representations. His book, although full of mistakes,[38] is interesting in so far as it grants a glimpse into the workshop methods of the time, the "Praxen" as Lencker puts it, and the use of plans and instruments in implementing these practical methods. Perspective, understood as diminution and drawing orthogonals to a central [vanishing] point, consists in constructing first a ground plan, the *perfetto* of the Italian artists, then deriving from it by means of compasses, strings and more compound instruments a foreshortened plan, the *degradato*, and finally obtaining the perspective of the solid by using an elevation (figure 8.8). The latter part of the process is described as easy going. The bulk of the book is dedicated to the techniques of drawing elevations and especially ground plans, and to foreshortening them ("gründe verjüngen").

---

36 Here is the original German text from Lautensack 1618, fol. $D_{iv}{}^v$: "Diese gründt die man also auffreist / werden die gründt in der Geometria oder Steinmetzengrund genennet / denn da sihet man wie alle ding im grundt kommen / so man es dann in die Perspectiff wil bringen / das man wisse wo ein jegliches ding sol im verjüngen hingebracht werden."

37 See Lencker 1571, fol. V^v: "So kan nun ein solches muster / oder was es sonst ist (doch nicht das es Corperlich sein müsse / sondern nur mit höhe / leng und breite / im sinn fürgenommen) auff einer vierung *a* ... auff zweierley gantz unterschiedliche form unnd gestalt gerichtet / unnd auß rechtem grund der Geometria / in die Perspectief fürgerissen werden."

38 See for instance the analysis of Elkins 1994, 96–101.

After having introduced in a first section the different instruments he needs, including templates of already foreshortened squares,[39] he describes in a second section how to prepare the plans, i.e. ground plan, "geometrischer Grund," called R throughout the book, and elevation, labelled P. His definition of both plans shows that

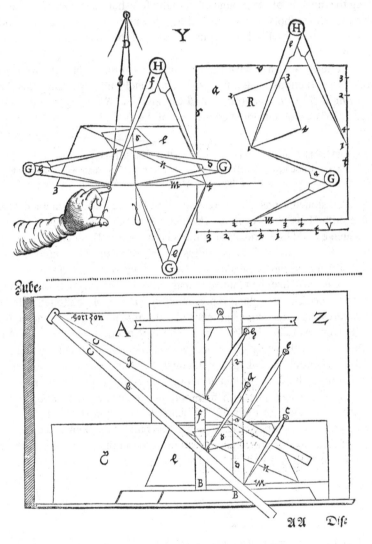

Figure 8.8. Foreshortening with the help of compasses and strings. (Lencker 1571 figure Y/Z.)

39 I study this aspect of Lencker's practical method in the Proceedings of the ESF Exploratory Workshop "Artists, work, and the challenge of perspective" in Rome (19–21 September 2002), to appear with the Ecole française de Rome.

we are far from a familiar projection technique. Lencker is appealing to the imagination and the experience of the apprentice. The ground plan located under the ground line, i.e. the intersection of the picture plane and the ground plane, is introduced as follows:

> all these plans you have to imagine them as the place or space straight from above which each thing (as a building, *corpora* or other, may it be polyhedral or circular, standing lying or oblique, largest at the bottom, in the middle or at the top, elevated at its two corners, at one or in the middle, or however you may imagine) would cover perpendiculariter on the horizontal plane, table or square, if it were a solid.[40]

He similarly defines an elevation as "the space or place which would be covered by some solid or building, by the rising points of its height and width (not perpendiculariter) but *à latre* (sic) on the sides like on a vertical wall."[41]

Two of Lencker's remarks deserve to be put forward in the context of this chapter. To bring the notions of plan and elevation closer to his reader, Lencker uses an analogy with the work of carpenters,[42] who first prepare floors and ceilings according to their length and width from their geometrical plans R. Even if they build columns, walls and vertical elements when laid down, all their thoughts are oriented towards how these elements will raise out of the ground plans (ligend gründ) according to plans P. Thus, it appears that, among craftsmen, carpenters are the ones who know best how to realize plans and to build in wood according to those plans. In Lencker's eyes painters are less informed. But once they have learned to draw the plans, it will be easy to obtain a perspective representation. The second remark points to architects, who already know how to draw plans. Lencker's perspective method will be convenient and particularly handy for them.[43] It is the method[44] called the circumscribed rectangle method by Elkins, the diagonal method by Kirsti Andersen and which I want to call the template method.

Taking a square and an already foreshortened square as a picture plane, this method allows the image of an arbitrary point of the geometrical square to be found

---

40 See Lencker 1571, fol. VII: "Und alle dise gründe unterthalb der Erdlinie / mustu dir eigentlich also für und einbilden / als den platz oder rhaum / welchen ein jedes ding (als Gebew / Corpora oder anders / es sey ecket oder rund / es stehe / lige / leine / es sey unten / mitten oder oben / am breitesten / es sey mit beiden orten / mit einem / oder nur mitten erhoben / oder wie es sonst erdacht werden mag) gerad von oben herab auff einem Estrich / tisch / oder vierung *a* Perpendiculariter bedecken würde / wans Cörperlich were / das ist sein rechter Geometrischer grund."

41 See Lencker 1571, fol. VII[r/v]: "Den grund aber oberhalb der linie *m* mustu verstehen gleich wie disen / als den rhaum oder platz / welchen ein jedes Corpus oder Gebew mit den auffsteigenden puncten seiner höhe unnd breite (doch nicht Perpendiculariter) sondern *à latre*, nach der seiten / als an einer auffrechten wand / bedecken würde…"

42 See Lencker 1571, fol. VII[v]: "Und haben dise beide gründe R und P gar ein ebens gleichnuß / mit dem werck der Zimmerleut / welche erstlich alle deck unnd böden nach der leng unnd breite auff jre Geometrische gründ richten / nach art des grundes R. Demnach ob sie wol alle auffrechte Gebew / als Seulen / Wend / unnd Gibel / auch niderligend zu werck ziehen / so stehen doch gleichwol alle jre gedancken dahin / wie sich hernach im auffrichten / solche Seulen / Wend / und Gibel / mit Düren und Fenstern / auff die ligenden gründ schicken werden / nach art des grundes P."

43 See Lencker 1571, fol. VIII[v]: "Bistu ein Architectus des Maßwercks und der Gebew / sampt der selben gründe / verstendig / so wird dir sonderlich und vor allen diser weg und gebrauch der Perspectief bequem / leicht / und sehr dienstlich dazu sein."

44 On this method, see Elkins 1994, 96–101; Ivins 1973; Peiffer 1995, 111–112; Schuritz 1919, 43–44; and Staigmüller 1891, 46.

in the foreshortened one. Piero della Francesca, Dürer, and Hirschvogel, for instance, used the diagonal to find the image. Lencker makes very clever and virtuosic use of compasses and strings. Once he has found the image point, he raises it to a height given by the plan P (an elevation).

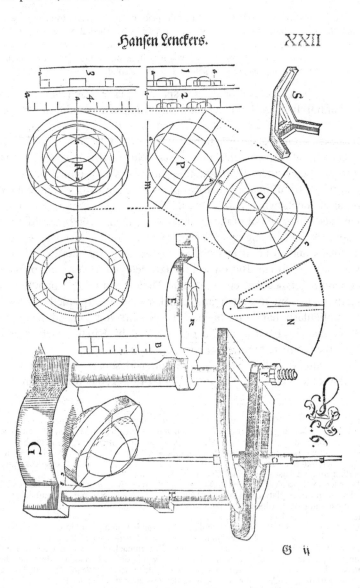

Figure 8.9. An instrument to draw plans and elevations. (Lencker 1571 fol. XXII, figure 6.)

As already mentioned, making the plans and elevations is considered by Lencker to be the core of the business. Seven of the ten figures of his *Perspectiva* are dedicated to obtaining various planes P and R. Lencker seems to have an optimal command of the technique. He combines the plans drawn in different ways such as to obtain different objects, or the same object in a different position. His examples include the sphere, polyhedra, a spiral staircase and an architectural setting, which closes and crowns the whole book. Thus he considers (see his figure 6) a sphere enclosed by a ring whose section is a square, and teaches how to draw it in two different positions: the ring stands vertically, the ring is oblique. His juggling with the different plans to obtain the desired result is quite impressive. For those who spare no expense, Lencker even conceives an instrument (G on his figure) to draw the ground plan and elevation of a given solid (figure 8.9). The description of this instrument is quite straightforward. In the half circle A made in brass, you fix half a diameter B pivoting around the screw F. In the socket (Hülslein) C moving along B, you fix a dry-point (or a lead pencil) D, which can be lifted and dropped. To obtain the ground plan R, you drop this point D onto the solid G, half of the ground plan of which you want to draw, turn the platform E into position H and transfer the coordinates of the point of the solid to the plan R. The heights of the different points of the solid are measured with the help of point D. Lencker prefers to work without this instrument, because the solid you want to draw must already be given, which is far from always being the case, especially if you are an architect and want to show a model of what you are going to realize. We should also add that the method described by Lencker is quite accurate for drawing the complex polyhedral representations, which were Nuremberg's specialty and even its trademark in the sixteenth century. Just think of Wentzel Jamnitzer's beautiful plates!

Johannes Faulhaber, 40 years later and in a quite different professional and theoretical context, includes Lencker's instrument in his *Newe Geometrische und Perspectivische Inventiones* (1610). Faulhaber,[45] a schoolmaster, "Rechenmeister" and "Modist" in the city of Ulm, was a prolific writer and earned his living in teaching mathematics. In his *Inventiones*, he is interested in military applications of perspective. He describes some instruments useful for easily drawing what he calls "planimetrisches Grundlegen" of fortifications, bastions, cities, and camps. To do so he places himself in a tradition of practical perspective, which he traces back to Albrecht Dürer. He follows this line through Gualterus Riff (sic), Heinrich Lautensack, Wentzel Jamnitzer, Daniele Barbaro, and Hans Lencker among others. For him, Dürer is at the beginning of a perspectival tradition based on the use of instruments. He was copied by Daniele Barbaro, and he inspired the so-called *Perspectivtisch* in which the lute was replaced by a simple plan, and which was included by Paul Pfinzing in his *Ein schöner kurtzer Extract der Geometriae unnd Perspectivae* (1599), where he claims to have seen it in Jamnitzer's house. Faulhaber improves Lencker's instrument which, according to him, is based on the same theoretical foundation as Dürer's. The figure

---

45 On Faulhaber, see Schneider 1993.

is inverted and from the point of vision the sight falls on the object perpendicularly; this is not the case in Dürer's lute woodcut, where the point of vision is fixed on the wall (figure 8.10).[46]

In my view there is another big difference between the two figures by Dürer and Lencker that is interesting for my purpose. Dürer offers a material model for the

Vnd damit günstiger lieber Herz will ich meinem schreyben end geben / vnd so mir Got genad ver=
leyeht die bücher so ich von menschlicher proporcion vñ anderen darzü gehörend geschrben hab mit
der zeyt in druck pringen/vnd darpey meniglich gewarnet haben/ob sich yemand vnder=
steen wurd mir diß außgangen büchlein wider nach zů drucken/das ich das
selb auch wider drucken will / vñ auß lassen geen mit meren vnd
grösserem zůsatz dañ ietz beschehen ist/darnach mag
sich ein yetlicher richtē/Got dem Herren
sey lob vnd eer ewigklich.

Ω iij

Figure 8.10. The painter with a lute. (Dürer 1983 figure IV.[64].)

46   See Faulhaber 1610, 35: "Weil er dardurch / so wol mit dem Perspectivtisch (doch auff die Maaß und
     Weiß / wie Albrecht Dürer / durch das Exempel / mit der Lauten angedeutet) als mit deß Lenckers
     Instrument / so alle Puncten fleissig observiret werden / zum fertigen Handgriff kommen kan / dann
     beyde Instrument / mit jhrem Gebrauch / auß einem Fundament zu demonstriren : Allein wird in deß
     Lenckers die Figur umbgekehrt / der gestallt dz der Augpunct *perpendiculariter* herab fällt. Mir ist der
     Perspectiv Tisch mit der Seyten umb etwas angenemer / dieweil der Augpunct stetigs an einem Orth
     bleibt / und nit verändert wird."

visual pyramid, the vertex of which is at the point of vision, which is not a human eye
in Dürer's representation, but a nail in the wall. The sight is materialized by a string,
cut by a transparent plane, the picture plane. The painter is doing nothing more than
manipulating strings and passively recording measures. While Dürer's famous woodcut embodies a theory, the Albertian definition of perspective, Lencker's instrument
and the way he represents it is first and foremost a technical device to draw plans.
Thus, in the practical workshop tradition of such craftsmen as the Nuremberg goldsmiths, carpenters, and architects, the descriptive and perspective methods developed
or described by Dürer are transformed into material devices, simple machines like
strings and templates, and into instruments designed to apply those techniques rather
mechanically.

3. THE GEOMETRY OF PROJECTION INVESTIGATED
BY MATHEMATICIANS IN THE SEVENTEENTH CENTURY

Orthographic projections, like those which the Nuremberg craftsmen learned to realize through some of the booklets we looked at in the preceding section, were used
systematically in architectural treatises of the late sixteenth century, especially in
Italy (Palladio, for instance). Plans and elevations, or horizontal and vertical orthographic projections, were combined, as in Piero della Francesca's *De prospectiva
pingendi*, in Dürer's *Underweysung der Messung*, and also much later, for instance in
Salomon de Caus' *La perspective avec la raison des ombres et miroirs*,[47] in order to
represent three-dimensional objects on a plane, either in perspective, i.e. as they are
perceived, or by methods that were later subsumed under descriptive geometry. Thus
two distinct mathematical techniques are involved, which led, when taken over by
mathematicians, to two distinct domains of geometry, namely projective geometry
founded in the seventeenth century by Desargues and Pascal, and descriptive geometry introduced systematically by Monge towards the end of the eighteenth century. In
this last section, we will have a rather quick look[48] at the way mathematicians up to
the seventeenth century handle orthographic projections and their combinations.
Then we will examine some of the known reactions of practical men, namely painters
and stonemasons, to the writings of professional mathematicians.

As Martin Kemp and J. V. Field have each emphasized, in the second half of the
sixteenth century in Italy, painters' perspective was taken over by humanists like
Daniele Barbaro and by professional mathematicians like Egnazio Danti, Giovanni
Battista Benedetti, and Guidobaldo del Monte. In their hands, perspective underwent
a significant change as different and more fundamental questions were asked, and
could slowly develop into a mathematical technique.

Daniele Barbaro, a Venetian patrician with a good humanist education, "a notable
patron"[49] who had Palladio and Veronese work for him, wrote a *Pratica della pers-*

---

47 London 1612. On Salomon de Caus, see Bessot 1991.
48 For this section we can rely on a rather important literature. See in particular Andersen 1991, Field
    1997, Flocon and Taton 1963, Kemp 1990, Le Goff 1994.
49 See Kemp 1990, 76.

pettiva published in Venice (1568). J. V. Field sees it as a "spin-off from Barbaro's work on Vitruvius."[50] Indeed, Barbaro prepared an Italian translation of Vitruvius' *De architectura* (Venice 1556) and published the original Latin text ten years later. His interest in the work of Vitruvius, which he amply annotated, is quite probably, according to Field, at least part of the reason for his interest in perspective. Piero della Francesca and Albrecht Dürer are the main sources for the methods displayed in his *Pratica della perspettiva*. He uses combined orthographic projections—plan and elevation—to give a perspective representation of objects, which are in most cases geometrical solids. He also offers a perspective machine inspired by Dürer's. Barbaro seems to take pleasure in the speculative opportunities offered by the study of mathematics. From the start, in his dedicatory letter, he claims that his mathematical studies "opened the route to high and subtle speculations"[51] and regrets the absence of demonstrations in the work of contemporary painters who "let themselves be led by simple practice."[52] In the first chapter of his treatise, introducing the principles and foundations of perspective, he offers a twofold approach to perspective: a practical one (as the title announces) and a more theoretical one referring to Euclid and Apollonius. His intention is to combine the experiments of art with the decrees of science. Being aware of the dependence of perspective on geometry and natural philosophy or physics, he treats perspective as a "*scienza subalterna*," subordinate to both sciences: from geometry, it receives the straight line, from physics the process of vision. The visual rays of perspective are simple lines without width if considered mathematically, but belong to the tangible world if they are to be visible. Even though Barbaro's treatise offers hardly any geometrical demonstration, it assigns the painters' technique a place as a science subalternate to physics and geometry in the classificatory tree of science.[53] The status of that technique is thus changed significantly.

Egnazio Danti,[54] a professional mathematician who translated Euclid's *Optics* into Italian (1573), considers perspective as an art depending on a science (optics) subordinated to geometry. To him, this is to say that it is impossible to proceed in perspective according to the rigor of the Euclidean method. Nevertheless, Danti attempts to provide geometrical proofs of the practical rules used by artists. Born in a Perugian family of artists, employed at the papal court in Rome, he is the author of an extensive commentary on the perspective rules of the architect Giacomo Barozzi da Vignola. He edited for publishing the latter's *Le due regole* (Rome 1583), making substantial additions printed in a different typeface and providing additional illustrations. Vignola's aim was to demonstrate that the two principal methods, i.e. the combined plan and elevation technique and the distance point technique, give the same result even if they look somewhat different in use. Showing the equivalence of two methods is a completely new objective absent from earlier treatises. Practitioners cre-

---

50 See Field 1997, 133.
51 Barbaro 1568 states: "ci è stata aperta la strada ad altissime e sottilissime speculationi" in the dedication to "Al molto magnifico et eccellente M. Matheo Macigni," which opens the book.
52 See Barbaro 1568, Proemio: "I Pittori dei nostri tempi ... si lasciano condurre da una semplice pratica."
53 On perspective as a mixed science, see Peiffer 2002a.
54 On Danti, see Dubourg–Glatigny 1999, Field 1997, 145–150 and Kemp 1990, 78–83.

ated new methods and applied them to the representation of single forms, rather than asking questions about the compatibility of existing methods.[55] Danti further develops this approach to perspective by focusing on geometrical proof and by offering different kinds of proofs for one proposition. Euclidean style proofs, which are sometimes beyond the understanding of most practitioners, are often followed by an instrument that makes the result tangible.

Mathematicians like Federico Commandino, Giovanni Battista Benedetti, and Guidobaldo del Monte moved further in this direction, as Kemp has shown convincingly. With their treatises, as he puts it, "we see the birth of projective geometry as a discipline in its own right, related to but increasingly separate from the painters' science in its means and ends."[56] The university-trained Commandino, well remembered for his editions and translations of ancient Greek mathematics, recognized that the projection methods used in Ptolemy's *Planisphere*, of which he gave an edition in 1558, are related to perspective,[57] of which he provides a mathematical treatment in Latin. As most commentators have underlined, Commandino gives here a piece of mathematics addressed to mathematicians. Giovanni Battista Benedetti too made "first-class original contributions to the high tradition of mathematics arising out of the problems conceived within the practical one."[58] Best remembered for its mathematical style and approach, however, are the *Perspectivae libri sex* (Pesaro 1600)[59] written by Guidobaldo del Monte, a nobleman trained by Commandino. His six books on perspective were to become "the main source of reference for anyone seriously interested in the underlying geometry of perspectival projection."[60] Right from the start, Guidobaldo states his conviction that perspective is a mathematical science: "I would like to make it quite clear that the proper and particular object of the perspectival science doesn't differ from the object of geometry on which it depends."[61] The mathematical approach to perspective, which was definitively his, appears even in the structure of the work. In his first book, he proves in a purely Euclidean style a corpus of propositions, which are to be applied to problem-solving in the following books.

Yet in the history of mathematics it is Girard Desargues who receives much attention as the founder of a new mathematical discipline, now called projective geometry, the methods of which easily can be connected from a mathematical point of view to those of perspective. His most important work, a short treatise, which contains concepts and methods later subsumed under the term of projective geometry, appeared in

---

55 See Elkins 1994, 89.
56 See Kemp 1990, 85.
57 For a clear exposé of Commandino's summary of perspective, see Field 1997, 150–161.
58 This is the conviction of J. V. Field 1997, 171. See also Field 1985, Kemp 1990, 86–89.
59 Guidobaldo del Monte has been studied in recent years especially by Rocco Sinisgalli 1984, who translated his *Six books on perspective* into Italian, and Christian Guipaud, who presented an as yet unpublished French translation. Guipaud 1991 focuses on the important result proved by Guidobaldo, namely that any set of parallels in the scene to be portrayed will appear in perspective as a set of lines converging to a point.
60 See Kemp 1990, 91.
61 "Hoc namque in primis praecognitum esse cupio, proprium, ac peculiare obiectum scientiae perspectivae nequaquam a subiecto geometriae, cui subalternatur, diversum esse." (Del Monte 1600, I 3; Sinisgalli 1984, 240).

1639 in 50 copies and was entitled: *Brouillon project d'une atteinte aux evenemens des rencontres du cone avec un plan.*[62] In this treatise you can find a projective theory of the conic sections considered as perspective images of a circle. Some properties of the circle can be transferred to the other conics—properties concerning intersection or contact for instance, but not measure—if you introduce new elements like points at infinity. That is exactly what Desargues did. The perspective images of parallel lines, as you know, intersect at a point of convergence, the vanishing point of the picture plane. If the parallel lines are considered as a pencil of intersecting lines (having their intersection point at infinity), then they are projected into a bundle of intersecting lines.

As early as 1636 Desargues had already published a leaflet of scarcely twelve pages on perspective: *Exemple de l'une des manieres universelles du S.G.D.L. touchant la pratique de la perspective sans emploier aucun tiers point, de distance ny d'autre nature, qui soit hors du champ de l'ouvrage.*[63] It shows a worked example, without proof, of Desargues' perspective method. The core of the method is the introduction of scales, which give the progressive diminution in the width and height of lengths seen at greater distances from the picture plane, the position of the eye remaining unchanged. Desargues speaks of "échelle des éloignements," which is a perspectival scale for reducing orthogonal lengths in the picture and "échelle des mesures," a scale of measures, which is geometrical and allows the measures of the depicted transversals to be read on the ground line according to how far away they are located. The treatise is explicitly addressed to practitioners, even though a twentieth-century scholar has called it "a parody of a practical treatise."[64] It is of course interesting to know the family, social and training background of such an author. Archival research done on the occasion of Desargues' 400th birthday shows that Desargues stemmed from a very wealthy family in Lyon and was perhaps better acquainted with the works of the learned mathematical tradition than his published work reflects. He also knew a great many of the practical traditions, but there is not much evidence that he was a working engineer or architect. The only architectural realizations that can be attributed to him date from the period after 1644 and were motivated by the necessity to prove the superiority of his methods, which had come under attack.

In the light of current historical reconstructions, taking for granted that the projective concepts introduced by Desargues, such as the point at infinity and involution, were rooted in his perspectival work, it is of course challenging to inquire about the connections between Desargues' work on perspective (1636) and his treatise on projective geometry (1639). This was investigated by Kirsti Andersen and Jan Hogendijk:[65]

---

62 *Rough draft for an essay on the results of taking plane sections of a cone* in the translation of Field and Gray 1987.
63 Example of one of S.G.D.L.'s general methods concerning drawing in perspective without using any third point, a distance point or any other kind, which lies outside the picture field.
64 See Field 1997, 195.
65 In a special issue of *Centaurus* devoted to Desargues, namely *Centaurus* 34(1991). For the following quotation, see Andersen 1991, 45.

> The mere fact that perspective and projective geometry are related has been taken as an evidence that the insights Desargues gained while working on perspective were essential for his new approach to geometry.

Jan Hogendijk has shown that there is a strong historical relationship between the *Brouillon project* and the *Conics* of Apollonius. Thanks to his introduction of points and lines at infinity, Desargues was able to derive most of the Apollonian theory of diameters and ordinates in a much easier way than Apollonius.[66] According to Hogendijk, Desargues took not only most of his initial motivation from the *Conics* but also more raw material than has hitherto been realized. Kirsti Andersen studies more specifically the connection in Desargues' writings between points at infinity and the vanishing points of perspective and obtains a more nuanced picture than the usual one. She discards vanishing points as a source of inspiration for Desargues' introduction of points at infinity.

Stereotomy and the Development of Descriptive Geometry

According to Joël Sakarovitch, who wrote an extensive history *Epures d'architecture* with the significant subtitle "From stonecutting to descriptive geometry. 16th to 18th centuries," stereotomy is to descriptive geometry as perspective is to projective geometry,[67] but he doesn't tell what precisely in his view perspective is to projective geometry. The writings referred to in the previous subsection show that from a historical perspective connections between the two are not straightforward but rather problematic. From the abstract mathematical point of view, it is true that the geometrical configuration underlying perspective, a visual cone cut by a plane, has an obvious link to conic sections interpreted as perspective images of a circle at the basis of the cone. It is also true that a vanishing point is a point at infinity, which one can add to a line. But from a historical perspective, things may be more complex, and it is not obvious that Desargues took his inspiration from the existence of vanishing points. Today, historical interpretations of the relations between painters' perspective and projective geometry are wide-ranging and divergent. Thus, at one end, the historian of art James Elkins claims that the connection is "virtually nonexistent." To him, "mathematics and perspective have developed in parallel, mutually isolated streams since the mid-sixteenth century."[68] At the other end, one can find the Husserlian thesis of Hubert Damisch,[69] interpreting perspective as an experimental ground for geometry before it took a new start in the seventeenth century and again adopted its

---

66 Hogendijk 1991, 2
67 In Sakarovitch 1994, 347.
68 See Elkins 1994, 3.
69 "Tout se passe en effet comme si, avant de prendre au xviie siècle un nouveau départ, la géométrie avait dû se donner, se constituer, comme elle l'aura fait à ses débuts, dans la Grèce antique, ce que le philosophe Edmund Husserl a décrit, dans son opuscule sur *L'Origine de la géométrie*, comme un sol préalable d'expérience: quitte pour ce faire à s'écarter pour un long temps des voies qui étaient les siennes et à parler un autre langage que strictement conceptuel, mais qui n'était pas non plus purement technique: celui des peintres et autres «perspecteurs», décorateurs, architectes ou tailleurs de pierre." This quotation is taken from Damisch's preface to Le Goff 1998, 5, but Damisch develops this thesis in Damisch 1987, which has been translated into English.

usual conceptual language. In between, one can find number of interpretations, most of them tying perspective to certain mathematical developments, some (as Kemp, Field and Gray, Andersen, Hogendijk) painting a more nuanced picture of these ties.

What Joël Sakarovitch shows in his *Epures d'architecture* is that methods belonging to what we call descriptive geometry were developed out of problems that stonecutters had to solve. The treatises of stereotomy, from Philibert de l'Orme's *Le premier tome de l'architecture* (1567) to Amédée-François Frézier's *La théorie et la pratique de la coupe des pierres* (1737–39), are seen as direct sources for Gaspard Monge's descriptive geometry. Monge himself was active for two years in the workshop for drawing and stonecutting at the *Ecole du génie* at Mézières, and knew quite well the problems and methods of that craft. His aim in *Géométrie descriptive* (1795) was to describe the mathematical principles underlying the complex graphical techniques, i.e. the method of projections, to build a unified theory that could easily be taught, and to apply it to such crafts as stonecutting and carpentry.

In the sixteenth century, various methods were traditionally applied to stonecutting, among which one, which used horizontal and vertical projections of the archstone to be cut out on the faces of the rough block of stone. Thus, geometrization in the form of orthographic projections—not necessarily laid down in drawings, but inscribed directly on the stone—was a necessary step in preparing stones before utilizing them in the planned building. It was at the origin of a large variety of graphical techniques, which were transmitted orally before they were collected and spread through the first treatises. The famous French architect Philibert de l'Orme included a treatise on stereotomy as Books III and IV of his *Le premier tome de l'architecture*. It is a kind of a catalogue of practical stonecutters' methods put together without establishing links between them, without any effort to single out common principles. From a geometrical point of view, most problems to be solved by stonecutters concern intersections of two different surfaces. The stone-dresser ("appareilleur")—the one who prepared the drawings allowing the stonecutter to obtain neatly the various surfaces bordering archstones—had to handle with sufficient precision complex surfaces such as cones, cylinders, ellipsoids, and spheres, which require a good command of methods for representing space. De l'Orme makes extensive use of the plan and elevation technique in the stereotomy books, while he refrains elsewhere in the treatise from using combined orthographic views. In describing the methods of stonecutters—which wasn't an easy task—, in putting into a written form procedures that were usually transmitted orally, in struggling to make them understood, in spreading them outside the realm of the craft, de l'Orme made a significant first move to single out the underlying geometrical theory.

While de l'Orme had described various isolated practical cases more or less familiar to his fellow architects, Girard Desargues, in his rough draft on stonecutting in architecture published in 1640, pretends to give a general method, as the title of his book claims: *Brouillon project d'une manière universelle du S.G.D.L. touchant la pratique du trait à preuves pour la coupe des pierres en l'architecture*. Like in his rough draft on perspective, he works out a single example. This example was not usual in the architectural practice of the time. Abstractly speaking, it involves a plane cutting a cylinder, i.e. a cylindrical vault penetrates a non-vertical wall and angles downward (a situation, which implies that the axis of the cylinder is not horizontal). The geometrical construction of Desargues is elegant, but the text is dense and obscure (as usual with Desargues). It is impossible to enter into technical detail here,[70] but it is interesting to report on the reactions Desargues' writing provoked among practitioners.

Practitioners' Responses to a Mathematician's Work

The practitioners violently rejected Desargues' geometrical approach to stereotomy. This led to a conflict between Desargues and Jacques Curabelle,[71] known as the best "appareilleur" (dresser) of his time. A series of pamphlets, leaflets and posters were printed in 1644 by the two protagonists, the aim of which was to show the falseness or the superiority of the stonecutters' traditional methods or of Desargues' proposal. To settle the question, Curabelle offered to organize a competition between two stonecutters, one following Curabelle's preparatory drawings, the other applying Desargues' procedures. The outcome of the competition was to decide, in his mind, which of the two methods was the more exact and efficient. For him, only a concrete piece of work could test the precision of the two competing models. Desargues didn't agree with this vision at all. What was important to him was the accuracy of the geometrical arguments. He asked that the two methods be submitted to a board of mathematicians able to evaluate and judge the theoretical foundations of both constructions. The working drawing was in his opinion sufficient to prove the exactitude of the construction; for Desargues there was no need to turn to a concrete realization in stone. On the contrary, the power of geometry is strong enough to legitimate the construction. As Desargues formulates it with some brutality: "geometers ... do not go to the school of stonemasons, on the contrary, masons ... go to the school and lessons of geometers, and thus the geometers are masters and the masons apprentices."[72] This intense polemic between Desargues and Curabelle is the expression of power struggles between mathematicians and practitioners for the control over the crafts.

While Curabelle had the whole profession behind him, Desargues came more and more under the attack of mathematicians and perspectivists such as Jean Dubreuil or Jean François Niceron. But not everybody turned against Desargues. In particular, the competent engraver Abraham Bosse spent much energy in explaining, defending and transmitting Desargues' ideas.[73] He is said to have followed courses by Desargues on geometry applied to drawing. A convinced defender of Desargues' methods, he wrote a treatise on stonecutting as well as a famous work on Desargues' perspective.

---

70 See Sakarovitch 1994 and Sakarovitch 1998, 149–179.
71 See Le Moël 1994 for biographical detail. See Desargues 1864, II 219–426 on the polemic.
72 Quoted by Sakarovitch 1998, 181, who gives a beautiful analysis of the polemic.
73 On Bosse and his relations to Desargues, see Heinich 1983, Kemp 1990, 120–123, Bottineau–Fuchs 1994, and Andersen 1991, 65.

Although not a full member of the *Académie royale de peinture et de sculpture,*
founded in 1648 and located in Paris, whose president was the painter Charles Le
Brun, Bosse was allowed to teach perspective at the Academy. He gave his first lesson
on 9 May 1648, but became increasingly involved in a harsh dispute on the role of
perspective in painting, which lead finally to his exclusion on 11 May 1661. Such an
exclusion seems to be unique in the history of the institution and one might wonder
what the motives behind this severe treatment were. They are probably to be found in
Bosse's notion of representation and in his allegiance to Desargues' methods.[74]
Bosse aims to represent things as they are (according to the geometrical laws of projection) and not as they are perceived subjectively: "You ought be careful not to draw
the relief or the natural as seen by the eye, but you should be able to reconstruct the
proper ground plan of a painting putting various objects together."[75] This is exactly
what Desargues' methods tend to provide. However, the notion of representation,
founded on strict geometrical laws, is violently rejected by the academicians who at
that time were seeking to discard such rigid laws.

For Abraham Bosse, the two practices of representing a solid by its ground plan or
by a perspective view are identical.[76] Bosse only repeats what Desargues had already
formulated in his 1640 rough draft on stonecutting: there is no

> difference between the way to draw, reduce or represent a thing in perspective and the
> way to draw, reduce or represent it by a ground plan, ground and perspective plans are
> thus two species of a same genre, and they can be described and proved together, with
> the same words.[77]

By identifying orthographic projections and perspective, Desargues shows that he
had understood clearly the link between cylindrical and central or conical projections. This is also obvious in his rough draft on conic sections, where he introduces
cone and cylinder by a single definition of a roll ("rouleau"). A line is said to move so
that one of its points is fixed at finite or infinite distance, and that another turns
around a circle. When the fixed point does not lie in the plane of the circle, the figure
obtained is called a roll. Thus in writings partially addressed to practical men,
Desargues rose to a level of abstractness that allowed him to treat horizontal and vertical orthographic projections (with the center of projection at infinity) in the same
way as perspective, i.e. central projection. Bosse made himself the champion of
Desargues' abstract ideas. One may doubt that such ideas were useful to the artists at
the Academy. But this does not explain the vehemence of the debates and the viru-
lence of the attacks against Bosse.[78] Painters and artists, whose activities had gained

74   This is at least the thesis of Nathalie Heinich's analysis of the bitter polemic at the Royal Academy of
     Painting and Sculpture. See Heinich 1993, 147–152.
75   Quoted by Heinich 1993, 148: "Il faut bien se garder de dessiner le relief ou naturel comme l'œil le voit,
     mais bien de [...] pouvoir remettre en son véritable géométral un tableau composé de divers objets
     perspectifs."
76   "ces deux pratiques de faire le trait de la représentation d'un corps en géométral, et en perspective, [...]
     ne sont qu'une même chose" (quoted by Heinich 1993, 148).
77   "[il n'y a] différence aucune entre la manière de figurer, réduire, ou représenter une quelconque chose
     en perspective & la manière de figurer, réduire, ou représenter en géométral, aussi le géométral & le
     perspectif ne sont-ils que deux espèces d'un mesme genre, & qui peuvent estre énoncées ou démontrées
     ensemble, en mesmes paroles." (Quoted by Saint-Aubin 1994, 364).

recognition as liberal arts and who were organized in an academic structure with royal patronage, were struggling for their autonomy by eliminating the control of geometry.

CONCLUSION

In this chapter, we have analyzed some of the complex drawings in Albrecht Dürer's *Underweysung der Messung* (1525), which historical scholarship since the beginning of the twentieth century places in the same line as Gaspard Monge's constructions. In these drawings, Dürer takes into account the laws of vision according to which these mathematical and namely architectural objects must be seen. These laws are those of Euclidean optics, a theory of appearances, and not the rules of perspective, although Dürer did include a short presentation of perspective constructions in his book. His immediate followers in the Nuremberg workshops drop the reference to Euclidean appearances. They invent various material devices and instruments in order to make Dürer's perspective methods more efficient. In the books describing these devices, they call the whole domain *Messung*, a term which translates in their eyes to the Latin *perspectiva*. Thus, *Messung* includes methods, like central projections, which do not preserve measure. Even if the Nuremberg craftsmen are not conscious of this fact, their main preoccupation, quite explicit in the treatises of Hirschvogel and Lencker, is the problem of obtaining the horizontal orthographic projection, the *Geometria* or ground plan, of the objects they want to portray. Having constructed the *Geometria*, or stonemasons' square as they call it, they of course know the real measures of the object. While the fabrique or the representation of the concrete object is at the center of their concern, once professional mathematicians, like Danti, Commandino or del Monte move in, the projective ideas underlying this kind of construction increasingly are studied for their own sake. In the seventeenth century Desargues clearly understands that measure is not preserved in projective transformations like those of perspective. Other properties are, if the Euclidean space is completed by some ideal elements. In his hands, a new geometrical discipline takes form, which is no longer concerned with measure and can no longer be called *Messung*. That's what Desargues is suggesting when he writes: "The words perspective, appearance, representation and portrait are the names of one and the same thing."[79] The thing, which remains unnamed by Desargues, is projective geometry.

---

78 See Desargues 1864, II 49–113, for some aspects of the polemic.
79 See Desargues 1636, 1: "Les mots perspective, apparence, représentation, & pourtrait, y sont chacun le nom d'une méme chose."

PART V
PRACTICE MEETS THEORY

In the period when the decisive steps toward classical mechanics were made (1500–1700), technical drawings played an important role in mediating between practical and theoretical knowledge. This much-neglected function of early modern engineering drawings is the topic of the last section of this volume.

Reflections and inquiries of a more theoretical nature were not at all beyond the horizon of early modern engineering. Design processes are themselves reflection processes of a specific kind and cannot be cordoned off clearly from reflections on issues connected with the working of machines that properly could be called theoretical. As already discussed at several places in this volume, particularly in the chapters by Marcus Popplow, David McGee, and Pamela Long, technical drawings played an important role in design processes. Naturally, the possibilities and restrictions that such drawings provide for and impose on these designing reflections mark the power and limits of drawings as means for general reflections about machines as well as their working principles. With respect to the use of drawings for such more theoretical purposes, the advantages and shortcomings of specific pictorial languages are of particular significance. It is not by chance that, as in the case of Leonardo's notebooks, drawings that doubtless served such general reflections often switched from a pictorial style of representation to the diagrammatic one that was common in treatises on mechanics. In this way, such reflections could make use of geometrical demonstrations and proofs that served as chief means of reasoning in early modern mechanics. However, these diagrams, whatever their advantages over more pictorial representations, shared some of the principal limits characteristic of engineers' models on paper. Their representational potency was overtaxed if one tried to represent on one and the same diagram both the spatial relations of a device and physical quantities such as its mass or forces acting upon it.

These limits of engineering drawings as means of theoretical reflections on machines, however, must not obscure another significant role these drawings played as mediators between practical and theoretical mechanics. As is now almost generally acknowledged, early modern mechanics developed along with the technological innovations of this period and hardly can be understood without this background. Almost all of the pioneers of preclassic mechanics were either engineers themselves, Tartaglia being probably the most prominent instance of such a theorizing engineer, or were occupied occasionally with engineering issues and tasks, as was the case with Galileo. The new technology provided a wealth of new subtle objects whose investigation advanced the understanding of the patterns and laws of the natural potencies of which these devices made use. Yet to become truly familiar with the advanced technology of the age, even in highly developed regions like the Padua-Venice area, personal experiences with real machinery of some sort must be complemented by knowledge gained through representations of machines, either three-dimensional models or technical drawings, and knowledge acquired elsewhere through the study of tracts. It even might be possible that, for investigations of machines in a theoretical perspective, drawings and other models were of even greater significance than real machines.

The main topic of Michael Mahoney's chapter is an exceptionally instructive instance of the power and limits of technical drawings for mediating between theoretical and practical mechanics. His example concerns diagrams by Christian Huygens that were at the interface between practical and theoretical mechanics in a twofold way. First, some of them served the communication between Huygens and Thuret, the Parisian clock-maker who translated Huygens' concept of an *Horologium Oscillatorium* into a working design, thereby proving how differently they could be read by a theoretician and by a practitioner. Second, some of them are highly intricate compounds of different layers of diagrams that represent entities from completely different worlds—from the practical world of machines, from the multidimensional world of physics, and from the ethereal world of mathematics. These diagrams demonstrate to which extremity technical drawings in combination with geometrical diagrams could be pushed when used in theoretical investigations and, at the same time, how impracticable this means of representation became for this employment.

9

DRAWING MECHANICS

MICHAEL S. MAHONEY

1. SETTING THE QUESTION

As the preceding chapters show, a variety of practitioners in the Renaissance drew machines for a variety of apparent reasons: to advertise their craft, to impress their patrons, to communicate with one another, to gain social and intellectual standing for their practice, to analyze existing machines and design new ones, and perhaps to explore the underlying principles by which machines worked, both in particular and in general. I say "perhaps," because this last point is least clear, both in extent and in nature. We lack a basis for judging. We have no corroborating evidence of anything resembling a theory or science of machines before the mid–sixteenth century, and what appeared then reached back to classical Antiquity through the newly recovered and translated writings of Aristotle, Archimedes, Hero, and Pappus, which came with their own illustrations of basic machine configurations.

The absence of a textual tradition to which the drawings themselves are linked, or to which we can link them, makes it difficult to know what to look for in them.[1] How does one know that one is looking at a visual representation of a mechanical principle? It will not do to invoke what is known from the science of mechanics that emerged over the course of the seventeenth century. That sort of "ante hoc, ergo gratia huius" identification of a valued feature of modern science or engineering begs the historical question of precisely what relationship, if any, the drawings bear to the emergence and development of that body of knowledge.[2] What did practitioners learn about the workings of machines from drawing them, and how did it inform the later theory?

That is actually several questions, which do not necessarily converge on the same result. What people learn depends on what they want to know, and why. What questions were the practitioners of the fifteenth and sixteenth centuries asking about the

---

1 Cf. Marcus Popplow's observation in his contribution to this volume (above, p. 42): "The analysis of early modern machine drawings is often confronted with such problems of interpretation. To narrow down the possibilities of interpretation, descriptive texts, text fragments on the drawing and textual documents preserved with the drawings again and again prove most useful. Where such additional material is missing—which is often the case due to the frequent separation of pictorial and textual material practised in a number of European archives some decades ago—interpretation is often confronted with considerable difficulties."

2 Popplow has pointed to the dangers of such retroactive identification: "Die wenigen Blicke, die von Seiten der Wissenschaftsgeschichte auf die Maschinenbücher geworfen wurden, suchten in erster Linie zu beurteilen, inwiefern ihre Diskussion mathematischer und mechanischer Prinzipien bereits auf die wissenschaftliche Revolution des 17. Jahrhunderts verweist. Wiederum ist damit die Tendenz zu erkennen, die Maschinenbücher also noch defiziente 'Vorläufererscheinung' späterer, wissenschaftlich exacterer Abhandlungen zu betrachten. Wie im folgenden deutlich werden wird, entsprechen die dieser Betwertung zugrundeliegenden Maßstäbe jedoch nicht den Intentionen hinter der Abfassung der Werke." (Popplow 1998a, 69) As Edgerton's example shows, it is not only the historians of science who have measured these works by later standards.

machines they were drawing? What sorts of answers were they seeking? What constituted an explanation for how a machine worked, and what could one do with that
explanation? How were the questions related to one another, and where did the
answers lead? What then (and only then) did these questions and answers have to do
with the science of mechanics as shaped by Galileo, Huygens, and Newton?

What follows addresses this historical inquiry and the contents of the foregoing
chapters only by contrast. It revisits an argument made almost 25 years ago in
response to a claim by Samuel Edgerton about the long-term theoretical importance
of Renaissance innovations in the depiction of machines.[3] In essence, he maintained
that ever more realistic pictures of machines led to the science of mechanics. Three
salient passages from the mechanical literature reveal the difficulties with that thesis.
The first is found in a letter written to Galileo Galilei in 1611 by Lodovico Cigoli,
who observed that "a mathematician, however great, without the help of a good drawing, is not only half a mathematician, but also a man without eyes."[4] The second
comes from Leonhard Euler, who in the preface of his *Mechanics, or the Science of
Motion Set Forth Analytically* lamented that the geometrical style of Newton's *Principia* hid more than it revealed about the mathematical structures underlying his
propositions.[5] Finally, capping a century's development of the subject, Joseph-Louis
Lagrange warned readers of his *Analytic Mechanics* that

> No drawings are to be found in this work. The methods I set out there require neither
> constructions nor geometric or mechanical arguments, but only algebraic operations sub-
> ject to a regular and uniform process. Those who love analysis will take pleasure in see-
> ing mechanics become a new branch of it and will be grateful to me for having thus
> extended its domain.[6]

---

3   Mahoney 1985. Cf. Edgerton 1980.
4   Quoted by Santillana 1955, 22, in a rather free translation. Cigoli made the remark in a letter to Galileo
    dated 11 August 1611 in a perplexed effort to understand why Christoph Clavius continued to resist
    Galileo's telescopic evidence of the moon's rough, earth-like surface (Galilei 1890–1909, XI 168):
    "Ora ci ò pensato et ripensato, nè ci trovo oltro ripiegho in sua difesa, se no che un matematico, sia
    grande quanto si vole, trovandosi senza disegnio, sia non solo un mezzo matematico, ma ancho uno
    huomo senza ochi." It is not clear just what Galileo's drawings of the moon's surface have to do with
    mathematics.
5   Euler 1736, Preface, [iv]: "Newton's *Mathematical Principles of Natural Philosophy,* by which the
    science of motion has gained its greatest increases, is written in a style not much unlike [the synthetic
    geometrical style of the ancients]. But what obtains for all writings that are composed without analysis
    holds most of all for mechanics: even if the reader be convinced of the truth of the things set forth,
    nevertheless he cannot attain a sufficiently clear and distinct knowledge of them; so that, if the same
    questions be the slightest bit changed, he may hardly be able to resolve them on his own, unless he
    himself look to analysis and evolve the same propositions by the analytic method." (Non multum
    dissimili quoque modo conscripta sunt Newtoni Principa Mathematica Philosophiae, quibus haec
    motus scientia maxima est adepta incrementa. Sed quod omnibus scriptis, quae sine analysi sunt
    composita, id potissimum Mechanicis obtingit, ut Lector, etiamsi de veritate eorum, quae proferuntur,
    convincatur, tamen non satis claram et distinctam eorum cognitionem assequatur, ita ut easdem
    quaestiones, si tantillum immutentur, proprio marte vix resolvere valeat, nisi ipse in analysin quirat,
    easdemque propositiones analytica methodo evolvat.)
6   Lagrange 1788, Avertissement: "On ne trouvera point de Figures dans cet Ouvrage. Les méthodes que
    j'y expose ne demandent ni constructions, ni raisonnemens géométriques ou mécaniques, mais
    seulement des opérations algébriques, assujetties à une marche régulier et uniforme. Ceux qui aiment
    l'Analyse, verront avec plaisir la Mécanique en devenir une nouvelle branche, et me sauront gré d'avoir
    étendu ainsi le domaine."

Over the short range, looking from Renaissance treatises on machines to the work of Galileo might support the notion that the science of mechanics emerged from ever more accurate modes of visual representation. However, looking beyond Galileo reveals that the long-range development of mechanics as a mathematical discipline directed attention away from the directly perceived world of three spatial dimensions and toward a multidimensional world of mass, time, velocity, force, and their various combinations. Practitioners ultimately found that the difficulties of encompassing those objects of differing dimensionality in a single, workable diagram reinforced a transition already underway in mathematics from a geometrical to an algebraic mode of expression and analysis.[7] One could not draw an abstract machine; one could not even make a diagram of it. But one could write its equation(s). Since the dynamics could find no place in the diagram, the diagram disappeared from the dynamics.[8] But that did not happen right away nor directly, and I want here to take a closer look at the process. In particular, I want to consider the role of drawings and diagrams in mediating between the real world of working devices and the abstract world of mathematical structures.

Edgerton insisted on the capacity of the new techniques to depict machines as they really appeared, rendering their structure in life-like detail. However, the science of mechanics was created as the science not of real, but of abstract machines. It took the form of the science of motion under constraint; as Newton put it in distinguishing between practical and rational mechanics,

> ... rational mechanics [is] the science, accurately set forth and demonstrated, of the motions that result from any forces whatever, and of the forces that are required for any sort of motions.[9]

The clock became a model for the universe not because the planets are driven by weights and gears, but because the same laws explain the clock and the solar system as instances of bodies moving under constraint with a regular, fixed (and measurable) period. Indeed, as Newton himself pointed out, his mathematical system allowed a multitude of possible physical worlds, depending on the nature of the particular forces driving them and on the nature of the initial conditions. Which of these corresponded to our world was an empirical, not a theoretical question.[10]

What made it an empirical question was that the entities and relationships of the mathematical system corresponded to measurable objects and behavior in the physi-

---

7   See Mahoney 1998, I, 702–55.

8   John J. Roche notes that Lagrange did not go unchallenged over the next century, pointing in particular to Louis Poinsot's complaint in 1834 that the "solutions are more lost in the analytical symbolism than the solution itself is hidden in the proposed question (Roche 1993, 225)."

9   Newton 1687, Praefatio ad lectorem, [I]. John Wallis had already spoken in similar terms at the beginning of Chapter 1 of Wallis 1670. In contrast to traditional definitions of mechanics that emphasized its artisanal origins or material focus, he insisted that, "We speak of *mechanics* in none of these senses. Rather we understand it as the part of geometry that treats of *motion* and investigates by geometrical arguments and apodictically by what force any motion is carried out (p. 2)." ("Nos neutro dictorum sensu *Mechanicen* dicimus. Sed eam Geometriae partem intelligimus, quae *Motum* tractat: atque Geometricis rationibus & ἀποδεικτικῶς inquirit, Quâ vi quique motus peragatur.")

10 See Newton 1687, the corollaries to Proposition 4 of Book I and the hypotheses (phaenomena in the 3rd edition) and opening propositions of Book III.

cal world. The fit was meant to be exact, and it grew ever more precise as theory and instrumentation developed in tandem. As will become clear below, that mapping of real to abstract, and the converse, emerged from drawings and diagrams in which representations of the two worlds were joined at an interface. But the drawings acted as more than a conceptual bridge. The principle that mathematical understanding be instantiated in specifiable ways in physical devices—in short, that knowledge work—entailed the need for a cognitive and social bridge between mathematical theory and technical practice and between the theoretician and the practitioner, and drawings played that role too. These issues emerge with particular clarity in the work of Christiaan Huygens, but they are rooted in earlier developments.

2. PUTTING MACHINES ON THE PHILOSOPHICAL AGENDA

Whatever body of principles might have been created by the designers and builders of machines, the science of mechanics as the mathematical theory of abstract machines was the work of seventeenth-century thinkers who considered themselves mathematicians and philosophers. Hence, before machines could become the subject of a science, they had to come to the attention of the people who made science at the time. That is, a body of artisanal practice had to attract the interest of the theory class. It has not always done so. Machines evidently did not impress medieval thinkers. The fourteenth-century philosophers who first compared the heavens to the recently invented mechanical clock lived in a society teeming with mills: windmills, watermills, floating mills, mills on streams and mills under bridges, grist-mills, sawmills, fulling-mills, smithing mills. And mechanically a clock is simply a version of a mill. Yet, no one before the late fifteenth century thought mills were worth writing about, much less suggested that the heavens might work along the same principles. Mills were of fundamental importance to medieval society, but as part of the mundane, workaday world they did not attract the interest of the natural philosophers.[11]

That had to change before a science of mechanics could even get onto the philosophical agenda. "It seems to me," says Salviati (speaking for Galileo) at the very beginning of the *Two New Sciences* (published in 1638 but essentially complete by 1609),

> that the everyday practice of the famous Arsenal of Venice offers to speculative minds a
> large field for philosophizing, and in particular in that part which is called "mechanics."[12]

That the everyday practice of mechanics should be the subject of philosophy is perhaps the most revolutionary statement in Galileo's famous work. Clearly, something had to have raised the intellectual standing of mechanics for Galileo to feel that the

---

11   On the place of the mill in medieval society, see Holt 1988, which significantly revises the long standard interpretation of Marc Bloch in his classic (Bloch 1935), at least for England. Boyer 1982 calls attention to the urban presence of mills and makes it all the more curious that medieval scholars could talk of the *machina mundi* without mentioning them.

12   Galilei 1638, 1: "Largo campo di filosofare à gl'intelletti specolativi parmi che porga la frequente pratica del famoso Arsenale di Voi Sig. Veneziani, & in particolare in quella parte che Mecanica si domanda." Cf., most recently, Renn and Valleriani 2001.

philosophical audience to whom he was addressing the *Two New Sciences* would continue reading past those first lines.[13] One may take the statement as an exhortation, but Galileo had to have grounds for believing it would receive a hearing and indeed enlist support.

The early machine literature appears to have been the start of that process. As several of the preceding chapters show, the "theaters of machines" were a form of advertising, through which engineers (in the original sense of machine-builders) sought to attract patronage and to enhance their social status. The pictures portrayed not so much actual working machines as mechanisms and the ingenious ways in which they could be combined to carry out a task. Depicting the machines in operation, often on mundane tasks in mundane settings, the engineer-authors of these books offered catalogues of their wares. Yet, in some cases, they pretended to more. The machines were also intended to serve a means of elevating their designers' status as intellectual workers. Claiming to be not mere mechanics but mathematicians, they purported to be setting out the principles, indeed even the mathematical principles, on which the machines were based.[14]

Yet it is hard to find much mathematics in these treatises. Contrary to Edgerton's claim of the Renaissance artist as quantifier, Popplow points out that it is precisely the dimensions that are missing.[15] In many, perhaps most, cases, these are not measured drawings; in some cases, the elements of the machines are not drawn to the same scale. Meant to show how machines worked and what kinds of machines might be built, the drawings were not intended to make it possible for someone actually to build a machine from its depiction, unless that someone already knew how to build machines of that sort. The need to persuade potential patrons of the desirability and feasibility of a device had to be balanced against the need to conceal essential details that would inform potential competitors.[16] The message seems clear: "These are the machines I can build. If you want one, I shall be glad to build it for you, bringing to bear the knowledge of dimensions, materials, and detailed structure that I have omitted from the pictures." If that is the case, then the drawings, however realistically crafted, were not headed toward a science of mechanics.

Dimensions include scale, and the absence of scale enables pictorial representation to mix the realistic with the fantastic without a clear boundary between the two. It is apparent in one of the pictures in Domenico Fontana's *Della trasportatione dell' obelisco vaticano* of 1590 (figure 9.1). Fontana's twin towers for lifting, lowering, and

---

13 The use of the vernacular in the first two days should not mislead us about Galileo's intended audience. The same Salviati who protested when Simplicio began to dispute in Latin in the *Two World Systems* slips unapologetically into the language of the schools in the third and fourth days of the *Two New Sciences* when setting out Galileo's new *scientia de motu*, a subject long part of the university curriculum. That part, at least, would not need translation to be understood by philosophers elsewhere in Europe.

14 In addition to the chapters in this volume by David McGee and Mary J. Henninger-Voss, see [Henninger-]Voss 1995 and Cuomo 1998.

15 Popplow 1998a, 72ff.

16 Popplow 1998a, 72: "Der Königsweg der Darstellung lag darin, auf der einen Seite die Umsetzbarkeit der vorgestellten Entwürfe sowie die dabei angewandten wissenschaftlichen Prinzipien überzeugend zu vermitteln und gleichzeitig auf der anderen Seite tatsächlich entscheidende Konstruktionsprinzipien zu verschweigen. Die Tendenz der Autoren der Maschinenbücher, 'unrealistische' Entwürfe zu präsentieren, scheint aus dieser Sicht durchaus vernünftig."

Figure 9.1. Schemes for moving the Vatican Obelisk. (Fontana 1590, fol. 8ʳ.)

raising the obelisk float above a collection of "eight designs, or models, that we consider among the best that were proposed" for carrying out the task.[17] What is striking
is that none of these others is even remotely feasible. All but possibly one fail on the
issue of scale. They are essentially human-sized devices that either cannot be made
larger or wouldn't make any sense if they were, because the humans necessary to
work them don't come in such large sizes. In some cases, if they were scaled up as
drawn, they would collapse under their own weight, much less that of the obelisk.
One can draw them, as one has done here, and one can make models of them, models
that might even work. But, scaled up to the dimensions of the obelisk, none of them
would constitute a working or workable machine.[18]

It is noteworthy that Galileo begins the *Two New Sciences* with precisely this
problem, aiming to provide a theoretical account of why machines do not scale up.
It's not merely a matter of geometry, though it may be demonstrated geometrically: at
a certain point, the internal stresses and strains of a material device cancel its
mechanical advantage.

> ... so that ultimately there is necessarily ascribed not only to all machines and artificial
> structures, but to natural ones as well, a limit beyond which neither art nor nature can
> transgress; transgress, I say, while maintaining the same proportions with the identical
> material.[19]

Several important things are going on in this statement. First, the task of a philosophical account, or a theory, of machines is to set limits *in principle*.[20] There are certain
things that machines cannot do, for both mathematical and physical reasons. Second,
machines in the hands of the natural philosopher have become part of nature, and
nature in turn has been made subject to the limits of machines. The same laws govern
both.

In *Diagrams and Dynamics* I pointed to the difficulty that Galileo had in adapting
his geometrical techniques from statics to kinematics. Taking his cue in statics first
from Archimedes, he transformed his figures while keeping them in equilibrium. For
example, in demonstrating the law of the lever, he took a realistically appearing beam
suspended at its midpoint and then cut it at various points, adding support at the midpoint of each segment (figure 9.2). The (geo)metrics of the situation coincided with
the object under consideration. In shifting then to the so-called "Jordanian" tradition
of medieval statics to treat the bent-arm balance, he could continue to work with a

---

17 Fontana 1590, fol. 7ᵛ.
18 A similar juxtaposition of real and fantastic occurs in Agricola 1556, in which the inventory of
    machines used in mining begins with the wholly practicable windlass, moves through ever more
    complicated combinations, and concludes with a water-driven crane that strains credulity once one
    takes into account the forces involved in reversing its direction in the times necessary to place loads at
    the desired level. At a certain point Agricola seems to be no longer reporting actual machines but rather
    imagining potential machines. Another example is Errard 1584. Few of the devices are drawn to scale,
    nor would they work if they were actually built, in some cases again because humans would not be able
    to drive them.
19 Galilei 1638, 3: "... si che ultimamente non solo di tutte le machine, e fabbriche artifiziali, mà delle
    naturali ancora sia un termine necesariamente ascritto, oltre al quale nè l'arte, nè la natura possa
    trapassare: trapassare dico con osservar sempre l'istesse proporzioni con l'identità della materia."
20 That is what conservation laws do, most notably in thermodynamics. In computer science, theory
    similarly sets limits on computability, decidability, and complexity.

picture of the apparatus by superimposing directly onto it the geometry of virtual dis-
placements, relying on the similarity of arcs traversed in the same time. Moving from
there to the inclined plane by way of the circle, he again could overlay the statical

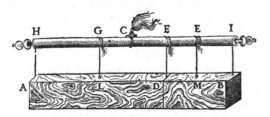

Figure 9.2. The law of the lever derived from a beam in
equilibrium. (Galilei 1638, 110.)

configuration on a picture of the
object. But, when it came to tracing the motion of a body accelerating down the plane, the spatial
configuration at first misled him
into thinking that the triangle
that was the picture of the plane
could also serve as the triangle
made up of the instantaneous
velocities of the body. Only by
separating the graph of motion from the physical trajectory could he get the mathematics to work.

It worked only for kinematics. As is well known, a mathematical account of the
dynamics of motion escaped him completely, as it did Descartes. Picking up where
they had left off, Christiaan Huygens devised a way of moving back and forth
between the physical and mathematical realms and thus to get some of the dynamics
into the diagram.

Before looking at specifics, it is perhaps worth emphasizing their context, if only
to make clear what the second part of this chapter has to do with the first. Huygens'
design of a pendulum clock in 1657 marked the beginning of a line of research that
continued until his death and that in many respects formed the central theme of his
scientific career.[21] In seeking to make his new clock accurate enough to serve for the
determination of longitude and durable enough to continue working aboard a ship at
sea, Huygens undertook a series of investigations in mechanics that led to fundamental results in the dynamics of moving and rigid bodies. In almost every case, those
results led in turn to practical mechanisms that improved either the accuracy or the
reliability of working timekeepers: the pendulum clock itself, the cycloidal pendulum, the conical pendulum, the sliding weight for adjusting the period, the balance-
spring regulator, the tricord pendulum, the "perfect marine balance." In the end, the
complete solution eluded him, in part because it was a matter of metallurgy rather
than mechanics. However, subsequent efforts picked up where he had left off, culminating in the success of John Harrison a half-century later.[22]

Huygens thus embodied the union of head and hand that is characteristic of the
new science of early modern Europe. His work on the clock and on the determination
of longitude at sea are a prime example of what happened when machines did attract
the attention of philosophers. In his hands, the clock constituted an interface between
the mathematical and the physical world, between theory and practice, and indeed
between the scientist and the artisan. Huygens not only derived and proved his results

21   See Mahoney 1980.
22   Andrewes 1996.

in theory, he also designed mechanisms to implement them in practice. He made his own sketches and, in some cases, built working models. But for the finished product he had to turn to master clockmakers and establish productive relations with them. In this he was less successful, in large part because of his inability or unwillingness to recognize the knowledge they brought to the collaboration along with their skill.

Three aspects of Huygens' work warrant closer attention: his use of diagrams in his mechanical investigations, his use of sketches in designing mechanisms, and his relations with his clockmakers. On close examination, his drawings reveal a subtle overlaying of three levels of pictorial representation and establish a visual interface between the physical and the mathematical. That same interface can be found in Newton's *Principia*, and its disappearance in later treatises marks the transition to an algebraic mode of analysis. Huygens' sketches range from the roughest outline to detailed plans, as they show him moving back and forth between theoretical inquiry and practical design. His famous dispute with his Parisian clockmaker, Isaac Thuret, shows what Huygens deemed to count as intellectual property and who could lay claim to possessing it.

3. THE PENDULUM AND THE CYCLOID

In December 1659 Huygens undertook to determine the period of a simple pendulum[23] or, as he put it, "What ratio does the time of a minimal oscillation of a pendulum have to the time of perpendicular fall from the height of the pendulum?" He

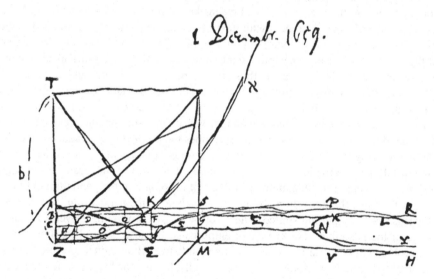

Figure 9.3. Huygens' original diagram. (Huygens 1888–1950, XVI 392.)

---

23 For a detailed account of what follows, see Mahoney 2000, 17–39.

began with a drawing of the pendulum with the bob displaced through an angle KTZ, which he described as "very small," but which he drew large to leave room for reasoning (figure 9.3). On that drawing of the physical configuration he then overlaid a semi-parabola ADΣ representing the increasing velocity of the pendulum as it swung down toward the center point. Next to the graph of motion, he then drew another curve ΣΣG representing the times inversely proportional to the speeds at each moment of the bob's fall. If the bob were falling freely, the area under that curve would represent the time over AZ. But the bob's motion is constrained along the arc KEZ of the circle, so that at, say, point E, one would have (from Galileo):[24]

$$\frac{\text{time at E}}{\text{time at B}} = \frac{\text{length at E}}{\text{length at B}} \times \frac{\text{speed at B}}{\text{speed at E}} = \frac{\text{TE}}{\text{BE}} \times \frac{\text{BF}}{\text{BD}} = \frac{\text{BG}}{\text{BE}} \times \frac{\text{BF}}{\text{BD}}$$

Huygens then constructed the curve of time over the arc by the relation

$\frac{\text{BX}}{\text{BF}} = \frac{\text{BG} \times \text{BF}}{\text{BE} \times \text{BD}}$, sketching it roughly as RLXNYH. Note that the curves of speeds

and times introduce non-spatial parameters into a drawing that began as a spatial configuration.

The mathematics of these "mixed" curves led Huygens then to the introduction of two important elements into the diagram. First, to simplify the expression, he drew on a mathematical result of unknown provenance: if BE were the ordinate to a parabola congruent to the one to which BD is an ordinate, then the product of BE and BD would be a constant times the ordinate BI to a semicircle of radius AZ drawn on the common base of the two parabolas. Huygens knew from earlier work on centrifugal force that a circle and a parabola with the right common parameters coincided in the immediate neighborhood of their point of mutual tangency, so he took the circular arc of the bob's trajectory to be a parabolic arc congruent to the graph of its speed and the intersecting curves as congruent, thus fixing crucial parameters, and drew in the semicircle. Second, from another source Huygens knew that the same circle also served the purpose of finding the area under the adjusted curve of times, and so he again shifted his gaze in the diagram and arrived at the result that the area, and hence the time of motion over the arc, varies as π times the product of the radius of the circle and the length of the pendulum. But, in comparing the time over the arc to the time of fall through the length of the pendulum, the radius of the circle cancels out, making the swing of the pendulum a function of the length of the pendulum only.

For the time, the derivation so far was already a mathematical *tour de force* but Huygens was only getting started. The result was only approximate over a minimal swing of the pendulum. He knew that empirically, because others had shown that, contrary to Galileo's claim, the period of a simple pendulum increases with increasing amplitude. But he now knew it mathematically, because he had explicitly made an approximation in deriving his result: he had taken BE as the ordinate to a parabola

---

24 Relying on the well known mean-speed theorem, Huygens takes the speed at B as the constant speed reached at Z, since the time over the interval AZ at that speed will be twice the time of uniform acceleration over the interval.

rather a circle. He now asked for what trajectory of the bob would BE in fact lie on that parabola? To determine the answer, he had to unpack his drawing to see how the mechanics of the body's motion would generate a parabola to match that of its speed of free fall. He appears to have found the answer by separating BE as ordinate to the physical trajectory from "another BE" as ordinate to the desired parabola, which was the mirror image of a mathematical (mechanical) curve. As he put it:

> ... I saw that, if we want a curve such that the times of descent through any of its arcs terminating at Z are equal, it is necessary that its nature be such that, if as the normal ET of the curve is to the applicate EB, so by construction a given straight line, say, BG is to another EB, point E falls on a parabola with vertex Z.

To determine that "other BE," he seems to have asked how the original BE entered into the expression $BE \times BD$. It came from the ratio $\dfrac{TE}{BE}$, which expresses the ratio of the "length" of point E on the circle to that of point B on the centerline (figure 9.4). Huygens had simply stated the relation without explanation and without drawing the lines in the diagram from which it derives. To see it, one draws the tangent at E, intersecting the centerline at, say, W. By similar triangles, $\dfrac{\text{length E}}{\text{length B}} = \dfrac{EW}{BW} = \dfrac{TE}{BE}$. Now, if one looks at the lines that Huygens did not draw but clearly had in mind in forming

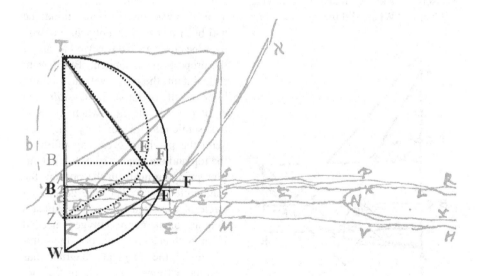

Figure 9.4. Unpacking the relation of physical trajectory and mechanical graph.

the ratio $\frac{\text{TE}}{\text{BE}}$, one more or less readily sees that $\frac{\text{TE}}{\text{BE}} = \frac{\text{TW}}{\text{EW}} = \frac{\text{EW}}{\text{BW}}$. If one takes TW

as a "given straight line" (Huygens' BG), then the first proportion maps the ratio $\frac{\text{TE}}{\text{BE}}$

into the form $\frac{k}{\text{EW}}$ (where EW is his "other EB"), and the second pair makes

$\text{EW}^2 = \text{TW} \times \text{BW}$. That is, if BE were extended to F such that BF = EW, then F

would lie on the parabola $\text{BF}^2 = \text{TW} \times \text{BW}$, with vertex at W and with latus rectum

TW. Now, if W coincided with Z, then $\text{BF}^2 = \text{TZ} \times \text{BZ}$. The osculating circle of
that parabola is precisely the circle on diameter TZ.

At this juncture, one needs a shift of focus. The original circle, centered at T, is no
longer of interest as the trajectory of the pendulum, but rather as a base for that trajectory. One is looking for another curve, with vertex at Z, the properties of whose normal and applicate can be mapped onto the relations just discussed in the circle of
diameter TZ. That is, if BE extended intersects the curve at H, one wants

$\frac{\text{normal at H}}{\text{BH}} = \frac{\text{TE}}{\text{BE}}$. That will be the case if the normal to the curve is parallel to TE,

and hence the tangent at H is parallel to ZE. But this last relation, as Huygens says, is
precisely, "the known method of drawing the tangent" to a cycloid.

Cycloid? Where did the cycloid come from? Well, it was on Huygens' mind; he
had been involved recently in a debate over the curve and so was well aware of the property of its tangent. But, I want to maintain, the curve was also before his eyes. It, or rather its *Gestalt*, had crept into his diagram when he drew that semicircle for auxiliary purposes. With the parabola streaming off from the top and the trajectory of the pendulum swinging up from the bottom, the semicircle now looked like generating circle of a cycloid in the then standard diagram of the curve. Huygens needed no more than a hint; note how the semicircle in the original diagram has become a generating circle in the diagram Huygens drew to show that the cycloid is indeed the curve in question (figure 9.5). Once he had the hint, the details quickly followed.

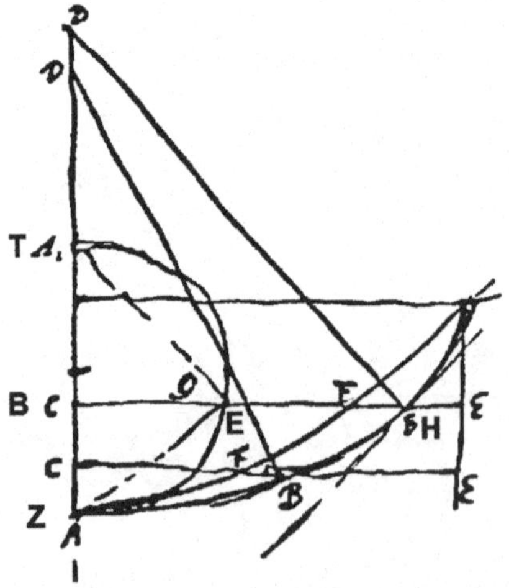

Figure 9.5. The revised diagram, with lettering added from figure IX.4. (Huygens 1888–1950, XVI 400.)

It will not have escaped notice, but I want to emphasize the composite nature of Huygens' main drawing here. It contains three spaces: the physical space of the pendulum, the mechanical space of the graphs of speeds and times, and the mathematical space of the auxiliary curves needed to carry out the quadrature of the curve of times. Huygens combined them without conflating them. That is how he was able so readily to modify the trajectory so as to make an approximate solution exact. He knew not only at what step he had made the approximation, but also in what space he had made it and how it was reflected in the other spaces. He needed a curve in physical space, the properties of whose normal and ordinate could be mapped by way of a mathematical curve so as to generate another mathematical curve congruent to a graph of velocity against distance.

4. PHYSICAL AND MATHEMATICAL SPACE

Huygens conjoined physical and mathematical space in another configuration in his work on the compound pendulum.[25] The problem itself, it should be noted, arose out of a quite practical concern, namely, that in the physical world pendulums have neither weightless cords nor point masses as bobs. Rather one is dealing with swinging objects whose weight is distributed over three dimensions. The task is to find a point in the pendulum at which it acts as if it were an ideal simple pendulum, its center of oscillation.[26]

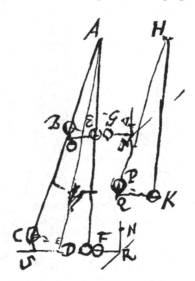

To determine that point, Huygens began with two bobs B and C joined by a common (weightless) rod AC and drew a simple pendulum HP swinging through the same angle in the same time (figure 9.6). Under the constraints of the pendulum, the speeds of B and C will be directly proportional to the speed of P at corresponding points of their swings. The speed of P can be measured by the square root of the height QP through which it falls to K, and that height is proportional to the heights BO and CS through which B and C fall toward E and D, respectively. But the speeds of B and C are constrained by their rigid connection and hence do not correspond directly to the heights through which they individually fall. To get a measure

Figure 9.6. The two-bob pendulum. (Huygens 1888–1950, XVI 415.)

of their speeds, Huygens imagines them impacting with equal bodies G and F, respectively, and imagines G and F then directed upward by reflection off planes inclined at

25 For details, see Mahoney 1980, 234–270.
26 The worknotes dated 1661 (Huygens 1888–1950, XVI 415–34) became Part IV, *De centro oscillationis*, of Huygens 1673.

45°. Each will climb to a height proportional to the square of its velocity, which can be expressed as a function of the height CS and of the ratio of the distance of the bob from A to the unknown length.[27]

At this point, Huygens invokes the principle that the center of gravity of G and F will rise to the same height as that of the compound pendulum at the beginning of its swing. Let me come back to the source of that principle in a moment and focus here on its application. It requires that Huygens move away from the geometrical configuration, which lacks the resources for determining the unknown length HK. That is, he cannot construct it directly by manipulation of the lines of the diagram, because the weights have no quantitative representation. Rather, he turns to algebra, translating the elements of the drawing into an algebraic equation in which the unknown is the length of the simple pendulum and the knowns the bobs and their distances from A: if HK = x, AB = b, AD = d, and B and D denote the weights of the respective bobs, then

the centers of gravity before and after will be $\dfrac{Bb + Dd}{(B + D)d}$ and $\dfrac{Bb^2 + Dd^2}{(B + D)xd}$, respec-

tively. That is, $x = \dfrac{Bb^2 + Dd^2}{Bb + Dd}$, which again cannot be exactly located on the dia-

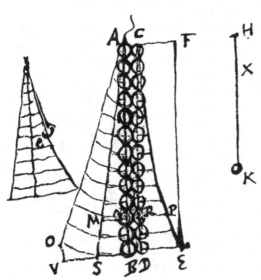

gram. The same thing happens when Huygens then turns to extend this result, by generalization of an n-body pendulum, to the oscillation of a uniform rod.[28] In the continuous case, he imagines the rod as consisting of contiguous small bodies, swinging down under the constraints of a rigid body and then freed to rise individually to the heights corresponding to their acquired speeds (figure 9.7). But, rather than measuring their heights vertically, Huygens draws them horizontally, thus forming a parabola. By reasoning *mutatis mutandis* from the case of two bodies, he shows

Figure 9.7. A rod resolved into contiguous elements. (Huygens 1888–1950, XVI 421.)

27 Huygens first used this technique in his derivation of the laws of elastic collision in *De motu ex percussione* in 1659. There, however, the bodies fall vertically from initial positions to acquire the speeds at which they collide, and then the speeds after collision are converted upward to resting positions. The basic principle is the same: the center of gravity of the system neither rises nor falls in the process. These drawings confirm what one suspects was the physical setup behind the diagrams in the earlier work, namely experiments on impact using pendulums.

28 The indeterminacy of n adds another reason for moving to algebra, where it can be treated operationally as a magnitude. There is no way to represent pictorially an indeterminate number of bodies.

that the equality of the heights of the centers of gravity before and after corresponds to the equality of areas of the triangle on the left (where BS = OV) and of the parabola on the right. The solution of the center of oscillation now comes down to the quadrature of the parabola. However, since the parameters of the parabola include the weight of the rod, that solution must again be couched in algebraic terms.[29]

Note that the triangle is simply an overlay on the physical picture of the pendulum, while the parabola is a mathematical configuration, graphing the height attained against the velocity as a function of the distance from the point of suspension. The centerline forms an interface between the two realms. The inclined planes that initially rendered that interface mechanically intelligible by directing the motion of the weights upward following collision disappear after the first construction. Thereafter, the physical configuration is pictured on the one side, the mathematical structure of the mechanics is pictured on the other. Transition from the one to the other takes place at the centerline by a transformation corresponding mathematically to Galileo's law relating velocity to height in free fall.

One finds similar configurations in Newton's *Principia*, for example in Proposition 41 of Book I:

> Assuming any sort of centripetal force, and *granting the quadrature of curvilinear figures*, required are both the trajectories in which the bodies move and the times of motions in the trajectories found.[30]

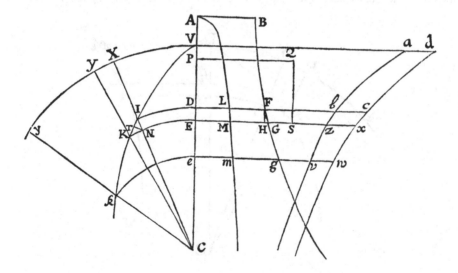

Figure 9.8. Newton's diagram. (Newton 1687, I.8, prop. xli probl. xxviii.)

---

29   Indeed, Huygens had to retain the algebra even in the finished geometrical form of [Huygens 1673].
30   For an extended discussion of Newton's mathematical methods, see Mahoney 1993, 183–205.

On the left is a picture of the orbit VIK of the body revolving about the center of force at C, together with a circle VXY superimposed as a measure of time; the angles in the drawing correspond to measurements that can be made by an observer (figure 9.8). On the right are a variety of curves, which represent the measures of various dynamic parameters such as force and velocity. They are mathematical structures with which one calculates, at least in principle, since in this diagram they are general curves drawn arbitrarily to demonstrate the structure of the problem, rather than any specific law of force. The lines connecting the two sets of curves at the centerline AC map areas under the mathematical curves on the right to sectors of the circle and orbit on the left, thereby determining the position of the planet on its orbit at any given time. The solution of the inverse problem of forces thus becomes a question of quadrature, of finding the areas under the curves on the right for particular laws of force.

For quite independent reasons, the reduced problem of quadrature took a new form with the development of the calculus. Geometry gave way to algebra as the language of analysis, and the construction of curves was supplanted by the manipulation of symbols. Pierre Varignon's adaptation of Newton's configuration shows the result (figure 9.9).[31] The left side remains the same, but complex of curves on the right is reduced to a single curve representing a general expression of the central force determining the orbit. Except for the two ordinates to the curve, there are no auxiliary constructions in the diagram to render graphically the transformations by which integration of the curve fixes the angle and radius of the corresponding position on the orbit. The curve itself thus plays no operational role in the argument, which

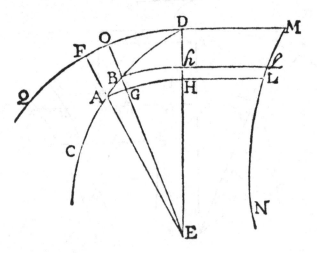

Figure 9.9. Varignon's diagram. (Varignon 1710.)

---

31 Varignon 1710; at 542. Varignon's solution of the problem of inverse central forces (given the force, to find the curve) rested on his earlier analysis of the direct problem (given the curve, to find the force) in Varignon 1700, which was based on a similar adaptation of Newton's diagram.

instead takes place entirely off the diagram in the Leibnizian notation of the infinites-imal calculus.[32] Indeed, the orbit is expressed in differential form, reflecting the gen-erality of the problem and its solution.

I come out here where I came out in *Diagrams and Dynamics*, albeit along a dif-ferent route. Here I have placed more emphasis on the importance (and the chal-lenges) of graphical modes of thought in the early development of the science of mechanics, even though they were later abandoned. It is important to see how draw-ings functioned for Huygens, if only to discern what was involved if one really set out to use pictorial means to analyze the workings of machines and to quantify the under-lying principles. Those principles were not to be found in the pictures, precisely because the pictures erased the boundary between the real and the fantastic; there has been no shortage of elegant pictures of perpetual motion machines.[33] What enabled Huygens to equate the areas on the right and the left of his diagram of the compound pendulum was a principle for which Torricelli is often given credit, but which surely predates him in the form of what I have referred to as a "maxim of engineering expe-rience."[34] It comes down to this: bodies do not rise of their own accord, or, as the author of the treatise attributed to Juanelo Turriano puts it for a particular case, "water cannot go upward on its own ... because of its heaviness and weight."[35] In the case of interest to Huygens, a swinging pendulum winds down, or at best it keeps swinging at the same rate. It certainly does not swing more widely. Huygens makes this a quantitative principle by applying Galileo's law of free fall to the center of gravity of a system of bodies, moving at first under constraint and then freed of con-straint. In doing so, he translates experience of the physical world into measurable behavior expressible in mathematical terms.

5. THEORY AND PRACTICE, KNOWLEDGE AND KNOW-HOW

What is particularly striking, and perhaps unusual, about Huygens' work on the clock is the close interplay between theory and practice. As noted above, the ultimate task was the reliable determination of longitude at sea, which is a matter of keeping time accurately. His abiding goal was to design a device accurate to within seconds a day and durable enough to withstand the rigors of service aboard ship under all condi-tions. In his own mind, the relation between theory and practice was seamless. A famous dispute surrounding his invention of a spring-regulated clock suggests other-

---

32 Indeed, Varignon's statement of the problem, while echoing Newton's proposition, reflected the shift of mathematical focus: "Problème: Les quadratures étant supposées, & la loy quelconque des Forces centrales f etant donnée à volonté en y & en constantes; Trouver en générale la nature de la courbe que ces forces doivent faire décrire au mobile pendent des tems ou des élemens de tems dt donnés aussi à volonté en y & en constantes multipliées par dx ou par dz variables ou non." (Varignon 1710, 536.)

33 David McGee makes a similar point (albeit eschewing the term "fantasy") in his contribution to this volume when he speaks of Taccola's style of drawing enabling him "to work in the absence of real physics." Indeed, he adds in a note, "the whole point of the drawing style is to *eliminate* the need for physical considerations on the part of the artist."

34 Mahoney 1998, 706–8.

35 Ps.–Juanelo Turriano, *Los veintiun libros de los ingenios y de las maquinas*, "el agua no puede ir de suyo para arriba ... por causa de su gravedad y peso." For an example of such maxims filtered through a reading of the *Mechanical Problems* attributed to Aristotle, see Keller 1976, 75–103.

wise and raises a question of interest in this context, namely, of what kinds of knowledge drawings contain and of how they serve as means of communication.

Shortly after publishing his *Horologium Oscillatorium* (Paris, 1673), Huygens uncovered the property of the cycloid that accounted for its tautochronicity: the accelerative force on a body sliding down the inverted curve is proportional to its displacement from the vertex at the bottom, the point of equilibrium. He quickly generalized the property into a principle he called *incitation parfaite décroissante:* in any situation in which the force acting on a body is proportional to its displacement from equilibrium, the body will oscillate with a period independent of its amplitude. By a series of experiments he then confirmed that the regular vibrations of springs rested on that principle and immediately sought to take advantage of the result.[36]

One of Huygens' worknotes shows that on 20 January 1675 he devised a mechanism for regulating a clock by means of a spiral spring.[37] Or rather I should say he sketched such a mechanism, for he did not build the mechanism himself, or even a model of it. Rather, he later related that on the 21st he sought out his clockmaker, Isaac Thuret, but did not find him until the morning of the 22nd, when he had Thuret construct a model of the mechanism while Huygens waited. Evidently, the model was completed by 3 p.m., and Huygens took it with him. The following day, Thuret built a model for himself and then on the 24th and 25th undertook to apply it to a watch. Soon thereafter, unbeknownst to Huygens, he displayed the watch to Colbert, presenting it in a such a manner as to suggest that it was his invention. Dismayed that Thuret would so violate a pledge of secrecy, Huygens vehemently rejected the claim, accused Thuret of violating his trust, and ended their longstanding collaboration.[38]

---

36 Worknotes and sketches in Huygens 1888–1950, XVIII 489–98.
37 Huygens 1888–1950, VII 408ff; on Huygens' invention of the spiral balance, see Leopold 1980, 221–233 and Leopold 1982.
38 Huygens appears to have had grounds for being angry. According to his version of the facts, which Thuret's supporters did not contest, Huygens had pledged Thuret to secrecy before explaining the mechanism he wished to have built. Over the following week, Huygens did more work on the design as he planned both its announcement to the scientific community and its presentation to Minister Colbert for the purpose of securing a *privilège* restricting its manufacture to those licensed by Huygens. On 30 January, he wrote Henry Oldenburg, secretary of the Royal Society, announcing a new mechanism for regulating watches accurately enough to determine longitude and encoding its basic principle in an anagram. The next day he visited Colbert. On 1 February, he learned that Thuret had already attended the Minister on the 24th to show him the second model and that people were now speaking of Thuret as the inventor. Thuret had said nothing of this to Huygens, even though the two had collaborated during the week. Thuret had made hints to Huygens that he desired a share in the credit. But Huygens rejected the idea, even as he pointed out that in enjoying Huygens' license to produce watches with the mechanism Thuret stood to reap the greater monetary gain. On learning of what he took as a betrayal of trust, Huygens cut off all relations with Thuret, excluding him from a license. As the leading clockmaker of Paris, Thuret enjoyed the protection of some powerful patrons, not least Madame Colbert and her son-in law, Charles Honoré d'Albert de Luynes, Duc de Chevreuse. Their intercession led to Thuret's written acknowledgment of Huygens' sole claim to the invention and expression of regret that he might have acted in any way to suggest otherwise. In return he received authorization to produce the new watches. But, then, so too did all clockmakers in Paris, as Huygens decided not to ask the Parlement de Paris to register his *privilège*. Although the two men eventually reconciled, and indeed Huygens recognized the superiority of Thuret's craftsmanship, they never resumed their active collaboration, which had constituted a powerful creative force in timekeeping. For details of the dispute, which dominated Huygens' attention for six months, see Huygens 1888–1950, VII 405–498.

Given the ensuing dispute, a question arises: How did Huygens express or record his invention of the balance spring? That question seems to depend on another, to wit, when did he make the sketches accompanied by "Eureka 20 Jan. 1675"? Was it on that day, or was it some two weeks later, when after learning of Thuret's preemptive visit to Colbert he felt the need to defend his ownership of the invention by means of a day-by-day account of what had transpired in the meantime? That the invention occurred on that day seems clear from the evidence. By Huygens' account, not contradicted at the time by any of several people in a position to do so, he had the idea on the 20th, spoke of it to Pierre Perrault on the morning of the 21st and described it to Isaac Thuret around midday on the 22nd. Rather, the question is how he described it to Thuret. Did he make one or more drawings, and, if so, are they the drawings bearing the date? As John H. Leopold has observed, a look at the manuscript itself suggests the answer.[39]

Surrounding the "eureka" are two drawings and some notes. One drawing shows only a coil spring attached to a dumbbell balance, seen in top view. The other is a side view of the dumbbell mounted on an escapement, with the spring in a cylindrical housing mounted underneath a mounting plate (figure 9.10).

At the top is a descriptive heading, "Watch balance regulated by a spring." To the left underneath the coil and balance are two notes:

> le ressort doit se tenir en l'air dans le tambour et estre rivè au costè et a l'arbre (the spring should be held in the air in the drum and be riveted at the side and at the arbor)

> le balancier en forme d'anneau comme aux montres ordinaires (the balance in the shape of a ring, as in ordinary watches)

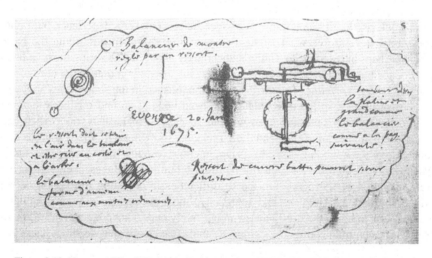

Figure 9.10. Huygens' "Eureka" sketch. (Leiden, University Library, Huygens Collection ms. E, fol. 35; reproduced in Leopold 1980 229.)

---

39 Leopold 1980, 228.

The note to the right of the drawing of the escapement reads:

le tambour dessus la platine et grand comme le balancier, comme a la pag. suivante (the
drum above the plate and as large as the balance, as on the following page (figure 9.11))

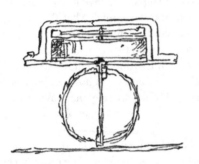

Finally, beneath the drawing, Huygens noted, "ressort de cuivre battu pourroit servir peutestre (spring of beaten copper could serve perhaps)."

Clearly, some of the notes describe not the drawings but rather changes to be wrought on the designs. Leopold suggests that we may have here the record of what transpired when Huygens visited Thuret. Huygens showed Thuret his sketches, described what he had in mind, and then jotted the notes as Thuret made suggestions based on his experience as a clockmaker. Except for the first note, that seems right. Thuret

Figure 9.11. The improved escapement. (Leiden, Museum Boerhaave, Huygens Collection ms. E, fol. 36; reproduced in Huygens 1888–1950, VII 409.)

looks at the dumbbell balance and says "Let's use an ordinary balance wheel, the same size as the spring." As to material, he thinks beaten copper might work. He looks at the escapement and suggests moving the spring from under the plate to above it, so as not to interfere with the 'scape wheel. Huygens makes a note, but waits until getting home to make a sketch, perhaps drawing it from the model he has brought home with him.

By contrast, the first note may not reflect the clockmaker's experience but rather have served to reinforce what Huygens was telling Thuret in describing the basic design, perhaps in response to Thuret's uncertainty about it. The spring must be fixed at both ends but move freely in between. Huygens insisted on that as the core of the design in both the *privilège* and the description published in the *Journal des Sçavans*. The idea of using a spring as regulator had been floating around for more than ten years, and several people had tried without success to make it work.[40] Most had worked from the model of the pendulum, replacing its swing with the vibrations of a metal strip or the bouncing of a coil spring, in either case leaving one end free. Quite apart from the difficulties of transmitting those vibrations to the escapement, such a direct quotation of the pendulum led to difficulties in adjusting the period of the spring's oscillations. Huygens had approached the spring as he had the pendulum. The goal was to leave the oscillator swinging freely and to keep it separately adjustable, while communicating its swings to the escapement and feeding back just

---

40 See, e.g., Huygens' reply of 18.IX.1665 (Huygens 1888–1950, V 486) to a report from Robert Moray (Huygens 1888–1950, V 427) that Robert Hooke had spoken of "applying a spring to the balance of a clock in place of a pendulum." Huygens said he recalled having heard of the idea on a visit to Paris in 1660 but did not think it would work, at least as proposed. Having himself made the idea work in 1675, he never adverted to those earlier suggestions and to what role, if any, they played in his thinking.

enough force to keep the oscillator from running down. In Huygens' design for the spring regulator, the balance constitutes the freely swinging, adjustable weight, held steady in its pivots and communicating its motion to the escapement, either directly by pallets on the arbor or indirectly by means of a pinion. The spring, independent of the escapement and subject only to the coiling force of the arbor, regulates the swing, and its tension may be separately adjusted.

Toward the end of the dispute with Thuret, Huygens reported having reminded Thuret of just this point of the design. In an effort to explain why he had sought some part of the credit for the invention, Thuret claimed to have been thinking of such a mechanism but to have held back from doing anything because he thought that lateral vibrations of the spring would vitiate its regular oscillations. "I responded," said Huygens,

> that what he said of the trouble with these vibrations was something contrived to make it appear that he knew something about the application of the spring, but that this itself showed that he had known nothing about it, because, if he had thought of attaching the spring by its two ends, he would have also easily seen that these vibrations were of no concern, occurring only when one knocked or beat against the clock and even then not undercutting the effect of the spring.[41]

Hence, the first note may account for something Huygens said on several occasions in his dispute with Thuret. "In explaining it to him," reported Huygens of his initial conversation with Thuret, "he said (as yet barely understanding it), 'I find that so beautiful that I still can't believe it is so.'"[42] As Huygens later reminded Thuret, he had said nothing about any investigations of his own prior to Huygens' visit. Yet, if Thuret had been thinking along the lines he later claimed, then he may well have expressed wonder about precisely the way in which the spring was mounted.

And it may well have been wonder born of having wrestled with the problem without seeing a solution. For Thuret evidently understood what Huygens showed him well enough both to make suggestions for its improvement and to build a working model from scratch in a couple of hours. Indeed, when Huygens then departed with the model, Thuret built another, evidently from memory (or did he make his own sketch?), and at once set about to incorporate a working version into a watch (figure 8.12). How much did he have to understand to do that? What did the second model look like, and what changes did Thuret make in adapting it to the watch? In this case, it is not a matter of scaling the model up, but rather of making it smaller and fitting it in with the other parts of the watch. Irrespective of whether Thuret had in fact been exploring the problem independently of Huygens, he might have thought that the know-how involved in that process alone entitled him to share in the privilege for the clock on the grounds that he had helped invent it. Huygens argued that Thuret had simply been following his sketch, which contained the essence of the invention.

---

41 Huygens 1888–1950, V 486.
42 "En la luy expliquant il dit, ne l'entendant encore qu'a peine, je trouve cela si beau que je me défie tousjours qu'il ne soit ainsi." Huygens 1888–1950, VII 410.

Figure 9.12. The escapement of a Thuret clock, 1675. (From Plomb 1999.)

Huygens' response to another challenge of his priority, this time from a gadfly named Jean de Hautefeuille, is revealing in this regard. Shortly after Huygens announced his invention to the Academy, Hautefeuille contested the claim both intellectually and legally, arguing that he had earlier proposed replacing the pendulum of a clock with a thin strip of steel, the uniform vibrations of which would have the same effect.[43] It would have the further advantage of working irrespective of the position of the clock and hence lend itself to use in watches. Hautefeuille admitted that he had not succeeded in getting his mechanism to work, but he insisted on having established the principle of using a vibrating spring as a regulator. The principle lay in the equal vibrations of a spring, whether straight, helical, or spiral. The particular shape and configuration of the spring was incidental, a matter to be determined by a workman *(ouvrier)*.

Huygens responded by noting that Hautefeuille was far from the first to suggest using a spring as a regulator. It had been proposed repeatedly, but none of the earlier designs had actually worked in practice. The trick lay in transforming the spring's equal vibrations into the uniform advance of an escapement, and more than the spring was involved. Thinking that one could simply replace a pendulum by a metal strip, even to the point of adding a weight at the end of the strip to strengthen the vibrations, reflected a basic misunderstanding of the problem.

---

43   J. de Hautefeuille, "Factum, touchant Les Pendules de Poche" (a petition to the Parlement de Paris to block the registration of Huygens' *privilège*, Huygens 1888–1950, VII 439–53; cf. his letter to the Académie des Sciences, 7 July 1674, describing his idea, Huygens 1888–1950, VII 458–60.

Huygens' claim to the invention of the balance-spring regulator rested on two different sorts of knowledge. First, he knew in principle that a spring, whatever its form, was a tautochronic oscillator. That meant more than knowing for a fact that springs vibrate at the same rate no matter how much they are stretched or compressed. It meant knowing why that was the case. His letter to the *Journal des Sçavans* announcing the invention spoke of the movement of the clock as being "regulated by a principle of equality, just as that of pendulums is corrected by the cycloid."[44] That piece of knowledge stemmed from his researches of 1673–74. As a result he knew that the relation between a spring and a cycloidal pendulum lay in their both instantiating the principle that the driving force is directly proportional to the distance from rest. They are different manifestations of the same kind of motion.

Second, Huygens had determined the particular arrangement that translated the principle into practice for the spring. As in the case of the pendulum, it was a matter of letting a weight swing freely, driven ever so slightly by the force of the driving weight or mainspring, communicated through the escapement. Hanging the pendulum by a cord attached to the frame, and connecting it to the escapement by means of a crutch, had been the key to using it to regulate the clock while using the clock to keep the pendulum swinging. In the case of the spring regulator, the weight was the balance wheel, pivoting about the arbor as an axis and governed by the winding and unwinding of a spiral spring attached at one end to the arbor and at the other to the plate or support on which the arbor was mounted. While his original drawings showed the arbor with pallets directly in contact with the crown wheel, the version for the *Journal des Sçavans* showed a pinion moving a rack wheel, which turned the pallets connected with the crown wheel. That design quoted the arrangement in his original clock of 1658.

What Hautefeuille thought was a mere "accident," a technicality to be left to artisan, Huygens considered essential. It was the weight of the balance wheel that held back the advance of the escapement. Since the wheel was swinging about its center of gravity, its motion was independent of its position in space. The spring maintained the tautochrony of that motion. Neither component could do the job alone, yet they could be independently adjusted for the force of the mainspring and the period of oscillation. The secret of a "regular, portable" timepiece lay in their combination.

For present purposes, the rights and wrongs of Huygens' dispute with Thuret are the least interesting aspect of the episode. What is more interesting is the intersection of craft knowledge and high science, of what one knew from building mechanisms and what one knew from analyzing their dynamics. Regrettably, only the latter knowledge was self-consciously set down in words; we have only Huygens' account of this affair. For the former, we must reason indirectly from the artifacts, something I must leave to the antique horologists who know enough about clockmaking to do it critically.[45] In the case of Renaissance machines we also lack the high science. But

---

44 *Journal des Sçavans*, 25.II.1675; repr. in Huygens 1888–1950, VII 424–5; at 424: "... leur [sc. les horologes] mouvement est reglé par un principe d'égalité, de même qu'est celuy des pendules corrigé par la Cycloïde."

we do have the drawings and must find ways to reconstruct from them and, with the help of modern craft expertise, from the artifacts, the conceptual framework of the artisans a half-millennium ago.

Also interesting is the direction Huygens took subsequently. The spring balance, it turned out, had its deficiencies. The practical goal of the enterprise was an accurate sea-going clock for determining longitude. While the spring was mechanically more stable than a pendulum, it was sensitive to changes in temperature and humidity to a degree that undermined its accuracy. So Huygens pursued other mechanisms, all of which had in common the underlying mechanical principle of the spring: the force driving them varied as the displacement from equilibrium. They were all what subsequently came to be called simple harmonic oscillators. Sketches for such mechanisms run for pages in his works, most accompanied by mathematical demonstrations of their workings. In some instances, it is not clear how they would be incorporated in a clock. In other cases, the designs were in fact realized. In the two best known instances, his tricord pendulum and his "perfect marine balance," Huygens built a model before turning to a clockmaker to produce a working timepiece.[46] In both cases, the model seems to have been a necessary proof of concept before attempting a full-scale clock.

What does this all have to do with the topic of this volume? Two things, I think. First, it shows that just drawing a picture of a device, however realistic the rendering, does not suffice to explain how it works. What it conveys depends to a large extent on what knowledge the viewer brings to it. As Huygens tells it, at least, Thuret built a model of the spring balance from Huygens' drawing without understanding how it worked. But having built one model and having heard Huygens' explanation, he evidently knew enough about the workings of the device to adapt it to a watch. That is, Thuret knew what he could change and what not; he knew how to scale it to his own needs. But he went beyond that. He claimed a share in the credit for the invention on the grounds that he had transformed Huygens' idea into a working device. If the interpretation above of what occurred between the two men is correct, Huygens' drawing did not suffice for that; it required a clockmaker's knowledge of the field of application. To Huygens' way of thinking, the invention lay in the idea. For all Thuret's expert knowledge, he had not understood at first why it worked.

---

45 On this point, see Mahoney 1996, 63–68. The several articles in that volume on the clocks of John Harrison are brilliant examples of such direct "readings" of the artifacts.
46 Huygens seems to have drawn a lesson from his experience with Thuret. In his "Application de Décembre 1683" he wrote: "Le 17 dec. 1683 j'ay porte [sic] a Van Ceulen l'horologer le modelle que j'avois fait de ce mouvement de Pendule Cylindrique, pour changer de cette facon les 2 horologes que je luy avois fait faire our la Compagnie des Indes Orientales. J'avois prié mon frere de Zeelhem de venir avec moy: parce que ledit horologer s'imaginoit d'avoir trouvé la mesme chose que moy, apres m'en avoir ouy dire quelque chose en gros. Mais ayant vu le modelle il avoua que ce qu'il avoit modelé n'y ressembloit nullement." (Huygens 1888–1950, XVIII 532) Here a comparison of models seems to have prevented confusion between what Huygens and Van Ceulen had in mind respectively and suggests that a drawing such as the one Huygens had earlier made for Thuret might have been too vague or, literally, sketchy to do so.

6. MATHEMATICAL MODELS

That difference of perspective leads to the second point. By the late seventeenth century, two kinds of people were emerging from the machine literature of the Renaissance, those concerned with machines and those concerned with (mathematical) mechanics. What Thuret, the craftsman, built was a model of a specific mechanism, a coiled-spring balance. From that point on, he was interested in the various ways in which it could be used in watches and clocks. As was the case for his counterparts in England, his clockmaking skills recommended him as an instrument maker for the new scientific institutions. In the following century, clockmakers built the first textile machinery. In short, artisans like Thuret became machine builders, for whom a new genre of machine literature would develop in the nineteenth century.[47]

What Huygens, the mathematician, showed Thuret was a model of a general principle, "perfect incitation." For Huygens the spring balance was one of a series of models that began with the cycloidal pendulum and included the vibrating string, the spring, and the variety of forms of the "perfect marine balance." Once he had discovered the principle, he began to look for the ways in which it was instantiated in physical systems. His drawings of those systems all aimed at bringing out the proportionality between the accelerative force and the displacement from equilibrium. In some cases, it emerged more or less directly, as in the case of chains being lifted from a surface or of cylinders being raised and lowered in containers of mercury. In other cases, it took some sophisticated and ingenious manipulation to relate the geometry of the mechanism to that of the cycloid. Later mechanicians would seek it symbolically by reducing the equation of the system to the form

$$F = m\frac{d^2S}{dt^2} = -kS.$$

To work that way is to build a model of another sort, namely a mathematical model, in which one seeks to map a physical system onto a deductive structure. As Newton showed, using the same mechanics as Huygens but positing another kind of *incitation*, one in which the accelerative force varies as the inverse square of the distance, unites Kepler's empirically derived laws of planetary motion in a mathematical structure, of which Galileo's laws of local motion are limiting cases. In a second set of "Queries" added to his *Opticks* in 1713, he turned his attention to the chemical and electrical properties of bodies and wondered rhetorically whether they might not be explained in terms of small particles of matter attracting and repelling one another by central forces of a different sort from gravity. "And thus Nature will be very conformable to her self," he mused, "and very simple, performing all the great Motions of the heavenly Bodies by the Attraction of Gravity which intercedes those Bodies, and almost all the small ones of their Particles by some other attractive and repelling

---

47   In England, the great clock and instrument makers could aspire to learned status, as Richard Sorrenson has shown in his dissertation [Sorrenson 1993]; for an example of one such clockmaker, see Sorrenson 1996.

Powers which intercede the particles."[48] His suggestion became the agenda of mathe-matical physics for the next two centuries, as practitioners sought to apply the increasingly sophisticated resources of analytic mechanics to mechanical models of natural phenomena. Following that agenda into the late twentieth century and to its encounter with complexity leads to a new problem of scaling and to what appears to be a redefinition of the relation of mathematics to nature. But that issue really would take this contribution beyond the question of the pictorial means of early modern engineering.

---

48   Newton 1730, 396.

APPENDIX

CONTRIBUTORS

Filippo Camerota
Istituto e Museo di Storia della Scienza
Florence, Italy

Mary Henninger-Voss
Independent scholar
Camp Hill, Pennsylvania, U.S.A.

Wolfgang Lefèvre
Max Planck Institute for the History of Science
Berlin, Germany

Rainer Leng
Institut für Geschichte
Universität Würzburg
Würzburg, Germany

Pamela O. Long
Independent scholar
Washington, D.C., U.S.A.

David McGee
Dibner Institute
Cambridge, Massachusetts, U.S.A.

Michael S. Mahoney
History of Science Department
Princeton University
Princeton, New Jersey, U.S.A.

Jeanne Peiffer
Centre Alexandre Koyré
Paris, France

Marcus Popplow
Lehrstuhl für Technikgeschichte
BTU Cottbus
Cottbus, Germany

REFERENCES

UNPUBLISHED SOURCES

Cologne. W* 232 – Konrad Kyeser 'Bellifortis'. Archiv der Stadt Köln.

Dresden. C 363 – Veitt Wolff von Senfftenberg 'hantbiechlin vnd ausszug von meinen erfindungen'. Staats- und Universitätsbibliothek Dresden.

Erlangen. Ms. B. 26 – Ludwig von Eyb 'Kriegsbuch'. Universitätsbibliothek Erlangen-Nürnberg.

Florence. Archivio Privato di D. Giov. De Medici, Filza 8. Archivio di Stato Firenze.

———. Ashburnham ms. 1212 – Matteo Zaccolini 'Della Descritione dell'ombre prodotte de'corpi opachi rettilinei (1617–1622)', 'Prospettiva lineale, De'colori' and 'Prospettiva del Colore'. Biblioteca Medicea Laurenziana.

———. ms. 2660 A. Gabinetto dei Disegni e delle Stampe.

———. MS Magliabechiana VII 380. Biblioteca Nazionale Centrale.

Frankfurt. Ms. germ. qu. 14 – Anonymous 'Rüst- und Feuerwerkbuch'. Stadt- und Universitätsbibliothek Frankfurt am Main.

———. Reichssachen Nachträge Nr. 741 – Anonymous 'Buchsenmeister – fewerwercker'. Bibliothek des Instituts für Stadtgeschichte Frankfurt am Main.

Gotha. Chart B 1032 – Hans Folz (?) 'Furibundi, Feuerwerk- und Büchsenmeisterbuch'. Universitäts- und Forschungsbibliothek Erfurt/Gotha.

Göttingen. Cod. Ms. philos. 63 – Konrad Kyeser 'Bellifortis'. Niedersächsische Staats- und Universitätsbibliothek Göttingen.

———. Cod. Ms. philos. 64 – Konrad Kyeser 'Bellifortis'. Niedersächsische Staats- und Universitätsbibliothek Göttingen.

———. Cod. Ms. philos. 64a – Konrad Kyeser 'Bellifortis'. Niedersächsische Staats- und Universitätsbibliothek Göttingen.

Heidelberg. cpg 126 – Philipp Mönch 'büch der stryt vnd buochsßen'. Universitätsbibliothek Heidelberg.

London. Add. ms codices 5218, 5228–5231 – Dürer's estate. British Library.

Madrid. ms. R 10 – Alonso de Valdelvira 'Libro de traças de cortes de piedras'. Biblioteca de la Escuela Técnica Superior de Arquitectura.

Munich. cgm 356 – Fireworkbook of 1420, Johannes Formschneider 'Armourers' book'. Bayerische Staatsbibliothek München.

———. cgm 599 – Anonymous 'Armourers' book', Fireworkbook of 1420, Martin Merz 'Kunst aus Büchsen zu schießen'. Bayerische Staatsbibliothek München.

———. cgm 600 – Anonymous 'Anleitung Schießpulver zu bereiten, Büchsen zu laden und zu beschießen'. Bayerische Staatsbibliothek München.

———. cgm 734 – Fireworkbook of 1420, Johannes Formschneider 'Armourers' book', Martin Merz (?) 'Armourers' book', etc. Bayerische Staatsbibliothek München.

———. cgm 973 – Christoph Sesselschreiber 'Von Glocken- und Stuckgießerei'. Bayerische Staatsbibliothek München.

———. Hs. 1949–258 – Johannes Formschneider 'Armourers' book' (fragment). Archiv des Deutschen Museums München.

Nuremberg. Hs. 25.801 – Anonymous 'Armourers' book'. Germanisches Nationalmuseum Nürnberg.

———. Hs. 28.893 – Berthold Holzschuher 'Buch der Erfindungen'. Germanisches Nationalmuseum Nürnberg.

Oxford. Canon Ital. 289. Bodleian Library.
Stuttgart. Cod. milit. 4° 31 – Johannes Formschneider 'Armourers' book' (fragment). Württembergische Landesbibliothek Stuttgart.
———. N 220 – Collection of documents and drawings of Heinrich Schickhardt. Hauptstaatsarchiv (HStA) Stuttgart.
Urbino. Fondo del Commune B. 77. Urbino Università.
Vatican. Cod. Vat. gr. 1164 (esp. fols. 88v.–135 r.) – Greek Poliorcetic Writings. Biblioteca Apostolica Vaticana.
Venice. Archivio proprio Contarini. Archivio di Stato di Venezia.
———. Cod. Cic. 3486/IV. Biblioteca Museo Correr.
———. Mat. Mist. Nob. Archivio di Stato di Venezia.
———. ms. Cl. 7 no. 2453 (10493). Biblioteca Nazionale Marciana.
———. Raccolta Terkuz n. 42. Archivio di Stato di Venezia.
———. Sen. Terr. Reg. 33. Archivio di Stato di Venezia.
———. SS. Reg. 83. Archivio di Stato di Venezia.
Vienna. cod. 3064 – Anonymous 'Visierbuch', Anonymous 'Armourers' book', Marcus Graecus 'Liber ignium', etc. Österreichische Nationalbibliothek Wien.
———. cod. 3069 – Anonymous 'Armourers' book'. Österreichische Nationalbibliothek Wien.
———. P 5014 – Anonymous 'Kriegs- und Pixenwerch'. Kunsthistorisches Museum Wien.
———. P 5135 – Anonymous 'Pixen, Kriegsrustung, Sturmzeug vnd Fewewerckh'. Kunsthistorisches Museum Wien.
Weimar. Fol. 328 – Anonymous 'Ingenieurkunst- und Wunderbuch'. Stiftung Weimarer Klassik, Anna Amalia Bibliothek.
———. Q 342 – Hanns Henntz (?) 'Rüst- und Büchsenmeisterbuch, Fireworkbook of 1420. Stiftung Weimarer Klassik, Anna Amalia Bibliothek.
Windsor. K[eele]/P[edretti] – Corpus of Leonardo's Anatomical Studies. Royal Library.
Wolfegg. no signature – Anonymous 'The Medieval Housebook'. Sammlung der Fürsten zu Waldburg Wolfegg.
Zurich. Ms. Rh. Hist. 33b – Anonymous 'Kriegs- und Befestigungskunde, Büchsen- und Pulvermacherei'. Zentralbibliothek Zürich.

PUBLISHED SOURCES AND LITERATURE

Accolti, Pietro. 1625. *Lo inganno de gli occhi – prospettiua pratica di Pietro Accolti*. Florence: Pietro Cecconcelli.
Ackerman, James S. 1949. "Ars sine scientia nihil est" – Gothic Theory of Architecture at the Cathedral of Milan. *The Art Bulletin* XXXI (2):84–111.
———. 1954. Architectural Practice in the Italian Renaissance. *Journal of the Society of Architectural Historians* XIII:3–11.
Adams, Nicholas. 1985. The Life and Times of Pietro dell'Abaco, a Renaissance Estimator from Siena (active 1457–1486). *Zeitschrift für Kunstgeschichte* XLVIII:384–396.
Adams, Nicholas, Daniela Lamberini, and Simon Pepper. 1988. Un disegno di spionaggio cinquecentesco, Giovanni Battista Belluzzi e il Rilievo delle difese di Siena ai tempi dell'assedio. *Mitteilungen des Kunsthistorischen Instituts in Florenz* XXXII (3):558–579.

Adams, Nicholas, and Simon Pepper. 1986. *Firearms and Fortifications*. Chicago: University of Chicago Press.

———. 1994. The Fortification Drawings. In *The Architectural Drawings of Antonio da Sangallo the Younger and his Circle. Vol. I*, edited by C. Frommel and N. Adams. Cambridge, MA: MIT Press.

Agricola, Georgius. 1556. *De re metallica libri XII*. Basel: Froben.

Aguilonius, François d'Aguillon (called). 1613. *Opticorum libri sex*. Antwerp: Moretus.

Aiken, Jane Andrews. 1994. Truth in Images: From the Technical Drawings of Ibn Al–Razzaz Al–Jazari, Campanus of Novara, and Giovanni de' Dondi to the Perspective Projections of Leon Battista Alberti. *Viator* XXV:325–59.

Akin, Omer. 1986. *Psychology of Architectural Design*. London: Pion.

Al–Hassan, Ahmed Y., and Donald Hill. 1986. *Islamic Technology: An Illustrated History*. Cambridge: Cambridge University Press.

Alberti, Leon Battista. 1966. *L'architectura (De re aedificatore): testo latino e traduzione*. Edited by G. Orlandi. Milan: Il Polifilo.

———. 1972. *On Painting and Sculpture: The Latin Texts of 'De Pictura' and 'De Sculptura'*. Edited by C. Grayson. London: Phaidon Press.

———. 1973. De pictura (1435–36). In *Leon Battista Alberti: Opere volgari*, edited by C. Grayson. Bari: Laterza.

———. 1975. *De re aedificatoria*. Munich: Prestel. Original edition, Florence 1485.

———. 1988. *On the Art of Building in Ten Books*. Edited by J. Rykwert, N. Leach and R. Tavernor. Cambridge M.A.: MIT Press.

———. 2000. *Das Standbild. Die Malkunst. Grundlagen der Malerei = De statua. De pictura. Elementa picturae*. Darmstadt: Wissenschaftliche Buchgesellschaft.

Alder, Ken. 1997. *Engineering the Revolution: Arms and Enlightenment in France, 1763–1815*. Princeton: Princeton University Press.

Alertz, Ulrich. 2001. Der Windwagen des Guido von Vigevano. *Technikgeschichte* LXVIII: 53–77.

Alghisi, Galasso. 1570. *Delle fortificatione*. Venice: n.p.

Andersen, Kirsti. 1991. Desargues' method of Perspective. *Centaurus* XXXIV:44–91.

———. 1992. Perspective and the Plan and Elevation Technique, in particular in the Work by Piero della Francesca. In *Amphora : Festschrift für Hans Wussing zu seinem 65. Geburtstag ; Festschrift for Hans Wussing on the occasion of his 65th birthday*, edited by S. Demidov, M. Folkerts, D. Rowe and C. Scriba, 1–23. Basel: Birkhäuser.

———. 1996. The Mathematical Treatment of Anamorphoses from Piero della Francesca to Nicéron. In *History of Mathematics. States of the Art*, edited by J. Dauben, M. Folkerts, E. Knobloch and H. Wussing. San Diego: Academic Press.

Andrewes, William J.H., ed. 1996. *The Quest for Longitude: The Proceedings of the Longitude Symposium, Harvard University, Cambridge, Massachusetts, November 4–6, 1993*. Cambridge, MA: Collection of Historical Scientific Instruments, Harvard University.

Anonymous. 1796. *Raccolta delle più belle vedute della Città e Porto di Livorno*. Livorno: Tommaso Masi.

———. 1912. Die alten Baurisse des Strassburger Münsters. *Strassburger Münsterblatt* VI (chap. 8).

Apianus, Petrus. 1524. *Cosmographicus liber*. Landshut: Weyssenburger.

Appelbaum, Stanley, ed. 1964. *The Triumpf of Maximilian I. 137 Woodcuts by Hans Burgkmair and Others. With a translation of descriptive text, introduction and notes by Stanley Appelbaum.* New York: Dover.

Arrighi, Gino. 1967. Un estratto dal 'De visu' di M° Grazia de' Castellani (dal Codice Ottoboniano latino 3307 della Biblioteca Apostolica Vaticana). *Atti della Fondazione Giorgio Ronchi* XXII:44–58.

Bacher, Jutta. 2000. Das Theatrum machinarum – Eine Schaubühne zwischen Nutzen und Vergnügen. In *Erkenntnis, Erfindung, Konstruktion. Studien zur Bildgeschichte von Naturwissenschaft und Technik vom 16. bis zum 19. Jahrhundert*, edited by H. Holländer, 255–297. Berlin: Mann.

Bachmann, A., and R. Forberg. 1959. *Technisches Zeichnen.* Leipzig: Teubner.

Baldi, Bernardino. 1998. *Le vite de' matematici. Edizione annotata e commentata della parte medievale e rinascimentale.* Edited by E. Nenci. Milan: Angeli.

Barbaro, Daniele. 1568. *La pratica della perspettiva.* Venice: Borgominieri.

Barnes, Carl F. 1982. *Villard de Honnecourt. The Artist and his Drawings: A Critical Bibliography.* Boston: G.K. Hall.

Bartoli, Cosimo. 1564. *Modo di misurar.* Venice: F. Franceschi.

Barzman, Karen–Eidis. 1989. The Florentine Accademia del Disegno: Liberal Education and the Renaissance Artist. In *Academies of Art between Renaissance and Romanticism*, edited by A. W. A. Boschloo, 14–30. 's-Gravenhage: SDU Uitgeverij.

Bätschmann, Oskar, and Sandra Giaufreda. 2002. Einleitung. In *Leon Battista Alberti. Della pittura. Über die Malkunst*, edited by O. Bätschmann and S. Giaufreda. Darmstadt: Wissenschaftliche Buchgesellschaft.

Battisti, Eugenio. 1981. *Filippo Brunelleschi: The Complete Work.* New York: Rizzoli.

Baxandall, Michael. 1972. *Painting and Experience in Fifteenth Century Italy.* Oxford: Oxford University Press.

Baynes, Ken, and Francis Pugh. 1981. *Art of the Engineer.* Woodstock, N.J.: Overlook Press.

Bechmann, Roland. 1993. *Villard de Honnecourt. La pensée technique au XIIIe siècle et sa communication.* Paris: Picard éditeur.

Belici, Giovanni Battista. 1598. *Nuova inventione di fabbricar fortezze.* Venice: T. Baglioni.

Berg, Theresia, and Udo Friedrich. 1994. Der 'Bellifortis' des Konrad Kyeser und das anonyme 'Feuerwerkbuch'. In *Der spätmittelalterliche Verschriftlichungsprozess am Beispiel Heidelberg im 15. Jahrhundert*, edited by J.-D. Müller, 233–288. Munich: Fink.

Besson, Jacques. 1569. *Theatrum instrumentorum et machinarum.* Orleans: n.p.

Bessot, Didier. 1991. Salomon de Caus (c.1576–1626): archaïque ou précurseur? In *Destins de l'art. Desseins de la science. Actes du colloque ADERHEM, Université de Caen, 24–29 octobre 1986*, edited by D. Bessot, Y. Helleguarch and J.-P. Le Goff. Caen: ADERHEM.

Betts, Richard J. 1977. On the Chronology of Francesco di Giorgio's Treatises: New Evidence from an Unpublished Manuscript. *Journal of the Society of Architectural Historians* XXXVI:3–14.

Binding, Günther. 1993. *Baubetrieb im Mittelalter.* Darmstadt: Wissenschaftliche Buchgesellschaft.

Bloch, Mark. 1935. Avènement et conquête du moulin à eau. *Annales d'histoire economique et sociale* XXXVI:583–663.

Blunt, Anthony. 1958. *Philibert de l'Orme.* London: Zwemmer.

Böckler, Georg Andreas. 1661. *Theatrum machinarum novum.* Nuremberg: Fürst and Gerhard.

Booker, Peter Jeffrey. 1963. *A History of Engineering Drawing.* London: Chatto and Windus.

Booz, Paul. 1956. *Der Baumeister der Gotik*. Munich and Berlin: Deutscher Kunstverlag.

Bosse, Abraham. 1647. *Manière universelle de M. Desargues pur practiquer la Perspective par petit pied par le Géometral*. Paris: Des-Hayes.

Bottineau–Fuchs, Yves. 1994. Abraham Bosse 'interprète' de Girard Desargues. In *Desargues en son temps*, edited by J. Dhombres and J. Sakarovitch, 371–388. Paris: A. Blanchard.

Bouvard, André. 2000. Un ingénieur à Montbéliard. Heinrich Schickhardt. Dessins et réalisations techniques (1593–1608). *Extrait des Bulletin et Mémoires de la Société d'Emulation de Montbéliard* CIL:1–92.

Boyer, Marjorie. 1982. Water Mills: A Problem for the Bridges and Boats of Medieval France. *History of Technology* VII:1–22.

Branca, Giovanni. 1629. *Le machine*. Rome: I. Mascardi.

Brancacio, Lelio. 1590. *I carichi militari*. Anversa: J. Trognesio.

Branner, Robert. 1958. Drawings from a Thirteenth-Century Architect's Shop: The Reims Palimpsest. *Journal of the Society of Architectural Historians* XVII (4): 9–21.

———. 1963. Villard de Honnecourt, Reims and the Origin of Gothic Architectural Drawing. *Gazette des Beau-Arts* LXI:129–146.

Bredekamp, Horst. 2000. *Sankt Peter in Rom und das Prinzip der produktiven Zerstörung – Bau und Abbau von Bramante bis Bernini*. Berlin: Wagenbach.

Broda, Werner, ed. 1996. *Dreiecks-Verhältnisse: Architektur- und Ingenieurzeichnungen aus vier Jahrhunderten, Ausstellungskataloge des Germanischen Nationalmuseums*. Nuremberg: Verlag des Germanischen Nationalmuseum.

Brusa, Giuseppe. 1994. L'orologio dei pianeti di Lorenzo della Volpaia. *Nuncius* IX:645–669.

Bruschi, Arnaldo, Corrado Maltese, Manfredo Tafuri, and Renato Bonelli, eds. 1978. *Scritti Rinascimentali di Architettura*. Milan: Il Polifilo.

Bucher, François, ed. 1979. *Architector – The Lodge Books and Sketchbooks of Mediaeval Architects*. Vol. I. New York: Abaris Books.

Büttner, Jochen, Peter Damerow, Jürgen Renn, and Matthias Schemmel. 2003. The Challenging Images of Artillery – Practical Knowledge at the Roots of the Scientific Revolution. In *The Power of Images in Early Modern Science*, edited by W. Lefèvre, J. Renn and U. Schoepflin, 3–27. Basel: Birkhäuser.

Calchini, Enrica. 1991. Glossario dei termini tecnici nel Trattato I (Ms. Saluzziano 148) di Francesco di Giorgio. In *Prima di Leonardo: Cultura delle macchine a Siena nel Rinascimento*, edited by P. Galluzzi, 452–470. Milan: Electa.

Camerota, Filippo. 1998. Misurare 'per perspectiva' : Geometria pratica e Prospectiva pingendi. In *La Prospettiva. Fondamenti teorici ed esperienze figurative dall'antichità al mondo moderno*, edited by R. Sinisgalli, 293–308. Fiesole: Cadmo.

———, ed. 2001. *Nel segno di Masaccio: l'invenzione della prospettiva*. Florence: Giunti.

Caramuel de Lobkowitz, Juan. 1678. *Architectura civil recta y obliqua, considerada y dibuxada en el Templo de Jerusalem*. Vigevano: Camillo Corrado.

Casali, Giovanna, and Ester Diana. 1983. *Bernardo Buontalenti e la burocrazia tecnica nella Toscana medicea*. Florence: Alinea.

Catalano, Franco. 1983. *Francesco Sforza*. Milan: dall'Oglio Editore.

Cataneo, Girolamo. 1584. *Dell'arte militare libri cinque*. Brescia: Mietro Maria Marchetti.

Cataneo, Pietro. 1554. *I quattro primi libri di architettura*. Venice: Figliuoli di Aldo.

———. 1567. *L'architettura di Pietro Cataneo Senese*. Venice: Aldus.

Caus, Salomon de. 1615. *Les raisons des forces mouuantes, auec diuerses machines tant vtilles que plaisantes*. Frankfurt: A. Pacquart.

Cecchini, Francesca. 1998a. Artisti, committenti e perspectiva in Italia alla fine del duecento. In *La Prospettiva. Fondamenti teorici ed esperienze figurative dall'antichità al mondo moderno*, edited by R. Sinisgalli, 56–74. Fiesole: Cadmo.

———. 1998b. 'Le misure secondo l'apparenza'. Ottica e illusionismo nella cultura del duecento : tracce figurative e testimonanzie letterarie. *Micrologus* VI:167–185.

Ceredi, Giuseppe. 1567. *Tre discorsi sopra il modo d'alzar acque da' luoghi bassi*. Parma: Viotti.

Chaboud, Marcel. 1996. *Girard Desargues, bourgeois de Lyon, mathématicien et architecte*. Lyon: IREM de Lyon/Aléas.

Chittolini, Giorgio, ed. 1989. *Gli Sforza, la chiesa lombardia, la corte di Roma: Strutture e pratiche beneficiarie nel ducato di Milano (1450–1535)*. Naples: Liguori.

Cigoli, Ludovico Cardi da. 1992. *Prospettiva pratica*. Edited by R. Profumo. Rome: Bonsignori.

Coenen, Ulrich. 1990. *Die spätgotischen Werkmeisterbücher in Deutschland – Untersuchung und Edition der Lehrschriften für Entwurf und Ausführung von Sakralbauten*. Munich: Scaneg.

Collado, Luys. 1606. *Prattica manuale dell'artiglieria*. Milan: G. Bordoni and P. Locarni.

Commandino, Federico. 1558. *Ptolomaei Planisphaerium, Jordani Planisphaerium, Federici Commandini in Planisphaerium Commentarius, in quo universa scenographices ratio quam brevissime traditur ad demonstrationibus confirmatur*. Venice: Aldus.

———. 1562. *Claudii Ptolemaei liber de analemmate*. Rome: C. Manutium.

Cousin, Jean. 1571. *La Vraye science de la portraicture*. Paris: G. Le Bé.

Cuomo, Serafina. 1997. Shooting by the Book. *History of Science* XXXV:157–188.

———. 1998. Niccolò Tartaglia, mathematics, ballistics and the power of possession of knowledge. *Endeavor: Review of the Progress of Science* XXII:31–35.

Da Costa Kaufmann, Thomas. 1975. Perspective of Shadows: the History of the Theory of Shadow Projection. *Journal of the Warburg and Courtald Institutes* XXXVIII (258–287).

———. 1993. *The Mastery of Nature: Aspects of Art, Science, and Humanism in the Renaissance*. Princeton: Princeton University Press.

Damisch, Hubert. 1987. *L'origine de la perspective*. Rev. ed. Paris: Flammarion.

Davis, Michael T. 2002. On the Drawing Board: Plans of the Clermont Cathedral Terrace. In *Ad Quadratum. The Practical Application of Geometry in Medieval Architecture*, edited by N. Y. Wu, 183–203. Aldershot: Ashgate.

De L'Orme, Philibert. 1567. *Le premier tome de l'architecture*. Paris: Frederic Morel.

De La Croix, Horst. 1963. The Literature on Fortification in Renaissance Italy. *Technology and Culture* IV:30–50.

De Rosa, Agostino. 1997. *Geometrie dell'ombra. Storia e simbolismo della teoria delle ombre*. Milan: Città Studi Edizioni.

Degenhart, Bernhard, and Annegrit Schmitt, eds. 1968. *Corpus der italienischen Zeichnungen 1300–1450, Teil I, Süd und Mittelitalien*. 4 vols. Berlin: Mann.

———, eds. 1980–82. *Corpus der italienischen Zeichnungen 1300–1450, Teil II, Venedig – Addenda zu Süd und Mittelitalien*. 4 vols. Berlin: Mann.

Del Monte, Guidobaldo. 1577. *Mechanicorvm liber*. Pesaro: H. Concordiam.

———. 1579. *Planisphaeriorum Universalium Theoricae*. Pesaro: H. Concordiam.

———. 1600. *Perspectivae libri sex*. Pesaro: H. Concordiam.

Desargues, Girard. 1636. *Exemple d'une desmanières universelles du S.G.D.L. touchant la pratique de la Perspective, sans employer aucun tiers poicts de distance ou d'autre qui soit hors du champ de l'ouvrage.* Paris: n.p.

———. 1640. *Brouillon projet d'exemple d'une manière universelle du S.G.D.L. touchant la pratique du trait à preuves pour la coupe de pierres en l'Architecture.* Paris: n.p.

———. 1864. *Oeuvres de Desargues.* Edited by N. G. Poudra. 2 vols. Paris: Leiber.

———. 1951. *L'oeuvre mathématique de G. Desargues.* Edited by R. Taton. Paris: J. Vrin.

Dhombres, Jean, ed. 1992. *L'École normale de l'an III. Leçons de mathématiques.* Paris: Vuibert.

Di Teodoro, Francesco Paolo. 1994. *Raffello, Baldassar Castiglione e la lettera a Leone X.* Bologna: La Nuova Alfa.

———. 2001. Piero della Francesca, il disegno, il disegno di architettura. In *Nel segno die Masaccio: l'invenzione della prospettiva,* edited by F. Camerota. Florence: Giunti.

———. 2002. Vitruvio, Piero della Francesca, Raffaello: note sulla teoria del disegno di architettura nel Rinascimento. *Annali di Architettura – Rivista del Centro Internationale di Architettura Andrea Palladio* XIV:35–54.

Diaz, Furio. 1976. *Il Granducato di Toscano.* Turin: Tosco.

Dohrn–van Rossum, Gerhard. 1990. Misura del tempo e ritmo di lavoro nei grandi cantieri medievali. In *Dalla torre di Babele al ponte di Rialto,* edited by J.–C. M. Vigueur and A. Paravicini Bagliani, 192–209. Palermo: Sellerio.

Dolza, Luisa, and Hélène Vérin. 2001. Introduction. In *Jacques Besson: Il Theatrum instrumentorum et machinarum (Lione, 1578),* edited by L. Dolza and H. Vérin, 227–254. Rome: Elefante.

Dubourg Glatigny, Pascal. 1999. Egnazio Danti 1536–1586 : discours scientifique et pratique artistique. Thèse de doctorat, Paris X – Nanterre.

Dupré, Sven. 2001. Galileo, Mathematical Instruments and Orthographic Projection. *Bulletin of the Scientific Instrument Society* LXIX:10–20.

Dürer, Albrecht. 1969. *Etliche underricht / zu befestigung der Stett / Schlosz / und flecken [Facsimile edition of the original edition Nuremberg 1527].* Unterschneidheim: Walter Uhl.

———. 1977. *The Painter's Manual – translated and with a Commentary by Walter L. Strauss.* New York: Abaris Books.

———. 1983. *Unterweisung der Messung [Facsimile edition of the original edition Nuremberg 1525].* Nördlingen: Alfons Uhl.

———. 1995. *Géométrie – Présentation, traduction de l'Allemand et notes par Jeanne Peiffer.* Paris: Le Seuil.

———. 1996. *Hierinn sind begriffen vier bücher von menschlicher Proportion [Facsimile edition of the original edition Nuremberg 1528].* 3 ed. Nördlingen: Alfons Uhl.

Edgerton, Samuel Y. 1980. The Renaissance Artist as Quantifier. In *The Reception of Pictures,* edited by M. A. Hagen. New York: Academic Press.

———. 1991. *The Heritage of Giotto's Geometry: Art and Science on the Eve of the Scientific Revolution.* Ithaca: Cornell University Press.

Egg, Erich, ed. 1969. *Ausstellung Maximilian I. Innsbruck 1. Juni bis 5. Oktober 1969.* Innsbruck: Land Tirol, Kulturreferat.

Ehrenberg, U. 1890. Ein finanz- und sozialpolitisches Projekt aus dem 16. Jahrhundert. *Zeitschrift für die gesamte Staatswissenschaft* XLVI:717–735.

Elkins, James. 1994. *The Poetics of Perspective.* Ithaca: Cornell University Press.

Errard, Jean. 1584. *Le premier livre des instruments mathematiques mechaniques*. Nancy: J. Janson.

Euler, Leonhard. 1736. *Mechanica sive motus scientia analytice exposita*. St. Petersburg: Typographia Academiae Scientiarum.

Evans, Robin. 1997. *Translations from Drawing to Building and Other Essays, AA Documents 2*. London: Architectural Association.

Falk, Tilman, ed. 1979. *Katalog der Zeichnungen des 15. und 16. Jahrhunderts im Kupferstich-kabinett Basel – Teil 1*. Basel: Schwabe.

Fanelli, Giovanni. 1980. *Brunelleschi*. Florence: Scala.

Fara, Amelio. 1988. *Bernardo Buontalenti: l'architettura, la guerra, e l'elemento geometrico*. Genoa: Sagep.

Farish, William. 1822. On Isometrical Perspective. *Cambridge Philosophical Society Transactions* I:1–20.

Faulhaber, Johann. 1610. *Newe Geometrische und Perspectivische Inventiones Etlicher sonderbahrer Instrument / die zum Perspectivischen Grundreissen der Pasteyen unnd Vestungen / wie auch zum Planimetrischen Grundlegen der Stätt / Feldläger und Landtschafften / deßgleichen zur Büchsenmeisterey sehr nützlich unnd gebrauchsam seynd*. Frankfurt: A. Hummen.

Federici Vescovini, Graziella. 1983. *'Arti' e filosofia nel secolo XIV. Studi sulla tradizione aristotelica e i moderni*. Florence: Vallecchi.

Ferguson, Eugene. 1992. *Engineering and the Mind's Eye*. Cambridge, MA: MIT Press.

Field, J.V. 1985. Giovanni Battista Benedetti on the Mathematics of Linear Perspective. *Journal of the Warburg and Courtauld Institutes* XLVIII:71–99.

———. 1997. *The Invention of Infinity – Mathematics and Art in the Renaissance*. Oxford, New York, Tokyo: Oxford University Press.

Field, J.V., and Jeremy Gray, eds. 1987. *The Geometrical Work of Girard Desargues*. New York: Springer.

Filarete (Averlino, Antonio). 1972. *Trattato di architettura*. Edited by A. M. Finoli and L. Grassi. 2 vols. Milan: Il Polifilo.

Filarete (Averlino, Antonio), and John R. Spencer (transl.). 1965. *Filarete's Treatise on Architecture*. 2 vols. New Haven: Yale University Press.

Fiocca, Alessandra. 1995. 'Libri d'Architetura et Matematicha' nella biblioteca di Giovan Battista Aleotti. *Bollettino di Storia delle Scienze Matematiche* XV:85–132.

———, ed. 1998. *Giambattista Aleotti e gli ingegneri del Rinascimento*. Florence: Olschki.

Fiore, Francesco Paolo. 1985. La traduzione da Vitruvio di Francesco di Giorgio; note ad una parziale trascrizione. *Architettura: Storia e documenti* I:7–30.

Fiore, Francesco Paolo, and Manfredo Tafuri, eds. 1993. *Francesco di Giorgio, architteto*. 2 vols. Milan: Electa.

Flocon, Albert, and René Taton. 1963. *La perspective, Que sais-je?* Paris: Presses Universitaires de France.

Fontana, Domenico. 1590. *Della trasportatione dell'Obelisco Vaticano, e delle fabriche di Nostro Signore Papa Sisto V*. Vol. I. Rome: D. Basso.

Francesco di Giorgio Martini. 1967. *Trattati di architettura ingegneria e arte militare*. Translated by L. Maltese Degrassi. Edited by C. Maltese. 2 vols. Milan: Il Polifilo.

Franci, Raffaella, and Laura Toti Rigatelli. 1989. La matematica nella tradizione dell'abaco nel XIV e XV secolo. In *La storia delle scienze*, edited by C. Maccagni and P. Freguglia, 68–94. Busto Arsizio: Bramante.

Frangenberg, Thomas. 1993. Optical correction in sixteenth-century theory and practice. *Renaissance Studies* VII:207–228.

French, Thomas E. 1947. *A Manual of Engineering Drawing for Students and Draftsmen.* New York and London: McGraw–Hill.

Friedrich, Udo. 1996. Herrscherpflichten und Kriegskunst. Zum intendierten Gebrauch früher 'Bellifortis'-Handschriften. In *Der Codex im Gebrauch,* edited by C. Meier, D. Hüpper and H. Keller. Munich: Fink.

Frommel, Christoph L. 1994a. Reflections on Early Architectural Drawings. In *The Renaissance from Brunelleschi to Michelangelo: the Representation of Architecture,* edited by H. A. Millon and V. M. Lampugnani. Milan: Bompiani.

———. 1994b. The Drawings of Antonio da Sangallo the Younger: History, Method, Function. In *The Architectural Drawings of Antonio da Sangallo the Younger and his Circle,* edited by C. L. Frommel and N. Adams. Cambridge, MA.: MIT Press.

Frommel, Christoph L., and Nicholas Adams, eds. 1994c. *The Architectural Drawings of Antonio da Sangallo the Younger and his Circle.* Cambridge, MA.: MIT Press.

Galilei, Galileo. 1638. *Discorsi e dimostrazioni matematiche intorno à due nuove scienze.* Leiden: Elzevier.

———. 1890–1909. *Le opere.* Edited by A. Favaro. 21 vols. Florence: Barbera.

———. 1968a. Trattato di fortificazione. In *Le Opere di Galileo Galilei,* II 77–146. Florence: Barbera.

———. 1968b. A proposito di una macchina con gravissimo pendolo adattato as un leva. In *Le Opere di Galileo Galilei,* VIII 571–581. Florence: Barbera.

Galluzzi, Paolo. 1987. The Career of a Technologist. In *Leonardo da Vinci: Engineer and Architect,* edited by P. Galluzzi, 41–109. Montreal: Montreal Museum of Fine Arts.

———, ed. 1991. *Prima di Leonardo. Cultura delle macchine a Siena nel Rinascimento.* Milan: Electa.

———. 1993. Portraits of Machines in Fifteenth-century Siena. In *Non-verbal Communication in Science Prior to 1900,* edited by R. Mazzolini, 53–90. Florence: Olschki.

———, ed. 1996a. *Mechanical Marvels. Invention in the Age of Leonardo.* Florence: Giunti.

———, ed. 1996b. *Renaissance Engineers, from Brunelleschi to Leonardo da Vinci.* Florence: Giunti.

———. 1999. *The Art of Invention: Leonardo and the Renaissance Engineers.* Florence: Istituto e Museo di Storia della Scienza.

Gambuti, Alessandro. 1972. Nuove ricerche sugli elementa picturae. *Studi e Documenti di Architettura* I:133–172.

Gärtner, Peter J. 1998. *Filippo Brunelleschi 1377–1446.* Cologne: Könemann.

Gemma Frisius, Reiner. 1556. *De astrolabo catholico.* Antwerp: Steelsius.

Germann, Georg. 1993. *Einführung in die Geschichte der Architekturtheorie.* Darmstadt: Wissenschaftliche Buchgesellschaft.

Gille, Bertrand. 1964. *Les ingénieurs de la Renaissance.* Paris: Hermann.

———. 1966. *Engineers of the Renaissance.* Cambridge, MA: MIT Press.

Goldthwaite, Richard. 1980. *The Building of Renaissance Florence.* Baltimore: Johns Hopkins University Press.

Goodman, Nelson. 1976. *Language of Art: An Approach to a Theory of Symbols.* Indianapolis: Hacket.

Goody, Jack. 1977. *The Domestication of the Savage Mind.* Cambridge: Cambridge University Press.

Grafton, Anthony. 2000. *Leon Battista Alberti: Master Builder of the Italian Renaissance.* New York: Hill and Wang.

Grassi, Giulio. 1996. Ein Kompendium spätmittelalterlicher Kriegstechnik aus einer Hand-schriften-Manufaktur (ZBZ, Ms. Rh. hist. 33b). *Technikgeschichte* LXIII:195–217.

Grayson, Cecil. 1964. L.B. Alberti's 'costruzione legittima'. *Italian Studies* XIX:14–27.

Guipaud, Christian. 1991. Les six livres de perspective de Guidobaldo del Monte. In *Destins de l'art. Desseins de la science. Actes du colloque ADERHEM, Université de Caen, 24–29 octobre 1986,* edited by D. Bessot, Y. Helleguarch and J.–P. Le Goff. Caen: ADERHEM.

———. 1998. De la représentation de la sphère céleste à la perspective dans l'œuvre de Guidobaldo del Monte. In *La prospettiva. Fondamenti teorici ed esperienze figurative dall'antichità al mondo moderno,* edited by R. Sinisgalli. Florence: Cadmo.

Günther, Siegmund. 1887. *Geschichte des mathematischen Unterrichts im deutschen Mittelalter bis zum Jahr 1525.* Berlin: Hofmann.

Gürtler, Eleonore. 1996. *Ruhm und Sinnlichkeit. Innsbrucker Bronzeguss 1500–1650 von Kaiser Maximilian I. bis Erzherzog Ferdinand Karl.* Innsbruck: Tiroler Landesmuseum Ferdinandeum.

Hahnloser, Hans R. 1972. *Villard de Honnecourt: Kritische Gesamtausgabe des Bauhüttenbu-ches ms. fr 19093 der Pariser Nationalbibliothek.* 2. ed. Graz: Akademische Druck- und Verlagsanstalt.

Hale, J.R. 1977. *Renaissance Fortification: Art or Engineering?* London: Thames and Hudson.

———. 1983a. The Early Development of the Bastion: An Italian Chronology c. 1450–1534. In *Renaissance War Studies,* edited by J. R. Hale, 1–30. London: The Hambledon Press.

———. 1983b. The First Fifty Years of a Venetian Magistry: the Provveditori alle Fortezze. In *Renaissance War Studies,* edited by J. R. Hale, 159–188. London: The Hambledon Press.

———. 1983c. The Military Education of the Officer Class in Early Modern Europe. In *Renaissance War Studies,* edited by J. R. Hale, 225–246. London: The Hambledon Press.

———. 1983d. Military Academies on the Venetian Terraferma in the Early Seventeenth Cen-tury. In *Renaissance War Studies,* edited by J. R. Hale, 285–308. London: The Hambledon Press.

Hale, J.R., and M.E. Mallet. 1984. *The Military Organization of a Renaissance State.* Cambridge: Cambridge University Press.

Hall, A. Rupert. 1976a. Guido's Texaurus, 1335. In *In On Pre-Modern Technology and Science. Studies in Honour of Lynn White, Jr,* edited by B. S. Hall and D. C. West, 11–51. Malibu: Undena.

Hall, Bert S. 1976b. The New Leonardo. *Isis* LXVII:463–475.

———. 1978. Giovanni de'Dondi and Guido da Vigevano: Notes Toward a Typology of Medieval Technological Writings. In *Machaut's World. Art in the 14'th Century,* edited by M. Pelner Cosman and B. Chandler, 127–142. New York: New York Academy of Sciences.

———. 1979a. Der Meister sol auch kennen schreiben und lesen: Writing about Technology ca. 1400 – ca. 1600 A.D. and their Cultural Implications. In *Early Technologies,* edited by D. Schmandt–Besserat, 47–58. Malibu: Undena.

———, ed. 1979b. *The technological illustrations of the so-called 'Anonymus of the Hussite Wars'. Codex Latinus Monacensis 197, Part I.* Wiesbaden: Reichert.

———. 1982a. Production et diffusion de certains traités techniques au moyen âge. In *Les arts mécaniques au moyen âge,* edited by G. H. Allard and S. Lusignan, 147–170. Montreal and Paris: Bellarmin.

————. 1982b. Editing Texts in the History of Early Technology. In *Editing Texts in the History of Science and Medicine*, edited by T. H. Levere, 69–100. New York and London: Garland.

————. 1982c. Guido da Vigevanos Texaurus regis franciae. In *Studies on medieval Fachliteratur*, edited by W. Eamon, 33–44. Brussels: Omirel.

————. 1996. The Didactic and the Elegant: Some Thoughts on Scientific and Technological Illustrations in the Middle Ages and Renaissance. In *Picturing Knowledge. Historical and Philosophical Problems Concerning the Use of Art in Science*, edited by B. S. Baigrie, 3–39. Toronto: Toronto University Press.

Hartwig, O. 1927. Christoph Sesselschreiber und sein Buch über Büchsenmeisterei. *Kultur des Handwerks* II:278–284.

Haselberger, Lothar. 1980. Werkzeichnungen am jüngeren Dydimeion. *Istanbuler Mitteilungen* XXX:191–215.

————. 1983. Bericht über die Arbeit am jüngeren Apollontempel. *Istanbuler Mitteilungen* XXXIII:90–123.

Hassenstein, Wilhelm, ed. 1941. *Das Feuerwerkbuch von 1420*. Munich: Verlag der Deutschen Technik.

Hecht, Konrad. 1997. *Maß und Zahl in der gotischen Baukunst*. 2 ed. Hildesheim: Olms.

Heinich, Nathalie. 1983. La perspective académique – peinture et tradition lettrée : la référence aux mathématiques dans les théories de l'art au 17e siècle. *Actes de la recherche en sciences sociales* IL:47–70.

————. 1987. Arts et sciences à l'âge classique – professions et institutions culturelles. *Actes de la recherche en sciences sociales* LXVI/LXVII:47–78.

————. 1993. *Du peintre à l'artiste. Artisans et académiciens à l'âge classique*. Paris: Les éditions de Minuit.

Heisel, Joachim P. 1993. *Antike Bauzeichnungen*. Darmstadt: Wissenschaftliche Buchgesellschaft.

Henninger–Voss, Mary. 2000. Working Machines and Noble Mechanics: Guidobaldo del Monte and the Translation of Knowledge. *Isis* XCI:233–259.

————. 2002. How the 'New Science' of Cannons Shook up the Aristotelian Cosmos. *Journal of the History of Ideas* LXIII (3):371–397.

[Henninger–]Voss, Mary. 1995. Between the Cannon and the Book. PhD-thesis, Johns Hopkins University.

Heron. 1589. *De gli automati*. Edited by B. Baldi. Venice: G. Porro.

Hess, O. 1920. Die fremden Büchsenmeister und Söldner in den Diensten der eidgen. Orte bis 1516. PhD-thesis, Zurich.

Heydenreich, Ludwig H., Bern Dibner, and Ladislao Reti. 1980. *Leonardo the Inventor*. New York et al.: McGraw–Hill.

Hill, Donald R. 1996. Engineering. In *Encyclopaedia of the History of Arabic Science*, edited by R. Rashed, 751–795. London: Routledge.

Hirschvogel, Augustin. 1543. *Ein aigentliche und gründtliche anweysung / in die Geometria …* Nuremberg: Vom Berg and Neuber.

Hogendijk, Jan. 1991. Desargues' Brouillon Project and the Conics of Apollonius. *Centaurus* XXXIV:1–43.

Holt, Richard. 1988. *The Mills of Medieval England*. Oxford: Oxford University Press.

Huygens, Christiaan. 1673. *Horologium oscillatorium*. Paris: F. Muguet.

————. 1888–1950. *Oeuvres complètes*. 22 vols. La Haye: Nijhoff.

Ianziti, Gary. 1988. *Humanistic Historiography under the Sforzas: Politics and Propaganda in Fifteenth-Century Milan*. Oxford: Clarendon Press.

Ivins, William M. Jr. 1973. *On the Rationalization of Sight*. New York: Da Capo Press.

Jazari, Ismail ibn al–Razzaz. 1974. *The book of knowledge of ingenious mechanical devices*. Edited by D. R. Hill. Dordrecht: Kluwer Academic Publishers.

Johnston, Marina Della Putta. 2000. Leonardo Da Vinci's Codex Madrid I: The Creation of the Self as Author. PhD-thesis, University of Pennsylvania.

Jolivet, Jean. 1997. Classifications des sciences. In *Histoire des sciences arabes III*, edited by R. Rashed, 255–270. Paris: Le Seuil.

Jones, Christopher. 1970. *Design Methods, Seeds of Human Futures*. London: Wiley Interscience.

Keller, Alex. 1976. A Manuscript Version of Jacques Besson's Book of Machines, with his Unpublished Principles of Mechanics. In *In On Pre-Modern Technology and Science. Studies in Honour of Lynn White, Jr*, edited by B. S. Hall and D. C. West, 75–95. Malibu: Undena.

———. 1976a. Mathematicians, Mechanics, and Experimental Machines in Northern Italy in the Sixteenth Century. In *The Emergence of Science in Western Europe*, edited by M. Crossland, 15–34. New York: Science History Pubs.

———. 1978. Renaissance Theaters of Machines (Review Article). *Technology and Culture* XIX:495–508.

Kemp, Martin. 1977. From 'Mimesis' to 'Fantasia': The Quattrocento Vocabulary of Creation, Inspiration and Genius in the Visual Arts. *Viator* VIII:347–398.

———. 1978. Science, Non-science and Nonsense: The Interpretation of Brunelleschi's Perspective. *Art History* II:134–161.

———. 1990. *The Science of Art – Optical Themes in Western Art from Brunelleschi to Seurat*. New Haven and London: Yale University Press.

———. 1991. *Leonardo da Vinci: The Marvellous Works of Nature and Man*. Cambridge, MA: Harvard University Press.

Kessler–Slotta, Elisabeth. 1994. Die Illustrationen in Agricolas "De re metallica". Eine Wertung aus kunsthistorischer Sicht. *Der Anschnitt* XLVI:55–67.

Keunecke, Hans–Otto. 1992–93. Ludwig von Eyb der Jüngere zum Hartenstein und sein Kriegsbuch. *Jahrbuch des Historischen Vereins für Mittelfranken* XCVI:21–36.

Kimpel, Dieter. 1983. Die Entfaltung der gotischen Baubetriebe – Ihre sozio-ökonomischen Grundlagen und ihre ästhetisch-künstlerischen Auswirkungen. In *Architektur des Mittelalters – Funktion und Gestalt*, edited by F. Möbius and E. Schubert, 246–272. Weimar: H. Böhlaus Nachfolger.

Klein, Ursula. 2003. *Experiments, Models, Paper Tools: Cultures of Organic Chemistry in the 19th Century*. Stanford, Calif.: Stanford University Press.

Kletzl, Otto. 1939. *Plan-Fragmente aus der Deutschen Dombauhütte von Prag in Stuttgart und Ulm, Veröffentlichungen des Archivs der Stadt Stuttgart*. Stuttgart: Felix Krais.

Knobloch, Eberhard. 1996. Technische Zeichnungen. In *Europäische Technik im Mittelalter 800–1400. Tradition und Innovation*, edited by U. Lindgren, 45–64. Berlin: Gebr. Mann.

Knoespel, Kenneth J. 1992. Gazing on Technology: Theatrum Mechanorum (!) and the Assimilation of Renaissance Machinery. In *Literature and Technology*, edited by M. C. Greenberg and L. Schachterle, 99–124. Bethlehem: Lehig University Press.

Koch, Peter. 1979. Berthold Hozschuher. *Neue Deutsche Biographie* IX:579f.

Koepf, Hans. 1969. *Die gotischen Planrisse der Wiener Sammlung.* Vienna: Hermann Böhlaus Nachf.

Kolb, Carolyn. 1988. The Francesco di Giorgio Material in the Zichy Codex. *Journal of the Society of Architectural Historians* XLVII:132–159.

Kostof, Spiro, ed. 1977. *The Architect: Chapters in the History of the Profession.* New York: Oxford University Press.

Kratzsch, Konrad. 1979. Das Weimarische Ingenieurkunst- und Wunderbuch und seine kulturgeschichtlichen Zeichnungen. *Marginalien* LXXIII:30–38.

———. 1981. Das Weimarische Ingenieurkunst- und Wunderbuch. Codex Wimariensis Fol. 328. *Studien zum Buch- und Bibliothekswesen* I:54–60.

Krautheimer, Richard. 1956. *Lorenzo Ghiberti.* Princeton: Princeton University Press.

Krieg, Luise. 1916. Die 'Erfindung' des Berthold Holzschuher. Eine Finanzreform des 16. Jahrhunderts. *Vierteljahrschrift für Sozial- und Wirtschaftsgeschichte* XIII:612–619.

Kühne, Andreas. 2002. Augustin Hirschvogel und sein Beitrag zur praktischen Mathematik. In *Verfasser und Herausgeber mathematischer Texte der frühen Neuzeit*, edited by R. Gebhardt. Annaberg-Buchholz: Adam-Ries-Bund.

Kunnert, Heinrich. 1965. Nürnberger Montanunternehmer in der Steiermark. *Mitteilungen des Vereins für die Geschichte der Stadt Nürnberg* LIII:229–258.

Kyeser, Konrad. 1967. *Bellifortis. Facsimile edition of Göttingen Cod. Ms. philos. 63.* Edited by G. Quarg. Düsseldorf: VDI.

———. 1995. *Bellifortis. Feuerwerkbuch. Microfiche edition of Göttingen Cod. Ms. philos. 64 and 64a.* Edited by U. Friedrich and F. Rädle. Munich: Lengenfelder.

La Penna, Pierlorenzo. 1997. *La fortezza et la citta.* Florence: Olschki.

Lagrange, Joseph–Louis. 1788. *Mécanique Analitique.* Paris: Desaint.

Laird, W.R. 1991. Patronage of Mechanics and Theories of Impact in Sixteenth-Century Italy. In *Patronage and Institutions: Science, Technology and Medicine at the European Court, 1500–1750*, edited by B. Moran, 51–66. Woodbridge: Bodell Press.

Lamberini, Daniela. 1986. The Military Architecture of Giovanni Battista Belluzzi. *Fort* XIV:4–16.

———. 1987. Practice and Theory in Sixteenth Century Fortifications. *Fort* XV:5–20.

———. 1990. *Il principe difeso. Vita e opere di Bernardo Puccini.* Florence: La Giuntina.

———. 2001. Disegno tecnico e rappresentazione prospettica delle macchine nel Rinascimento italiano. *Bolletino degli ingegneri* VII:3–10.

———. (in press). 'Machines in Perspective': Technical Drawings in Unpublished Treatises and Notebooks of the Italian Renaissance. In *The Treatise on Perspective: Published and Unpublished (Symposium, 7–8 November 1997, National Gallery of Art, Center for Advanced Studies in the Visual Arts).* Washington.

Lanteri, Giacomo. 1557. *Due dialoghi ... del modo di designare le piante delle fortezze secondo Euclide.* Venice: V. Valgrisi and B. Costantini.

Lautensack, Heinrich. 1618. *Deß Circkelß und Richtscheyts / auch der Perspectiva / und Proportion der Menschen und Rosse / kurtze / doch gründtliche underweysung / deß rechten gebrauchs [1563].* 2nd ed. Frankfurt: Schamberger.

Lawson, Bryan. 1980. *How Designers Think: The Design Process Demystified.* London: Architectural Press.

Le Goff, Jean–Pierre. 1994. Desargues et la naissance de la géométrie projective. In *Desargues en son temps*, edited by J. Dhombres and J. Sakarovitch, 157–206. Paris: A. Blanchard.

Le Moël, Michel. 1994. Jacques Curabelle et le monde des architectes parisiens. In *Desargues en son temps*, edited by J. Dhombres and J. Sakarovitch, 389–392. Paris: A. Blanchard.

Lefèvre, Wolfgang. 2002. Drawings in Ancient Treatises on Mechanics. In *Homo Faber: Studies in Nature, Technology and Science at the Time of Pompeii*, edited by J. Renn and G. Castagnetti, 109–120. Rome: Bretschneider.

———. 2003 The Limits of Pictures – Cognitive Functions of Images in Practical Mechanics, 1400 to 1600. In *The Power of Images in Early Modern Science*, edited by W. Lefèvre, J. Renn and U. Schoepflin, 69–89. Basel: Birkhäuser.

Lencker, Hans. 1571. *Perspectiva*. Nuremberg: D. Gerlatz.

Leng, Rainer. 1996. getruwelich dienen mit Buchsenwerk. Ein neuer Beruf im späten Mittelalter: Die Büchsenmeister. In *Strukturen der Gesellschaft im Mittelalter*, edited by D. Rödel and J. Schneider, 302–321. Wiesbaden: Reichert.

———. 1997. Burning, killing, treachery everywhere / Stabbing, slaying in fiercest war. War in the Medieval Housebook. In *The Medieval Housebook. Facsimile and commentary volume*, edited by C. Waldburg, 145–161. Munich: Prestel.

———, ed. 2000. *Anleitung Schießpulver zu bereiten, Büchsen zu beladen und zu beschießen. Eine kriegstechnische Bilderhandschrift im cgm 600 der Bayerischen Staatsbibliothek München*. Wiesbaden: Reichert.

———. 2001. *Franz Helm und sein 'Buch von den probierten Künsten'. Ein handschriftlich verbreitetes Büchsenmeister-buch in der Zeit des frühen Buchdrucks*. Wiesbaden: Reichert.

———. 2002. *Ars belli. Deutsche taktische und kriegstechnische Bilderhandschriften und Traktate im 15. und 16. Jahrhundert*. 2 vols. Wiesbaden: Reichert.

Leonardo, da Vinci. 1973–80. *Il Codice Atlantico della Biblioteca Ambrosiana di Milano*. Edited by A. Marinoni. 24 vols. Florence: Giunti-Barbèra.

———. 1974. *The Madrid Codices*. Translated by L. Reti. 5 vols. New York: McGraw-Hill.

———. 1986. *I manoscritti dell'Istitut de France*. Edited by A. Marinoni. Florence: Giunti-Barbèra.

———. 1987. *Scritti letterari*. Edited by A. Marinoni. 3 ed. Milan: Rizzoli.

———. 1995. *Libro di pittura*. Edited by C. Pedretti and C. Vecce. 2 vols. Florence: Giunti-Barbèra.

Leopold, J.H. 1980. Christiaan Huygens and his Instrument Makers. In *Studies on Cristiaan Huygens*, edited by H. J. M. Bos and others. Lisse: Swets.

———. 1982. L'invention par Christiaan Huygens du ressort spiral réglant pour les montres. In *Huygens et la France: table ronde du Centre national de la recherche scientifique, Paris, 27–29 mars 1979*. Paris: Vrin.

Lepik, Andres. 1994. *Das Architekturmodell in Italien 1335–1550*. Worms: Werner.

Lindberg, David C. 1976. *Theories of Vision from al–Kindi to Kepler*. Chicago: The University of Chicago Press.

Lindgren, Uta. 2000. Technische Enzyklopädien des Spätmittelalters. Was ist daran technisch? *Patrimonia* CXXXVII:9–20.

Long, Pamela O. 1985. The Contribution of Architectural Writers to a 'Scientific' Outlook in the Fifteenth and Sixteenth Centuries. *Journal of Medieval and Renaissance Studies* XV (2):265–289.

———. 1997. Power, Patronage, and the Authorship of Ars: From Mechanical Know-How to Mechanical Knowledge in the Last Scribal Age. *Isis* LXXXVIII:1–41.

———. 2001. *Openness, Secrecy, Authorship: Technical Arts and the Culture of Knowledge from Antiquity to the Renaissance*. Baltimore et al.: Johns Hopkins University Press.

Loria, Gino. 1921. *Storia della geometria descrittiva delle origini, sino ai giorni nostri*. Milan: U. Hoepli.

Lorini, Buonaiuto. 1597. *Delle fortificationi libri cinque*. Venice: G. A. Rampazetto.

Lotz, Wolfgang. 1977. The Rendering of the Interior in Architectural Drawings of the Renaissance. In *Studies in Italian Renaissance Architecture*, edited by W. Lotz, 1–65. Cambridge, M.A.: MIT Press.

Lubkin, Gregory. 1994. *A Renaissance Court: Milan under Galeazzo Maria Sforza*. Berkeley and Los Angeles: University of California Press.

Ludwig, Karl–Heinz. 1997. Historical and Metallurgical Observations on the Medieval Housebook. In *The Medieval Housebook. Facsimile and commentary volume*, edited by C. Waldburg, 127–134. Munich: Prestel.

Lupicini, Antonio. 1582. *Discorso sopra la fabrica e uso della nuove verghe astronomiche*. Florence: G. Marescotti.

Maggi, Girolamo, and Jacopo Castriotto. 1564. *Della Fortificatione della Città ... libri tre*. Venice: R. Borgominiero.

Mahoney, Michael. 1980. Christiaan Huygens, The Measurement of Time and Longitude at Sea. In *Studies on Cristiaan Huygens*, edited by H. J. M. Bos and others. Lisse: Swets.

———. 1985. Diagrams and Dynamics: Mathematical Perspectives on Edgerton's Thesis. In *Science and the Arts in the Renaissance*, edited by J. W. Shirley and F. D. Honiger, 198–220. Washington: The Folger Shakespeare Library.

———. 1993. Algebraic vs. Geometric Techniques in Newton's Determination of Planetary Orbits. In *Action and Reaction: Proceedings of a Symposium to Commemorate the Tercentenary of Newton's Principia*, edited by P. Theerman and A. F. Seeff. Newark: University of Delaware Press.

———. 1996. Longitude in the History of Science. In *The Quest for Longitude*, edited by W. J. H. Andrewes. Cambridge, MA: Collection of Historical Scientific Instruments, Harvard University.

———. 1998. The Mathematical Realm of Nature. In *Cambridge History of Seventeenth-Century Philosophy*, edited by D. Garber and M. Ayers. Cambridge: Cambridge University Press.

———. 2000. Huygens and the Pendulum: From Device to Mathematical Relation. In *The Growth of Mathematical Knowledge*, edited by H. Breger and E. Grosholz. Amsterdam: Kluwer Academic Publishers.

Mancini, Girolamo. 1916. L'opera 'De corporibus regolaribus' di Pietro Franceschi detto Della Francesca, ursupata da Fra Luca Pacioli. *Atti della R. Accademia dei Lincei* XIV:441–580.

Manetti, Antonio di Tuccio. 1970. *The Life of Brunelleschi*. Translated by C. Enggass. Edited by H. Saalman. London: The Pennsylvania State University Press.

———. 1976. *Vita di Filippo Brunelleschi, preceduta da la novella del grasso*. Edited by D. De Robertis. Milan: Il Polifilo.

Manno, Antonio. 1987. Giulio Savorgnan: Machinatio e ars fortificatoria a Venezia. In *Atti del Convegno Internazionale di Studio Giovan Battista Benedetti e il suo Tempo*, 227–245. Venice: Istituto Veneto di Scienze, Lettere ed Arti.

Marani, Pietro C. 1991. Francesco di Giorgio a Milano e a Pavia: consequenze e ipotesi. In *Prima di Leonardo: Cultura delle macchine a Siena nel Rinascimento*, edited by P. Galluzzi, 93–104. Milan: Electa.

Marchis, Vittorio. 1991. Nuove dimensioni per l'energia: le macchine di Francesco di Giorgio. In *Prima di Leonardo: Cultura delle macchine a Siena nel Rinascimento*, edited by P. Galluzzi, 113–120. Milan: Electa.

Marchis, Vittorio, and Luisa Dolza. 1998. L'acqua, i numeri, le machine: Al sorgere dell'ingegneria idraulica moderna. In *Giambattista Aleotti e gli ingegneri del Rinascimento*, edited by A. Fiocca, 207–222. Florence: Olschki.

Maschat, Herbert. 1989. *Leonardo da Vinci und die Technik der Renaissance*. Munich: Profil.

McGee, David. 1999. From Craftsmanship to Draftsmanship: Naval Architecture and the three traditions of Early Modern Design. *Technology and Culture* XL (2):209–236.

Mende, Matthias. 1971. *Dürer-Bibliographie*. Wiesbaden: Harrassowitz.

Metzger, Wolfgang. 2001. Ein Bildzyklus des Spätmittelalters zwischen Hofkunst und 'Magia naturalis'. In *Opere e Giorni. Studi su mille anni di arte europea dedicati a Max Seidel*, edited by G. Bonsanti and K. Bergdoldt, 253–264. Venice: Marsilio.

Millon, Henry. 1958. The Architectural Theory of Francesco di Giorgio. *Art Bulletin* XL:257–261.

Morachiello, Paolo. 1988. Da Lorini e de Ville: per una scienza e per una statuto dell'Ingegnero. In *L'architettura militare veneta del Cinquecento*, edited by Centro internazionale di studi di architettura 'Andrea Palladio' di Vicenza, 45–47. Milan: Electa.

Müller, Werner. 1990. *Grundlagen gotischer Bautechnik*. Munich: Deutscher Kunstverlag.

Müller, Winfried. 2000. Die Wasserkunst des Schlosses Hellenstein bei Heidenheim. In *Die Wasserversorgung in der Renaissancezeit*, edited by Frontinus-Gesellschaft, 260–266. Mainz: von Zabern.

Mussini, Massimo. 1991a. *Il Trattato di Francesco di Giorgio Martini e Leonardo: Il Codice Estense restituito*. Parma: Universita di Parma.

———. 1991b. Un frammento del Trattato di Francesco di Giorgio Martini nell'archivio di G. Venturi alla Biblioteca Municipale di Reggio Emilia. In *Prima di Leonardo: Cultura delle macchine a Siena nel Rinascimento*, edited by P. Galluzzi, 81–92. Milan: Electa.

———. 1993. La trattatistica di Francesco di Giorgio: un problema critico aperto. In *Francesco di Giorgio architetto*, edited by F. P. Fiore and M. Tafuri, 358–379. Milan: Electa.

Neuhaus, August. 1940. Der Kampfwagen des Berthold Holzschuher. *Nürnberger Schau*:187–191.

Newton, Isaac. 1687. *Philosophiae naturalis principia mathematica*. London: J. Streater.

———. 1730. *Opticks, or a Treatise of the Reflections, Refractions, Inflections & Colours of Light*. 4th ed. London: W. Innys.

Noble, David. 1984. *Forces of Production: A Social History of Industrial Automation*. New York: Knopf.

Oddi, Mutio. 1625. *Dello Squadro*. Milan: B. Fobella.

———. 1633. *Fabrica et Uso del Compasso Polimetro*. Milan: B. Fobella.

Pacioli, Luca. 1494. *Summa de arithmetica, geometria, proportioni et proportionalità*. Venice: P. de Paganini.

Panofsky, Erwin. 1955. *The Life and Art of Albrecht Dürer*. Princeton: Princeton University Press.

Pappus. 1588. *Mathematicae collectiones*. Edited by F. Commandinus. Pesaro: Concordiam.

Parsons, William Barclay. 1968. *Engineers and Engineering in the Renaissance*. Cambridge, MA: MIT Press.

Pause, Peter. 1973. Gotische Architekturzeichnungen in Deutschland. PhD-thesis, Bonn.

Pedretti, Carlo. 1985. Excursus 1: Architectural Studies under Ultraviolet Light: New Documentation on Leonardo's Work at Pavia; Excursus 3: Francesco di Giorgio, 'Trattato di Architettura Civile e Militare'. In *Leonardo Architect*, 19–27; 196–204. New York: Rizzoli.

———. 1987. Introduction. In *Leonardo da Vinci: Engineer and Architect*, edited by P. Galluzzi, 1–21. Montreal: Montreal Museum of Fine Arts.

Peiffer, Jeanne. 1995. Dürer Géomètre. In *Albrecht Dürer: Géométrie*, edited by J. Peiffer, 17–131. Paris: Seuil.

———. 2002a. La perspective, une science mêlée. *Nouvelle revue du seizième siècle* XX:97–121.

———. 2002b. L'art de la mesure chez Dürer. In *Mesure & grands chantiers. 4000 ans d'histoire*, edited by J. Sakarovitch, 22–27. Paris: Publi-Topex.

———. (in press). Adapter, simplifier et mettre en pratique la perspective. Les Kunstbücher du XVIe siècle. Paper read at ESF Exploratory Workshop "Artists, work, and the challenge of perspective", at Rome.

———. (in press). L'optique euclidienne dans les pratiques artisanales de la Renaissance. In *Oriens-Occidens. Sciences, mathématiques et philosopie de l'Antiquité à l'Âge classique*. Cahiers du Centre d'histoire des sciences et des philosophies arabes et médiévales.

Pellicanò, Antonio. 2000. *Del periodo giovanile di Galileo Galilei*. Rome: Gangemi.

Pérouse de Montclos, Jean–Marie. 2000. *Philibert De l'Orme, architecte du roi (1514–1570)*. Paris: Mengès.

Pevsner, Nikolaus. 1942. The Term 'Architect' in the Middle Ages. *Speculum* XVII (4):549–562.

Pfintzing von Henfenfeld, Paul. 1599. *Ein schöner kurtzer Extract der Geometriae unnd Perspectivae*. Nuremberg: Fuhrmann.

Piero della Francesca. 1984. *De prospectiva pingendi – Edizione critica*. Edited by G. Nicco-Fasola. Florence: Casa editrice le lettere.

———. 1995. *Libellus de quinque corporibus regularibus*. Edited by F. P. di Teodoro. Florence: Giunti.

———. 1998. *De la perspective en peinture*. Translated by J.–P. Le Goff. Edited by J.–P. Le Goff. Paris: In Medias Res.

Plomb, Reinier. 1999. A Longitude Timekeeper by Isaac Thuret with the Balance Spring invented by Christiaan Huygens. *Annals of Science* LVI:379–394. Online version at *http://www.antique-horology.org/_Editorial/thuretplomp/thuretplomp.htm*.

Pollak, Martha. 1991. *Turin: 1564–1680: Urban Design, Military Culture and the Creation of the Absolutist Capital*. Chicago: The University of Chicago Press.

Poni, Carlo. 1985. Introduction. In *Vittorio Zonca: Novo Teatri di machine et edificii*, edited by C. Poni. Milan: Polifilo.

Popplow, Marcus. 1998a. *Neu, nützlich und erfindungsreich: die Idealisierung von Technik in der frühen Neuzeit*. Münster: Waxmann.

———. 1998b. Protection and Promotion – Privileges for Inventions and Books of Machines in the Early Modern Period. *History of Technology* XX:103–124.

———. 1998c. Books of Machines in the Early Modern Period. *History of Technology* XX:103–124.

———. 1999. Heinrich Schickhardt als Ingenieur. In *Heinrich Schickhardt: Baumeister der Renaissance – Leben und Werk des Architekten, Ingenieurs und Städteplaners*, edited by S. Lorenz and W. Setzler, 75–82. Stuttgart: DRW-Verlag.

————. 2001. Militärtechnische Bildkataloge des Spätmittelalters. In *Krieg im Mittelalter*, edited by H.–H. Kortüm. Berlin: Akademie-Verlag.

————. 2002. Heinrich Schickhardts Maschinenzeichnungen. Einblicke in die Praxis frühneuzeitlicher Ingenieurtechnik. In *Neue Forschungen zu Heinrich Schickhardt*, edited by R. Kretzschmar, 145–170. Stuttgart: Kohlhammer.

————. (in press). Models of Machines. A 'Missing Link' between Early Modern Engineering and Mechanics? In *A Reappraisal of the Zilsel Thesis*, edited by D. Raven and W. Krohn. Dordrecht: Kluwer Academic Publishers.

Post, Robert C. 2001. *High Performance: The Culture and Technology of Drag Racing*. Rev. ed. Baltimore: Johns Hopkins University Press.

Potié, Philippe. 1996. *Philibert De L'Orme. Figures de la pensée constructive*. Marseille: Editions Parenthèses.

Promis, Carlo. 1874. *Biografie di ingegneri militari italiani dal secolo xiv alla metà xviii*. Turin: Fratelli Bocca.

Ptolemaeus, Claudius. 1525. *Claudii Ptolemaei Geographicae enarrationis libri octo*. Strasbourg: Koberger.

Puppi, Lionello. 1981. La Teoria Artistica nel Cinquecento. In *Storia della cultura veneta III/3*, edited by G. Arnaldi and M. Pastore Stocchi. Vicenza: Pozza.

Ramelli, Agostino. 1588. *Le diverse et artificiose machine*. Paris: In casa del Autore.

Rauck, Max J.B. 1964. Als die Automobile noch keine Motoren hatten. *Das Schnauferl* XII (3ff.):4–7, 6–12, 6–8, 6–8.

Recht, Roland, ed. 1989. *Les Batisseurs des Cathedrales Gothiques*. Strasbourg: Editions les Musees de la Ville de Strasbourg.

Renn, Jürgen, and Matteo Valleriani. 2001. Galileo and the Challenge of the Arsenal. In *Preprint 179 of the Max Planck Institute for the History of Science*. Berlin.

Reti, Ladislao. 1969. Leonardo Da Vinci the Technologist: The Problem of Prime Movers. In *Leonardo Technologist*, edited by B. Dibner and L. Reti. Norwalk: Burndy Library.

Ricossa, Sergio, and Pier Luigi Bassignana, eds. 1988. *Le Maccine di Valturio: nei documenti dell'Archivio storico Amma*. Turin: Allemandi.

Rising, James S., and Maurice W. Amfeldt. 1964. *Engineering Graphics*. 3rd ed. Dubuque, Iowa: WM. C. Brown Book Company.

Roche, John J. 1993. The Semantics of Graphs in Mathematical Natural Philosophy. In *Non-Verbal Communication in Science Prior to 1900*, edited by R. G. Mazzolini. Firenze: Olschki.

Rodakiewicz, Erla. 1940. The Editio princeps of Roberto Valturio's 'De re militari' in relation to the Dresden and Munich manuscripts. *Maso Finiguerra* V (1/2):15–82.

Rojas Sarmiento, Juan de. 1550. *Commentariorum in astrolabium*. Paris: Vascosanus.

Romano, Bartelomeo. 1595. *Proteo Militare*. Naples: G.C. Carlino and A. Pace.

Roriczer, Matthäus. 1486. *Das Büchlein von der Fialen Gerechtigkeit*. Regensburg: Roriczer.

Rose, Paul Laurence. 1970. Renaissance Italian Methods of Drawing Ellipse and Related Curves. *Physis* XII:371–404.

Rosenfeld, Myra Nan. 1989. From Drawn to Printed Model Book: Jacques Androuet Du Cerceau and the Transmission of Ideas from Designer to Patron, Master Mason and Architect in the Renaissance. *RACAR* XVI (2):131–145.

Rossi, Massimo. 1998. La cartografia aleottiana. In *Giambattista Aleotti e gli ingegneri del Rinascimento*, edited by A. Fiocca, 161–188. Florence: Olschki.

Rowe, Peter G. 1987. *Design Thinking*. Cambridge, MA: MIT Press.

Rowland, Ingrid D. 1998. *The Culture of the High Renaissance: Ancients and Moderns in Sixteenth-Century Rome*. Cambridge: Cambridge University Press.

Rupprich, Hans, ed. 1956–1969. *Dürer: Schriftlicher Nachlaß*. 3 vols. Berlin: Deutscher Verein für Kunstwissenschaft.

Ruscelli, Girolamo. 1568. *Precetti della militia moderna*. Venice: M. Sessa.

Ryff, Walther. 1547. *Der furnembsten, notwendigsten, der gantzen Architectur angehörigen mathematischen und mechanischen Künst eygentlicher Bericht [...] f.XI–LII, Als der neuen Perspectiva das I. buch*. Nuremberg: J. Petreius.

———. 1558. *Der Architectur fürnembsten/ notwendigsten, angehörigen Mathematischen und Mechanischen künst/ eygentlicher Bericht/ und verstendliche unterrichtung zu rechtem verstandt der lehr Vitruvii : in drey fürneme Bücher abgetheilet ... / Durch Gualtherum H. Rivium*. Nuremberg: Heyn.

Saalman, Howard. 1959. Early Renaissance Architectural Theory and Practice in Antonio Filaretes Trattato di Architettura. *Art Bulletin* XLI (1):89–106.

———. 1980. *Filippo Brunelleschi : the cupola of Santa Maria del Fiore*. London: Zwemmer.

Saint–Aubin, Jean–Paul. 1994. Les enjeux architecturaux de la didactique stéréotomique de Desargues. In *Desargues en son temps*, edited by J. Dhombres and J. Sakarovitch. Paris: A. Blanchard.

Sakarovitch, Joël. 1992. La coupe des pierres et la géométrie descriptive. In *L'Ecole normale de l'an III. Leçons de Mathématiques, Laplace-Lagrange-Monge*, edited by J. Dhombres, 530–540. Paris: Dunod.

———. 1994. Le fascicule de stéréotomie, entre savoir et métiers, la fonction de l'architecte. In *Desargues en son temps*, edited by J. Dhombres and J. Sakarovitch, 347–362. Paris: A. Blanchard.

———. 1998. *Epures d'architecture: De la coupe des pierres à la géométrie descriptive – XVIe–XIXe siècles*. Basel: Birkhäuser.

Salmi, Mario. 1947. Disegni di Francesco di Giorgio nella collezione Chigi Saracini. *Quaderni dell'Accademia Chigiana* XI:7–45.

Sanmarino, Giovan Battista Belici (called). 1598. *Nuova inventione di fabbricar fortezze, di varie forme*. Vencice: R. Meietti.

Santillana, Giorgio di. 1955. *The Crime of Galileo*. Chicago: The University of Chicago Press.

Scaglia, Gustina. 1985. *Il 'Vitruvio Magliabechiano' di Francesco di Giorgio Martini, Documenti inediti di cultura toscana VI*. Florence: Gonelli.

———. 1992. *Francesco di Giorgio: Checklist and History of Manuscripts and Drawings in Autographs and Copies from ca. 1470 to 1687 and Renewed Copies (1764–1839)*. Bethlehem: Lehigh University Press.

Scheller, Robert W. 1995. *Exemplum: Model-Book Drawings and the Practice of Artistic Transmission in the Middle Ages*. Translated by M. Hoyle. Amsterdam: Amsterdam University Press.

Schickhardt, Heinrich. 1902. *Handschriften und Handzeichnungen des herzoglich württembergischen Baumeisters Heinrich Schickhardt*. Edited by W. Heyd. Stuttgart: Kohlhammer.

Schmidtchen, Volker. 1977. *Bombarden, Befestigungen, Büchsenmeister. Von den ersten Mauerbrechern des Spätmittelalters zur Belagerungsartillerie der Renaissance. Eine Studie zur Entwicklung der Militärtechnik*. Düsseldorf: Droste.

————. 1982. Militärische Technik zwischen Tradition und Innovation am Beispiel des Antwerks. Ein Beitrag zur Geschichte des mittelalterlichen Kriegswesens. In *gelêrter der arzenîe, ouch apoteker. Beiträge zur Wissenschaftsgeschichte. Festschrift zum 70. Geburtstag von Willem F. Daems*, edited by G. Keil, 91–195. Pattensen: Wellm.

————. 1990. *Kriegswesen im späten Mittelalter. Technik, Taktik, Theorie.* Weinheim: VCH.

————. 1997. Technik im Übergang vom Mittelalter zur Neuzeit zwischen 1350 und 1600. In *Metalle und Macht 1000 – 1600*, edited by K.–H. Ludwig and V. Schmidtchen, 209–598. Berlin: Propyläen.

Schmuttermayer, Hans. 1489. *[Fialenbüchlein].* Nuremberg: n.p.

Schneider, Ivo. 1993. *Johannes Faulhaber 1580–1635, Rechenmeister in einer Welt des Umbruchs, Vita Mathematica 7.* Basel: Birkhäuser.

Schneider, Karl. 1868. Zusammenstellung und Inhalts-Angabe der artilleristischen Schriften und Werke in der Bibliothek Seiner Exellenz des Herrn Feldzeugmeisters Ritter v. Hauslab. In *Mittheilungen über Gegenstände der Artillerie- und Kriegs-Wissenschaften*, edited by K. K. Artillertie-Comitté, 125–211. Vienna.

Schneider, Rudolf. 1912. Griechische Poliorketiker. *Abhandlungen der Königlichen Gesellschaft der Wissenschaften zu Göttingen, Philologische-historische Klasse* NF XII:1–87 and plates.

Schnitter, Niklaus. 2000. Hanns Zschans Plan der Basler Wasserversorgung. In *Die Wasserversorgung in der Renaissancezeit*, edited by Frontinus-Gesellschaft, 213–216. Mainz: von Zabern.

Schoeller, Wolfgang. 1980. Eine mittelalterliche Architekturzeichnung im südlichen Querhausarm der Kathedrale von Soissons. *Zeitschrift für Kunstgeschichte* XLIII:196–202.

————. 1989. Ritzzeichnungen – Ein Beitrag zur Geschichte der Architekturzeichnung im Mittelalter. *Architectura* XIX:36–61.

Schofield, Richard. 1991. Leonardo's Milanese Architecture: Career, Sources and Graphic Techniques. *Achademia Leonardo Vinci* IV:111–157.

Schuritz, Hans. 1919. *Die Perspektive in der Kunst Albrecht Dürers.* Frankfurt: Heinrich Keller.

Scolari, Massimo. 1984. Elementi per una storia dell'axonometria. *Casabella* D: 42–49.

Serlio, Sebastiano. 1545. *Il primo libro d'architettura [Geometria]; Il Secondo libro di perspettia.* Paris: Barbé.

Settle, Thomas B. 1978. Brunelleschi's Horizontal Arches and Related Devices. *Annali dell'Istituto e Museo di Storia della Scienza di Firenze* III:65–80.

Severini, Giancarlo. 1994. *Progetto e disegno nei trattati di architettura militare del '500.* Pisa: Pacini.

Shelby, Lon R. 1971. Mediaeval Masons Templates. *Journal of the Society of Architectural Historians* XXX:140–154.

————. 1972. The Geometrical Knowledge of Mediaeval Master Masons. *Speculum* IV:395–421.

————, ed. 1977. *Gothic Design Techniques – The Fifteenth-Century Design Booklets of Mathes Roriczer and Hanns Schmuttermayer.* London and Amsterdam: Southern Illinois University Press.

Simi, Annalisa. 1996. Celerimensura e strumenti in manoscritti dei secoli XIII–XV. In *Itinera mathematica. Studi in onore di Gino Arrighi per il suo 90° compleanno*, edited by R. Franci, P. Pagli and L. Toti Rigatelli, 71–122. Siena: Universitá di Siena.

Simmern, Johann von. 1546. *Eyn schön nützlich büchlin und underweysung der kunst des Messens / mit dem Zirckel / Richtscheidt oder Linial (1531)*. 2nd ed. Frankfurt: Jacob.

Sinisgalli, Rocco. 1984. *I sei libri della Prospettiva di Guidobaldo dei Marchesi del Monte dal latino tradotti interpretati e commentati*. Rome: L'Erma di Bretschneider.

Smith, Merrit Roe. 1977. *The Harpers Ferry Armory and the 'New Technology' in America, 1794–1854*. Ithaca: Cornell University Press.

Sorrenson, Richard. 1993. Scientific Instrument Makers at the Royal Society of London, 1720–1780. PhD-thesis, Princeton.

———. 1996. George Graham, Visible Technician. *British Journal for the History of Science* XXXII:203–221.

Spies, Gerd. 1992. *Technik der Steingewinnung und der Flußschiffahrt im Harzvorland in früher Neuzeit*. Braunschweig: Waisenhaus-Druckerei.

Staigmüller, H. 1891. Dürer als Mathematiker. In *Programm des Königlichen Gymnasiums in Stuttgart 1890/91*. Stuttgart: Zu Guttenberg.

Steck, Max. 1948. *Dürers Gestaltlehre der Mathematik und der bildenden Künste*. Halle: Niemeyer.

Steinmann, Martin. 1979. Regiomontan und Dürer – eine Handschrift mit berühmten Vorbesitzern. *Basler Zeitschrift für Geschichte und Altertumskunde* LXXIX:81–89.

Stevin, Simon. 1955. De weeghdaet. In *The principal works of Simon Stevin*, edited by E. Crone, E. J. Dijksterhuis, R. J. Forbes, M. G. J. Minnaert and A. Pannekoek, I 287–373. Amsterdam: Swets & Zeitlinger.

———. 1955–1966. *The principal works of Simon Stevin*. Edited by E. Crone, E. J. Dijksterhuis, R. J. Forbes, M. G. J. Minnaert and A. Pannekoek. 6 vols. Amsterdam: Swets & Zeitlinger.

Strada, Jacopb de. 1617. *Kunstliche Abriß allerhand Wasser- Wind- Ross- und Handt Mühlen*. Frankfurt: Jacobi.

Straker, Stephen. 1976. The eye made 'other' : Dürer, Kepler and the mechanization of light and vision. In *Science, Technology and Culture in Historical Perspective*, edited by L. A. Knafla, M. S. Staum and T. H. E. Travers, 7–25. Calgary: University of Calgery.

Strauss, Walter L., ed. 1974. *The Complete Drawings of Albrecht Dürer 6 : Architectural Studies*. New York: Abaris Books.

Summers, David. 1987. *The Judgment of Sense: Renaissance Naturalism and the Rise of Aesthetics*. Cambridge: Cambridge University Press.

Taccola, Mariano di Jacopo. 1969. *Liber tertius de ingeneis ac edifitiis non usitatis*. Edited by J. H. Beck. Milan: Il Polifilo.

———. 1971. *De machinis: The Engineering Treatise of 1449*. Translated by G. Scaglia. Edited by G. Scaglia. 2 vols. Wiesbaden: Reichert.

———. 1972. *Mariano Taccola and his Book 'De Ingeneis'*. Edited by F. D. Prager and G. Scaglia. Cambridge, MA: MIT Press.

———. 1984a. *De ingeneis: Liber primus leonis, liber secundus draconis, addenda; Books I and II, On Engines and Addenda (The Notebook)*. Translated by G. Scaglia. Edited by G. Scaglia, F. D. Prager and U. Montag. 2 vols. Wiesbaden: Reichert.

———. 1984b. *De rebus militaribus (De machinis, 1449)*. Edited by E. Knobloch. Baden-Baden: Koerner.

Tafuri, Manfredo. 1990. Science, Politics, and Architecture. In *Venice and the Renaissance*, edited by M. Tafuri. Cambridge, MA: MIT Press.

Tapia, Nicolás García. 1991. *Patentes de invencion españolas en el siglo de oro*. Madrid: Ministerio de Industria y Energia.

Tartaglia, Niccolò. 1537. *Nova Scientia*. Venice: S. Sabio.

————. 1546. *Quesiti et Inventioni*. Venice: V. Ruffinelli.

Thoenes, Christoph. 1993. Vitruv, Alberti, Sangallo. Zur Theorie der Architekturzeichnung in der Renaissance. In *Hülle und Fülle. Festschrift für Tilmann Buddensieg*, edited by A. Beyer, V. Lampugnani and G. Schweikhart, 565–584. Alfter: VDG.

————. 1998. Vitruvio, Alberti, Sangallo. La teoria del disegno architettonico nel Rinascimento. In *Sostegno e adornamento, Saggi sull' architettura del rinascimento: disegni, ordini, magnificenza*, edited by C. Thoenes. Milan: Electa.

Tigler, Peter. 1963. *Die Architekturtheorie des Filarete*. Berlin: De Gruyter.

Tocci, Luigi Michelini. 1962. Disegni e appunti autografi di Francesco di Giorgio in un codice del Taccola. In *Scritti di Storia dell'Arte in onore di Mario Salmi*, 203–212. Rome: De Luca.

Toldano, Ralph. 1987. *Francesco di Giorgio Martini: Pittore e scultore*. Milan: Electa.

Torlontano, Tossana. 1998. Aleotti trattatista di artiglieria e fortificationi. In *Giambattista Aleotti e gli ingegneri del Rinascimento*, edited by A. Fiocca, 117–133. Florence: Olschki.

Truesdell, Clifford A. 1982. Fundamental Mechanics in the Madrid Codices. In *Leonardo e l'età della ragione*, edited by E. Bellone and P. Rossi, 309–324. Florence: Scientia.

Turriano, Juanelo. 1996. *Los veintiún libros de los ingenios y máquinas de Juanelo Turriano*. 5 vols. Aranjuez: Fundación Juanelo Turriano.

Valdelvira, Alonso de. 1977. *El tratado de Arquitectura (ms. R 10 Biblioteca de la Escuela Técnica Superior de Arquitectura Madrid)*. Edited by G. Barbé-Coquelin de Lisle. 2 vols. Albacete: Caja de Ahorros Provincial de Albacete.

Valturio, Roberto. 1472. *De re militari libri XII*. Verona: G. di Niccolò.

Van Egmond, Warren. 1980. Practical Mathematics in the Italian Renaissance: a Catalog of the Italian Abbacus Manuscripts and Printed Books to 1600. *Annali dell'Istituto e Museo di Storia della Scienza* I:Supplement.

Varignon, Pierre. 1700. Des forces centrales, ou des pesanteurs necessaires aux Planetes pour leur fair décrire les orbes qu'on leur a supposés jusqu'icy. *Mémoires de l'Académie Royale des Sciences* 1700:218–237.

————. 1710. Des forces centrales inverses. *Mémoires de l'Académie Royale des Sciences* 1710:533–544.

Vasari, Giorgio. 1906. *Le opere di Giorgio Vasari*. Edited by G. Milanesi. 9 vols. Florence: Sansoni.

————. 1963. *The Lives of the Painters, Sculptors and Architects*. Vol. 1. London and New York: Everyman's Library.

Vegetius Renatus, Flavius. 2002. *Von der Ritterschaft. Aus dem Lateinischen übertragen von Ludwig Hohenwang in der Ausgabe Augsburg, Johann Wiener, 1475/76*. Edited by F. Fürbeth and R. Leng. Munich: Lengenfelder.

Veltman, Kim. 1979. Military Surveying and Topography: The Practical Dimension of Renaissance Linear Perspective. *Revista da Universidade de Coimbra* XXVII:329–368.

Ventrice, Pasquale. 1998. Architettura militare e ingegneria tra XVI and XVII secolo a Venezia. In *Giambattista Aleotti e gli ingegneri del Rinascimento*, edited by A. Fiocca, 309–330. Florence: Olschki.

Viator (Pelerin), Jean. 1505. *De artificiali perspectiva*. Toul: P. Jacobus.

Vigevano, Guido. 1993. *Le macchine del re. Il texaurus regis Francie di Guido Vigevano.* Edited by G. Ostuni. Vigevano: Diakronia.

Vignola, Giacomo Barozzi da. 1583. *Le due regole della prospettiva pratica.* Edited by E. Danti. Rome: F. Zannetti.

Vitruvius. 1521. *De architectura libri dece.* Edited by C. Cesarino. Como: G. De Ponte.

————. 1567. *I dieci libri dell'architettura di M. Vitruvio tradotti e commentati da Daniele Barbaro.* Venice: F. de Franceschi Senese & G. Chrieger Alemano.

————. 1975. *Vitruvio e Raffaello. Il 'De architectura' di Vitruvio nella traduzione inedita di Fabio Calvo Ravennate.* Edited by V. Fontana and P. Morachiello. Rome: Officina.

Waldburg, Christoph, ed. 1997. *The Medieval Housebook. Facsimile and commentary volume.* Munich: Prestel.

Wallis, John. 1670. *Mechanica, sive de motu.* London: Godbid.

Ward, J.D., and F.A.B.H. Ward, eds. 1974. *The planetarium of Giovanni de Dondi, citizen of Padua : a manuscript of 1397 ... with additional material from another Dondi manuscript.* London: Antiquarian Horological Society.

Ward, Mary Ann Jack. 1972. The Accademia del Disegno in Sixteenth Century Florence: A Study of an Artists' Institution. PhD-thesis, University of Chicago.

Wazbinski, Zygmunt. 1987. *L'Accademia Medicea del Disegno a Firenze nel Cinquecento.* 2 vols. Florence: Olschki.

Welch, Evelyn. 1996. *Art and Authority in Renaissance Milan.* New Haven: Yale University Press.

Wilkinson, Catherine. 1988. Renaissance Treatises on Military Architecture and the Science of Mechanics. In *Les traités d'architecture de la Renaissance*, edited by J. Guillaume, 466–475. Paris: Picard.

Zanchi, Giovan Battista. 1554. *Del modo di fortificar la citta.* Venice: P. Pietrasanta.

Zeising, Heinrich. 1607. *Theatri machinarum.* Leipzig: Gross.

Zonca, Vittorio. 1985. *Novo Teatri di machine et edificii (1607).* Edited by C. Poni. Milan: Polifilo.

Zwijnenberg, Robert. 1999. *The Writings and Drawings of Leonardo da Vinci: Order and Chaos in Early Modern Thought.* Translated by C. A. van Eck. Cambridge: Cambridge University Press.

NAME INDEX

Accolti, Pietro (1579–1642), 182–3, 190–2, 195, 233
Agricola, Georgius (1494–1555), 9, 14, 26, 40–1, 287
De re metallica, 9, 26, 40–1
Aguilonius (d'Aguillon), François (1566–1617), 187, 195
Alberti, Leon Battista (1404–1472), 108, 131, 139, 151, 158, 175–80, 192–3, 196, 212, 219, 233–4, 259, 267
De re aedificatoria, 196, 212
De pictura, 176–7, 192, 233
Elementa picturae, 176
Alcamenes (Alkamenes) (5th c. B.C.), 257
Aleotti, Giambattista (1546–1636), 40
Al-Farabi (870–950), 259
Alfonso II (Alfonso da Calabria), King of Naples (1448–1495), 120
"Anonymous of the Hussite Wars" (fl. 1470–1490), 14, 21, 96
Apianus, Petrus (1495–1552), 180
Apollonius (Appollonios) of Perga (c. 262–c. 190 B.C.), 268, 271
Archimedes (287–212 B.C.), 43, 120, 151, 281, 287
Aristotle (Aristoteles) (384–322 B.C.), 43, 281, 297
Averlino (Filarete), Antonio (1400–1469), 131, 139, 175, 219
Trattato di Architettura, 219
Ayanz, Jerónimo de (1553–1613), 28

Baldi, Bernardino (1553–1617), 184
Barbaro, Daniele (1513–1570), 182, 184, 186, 197, 257, 265, 267–8
La pratica della perspettiva, 257, 267–8
Bartoli, Cosimo (1503–1572), 21–2, 156–7
Belici, Giovani Battista (fl. 1598), 184
Bellucci, Giovanni Battista, *see* Belici
Benedetti, Giovanni Battista (1530–1590), 267, 269
Perspectivae libri sex, 269
Besson, Jacques (c. 1540–1575), 14, 21, 28
Theatrum instrumentorum et machinarum, 21
Boncompagni, Giacomo (1548–1612), 182
Bosse, Abraham (1602–1676), 245, 273–4

Bramante, Donato di Pascuccio di Antonio, called (1444–1514), 205, 227, 235
Branca, Giovanni (1571–1645), 14, 21
Brancaccio, Lelio (1560–1637), 156, 160
Brunelleschi, Filippo (1377–1446), 108, 148, 150–1, 176–7, 179, 181–2, 193
Buontalenti, Bernardo (1536–1608), 149–50, 153

Caramuel de Lobkowitz, Juan (1606–1682), 205
Castellani, Grazia de' (?–1401), 181
Castiglione, Baldassare (1478–1529), 212
Castriotto, Jacopo Fausto (1500–1562), 168, 184–5
Cataneo, Girolamo (fl. 1540–1584), 105, 158, 167
Cataneo, Pietro (?–1569), 146, 160
Caus, Salomon de (1576–1626), 14, 21, 267
Ceredi, Giuseppe (fl. 1567), 14, 46
Charles IX, King of France (1560–1574), 28
Charles V, Holy Roman Emperor (1500–1558), 158
Cigoli, Lodovico (1559–1613), 152, 192, 195–6, 282
Colbert, Jean Baptiste (1619–1683), 298–9
Commandino, Federico (1509–1575), 184, 186–7, 269, 275
Contarini, Giacomo (1536–1595), 157
Cousin (the Younger), Jean (1522–1594), 190
Curabelle, Jacques (fl. 1644), 273

Danti, Egnazio (1536–1586), 180–2, 245, 267–9, 275
De L'Orme, Philibert (1515?–1570), 201–8, 272
Le premier tome de l'architecture, 201–2, 272
De Marchi, Francesco (1490–1574), 168
Del Monte, Guidobaldo (1545–1607), 14, 43, 48, 164, 182, 184, 186–7, 190–1, 195–6, 245, 267, 269, 275
Perspectivae libri sex, 195, 269
Desargues, Girard (1591–1661), 174, 191, 195, 208, 245, 267, 269–75
Brouillon projet, 270–2

SUBJECT INDEX

Printed in the United States
by Baker & Taylor Publisher Services